SOCIOLOGY
IN OUR TIMES

SOCIOLOGY
IN OUR TIMES
11E

Diana Kendall
Baylor University

CENGAGE
Learning·

Australia • Brazil • Mexico • Singapore • United Kingdom • United States

Sociology in Our Times, **Eleventh Edition**
Diana Kendall

Product Director: Marta Lee-Perriard

Product Manager: Libby Beiting-Lipps

Content Developer: Lin Gaylord

Product Assistant: Chelsea Meredith

Marketing Manager: Kara Kindstrom

Content Project Manager: Cheri Palmer

Art Director: Vernon Boes

Manufacturing Planner: Judy Inouye

Production Service: Greg Hubit Bookworks

Photo/Text Researcher: Lumina Datamatrics Ltd

Copy Editor: Donald Pharr

Proofreader: Debra Nichols

Compositor and Illustrator: MPS Limited

Text and Cover Designer: Terri Wright

Cover Credits: Cityscape © Aaron Morris/
Getty Images; Genome © CC by Biggs et al.,
adapted. DOI: 10.1371/journal.pone.0027121,
PLOS Collections; People © Caiaimage/Martin
Barraud/Getty Images; Mandala © iStock/
Getty Images Plus; Globe CC BY-SA 2.0 Satellite
imagery, NASA (Blue Marble Project)

For product information and technology assistance, contact us at
Cengage Learning Customer & Sales Support, 1-800-354-9706.

For permission to use material from this text or product,
submit all requests online at **www.cengage.com/permissions.**
Further permissions questions can be e-mailed to
permissionrequest@cengage.com.

Library of Congress Control Number: 2015949918

Student Edition:
ISBN: 978-1-305-50309-0

Loose-leaf Edition:
ISBN: 978-1-305-67712-8

Cengage Learning
20 Channel Center Street
Boston, MA 02210
USA

Cengage Learning is a leading provider of customized learning solutions
with employees residing in nearly 40 different countries and sales in more
than 125 countries around the world. Find your local representative at
www.cengage.com.

Cengage Learning products are represented in Canada by Nelson Education, Ltd.

To learn more about Cengage Learning Solutions, visit **www.cengage.com.**

Purchase any of our products at your local college store or at our preferred
online store **www.cengagebrain.com.**

Printed in Mexico
Print Number: 03   Print Year: 2019

BRIEF CONTENTS

BRIEF CONTENTS

CONTENTS

CHAPTER 3 Culture   56

CHAPTER 4 Socialization   84

PART 2: Social Groups and Social Control

PART 3: Social Inequality

CHAPTER 8 Class and Stratification in the United States 208

CHAPTER 9 Global Stratification 240

CHAPTER 10 Race and Ethnicity 264

CHAPTER 11 Sex, Gender, and Sexuality 300

FEATURES

CHAPTER 12 Aging and Inequality Based on Age 336

FEATURES

PART 4: Social Institutions

CHAPTER 15 Families and Intimate Relationships   426

CHAPTER 16 Education   456

FEATURES

CHAPTER 17 Religion 484

FEATURES

CHAPTER 18 Health, Health Care, and Disability 514

FEATURES

PART 5: Social Dynamics and Social Change

FEATURES

Sociology & Everyday Life

Sociology in Global Perspective

Sociology & Social Policy

You Can Make a Difference

PREFACE

Hello and welcome to the eleventh edition of *Sociology in Our Times*! I think you will find that this best-selling text lives up to the timeliness in its name. Each edition is thoroughly revised and updated to reflect the newest sociological insights and statistical data. It also looks at contemporary social life and pressing societal problems through a sociological lens. Topics include injustice and inequality, family-related issues, educational and health problems, crime and gun violence, racism and hate crimes, terrorism and war, and environmental degradation.

What would you like to know more about in the social world where you live? As you know, we face unprecedented challenges and opportunities in the twenty-first century. By gaining new sociological insights on pressing social issues, you can enhance your perspective on the world and envision new ways in which you might make a difference for yourself and for future generations. The issues of social injustice and inequality are explored in depth in all chapters because these factors may hamper social change or contribute to it, particularly through collective behavior and social movements in which people demand that change occur. By studying sociology and reading *Sociology in Our Times,* you will gain a better understanding of why people seek stability within social institutions—including family, religion, education, politics, and government—even if they believe that these institutions might benefit from certain changes.

Like previous editions, the eleventh edition of *Sociology in Our Times* highlights the relevance of sociology to enable you to connect with the subject and the full spectrum of topics and issues that it encompasses. It achieves this connection by providing a meaningful, concrete context within which to learn. Specifically, it presents the comments and stories—the *lived experiences*—of real individuals in Sociology and Everyday Life features that describe the social issues they face, within the context of discussing classical and contemporary social theory and examining interesting and relevant research. The first-person commentaries and real-life examples that begin each chapter show how sociology can help you and other students understand the important questions we face and how these concerns are embedded within the larger culture and social world in which we live. These opening experiences also provide a framework that helps organize and highlight the chapter material and makes it easier for you to understand the new concepts, theories, and research that are introduced in the chapter.

Why is this text different? *Sociology in Our Times* includes the best of classical and contemporary sociologists, and it weaves an inclusive treatment of all people—across lines of race/ethnicity, class, gender, sexual identity and orientation, age, ability/disability, and other important attributes—into the examination of sociology. It does not oversimplify or water down the sociology! While helping you appreciate how sociology provides a better understanding of the world, this text gives you insights on your personal role as a *member of your various groups, organizations, and communities.* As a result, you will see that sociology is more than a collection of concepts and theories: Sociology is an academic area of study that can make a significant difference in your life! I invite you to join me on this exciting journey and to let me know what you think as we move forward.

What's New to the Eleventh Edition?

The eleventh edition builds on the best of previous editions but places more emphasis on social change while offering new cutting-edge insights, helpful learning tools, and fresh opportunities to apply the content of each chapter to relevant sociological issues and major concerns of the twenty-first century. As it is my goal to make each edition better than the previous one, I have revised all chapters to reflect the latest in sociological theory and research, and have updated examples throughout. Additionally, all statistics, such as data relating to crime, demographics, health, and the economy, are the latest available at the time of this writing. In sum, I have done more than simply revise existing materials and features. The page layouts have been refreshed and are very easy to follow: Boxed inserts have been reduced to provide you with a straightforward, more concise text that is highly relevant to your interests and makes studying for exams easier.

One feature of the eleventh edition that you will notice immediately is that learning objectives are restated at relevant locations throughout each chapter to help you with your reading and note taking. All chapter learning objectives are listed at the beginning of each chapter, and then they are listed individually when that specific topic is introduced in the chapter, and finally the Chapter Review provides a brief synopsis of all the learning objectives and key points.

Changes by Chapter

CHAPTER 1, The Sociological Perspective

- Throughout the chapter, more discussion on applied sociology and on how social change affects sociological theory and students' use of what they are learning in the course
- Revised "Sociology & Everyday Life" feature about the sociological aspects of consumer debt and student credit card debt
- New discussion of the American Sociological Association's organizational section on consumers and consumption
- Expanded discussion of consumption and consumerism and why these subjects should be important to students studying sociology
- New research on how people in similar income brackets spend money differently by city and region
- Shorter and more concise section on "The Development of Sociological Thinking"
- Revised discussion of "The Development of Modern Sociology"
- Information from "Sociology Works!" feature moved into discussion of Karl Marx
- Information from "Census Profiles" on consumer spending moved into body of text
- Expanded final section on "Looking Ahead: Are Theory and/or Practice in Your Future?" regarding sociological practice and public sociology

CHAPTER 2, Sociological Research Methods

- New opening "Sociology & Everyday Life" lived experience on bullying, social media abuse, and suicide among young people
- Updated "How Much Do You Know About Suicide?"
- "Sociology Works!" moved into "Sociology in Global Perspective" to relate Durkheim's work to contemporary suicide among young people in India
- Updated "Understanding Statistical Data Presentations" to reflect latest data on homicide, suicide, and firearm-related deaths of youths ages 15–19 by gender
- Revised "Sociology & Social Policy" to provide latest data on suicides among U.S. military personnel and efforts being made to reduce rates
- "Census Profiles" deleted to streamline chapter
- Updated "Statistics: What We Know (and Don't Know)"
- Deleted "Framing Suicide in the Media"
- Deleted dated study in "The Humphreys Research" section on "Ethical Issues in Sociological Research" because it potentially casts aspersion on groups based on sexual identity and orientation
- Revised and expanded "Looking Ahead: Research, Social Change, and Your Future" to discuss how computational social science has transformed research and how methods such as social geospatial modeling (GIS) and social network analysis have brought about rapid change in sociological studies

CHAPTER 3, Culture

- New opening "Sociology & Everyday Life" lived experience on how culture may be spread through food trucks
- Updated map showing "States with Official English Laws"
- Deleted "Census Profiles"
- Updated "Sociology in Global Perspective" to show how old the cultural norms about drinking behavior are in the Republic of Georgia
- Deleted "Framing Culture in the Media"
- Revised "Looking Ahead: Culture, Social Change, and Your Future" to emphasize ways in which culture is affected by technology and why the study of culture helps students understand their own social world while becoming more aware of how other people live

CHAPTER 4, Socialization

- New opening "Sociology & Everyday Life" lived experience on "Class Attendance in Higher Education" about digital-age methods of reducing skipping classes
- Updated figure on "Types of Maltreatment Among Children Under Age 18"
- Deleted "Sociology Works!" and "Media Framing" to make material more concise for students
- Revised and updated "Sociology in Global Perspective: Open Doors: Study Abroad and Global Socialization"
- New discussion of the effects of social isolation and loneliness, particularly among older individuals
- New final section on "Looking Ahead: Socialization, Social Change, and Your Future" discussing digital natives and digital immigrants

CHAPTER 5, Social Structure and Interaction in Everyday Life

- Revised figure on "Causes of Family Homelessness in 25 Cities"
- Deleted "Framing Homelessness in the Media" and incorporated some of the information into the main text
- Revised figure: "Who Are the Homeless?"
- Deleted photo essay
- Deleted "Census Profiles"
- Revised and updated "Sociology & Social Policy: What's Going on in 'Paradise'?—Homeless Rights Versus Public Space"
- Deleted "Sociology Works!"
- Updated "You Can Make a Difference" feature

CHAPTER 6, Groups and Organizations

- Deleted "Framing Community in the Media" and moved some information into the text
- Deleted "Sociology Works!" and moved some of the content into "Sumner's Ingroups and Outgroups"
- Revised and updated "Sociology & Social Policy: Technology and Social Change in the Workplace: BYOD?"
- Revised "Looking Ahead: Social Change and Organizations in the Future"

CHAPTER 7, Deviance and Crime

- New opening "Sociology & Everyday Life" lived experience, "When the Unspeakable Happens," about the final report on the Sandy Hook school killings
- Deleted "Sociology Works!" and moved some of the content into discussion about deviance
- Updated discussion and examples throughout sections on crime theories
- Updated crime statistics throughout chapter
- Revised and updated all figures pertaining to crime statistics
- Deleted "Framing Violent Crime in the Media" and moved some content into discussion about violent crime
- Updated discussion about terrorism and crime to include violence in France
- Updated statistics on the criminal justice system
- New "Sociology in Global Perspective: A Wider Perspective on Gangs: Look and Listen Around the World"
- Revised and expanded section on Internet crime

CHAPTER 8, Class and Stratification in the United States

- Updated statistics on income, poverty, health insurance, and other issues pertaining to inequality throughout chapter
- Updated models and figures for the U.S. class structure
- Revised figure: "Distribution of Pretax Income in the United States, 2013"
- Revised figure: "Median Household Income in the United States"
- Revised figure: "Median Household Income by Race/Ethnicity in the United States"
- Revised figure: "Racial Divide in Net Worth, 2013"
- Revised figure: "Rate of Uninsurance by Household Income, 2013"
- Deleted photo essay
- Revised and updated map: "Percentage of People in Poverty in the Past 12 Months by State, 2013"
- Revised and updated figure: "U.S. Poverty Rate by Age, 1959–2013"
- Deleted "Sociology Works!"
- Updated "You Can Make a Difference: Students Helping Others Through Campus Kitchen"

CHAPTER 9, Global Stratification

- Revised and updated information in "Sociology & Everyday Life: How Much Do You Know About Global Wealth and Poverty?"
- Revised "Classifications of Economies by Income" and map of "High-, Middle-, and Low-Income Economies in Global Perspective"
- Deleted "Framing Child Labor Issues in the Media"
- Revised and updated "Global Poverty and Human Development Issues," including life expectancy, per capita gross national income, and accompanying figures
- Updated information in "Education and Literacy" section
- Updated discussion of multidimensional poverty index
- Updated information on maquiladora plants

- Deleted "Sociology Works!" and incorporated some information into the main body of the text
- Revised and updated "Looking Ahead: Global Inequality in the Future"

CHAPTER 10, Race and Ethnicity

- New opening "Sociology & Everyday Life" lived experience about Selma, Alabama, fifty years after the Civil Rights March
- Revised and updated information in "Sociology & Everyday Life: How Much Do You Know About Race, Ethnicity, and Sports?"
- Deleted "Census Profiles" and included information within the chapter
- Replaced "Sociology in Global Perspective" box with new "Sociology & Social Policy" box on "Racist Hate Speech on Campus Versus First Amendment Right to Freedom of Speech"
- Deleted "Sociology Works!" and "Framing Sports in the Media" in order to expand discussion of current racial/ethnic strife in the United States
- Updated all information on "Racial and Ethnic Groups in the United States"
- Revised final section: "Looking Ahead: The Future of Global Racial and Ethnic Inequality"

CHAPTER 11, Sex, Gender, and Sexuality

- New opening "Sociology & Everyday Life" lived experience on "When Gender, Sexual Orientation, and Weight Bias Collide"
- New "Sociology & Everyday Life" quiz on "How Much Do You Know About Gender, Sexual Orientation, and Weight Bias?"
- Revised and updated discussion about LGBTQ issues and made extensive changes to "Intersex and Transgender Persons" section
- Deleted "Sociology Works!" and "Census Profiles" to provide more space for expanded discussion of sexual identity, sexual orientation, and the U.S. Supreme Court decision on same-sex marriage
- Updated section on "Gender and Socialization"
- Deleted "Framing Gender in the Media" and expanded discussion of "Mass Media and Gender Socialization" to include more on social media
- Revised and updated "Contemporary Gender Inequality" section, especially "Gendered Division of Paid Work in the United States"
- Updated table: "Percentage of the Workforce Represented by Women, African Americans, Hispanics, and Asian Americans in Selected Occupations"
- Updated figure: "The Wage Gap, 2013"
- Updated figure: "Women's Wages as a Percentage of Men's in Each Racial–Ethnic Category"
- Updated map: "Women's Earnings as a Percentage of Men's Earnings by State and Puerto Rico, 2013"
- New "Sociology in Global Perspective" box: "Women's Body Size and the Globalization of 'Fat Stigma'"

- Updated "You Can Make a Difference" box on the "Love Your Body" campaign
- Deleted photo essay

CHAPTER 12, Aging and Inequality Based on Age

- Updated "Sociology & Everyday Life" on "Facing Obstacles to Living a Long, Full Life"
- Revised statistics on aging throughout chapter
- Updated discussion of "Age and the Life Course in Contemporary Society"
- Revised map: "Median Age by State, 2013"
- Deleted "Sociology Works!"
- Updated "Sociology & Social Policy" box on elderly drivers
- Deleted "Framing Aging in the Media"
- Updated figure: "Percentage of Persons Age 65+ Below the Poverty Level"
- Updated figure: "Living Arrangements of the Population Ages 65 and Over, by Sex and Race and Hispanic Origin, 2012"
- Updated discussion of nursing homes

CHAPTER 13, The Economy and Work in Global Perspective

- Updated opening "Sociology & Everyday Life" lived experience on defining the twenty-first-century workplace and updated "How Much Do You Know About Work in the United States in the 2010s?"
- Deleted "Census Profiles"
- Updated figure: "Top Ten Fastest-Growing Occupations, 2012–2022"
- Updated table: "Revenues of the World's 20 Largest Public and Private Corporations (2014)"
- Updated table: "The Music Industry's Big Three"
- Updated figure: "The General Motors Board of Directors"
- Updated discussion of socialism
- New "Sociology in Global Perspective" box on the "Lopsided Job Market in China: A Mismatch Between Workers and Jobs"
- Deleted "Sociology Works!"
- Updated figure: "Selected Highest-Paying Occupations, 2014"
- Updated figure: "SAT Scores by Parents' Income and Education, 2014"
- Updated statistics in "Unemployment" section
- Deleted "Framing Luxury Consumption in the Media"
- Updated map: "U.S. Unemployment Rates by State, 2015"
- Revised and updated discussion of "Worker Resistance and Activism," particularly labor union statistics
- Updated figure: "Major Work Stoppages in the United States, 1960–2012"
- Updated "You Can Make a Difference"
- Revised "Looking Ahead: The Global Economy and Work in the Future"
- Updated "Sociology & Social Policy" box on "How Globalization Changes the Nature of Social Policy"

CHAPTER 14, Politics and Government in Global Perspective

- Updated "Sociology & Everyday Life: How Much Do You Know About Politics and the Media?"
- Revised discussion of the 2012 presidential election and midterm elections throughout the chapter
- Revised and updated discussion of super PACs
- Deleted "Sociology Works!" box
- Revised "Voter Participation and Voter Apathy"
- Revised discussion of federal bureaucracy
- Revised figure: "Categories and Percentages of U.S. Federal Spending in Fiscal Year 2015"
- Updated discussion of terrorism and war
- Deleted "Framing Politics in the Media"
- Revised and updated "You Can Make a Difference: Keeping an Eye on the Media"

CHAPTER 15, Families and Intimate Relationships

- Revised chapter opening "Sociology & Everyday Life" lived experience on "Diverse Family Landscapes in the Twenty-First Century" and updated "How Much Do You Know About Contemporary Trends in U.S. Family Life?"
- Revised statistics on families throughout chapter
- Deleted "Census Profiles" and added new research to section on "Love and Intimacy"
- Updated data on cohabitation and domestic partnerships
- Added figure: "Estimated Number of Opposite-Sex Couples Cohabiting in the United States in Selected Years, 1996–2014"
- Deleted map: "Percentage of All Households Reporting as Same-Sex Couple Households in 2010 Census"
- Updated statistics on "Marriage," "Same-Sex Marriages," and "Housework and Child-Care Responsibilities"
- Updated statistics on "Deciding to Have Children"
- Deleted "Sociology Works!"
- Updated statistics on "Adoption," "Teenage Childbearing," and "Single-Parent Households"
- Deleted "Framing Teen Pregnancy in the Media"
- Updated figure: "Living Arrangements of Children Under 18 Years Old for Selected Years, 1970–2014"
- Updated figure: "Marital Status of U.S. Population Ages 18 and Over by Race/Ethnicity"
- Updated figure: "U.S. Divorce Rate by State, 1990–2012"
- Updated statistics on divorce and remarriage
- Revised "Looking Ahead: Family Issues in the Future"

CHAPTER 16, Education

- Reworked data throughout chapter to provide latest information on education
- Updated "Sociology & Everyday Life: How Much Do You Know About U.S. Education?"
- Deleted "Sociology Works!"
- Deleted "Framing Education in the Media"
- Updated figure: "Percentage Distribution of Total Public Elementary–Secondary School System Revenue, 2014–2015"

- Revised figure: "Status Dropout Rates for 16- to 24-Year-Olds, by Race/Ethnicity, Gender, and Region"
- Updated discussion of "School Safety and Violence at All Levels" to include the issue of concealed carry of guns on college and university campuses
- Updated discussion of community colleges and tribal colleges
- Updated "The High Cost of a College Education"
- Updated "Slashed Budgets at State Colleges and Universities"
- Updated discussion of racial and ethnic diversity in student and faculty populations
- Deleted "Census Profiles" and moved figure on "Educational Attainment of Persons Ages 25 and Over" into the body of the text
- Revised and updated "Looking Ahead: Future Trends in Education"

CHAPTER 17, Religion

- Deleted "Sociology Works!" and incorporated ideas of Peter Berger into text
- Updated table: "Major World Religions"
- Updated figure: "World Religions by Percentage of Adherents"
- Updated "Sociology & Social Policy" on the issue of separation of church and state
- Deleted table on "Top 25 U.S. Denominations That Self-Identify as Christian" because newer, comparable data are not available
- Updated figure: "U.S. Religious Traditions' Membership"
- Deleted "Framing Religion in the Media"
- Updated "Looking Ahead: Religion in the Future"

CHAPTER 18, Health, Health Care, and Disability

- Updated discussion and data for illness and health care throughout the chapter.
- Deleted "Framing Health Issues in the Media" and incorporated this material in the text
- Updated statistics on health-related data such as life expectancy, infant mortality, and racial/ethnic and class differences.
- Updated "Sociology in Global Perspective" on "Medical Crises in the Aftermath of Disasters: From Oklahoma to Nepal" to include earthquake in Nepal
- Updated information on sexually transmitted diseases
- Updated figure: "Chlamydia: Rates by Age and Sex, United States, 2012"
- Updated map: "Adult Obesity in the United States: 2013"
- Updated discussion on implementation and legal issues associated with the Affordable Care Act
- Updated figure: "Increase in Cost of Health Care, 1993–2014"
- Revised discussion of "Private Health Insurance," "Public Health Insurance," "Managed Care," and "The Uninsured"
- Updated figure: "Uninsured Children by Poverty Status, Household Income, Age, Race, Hispanic Origin, and Nativity, 2013"

- Deleted "Sociology Works!"
- Updated discussion and data on disability

CHAPTER 19, Population and Urbanization

- Updated figure: "Growth in the World's Population, 2014"
- Revised demography discussion and included new data for fertility, mortality, and migration
- New "Sociology in Global Perspective" on "Problems People Like to Ignore: Global Diaspora and the Migrant Crisis"
- Updated table: "The Ten Leading Causes of Death in the United States, 1900 and 2014"
- Updated figure: "Population Pyramids for Mexico, Iran, the United States, and France, 2014"
- Deleted "Framing Immigration in the Media," "Photo Essay," and "Sociology Works!" to provide room for more information on population and urbanization
- Updated "Problems in Global Cities"
- Updated figure: "The World's Fifteen Largest Agglomerations"
- Updated figure: "Increase in the World's Population in Billions of People"
- Revised "Looking Ahead: Population and Urbanization in the Future."

CHAPTER 20, Collective Behavior, Social Movements, and Social Change

- New opening "Sociology & Everyday Life" lived experience on "Collective Behavior and Environmental Issues"
- Updated discussion of collective behavior, social movements, and social change
- Revised table: "Top 15 Policy Priorities of the U.S. Public, 2015"
- Deleted "Sociology Works!" to provide more space to revise and update discussion of environmental issues and social movements
- Added more-contemporary environmental activists and issues, including divestment in fossil-fuel industries by universities and other institutional investors

Overview of the Text's Contents

Sociology in Our Times, eleventh edition, contains twenty high-interest, up-to-date, clearly organized chapters to introduce students to the best of sociological thinking. The length of the text has been streamlined and carefully organized to make full coverage of the book possible in the time typically allocated to the introductory course.

Sociology in Our Times is divided into five parts. **Part 1** establishes the foundation for studying society and social life. **Part 2** examines social groups and social control. **Part 3** focuses on social inequality, looking at issues of class, race/ethnicity, sex/gender/sexuality, and age discrimination. **Part 4** offers a systematic discussion of social institutions, building students' awareness of the importance of these foundational elements of society and showing how a problem in one often has a significant influence on others. **Part 5** surveys social dynamics and social change.

Part 1

Chapter 1 introduces students to the sociological imagination and traces the development of sociological thinking. The chapter sets forth the major theoretical perspectives used by sociologists in analyzing compelling social issues, such as the problem of credit card abuse and hyperconsumerism among college students and others.

Chapter 2 focuses on sociological research methods and shows students how sociologists conduct research. This chapter provides a thorough description of both quantitative and qualitative methods of sociological research. Throughout the chapter, new updates concentrate on various factors that influence suicide rates.

In **Chapter 3**, culture is spotlighted as either a stabilizing force or a force that can generate discord, conflict, and even violence in societies. Cultural diversity is discussed as a contemporary issue, and unique coverage is given to popular culture and leisure and to divergent perspectives on popular culture.

Chapter 4 looks at the positive and negative aspects of socialization, including opening lived experiences about learning socialization cues in college and medical school. This chapter presents an innovative analysis of gender and racial–ethnic socialization and of issues associated with recent immigration.

Part 2

Chapter 5 applies the sociological imagination to an examination of social structure and social interaction, using homelessness as a sustained example of the dynamic interplay of structure and interaction in society. Unique to this chapter are discussions of the sociology of emotions and of personal space as viewed through the lenses of race, class, gender, and age.

Chapter 6 analyzes groups and organizations, including innovative forms of social organization and ways in which organizational structures may differentially affect people based on race, class, gender, and age. The opening narrative discusses an MIT professor's experience with students using digital technology in the classroom.

Chapter 7 examines how deviance and crime emerge in societies, using diverse theoretical approaches to describe the nature of deviance, crime, and the criminal justice system. Key issues are dramatized for students through an analysis of recent mass shootings and the consequences of violence on individuals and society.

Part 3

Chapter 8 focuses on class and stratification in the United States, analyzing the causes and consequences of inequality and poverty, including a discussion of the ideology and accessibility of the American Dream.

Chapter 9 addresses the issue of global stratification and examines differences in wealth and poverty in rich and poor nations around the world. Explanations for these differences are discussed.

The focus of **Chapter 10** is race and ethnicity, which includes an illustration of the historical relationship (or lack of it) between sports and upward mobility by persons from diverse racial–ethnic groups. A thorough analysis of prejudice, discrimination, theoretical perspectives, and the experiences of diverse racial and ethnic groups is presented, along with global racial and ethnic issues.

Chapter 11 examines sex, gender, and sexuality, with special emphasis on gender stratification in historical perspective. Linkages between gender socialization and contemporary gender inequality are described and illustrated by lived experiences and perspectives on body image.

Chapter 12 provides a cutting-edge analysis of aging, including theoretical perspectives and inequalities experienced by people across the life course. This chapter has thorough discussions of adolescence, young adulthood, middle adulthood, and late adulthood.

Part 4

The economy and work are explored in **Chapter 13**, including the different types of global economic systems, the social organization of work in the United States, unemployment, and unions. The chapter has been extensively revised to include issues pertaining to the aftermath of the "Great Recession," including job loss, higher rates of unemployment, and the gradual economic recovery during the second decade of the twenty-first century.

Chapter 14 discusses the intertwining nature of politics, government, and the media. Political systems are examined in global perspective, and politics and government in the United States are analyzed with attention to governmental bureaucracy and the military–industrial complex.

Chapter 15 focuses on families in global perspective and on the diversity found in U.S. and global families today. The latest figures on family-related issues such as family violence, foster care, and teenage pregnancy are included.

Chapter 16 investigates education in the United States and other nations. In the process the chapter highlights issues of race, class, and gender inequalities in current U.S. education.

In **Chapter 17**, religion is examined from a global perspective, including a survey of world religions and an analysis of how religious beliefs affect other aspects of social life. Current trends in U.S. religion are also explored, including various sociological explanations of why people look to religion to find purpose and meaning in life.

Chapter 18 analyzes health, health care, and disability from both U.S. and global perspectives. Among the topics included are social epidemiology, lifestyle factors influencing health and illness, health care organization in the United States and other nations, social implications of advanced medical technology, and holistic and alternative medicine. This chapter is unique in that it contains a thorough discussion of the sociological perspectives on disability and of social inequalities based on disability. The Affordable Care Act is explored in detail.

Part 5

Chapter 19 examines population and urbanization, looking at demography, global population change, and the process and consequences of urbanization. Special attention is given to race- and class-based segregation in urban areas and the crisis in health care in central cities.

Chapter 20 concludes the text with an innovative analysis of collective behavior, social movements, and social change. The need for persistence in social movements, such as the continuing work of environmental activists over the past fifty years, is used as an example to help students grasp the importance of collective behavior and social movements in producing social change.

Distinctive, Classroom-Tested Features

The following special features are specifically designed to demonstrate the relevance of sociology in our lives, as well as to support students' learning. As the preceding overview of the book's contents shows, these features appear throughout the text, some in every chapter, others in selected chapters.

Unparalleled Coverage of and Attention to Diversity

From its first edition, I have striven to integrate diversity in numerous ways throughout this book. The individuals portrayed and discussed in each chapter accurately mirror the diversity in society itself. As a result, this text speaks to a wide variety of students and captures their interest by taking into account their concerns and perspectives. Moreover, the research used includes the best work of classical and established contemporary sociologists—including many women and people of color—and it weaves an inclusive treatment of *all* people into the examination of sociology in *all* chapters. Therefore, this text helps students consider the significance of the intersectionality of class, race, gender, and age in all aspects of social life.

Personal Narratives That Highlight Key Issues

Authentic first-person commentaries serve as the vignettes that open each chapter and personalize the issue that unifies the chapter's coverage. These lived experiences provide opportunities for students to examine social life beyond their own experiences and for instructors to systematically incorporate into lectures and discussions an array of interesting and relevant topics that help demonstrate to students the value of applying sociology to their everyday lives. New topics include "Class Attendance in Higher Education," "When Gender, Sexual Orientation, and Weight Bias Collide," and "Collective Behavior and Environmental Issues."

Focus on the Relationship Between Sociology and Everyday Life

Each chapter has a brief quiz in the opening "Sociology & Everyday Life" feature that relates the sociological perspective to the pressing social issues presented in the opening vignette. (Answers are provided at the end of the chapter.)

Emphasis on the Importance of a Global Perspective

The global implications of all topics are examined throughout each chapter and in the "Sociology in Global Perspective" boxes, which highlight our interconnected world and reveal how the sociological imagination extends beyond national borders.

Emphasis on Social and Global Change

The eleventh edition also strives to relate the importance of social and global change in its many forms and how this affects not only our everyday lives but also our communities and the entire nation and world.

Applying the Sociological Imagination to Social Policy

The "Sociology & Social Policy" boxes in selected chapters help students understand the connection between law and social policy issues in society.

Focus on Making a Difference

Designed to help students learn how to become involved in their communities, the "You Can Make a Difference" boxes look at ways in which students can address, on a personal level, issues raised by the chapter themes.

Effective Study Aids

In addition to basic reading and study aids such as learning objectives, chapter outlines, key terms, a running glossary, and our popular online study system, *Sociology in Our Times* includes the following pedagogical aids to enhance students' mastery of course content:

- **Concept Quick Review.** These tables categorize and contrast the major theories or perspectives on the specific topics presented in a chapter.
- **Questions for Critical Thinking.** Each chapter concludes with "Questions for Critical Thinking" to encourage students to reflect on important issues, to develop their own critical-thinking skills, and to highlight how ideas presented in one chapter often build on those developed previously.
- **Feature-Concluding Reflect & Analyze Questions.** From activating prior knowledge related to concepts and themes, to highlighting main ideas and reinforcing

diverse perspectives, this text's questions consistently contribute to student engagement.

- **End-of-Chapter Summaries.** Connected to the learning objectives, chapter summaries provide a built-in review for students by reexamining material covered in the chapter in an easy-to-read question-and-answer format to review, highlight, and reinforce the most important concepts and issues discussed in each chapter. Each element in the chapter summaries is related to one of the learning objectives introduced at the beginning of the chapter.

Comprehensive Supplements Package

The eleventh edition of *Sociology in Our Times* is accompanied by a wide array of supplements developed to create the best teaching and learning experience inside as well as outside the classroom. All of the continuing supplements have been thoroughly revised and updated, and some new supplements have been added. Cengage Learning prepared the following descriptions, and I invite you to start taking full advantage of the teaching and learning tools available to you by reading this overview.

Products for Blended and Online Courses

MindTap™: The Personal Learning Experience

MindTap Sociology for Kendall's *Sociology in Our Times*, eleventh edition, from Cengage Learning represents a new approach to a highly personalized, online learning platform. A fully online learning solution, MindTap combines all of a student's learning tools—readings, multimedia, activities, and assessments—into a "Learning Path" that guides the student through the introduction to sociology course. Instructors personalize the experience by customizing the presentation of these learning tools to their students, even seamlessly introducing their own content into the Learning Path via apps that integrate into the MindTap platform. Learn more at **www.cengage.com/mindtap**.

MindTap for Kendall's *Sociology in Our Times*, eleventh edition, is easy to use and saves instructors time by allowing them to

- Seamlessly deliver appropriate content and technology assets from a number of providers to students, as needed.
- Break course content down into movable objects to promote personalization, encourage interactivity, and ensure student engagement.
- Customize the course—from tools to text—and make adjustments "on the fly," making it possible to intertwine breaking news into their lessons and incorporate today's teachable moments.
- Bring interactivity into learning through the integration of multimedia assets (apps from Cengage Learning and other providers) and numerous in-context exercises and supplements; student engagement will increase, leading to better student outcomes.

- Track students' use, activities, and comprehension in real time, which provides opportunities for early intervention to influence progress and outcomes. Grades are visible and archived so students and instructors always have access to current standings in the class.
- Assess knowledge throughout each section: after readings and in activities, homework, and quizzes.
- Automatically grade all homework and quizzes.

MindTap Sociology for Kendall's *Sociology in Our Times,* eleventh edition, features Aplia assignments, which help students learn to use their sociological imagination through compelling content and thought-provoking questions. Students complete interactive activities that encourage them to think critically in order to practice and apply course concepts. These valuable critical-thinking skills help students become thoughtful and engaged members of society. Aplia for Kendall's *Sociology in Our Times,* eleventh edition, is also available as a stand-alone product. Log in to **CengageBrain.com** for access.

CourseReader for Sociology CourseReader for Sociology allows you to create a fully customized online reader in minutes. You can access a rich collection of thousands of primary and secondary sources, readings, and audio and video selections from multiple disciplines. Each selection includes a descriptive introduction that puts it into context, and every selection is further supported by both critical-thinking and multiple-choice questions designed to reinforce key points. This easy-to-use solution allows you to select exactly the content you need for your courses, and it is loaded with convenient pedagogical features, such as highlighting, printing, note taking, and downloadable MP3 audio files for each reading. You have the freedom to assign and customize individualized content at an affordable price. CourseReader is the perfect complement to any class.

Resources for Customizing Your Textbook

Cengage Learning is pleased to offer three modules that help you tailor *Sociology in Our Times,* eleventh edition, to your course. The modules present topics not typically covered in most introductory texts but often requested by instructors. In addition, you can choose to add your own materials or reorganize the table of contents. Work with your local Cengage Learning consultant to find out more.

Careers in Sociology Module Written by leading author Joan Ferrante, Northern Kentucky University, the Careers in Sociology module offers the most extensive and useful information on careers that is available. This module provides six career tracks, each of which has a "featured employer," a job description, and a letter of recommendation (written by a professor for a sociology student) or application (written by a sociology student). The module also includes résumé-building tips on how to make the most out of being a sociology major and offers specific course suggestions along with the transferable skills gained by taking these courses. As part of Cengage Learning's Add-a-Module Program, Careers in Sociology can be purchased separately, bundled, or customized with any of our introductory texts.

Sociology of Sports Module   The Sociology of Sports module, authored by Jerry M. Lewis, Kent State University, examines why sociologists are interested in sports, mass media and sports, popular culture and sports (including feature-length films on sports), sports and religion, drugs and sports, and violence and sports. As part of Cengage Learning's Add-a-Module Program, Sociology of Sports can be purchased separately, bundled, or customized with any of our introductory texts.

Rural Sociology Module   The Rural Sociology module, authored by Carol A. Jenkins, Glendale Community College–Arizona, presents the realities of life in rural America. Many people imagine a rural America characterized by farming, similar cultures, and close-knit communities. However, rural Americans and rural communities are extremely diverse—demographically, culturally, socially, economically, and environmentally. The module presents these characteristics of rural life in a comprehensive and accessible format for introductory sociology students. As part of Cengage Learning's Add-a-Module program, Rural Sociology can be purchased separately, bundled, or customized with any of our introductory sociology texts.

Teaching Aids for Instructors

A broad array of teaching aids is available to make course planning faster and easier, giving you more time to focus on your students. All of these resources can be accessed with a single account. Go to **login.cengage.com** to log in.

Online Instructor's Resource Center

Online Instructor's Resource Manual   Prepared by Josh Packard of the University of Northern Colorado, this text is designed to maximize the effectiveness of your course preparation. It offers you Quick Start questions to launch your lecture, brief chapter outlines, key terms, and student learning objectives, plus extensive chapter lecture outlines mapped to the learning objectives, lecture ideas, questions for discussion, Internet activities, student activities, and creative lecture and teaching suggestions, including video suggestions.

Online Test Bank   Prepared by Josh Packard of the University of Northern Colorado, the eleventh edition's test bank consists of revised and updated true/false and multiple-choice questions for every chapter of the text, along with an answer key and page references for each question. Each multiple-choice item has the question type (fact, concept, or application) indicated. Also included are short-answer and essay questions for every chapter.

Online PowerPoint® Slides   Helping you make your lectures more engaging while effectively reaching your visually oriented students, these handy Microsoft® PowerPoint slides outline the chapters of the main text in a classroom-ready presentation that include tables, selected figures, a Quick Quiz, and now more photos as "Consider This" slides. The PowerPoint slides are updated to reflect the content and organization of the new edition of the text.

The Sociology Video Library Vols. I–IV   These DVDs drive home the relevance of course topics through short, provocative clips of current and historical events. Perfect for enriching lectures and engaging students in discussion, many of the segments on this volume have been gathered from BBC Motion Gallery. Ask your Cengage Learning representative for a list of contents.

Cengage Learning Testing Powered by Cognero
This is a flexible, online system that allows you to

- import, edit, and manipulate test bank content from the *Sociology in Our Times* test bank or elsewhere, including your own favorite test questions
- create multiple test versions in an instant
- deliver tests from your LMS, your classroom, or wherever you want

Acknowledgments

Sociology in Our Times, eleventh edition, would not have been possible without the insightful critiques of these colleagues, who have reviewed some or all of this book. My profound thanks to each one for engaging in this time-consuming process:

Virginia Chase, University of Toledo
Anne Eisenberg, SUNY–Geneseo
Dina Giovanelli, Century College
Aaron Major, University at Albany–SUNY
Amanda Miller, University of Indianapolis
Carla Pfeffer, Purdue University North Central
Melissa Russiano, Mercyhurst University
Megan Swindal, University of Alabama

I deeply appreciate the energy, creativity, and dedication of the many people responsible for the development and production of this edition of *Sociology in Our Times*. I wish to thank Cengage Learning's Elizabeth Beiting-Lipps, Lin Gaylord, John Chell, and Kara Kindstrom for their enthusiasm and insights throughout the development of this text. Many other people worked hard on the production of the eleventh edition of *Sociology in Our Times*, especially Cheri Palmer, Greg Hubit, Donald Pharr, and Debra Nichols. I am extremely grateful to them.

I invite you to send your comments and suggestions about this book to me in care of:

Sociology Team
Cengage Learning
500 Terry A. Francois Blvd., 2nd Floor
San Francisco, CA 94158

ABOUT THE AUTHOR

DIANA KENDALL is Professor of Sociology at Baylor University, where she was named an Outstanding University Professor. Dr. Kendall has taught a variety of courses, including Introduction to Sociology; Sociological Theory (undergraduate and graduate); Sociology of Medicine; Sociology of Law; and Race, Class, and Gender. Previously, she enjoyed many years of teaching sociology and serving as chair of the Social and Behavioral Science Division at Austin Community College.

Diana Kendall received her Ph.D. from the University of Texas at Austin, where she was invited to membership in Phi Kappa Phi Honor Society. Her areas of specialization and primary research interests are sociological theory and the sociology of medicine. She is the author of *Sociology in Our Times: The Essentials,* tenth edition; *The Power of Good Deeds: Privileged Women and the Social Reproduction of the Upper Class* (Rowman & Littlefield, 2002); *Members Only: Elite Clubs and the Process of Exclusion* (Rowman & Littlefield, 2008); and *Framing Class: Media Representations of Wealth and Poverty in America,* second edition (Rowman & Littlefield, 2012). Professor Kendall is a member of numerous sociological associations, including the American Sociological Association, the Society for the Study of Social Problems, and the Southwestern Sociological Association.

SOCIOLOGY
IN OUR TIMES

THE SOCIOLOGICAL PERSPECTIVE

© iStockphoto.com/Rory Vanucci

LEARNING OBJECTIVES

1 **Define** sociology and explain how it can contribute to our understanding of social life.

2 **Identify** what is meant by the sociological imagination.

3 **Describe** how we can develop a global sociological imagination.

4 **Describe** the historical context in which sociological thinking developed.

5 **Discuss** why early social thinkers were concerned with social order and stability.

6 **Identify** reasons why many later social thinkers were concerned with social change.

7 **Discuss** how industrialization and urbanization influenced the theories of Max Weber and Georg Simmel.

8 **Compare** and contrast contemporary functionalist and conflict perspectives on social life.

9 **Identify** key differences in contemporary symbolic interactionism and postmodernist perspectives on social life.

SOCIOLOGY & EVERYDAY LIFE

College Life and the Consumer Society

What I enjoyed about college was that I was able to walk away with a degree and go find a job, but what I regret most is getting a credit card, racking it up and getting multiple credit cards and doing the same thing, 'cause now I have to deal with it and I'm paying it off now and it's kind of hard to deal with. Things that I charged on my credit card in college were those spring break vacations, going out to eat with friends numerous times. Other things were like materialistic things like clothes, accessories, makeup—all that good stuff—trying to keep up with everyone else. [Slight laugh.] I wish I could do those things now. Now, I can't have those things; I have to do with what I've got. . . . I can't enjoy the things I enjoyed in college because I enjoyed them in college. I guess when I was making the purchases in college with my credit card saying, "Oh, I can just pay that off later," I figured I would be making more money than what I was given through financial aid and through my parents, [but] in reality, you're not. You have to

compensate for other things like tax being taken out of your salary, groceries, gas is something I didn't even think about because my parents always paid it. I mean, all those little things: They will add up!

—Robyn Beck (2014) describing her experience of struggling to pay off $7,000 in credit card debt that she ran up during her college years

Young people who run up credit card debt may find that paying off the debt can take decades.

Like millions of college students in the United States and other high-income nations, Robyn Beck quickly learned both the liberating and constraining aspects of living in a "consumer society" where many of us rely on our credit cards to pay for the goods or services that we want or need. For many years companies targeted college students, trying to get you to apply for a credit card regardless of whether you had the financial ability to pay off your balance. Since 2009, the Credit Card Accountability, Responsibility, and Disclosure (CARD) Act has banned credit card approval for anyone under age twenty-one unless the person can prove that he or she has sufficient income to pay their bill or unless someone over age twenty-one agrees to cosign the application. However, this has not ended the problems related to students' credit card debt. Are you aware, for example, that most college students who own a credit card have an unpaid balance of at least $500 on that card? The companies that issue credit cards are concerned because students have little, if any, credit card history and have the potential to be less "creditworthy." As a result, these companies frequently charge the highest annual interest rates for student-issued cards. These rates were at about 21.4 percent in 2014. Think about this: If you charge $1,000 on a credit card with a 21.4 percent interest rate and pay the minimum amount due each month, eventually you will pay $1,941 for the $1,000 you initially borrowed. If you add no additional

charges beyond the initial $1,000 and pay only the minimum amount due each month, you will need 7.6 years to pay off the debt (Lowry, 2014). Not only on college campuses but also across nations and the entire world, consumerism is an important aspect of social life in the twenty-first century.

As sociologists in training, why should you be interested in studying consumerism? According to the American Sociological Association's "Section on Consumers and Consumption" (2014), sociologists who specialize in this area do research to show "the pervasiveness of consumer goods and consumerism in shaping our everyday lives, the social structure, and the contemporary social, political, economic, and environmental problems that we face as a global society." From this explanation, you can see that what we buy is much more than just acquiring goods or services; it also strongly influences how we live our everyday lives and how social change occurs locally and globally. Consumer goods and consumerism are deeply intertwined with the social structure of our nation and the larger global society, including significant global issues such as environmental problems, widening gaps in wealth and other resources, and greater inequalities based on gender, race/ethnicity, and class hierarchies (ASA Section on Consumers and Consumption, 2014).

When you study the *consumer society*—a society in which discretionary consumption is a mass phenomenon among people across diverse income categories—you will

How Much Do You Know About Consumption and Credit Cards?

TRUE	FALSE		
T	F	1	The average U.S. household owes more than $15,000 in credit card debt.
T	F	2	It has become easier in the 2010s for undergraduate college students to get a credit card in their own name.
T	F	3	Among college students who have their own credit card, the average unpaid balance is slightly less than $500.
T	F	4	Millennials (people ages 18–34) spend more money online in a given year than people in any other age category.
T	F	5	Credit cards are not a financial safety net for low-income families because people with low and middle incomes have greater difficulty acquiring credit.
T	F	6	Fewer individuals filed for bankruptcy in the 2010s because more people are better off financially today than in the past.
T	F	7	The total U.S. outstanding revolving debt on credit cards alone is over $800 billion.
T	F	8	Overspending is primarily a problem for people in higher-income brackets in the United States and other affluent nations.

Answers can be found at the end of the chapter.

gain important insights into all aspects of social life and social change. However, we must distinguish between *consumption* and *consumerism* for a clearer sociological understanding. When we study *consumption*, we are looking at *behavior* that people around the world routinely engage in—the selection, purchase, and use of products and services that are available in a society. By contrast, *consumerism* refers to social and economic *beliefs* and *structures* that encourage people to acquire goods and services in an ever-increasing manner. Consumerism refers to the characteristics of a larger society and an ideology that frames people's worldviews, values, relationships, identities, and behaviors.

From a sociological perspective, we want to know more about the influences, experiences, and social relationships that produce specific patterns of consumption. When we examine consumption from a larger structural perspective, we see that it is linked to nationality, region, race/ethnicity, culture, gender, sexuality, age, ability/disability, and other social attributes and characteristics. Learning to study how people think, live, work and play is at the core of all sociological inquiry. This investigation is closely linked to social change and a better understanding of how social differences and inequalities originate, persist, and sometimes change over time and place. For example, recent studies have found that higher-income people in the United States spend money on different things. Even among those with similar wealth,

consumer patterns vary by region and by city. In Manhattan affluent people spend more money on luxury watches and shoes than in cities such as Boston, where people with similar economic resources are more likely to spend their money on private-school tuition and elite college education (Currid-Halkett, 2014). To explain this, some social analysts have said, "Geography is consumer destiny," meaning that where individuals live affects how they will spend money.

How can we analyze spending patterns by city and region to learn what people value? What can we learn about how people attempt to secure or enhance their social position through what they buy? (Currid-Halkett, 2014). Think about your own city: How do the spending patterns compare for high-, middle-, and low-income persons in relation to clothing, entertainment, and transportation? Do spending patterns reflect economic and social conditions in your community?

Although excessive consumerism may contribute to the individual's personal financial problems, larger economic conditions are linked to national and global instability. In this chapter you will see how the sociological perspective helps us examine complex questions such as this, as well as learning how sociologists look at difficulties associated with studying human behavior generally. Before reading on, take the "Sociology and Everyday Life" quiz, which examines a number of commonsense notions about consumption and credit card debt.

Putting Social Life into Perspective

Sociology is the systematic study of human society and social interaction. It is a *systematic* study because sociologists apply both theoretical perspectives and research methods (or orderly approaches) to examinations of social behavior. Sociologists study human societies and their social interactions to develop theories of how human behavior is shaped by group life and how, in turn, group life is affected by individuals.

To better understand the scope of sociology, you can compare it to other social sciences, such as anthropology, psychology, economics, and political science. Like anthropology, sociology studies many aspects of human behavior; however, sociology is particularly interested in contemporary social organization, relations, and social change. Anthropology primarily concentrates on the study of humankind in all times and spaces. It focuses on both traditional and contemporary societies and the development of diverse cultures. Closest to sociology is cultural anthropology—the comparative study of cultural similarities and differences that looks for patterns in human behavior, beliefs, and practices that are typical in groups of people. By contrast, psychology primarily focuses on *internal* factors relating to the individual in its explanations of human behavior and mental processes, such as how the human mind thinks, remembers, and learns. Social psychology is most akin to sociology because social psychologists examine how we perceive ourselves in relation to the rest of the world and how this affects our choices, behavior, and beliefs. Sociology specifically focuses on *external* social factors, such as the effects of groups, organizations, and social institutions on individuals and social life. Although sociology examines all major social institutions, including the economy and politics, the fields of economics and political science concentrate primarily on a single institution—the economy or the political

system. Topics of mutual interest to economics and sociology include issues such as consumerism and debt, which can be analyzed at global, national, and individual levels. As you can see from these examples, sociology shares similarities with other social sciences but offers a distinct approach for gaining greater understanding of our social world.

Why Should You Study Sociology?

Sociology helps you gain a better understanding of yourself and the social world. It enables you to see how the groups to which you belong and the society in which you live largely shape behavior. A *society* is a large social grouping that shares the same geographical territory and is subject to the same political authority and dominant cultural expectations, such as the United States, Mexico, or Nigeria. Many changes are occurring in the twenty-first century. Many societies have not only dominant cultural groupings and expectations but also many smaller groupings that have their own unique cultural identities. Migration and interdependence have shifted the meaning of society in the twenty-first century.

Examining the world order helps us understand that we are all affected by *global interdependence*—a relationship in which the lives of all people everywhere are intertwined closely and any one nation's problems are part of a larger global problem. Environmental problems are an example. People throughout the world share the same biosphere. When environmental degradation, such as removing natural resources or polluting the air and water, takes place in one region, it may have an adverse effect on people around the globe.

You can make use of sociology on a more personal level. Sociology enables us to move beyond established ways of thinking, thus allowing us to gain new insights into ourselves and to develop a greater awareness of the connection between our own "world" and that of other people. Sociology provides new ways of approaching problems and making decisions in everyday life. For this reason, people with knowledge of sociology are employed in a variety of fields that apply sociological insights to everyday life (see ● Figure 1.1).

Health and Human Services	Business	Communication	Academia	Law
Medicine Nursing Physical Therapy Occupational Therapy Counseling Education Social Work	Advertising Labor Relations Management Marketing	Broadcasting Public Relations Journalism Social Media	Anthropology Economics Geography History Information Studies Media Studies/ Communication Political Science Psychology Sociology	Law Criminal Justice Mediation Conflict Resolution

FIGURE 1.1   Fields That Use Social Science Research

In many careers, including jobs in health and human services, business, communication, academia, and law, the ability to analyze social science research is an important asset.

Source: Based on Katzer, Cook, and Crouch, 1991.

Sociology promotes understanding and tolerance by enabling each of us to look beyond intuition, common sense, and our personal experiences. Many of us rely on intuition or common sense gained from personal experience to help us understand our daily lives and other people's behavior. *Commonsense knowledge* guides ordinary conduct in everyday life. However, many commonsense notions are actually myths. A *myth* is a popular but false notion that may be used, either intentionally or unintentionally, to perpetuate certain beliefs or "theories" even in the light of conclusive evidence to the contrary.

By contrast, sociologists strive to use scientific standards, not popular myths or hearsay, in studying society and social interaction. They use systematic research techniques and are accountable to the scientific community for their methods and the presentation of their findings. Whereas some sociologists argue that sociology must be completely value free—free from distorting subjective (personal or emotional) bias—others do not think that total objectivity is an attainable or desirable goal when studying human behavior. However, all sociologists attempt to discover patterns or commonalities in human behavior. When they study consumerism, such as regional spending habits or credit card abuse, for example, they look for recurring patterns of behavior in individuals and groups. Consequently, we seek the multiple causes and effects of social issues and analyze the impact of the problem not only from the standpoint of the people directly involved but also from the standpoint of the effects of such behavior on all people.

LO2 Identify what is meant by the sociological imagination.

The Sociological Imagination

Do you wonder how your daily life compares to what other people are doing? Our interest in Facebook, Instagram, Twitter, and other social media sites reflects how fascinated we are by what other people are thinking and doing. But how can you really link your personal life with what is going on with other people in the larger social world? You can make an important linkage known as the sociological imagination.

Sociological reasoning is often referred to as the *sociological imagination*—the ability to see the relationship between individual experiences and the larger society (Mills, 1959b). The sociological imagination is important to each of us because having this awareness enables us to understand the link between our personal experiences and the social contexts in which they occur. Each of us lives in a society, and we live out a biography within some historical setting. Throughout your life, you contribute to the shaping of society and to its history, even as you are made by society and the historical events that take place during your lifetime. The sociological imagination will enable you to grasp the relationship between history at the societal level

and your own biography at the individual level. It also helps you distinguish between personal troubles and social (or public) issues. *Personal troubles* are private problems that affect individuals and the networks of people with whom they associate regularly. As a result, individuals within their immediate social settings must solve those problems. For example, one person being unemployed or having a high level of credit card debt may be a personal trouble. *Public issues* are problems that affect large numbers of people and often require solutions at the societal level (Mills, 1959b). Widespread unemployment and extensive consumer debt are public issues. The sociological imagination helps us place seemingly personal troubles into a larger social context, where we can distinguish whether and how personal troubles may be related to public issues. Let's compare the two perspectives by looking at overspending.

Overspending as a Personal Trouble Have you heard someone say, "He has no one to blame but himself" for some problem? In everyday life we often blame people for "creating" their own problems. Although individual behavior can contribute to social problems, our individual experiences are often largely beyond our own control. They are determined by society as a whole—by its historical development and its organization. If a person sinks into debt because of overspending or credit card abuse, other people often consider the problem to be the result of the individual's personal failings. However, thinking about it this way overlooks debt among people in low-income brackets who have no other way than debt to acquire basic necessities of life such as food, clothing, and housing. By contrast, at middle- and upper-income levels, overspending takes on a variety of meanings typically dictated by what people think of as essential for their well-being and associated with the so-called "good life" that is so heavily marketed and flaunted by high-end consumers. But across income and wealth levels, larger-scale economic, political, and social problems may affect the person's ability to pay for consumer goods and services (● Figure 1.2).

Overspending as a Public Issue Let's apply the sociological imagination to the problem of overspending and credit card debt by looking at it first as a public issue—a societal problem. In 2014 consumer debt in the United States added up to more than $3.24 trillion for student loans, car loans, and revolving debt (amounts that are not paid off in full each month) on credit cards. Debt on credit card balances alone was more than $880.5 billion (Currid-Halkett, 2014).

sociology
the systematic study of human society and social interaction.

society
a large social grouping that shares the same geographical territory and is subject to the same political authority and dominant cultural expectations.

sociological imagination
C. Wright Mills's term for the ability to see the relationship between individual experiences and the larger society.

FIGURE 1.2 Because of an overreliance on credit, many Americans now owe more than they can pay back. This couple is signing up for debt consolidation, a somewhat controversial process that may help them avoid bankruptcy.

Sociologically speaking, why do you think this is considered to be a public issue? Consumerism is a way of life in the United States and most nations around the world. Through media and popular culture, people are continually encouraged to buy goods and services that they do not necessarily need. They are bombarded with more and more choices, and many items they own, such as smartphones and tablets, are declared outdated and obsolete not long after they purchase them. A shiny new device is available for them to purchase, and the rapid cycle of buy-and-replace continues into the future. Marketing and advertising encourages people to "buy now and pay later." The credit card industry stands ready to help people buy what they want and sometime more than they can pay off. Card users are encouraged by credit card companies to carry large balances on their cards, pay high interest rates month after month, and transfer balances to newly issued cards. Spending and overspending are encouraged by the values of society. Similarly, government policies and laws may favor credit card issuers over individual credit card holders when it comes to rights pertaining to debt collection and the handling of lawsuits.

The sociological imagination is useful for examining such issues because it integrates microlevel (individual and small-group) troubles with macrolevel (larger social institutions and social forces) issues on a global basis.

 LO3 Describe how we can develop a global sociological imagination.

The Importance of a Global Sociological Imagination

How is it possible to think globally when you live in one location and have been taught to think a certain way? Although we live in one country and rely heavily on Western sociological theory and research, we can access the world beyond the United States and learn to develop a more comprehensive *global* approach for the future. One way we can do this is to reach beyond studies that have focused primarily on the United States to look at the important challenges that we face in a rapidly changing world. These issues range from political and economic instability to environmental concerns, natural disasters, and terrorism. We can also examine the ways in which nations are not on equal footing when it comes to economics and politics (see ● Figure 1.3).

The world's *high-income countries* are nations with highly industrialized economies; technologically advanced industrial, administrative, and service occupations; and relatively high levels of national and personal income. Examples include the United States, Canada, Australia, New Zealand, Japan, and the countries of Western Europe.

As compared with other nations of the world, many high-income nations have a high standard of living and a lower death rate because of advances in nutrition and medical technology. However, everyone living in a so-called high-income country does not necessarily have these advantages.

In contrast, *middle-income countries* are nations with industrializing economies, particularly in urban areas, and moderate levels of national and personal income. Examples of middle-income countries include the nations of Eastern Europe and many Latin American countries.

Low-income countries are primarily agrarian nations with little industrialization and low levels of national and personal income. Examples of low-income countries include many of the nations of Africa and Asia, particularly the People's Republic of China and India, where people typically work the land and are among the poorest in the world. However, generalizations are difficult to make because there are wide differences in income and standards of living within many nations (see Chapter 9, "Global Stratification").

Throughout this text we will continue to develop our sociological imaginations by examining social life in the United States and other nations. The future of our nation is deeply intertwined with the future of all other nations of the world on economic, political, environmental, and humanitarian levels.

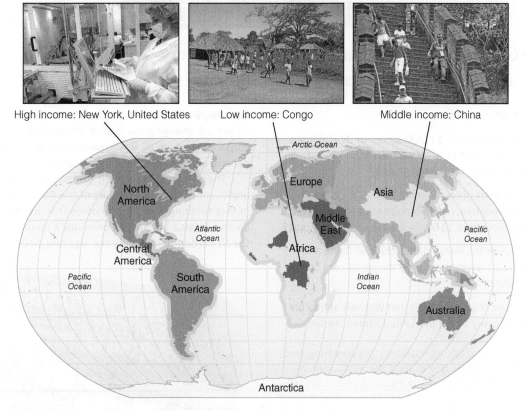

High income: New York, United States Low income: Congo Middle income: China

FIGURE 1.3 The World's Economies in the Twenty-First Century
High-income, middle-income, and low-income countries.
Photos: © Cengage Learning. *Photos, left to right:* John Berry/Syracuse Newspapers/The Image Works; Gable/Alamy; philipbigg/Alamy.

Whatever your race/ethnicity, class, sex, or age, are you able to include in your thinking the perspectives of people who are quite different from you in experiences and points of view? Before you answer this question, a few definitions are in order. *Race* is a term used by many people to specify groups of people distinguished by physical characteristics such as skin color. *Ethnicity* refers to the cultural heritage or identity of a group and is based on factors such as language or country of origin. *Class* is the relative location of a person or group within the larger society, based on wealth, power, prestige, or other valued resources. *Sex* refers to the biological and anatomical differences between females and males. By contrast, *gender* refers to the meanings, beliefs, and practices associated with sex differences, referred to as *femininity* and *masculinity*. Although these terms sound very precise, they often do not have a precise meaning and are, instead, social constructions that people use to justify social inequalities. When we refer to something as a "social construction," we mean that race, ethnicity, class, and gender do not really indicate anything apart from the social meaning that people in a given society confer on them. However, the result is that we may—either intentionally or unintentionally—privilege some categories of people over others who are placed in disadvantaged or subordinate positions. In sum, a "social construction of reality" occurs when large

numbers of people act and respond as if these categories exist in reality instead of being socially created.

 LO4 Describe the historical context in which sociological thinking developed.

The Development of Sociological Thinking

Throughout history, social philosophers and religious authorities have made countless observations about human behavior. However, the idea of observing how people lived, finding out what they thought, and doing so in a systematic

high-income countries
(sometimes referred to as **industrial countries**) nations with highly industrialized economies; technologically advanced industrial, administrative, and service occupations; and relatively high levels of national and personal income.

middle-income countries
(sometimes referred to as **developing countries**) nations with industrializing economies, particularly in urban areas, and moderate levels of national and personal income.

low-income countries
(sometimes referred to as **underdeveloped countries**) primarily agrarian nations with little industrialization and low levels of national and personal income.

SOCIOLOGY in GLOBAL PERSPECTIVE

Global Walmartization: From Big-Box Stores to Online Supermarkets in China

Did you know that:

- Walmart has more than 11,200 stores in 27 countries and that more than half of all Walmart stores worldwide are located outside the United States?

- In China, Walmart operates more than 400 stores, including supercenters, neighborhood markets, and Sam's Clubs, and more than 110 new facilities are being built?

- Walmart serves the rising middle class in China and continues to be a major player in the credit card business in China?

Although most of us are aware that Walmart stores are visible in virtually every city in the United States, we are

An exciting aspect of studying sociology is comparing our own lives with those of people around the world. Global consumerism, as evidenced by the opening of a Walmart Supercenter in Shanghai, China, provides a window through which we can observe how issues such as shopping and credit affect all of us. Which aspects of this photo reflect local culture? Which aspects reflect a global cultural phenomenon?

less aware of the extent to which Walmart and other big-box stores are changing the face of the world economy as mega-corporations such as this expand their operations into other nations and into the credit card business.

The strategic placement of Walmart stores both here and abroad accounts for part of the financial success of this retailing giant, but another U.S. export—credit cards—is also part of the company's business plan. Credit cards are changing the way that people shop and

Top Photo Corporation/Alamy

manner that could be verified did not take hold until the nineteenth century and the social upheaval brought about by industrialization and urbanization.

Industrialization is the process by which societies are transformed from dependence on agriculture and handmade products to an emphasis on manufacturing and related industries. This process occurred first during the Industrial Revolution in Britain between 1760 and 1850, and was soon repeated throughout Western Europe. By the mid-nineteenth century, industrialization was well under way in the United States (• Figure 1.4). Massive economic, technological, and social changes occurred as machine technology and the factory system shifted the economic base of these nations from agriculture to manufacturing: textiles, iron smelting, and related industries. Many people who had labored on the land were forced to leave their tightly knit rural communities and sacrifice well-defined social relationships to seek employment as factory workers in the emerging cities, which became the centers of industrial work.

Urbanization is the process by which an increasing proportion of a population lives in cities rather than in rural areas. Although cities existed long before the Industrial Revolution, the development of the factory system led to a rapid increase in both the number of cities and the size of their

Library of Congress Prints and Photographs Division [LC-DIG-nclc-02949]

FIGURE 1.4   As the Industrial Revolution swept through the United States beginning in the nineteenth century, children being employed in factories became increasingly common. Soon social thinkers began to explore such new social problems brought about by industrialization.

how they think about spending money in emerging nations such as China. For example, Walmart China continues to aggressively seek both shoppers and credit card holders. By encouraging people to spend money now rather than save it for later, corporations such as Walmart gain in two ways: (1) people buy more goods than they would otherwise, thus increasing sales; and (2) the corporation whose "brand" is on the credit card increases its earnings as a result of the interest the cardholder pays on credit card debt.

The motto for the original Walmart credit card in China was "Maximizing value, enjoying life," and this idea encouraged a change in attitude from the past, when—regardless of income level—most residents of that country did not purchase items on credit or possess a credit card. Introduction of the credit card brought a corresponding surge in debt in China. This change has been partly attributed to aggressive marketing of goods and services by transnational retailers, but it also relates to credit card companies encouraging consumers to "buy now, pay later."

Throughout this course, you will see that many issues we discuss, such as consumerism and globalization, have positive and negative effects on people's daily lives. Global consumerism, whether in big-box stores or through credit cards or electronic commerce, provides a window through which we can observe how an issue such as shopping affects all of us. Among the poor and those who have been most hard-hit by difficult economic times, the lack of ability to purchase basic necessities is a central litmus test for analyzing quality of life and social inequality. Among persons in the middle classes, purchasing power is often used to determine social mobility (the ability to move into) or social stability (the ability to stay on) the middle rungs of a society's ladder of income and wealth. Among persons in the upper classes, high rates of luxury consumerism are often seen as an outward sign of "having it all," but such behaviors reveal much more than this from a sociological perspective. As you look around you, what important social issues do you think are revealed by *what* people purchase? Or *why* they buy specific items and not others?

populations. People from very diverse backgrounds worked together in the same factory. At the same time, many people shifted from being *producers* to being *consumers.* For example, families living in the cities had to buy food with their wages because they could no longer grow their own crops to eat or to barter for other resources. Similarly, people had to pay rent for their lodging because they could no longer exchange their services for shelter.

These living and working conditions led to the development of new social problems: inadequate housing, crowding, unsanitary conditions, poverty, pollution, and crime. Wages were so low that entire families—including very young children—were forced to work, often under hazardous conditions and with no job security. As these conditions became more visible, a new breed of social thinkers tried to understand why and how society was changing.

The Development of Modern Sociology

At the same time that urban problems were growing worse, natural scientists had been using reason, or rational thinking, to discover the laws of physics and the movement of the planets. Social thinkers started to believe that by applying the methods developed by the natural sciences, they might discover the laws of human behavior and apply these laws to solve social problems. Historically, the time was ripe because the Age of Enlightenment had produced a belief in reason and humanity's ability to perfect itself.

 LO5 **Discuss** why early social thinkers were concerned with social order and stability.

Early Thinkers: A Concern with Social Order and Stability

Early social thinkers—such as Auguste Comte, Harriet Martineau, Herbert Spencer, and Emile Durkheim—were interested in analyzing social order and stability, and many of their ideas have had a dramatic and long-lasting

industrialization
the process by which societies are transformed from dependence on agriculture and handmade products to an emphasis on manufacturing and related industries.

urbanization
the process by which an increasing proportion of a population lives in cities rather than in rural areas.

influence on modern sociology. The first of these, Auguste Comte, focused primarily on continuity in societies, but his theorizing also highlights how societies contain forces for change as well.

Auguste Comte

The French philosopher Auguste Comte (1798–1857) coined the term *sociology* from the Latin *socius* ("social, being with others") and the Greek *logos* ("study of") to describe a new science that would engage in the study of society. Even though he never actually conducted sociological research, Comte is considered by some to be the "founder of sociology." Comte's theory that societies contain *social statics* (forces for social order and stability) and *social dynamics* (forces for conflict and change) continues to be used in contemporary sociology. In fact, we can trace the origins of applied sociology (which focuses on social change and intervention) to the 1850s, when Comte divided sociology into two areas: theories of stability (social statics) and the practice of social interventionism (social progress and development). Although many contemporary social theorists and researchers participate in academic studies and influence public debate through their writing and presentations, *applied sociologists* are practitioners and social activists who adapt sociological thinking to real-life situations (typically outside academic settings) and help formulate social policy that may promote social change (Perlstadt, 2007).

Comte stressed that the methods of the natural sciences should be applied to the objective study of society. He sought to unlock the secrets of society so that intellectuals like him could become the new secular (as contrasted with religious) "high priests" of society. For Comte, the best policies involved order and authority. He envisioned that a new consensus would emerge on social issues and that the new science of sociology would play a significant part in the reorganization of society.

Comte's philosophy became known as *positivism*—a belief that the world can best be understood through scientific inquiry. He believed that positivism had two dimensions: (1) methodological—the application of scientific knowledge to both physical and social phenomena—and (2) social and political—the use of such knowledge to predict the likely results of different policies so that the best one could be chosen.

Social analysts have praised Comte for his advocacy of sociology and his insights regarding linkages between the social structural elements of society (such as family, religion, and government) and social thinking in specific historical periods. However, a number of contemporary sociologists argue that Comte contributed to an overemphasis on the "natural science model" and focused on the experiences of a privileged few, to the exclusion of all others.

Harriet Martineau

Comte's works were made more accessible for a wide variety of scholars through the efforts of the British sociologist Harriet Martineau (1802–1876). Until fairly recently, Martineau received no recognition in the field of sociology, partly because she was a woman in a male-dominated discipline and society. Not only did she translate and condense Comte's works, but she was also an active sociologist in her own right. Martineau studied the social customs of Britain and the United States, analyzing the consequences of industrialization and capitalism. In *Society in America* (1962/1837), she examined religion, politics, child rearing, slavery, and immigration, paying special attention to social distinctions based on class, race, and gender. Her works explore the status of women, children, and "sufferers" (persons who are considered to be criminal, mentally ill, handicapped, poor, or alcoholic).

Martineau was also an advocate of social change, encouraging greater racial and gender equality. She was also committed to creating a science of society that would be grounded in empirical observations and be widely accessible to people. She argued that sociologists should be impartial in their assessment of society but that it is entirely appropriate to compare the existing state of society with the principles on which it was founded. Martineau believed that a better society would emerge if women and men were treated equally, enlightened reform occurred, and cooperation existed among people in all social classes (but led by the middle class).

Herbert Spencer

Unlike Comte, who was strongly influenced by the upheavals of the French Revolution, the British social theorist Herbert Spencer (1820–1903) was born in a more peaceful and optimistic period in his country's history. Spencer's major contribution to sociology was an evolutionary perspective on social order and social change. Evolutionary theory helps to explain how organic and/or social change occurs in societies. According to Spencer's Theory of General Evolution, society, like a biological organism, has various interdependent parts (such as the family, the economy, and the government) that work to ensure the stability and survival of the entire society.

Spencer believed that societies develop through a process of "struggle" (for existence) and "fitness" (for survival), which he referred to as the "survival of the fittest." Because this phrase is often attributed to Charles Darwin, Spencer's view of society is known as *social Darwinism*—the belief that those species of animals, including human beings, best adapted to their environment survive and prosper, whereas those poorly adapted die out. Spencer equated this process of *natural selection* with progress because only the "fittest" members of society would survive the competition.

Critics believe that Spencer's ideas are flawed because societies are not the same as biological systems; people are able to create and transform the environment in which they live. Moreover, the notion of the survival of the fittest can easily be used to justify class, racial–ethnic, and gender inequalities.

Emile Durkheim   French sociologist Emile Durkheim (1858–1917) stressed that people are the product of their social environment and that behavior cannot be understood fully in terms of *individual* biological and psychological traits. He believed that the limits of human potential are *socially* based, not *biologically* based.

© Bettmann/Corbis

In his work *The Rules of Sociological Method* (1964a/1895), Durkheim set forth one of his most important contributions to sociology: the idea that societies are built on social facts. *Social facts* are patterned ways of acting, thinking, and feeling that exist outside any one individual but that exert social control over each person. Durkheim believed that social facts must be explained by other social facts—by reference to the social structure rather than to individual attributes.

Durkheim observed that rapid social change and a more specialized division of labor produce *strains* in society. These strains lead to a breakdown in traditional organization, values, and authority and to a dramatic increase in *anomie*—a condition in which social control becomes ineffective as a result of the loss of shared values and of a sense of purpose in society. According to Durkheim, anomie is most likely to occur during a period of rapid social change. In *Suicide* (1964b/1897), he explored the relationship between anomic social conditions and suicide, a concept that remains important in the twenty-first century.

Durkheim's contributions are so significant that he is considered to be one of the crucial figures in the development of sociology as an academic area of study. He is one of the founding figures in the functionalist theoretical tradition, but he also made important contributions to other perspectives, particularly symbolic interactionism. Later in this chapter, we look at these theoretical approaches.

Although critics acknowledge Durkheim's important contributions, some argue that his emphasis on societal stability, or the "problem of order"—how society can establish and maintain social stability and cohesiveness—obscured the *subjective meanings* that individuals give to religion, work, and suicide. From this view, overemphasis on *structure* and the determining power of "society" resulted in a corresponding neglect of *agency* (the beliefs and actions of the actors involved) in much of Durkheim's theorizing.

LO6   **Identify** reasons why many later social thinkers were concerned with social change.

Differing Views on the Status Quo: Stability or Change?

Together with Karl Marx, Max Weber, and Georg Simmel, Durkheim established the direction of modern sociology. We will look first at Marx's and Weber's divergent thoughts about conflict and social change in societies and then at Georg Simmel's microlevel analysis of society.

Karl Marx   In sharp contrast to Durkheim's focus on the stability of society, German economist and philosopher Karl Marx (1818–1883) stressed that history is a continuous clash between conflicting ideas and forces. He believed that conflict—especially class conflict—is necessary in order to produce social change and a better society. For Marx, the most important changes are economic. He concluded that the capitalist economic system was responsible for the overwhelming poverty that he observed in London at the beginning of the Industrial Revolution (Marx and Engels, 1967/1848).

North Wind/North Wind Picture Archives

In the Marxian framework, *class conflict* is the struggle between the capitalist class and the working class. The capitalist class, or *bourgeoisie,* is those who own and control the means of production—the tools, land, factories, and money for investment that form the economic basis of a society. The working class, or *proletariat,* is those who must sell their labor because they have no other means to earn a livelihood. From Marx's viewpoint, the capitalist class controls and exploits the masses of struggling workers by paying less than the value of their labor. This exploitation results in the workers' *alienation*—a feeling of powerlessness and

positivism
a term describing Auguste Comte's belief that the world can best be understood through scientific inquiry.

social Darwinism
Herbert Spencer's belief that those species of animals, including human beings, best adapted to their environment survive and prosper, whereas those poorly adapted die out.

social facts
Emile Durkheim's term for patterned ways of acting, thinking, and feeling that exist *outside* any one individual but that exert social control over each person.

anomie
Emile Durkheim's designation for a condition in which social control becomes ineffective as a result of the loss of shared values and of a sense of purpose in society.

Online Shopping and Your Privacy

Motorcycle jacket for kid brother on the Internet—$300
Monogrammed golf balls for dad on the Internet—$50
Vintage smoking robe for husband on the Internet—$80
Not having to hear "attention shoppers"—not even once—priceless.

The way to pay on the Internet and everywhere else you see the MasterCard logo: MasterCard.

—MasterCard advertisement (qtd. in Manning, 2000: 114)

Clearly, this older advertisement for MasterCard taps into a vital source of revenue for companies that issue credit cards: online customers. Earlier, we mentioned that industrialization and urbanization were important historical factors that brought about significant changes in social life. Today, social life has changed as the Internet has become an integral part of our daily lives, including how we gather information, communicate with others, shop, and view our privacy.

Shopping online raises important questions: Who is watching your online activity? How far do companies go in "snooping" on those who visit their websites? Companies that sell products or services on the Internet are not required to respect the privacy of shoppers. According to the American Bar Association (2012), "When you buy something online, that company collects information about you. The information it collects is not necessarily limited to what the company needs to process your order. . . ." This means the seller may collect data on which site pages you visit, which products you buy,

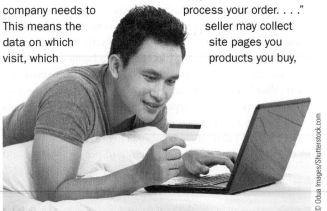

when you buy them, and where you ship them. Then the seller may share the information with other companies or sell it to them. Some websites have privacy policies but still insert cookies, data stored on a user's computer that tell the site's owner where you go and what you do on the site. Sometimes, the site owner records your e-mail address and begins sending you messages about that company's products, whether you have asked to receive them or not. It is possible, but sometimes not easy, to "unsubscribe" from these mailings.

To offset people's fears of invasion of privacy or abuse of their credit card information, corporations reassure customers that they are not being tracked and that it is safe to give out personal information online. However, the American Bar Association (ABA) advises caution in Internet interactions. According to the ABA, consumers using a credit card for an online purchase should find out if their credit card number will be kept on file by the seller for automatic use in future orders. Many online merchants offer the option to "keep" or "not keep" such information in their database, but they generally encourage consumers to let them keep it because it allegedly makes future transactions so much faster and easier. Online shoppers should find out what information the seller is gathering about them, how this information will be used, and whether they can "opt out" of having this information gathered (American Bar Association, 2012). If you think that you are being "watched" when you browse or make purchases online, you probably are!

Do you feel comfortable shopping online? Do you care if retailers use your private information for their own purposes? Why or why not?

REFLECT & ANALYZE

Are you responsible for protecting your own privacy online, or should federal law require that companies obtaining information about you let you know exactly what data they are collecting and why? How can sociology make us more aware of key social policy issues—such as this—that affect our everyday life?

Source: Based on American Bar Association, 2012.

---

estrangement from other people and from oneself. Marx predicted that the working class would become aware of its exploitation, overthrow the capitalists, and establish a free and classless society.

Overall, Marx's ideas are still influential in contemporary societies where alienation is viewed as a pressing problem and where social movements call people's attention to large economic disparities brought about by the emergence of global capitalism. Marx specifically linked alienation to social relations that are inherent in the *production* side of capitalism. Because of his emphasis on the negative effects of production under capitalism, social scientists for many years focused primarily on problems associated with production and social organization in industrial societies. However, some theorizing and research have shifted to the issue of rampant global consumerism. The effects of consumerism are examined on both the macrolevel and the microlevel. At the macrolevel, there is concern about issues such as environmental degradation and national debt. At the microlevel, studies often focus on why individuals and families spend more money than they can afford in hopes of finding personal happiness, gaining approval of others, and elevating their own social importance. Industrialization and urbanization were not only important influences in production but also in how societies consume, and this is reflected in the works of Max Weber and Georg Simmel.

LO7 **Discuss** how industrialization and urbanization influenced the theories of Max Weber and Georg Simmel.

Max Weber German social scientist Max Weber (pronounced VAY-ber) (1864–1920) was also concerned about the changes brought about by the Industrial Revolution. Although he disagreed with Marx's idea that economics is *the* central force in social change, Weber acknowledged that economic interests are important in shaping human action. Even so, he thought that economic systems were heavily influenced by other factors in a society.

Unlike many early analysts who believed that values could not be separated from the research process, Weber emphasized that sociology should be *value free*—research should be done scientifically, excluding the researcher's personal values and economic interests. However, Weber realized that social behavior cannot be analyzed by purely objective criteria, so he stressed that sociologists should employ *verstehen* (German for "understanding" or "insight") to gain the ability to see the world as others see it. In contemporary sociology, Weber's idea is incorporated into the concept of the sociological imagination (discussed earlier in this chapter).

Weber was also concerned that large-scale organizations (bureaucracies) were becoming increasingly oriented toward routine administration and a specialized division of labor, which he believed were destructive to human vitality and freedom. According to Weber, rational bureaucracy, rather than class struggle, is the most significant factor in determining the social relations between people in industrial societies. From this view, bureaucratic domination can be used to maintain powerful (capitalist) interests in society. As discussed in Chapter 6 ("Groups and Organizations"), Weber's work on bureaucracy has had a far-reaching impact.

Weber also provided important insights on the process of rationalization, on religion, and on many other topics. In his writings, Weber was more aware of women's issues than many of the scholars of his day. Perhaps his awareness at least partially resulted from the fact that his wife, Marianne Weber, was an important figure in the women's movement in Germany.

Georg Simmel At about the same time that Durkheim was developing the field of sociology in France, the German sociologist Georg Simmel (pronounced ZIM-mel) (1858–1918) was theorizing about the importance of social change in his own country and elsewhere. Simmel was also focusing on how society is a web of patterned interactions among

people (• Figure 1.5). In *The Sociology of Georg Simmel* (1950/1902–1917), he described how social interactions are different based on the size of the social group. According to Simmel, interaction patterns differ between a *dyad* (a social group with two members) and a *triad* (a group with three members) because the presence of an

FIGURE 1.5 According to the sociologist Georg Simmel, society is a web of patterned interactions among people. If we focus on the behavior of individuals only, we miss the underlying forms that make up the "geometry of social life."

additional person often changes the dynamics of communication and the overall interaction process. Simmel also developed *formal sociology,* an approach that focuses attention on the universal social forms that underlie social interaction. He referred to these forms as the "geometry of social life."

Like the other social thinkers of his day, Simmel analyzed the impact of industrialization and urbanization on people's lives. He concluded that class conflict was becoming more pronounced in modern industrial societies. He also linked the increase in individualism, as opposed to concern for the group, to the fact that people now had many cross-cutting "social spheres"—membership in a number of organizations and voluntary associations—rather than the singular community ties of the past.

Simmel's contributions to sociology are significant. He wrote more than thirty books and numerous essays on diverse topics, leading some critics to state that his work is fragmentary and piecemeal. However, his thinking has influenced a wide array of sociologists, including the members of the "Chicago School" in the United States.

The Beginnings of Sociology in the United States

From Western Europe, sociology spread in the 1890s to the United States, where it thrived as a result of the intellectual climate and the rapid rate of social change. The first departments of sociology in the United States were located at the University of Chicago and at Atlanta University, then an African American school.

The Chicago School   The first department of sociology in the United States was established at the University of Chicago, where the faculty was instrumental in starting the American Sociological Society (now known as the American Sociological Association). Robert E. Park (1864–1944), a member of the Chicago faculty, asserted that urbanization has a disintegrating influence on social life by producing an increase in the crime rate and in racial and class antagonisms that contribute to the segregation and isolation of neighborhoods (Ross, 1991). George Herbert Mead (1863–1931), another member of the faculty at Chicago, founded the symbolic interaction perspective, which is discussed later in this chapter.

Jane Addams   Jane Addams (1860–1935) is one of the best-known early women sociologists in the United States because she founded Hull House, one of the most famous settlement houses, in an impoverished area of Chicago. Throughout her career, she was actively engaged in sociological endeavors: She lectured at numerous colleges, was a charter member of the American Sociological Society, and published a number of articles and books. Addams was one of the authors of *Hull-House*

Maps and Papers, a groundbreaking book that used a methodological technique employed by sociologists for the next forty years. She was also awarded a Nobel Prize for her assistance to the underprivileged. In recent years, Addams has received greater recognition from contemporary sociologists because of her role as an early theorist of social change who influenced later feminist theorists and activists.

W. E. B. Du Bois and Atlanta University   The second department of sociology in the United States was founded by W. E. B. Du Bois (1868–1963) at Atlanta University. He created a laboratory of sociology, instituted a program of systematic research, founded and conducted regular sociological conferences on research, founded two journals, and established

a record of valuable publications. His classic work, *The Philadelphia Negro: A Social Study* (1967/1899), was based on his research into Philadelphia's African American community and stressed the strengths and weaknesses of a community wrestling with overwhelming social problems. Du Bois was one of the first scholars to note that a dual heritage creates conflict for people of color. He called this duality *double-consciousness*—the identity conflict of being both a black and an American. Du Bois pointed out that although people in this country espouse such values as democracy, freedom, and equality, they also accept racism and group discrimination. African Americans are the victims of these conflicting values and the actions that result from them. The influence of Du Bois continues to grow in contemporary studies of inequality, social justice, and the need for change in race/ethnic and class relations in the United States and worldwide.

LO**8** **Compare** and contrast contemporary functionalist and conflict perspectives on social life.

Contemporary Theoretical Perspectives

Given the many and varied ideas and trends that influenced the development of sociology, how do contemporary sociologists view society? Some see it as basically a stable and ongoing entity; others view it in terms of many groups competing for scarce resources; still others describe it based on the everyday, routine interactions among individuals. Each of these views represents a method of examining the same phenomena. Each is based on general ideas about how social life is organized and represents an effort to link specific observations in a meaningful way. Each uses a *theory*—a set of logically interrelated statements that attempts to describe, explain, and (occasionally) predict social events. Each theory helps interpret reality in a distinct way by providing a framework in which observations may be logically ordered. Sociologists refer to this theoretical framework as a *perspective*—an overall approach to or viewpoint on some subject. Three major theoretical perspectives have been predominant in U.S. sociology: the functionalist, conflict, and symbolic interactionist perspectives. Other perspectives, such as postmodernism and globalization, have emerged and gained acceptance among social thinkers. Before turning to the specifics of these perspectives, we should note that some theorists and theories do not fit neatly into any of these perspectives. Although the categories may be viewed as oversimplified by some people, most of us organize our thinking into categories and find it easier for us to compare and contrast ideas if we have a basic outline of key characteristics associated with each approach (see the Concept Quick Review at the end of this section).

Functionalist Perspectives

Also known as *functionalism* and *structural functionalism*, **functionalist perspectives** are based on the assumption that society is a stable, orderly system. This stable system is characterized by *societal consensus*, whereby the majority of members share a common set of values, beliefs, and behavioral expectations. According to this perspective, a society is composed of interrelated parts, each of which serves a function and (ideally) contributes to the overall stability of the society. Societies develop social structures, or institutions that persist because they play a part in helping society survive. These institutions include the family, education, government, religion, and the economy. If anything adverse happens to one of these institutions or parts, all other parts are affected, and the system no longer functions properly.

Talcott Parsons and Robert K. Merton Talcott Parsons (1902–1979), perhaps the most influential contemporary advocate of the functionalist perspective, stressed that all societies must meet social needs in order to survive. Parsons (1955) suggested, for example, that a division of labor (distinct, specialized functions) between husband and wife is essential for family stability and social order. The husband/father performs the *instrumental tasks,* which involve leadership and decision-making responsibilities in the home and employment outside the home to support the family. The wife/mother is responsible for the *expressive tasks,* including housework, caring for the children, and providing emotional support for the entire family. Parsons believed that other institutions, including school, church, and government, must function to assist the family and that all institutions must work together to preserve the system over time (Parsons, 1955).

Pictorial Parade/Archive Photos/Getty Images

Functionalism was refined further by Robert K. Merton (1910–2003), who distinguished between manifest and latent functions of social institutions. **Manifest functions** are intended and/or overtly recognized by the participants in a social unit. In contrast, **latent functions** are unintended functions that are hidden and remain unacknowledged by participants (● Figure 1.6). For example, a manifest function of education is the transmission of knowledge and skills from one generation to the next; a latent function is the establishment of social relations and networks. Merton noted that all features of a social system may not be functional at all times; *dysfunctions* are the undesirable consequences of any element of a society. A dysfunction of education in the United States is the perpetuation of gender, racial, and class inequalities.

theory
a set of logically interrelated statements that attempts to describe, explain, and (occasionally) predict social events.

functionalist perspectives
the sociological approach that views society as a stable, orderly system.

manifest functions
functions that are intended and/or overtly recognized by the participants in a social unit.

latent functions
unintended functions that are hidden and remain unacknowledged by participants.

FIGURE 1.6 Shopping malls are a reflection of a consumer society. A manifest function of a shopping mall is to sell goods and services to shoppers; however, a latent function may be to provide a communal area in which people can visit friends and enjoy an event.

Such dysfunctions may threaten the capacity of a society to adapt and survive.

Applying a Functional Perspective to Shopping and Consumption How might functionalists analyze shopping and consumption? When we examine the part-to-whole relationships of contemporary society in high-income nations, it immediately becomes apparent that each social institution depends on the others for its well-being. For example, a booming economy benefits other social institutions, including the family (members are gainfully employed), religion (churches, mosques, synagogues, and temples receive larger contributions), and education (school taxes are higher when property values are higher). A strong economy also makes it possible for more people to purchase more goods and services. By contrast, a weak economy has a negative influence on people's opportunities and spending patterns. For example, if people have "extra" money to spend and can afford leisure time away from work, they are more likely to dine out, take trips, and purchase things they might otherwise forgo. However, in difficult economic times, people are more likely to curtail family outings and some purchases.

Clearly, a manifest function of shopping and consumption is purchasing necessary items such as food, clothing,

household items, and sometimes transportation. But what are the latent functions of shopping? Consider, shopping malls, for example: Many young people go to the mall to "hang out," visit with friends, and eat lunch at the food court. People of all ages go shopping for pleasure, relaxation, and perhaps to enhance their feelings of self-worth. ("If I buy this product, I'll look younger/beautiful/handsome/sexy, etc.!") However, shopping and consuming may also produce problems or dysfunctions. Some people are "shopaholics" or "credit card junkies" who cannot stop spending money; others are kleptomaniacs, who steal products rather than pay for them.

The functionalist perspective is useful in analyzing consumerism because of the way in which it examines the relationship between part-to-whole relationships. How the economy is doing affects individuals' consumption patterns, and when the economy is not doing well, political leaders often encourage us to spend more to help the national economy and keep other people employed.

Conflict Perspectives

According to *conflict perspectives*, groups in society are engaged in a continuous power struggle for control of scarce resources. Conflict may take the form of politics, litigation, negotiations, or family discussions about financial

FIGURE 1.7 This multimillion-dollar penthouse is an example of conspicuous consumption. What examples of conspicuous consumption do you see in your community?

inevitable and as a prime source of social change. A second branch focuses on racial–ethnic inequalities and the continued exploitation of members of some racial–ethnic groups. A third branch is the feminist perspective, which focuses on gender issues.

The Feminist Approach A feminist theoretical approach (or "feminism") directs attention to women's experiences and the importance of gender as an element of social structure. This approach is based on a belief in the equality of women and men and the idea that all people should be equally valued and have equal rights. According to feminist theorists, we live in a *patriarchy*, a system in which men dominate women and in which things that are considered to be "male" or "masculine" are more highly valued than those considered to be "female" or "feminine." The feminist perspective assumes that gender is socially created and that change is essential in order for people to achieve their human potential without limits based on gender. Some feminists argue that women's subordination can end only after the patriarchal system becomes obsolete. However, feminism is not one single, unified approach; there are several feminist perspectives, which are discussed in Chapter 11 ("Sex, Gender, and Sexuality").

Applying Conflict Perspectives to Shopping and Consumption How might advocates of a conflict approach analyze the process of shopping and consumption? A contemporary conflict analysis of consumption might look at how inequalities based on racism, sexism, and income differentials affect people's ability to acquire the things they need and want. It might also look at inequalities regarding the issuance of credit cards and access to "cathedrals of consumption" such as mega shopping malls and tourist resorts (see Ritzer, 1999: 197–214).

However, one of the earliest social theorists to discuss the relationship between social class and consumption patterns was the U.S. social scientist Thorstein Veblen

matters. Simmel, Marx, and Weber contributed significantly to this perspective by focusing on the inevitability of clashes between social groups. Today, advocates of the conflict perspective view social life as a continuous power struggle among competing social groups.

Max Weber and C. Wright Mills As previously discussed, Karl Marx focused on the exploitation and oppression of the proletariat by the bourgeoisie. Max Weber recognized the importance of economic conditions in producing inequality and conflict in society, but he added *power* and *prestige* as other sources of inequality. Weber (1968/1922) defined *power* as the ability of a person within a social relationship to carry out his or her own will despite resistance from others, and *prestige* as a positive or negative social estimation of honor (Weber, 1968/1922).

C. Wright Mills (1916–1962), a key figure in the development of contemporary conflict theory, encouraged sociologists to get involved in social reform. Mills encouraged everyone to look beneath everyday events in order to observe the major resource and power inequalities that exist in society (• Figure 1.7). He believed that the most important decisions in the United States are made largely behind the scenes by the *power elite*—a small clique of top corporate, political, and military officials. Mills's power elite theory is discussed in Chapter 14 ("Politics and Government in Global Perspective").

The conflict perspective is not one unified theory but one with several branches. One branch is the neo-Marxist approach, which views struggle between the classes as

conflict perspectives
the sociological approach that views groups in society as engaged in a continuous power struggle for control of scarce resources.

YOU CAN MAKE A DIFFERENCE

Thinking Less About Things and More About People

We must rapidly begin the shift from a thing-oriented society to a person-oriented society. When machines and computers, profit motives and property rights are considered more important than people, the giant triplets of racism, militarism and economic exploitation are incapable of being conquered.

—Dr. Martin Luther King Jr., April 1967
(qtd. in Postman, 2011)

Almost fifty years ago, Dr. King encouraged people to find fulfillment in social relationships with other people rather than in new technologies or in making more money. Since King's era, the United States has had periods of economic boom and bust, accompanied by unparalleled consumerism. Many analysts believe that consumerism and constant pressures to "buy, buy, buy more!" have created financial havoc for many individuals and families. We are continually surrounded by advertisements, shopping malls, and online buying opportunities that set in front of us a veritable banquet of merchandise to buy. However, shopping that gets out of hand is a serious habit that may have lasting psychological and economic consequences. If we are aware of these problems, we may be able to help ourselves or others avoid hyperconsumerism.

Do you know the symptoms of compulsive overspending and debt dependency? Consider these questions:

Stan Rohrer/Alamy

These students are spending a day of service, helping to build a home. Many colleges and universities have similar service days to help their communities. What projects could you and your peers undertake?

- Do you or someone you know spend large amounts of time shopping or thinking about going shopping?

- Do you or someone you know rush to the store or to the computer for online shopping when feeling frustrated, depressed, or otherwise "out of sorts"?

(1857–1929). In *The Theory of the Leisure Class* (1967/1899), Veblen described early wealthy U.S. industrialists as engaging in *conspicuous consumption*—the continuous public display of one's wealth and status through purchases such as expensive houses, clothing, motor vehicles, and other consumer goods. According to Veblen, the leisurely lifestyle of the upper classes typically does not provide them with adequate opportunities to show off their wealth and status. In order to attract public admiration, the wealthy often engage in consumption and leisure activities that are both highly visible and highly wasteful. Examples of conspicuous consumption range from Cornelius Vanderbilt's 8 lavish mansions and 10 major summer estates in the Gilded Age to a contemporary $85 million mansion formerly owned by the Spelling family in Los Angeles, which has 56,500 square feet, 14 bedrooms, 27 bathrooms, and a two-lane bowling alley, among many other luxurious amenities. By contrast, however, some of today's wealthiest people engage in *inconspicuous consumption,* perhaps to maintain a low public profile or out of fear for their own safety or that of other family members.

Conspicuous consumption has become more widely acceptable at all income levels, and many families live on credit in order to purchase the goods and services that they would like to have. According to conflict theorists, the economic gains of the wealthiest people are often at the expense of those in the lower classes, who may have to struggle (sometimes unsuccessfully) to have adequate food, clothing, and shelter for themselves and their children. Chapter 8 ("Class and Stratification in the United States") and Chapter 9 ("Global Stratification") discuss contemporary conflict perspectives on class-based inequalities.

 LO9 **Identify** key differences in contemporary symbolic interactionism and postmodernist perspectives on social life.

Symbolic Interactionist Perspectives

The conflict and functionalist perspectives have been criticized for focusing primarily on macrolevel analysis. A **macrolevel analysis** examines whole societies, large-scale social structures, and social systems instead of looking at important social dynamics in individuals' lives. Our third perspective, symbolic interactionism, fills this void by examining people's day-to-day interactions and their behavior in groups. Thus, symbolic interactionist approaches are based on a **microlevel analysis**, which focuses on small groups rather than large-scale social structures.

- Do you or someone you know routinely argue with parents, friends, or partners about spending too much money or overcharging on credit cards?

- Do you or someone you know hide purchases or make dishonest statements—such as "It was a gift from a friend"—to explain where new merchandise came from?

According to economist Juliet Schor (1999), who has extensively studied the problems associated with excessive spending and credit card debt, each of us can empower ourselves and help others as well if we follow simple steps in our consumer behavior. Among these steps are *controlling desire* by gaining knowledge of the process of consumption and its effect on people, *helping to make exclusivity uncool* by demystifying the belief that people are "better" simply because they own excessively expensive items, and *discouraging competitive consumption* by encouraging our friends and acquaintances to spend less on presents and other purchases. Finally, Schor suggests that we should *become educated consumers* and *avoid use of shopping as a form of therapy*. By following Schor's simple steps and encouraging our friends and relatives to do likewise, we may be able to free ourselves from the demands of a hyperconsumer society that continually bombards us with messages indicating that we should spend more and go deeper in debt on our credit cards.

How might we think more about people? Some analysts suggest that we should make a list of things that are more important to us than money and material possessions. These might include our *relationships* and *experiences* with family, friends, and others whom we encounter in daily life. Are we so engrossed in our own life that we fail to take others into account? Around school, are we so busy texting or talking on our cell phone that we fail to speak to others? During holidays and special occasions, do we make time for friends and loved ones even if we think we have "better" things to do?

Other suggestions for thinking about others might include *looking for ways, small and large, to help others.* Small ways to help others might be opening a door for someone whose hands are full, letting someone go before us in a line or while driving in traffic, or any one of a million small kindnesses that might brighten someone else's day as well as our own. Large ways of helping others would include joining voluntary organizations that assist people in the community, including older individuals, persons with health problems, children who need a tutor or mentor, or many others you might learn of from school organizations, social service agencies, or churches in your area. Are we up to the challenge? Many who have tried thinking less about things and more about people highly recommend this as a life-affirming endeavor for all involved.

Source: Schor, 1999.

We can trace the origins of this perspective to the Chicago School, especially George Herbert Mead and the sociologist Herbert Blumer (1900–1986), who is credited with coining the term *symbolic interactionism*. According to **symbolic interactionist perspectives**, society is the sum of the interactions of individuals and groups. Theorists using this perspective focus on the process of *interaction*—defined as immediate reciprocally oriented communication between two or more people—and the part that symbols play in communication. A *symbol* is anything that meaningfully represents something else. Examples include signs, gestures, written language, and shared values. Symbolic interaction occurs when people communicate through the use of symbols—for example, a ring to indicate a couple's engagement. But symbolic communication occurs in a variety of forms, including facial gestures, posture, tone of voice, and other symbolic gestures (such as a handshake or a clenched fist).

Symbols are instrumental in helping people derive meanings from social interactions (● Figure 1.8). In social encounters each person's interpretation or definition of a given situation becomes a *subjective reality* from that person's viewpoint. We often assume that what we consider to be "reality" is shared by others; however,

this assumption is often incorrect. Subjective reality is acquired and shared through agreed-upon symbols, especially language. If a person shouts "Fire!" in a crowded movie theater, for example, that language produces the same response (attempting to escape) in all of those who hear and understand it. When people in a group do not share the same meaning for a given symbol, however, confusion results: People who do not know the meaning of the word *fire* will not know what the commotion is about. How people *interpret* the messages they receive and the situations they encounter becomes their subjective reality and may strongly influence their behavior.

macrolevel analysis
an approach that examines whole societies, large-scale social structures, and social systems instead of looking at important social dynamics in individuals' lives.

microlevel analysis
sociological theory and research that focus on small groups rather than on large-scale social structures.

symbolic interactionist perspectives
the sociological approach that views society as the sum of the interactions of individuals and groups.

Applying Symbolic Interactionist Perspectives to Shopping and Consumption Sociologists applying a symbolic interactionist framework to the study of shopping and consumption would primarily focus on a microlevel analysis of people's face-to-face interactions and the roles that people play in society. In our efforts to interact with others, we define any situation according to our own subjective reality. This theoretical viewpoint applies to shopping and consumption just as it does to other types of conduct. For example, when a customer goes into a store to make a purchase and offers a credit card to the cashier, what meanings are embedded in the interaction process that takes place between the two of them? The roles that the two people play are based on their histories of interaction in previous situations. They bring to the present encounter symbolically charged ideas, based on previous experiences. Each person also has a certain level of emotional energy available for each interaction. When we are feeling positive, we have a high level of emotional energy, and the opposite is also true. Every time we engage in a new interaction, the situation has to be negotiated all over again, and the outcome cannot be known beforehand.

In the case of a shopper–cashier interaction, how successful will the interaction be for each of them? The answer to this question depends on a kind of social marketplace in which such interactions can either raise or lower one's emotional energy. If the customer's credit card is rejected, he or she may come away with lower emotional energy. If the customer is angry at the cashier, he or she may attempt to "save face" by reacting in a haughty manner regarding the rejection of the card. ("What's wrong with you? Can't you do anything right? I'll never shop here again!") If this type of encounter occurs, the cashier may also come out of the interaction with a lower level of emotional energy, which may affect the cashier's interactions with subsequent customers. Likewise, the next time the customer uses a credit card, he or she may say something

FIGURE 1.8 Sporting events are a prime location for seeing how college students use symbols to convey shared meanings. The colors of clothing and the display of the school logo emphasize these students' pride in their school.

like "I hope this card isn't over its limit; sometimes I lose track," even if the person knows that the card's credit limit has not been exceeded. This is only one of many ways in which the rich tradition of symbolic interactionism might be used to examine shopping and consumption. Other areas of interest might include the social nature of the shopping experience, social interaction patterns in families regarding credit card debts, and why we might spend money to impress others.

Postmodern Perspectives

According to **postmodern perspectives**, existing theories have been unsuccessful in explaining social life in contemporary societies that are characterized by postindustrialization, consumerism, and global communications. Postmodern social theorists reject the theoretical perspectives we have previously discussed, as well as how those theories were created (Ritzer, 2011).

Postmodern theories are based on the assumption that large-scale and rapid social change, globalization, and technology are central features of the postmodern era. Moreover, these conditions tend to have a harmful effect on people because they often result in ambiguity and chaos. One evident change is a significant decline in the influence of social institutions such as the family, religion, and education on people's lives. Those who live in postmodern societies typically pursue individual freedom and do not want the structural constraints that are imposed by social institutions. As social inequality and class differences increase, people are exposed to higher levels of stress that produce depression, fear, and ambivalence. Problems such as these are found in nations throughout the world.

Postmodern (or "postindustrial") societies are characterized by an *information explosion* and an economy in which large numbers of people either provide or apply information, or are employed in professional occupations (such as attorneys and physicians) or service jobs (such as fast-food servers and health care workers). There is a corresponding *rise of a consumer society* and the emergence of a *global village* in which people around the world instantly communicate with one another.

Jean Baudrillard, a well-known French social theorist, has extensively explored how the shift from production of goods to consumption of information, services, and products has created a new form of social control. According to Baudrillard's approach, capitalists strive to control people's shopping habits, much like the output of factory workers in industrial economies, to enhance their profits and to keep everyday people from rebelling against social inequality (1998/1970). How does this work? When consumers are encouraged to purchase more than they need or can afford, they often sink deeper in debt and must keep working to meet their monthly payments. Consumption comes to be based on factors such as our "wants" and our need to distinguish ourselves from others. We will return to Baudrillard's general ideas on postmodern societies in Chapter 3 ("Culture").

Postmodern theory opens up broad new avenues of inquiry by challenging existing perspectives and questioning current belief systems. However, postmodern theory has also been criticized for raising more questions than it answers.

Applying Postmodern Perspectives to Shopping and Consumption According to some social theorists, the postmodern society is a consumer society. The focus of the capitalist economy has shifted from production to consumption: The emphasis is on getting people to consume more and to own a greater variety of things. As previously discussed, credit cards may encourage people to spend more money than they should, and often more than they can afford (Ritzer, 1998). Television shopping networks, online shopping, and mobile advertising and shopping devices make it possible for people to shop around the clock without having to leave home or encounter "real" people. As Ritzer (1998: 121) explains, "So many of our interactions in these settings . . . are simulated, and we become so accustomed to them, that in the end all we have are simulated interactions; there are no more 'real' interactions. The entire distinction between the simulated and the real is lost; simulated interaction *is* the reality" (see also Baudrillard, 1983).

For postmodernists, social life is not an objective reality waiting for us to discover how it works. Rather, what we experience as social life is actually nothing more or less than how we think about it, and there are many diverse ways of doing that. According to a postmodernist perspective, the Enlightenment goal of intentionally creating a better world out of some knowable truth is an illusion. Although some might choose to dismiss postmodern approaches, they do give us new and important questions to think about regarding the nature of social life.

The Concept Quick Review reviews all four of the major sociological perspectives. Throughout this book we will be using these perspectives as lenses through which to view our social world.

Looking Ahead: Are Theory and/or Practice in Your Future?

One of the themes of *Sociology in Our Times* is that we live in a world that is constantly changing. Some people might argue that sociological theory has not changed sufficiently to adequately describe these social changes. Others believe that newer perspectives, such as postmodernism and globalization, adequately address the changes and continuities we are seeing today. Now it is time for you to consider how useful sociology and its theories might be to you not only in this course but in the future.

As previously discussed, the sociological imagination will help you gain a better understanding of yourself, other people, and the larger social world. We have discussed consumerism as an example of sociological inquiry in this chapter because it is a reflection of both social continuity and change in our nation and world. Consumerism is underpinned by power relationships, and studying it reveals how economic structures

postmodern perspectives
the sociological approach that attempts to explain social life in contemporary societies that are characterized by postindustrialization, consumerism, and global communications.

such as capitalism have an inescapable influence on all aspects of our lives. Today, consumerism is a rapidly growing area of inquiry in both academic and applied sociology. Jobs in applied sociology include project managers, urban development specialists, human rights officers, case managers, health planners, research coordinators, and a myriad of other careers in which sociological thinking and research can help people gain a better understanding of their work, clients, and pressing social issues facing their organizations.

Studying consumerism reveals something very important about sociology: Theory for its own sake is useless. We need to explore ways that theory can provoke thought and debate. What are the practical implications for theory in the social world? Practical sociological knowledge can be divided into five roles, one or more of which you might have in the future:

1. The *decision maker* who uses social science to shape public policy decisions.

2. The *educator* who teaches sociology to students.

3. The *commentator and social critic* who writes for the public through books, articles, and blogs and other social media.

4. The *researcher for clients* who works with public or private organizations (such as health care institutions or mental health groups).

5. The *consultant* who works for specific clients to answer questions or solve problems that are of interest to that client. (based on Zetterberg, 2002/1964)

People in roles 4 and 5 are classified as applied sociologists.

The "pure" sociology of the past has been joined by those who engage in applied sociology and public sociology. In the twenty-first century, *public sociology* aims to engage nonacademic audiences—such as people in grassroots environmental organizations or neighborhood activist groups—in informed public discussions through which both sides gain a better understanding of public issues (Burawoy, 2005).

If you are thinking about how to use sociology in your own life, it is exciting to realize that there are a variety of paths you might follow, using academic and nonacademic applications of what you are learning in this course. The future of the sociological perspective may be linked to your future as well!

CHAPTER REVIEW

LO1 What is sociology, and how can it contribute to our understanding of social life?

Sociology is the systematic study of human society and social interaction. We study sociology to understand how human behavior is shaped by group life and, in turn, how group life is affected by individuals. Our culture tends to emphasize individualism, and sociology pushes us to consider more-complex connections between our personal lives and the larger world.

LO2 What is meant by the sociological imagination, and how can it be used?

The sociological imagination helps us understand how seemingly personal troubles, such as debt or unemployment, are actually related to larger social forces. It is the ability to see the relationship between individual experiences and the larger society.

LO3 How can we develop a global sociological imagination?

We must reach beyond past studies that have focused primarily on the United States to develop a more comprehensive global approach for the future. It is important to have a global sociological imagination because the future of this nation is deeply intertwined with the future of all nations of the world on economic, political, and humanitarian levels.

LO4 What was the historical context in which sociological thinking developed?

The origins of sociological thinking as we know it today can be traced to the beginnings of industrialization and urbanization, trends that increased rapidly in the late eighteenth century and attracted the attention of social thinkers.

LO5 Why were many early social thinkers concerned with social order and stability?

Early social thinkers—such as Auguste Comte, Harriet Martineau, Herbert Spencer, and Emile Durkheim—were interested in analyzing social order and stability because they were concerned about the future of the nations in which they lived. Order and stability seemed functional for everyone's well-being in a rapidly changing world, and as such, many of these early sociologists' ideas had a dramatic influence on consensus and order perspectives in contemporary sociology. Auguste Comte coined the term *sociology* to describe a new science that would engage in the study of society. Comte's works were made more accessible for a wide variety of scholars through the efforts of the British sociologist Harriet Martineau. Herbert Spencer's major contribution to sociology was an evolutionary perspective on social order and social change. Durkheim argued that societies are built on social facts, that rapid social change produces strains in society, and that the loss of shared values and purpose can lead to a condition of anomie.

LO6 Why were many later social thinkers concerned with social change?

In sharp contrast to Durkheim's focus on the stability of society, German economist and philosopher Karl Marx stressed that history is a continuous clash between conflicting ideas and forces. He believed that conflict—especially class conflict—is necessary in order to produce social change and a better society. Although he disagreed with Marx's idea that economics is *the* central force in social change, German social scientist Max Weber acknowledged that economic interests are important in shaping human action.

LO7 How did industrialization and urbanization influence theorists such as Weber and Simmel?

Weber was concerned about the changes brought about by the Industrial Revolution and the influences these changes had on human behavior. In particular, Weber was concerned that large-scale organizations were becoming increasingly oriented toward routine administration and a specialized division of labor, which he believed were destructive to human vitality and freedom. Whereas other sociologists primarily focused on society as a whole, Simmel explored small social groups and social life in urban areas, arguing that society is best seen as a web of patterned interactions among people.

LO8 What are key differences in contemporary functionalist and conflict perspectives on social life?

Functionalist perspectives assume that society is a stable, orderly system characterized by societal consensus. Conflict perspectives argue that society is a continuous power struggle among competing groups, often based on class, race, ethnicity, or gender.

LO9 What are key differences in contemporary symbolic interactionism and postmodernist perspectives on social life?

Interactionist perspectives focus on how people make sense of their everyday social interactions, which are made possible by the use of mutually understood symbols. From an alternative perspective, postmodern theorists believe that entirely new ways of examining social life are needed and that it is time to move beyond functionalist, conflict, and interactionist approaches.

KEY TERMS

anomie 13

conflict perspectives 18

functionalist perspectives 17

high-income countries 8

industrialization 10

latent functions 17

low-income countries 8

macrolevel analysis 20

manifest functions 17

microlevel analysis 20

middle-income countries 8

positivism 12

postmodern perspectives 22

social Darwinism 12

social facts 13

society 6

sociological imagination 7

sociology 6

symbolic interactionist perspectives 21

theory 17

urbanization 10

QUESTIONS for CRITICAL THINKING

1 What does C. Wright Mills mean when he says the sociological imagination helps us "to grasp history and biography and the relations between the two within society"? (Mills, 1959b: 6). How might this idea be applied to today's consumer society?

2 As a sociologist, how would you remain objective yet see the world as others see it? Would you make subjective decisions when trying to understand the perspectives of others?

3 Early social thinkers were concerned about stability in times of rapid change. In our more-global world, is stability still a primary goal? Or is constant conflict important for the well-being of all humans? Use the conflict and functionalist perspectives to bolster your analysis.

4 Some social analysts believe that college students relate better to commercials, advertising culture, and social media than they do to history, literature, or probably anything else. How would you use the various sociological perspectives to explore the validity of this assertion in regard to students on your college campus?

ANSWERS to the SOCIOLOGY QUIZ

ON CONSUMPTION AND CREDIT CARDS

1	**True**	The credit card debt owed by the average U.S. household (that had credit card debt) in 2014 was $15,611.
2	**False**	Since the 2009 Credit CARD Act, it has become more difficult for college students to get a credit card in their name unless an adult cosigns for the card or students can prove they have sufficient income to pay the bills.
3	**True**	The average balance among college students with a credit card in 2014 was about $499. More students are now using debit cards than in the past.
4	**True**	Millennials are the biggest spenders online. Despite their typically lower incomes, they spend about $2,000 in online shopping each year.
5	**False**	Although the credit card industry has cracked down on companies issuing cards to people in lower-income brackets, many lower- and middle-income families *do* use credit cards for their financial safety net as they purchase necessities such as food, housing, and utilities because they lack adequate cash to pay for these goods and services.
6	**False**	In the United States, approximately 900,000 people filed for bankruptcy in 2014 (as compared with more than one million each year in previous years). However, this change is a partial result of new regulations that made it more difficult to discharge (get rid of) debt by bankruptcy and not because people are now better off financially.
7	**True**	The total U.S. outstanding revolving debt on credit cards alone was over $880.5 billion in 2014.
8	**False**	People in all income brackets in the United States have problems with overspending, partly as a result of excessive use of credit and not paying off debts when they are due.

Sources: Adapted from Holmes and Ghahremani, 2014; Smith, 2014; and U.S. Courts.gov, 2014.

SOCIOLOGICAL RESEARCH METHODS

Richard G. Bingham II/Alamy

LEARNING OBJECTIVES

1 **Explain** why sociological research is necessary and how it challenges our commonsense beliefs about pressing social issues such as suicide.

2 **Compare** deductive and inductive approaches in the theory and research cycle.

3 **Distinguish** between quantitative research and qualitative research.

4 **List** and briefly describe the steps in the quantitative research model.

5 **List** and briefly describe the steps in a qualitative research model.

6 **Explain** what is meant by survey research and briefly discuss three types of surveys.

7 **Compare** research methods used in secondary analysis of existing data, field research, experiments, and triangulation.

8 **Discuss** ethical issues in research and identify professional codes that protect research participants.

SOCIOLOGY & EVERYDAY LIFE

The Sociology of Suicide Trends Today

"Why are you alive?"

"You're ugly."

"You should die."

"Why don't you go kill yourself."

"Can u die please?"

—After more than a year of receiving online bullying comments like these, Rebecca Ann Sedwick, a twelve-year-old Tampa, Florida, girl, jumped from a concrete silo tower to her death at an abandoned cement plant. Prior to this terrible event, Rebecca changed her online screen name to "That Dead Girl." After her death, law enforcement officials stated that she was "absolutely terrorized on social media" (qtd. in Alvarez, 2013), while psychologists pointed out that she had been a victim of the "cool to be cruel" cyber culture (Ng, 2013).

" . . . go die, evry1 wud be happy."

"Drink bleach."

—London, England, resident Hanna Smith, age fourteen, repeatedly had hostile comments like these posted about her on ASK.fm prior to her suicide (Dolan and Robinson, 2013).

"You think you want to die, but in reality you just want to be saved."

—Hanna Smith posted a picture of this statement written in a spiral notebook to her Facebook page less than twenty-four hours prior to killing herself (Dolan and Robinson, 2013).

Polaris/Newscom

Rebecca Ann Sedwick

Suicides committed by young people who have been the victims of online bullying deeply touch the lives of their families, friends, and others who have not met them. Although we will never know the full story of what happened to Rebecca Ann Sedwick and the others described, these tragic occurrences bring us to larger sociological questions: Why does anyone commit suicide? Is suicide purely an individual phenomenon, or is it related to our social interactions and the social environment and society in which we live? How have technologies such as smartphones and social media affected our communication—both positively and negatively—with others?

As you are well aware, social media use among teens and college students continues to grow rapidly. You are engulfed by smartphones, tablets, and computers. Facebook, Twitter, YouTube, and other Internet-based social networking sites are taken for granted. You enjoy the positive effects of

social media, but the digital age may also produce harmful outcomes, particularly when some people harass others, causing psychological and physical harm and sometimes even bullying them into suicide.

Although suicide may seem like a "downer" for your study of sociology, I have chosen this topic because it is one of the first social issues that early sociologists studied. These thinkers believed that identifying the *social causes* of such behavior sets sociology apart from psychology, philosophy, and other areas of inquiry.

In this chapter we examine how sociological theories and research can help you understand social life, including seemingly individualistic acts such as committing suicide. You will see how sociological theory and research methods might be used to answer complex questions, and you will be able to wrestle with some of the difficulties that sociologists experience as they study human behavior.

"Goodbye forever my good friends, goodbye, I regret nothing. I have chosen to go with 3 people's advice and kill myself. I just wish it was faster."

—Prior to his suicide, Bart Palosz, a fifteen-year-old native of Poland who resided in Greenwich, Connecticut, for seven years, used his Google Plus page to express his suicidal musings during patterns of constant bullying by other boys at his school (Hussey and Leland, 2013).

How Much Do You Know About Suicide?

TRUE	FALSE		
T	F	1	After cancer and heart disease, suicide accounts for more years of life lost than any other cause of death.
T	F	2	Young people between the ages of 15 and 24 have the highest suicide rate in the United States.
T	F	3	White males accounted for 65 percent of all U.S. suicides in 2012.
T	F	4	Although females are more likely to attempt suicide, males are more likely to complete suicide (take their own life).
T	F	5	Each year, about 500,000 suicide deaths occur worldwide.
T	F	6	Firearms are the most commonly used method of suicide among males and females.
T	F	7	Alcohol intoxication is present in nearly one-fourth of all suicide deaths in the United States.
T	F	8	Government agencies such as the U.S. Centers for Disease Control and Prevention have developed public health programs to prevent suicidal behavior and now emphasize that promoting individual, family, and community connectedness is an important way to prevent such behavior.

Answers can be found at the end of the chapter.

 LO1 **Explain** why sociological research is necessary and how it challenges our commonsense beliefs about pressing social issues such as suicide.

Why Sociological Research Is Necessary

Most of us rely on our own experiences and personal knowledge to help us form ideas about what happens in everyday life and how the world works. However, there are many occasions when our personal knowledge is not enough to provide us with a thorough understanding of what is going on around us. This is why sociologists and other social scientists learn to question ordinary assumptions and to use specific research methods to find out more about the social world.

Sociologists obtain their knowledge of human behavior through research, which results in a body of information that helps us move beyond guesswork and common sense

in understanding society. The sociological perspective incorporates theory and research to arrive at a more accurate understanding of the "hows" and "whys" of human social interaction. Once we have an informed perspective about social issues, we are in a better position to find solutions and make changes. Social research, then, is a key part of sociology.

Common Sense and Sociological Research

How does sociological research challenge your commonsense beliefs about an issue? Consider suicide, for example. Most of us have commonsense ideas about suicide. Common sense may tell us that people who threaten suicide will not commit suicide. However, sociological research indicates that this assumption is frequently incorrect: People who threaten to kill themselves are often sending messages to others and may indeed attempt suicide. Common sense

SOCIOLOGY in GLOBAL PERSPECTIVE

Durkheim's Classical Study of Suicide Applied to Twenty-First-Century Young People in India

The bond attaching [people] to life slackens because the bond which attaches [them] to society is itself slack.

—Emile Durkheim, *Suicide* (1964b/1897)

Although this statement described social conditions accompanying the high rates of suicide found in late nineteenth-century France, Durkheim's words ring true today as we look at contemporary suicide rates for young people in cities such as New Delhi, India. Suicide rates in India are highest in the 15–29 age category and are especially great among those living in the wealthier and more-educated regions of the nation (NDTV.com, 2012; *Lancet*, 2012).

Why might those living in large cities in India be more prone to suicide than those living in rural areas?

Doesn't this seem unlikely? Many people think rural farmers facing poor harvests and high debt would have

may also tell us that suicide is caused by despair or depression. However, research suggests that suicide is sometimes used as a means of lashing out at friends and relatives because of real or imagined wrongs. Research also shows that some younger people commit suicide because they believe there is no way out of their problems, particularly when they are continually harassed or bullied by individuals whom they encounter daily.

Historically, the commonsense view of suicide was that it was a sin, a crime, and a mental illness. French sociologist Emile Durkheim refused to accept these explanations. In what is probably the first sociological study to use scientific research methods, he related suicide to the issue of cohesiveness (or lack of cohesiveness) in society instead of viewing suicide as an isolated act that could be understood only by studying individual personalities or inherited tendencies. In *Suicide* (1964b/1897), Durkheim documented his contention that a high suicide rate was symptomatic of large-scale societal problems. In the process he developed an approach to research that continues to influence researchers and help us look at suicide worldwide (see "Sociology in Global Perspective"). As we discuss sociological research, we will use the problem of suicide to demonstrate the research process.

Because much of sociology deals with everyday life, we might think that common sense, our own personal experiences, and the media are the best sources of information. However, our personal experiences are subjective, and much of the information provided by the media comes from sources seeking support for a particular point of view.

The content of the media is also influenced by the necessity of selling advertisements based on readership, audience ratings, or the number of hits that a website receives.

We need to evaluate the information we receive because the quantity—but, in some instances, not the quality—of information available has grown dramatically as a result of the information explosion brought about by corporate and social media.

Sociology and Scientific Evidence

In taking this course, you will be studying social science research and may be asked to write research reports or read and evaluate journal articles. If you attend graduate or professional school in fields that use sociological research, you will be expected to evaluate existing research and perhaps do your own. Hopefully, you will find that social research is relevant to the practical, everyday concerns of the real world.

Sociology involves *debunking*—the unmasking of fallacies (false or mistaken ideas or opinions) in everyday life, as well as official interpretations of society (Mills, 1959b). Because problems such as suicide involve threats to existing societal values, we cannot analyze these problems without acknowledging what values are involved. For example, we might ask a question like "Should assisted suicide be legal for terminally ill patients who wish to die?" We often answer questions such as this by using either a normative or empirical approach. The *normative approach* relies on religion, customs, habits, traditions, and law to answer questions.

the highest risk of suicide; however, this has not proven true in India. At first glance, we might think that economic success and a good education would provide insurance against suicide because of the greater happiness and job satisfaction among individuals in cities such as New Delhi, as these individuals have gained new opportunities and higher salaries in recent years. However, this economic boom—including the more-open markets of India in the past twenty years—has not only created new opportunities for people; these changes have also contributed to rapid urbanization and weakened social ties. The result? Intensified job anxiety, higher expectations, and more pressure for individual achievement. Social bonds have been weakened or dissolved as people move away from their families and their community. Ironically, newer technologies such as cell phones and social networking sites have contributed to the breakdown of traditional family units as communication has become more impersonal and fragmented.

In addition, life in the cities moves at a much faster pace than in the rural areas, and many individuals experience loneliness, sleep disorders, family discord, and

major health risks such as heart disease and depression (Mahapatra, 2007). In the words of Ramachandra Guha (2004), a historian residing in India, Durkheim's sociology of suicide remains highly relevant to finding new answers to this challenging problem: "The rash of suicides in city and village is a qualitatively new development in our history. We sense that tragedies are as much social as they are individual. But we know very little of what lies behind them. What we now await, in sum, is an Indian Durkheim."

REFLECT & ANALYZE

How does sociology help us examine seemingly private acts such as suicide within a larger social context? Why are some people more inclined to commit suicide if they are not part of a strong social fabric and have, at the same time, high job anxiety and intensive pressure to achieve?

It is based on strong beliefs about what is right and wrong and what "ought to be" in society. However, some sociologists discourage the use of the normative approach and advocate instead using an empirical approach. The *empirical approach* attempts to answer questions through conclusions drawn from systematic collection and analysis of data. This approach is referred to as the conventional model, or the "scientific method," and is based on the assumption that knowledge is best gained by direct, systematic observation. Two basic scientific standards must be met: (1) scientific beliefs should be supported by good evidence or information, and (2) these beliefs should be open to public debate and critiques from other scholars, with alternative interpretations being considered (Cancian, 1992).

 Compare deductive and inductive approaches in the theory and research cycle.

The Theory and Research Cycle

The relationship between theory and research has been referred to as a continuous cycle, as shown in ● Figure 2.1 (Wallace, 1971). You will recall that a *theory* is a set of logically interrelated statements that attempts to describe, explain, and (occasionally) predict social events. A theory attempts to explain why something is the way it is. *Research* is the process of systematically collecting information for the purpose of testing an existing theory or generating a new one.

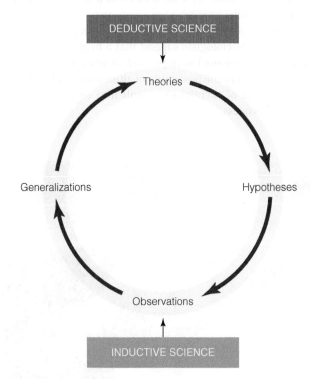

FIGURE 2.1 The Theory and Research Cycle
The theory and research cycle can be compared to a relay race; although all participants do not necessarily start or stop at the same point, they share a common goal—to examine all levels of social life.

Source: Adapted from Walter Wallace, *The Logic of Science in Sociology*. New York: Aldine de Gruyter, 1971.

The theory and research cycle consists of deductive and inductive approaches. In the *deductive approach* the researcher begins with a theory and uses research to test the theory. This approach proceeds as follows: (1) theories generate hypotheses, (2) hypotheses lead to observations (data gathering), (3) observations lead to the formation of generalizations, and (4) generalizations are used to support the theory, to suggest modifications to it, or to refute it. To illustrate, if we use the deductive method to determine why people commit suicide, we start by formulating a theory about the "causes" of suicide and then test our theory by collecting and analyzing data (for example, vital statistics on suicides or surveys to determine whether adult church members view suicide differently from nonmembers).

In the *inductive approach* the researcher collects information or data (facts or evidence) and then generates theories from the analysis of that data. Under the inductive approach, we would proceed as follows: (1) specific observations suggest generalizations, (2) generalizations produce a tentative theory, (3) the theory is tested through the formation of hypotheses, and (4) hypotheses may provide suggestions for additional observations (see ● Figure 2.2). Using the inductive approach to study suicide, we might start by simultaneously collecting and analyzing data related to suicidal behavior and then generate a theory. Researchers may break into the cycle at different points depending on what they want to know and what information is available.

Theory gives meaning to research; research helps support theory. Sociologists suggest that a healthy skepticism (a feature of science) is important in research because it keeps us open to the possibility of alternative explanations. Some degree of skepticism is built into each step of the research process.

FIGURE 2.2 Why do older African American men have a lower rate of suicide than white males of similar ages? Questions such as this often serve as the foundation for explanatory studies as sociologists attempt to understand and describe certain cause-and-effect relationships.

LO3 **Distinguish** between quantitative research and qualitative research.

The Sociological Research Process

Not all sociologists conduct research in the same manner. Some researchers primarily engage in quantitative research, whereas others engage in qualitative research. With *quantitative research*, the goal is scientific objectivity, and the focus is on data that can be measured numerically. Quantitative research typically emphasizes complex statistical techniques. Most sociological studies on suicide have used quantitative research. They have compared rates of suicide with almost every conceivable variable, including age, sex, race/ethnicity, education, and even sports participation. "Understanding Statistical Data Presentations" explains how to read a table, interpret the data, and draw conclusions.

With *qualitative research*, interpretive description (words) rather than statistics (numbers) is used to analyze underlying meanings and patterns of social relationships. An example of qualitative research is a study in which the researcher systematically analyzed the contents of the notes of suicide victims to determine recurring themes, such as a feeling of despair or failure. Through this study the researcher hoped to determine whether any patterns could be found that would help in understanding why people might kill themselves.

LO4 **List** and briefly describe the steps in the quantitative research model.

The Quantitative Research Model

Research models are tailored to the specific problem being investigated and the focus of the researcher. Both quantitative research and qualitative research contribute to our knowledge of society and human social interaction, and both involve a series of steps, as shown in ● Figure 2.3. We will now trace the steps in the "conventional" research model, which focuses on quantitative research. Then we will describe an alternative model that emphasizes qualitative research.

1. *Select and define the research problem.* When you engage in research, the first step is to select and clearly define the research topic. Sometimes, a specific experience such as having known someone who committed suicide can trigger your interest in a topic. Other times, you might select topics to fill gaps or challenge misconceptions in existing research or to test a specific theory (Babbie, 2013). Emile Durkheim selected suicide because he wanted to demonstrate the importance of *society* in

Yellow Dog Productions/Stone/Getty Images

UNDERSTANDING

Statistical Data Presentations

Are young males or females more likely to die violently? How do homicide, suicide, and firearm death rates (per 100,000 population) compare for males and females ages 15 to 19 in the United States? Sociologists use statistical tables as a concise way to present data in a relatively small space;
• Table 2.1 gives an example. To understand a table, follow these steps:

1. *Read the title.* From the title, "Rates (per 100,000 U.S. Population) for Homicide, Suicide, and Firearm-Related Deaths of Youths Ages 15–19, by Gender, 2012," we learn that the table shows relationships between two variables: gender and three causes of violent deaths among young people in a specific age category.
2. *Check the source and other explanatory notes.* In this case the first source is the Centers for Disease Control and Prevention (2014), a site that allows researchers to put in online inquiries for specific information from the CDC's database. The second source is Child Trends Data Bank (2014). This data bank is a nonprofit research and policy center that researches issues pertaining to children and young people. The explanatory note in this table states that firearm deaths, which constitute a majority of teen homicides and suicides, may also include *accidental* deaths that are firearm related. This distinction is made in Table 2.1 because it is possible for "firearm-related death" to occur accidentally. However, firearms were the method of death in 88 percent of teen homicides and 42 percent of teen suicides in 2012.
3. *Read the headings for each column and each row.* The main column headings in Table 2.1 are "Method," "Males," and "Females." The columns present information (usually numbers) arranged vertically. The rows present information horizontally. Here, the row headings indicate homicide, suicide, and firearm-related death. Based on the explanation above regarding firearm-related death, we know that some overlap exists between the first two categories—homicide and suicide—and the third, deaths that are firearm related.
4. *Examine and compare the data.* To examine the data, determine what units of measurement have been used. In Table 2.1 the figures are rates per 100,000 males or females in a specific age category. For example, the suicide rate is 12.5 per 100,000 population of males between the ages of 15 and 19 as compared with only 3.9 per 100,000 population of females in the same age category.

TABLE 2.1

Rates (per 100,000 U.S. Population) for Homicide, Suicide, and Firearm-Related Deaths of Youths Ages 15–19, by Gender, 2012		
Method	**Males**	**Females**
Homicide	12.8	2.0
Suicide	12.5	3.9
Firearm-Related Deaths[a]	18.5	2.4

[a]Firearm deaths, which constitute a majority of teen homicides and suicides, also include accidental deaths that are firearm related.

Sources: U.S. Centers for Disease Control and Prevention, 2014; Child Trends Data Bank, 2014.

5. *Draw conclusions.* By looking for patterns, some conclusions can be drawn from Table 2.1:
   a. *Determining differences by gender.* Males between the ages of 15 and 19 are more than three times more likely than females to die from suicide (12.5 compared with 3.9 per 100,000 in 2012). Males in this age category are also more than six times more likely to die from homicide (12.8 compared with 2.0 per 100,000). And even more noteworthy, males are nearly eight times more likely to die from any firearm-related incident (either intentional or unintentional) than females of this age. As shown in Table 2.1, 18.5 per 100,000 males ages 15–19 died by firearms in 2012, compared with 2.4 per 100,000 females in that same age category.
   b. *Drawing appropriate conclusions.* Males between the ages of 15 and 19 are much more likely than females in their age category to die violently, and many of those deaths are firearm related. Although not indicated in this table, it is important to note that differences by race and Hispanic origin are also significant. In 2012 the homicide rate for African American male teens was 50.9 per 100,000, more than 20 times higher than the rate for white male teens. The highest rate of suicide in the 15–19 age category was among American Indian males at 22.2 per 100,000. For more information, visit the Child Trends DataBank website.

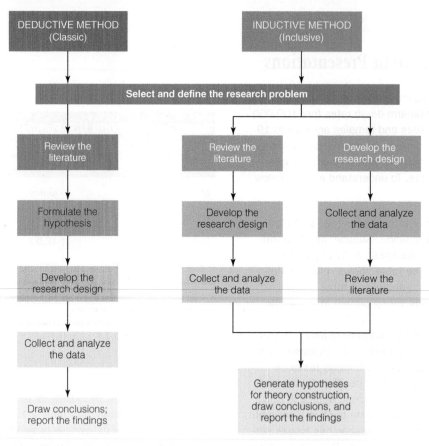

FIGURE 2.3 **Steps in Sociological Research**

situations that might appear to be arbitrary acts by individuals: In his time, suicide was widely believed to be a uniquely individualistic act. Durkheim emphasized that *suicide rates* provide better explanations of suicide than do *individual acts* of suicide. He reasoned that if suicide were purely an individual act, then the rate of suicide (the relative number of people who kill themselves each year) should be the same for every group regardless of culture and social structure. Durkheim wanted to know why there were different rates of suicide—whether factors such as religion, marital status, sex, and age had an effect on social cohesion.

2. *Review previous research.* Before you begin your research, it is important to review the literature to see what others have written about the topic. Analyzing what previous researchers have found helps to clarify issues and focus the direction of your own research. But when Durkheim began his study, very little sociological literature existed for him to review other than the works of Henry Morselli (1975/1881), who concluded that suicide was a part of an evolutionary process whereby "weak-brained" individuals were sorted out by insanity and voluntary death.

3. *Formulate the hypothesis (if applicable).* You may formulate a **hypothesis**—a tentative statement of the relationship between two or more concepts. Concepts

are the abstract elements representing some aspect of the world in simplified form (such as "social integration" or "loneliness"). As you formulate your hypothesis about suicide, you may need to convert concepts to variables. A *variable* is any concept with measurable traits or characteristics that can change or vary from one person, time, situation, or society to another. Variables are the observable and/or measurable counterparts of concepts. For example, "suicide" is a concept; the "rate of suicide" is a variable.

The most fundamental relationship in a hypothesis is between a dependent variable and one or more independent variables (see ● Figure 2.4). The **independent variable** is presumed to cause or determine a dependent variable. Age, sex, race, and ethnicity are often used as independent variables. The **dependent variable** is assumed to depend on or be caused by the independent variable(s) (Babbie, 2013). Durkheim used the degree of social integration in society as the independent variable to determine its influence on the dependent variable, the rate of suicide.

Whether a variable is dependent or independent depends on the context in which it is used. To use variables in the contemporary research process, sociologists create operational definitions. An *operational definition* is an explanation of an abstract concept in terms of observable features that are specific enough

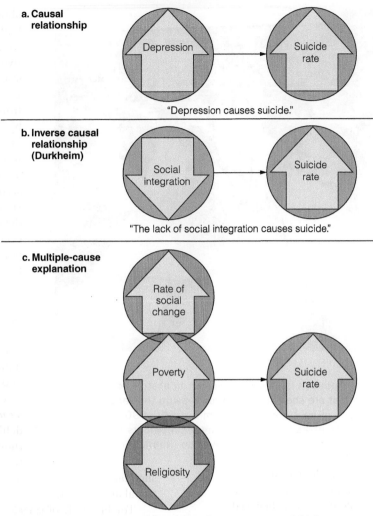

a. Causal
   relationship

Depression → Suicide rate

"Depression causes suicide."

b. Inverse causal
   relationship
   (Durkheim)

Social integration → Suicide rate

"The lack of social integration causes suicide."

c. Multiple-cause
   explanation

Rate of social change

Poverty → Suicide rate

Religiosity

"Many factors interact to cause suicide."

FIGURE 2.4   Hypothesized Relationships Between Variables
A causal hypothesis connects one or more independent (causal) variables
with a dependent (affected) variable. The diagram illustrates three hypotheses about the causes of suicide. To test these hypotheses, social scientists
would need to operationalize the variables (define them in measurable terms)
and then investigate whether the data support the proposed explanation.

to measure the variable. For example, suppose that your goal is to earn an *A* in this course (● Figure 2.5). Your professor may have created an operational definition by defining an *A* as earning an exam average of 90 percent or above (Babbie, 2013).

Events such as suicide are too complex to be caused by any one variable. Therefore, they must be explained in terms of *multiple causation*—that is, an event occurs as a result of many factors operating in combination. What *does* cause suicide? Social scientists cite multiple causes, including rapid social change, economic conditions, hopeless poverty, and lack of religiosity (the degree to which an individual or group feels committed to a particular system of religious beliefs).

4. *Develop the research design.* In developing the research design, you must first consider the units of analysis

and the time frame of the study. A *unit of analysis* is *what* or *whom* is being studied (Babbie, 2013). In social science research, individuals are the most typical unit of analysis. Social groups (such as families, cities, or geographic regions), organizations (such as clubs, labor unions, or political parties), and social artifacts

hypothesis
in research studies, a tentative statement of the relationship between two or more concepts.

independent variable
a variable that is presumed to cause or determine a dependent variable.

dependent variable
a variable that is assumed to depend on or be caused by the independent variable(s).

FIGURE 2.5 An operational definition is an explanation of an abstract concept in terms of observable features that are specific enough to measure the variable. For example, the operational definition of an *A* may be an exam average of 90 percent or above. After college professors have established the grading requirements for a course, students seek to meet those expectations by performing well on examinations.

(such as books, paintings, or weddings) may also be units of analysis. Durkheim's unit of analysis was social groups, not individuals, because he believed that the study of individual cases of suicide would not explain the rates of suicide.

After determining the unit of analysis for your study, you must select a time frame for study: cross-sectional or longitudinal. *Cross-sectional studies* are based on observations that take place at a single point in time; these studies focus on behavior or responses at a specific moment. *Longitudinal studies* are concerned with what is happening over a period of time or at several different points in time; they focus on processes and social change. Some longitudinal studies are designed to examine the same set of people each time, whereas others look at trends within a general population. Using longitudinal data, Durkheim was able to compare suicide rates over a period of time in France and other European nations.

5. *Collect and analyze the data.* Your next step is to collect and analyze data. You must decide which *population*—persons about whom we want to be able to draw conclusions—will be observed or questioned. Then it is necessary to select a sample of people from the larger population to be studied. It is important that the sample accurately represents the larger population. For example, if you arbitrarily selected five students from your class to interview

about the problem of bullying, they probably would not be representative of your school's total student body. However, if you selected five students from the total student body by a random sample, they might be closer to being representative (although a random sample of five students would be too small to yield much useful data). In *random sampling*, every member of an entire population being studied has the same chance of being selected. You would have a more representative sample of the total student body, for example, if you placed all the students' names in a rotating drum and conducted a drawing. By contrast, in *probability sampling*, participants are deliberately chosen because they have specific characteristics, possibly including such factors as age, sex, race/ethnicity, and educational attainment.

For his study of suicide, Durkheim collected data from vital statistics for approximately 26,000 suicides. He classified them separately according to age, sex, marital status, presence or absence of children in the family, religion, geographic location, calendar date, method of suicide, and a number of other variables. As Durkheim analyzed his data, four distinct categories of suicide emerged: egoistic, altruistic, anomic, and fatalistic. *Egoistic suicide* occurs among people who are isolated from any social group. For example, Durkheim concluded that suicide rates were relatively high in Protestant countries in Europe because Protestants believed in individualism and were more loosely tied to the church than were Catholics. Single people had proportionately higher suicide rates than married persons because they had a low degree of social integration, which contributed to their loneliness. In contrast, *altruistic suicide* occurs among individuals who are excessively integrated into society. An example is military leaders who kill themselves after defeat in battle because they have so strongly identified themselves with their cause that they believe they cannot live with defeat. Today, other factors such as family conflict at home or extended periods of military service may also contribute to relatively high rates of suicide among U.S. military personnel (see "Sociology and Social Policy"). According to Durkheim, people are more likely

SOCIOLOGY & SOCIAL POLICY

Establishing Policies to Help Prevent Military Suicides

Grandpa, I just wanted to give you my thanks for being a great influence in my life.

—Marine Corporal Daniel O'Brien, who completed two tours of duty in Iraq, sent a note containing this statement to his grandfather prior to taking his own life (qtd. in Roberts, 2011).

We think that we are seeing a societal problem [in regard to suicides by military personnel], and frankly the Army is the canary in the mine shaft here.

—General Ray Carpenter, commander of the Army National Guard, discussing how the incidence of suicide has continued to be a problem in most branches of the U.S. military (qtd. in Roberts, 2011)

Suicide is a nationwide problem, but it has grown as a special problem in the military, where it is now the leading cause of death, after accidents, among U.S. service members (Corr, 2014). Most suicides and suicide attempts occurred among service members stationed in the United States, and most suicides involved the use of a firearm.

Shocked by relatively high rates of both suicide and suicide attempts among members of the Air Force, Army, Navy, and Marine Corps, the U.S. Department of Defense and the executive branch of the U.S. government have encouraged all branches of the military to learn more about the sociological causes of suicide and to develop comprehensive

Fabrizio Bensch/Reuters

What unique social conditions do military personnel face that might contribute to suicide or other conditions such as depression and alcoholism?

suicide-prevention initiatives to help reduce the problem and support military service members around the globe. Efforts are being made to educate military leaders, service members, and family members about the risk factors for self-harm. Policies are being implemented to increase the availability of, and access to, mental health resources for individuals who cope with suicidal ideation and other psychological distress. Special initiatives have been established to destigmatize the process of seeking help from mental health professionals. Branches of the military, such as the Navy, are advising commanders to ask military personnel thought to be at risk of harming themselves to voluntarily turn over their personal firearms for temporary safekeeping (Reilly, 2014).

Sociological issues are apparent in social policy pertaining to dealing with suicidal behavior in the military. According to a U.S. Army report (2012: 51),

Each potential suicide or attempted suicide is different with respect to contributing factors and triggering events. Each victim responds differently to pre-suicide stressors based on protective factors such as personal resilience, coping skills, and whether or not they are help-seeking. . . . To be sure, the Army has investigated numerous suicide cases that, in hindsight, seemed to present a clear trail of behavioral indicators that may have afforded leaders or others in the social circle an opportunity to respond.

Although no single cause can be identified for suicide, factors such as financial worries, relationship issues, legal trouble, substance abuse, medical problems, and posttraumatic stress are all thought to be associated with suicidal behavior among military personnel. Perhaps greater awareness and new social policies implemented by the military will help reduce this problem over time.

REFLECT & ANALYZE

How might lengthy wars and deployments away from home contribute to suicidal behavior among troops? Do you anticipate that military suicide rates might decrease in peacetime? Why or why not?

to kill themselves when social cohesion is either very weak or very strong, and/or when nations experience rapid social change.

We have traced the steps in the "conventional" research process (based on quantitative research). But what steps do you think might be taken in an alternative approach based on qualitative research?

random sampling
a study approach in which every member of an entire population being studied has the same chance of being selected.

probability sampling
choosing participants for a study on the basis of specific characteristics, possibly including such factors as age, sex, race/ethnicity, and educational attainment.

LO5 List and briefly describe the steps in a qualitative research model.

A Qualitative Research Model

Although the same underlying logic is involved in both quantitative and qualitative sociological research, the *styles* of these two models are very different. As previously stated, qualitative research is more likely to be used when the research question does not easily lend itself to numbers and statistical methods (• Figure 2.6). As compared to a quantitative model, a qualitative approach often involves a different type of research question and a smaller number of cases.

Although the qualitative approach follows the conventional research approach in presenting a problem, asking a question, collecting and analyzing data, and seeking to answer the question, it also has several unique features:

1. *The researcher begins with a general approach rather than a highly detailed plan.* Flexibility is necessary

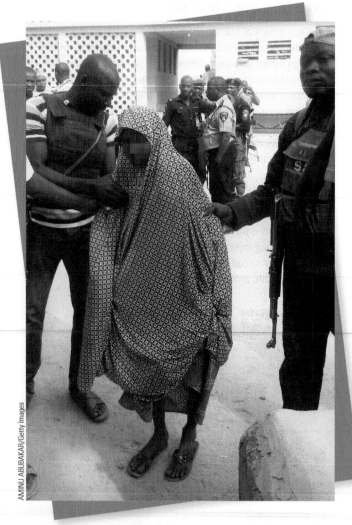

FIGURE 2.6 Sociological research has begun to look at issues such as what social factors might motivate individuals to take their own life in the process of committing a terrorist act.

because of the nature of the research question. The topic needs to be explored so that we can know "how" or "what" is going on, but we may not be able to explain "why" a particular social phenomenon is occurring.

2. *The researcher has to decide when the literature review and theory application should take place.* Initial work may involve redefining existing concepts or conceptualizing how existing studies have been conducted. The literature review may take place at an early stage, before the research design is fully developed, or it may occur after the development of the research design and after the data collection has already occurred.

3. *The study presents a detailed view of the topic.* Qualitative research usually involves a smaller number of cases and many variables, whereas quantitative researchers typically work with a few variables and many cases.

4. *Access to people or other resources that can provide the necessary data is crucial.* Unlike the quantitative researcher, who often uses existing databases, many qualitative researchers generate their own data. As a result, it is necessary to have access to people and build rapport with them.

5. *Appropriate research method(s) are important for acquiring useful qualitative data.* Qualitative studies are often based on field research such as observation, participant observation, case studies, ethnography, and unstructured interviews, as discussed in the "Research Methods" section of this chapter.

How might qualitative research be used to study suicidal behavior? In studying different rates of suicide among women and men, for example, the social psychologist Silvia Canetto (1992) questioned whether existing theories and quantitative research provided an adequate explanation for gender differences in suicidal behavior and decided that she would explore alternate explanations. Analyzing previous research, Canetto learned that most studies linked suicidal behavior in women to problems in their personal relationships, particularly with members of the opposite sex. By contrast, most studies of men's suicides focused on their performance and found that men are more likely to be suicidal when their self-esteem and independence are threatened. According to Canetto's analysis, gender differences in suicidal behavior are more closely associated with beliefs about and expectations for men and women rather than purely interpersonal crises.

As in Canetto's study, researchers using a qualitative approach may engage in *problem formulation* to clarify the research question and to develop questions of concern and interest to the research participants. To create a research design for Canetto's study, you might start with the proposition that most studies may have attributed women's and men's suicidal behavior to the wrong causes. Next, you might decide to interview people who have attempted suicide by using a collaborative approach in which the

AMINU ABUBAKAR/Getty Images

participants suggest avenues of inquiry that the researcher should explore.

Although Canetto did not gather data in her study, she reevaluated existing research, concluding that alternative explanations of women's and men's suicidal behavior are justified from existing data.

In a qualitative approach the next step is collecting and analyzing data to assess the validity of the starting proposition. Data gathering is the foundation of the research. Researchers pursuing a qualitative approach tend to gather data in natural settings, such as where the person lives or works, rather than in a laboratory or other research setting. Data collection and analysis frequently occur at the same time, and the analysis draws heavily on the language of the persons being studied, not the researcher.

Research Methods

How do sociologists know which research method to use? Are some approaches better than others? Which method is best for a particular problem? **Research methods** are specific strategies or techniques for systematically conducting research. We will look at four of these methods.

 LO6 **Explain** what is meant by survey research and briefly discuss three types of surveys.

Survey Research

A **survey** is a poll in which the researcher gathers facts or attempts to determine the relationships among facts. Surveys are the most widely used research method in the social sciences because they make it possible to study things that are not directly observable—such as people's attitudes and beliefs—and to describe a population too large to observe directly (Babbie, 2013). Researchers frequently select a representative sample (a small group of respondents) from a larger population (the total group of people) to answer questions about their attitudes, opinions, or behavior. **Respondents** are persons who provide data for analysis through interviews or questionnaires. The Gallup, Harris, Roper, and Pew polls are among the most widely known large-scale surveys; however, government agencies such as the U.S. Census Bureau conduct a variety of surveys as well.

Unlike many polls that use various methods of gaining a representative sample of the larger population, the Census Bureau attempts to gain information from all persons in the United States. The decennial census occurs every 10 years, in the years ending in "0." The purpose of this census is to count the population and housing units of the entire United States. The population count determines how seats in the U.S. House of Representatives are apportioned; however, census figures are also used in formulating public policy and in planning and decision making in the private sector. Statistics from the Census Bureau provide information that sociologists use in their research.

Let's take a brief look at the most frequently used types of surveys.

Types of Surveys Survey data are collected by using self-administered questionnaires, face-to-face interviews, and telephone or computer surveys. A **questionnaire** is a printed research instrument containing a series of items to which subjects respond. Items are often in the form of statements with which the respondent is asked to "agree" or "disagree." Questionnaires may be administered by interviewers in face-to-face encounters or by telephone, but the most commonly used technique is the *self-administered questionnaire*. The questionnaires are typically mailed or delivered to the respondents' homes; however, they may also be administered to groups of respondents gathered at the same place at the same time. For example, the sociologist Kevin E. Early (1992) used survey data collected through questionnaires to test his hypothesis that suicide rates are lower among African Americans than among white Americans because of the influence of the black church. Data from questionnaires filled out by members of six African American churches in Florida supported Early's hypothesis that the church buffers some African Americans against harsh social forces, such as racism, that might otherwise lead to suicide. Studies subsequent to Early's have also found that religious involvement and thoughts of looking to God for strength, comfort, and guidance are largely protective of suicidal thoughts and attempts among African Americans (see Taylor, Chatters, and Joe, 2011, for an example).

Survey data may also be collected by interviews (● Figure 2.7). An **interview** is a data-collection encounter in which an interviewer asks the respondent questions and records the answers. Survey research often uses *structured interviews,* in which the interviewer asks questions from a standardized questionnaire. Structured interviews tend to produce uniform or replicable data that can be elicited time after time by different interviewers. For example, in addition to surveying congregation members, Early (1992) conducted interviews with pastors of African American churches to determine the pastors' opinions about the extent to which the African American church reinforces values and beliefs that discourage suicide.

research methods
specific strategies or techniques for systematically conducting research.

survey
a poll in which the researcher gathers facts or attempts to determine the relationships among facts.

respondents
persons who provide data for analysis through interviews or questionnaires.

questionnaire
a printed research instrument containing a series of items to which subjects respond.

interview
a data-collection encounter in which an interviewer asks the respondent questions and records the answers.

Paul Conklin/PhotoEdit

FIGURE 2.7 Conducting surveys and polls is an important means of gathering data from respondents. Some surveys take place on street corners; increasingly, however, such surveys are done by telephone, the Internet, or other means.

Interviews have specific advantages. They are usually more effective in dealing with complicated issues and provide an opportunity for face-to-face communication between the interviewer and the respondent. Although interviews provide a wide variety of useful information, a major disadvantage is the cost and time involved in conducting the interviews and analyzing the results. A quicker method of administering questionnaires is the *telephone* or *computer survey.* Telephone and computer surveys give greater control over data collection and provide greater personal safety for respondents and researchers than do personal encounters. In *computer-assisted telephone interviewing* (sometimes called CATI), the interviewer uses a computer to dial random telephone numbers, reads the questions shown on the video monitor to the respondent, and then types the responses into the computer terminal (• Figure 2.8). The answers are immediately stored in the central computer, which automatically prepares them for data analysis. However, the respondent must answer the phone before the interview can take place, and many people screen their phone calls. In the second decade of the twenty-first century, online survey research has increased dramatically as software packages and online survey services have made this type of research easier to conduct. Online research makes it possible to study virtual communities, online relationships, social media interactions, and other types of computer-mediated communications networks around the world.

Strengths and Weaknesses of Surveys Survey research is useful in describing the characteristics of a large population without having to interview each person in that population. In recent years, computer technology has enhanced researchers' ability to do *multivariate analysis*—research involving more than two independent variables. For example, to assess the influence of religion on suicidal behavior among African Americans, a researcher might look at the effects of age, sex, income level, and other variables all at once to determine which of these independent variables influences suicide the most or least and how influential each variable is relative to the others. However, a weakness of survey research is the use of standardized questions; this approach tends to force respondents into categories in which they may or may not belong. Moreover, survey research relies on self-reported information, and some people may be less than truthful, particularly on emotionally charged issues such as suicide. Some scholars have also criticized the way that survey data are used. They believe that survey data do not always constitute the "hard facts" that other analysts may use to justify changes in public policy or law. For example, survey statistics may overestimate or underestimate the extent of a problem and work against some categories of people more than others, as shown in • Table 2.2.

LO7 **Compare** research methods used in secondary analysis of existing data, field research, experiments, and triangulation.

Secondary Analysis of Existing Data

In ***secondary analysis,*** researchers use existing material and analyze data that were originally collected by others. Existing data sources include public records, official reports of organizations and government agencies, and surveys conducted by researchers in universities and private corporations. Research data gathered from studies are available in data banks, such as the Pew Research Center, the Inter University Consortium for Political and Social Research, the National Opinion Research Center (NORC), and the Roper Public Opinion Research Center. Today, many researchers studying suicide use data compiled by the U.S. Centers for Disease Control and Prevention (CDC) and the National Center for Injury Prevention and Control (see • Figure 2.9). Other sources of data for secondary analysis are books, magazines, newspapers, radio and television programs, websites, and personal documents. Secondary analysis is referred to as *unobtrusive research* because it has

FIGURE 2.8 Computer-assisted telephone interviewing is an easy and cost-efficient method of conducting research. However, the widespread use of answering machines, voicemail, and caller ID may make this form of research more difficult in the twenty-first century.

TABLE 2.2

Statistics: What We Know (and Don't Know)		
Topic	**Homelessness in the United States**	**Suicide in the United States**
Research Finding	At least 610,000 people are officially identified as homeless in the United States.	More than 40,600 suicides occurred in this country in 2012.
Possible Problems	How do we count "homelessness"? What problems do "homeless" individuals and families face?	Are suicide rates different for some categories of U.S. residents?
Explanation	The homeless are difficult to count, frequently attempting to avoid encounters with public officials and census takers.	It is difficult to know about racial and ethnic disparities in suicide rates: Census data tend to shift categories (such as Hispanic origin and white [non-Hispanic]) over time, making comparisons difficult, if not impossible.

As the examples in this table show, statistics provide certain insights into the prevalence of social issues such as homelessness and suicide but do not always provide the *answer* regarding the nature and extent of the problem. What other difficulties do researchers encounter when gathering data on people?

no impact on the people being studied. In Durkheim's study of suicide, for example, his analysis of existing statistics on suicide did nothing to increase or decrease the number of people who *actually* committed suicide.

Analyzing Existing Statistics Secondary analysis may involve obtaining *raw data* collected by other researchers and undertaking a statistical analysis of the data, or it may involve the use of other researchers' existing statistical analyses. In analysis of existing statistics, the unit of analysis

secondary analysis
a research method in which researchers use existing material and analyze data that were originally collected by others.

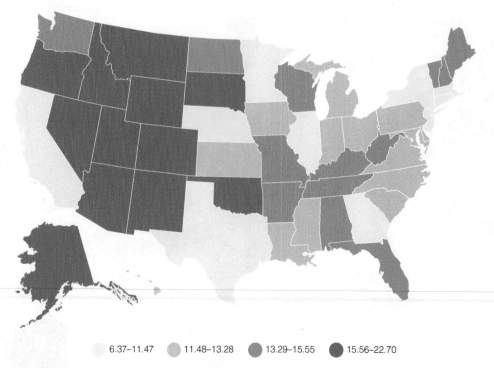

6.37–11.47　　11.48–13.28　　13.29–15.55　　15.56–22.70

FIGURE 2.9　**National Suicide Statistics at a Glance**
Age-adjusted suicide rate per 100,000 population per state, 2000–2006, United States.

is often *not* the individual. Most existing statistics are *aggregated:* They describe a group. For example, Durkheim used vital statistics (death records) that were originally collected for other purposes to examine the relationship among variables such as age, marital status, and the circumstances surrounding the suicides of a large number of individuals.

In one study, K. D. Breault (1986) analyzed secondary data collected by government agencies to test Durkheim's hypothesis that religion and social integration provide protection from suicide. Using suicide as the dependent variable and church membership, divorce, unemployment, and female labor-force participation as several of his independent variables, Breault performed a series of sophisticated statistical analyses and concluded that the data supported Durkheim's views on social integration and his theory of egoistic suicide.

Analyzing Content　*Content analysis* is the systematic examination of cultural artifacts or various forms of communication to extract thematic data and draw conclusions about social life. *Cultural artifacts* are products of individual activity, social organizations, technology, and cultural patterns. Among the materials studied are written records (such as diaries, love letters, poems, books, and graffiti), narratives and visual texts (such as movies, television programs, websites, advertisements, and greeting cards), and material culture (such as music, art, and even garbage). Researchers may look for regular patterns, such as frequency of suicide as a topic on television talk shows.

Content analysis provides objective coding procedures for analyzing written material. It also allows for the counting and arranging of data into clearly identifiable categories (manifest coding) and provides for the creation of analytically developed categories (latent or open coding). Using latent or open coding, it is possible to identify general themes, create generalizations, and develop "grounded theoretical" explanations (Glaser and Strauss, 1967). As this explanation suggests, researchers use both qualitative and quantitative procedures in content analysis.

How might a social scientist use content analysis in research on why people commit suicide? Suicide notes and diaries are useful forms of cultural artifacts. Suicide notes have been subjected to extensive analysis because they are "ultrapersonal documents" that are not solicited by others and are frequently written just before the person's death (Leenaars, 1988: 34). Many notes provide new levels of meaning regarding the *individuality* of the person who committed or attempted suicide. For example, some notes indicate that people may want to get revenge and make other people feel guilty or responsible for their suicide (Leenaars, 1988). Thus, suicide notes may be a valuable starting point for finding patterns of suicidal behavior and determining the characteristics of people who are most likely to commit suicide (Leenaars, 1988).

Strengths and Weaknesses of Secondary Analysis
One strength of secondary analysis is that data are readily available and inexpensive. Another is that, because the researcher often does not collect the data personally, the chances of bias may be reduced. In addition, the use of existing sources makes it possible to analyze longitudinal data (things that take place over a period of time or at several different points in time) to provide a historical context within

which to locate original research. However, secondary analysis has inherent problems. For one thing, the researcher does not always know if the data are incomplete, unauthentic, or inaccurate.

Field Research

Field research is the study of social life in its natural setting: observing and interviewing people where they live, work, and play. Some kinds of behavior can be studied best by "being there"; a fuller understanding can be developed through observations, face-to-face discussions, and participation in events. Researchers use these methods to generate *qualitative* data: observations that are best described verbally rather than numerically (• Figure 2.10).

Participant Observation Sociologists who are interested in observing social interaction as it occurs may use *participant observation*—the process of collecting systematic observations while being part of the activities of the group that the researcher is studying. Participant observation generates more "inside" information than simply asking questions or observing from the outside. For example, to learn more about how coroners make a ruling of "suicide" in connection with a death and to analyze what (if any) effect such a ruling has on the accuracy of official suicide statistics, the sociologist Steve Taylor (1982) engaged in participant observation at a coroner's office over a six-month period. As he followed a number of cases from the initial report of death through the various stages of investigation, Taylor learned that it was important to "be around" so that he could listen to discussions and ask the coroners questions because intuition and guesswork play a large part in some decisions to rule a death as a suicide.

Case Studies Most participant observation research takes the form of a *case study*, which is often an in-depth, multifaceted investigation of a single event, person, or social grouping. However, a case study may also involve multiple

FIGURE 2.10 Field research takes place in a wide variety of settings. For example, how might sociologists study the ways in which parents and their college-age children cope with change when the students first leave home and move into college housing?

cases and is then referred to as a *collective case study*. Whether the case is single or collective, most case studies require detailed, in-depth data collection involving multiple sources of rich information such as documents and records and the use of methods such as participant observation, unstructured or in-depth interviews, and life histories. As they collect extensive amounts of data, the researchers seek to develop a detailed description of the case, to analyze the themes or issues that emerge, and to interpret or create their own assertions about the case. For example, the anthropologist Elliot Liebow "backed into" his study of single, homeless women living in emergency shelters by becoming a volunteer at a shelter. As he got to know the women, Liebow became fascinated with their lives and survival strategies and spent four years engaged in participant observation research that culminated in his book *Tell Them Who I Am* (1993).

Ethnography An *ethnography* is a detailed study of the life and activities of a group of people by researchers who may live with that group over a period of years. Unlike participant observation, ethnographic studies usually take place over a longer period of time. For example, the sociologist Sudhir Venkatesh (2013) conducted more than a decade

content analysis
the systematic examination of cultural artifacts or various forms of communication to extract thematic data and draw conclusions about social life.

field research
the study of social life in its natural setting: observing and interviewing people where they live, work, and play.

participant observation
a research method in which researchers collect systematic observations while being part of the activities of the group being studied.

ethnography
a detailed study of the life and activities of a group of people by researchers who may live with that group over a period of years.

FIGURE 2.11 The research of Sudhir Venkatesh (left) resulted in the publication of a book titled *Floating City: A Rogue Sociologist Lost and Found in New York's Underground Economy* (2013).

of research on New York City's underground economy before writing his book *Floating City: A Rogue Sociologist Lost and Found in New York's Underground Economy* (● Figure 2.11). He observed and interviewed a wide array of people, many of whom were involved in illegal transactions. Among his participants were poor immigrants, drug bosses and dealers, prostitutes, businesspeople, socialites, and academics, all of whom had their own stories to tell. Based on his extensive ethnographic research, Venkatesh was able to describe in great detail how a vast underground economy serves to unite the wealthy and poor in New York and other large cities.

Unstructured Interviews An *unstructured interview* is an extended, open-ended interaction between an interviewer and an interviewee. This type of interview is referred to as *unstructured* because few predetermined or standardized procedures are established for conducting it. Because many decisions have to be made during the interview, this approach requires that the researcher have a high level of skill in interviewing and extensive knowledge regarding the interview topic. Unstructured interviews are essentially conversations in which interviewers establish the general direction by asking open-ended questions, to which interviewees

may respond flexibly, and then interviewers may "shift gears" to pursue specific topics raised by interviewees.

Before conducting in-depth interviews, researchers must make a number of decisions, including how the people to be interviewed will be selected. Respondents are often chosen by "snowball sampling"—a method in which the researcher interviews a few individuals who possess a certain characteristic, and then these interviewees are asked to supply the names of others with the same characteristic (such as persons who are members of the same social organization). This process continues until the sample has "snowballed" into an acceptable size and no new information of any significance is being gained by the researchers.

Interviews and Theory Construction In-depth interviews, along with participant observation and case studies, are frequently used to develop theories through observation. The term *grounded theory* was developed by sociologists Barney Glaser and Anselm Strauss (1967) to describe this inductive method of theory construction. Researchers who use grounded theory collect and analyze data simultaneously. For example, after in-depth interviews with 106 suicide attempters, researchers in one study concluded that half of the individuals who attempted suicide wanted *both* to live *and* to die at the time of their attempt. From these unstructured interviews it became obvious that ambivalence led about half of "serious" suicidal attempters to "literally gamble with death" (Kovacs and Beck, 1977, qtd. in Taylor, 1982: 144). After asking their initial unstructured questions of the interviewees, Kovacs and Taylor decided to widen the research question from "Why do people kill themselves?" to a broader question: "Why do people engage in acts of self-damage which may result in death?" In other words, uncertainty of outcome is a common feature of most suicidal acts.

Strengths and Weaknesses of Field Research
Participant observation research, case studies, ethnography, and unstructured interviews provide opportunities for researchers to view from the inside what may not be obvious to an outside observer. They are useful when attitudes and behaviors can be understood best within their natural setting or when the researcher wants to study social processes and change over a period of time. They provide a wealth of information about the reactions of people and give us an opportunity to generate theories from the data collected.

A weakness of field research is the inability to generalize what is learned from a specific group or community to a larger population. Data collected in natural settings are descriptive and do not lend themselves to precise measurement. To counteract these criticisms, some qualitative researchers use computer-assisted qualitative-data-analysis programs that make it easier for the researchers to enter, organize, annotate, code, retrieve, count, and analyze data.

Experiments

An *experiment* is a carefully designed situation in which the researcher studies the impact of certain variables on subjects' attitudes or behavior. Experiments are designed to create "real-life" situations, ideally under controlled circumstances, in which the influence of different variables can be modified and measured. Conventional experiments require that subjects be divided into two groups: an experimental group and a control group. The *experimental group* contains the subjects who are exposed to an independent variable (the experimental condition) to study its effect on them. The *control group* contains the subjects who are not exposed to the independent variable. The members of the two groups are matched for similar characteristics so that comparisons may be made between the groups. The experimental and control groups are then compared to see if they differ in relation to the dependent variable, and the hypothesis stating the relationship of the two variables is confirmed or rejected. For example, the sociologist Arturo Biblarz and colleagues (1991) examined the effects of media violence and depictions of suicide on attitudes toward suicide by showing one group of subjects (an experimental group) a film about suicide, while a second (another experimental group) saw a film about violence, and a third (the control group) saw a film containing neither suicide nor violence. The research found

some evidence that people exposed to suicidal acts or violence in the media may be more likely to demonstrate an emotional state favorable to suicidal behavior, particularly if they are already "at risk" for suicide. If we were able to replicate this study today, do you believe that we would find similar results? Why or why not?

Researchers may use experiments when they want to demonstrate that a cause-and-effect relationship exists between variables (• Figure 2.12). In order to show that a change in one variable causes a change in another, these three conditions must be satisfied:

1. *You must show that a correlation exists between the two variables. Correlation* exists when two variables are associated more frequently than could be expected by chance. For example, suppose that you wanted to test the hypothesis that the availability of a crisis intervention center with a twenty-four-hour counseling "hotline" on your campus causes a change in students' attitudes toward suicide. To demonstrate correlation you would need to show that the students had different attitudes toward committing suicide depending on whether they had any experience with the crisis intervention center.

2. *You must ensure that the independent variable preceded the dependent variable.* If differences in students' attitudes toward suicide were evident before the students were exposed to the intervention center, exposure to the center could not be the cause of these differences.

3. *You must make sure that any change in the dependent variable was not because of an extraneous variable*—one outside the stated hypothesis. If some of the students receive counseling from off-campus psychiatrists, any change in attitude that they experience could be because of this third variable, not the hotline. This is referred to as a *spurious correlation*—the association of two variables that is actually caused by a third variable and does not demonstrate a cause-and-effect relationship (see • Figure 2.13).

---

unstructured interview
an extended, open-ended interaction between an interviewer and an interviewee.

experiment
a carefully designed situation in which the researcher studies the impact of certain variables on subjects' attitudes or behavior.

experimental group
in an experiment, the group that contains the subjects who are exposed to an independent variable (the experimental condition) to study its effect on them.

control group
in an experiment, the group that contains the subjects who are not exposed to the independent variable.

correlation
a relationship that exists when two variables are associated more frequently than could be expected by chance.

FIGURE 2.12 Do extremely violent video games cause an increase in violent tendencies in their users? Experiments are one way to test this hypothesis.

a. Observed correlation

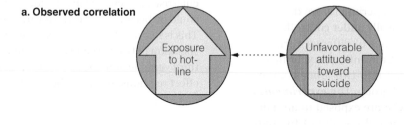

b. Possible causal explanation

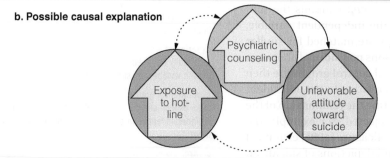

FIGURE 2.13 Correlation Versus Causation
A study might find that exposure to a suicide hotline is associated (correlated) with a change in attitude toward suicide. But if some of the people who were exposed to the hotline also received psychiatric counseling, the counseling may be the "hidden" cause of the observed change in attitude. In general, correlations alone do not prove causation.

Strengths and Weaknesses of Social Research Methods

Research Method	Strengths	Weaknesses
Experiments (laboratory, field, natural)	Control over research	Artificial by nature
	Ability to isolate experimental factors	Frequent reliance on volunteers or captive audiences
	Relatively little time and money required	Ethical questions of deception
	Replication possible, except for natural experiments	
Survey Research (questionnaire, interview, telephone survey)	Useful in describing features of a large population without interviewing everyone	Potentially forced answers
	Relatively large samples possible	Respondent untruthfulness on emotional issues
	Multivariate analysis possible	Data that are not always "hard facts" presented as such in statistical analyses
Secondary analysis of existing data (existing statistics, content analysis)	Data often readily available, inexpensive to collect	Difficulty in determining accuracy of some of the data
	Longitudinal and comparative studies	Failure of data gathered by others to meet goals of current research
	Replication possible	Questions of privacy when using diaries and other personal documents
Field research (participant observation, case study, ethnography, unstructured)	Opportunity to gain insider's view	Problems in generalizing results to a larger population
	Useful for studying attitudes and behavior in natural settings	Imprecise data measurements
	Longitudinal/comparative studies possible	Inability to demonstrate cause/effect relationships or test theories
	Documentation of important social problems of excluded groups possible	Difficult to make comparisons because of lack of structure
	Access to people's ideas in their words	Not a representative sample
	Forum for previously excluded groups	
	Documentation of need for social reform	

Strengths and Weaknesses of Experiments The major advantage of an experiment is the researcher's control over the environment and the ability to isolate the experimental variable. Because many experiments require relatively little time and money and can be conducted with limited numbers of subjects, it is possible for researchers to replicate an experiment several times by using different groups of subjects. Perhaps the greatest limitation of experiments is that they are artificial. Social processes that are set up by researchers or that take place in a laboratory setting are often not the same as real-life occurrences.

Another potential concern with experiments relates to this question: What happens when people know

that they are being studied? This problem is known as *reactivity*—the tendency of subjects to change their behavior in response to the researcher or to the fact that they know they are being studied. Social psychologist Elton Mayo first noticed this problem in a study conducted between 1927 and 1932 to determine how worker productivity and morale could be improved at Western Electric's Hawthorne plant. To identify variables that increase worker productivity, Mayo separated one group of women (the experimental group) from other workers and systematically varied factors in that group's work environment, while the working conditions for the other workers (the control group) were not changed. The researchers tested a number of factors, including an increase in the amount of lighting to see if more light would raise the workers' productivity. Much to the researchers' surprise, the workers' productivity increased not only when the light was brighter but also when it was dimmed. In fact, all of the changes increased productivity, leading Mayo to conclude that the Hawthorne subjects were trying to please the researchers because interest was being shown in the workers (Roethlisberger and Dickson, 1939). Thus, the *Hawthorne effect* refers to changes in a subject's behavior caused by the researcher's presence or by the subject's awareness of being studied.

Multiple Methods: Triangulation

What is the best method for studying a topic? The Concept Quick Review compares the various social research methods. There is no one best research method because social reality is complex and all research methods have limitations. Many sociologists believe that triangulation—the use of multiple methods in one study—is the solution to this problem. Triangulation refers not only to research methods but also to multiple data sources, investigators, and theoretical perspectives in a study. For example, in a study of more than 700 homeless people in Austin, Texas, the sociologists David Snow and Leon Anderson (1991: 158) used as their primary data sources "the homeless themselves and the array of settings, agency personnel, business proprietors, city officials, and neighborhood activities relevant to the routines of the homeless." Snow and Anderson gained a detailed portrait of the homeless and their experiences and institutional contacts by tracking homeless individuals through a network of seven institutions with which they had varying degrees of contact. The study used a variety of methods, including "participant observation and informal, conversational interviewing with the homeless; participant and nonparticipation observation, coupled with formal and informal interviewing in street agencies and settings; and a systematic survey of agency records" (Snow and Anderson, 1991: 158–169). This study is discussed in depth in Chapter 5 ("Social Structure and Interaction in Everyday Life").

Multiple methods and approaches provide a wider scope of information and enhance our understanding of critical issues (• Figure 2.14). Many researchers also use multiple methods to validate or refine one type of data by use of another type.

Michael Wheatley/PhotoLibrary

FIGURE 2.14 Multiple research methods are often used to gain information about important social concerns. Which methods might be most effective in learning more about the problems of the homeless, such as these street people warming themselves on a heated grate in Moscow, Russia?

 LO8 **Discuss** ethical issues in research and identify professional codes that protect research participants.

Ethical Issues in Sociological Research

The study of people ("human subjects") raises vital questions about ethical concerns in sociological research. Researchers are required to obtain written "informed consent" statements from the persons they study—but what constitutes "informed consent"? And how do researchers protect the identity and confidentiality of their sources?

The ASA *Code of Ethics*

The American Sociological Association (ASA) *Code of Ethics* (2008/1999) sets forth certain basic standards that sociologists must follow in conducting research. This code is available on the ASA website and includes the following:

1. Researchers must endeavor to maintain objectivity and integrity in their research by disclosing their research findings in full and including all possible interpretations of the data (even those interpretations that do not support their own viewpoints).

2. Researchers must safeguard the participants' right to privacy and dignity while protecting them from harm.

3. Researchers must protect confidential information provided by participants, even when this information is not considered to be "privileged" (legally protected, as is the case between doctor and patient and between attorney and client) and legal pressure is applied to reveal this information.

4. Researchers must acknowledge research collaboration and assistance they receive from others and disclose all sources of financial support.

Sociologists are obligated to adhere to this code and to protect research participants; however, many ethical issues arise that cannot be easily resolved. Ethics in sociological research is a difficult and often ambiguous topic. But ethical issues cannot be ignored by researchers, whether they are sociology professors, graduate students conducting investigations for their dissertations, or undergraduates conducting a class research project. Sociologists have a burden of "self-reflection"—of seeking to understand the role they play in contemporary social processes while at the same time assessing how these social processes affect their findings.

How honest do researchers have to be with potential participants? Where does the "right to know" end and the "right to privacy" begin in these situations? We will look at a historical case in which the researcher's ethics were challenged to see how cases such as this one have contributed to contemporary values and ethics in sociological research.

The Zellner Research

Sociologist William Zellner (1978) wanted to look at fatal single-occupant automobile accidents to determine whether some drivers were actually committing suicide. To examine this issue further, he sought to interview the family, friends, and acquaintances of persons killed in single-car crashes to determine whether the deaths were possibly intentional. To recruit respondents, Zellner told them that he hoped the research would reduce the number of automobile accidents in the future. He did not mention that he suspected "autocide" might have occurred in the case of their friend or loved one. From his data, Zellner concluded that at least 12 percent of the fatal single occupant crashes were suicides—and that those crashes sometimes also killed or critically injured other people as well. However, Zellner's research raised important research questions: Was his research unethical? Did he misrepresent the reasons for his study?

In this chapter we have looked at the research process and the methods used to pursue sociological knowledge. The important thing to remember is that research is the "lifeblood" of sociology: Theory provides the framework for analysis, and research takes us beyond common sense and provides opportunities for us to use our sociological imagination to generate new knowledge. As we have seen in this chapter, for example, suicide cannot be explained by common sense or a few isolated variables. We have to take into account many aspects of personal choice and social structure that are related to one another in extremely complex ways. Research can help us unravel the complexities of social life if sociologists observe, talk to, and interact with people in real-life situations.

Our challenge today is to find new ways to integrate knowledge and action and to include all people in the research process in order to help fill the gaps in our existing knowledge about social life and how it is shaped by gender, race, class, age, and the broader social and cultural contexts in which everyday life occurs. Each of us can and should find new ways to integrate knowledge and action into our daily lives (see "You Can Make a Difference").

Looking Ahead: Research, Social Change, and Your Future

Rapid changes have taken place in the United States and worldwide since the 1970s, when computers were introduced in a format that made them accessible to larger numbers of researchers who conduct wide-scale systematic social science research. By the 1980s and 1990s, with improved software, the availability of the Internet and the World Wide Web, and the introduction of browsers and search engines such as Yahoo and Google, research methods and data-gathering practices in sociology were forever changed.

New technologies produced new data sources and innovative methods for working with that data. From networking sites such as Facebook, YouTube, Twitter, and Instagram, social media researchers have generated vast amounts of data for studying interpersonal relationships. However, these changes also brought new challenges, such as how to build intersections among social science disciplines and how to reconcile the legal and ethical issues associated with collecting vast amounts of information.

What do these changes have to do with your future? They have opened up new avenues of inquiry for you in sociology and beyond. Computational social science (CSS) is an important and rapidly growing area of study. CSS looks at complex social systems through computational

Hawthorne effect
a phenomenon in which changes in a subject's behavior are caused by the researcher's presence or by the subject's awareness of being studied.

Responding to a Cry for Help

Chad felt that he knew Frank quite well. After all, they had been roommates for two years at State U. As a result, Chad was taken aback when Frank became very withdrawn, sleeping most of the day and mumbling about how unhappy he was. One evening, Chad began to wonder whether he needed to do something because Frank had begun to talk about "ending it all" and saying things like "the world will be better off without me." If you were in Chad's place, would you know the warning signs that you should look for? Do you know what you might do to help someone like Frank?

The American Foundation for Suicide Prevention, a national nonprofit organization dedicated to funding research, education, and treatment programs for depression and suicide prevention, suggests that each of us should be aware of these warning signs of suicide:

Can a suicide crisis center prevent a person from committing suicide? People who understand factors that contribute to suicide may be able to better counsel those who call for help.

- *Talking about death or suicide.* Be alert to such statements as "Everyone would be better off without me." Sometimes, individuals who are thinking about suicide speak as if they are saying goodbye.

- *Making plans.* The person may do such things as giving away valuable items, paying off debts, and otherwise "putting things in order."

- *Showing signs of depression.* Although most depressed people are not suicidal, most suicidal people are depressed.

Serious depression tends to be expressed as a loss of pleasure or withdrawal from activities that a person has previously enjoyed. It is especially important to note whether five of the following symptoms are present almost every day for several weeks: change in appetite or weight, change in sleeping patterns, speaking or moving with unusual speed or slowness, loss of interest in usual activities, decrease in sexual drive, fatigue, feelings of guilt or worthlessness, and indecisiveness or inability to concentrate.

The possibility of suicide must be taken seriously: Most people who commit suicide give some warning to family members or friends. Instead of attempting to argue the person out of suicide or saying "You have so much to live for," let the person know that you care and understand, and that his or her problems can be solved. Urge the person to see a school counselor, a physician, or a mental health professional immediately. If you think the person is in imminent danger of committing suicide, you should take the person to an emergency room or a walk-in clinic at a psychiatric hospital. It is best to remain with the person until help is available.

For more information about suicide prevention, contact the following organizations:

- American Foundation for Suicide Prevention is a leading not-for-profit organization dedicated to understanding and preventing suicide through research and education.

- Suicide Awareness Voices of Education is a resource index with links to other valuable resources.

- Befrienders Worldwide is a website providing information for anyone feeling depressed or suicidal or who is worried about a friend or relative who feels that way. It includes a directory of suicide and crisis help lines.

---

modeling and related techniques. It spans anthropology, communications, economics, linguistics, organizational science, political science, sociology, and social psychology, among other disciplines. CSS includes social geospatial modeling methods, social network analysis, automated information and content analysis methods, and other computational methods to investigate the social world in the twenty-first century.

Let's look briefly at two of these approaches: social network analysis and social geospatial modeling methods. The idea of *social network analysis* can be traced to early sociologists such as Emile Durkheim and Georg Simmel, who pointed out the importance of studying the patterns of relationships that connected social actors. Today, computer-assisted social network analysis investigates complex relationships such as friendship ties among

Facebook users. Social network analysis can also be used to understand negative effects of social relationships, such as how having had family members or friends who attempt suicide may be associated with individuals' thoughts of suicide or even suicide attempts (Mueller, Abrutyn, and Stockton, 2015). Some new research integrates knowledge from Durkheim's important work on social integration with social network analysis to show the negative impact that social ties can have on the health, well-being, and likelihood of reporting suicidal thoughts or attempts, such as in the case of adolescent bullying and suicide (Mueller, Abrutyn, and Stockton, 2015).

Some types of social network analysis rely on network theory and have displays of *nodes* that represent individuals within the network and *ties* that represent the relationships between those individuals. Social network analysis is used in many settings, including higher education and business, where it is important to study customer relations, marketing, and other social interactions. Similarly, social network analysis is used in public-sector work to study crime and environmental issues, elections, media representations of social problems, and city- and community-based problem solving.

Turning to our second example of CSS, *social geospatial modeling methods (GIS)* is a computer system designed to capture, store, manipulate, analyze, manage, and present all types of spatial or geographical data. User-created GIS searches generate vast amounts of information using space–time location as the key variable. GIS can also be used to analyze spatial information, edit data on maps, and present the results. Originally, digitization involved the transfer of a printed map or survey plan into a digital medium through use of a computer-assisted program.

Today, powerful computers and the Internet have greatly improved GIS by using satellite and aerial sources.

GIS has many applications in a wide variety of fields, ranging from urban planning to mental health care. For example, the location and provision of mental health care and advocacy can be mapped by GIS to show both the geographic availability of mental health services, such as suicide prevention and counseling facilities, and the level of satisfaction of those who avail themselves of the services. GIS also makes it possible to evaluate inequalities in access and observe spatial gaps in existing services and variation in the use of all types of health care resources. Research such as this makes it possible for public officials and mental health workers to bring about social change that will be beneficial for many individuals and families. GIS can produce maps that show areas without adequate access to mental health services—not only in terms of available care but also in regard to accessibility through public transportation, payment methods, and similar issues. What can you do with GIS? The effective use of GIS and the application of the skills associated with it can be applied to careers in social work, criminology, health care, community involvement, and government employment.

Whether or not you choose to engage in advanced sociological research, some ideas that you gain from this course will probably be useful in the future as you address pressing social problems of your times and work to bring positive changes to your home, workplace, community, and nation. We have research tools available to produce social policy suggestions and find new ways to reduce pressing social problems such as bullying and suicide. Do you see yourself contributing to social change in the future? How might you go about this?

CHAPTER REVIEW Q & A

LO1 Why is sociological research necessary, and how does it challenge our commonsense beliefs about pressing social issues such as suicide?

Sociological research provides a factual and objective counterpoint to commonsense knowledge and ill-informed sources of information. It is based on an empirical approach that answers questions through a direct, systematic collection and analysis of data.

LO2 How do the deductive and inductive approaches in the theory and research cycle compare?

Theory and research form a continuous cycle that encompasses both deductive and inductive approaches. With the deductive approach, the researcher begins with a theory and then collects and analyzes research to test it. With the inductive approach, the researcher collects and analyzes data and then generates a theory based on that analysis.

LO3 How does quantitative research differ from qualitative research?

Quantitative research focuses on data that can be measured numerically (comparing rates of suicide, for example). Qualitative research focuses on interpretive description (words) rather than statistics to analyze underlying meanings and patterns of social relationships.

LO4 What are the key steps in the quantitative research process?

A conventional research process based on deduction and the quantitative approach has these key steps: (1) selecting and defining the research problem; (2) reviewing previous research; (3) formulating the hypothesis, which involves constructing variables; (4) developing the research design; (5) collecting and analyzing the data; and (6) drawing conclusions and reporting the findings.

LO5 What steps are often taken by researchers using the qualitative approach?

A researcher taking the qualitative approach might (1) formulate the problem to be studied instead of creating a hypothesis, (2) collect and analyze the data, and (3) report the results.

LO6 What is survey research, and what are the three types of surveys?

The main types of research methods are surveys, secondary analysis of existing data, field research, and experiments. Surveys are polls used to gather facts about people's attitudes, opinions, or behaviors; a representative sample of respondents provides data through questionnaires or interviews. Survey data are collected by using self-administered questionnaires, face-to-face interviews, and telephone or computer surveys.

LO7 How do the following compare: research methods used in secondary analysis of existing data, field research, experiments, and triangulation?

In secondary analysis, researchers analyze existing data, such as a government census, or cultural artifacts, such as a diary. In field research, sociologists study social life in its natural setting through participant observation, case studies, unstructured interviews, and ethnography. Through experiments, researchers study the impact of certain variables on their subjects. Triangulation is the use of multiple methods in one study—not only research methods but also multiple data sources, investigators, and theoretical perspectives in a study.

LO8 What ethical issues are involved in sociological research, and what professional codes protect research participants?

Because sociology involves the study of people ("human subjects"), researchers are required to obtain the informed consent of the people they study; however, in some instances what constitutes "informed consent" may be difficult to determine. The American Sociological Association (ASA) *Code of Ethics* sets forth certain basic standards that sociologists must follow in conducting research.

KEY TERMS

content analysis 44
control group 47
correlation 47
dependent variable 36
ethnography 45
experiment 47
experimental group 47

field research 45
Hawthorne effect 50
hypothesis 36
independent variable 36
interview 41
participant observation 45
probability sampling 38

questionnaire 41
random sampling 38
research methods 41
respondents 41
secondary analysis 42
survey 41
unstructured interview 46

QUESTIONS for CRITICAL THINKING

1 The agency that funds the local suicide clinic has asked you to study the clinic's effectiveness in preventing suicide. What would you need to measure? What can you measure? What research method(s) would provide the best data for analysis?

2 Some classical studies have suggested that groups with high levels of *suicide acceptability* (holding the belief that suicide is an acceptable way to end one's life under certain circumstances) tend to have a higher-than-average suicide risk (Stack and Wasserman, 1995; Stack, 1998). What implications might such findings have on public policy issues today such as the legalization of physician-assisted suicide and euthanasia? What implications might the findings have on an individual who is thinking about committing suicide? Analyze your responses using a sociological perspective.

3 In high-income nations, computers have changed many aspects of people's lives. Thinking about the various research methods discussed in this chapter, which approaches do you believe would be most affected by greater reliance on computers for collecting, organizing, and analyzing data? What are the advantages and limitations of conducting sociological research via the Internet?

ANSWERS to the SOCIOLOGY QUIZ
ON SUICIDE

1	**True**	Suicide was the number-one cause of injury-related death in the United States in 2012 (the latest year for which data are available).
2	**False**	According to data from the U.S. Centers for Disease Control and Prevention, the highest suicide rate (19.88 per 100,000 people) in the United States is among people 45 to 59 years old. In 2012 the rate for people aged 15 to 24 was 10.9 per 100,000 people in the population.
3	**True**	White males accounted for 65 percent of all U.S. suicides in 2012.
4	**True**	Females attempt suicide three times as often as males, but males are four times more likely than females to die by suicide.
5	**False**	The World Health Organization estimates that more than 800,000 suicide deaths occur worldwide each year. For 2012 the number was estimated to be 804,000, but WHO officials believe that many more occurred but were not reported as such.
6	**False**	Although firearms are the most commonly used method of suicide among males (56.1 percent), poisoning is the most common method of suicide among females (38 percent).
7	**True**	Alcohol intoxication is present in 24 percent (nearly one-fourth) of all suicide deaths in the United States.
8	**True**	Government organizations such as the U.S. Centers for Disease Control and Prevention have developed these programs and now emphasize that promoting individual, family, and community connectedness is an important way to prevent such behavior.

Sources: Based on American Foundation for Suicide Prevention, 2014; U.S. Centers for Disease Control and Prevention, 2014; and World Health Organization, 2014.

CULTURE

LEARNING OBJECTIVES

1 **Define** *culture* and explain why it is important in helping people in their daily life.

2 **Analyze** material and nonmaterial cultures, and give examples of each.

3 **Explain** what is meant by the term *cultural universal* and provide three recent examples.

4 **Discuss** how symbols and language reflect cultural values.

5 **Explain** the differences among folkways, mores, and laws, and provide at least one example of each.

6 **Distinguish** ways in which technological changes affect culture in a single nation and throughout the world.

7 **Explain** the concepts of *culture shock, ethnocentrism,* and *cultural relativism,* and provide one example of each.

8 **Compare** and contrast functionalist, conflict, symbolic interactionist, and postmodernist perspectives on society and culture.

SOCIOLOGY & EVERYDAY LIFE

Spreading Culture Through Food Trucks?

Food brings people together in ways that words cannot, as it provides a universal outlet of creativity through which we can express ourselves. . . . We can pour culture and inspiration into our food, a subtle form of communication that is understood by all. The recent food truck craze in America has made this ability to bring people together through food even easier, and has indirectly spread cultural awareness with food as the medium.

—Veronica Werhane (2011), senior arts and culture editor for @neontommy, USC Annenberg School for Communication and Journalism, explaining how food trucks have become vehicles for spreading culture through food

In a Yale University Introduction to Public Humanities class taught by Dean Ryan Brasseaux in fall 2013, students examined how New Haven, Connecticut, became a "food truck town" (*Yale News*, 2013). The students specifically wanted to find out how the local food culture was intertwined with New Haven's history, particularly in regard to racial and ethnic diversity. Using ethnographic filming, interactive

How is the food that we consume linked to our identity and the larger culture of which we are a part? Do people who identify with more than one culture face more-complex issues when it comes to food preferences?

mapping, and oral histories from the owners of food trucks ranging from the Cannoli Truck and Joe Grate's BBQ to Peking Edo Cart and Ricky D's Rib Shack, the students created a "food routes" project. The proprietors were asked questions, such as why they chose to enter the food truck business, how they determined their routes, what their relationship was to the cuisine they served, and where they saw themselves and their trucks fitting into the bigger picture of New Haven life. Then the students created maps that reflected patterns of cultural diversity and food consumption in New Haven.

© Blulz60/Shutterstock.com

Have you eaten at a mobile food truck or a street vendor's cart? Many of us think of these as new inventions as they have recently become popular on TV programs such as seen on Food Network as participants compete to provide judges with the best food and service. However, food trucks and street carts have sold eats and drinks on the streets of U.S. cities for many years. Dating at least back to the 1960s, *loncheros,* or taco trucks, made daily stops at multiple locations in cities such as Los Angeles. The drivers slid open the shiny metal sides of their vehicles to display a variety of foods available for purchase at reasonable prices by workers from nearby offices or construction sites. Also, street vendors in New York City have sold food to hungry passersby for many decades. Everything from a banana and muffin for breakfast to a full steak dinner has been available from some of these vendors.

By the second decade of the twenty-first century, dramatic shifts have occurred in the mobile food industry as elaborately painted food trucks are a common sight on public streets and college campuses throughout the nation. These newer trucks offer customers a wider variety of ethnic foods and even haute cuisine prepared from the menu of a famous chef. Many people previously ridiculed food trucks, calling them "roach coaches," suggesting a lack of cleanliness, but now they refer to the elaborate trucks as a new cultural phenomenon and order foods such as Vietnamese bánh mi (a baguette with ingredients such as pork, cucumbers, fresh cilantro, sweet pickled carrots, daikon radish, cilantro, and chili peppers), Korean short-rib tacos and kimchi dogs, Brazilian barbeque, Greek sausages, Indian dosas, baklava (a popular dessert in Greece and Albania), and many other food offerings seemingly reflecting cultural diversity.

Although food trucks gain popularity in difficult times, economic problems are only one reason for the success of these mobile ventures. Other important cultural trends are also at play in the United States: The population is more culturally diverse, younger generations are willing to experiment with a wider variety of tastes and textures in their food choices, and the explosion of social media

According to the students, the food truck movement helps to "illustrate the importance of migration to the city of New Haven and the various cultural and social networks that exist throughout the city" (*Yale News,* 2013).

What do you think you might learn about food and culture if you participated in a similar project in your own city? Has the food truck trend helped to spread cultural awareness? Why or why not?

How Much Do You Know About Global Food and Culture?

TRUE	FALSE		
T	F	1	Cheese is a universal food enjoyed by people of all nations and cultures.
T	F	2	Giving round-shaped foods to the parents of new babies is considered to be lucky in some cultures.
T	F	3	Wedding cakes are made of similar ingredients in all countries, regardless of culture or religion.
T	F	4	Food is an important part of religious observance for many different faiths.
T	F	5	In authentic Chinese cuisine, cooking methods are divided into "yin" and "yang" qualities.
T	F	6	Because of the fast pace of life in the United States, virtually everyone relies on mixes and instant foods at home and fast foods when eating out.
T	F	7	Potatoes are the most popular mainstay in the diet of first- and second-generation immigrants who have arrived in the United States over the past 40 years.
T	F	8	According to sociologists, individuals may be offended when a person from another culture does not understand local food preferences or the cultural traditions associated with eating even if the person is obviously an "outsider" or a "tourist."

Answers can be found at the end of the chapter.

makes it possible for potential customers to know the location and daily offerings of their favorite food trucks. Finally, people are more aware that food is intricately intertwined with culture and with personal identities pertaining to race, ethnicity, religion, class, gender, age, and other social attributes.

For all of us, the food we consume is linked to our individual identity and to the larger culture of which we are a part. To some people, food consumption is nothing more than how we meet a basic biological need; however, many sociologists are interested in food and eating because of this subject's cultural significance in our lives.

In this chapter we examine society and culture, with special attention given to how our material culture, including the food we eat, is related to our beliefs, values, and actions. We also analyze culture from functionalist, conflict, symbolic interactionist, and postmodern perspectives. Before reading on, test your knowledge of the relationship among food, its distribution, and the culture in which we live by answering the questions in "Sociology and Everyday Life."

 LO1 **Define** *culture* **and explain why it is important in helping people in their daily life.**

Culture and Society in a Changing World

What is culture? **Culture** is the knowledge, language, values, customs, and material objects that are passed from person to person and from one generation to the next in a human group or society. As previously defined, a *society* is a large social grouping that occupies the same geographic territory and is subject to the same political authority and dominant cultural expectations. Whereas a society is composed of people, a culture is composed of ideas, behavior, and material possessions. Society and culture are interdependent; neither could exist without the other.

culture
the knowledge, language, values, customs, and material objects that are passed from person to person and from one generation to the next in a human group or society.

How important is culture in determining how people think and act on a daily basis? Simply stated, culture is essential for our individual survival and our communication with other people. We rely on culture because we are not born with the information we need to survive. We do not know how to take care of ourselves, how to behave, how to dress, what to eat, which gods to worship, or how to make or spend money. We must learn about culture through interaction, observation, and imitation in order to participate as members of the group. Sharing a common culture with others simplifies day-to-day interactions. However, we must also understand other cultures and the worldviews therein.

Just as culture is essential for individuals, it is also fundamental for the survival of societies. Culture has been described as the common foundation or core that makes individuals understandable to the larger group of which they are a part. Some system of rulemaking and enforcing necessarily exists in all societies. What would happen, for example, if *all* rules and laws in the United States suddenly disappeared? At a basic level, we need rules in order to navigate our bicycles and cars through traffic. At a more abstract level, we need rules to establish and protect our rights.

In order to survive, societies need rules about civility and tolerance. We are not born knowing how to express certain kinds of feelings toward others. When a person shows kindness or hatred toward another individual, some people may say "Well, that's just human nature" when explaining this behavior. Such a statement is built on the assumption that what we do as human beings is determined by *nature* (our biological and genetic makeup) rather than *nurture* (our social environment)—in other words, that our behavior is instinctive. An *instinct* is an unlearned, biologically determined behavior pattern common to all members of a species that predictably occurs whenever certain environmental conditions exist. For example, spiders do not learn to build webs. They build webs because of instincts that are triggered by basic biological needs such as protection and reproduction.

Culture is similar to instincts in animals because it helps us deal with everyday life. Although people may have some instincts, what we most often think of as instinctive behavior can actually be attributed to reflexes and drives. A *reflex* is an unlearned, biologically determined involuntary response to some physical stimuli (such as a sneeze after breathing some pepper in through the nose or the blinking of an eye when a speck of dust gets in it). *Drives* are unlearned, biologically determined impulses common to all members of a species that satisfy needs such as those for sleep, food, water, or sexual gratification. Reflexes and drives do not determine how people will behave in human societies; even the expression of these biological characteristics is channeled by culture. For example, we may be taught that the "appropriate" way to sneeze (an involuntary response) is to use a tissue

or turn our head away from others (a learned response). Similarly, we may learn to sleep on mats or in beds. Most contemporary sociologists agree that culture and social learning, not nature, account for virtually all of our behavior patterns.

Because humans cannot rely on instincts in order to survive, culture is a "tool kit" for survival (Swidler, 1986). From this approach, culture serves as a tool kit full of abstract things such as our beliefs and rituals, symbols, personal narratives, and overall perspectives on the world. People use these in a variety of configurations to help solve the problems they face. The tools we choose vary according to our own personality and the situations we face. We are not puppets on a string; we make choices from among the items in our own "tool kit."

LO2 Analyze material and nonmaterial cultures, and give examples of each.

Material Culture and Nonmaterial Culture

Our cultural tool kit is divided into two major parts: material culture and nonmaterial culture. **Material culture** consists of the physical or tangible creations that members of a society make, use, and share. Initially, items of material culture begin as raw materials or resources such as ore, trees, and oil. Through technology, these raw materials are transformed into usable items, ranging from books and computers to guns and tanks. Technology is both concrete and abstract. For example, technology includes computers, smartphones, iPads and other tablets, and the knowledge and skills necessary to use them. At the most basic level, material culture is important because it is our buffer against the environment. For example, we create shelter to protect ourselves from the weather and to give ourselves privacy. Beyond the survival level, we make, use, and share objects that are interesting and important to us. Why are you wearing the particular clothes you have on today? Perhaps you're communicating something about yourself, such as where you attend school, what kind of music you like, where you went on vacation, or something that you saw and liked online.

Nonmaterial culture consists of the abstract or intangible human creations of society that influence people's behavior. Language, beliefs, values, rules of behavior, family patterns, and political systems are examples of nonmaterial culture. Even the gestures that we use in daily conversation are part of the nonmaterial culture in a society. As many international travelers and businesspeople have learned, it is important to know what gestures mean in various nations (see • Figure 3.1). Although the "hook 'em Horns" sign—the pinky and index finger raised up and the middle two fingers folded down—is

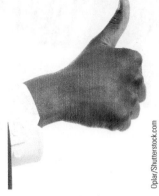

a. HORNS: "Hook'em Horns" or "Your spouse is unfaithful."

b. CIRCLE: "OK" (absolutely fine) or "I'll kill you."

c. THUMBS UP: "Great" or an obscenity.

FIGURE 3.1   Hand Gestures with Different Meanings
As international travelers and businesspeople have learned, hand gestures may have very different meanings in different cultures.

used by fans to express their support for University of Texas at Austin sports teams, for millions of Italians the same gesture means "Your spouse is being unfaithful." In Argentina, rotating one's index finger around the front of the ear means "You have a telephone call," but in the United States it usually suggests that a person is "crazy." Similarly, making a circle with your thumb and index finger indicates "OK" in the United States, but in Tunisia it means "I'll kill you!"

As the example of hand gestures shows, a central component of nonmaterial culture is *beliefs*—the mental acceptance or conviction that certain things are true or real. Beliefs may be based on tradition, faith, experience, scientific research, or some combination of these. Faith in a Supreme Being and trust in another person are examples of beliefs. We may also have a belief in items of material culture. When we travel by airplane, for instance, we believe that it is possible to fly at 33,000 feet and to arrive at our destination even though we know that we could not do this without the airplane itself.

 LO3 **Explain** what is meant by the term *cultural universal* and provide three recent examples.

Cultural Universals

Because all humans face the same basic needs (such as for food, clothing, and shelter), we engage in similar activities that contribute to our survival. Anthropologist George Murdock (1945: 124) compiled a list of more than seventy *cultural universals*—customs and practices that occur across all societies. His categories included appearance (such as clothing and hairstyles), activities (such as sports, dancing, games, joking, and visiting), social institutions (such as family, law, and religion), and customary

practices (such as cooking, folklore, gift giving, and hospitality). Whereas these general customs and practices may be *present* in all cultures, their specific *forms* vary from one group to another and from one time to another within the same group. For example, although telling jokes may be a universal practice, what is considered to be a joke in one society may be an insult in another.

How do sociologists view cultural universals? In terms of their functions, cultural universals are useful because they ensure the smooth and continual operation of society. A society must meet basic human needs by providing food, shelter, and some degree of safety for its members so that they will survive (• Figure 3.2). Children and other new members (such as immigrants) must be taught the ways of the group. A society must also settle disputes and deal with people's emotions. All the while, the self-interest of individuals must be balanced with the needs of society as a whole. Cultural universals help fulfill these important functions of society.

From another perspective, however, cultural universals are not the result of functional necessity; members of one society may have *imposed* these practices on members of another. Similar customs and practices do not necessarily constitute cultural universals. They may be an indication that a conquering nation used its power

material culture
the physical or tangible creations that members of a society make, use, and share.

nonmaterial culture
the abstract or intangible human creations of society that influence people's behavior.

beliefs
the mental acceptance or conviction that certain things are true or real.

cultural universals
customs and practices that occur across all societies.

to enforce certain types of behavior on those who were defeated. Sociologists might ask questions such as "Who determines the dominant cultural patterns?" For example, although religion is a cultural universal, traditional religious practices of indigenous peoples (those who first live in an area) have often been repressed and even stamped out by subsequent settlers or conquerors who possess political and economic power over them. However, many people believe there is cause for optimism in the United States because the democratic ideas of this nation provide more guarantees of religious freedom than do some other nations.

 LO4 **Discuss** how symbols and language reflect cultural values.

Components of Culture

Even though the specifics of individual cultures vary widely, all cultures have four common nonmaterial cultural components: symbols, language, values, and norms. These components contribute to both harmony and strife in a society.

Symbols

A *symbol* is anything that meaningfully represents something else. Culture could not exist without symbols because there would be no shared meanings among people. Symbols can simultaneously produce loyalty and animosity, and love and hate. They help us communicate ideas such as love or patriotism because they express abstract concepts with visible objects (• Figure 3.3). For example, flags can stand for patriotism, nationalism, school spirit, or religious beliefs held by members of a group or society. Symbols can stand for love (a heart on a valentine), peace (a dove), or hate (a Nazi swastika), just as words can be used to convey these meanings. Symbols can also transmit other types of ideas. A siren is a symbol that denotes an emergency situation and sends the message to clear the way immediately. Gestures are also a symbolic form of communication—a movement of the head, body, or hands can express our ideas or feelings to others. For example, in the United States, pointing toward your chest with your thumb or finger is a symbol for "me."

FIGURE 3.2   Food is a universal type of material culture, but what people eat and how they eat it vary widely, as shown in these cross-cultural examples from the United Arab Emirates (top), Holland (middle), and China (bottom). What might be some reasons for the similarities and differences that you see in these photos?

FIGURE 3.3 The customs and rituals associated with weddings are one example of nonmaterial culture. What can you infer about beliefs and attitudes concerning marriage in the societies represented by these photographs?

Symbols affect our thoughts about class. For example, how a person is dressed or the kind of car that he or she drives is often at least subconsciously used as a measure of that individual's economic standing or position (•Figure 3.4). With regard to clothing, although many people wear casual clothes on a daily basis, where the clothing was purchased is sometimes used as a symbol of social status. Were the items purchased at Walmart, Old Navy, Forever 21, or Saks? What indicators are there on the items of clothing—such as the Nike *swoosh*, some other logo, or a brand name—that indicate something about the status of the product? Automobiles and their logos are also symbols that have cultural meaning beyond the shopping environment in which they originate.

Finally, symbols may be specific to a given culture and have special meaning to individuals who share that culture but not necessarily to other people. Consider, for example, the use of certain foods to celebrate the Chinese New Year: Bamboo shoots and black-moss seaweed both represent wealth, peanuts and noodles symbolize a long life, and tangerines represent good luck. What foods in other cultures represent "good luck" or prosperity? In countries throughout the world, food and drink are powerful symbols of the history and cultural identity of people residing in the area.

Language

Language is a set of symbols that expresses ideas and enables people to think and communicate with one another. Verbal (spoken) language and nonverbal (written or gestured) language help us describe reality. One of our most important human attributes is the ability to use language to share our experiences, feelings, and knowledge with others. Language can create visual images in our head, such as describing small white kittens as looking like "snowballs." Language also allows people to distinguish themselves from outsiders and maintain group boundaries and solidarity.

Language is not solely a human characteristic. Other animals use sounds, gestures, touch, and smell to communicate with one another, but they use signals with fixed meanings that are limited to the immediate situation (the present) and that cannot encompass past or future situations. For example, chimpanzees can use elements of American Sign Language and manipulate physical objects to make "sentences," but they are not physically endowed with the vocal apparatus needed to form the consonants required for oral language. As a result, nonhuman animals cannot transmit the more-complex aspects of culture to their offspring. Humans have a unique ability to manipulate symbols to express

symbol
anything that meaningfully represents something else.

language
a set of symbols that expresses ideas and enables people to think and communicate with one another.

FIGURE 3.4 Would you expect the user of this device to be impoverished or affluent? What do possessions indicate about their owner's social class?

abstract concepts and rules and thus to create and transmit culture from one generation to the next.

Language and Social Reality Does language *create* reality or simply *communicate* reality? Anthropological linguists Edward Sapir and Benjamin Whorf have suggested that language not only expresses our thoughts and perceptions but also influences our perception of reality. According to the **Sapir–Whorf hypothesis**, language shapes the view of reality of its speakers (Whorf, 1956; Sapir, 1961). If people are able to think only through language, then language must precede thought. If language actually shapes the reality that we perceive and experience, then some aspects of the world are viewed as important and others are virtually neglected because people know the world only in terms of the vocabulary and grammar of their own language.

If language does create reality, does our language trap us? Many social scientists agree that the Sapir–Whorf hypothesis overstates the relationship between language and our thoughts and behavior patterns. Although they acknowledge that language has many subtle meanings and that words used by people reflect their central concerns, most sociologists contend that language may *influence* our behavior and interpretation of social reality but does not *determine* it.

Language and Gender How are language and gender related? What cultural assumptions about women and men does language reflect? Scholars have suggested several ways in which language and gender are intertwined:

- The English language ignores women by using the masculine form to refer to human beings in general. For example, the word *man* is used generically in words such as *chairman* and *mankind,* which allegedly include both men and women.

- Use of the pronouns *he* and *she* affects our thinking about gender. Pronouns show the gender of the person we *expect* to be in a particular occupation. For instance, nurses, secretaries, and schoolteachers are usually referred to as *she,* but doctors, engineers, electricians, and presidents are usually referred to as *he.*

- Words have positive connotations when relating to male power, prestige, and leadership; when relating to women, they carry negative overtones of weakness, inferiority, and immaturity. • Table 3.1 shows how gender-based language reflects the traditional acceptance of men and women in certain positions, implying that the jobs are different when filled by women rather than men.

- A language-based predisposition to think about women in sexual terms reinforces the notion that women are sexual objects. Terms such as *fox, broad, bitch,*

TABLE 3.1

Language and Gender		
Male Term	**Female Term**	**Neutral Term**
Teacher	Teacher	Teacher
Chairman	Chairwoman	Chair, chairperson
Congressman	Congresswoman	Representative
Policeman	Policewoman	Police officer
Fireman	Female fireman	Firefighter
Airline steward	Airline stewardess	Flight attendant
Race car driver	Female race car driver	Race car driver
Professor	Teacher/female professor	Professor
Doctor	Female doctor	Doctor
Bachelor	Bachelorette	Single person
Male prostitute	Prostitute	Prostitute
Welfare recipient	Welfare mother	Welfare recipient
Worker/employee	Working mother	Worker/employee
Janitor/maintenance man	Maid/cleaning lady	Custodial attendant

Sources: Adapted from Korsmeyer, 1981: 122; and Miller and Swift, 1991.

babe, and *doll* often describe women, which ascribe childlike or even pet-like characteristics to them. By contrast, men have performance pressures placed on them by being defined in terms of their sexual prowess, such as *dude, stud,* and *hunk.*

Gender in language has been debated and studied extensively for many years now, and some changes have occurred. The preference of many women to be called *Ms.* (rather than *Miss* or *Mrs.* in reference to their marital status) has received a degree of acceptance in public life and the media. Many organizations and publications have established guidelines for the use of nonsexist language and have changed titles such as *chairman* to *chair* or *chairperson.* To develop a more inclusive and equitable society, many analysts suggest that a more inclusive language is needed.

Language, Race, and Ethnicity Language may create and reinforce our perceptions about race and ethnicity by transmitting preconceived ideas about the superiority of one category of people over another. Let's look at a few images conveyed by words in the English language in regard to race/ethnicity:

- Words may have more than one meaning and create and/or reinforce negative images. In the nineteenth and twentieth centuries, terms such as *black-hearted* (malevolent) and expressions such as a *black mark* (a detrimental fact) and *Chinaman's chance of success* (unlikely to succeed) were used to associate the words *black* and *Chinaman* with derogatory imagery. Although these terms are seldom used today, they are occasionally referenced in popular culture and film.

- Overtly derogatory terms such as *nigger, kike, gook, honky, chink, spic,* and other racial–ethnic slurs have been "popularized" in movies, music, comedy routines, and so on. Such derogatory terms are often used in conjunction with physical threats against persons.

- Words are frequently used to create or reinforce perceptions about a group. For example, Native Americans have been referred to as "savage" and "primitive," and African Americans have been described as "uncivilized," "cannibalistic," and "pagan."

- The "voice" of verbs may minimize or incorrectly identify the activities or achievements of people of color. For example, the use of the passive voice in the statement "African Americans *were given* the right to vote" ignores how African Americans *fought* for that right. Active-voice verbs may also inaccurately attribute achievements to people or groups. Some historians argue that cultural bias is shown by the very notion that "Columbus discovered America"—given that America

was already inhabited by people who later became known as Native Americans.

In addition to these concerns about the English language, problems also arise when more than one language is involved. Across the nation, the question of whether the United States should have an "official" language continues to arise. Some people believe that there is no need to designate an official language; other people believe that English should be designated as the official language and that the use of any other language in official government business should be discouraged or negatively sanctioned. By 2014, thirty-one states (see ● Figure 3.5) had passed laws which require that all public documents, records, legislation, and regulations, as well as hearings, official ceremonies, and public meetings, be written or conducted solely in English.

Are deep-seated social and cultural issues embedded in social policy decisions such as these? Although the United States has always been a nation of immigrants, in recent decades this country has experienced rapid changes in population that have brought about greater diversity in languages and cultures. Recent data gathered by the U.S. Census Bureau indicate that although 79 percent of the people in this country speak only English at home, 21 percent speak a language other than English. Spanish is the language most frequently used at home by non-English speakers (see ● Figure 3.6).

Language is an important means of cultural transmission. Through language, children learn about their cultural heritage and develop a sense of personal identity in relationship to their group. Latinos/as in New Mexico and south Texas use *dichos*—proverbs or sayings that are unique to the Spanish language—as a means of expressing themselves and as a reflection of their cultural heritage. Examples of *dichos* include *"Anda tu camino sin ayuda de vecino"* ("Walk your own road without the help of a neighbor") and *"Amor de lejos es para pendejos"* ("A long-distance romance is for fools"). *Dichos* are passed from generation to generation as a priceless verbal tradition whereby people can give advice or teach a lesson (Gandara, 1995).

Language is also a source of power and social control; language perpetuates inequalities between people and between groups because words are used (intentionally or not) to "keep people in their place." As the linguist Deborah Tannen (1993: B5) has suggested, "The devastating group hatreds that result in so much suffering in our own country and around the world are related in origin to the small intolerances in our everyday conversations—our readiness to attribute good intentions to ourselves and bad intentions to others." Language, then, is a reflection of our feelings and values.

Sapir–Whorf hypothesis
the proposition that language shapes the view of reality of its speakers.

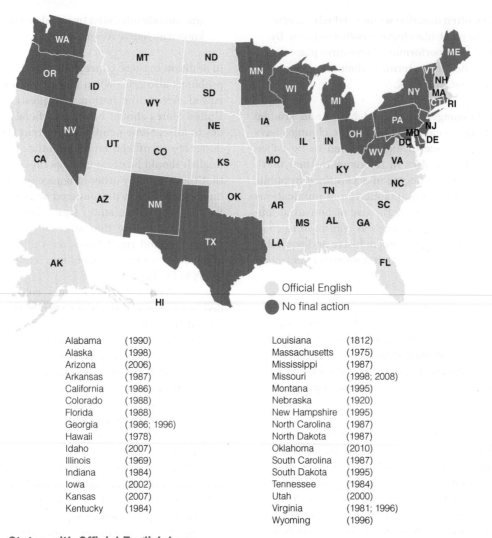

Official English

No final action

Alabama	(1990)	Louisiana	(1812)
Alaska	(1998)	Massachusetts	(1975)
Arizona	(2006)	Mississippi	(1987)
Arkansas	(1987)	Missouri	(1998; 2008)
California	(1986)	Montana	(1995)
Colorado	(1988)	Nebraska	(1920)
Florida	(1988)	New Hampshire	(1995)
Georgia	(1986; 1996)	North Carolina	(1987)
Hawaii	(1978)	North Dakota	(1987)
Idaho	(2007)	Oklahoma	(2010)
Illinois	(1969)	South Carolina	(1987)
Indiana	(1984)	South Dakota	(1995)
Iowa	(2002)	Tennessee	(1984)
Kansas	(2007)	Utah	(2000)
Kentucky	(1984)	Virginia	(1981; 1996)
		Wyoming	(1996)

FIGURE 3.5   States with Official English Laws
Why do some states have official English laws while others do not? How does the composition of the population in each state affect the passage of laws regarding language? Do you see any similarities in the states that have official English laws versus those that don't? What conclusions can you draw from this map?
Source: U.S. English, 2013.

Values are collective ideas about what is right or wrong, good or bad, and desirable or undesirable in a particular culture. Values do not dictate which behaviors are appropriate and which ones are not, but they provide us with the criteria by which we evaluate people, objects, and events. Values typically come in pairs of positive and negative values, such as being brave or cowardly, hardworking or lazy. Because we use values to justify our behavior, we tend to defend them staunchly.

Core American Values   Do we have shared values in the United States? Sociologists disagree about the extent to which all people in this country share a core set of values. Functionalists tend to believe that shared values are essential for the maintenance of a society, and scholars using a functionalist approach have conducted most of the research on core values. Analysts who focus on the

importance of core values maintain that the values identified by sociologist Robin M. Williams Jr., in 1970 are important to people in the United States but that they have changed somewhat in recent years. How important do you think the following ten values are?

1. *Individualism.* People in the United States are responsible for their own success or failure. Those who do not succeed have only themselves to blame because of their lack of ability, laziness, immorality, or other character defects.

2. *Achievement and success.* Personal achievement results from successful competition with others. Individuals are encouraged to do better than others in school and to work in order to gain wealth, power, and prestige. Material possessions are seen as a sign of personal achievement.

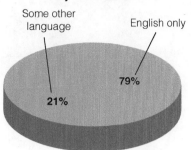

What percent speak English only in the home?

Some other language

English only

79%

21%

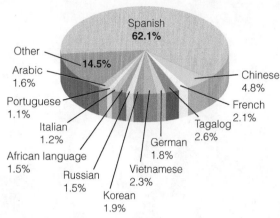

Among those who speak a language other than English, what do they speak?

Spanish
62.1%

Other
14.5%

Arabic
1.6%

Portuguese
1.1%

Italian
1.2%

African language
1.5%

Russian
1.5%

Korean
1.9%

Vietnamese
2.3%

German
1.8%

Tagalog
2.6%

French
2.1%

Chinese
4.8%

*Note: Percentages do not total 100% because of rounding.

FIGURE 3.6   Languages Spoken at Home, Other than English

[a]In Census Bureau terminology, a household consists of people who occupy a housing unit.
[b]Includes Native Hawaiian, Guamanian or Chamarro, Samoan, and other Pacific Islanders.
[c]Includes Chinese, Filipino, Japanese, Asian, Indian, Korean, Vietnamese, and other Asians.
Source: U.S. Census Bureau, 2012b.

3. *Activity and work.* People who are industrious are praised for their achievement; those perceived as lazy are ridiculed. From the time of the early Puritans, work has been viewed as important. Even during their leisure time, many people "work" in their play. For example, think of all the people who run in marathons, repair or restore cars, and so on in their spare time (● Figure 3.7).

4. *Science and technology.* People in the United States have a great deal of faith in science and technology. They expect scientific and technological advances ultimately to control nature, the aging process, and even death.

5. *Progress and material comfort.* The material comforts of life include not only basic necessities (such as adequate shelter, nutrition, and medical care) but also the goods and services that make life easier and more pleasant.

6. *Efficiency and practicality.* People want things to be bigger, better, and faster. As a result, great value is placed on efficiency ("how well does it work?").

7. *Equality.* Since colonial times, overt class distinctions have been rejected in the United States. However, "equality" has been defined as "equality of *opportunity*"—an assumed equal chance to achieve success—not as "equality of *outcome*."

8. *Morality and humanitarianism.* Aiding others, especially following natural disasters (such as fires, hurricanes, and other natural disasters), is seen as a value. The notion of helping others was originally a part of religious teachings and tied to the idea of morality. Today, people engage in humanitarian acts without necessarily perceiving that it is the "moral" thing to do.

9. *Freedom and liberty.* Individual freedom is highly valued in the United States. The idea of freedom includes the right to private ownership of property, the ability to engage in private enterprise, freedom of the press, and other freedoms considered to be "basic" rights.

10. *Ethnocentrism and group superiority.* People value their own racial or ethnic group above all others. Such feelings of pride and superiority may lead to discrimination; in the past, slavery and segregation laws were an example. Many people also believe in the superiority of this country and that "the American way of life" is best.

Are these values still important today? According to a report by the Pew Research Center Global Attitudes Project (2011), people in the United States remain more individualistic than people residing in Western Europe. However, Americans appear to be less inclined to view their culture and way of life as superior to others than they have in the past. Overall, it appears that core values are an important component of culture in all societies but that over time, they tend to shift based on economic conditions, social

values
collective ideas about what is right or wrong, good or bad, and desirable or undesirable in a particular culture.

FIGURE 3.7 Even during their leisure time, many people "work" in their play.

trends, religious beliefs, and other factors that arise in those nations and around the globe.

Value Contradictions All countries, including the United States, have value contradictions. *Value contradictions* are values that conflict with one another or are mutually exclusive (meaning that achieving one value makes it difficult, if not impossible, to achieve another). Core values of morality and humanitarianism may conflict with values of individual achievement and success. For example, humanitarian values reflected in welfare and other government-aid programs for people in need can come into conflict with values emphasizing hard work and personal achievement.

Ideal Culture Versus Real Culture What is the relationship between values and human behavior? Sociologists stress that a gap always exists between ideal culture and real culture in a society. *Ideal culture* refers to the values and standards of behavior that people in a society profess to hold. *Real culture* refers to the values and standards of behavior that people actually follow. For example, we may claim to be law-abiding (ideal cultural value) but may regularly drive over the speed limit (real cultural behavior) and still think of ourselves as "good citizens."

We may believe in the value of honesty and, at the same time, engage in deception.

 LO5 **Explain** the differences among folkways, mores, and laws, and provide at least one example of each.

Norms

Values provide ideals or beliefs about behavior but do not state explicitly how we should behave. Norms, on the other hand, do have specific behavioral expectations. *Norms* are established rules of behavior or standards of conduct. *Prescriptive norms* state what behavior is appropriate or acceptable. For example, persons making a certain amount of money are expected to file a tax return and pay any taxes they owe. Norms based on custom direct us to open a door for a person carrying a heavy load or give up our seat for an elderly person on a bus. By contrast, *proscriptive norms* state what behavior is inappropriate or unacceptable. Laws that prohibit us from driving over the speed limit and "good manners" that preclude you from texting during class are examples. Prescriptive and proscriptive norms operate at all levels of society, from our everyday actions to the formulation of laws.

Formal and Informal Norms Not all norms are of equal importance; those that are most crucial are formalized. *Formal norms* are written down and involve specific punishments for violators. Laws are the most common type of formal norms; they have been codified and may be enforced by sanctions. **Sanctions** are rewards for appropriate behavior or penalties for inappropriate behavior. Examples of *positive sanctions* include praise, honors, or medals for conformity to specific norms. *Negative sanctions* range from mild disapproval to the death penalty.

Norms that are considered to be less important are referred to as *informal norms*—unwritten standards of behavior understood by people who share a common identity. When individuals violate informal norms, other people may apply informal sanctions. *Informal sanctions* are not clearly defined and can be applied by any member of a group (such as frowning at someone or making a negative comment or gesture). Around the world, people of all nations have formal and informal norms that are unique to their culture and way of life (see "Sociology in Global Perspective.")

Folkways Norms are also classified according to their relative social importance. **Folkways** are informal norms or everyday customs that may be violated without serious consequences within a particular culture (Sumner, 1959/1906). They provide rules for conduct but are not considered to be essential to society's survival (• Figure 3.8). In the United States, folkways include using underarm deodorant, brushing our teeth, and wearing appropriate clothing for a specific occasion. Often, folkways are not

enforced; when they are enforced, the resulting sanctions tend to be informal and relatively mild.

Mores Other norms are considered to be highly essential to the stability of society even though they are not laws. **Mores** are strongly held norms with moral and ethical connotations that may not be violated without serious consequences in a particular culture. Because mores (pronounced MOR-ays) are based on cultural values and are considered to be crucial to the well-being of the group, violators are subject to more-severe negative sanctions (such as ridicule or loss of employment) than are those who fail to adhere to folkways. The strongest mores are referred to as taboos. **Taboos** are mores so strong that their violation is considered to be extremely offensive and even unmentionable. Violation of taboos is punishable by the group or even, according to certain belief systems, by a supernatural force. The incest taboo, which prohibits sexual or marital relations between certain categories of kin, is an example of a nearly universal taboo. In the United States, mores are considered to be informal norms unless they are officially made into law.

Laws **Laws** are formal, standardized norms that have been enacted by legislatures and are enforced by formal sanctions. Laws may be either civil or criminal. *Civil law* deals with disputes among persons or groups. Persons who lose civil suits may encounter negative sanctions such as having to pay compensation to the other party or being ordered to stop certain conduct. *Criminal law,* on the other hand, deals with public safety and well-being. When criminal laws are violated, fines and prison sentences are the most likely negative sanctions, although in some states the death penalty is handed down for certain major offenses.

FIGURE 3.8 Folkways—such as how to behave in a crowded elevator—provide rules for conduct but are not considered to be essential to society's survival.

Nico Kai/The Image Bank/Getty Images

value contradictions
values that conflict with one another or are mutually exclusive.

norms
established rules of behavior or standards of conduct.

sanctions
rewards for appropriate behavior or penalties for inappropriate behavior.

folkways
informal norms or everyday customs that may be violated without serious consequences within a particular culture.

mores
strongly held norms with moral and ethical connotations that may not be violated without serious consequences in a particular culture.

taboos
mores so strong that their violation is considered to be extremely offensive and even unmentionable.

laws
formal, standardized norms that have been enacted by legislatures and are enforced by formal sanctions.

SOCIOLOGY in GLOBAL PERSPECTIVE

What Do Cultural Norms Say About Drinking Behavior?

In the United States, most of us are familiar with norms and rituals associated with alcohol consumption and/or drinking behavior. Some norms are formal, such as laws that set forth the legal age at which young people may consume alcoholic beverages. Other drinking norms are informal, such as whether champagne, wine, or beer should be served at a wedding celebration. Let's take a brief look at the Republic of Georgia, where highly formalized norms have been established regarding the consumption of wine at meals. And many of these norms date back more than 7,000 years ago in this ancient wine region (Salcito, 2014).

In the Republic of Georgia, people participate in celebratory and memorial occasions known as *supras*—feasts or banquets. More-formal *supras* take place at holidays or significant events, such as birthdays, weddings, baptisms, or funerals. Less-formal *supras* mark the gathering of friends and family to enjoy food, music, and dancing while engaging in a ritualized

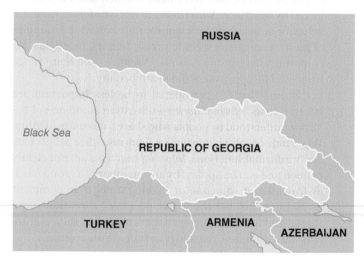

The Republic of Georgia

LO6 **Distinguish** ways in which technological changes affect culture in a single nation and throughout the world.

Technology, Cultural Change, and Diversity

Cultures do not generally remain static. There are many forces working toward change and diversity. Some societies and individuals adapt to this change, whereas others suffer culture shock and succumb to ethnocentrism.

Cultural Change

Societies continually experience cultural change at both material and nonmaterial levels. Changes in technology continue to shape the material culture of society. *Technology* refers to the knowledge, techniques, and tools that allow people to transform resources into usable forms, and the knowledge and skills required to use them after they are developed. Although most technological changes are primarily modifications of existing technology, *new technologies* refers to changes that make a significant difference in many people's lives. Examples of new technologies include the introduction of the printing press more than 500 years ago and the advent of computers and electronic communications in the twentieth century. The pace of technological change has increased rapidly in the past 150 years, as contrasted with the 4,000 years prior to that, during which humans advanced from digging sticks and hoes to the plow.

All parts of culture do not change at the same pace. When a change occurs in the material culture of a society, nonmaterial culture must adapt to that change. Frequently, this rate of change is uneven, resulting in a gap between the two. Sociologist William F. Ogburn (1966/1922) referred to this disparity as ***cultural lag***, which is a gap between the technical development of a society and its moral and legal institutions (• Figure 3.9). In other words, cultural lag occurs when material culture changes faster than nonmaterial culture, thus creating a lag between the two cultural components. For example, at the material cultural level the personal computer and electronic coding have made it possible to create a unique health identifier for each person in the United States. Based on available technology (material culture), it is possible to create a national data bank that would include everyone's individual medical records from birth to death. Using this identifier, health providers and insurance companies could rapidly transfer medical records around the globe, and researchers could access unlimited data on people's diseases, test results, and treatments. However, the availability of this technology does not mean that it will be accepted by people who believe (nonmaterial culture) that such a national data bank would constitute an invasion of privacy and could easily be abused by others. The failure of nonmaterial culture to keep pace with material culture is linked to social conflict and societal problems. As in the above example, such changes are often set in motion by discovery, invention, and diffusion.

Discovery is the process of learning about something previously unknown or unrecognized. Historically,

wine drinking that marks one's personal and national identity, as well as his or her place in the social hierarchy.

The norms of eating are more informal at the *supra*, but the drinking norms are strictly formal, starting with the first toast, which sets the character, structure, and meaning of what is being celebrated or mourned (Muehlfried, 2007). Georgian wine cannot be consumed without first having an eloquent speech given by the *tamada*, or toastmaster, who must adhere to strict rules of etiquette to ensure that honor is appropriately distributed to the guests. The *tamada* must be able to give extensive toasts, share humorous stories, and state and control the order and duration of each part of the evening. It is also important who drinks when and how much because such details demonstrate people's status. When a toast is given to women or the deceased, for example, women and children remain seated, but boys of a certain age must stand up to show that they have become men. Older men who no longer participate in drinking and toasting typically lose their status as head of the family (Muehlfried 2007).

Why do Georgians maintain these cultural norms governing *supras* across centuries and generations? In the past, such norms may have helped Georgians hold on to their cultural identity when they were overtaken by other nations, such as Russia. Today, some social scientists believe that these norms and rituals continue to help people in Georgia maintain their identity and cultural heritage in the face of rapid social change and globalization (Muehlfried, 2007).

REFLECT **& ANALYZE**

Can you think of eating or drinking norms that are specific to your family's national, religious, or racial–ethnic traditions? What about those of your friends or acquaintances? Do these norms contribute to your identity and social interaction patterns?

FIGURE 3.9 Facebook and other social networking companies are examples of cultural lag—a gap between technology and a society's morals and laws. Who owns what you have posted on your social networking page?

discovery involved unearthing natural elements or existing realities, such as "discovering" fire or the true shape of the Earth. Today, discovery most often results from scientific research. For example, the discovery of a polio vaccine virtually eliminated one of the major childhood diseases. A future discovery of a cure for cancer or the common cold could result in longer and more productive lives for many people.

As more discoveries have occurred, people have been able to reconfigure existing material and nonmaterial cultural items through invention. *Invention* is the process of reshaping existing cultural items into a new form. Guns, airplanes, TV sets, and digital devices are examples of inventions that positively or negatively affect our lives today.

When diverse groups of people come into contact, they begin to adapt one another's discoveries, inventions, and ideas for their own use. *Diffusion* is the transmission of cultural items or social practices from one group or society to

technology
the knowledge, techniques, and tools that allow people to transform resources into usable forms, and the knowledge and skills required to use them after they are developed.

cultural lag
William Ogburn's term for a gap between the technical development of a society (material culture) and its moral and legal institutions (nonmaterial culture).

another through such means as exploration, war, the media, tourism, and immigration. Today, cultural diffusion moves at a very rapid pace in the global economy.

Cultural Diversity

Cultural diversity refers to the wide range of cultural differences found between and within nations. Cultural diversity between countries may be the result of natural circumstances (such as climate and geography) or social circumstances (such as level of technology and composition of the population). Some nations—such as Sweden—are referred to as *homogeneous societies,* meaning that they include people who share a common culture and who are typically from similar social, religious, political, and economic backgrounds (though this is changing in Sweden and other countries as they become more diverse). By contrast, other nations—including the United States—are referred to as *heterogeneous societies,* meaning that they include people who are dissimilar in regard to social characteristics such as religion, income, or race/ethnicity (see • Figure 3.10).

Immigration contributes to cultural diversity in a society. Throughout its history, the United States has been a nation of immigrants (see • Figure 3.11). Over the past 200 years, more than 60 million "documented" (legal) immigrants have arrived here; innumerable people have also entered the country as undocumented immigrants. Immigration can cause feelings of frustration and hostility, especially in people who feel threatened by the changes that large numbers of immigrants may produce. Often, people are intolerant of those who are different from themselves. When societal tensions rise, people may look for others on whom they can place blame—or single out persons because they are the "other," the "outsider," the one who does not "belong."

Do you believe that people can overcome these feelings in a culturally diverse society such as the United States? Some analysts believe that it is possible to communicate with others despite differences in race, ethnicity, national origin, age, sexual orientation, religion, social class, occupation, leisure pursuits, regionalism, and so on. People who differ from the dominant group may also find reassurance and social support in a subculture.

Subcultures   A **subculture** is a category of people who share distinguishing attributes, beliefs, values, and/or norms that set them apart in some significant manner from the dominant culture. This concept has been applied to distinctions ranging from ethnic, religious, regional, and age-based categories to those categories presumed to be "deviant" or marginalized from the larger society. In the broadest use of the concept, members of thousands of categories of people residing in the United States might be classified as participants in one or more subcultures, including Lady Gaga fans, Muslims,

Religious Affiliation

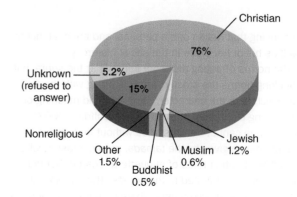

Household Income[a]

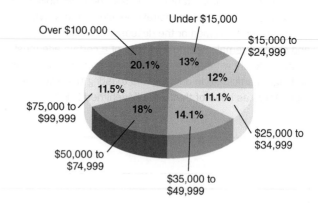

Race and Ethnic Distribution*

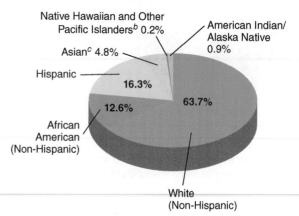

*May not total to 100% due to rounding.

[a]In Census Bureau terminology, a household consists of people who occupy a housing unit.
[b]Includes Native Hawaiian, Guamanian or Chamorro, Samoan, and other Pacific Islanders.
[c]Includes Chinese, Filipino, Japanese, Asian, Indian, Korean, Vietnamese, and other Asians.

FIGURE 3.10   Heterogeneity of U.S. Society
Throughout history, the United States has been heterogeneous. Today, we represent a wide diversity of social categories, including our religious affiliations, income levels, and racial-ethnic categories.
Sources: U.S. Census Bureau, 2011g; Humes, Jones, and Ramirez, 2011.

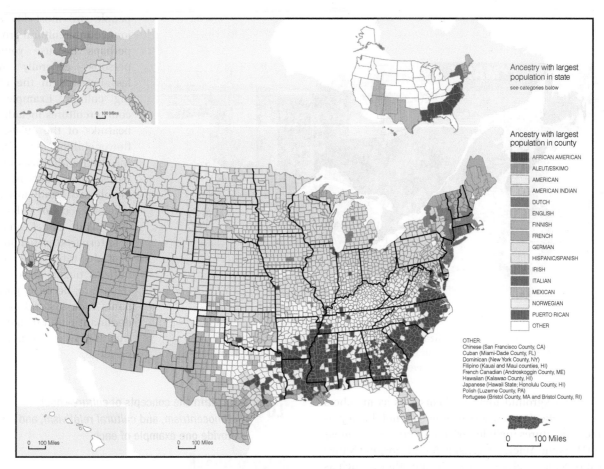

FIGURE 3.11   Cultural Diversity: A Nation of Immigrants
Source: U.S. Census Bureau, American Fact Finder, 2012.

Ancestry with largest population in state
see categories below

Ancestry with largest population in county

AFRICAN AMERICAN
ALEUT/ESKIMO
AMERICAN
AMERICAN INDIAN
DUTCH
ENGLISH
FINNISH
FRENCH
GERMAN
HISPANIC/SPANISH
IRISH
ITALIAN
MEXICAN
NORWEGIAN
PUERTO RICAN
OTHER

OTHER:
Chinese (San Francisco County, CA)
Cuban (Miami-Dade County, FL)
Dominican (New York County, NY)
Filipino (Kauai and Maui counties, HI)
French Canadian (Androskoggin County, ME)
Hawaiian (Kalawao County, HI)
Japanese (Hawaii State; Honolulu County, HI)
Polish (Luzerne County, PA)
Portugese (Bristol County, MA and Bristol County, RI)

and motorcycle enthusiasts. However, many sociological studies of subcultures have limited the scope of inquiry to more-visible, distinct subcultures such as the Old Order Amish and ethnic enclaves such as Little Saigon or Little Gaza to see how participants in subcultures interface with the dominant U.S. culture. Let's take a brief look at the Old Order Amish, a classic sociological example.

The Old Order Amish   Having arrived in the United States in the early 1700s, members of the Old Order Amish have fought to maintain their distinct identity. Today, most of the Amish live in Pennsylvania, Ohio, and Indiana, where they practice their religious beliefs and remain in a relatively closed social network. According to sociologists, this religious community is a subculture because its members share values and norms that differ from those of people who primarily identify with the dominant culture. The Amish have a strong faith in God and reject worldly concerns. Their core values include the joy of work, the primacy of the home, faithfulness, thriftiness, tradition, and humility. The Amish hold a conservative view of the family, believing that women are subordinate to men, birth control is unacceptable,

and wives should remain at home. Children (about seven per family) are cherished and seen as an economic asset: They help with the farming and other work. Many of the Old Order Amish speak Pennsylvania Dutch (a dialect of German) as well as English. They dress in traditional clothing, live on farms, and rely on the horse and buggy for transportation (• Figure 3.12).

The Amish are aware that they share distinctive values and look different from other people; these differences provide them with a collective identity and make them feel close to one another. The belief system and group cohesiveness of the Amish remain strong despite the intrusion of corporations and tourists, the vanishing farmlands, and increasing levels of government regulation in their daily lives (Schaefer and Zellner, 2010).

Ethnic Subcultures   Some people who have unique shared behaviors linked to a common racial, language, or

subculture
a category of people who share distinguishing attributes, beliefs, values, and/or norms that set them apart in some significant manner from the dominant culture.

FIGURE 3.12 Although modernization and consumerism have changed the way of life of some subcultures, groups such as the Old Order Amish have preserved some of their historical practices, including traveling by horse-drawn carriage.

people are most likely to join countercultural groups, perhaps because younger persons generally have less invested in the existing culture. Examples of countercultures include the beatniks of the 1950s, the flower children of the 1960s, the drug enthusiasts of the 1970s, and contemporary members of nonmainstream religious sects or cults. Occupy Wall Street and its counterparts throughout the United States and other nations began as a counterculture because participants took a stand against dominant cultural values of wealth, power, and political privilege.

nationality background identify themselves as members of a specific subculture, whereas others do not. Examples of ethnic subcultures include African Americans, Latinos/Latinas (Hispanic Americans), Asian Americans, and Native Americans. Some analysts include "white ethnics" such as Irish Americans, Italian Americans, and Polish Americans. Others also include Anglo Americans (Caucasians).

Although people in ethnic subcultures are dispersed throughout the United States, a concentration of members of some ethnic subcultures is visible in many larger communities and cities. For example, Chinatowns, located in cities such as San Francisco, Los Angeles, and New York, are one of the more visible ethnic subcultures in the United States. By living close to one another and retaining their original customs and language, first-generation immigrants can survive the abrupt changes they experience in material and nonmaterial cultural patterns. In New York City, for example, Korean Americans and Puerto Rican Americans constitute distinctive subcultures, each with its own food, music, and personal style. In San Antonio, Mexican Americans enjoy different food and music than do Puerto Rican Americans or other groups.

Subcultures provide opportunities for the expression of distinctive lifestyles, as well as sometimes helping people adapt to abrupt cultural change. Subcultures can also serve as a buffer against the discrimination experienced by many ethnic or religious groups in the United States. However, some people may be forced by economic or social disadvantage to remain in such ethnic enclaves.

Countercultures A *counterculture* is a group that strongly rejects dominant societal values and norms and seeks alternative lifestyles (Yinger, 1960, 1982). Young

  Explain the concepts of *culture shock, ethnocentrism*, and *cultural relativism*, and provide one example of each.

Culture Shock

Culture shock is the disorientation that people feel when they encounter cultures radically different from their own and believe they cannot depend on their own taken-for-granted assumptions about life. When people travel to another society, they may not know how to respond to that setting. For example, Napoleon Chagnon (1992) described his initial shock at seeing the Yanomamö (pronounced yahnoh-MAH-mah) tribe of South America on his first trip in 1964 (● Figure 3.13).

The Yanomamö (also referred to as the "Yanomami") are a tribe of about 20,000 South American Indians who live in the rain forest. Although Chagnon traveled in a small aluminum motorboat for three days to reach these people, he was not prepared for the sight that met his eyes when he arrived:

I looked up and gasped to see a dozen burly, naked, sweaty, hideous men staring at us down the shafts of their drawn arrows. Immense wads of green tobacco were stuck between their lower teeth and lips, making them look even more hideous, and strands of dark-green slime dripped from their nostrils—strands so long that they reached down to their pectoral muscles or drizzled down their chins and stuck to their chests and bellies. We arrived as the men were blowing ebene, a hallucinogenic drug, up their noses. . . . I was horrified. What kind of welcome was this for someone who had come to live with these people and learn their way of life—to become friends with them? But when they recognized Barker [a guide], they put their weapons down and returned to their chanting, while keeping a nervous eye on the village entrances. (Chagnon, 1992: 12–14)

The Yanomamö have no written language, system of numbers, or calendar. They lead a nomadic lifestyle, carrying everything they own on their backs. They wear no clothes and paint their bodies; the women insert slender sticks through holes in the lower lip and through the pierced nasal septum. In other words, the Yanomamö—like the members of thousands of other cultures around the world—live in a culture very different from that of the United States.

Ethnocentrism and Cultural Relativism

When observing people from other cultures, many of us use our own culture as the yardstick by which we judge their behavior. Sociologists refer to this approach as **ethnocentrism**—the practice of judging all other cultures by one's own culture. Ethnocentrism is based on the assumption that one's own way of life is superior to all others. For example, most schoolchildren are taught that their own school and country are the best (● Figure 3.14). The school song, the pledge to the flag, and the national anthem are forms of *positive ethnocentrism.* However, *negative ethnocentrism* can also result from constant emphasis on the superiority of one's own group or nation. Negative ethnocentrism is manifested in derogatory stereotypes that ridicule recent immigrants whose customs, dress, eating habits, or religious beliefs are markedly different from those of dominant-group members. Long-term U.S. residents who are members of racial and ethnic minority groups, such as Native Americans, African Americans, Asian Americans, and Latinas/os, have also been the target of ethnocentric practices by other groups.

An alternative to ethnocentrism is *cultural relativism*—the belief that the behaviors and customs of any culture must be viewed and analyzed by the culture's own standards. For

FIGURE 3.13   Even as global travel and the media make us more aware of people around the world, the distinctiveness of the Yanomamö in South America remains apparent. Are people today more or less likely than those in the past to experience culture shock upon encountering diverse groups of people such as these Yanomamö?

Gavriel Jecan/Danita Delimont/Alamy

example, the anthropologist Marvin Harris (1974, 1985) uses cultural relativism to explain why cattle, which are viewed as sacred, are not killed and eaten in India, where widespread hunger and malnutrition exist. From an ethnocentric viewpoint, we might conclude that cow worship is the cause of the hunger and poverty in India. However, according to Harris, the Hindu taboo against killing cattle is very important to the Indian economic system. Live cows are more valuable than dead ones because they have more important uses than as a direct source of food. As part of the ecological system, cows consume grasses of little value to humans. Then they produce two valuable resources—oxen (the neutered offspring of cows) to power the plows and manure (for fuel and fertilizer)—as well as milk for children. As Harris's study reveals, culture must be viewed from the standpoint of those who live in a particular society.

Cultural relativism also has a downside. It may be used to excuse customs and behavior (such as cannibalism) that may violate basic human rights. Cultural relativism is a part of the sociological imagination; researchers must be aware of the customs and norms of the society they are studying and then spell out their background assumptions so that others can spot possible biases in their studies. However, according to some social scientists, issues surrounding ethnocentrism and cultural relativism may become less distinct in the future as

counterculture
a group that strongly rejects dominant societal values and norms and seeks alternative lifestyles.

culture shock
the disorientation that people feel when they encounter cultures radically different from their own and believe they cannot depend on their own taken-for-granted assumptions about life.

ethnocentrism
the practice of judging all other cultures by one's own culture.

cultural relativism
the belief that the behaviors and customs of any culture must be viewed and analyzed by the culture's own standards.

FIGURE 3.14 These children from Norway wave their country's flag on May 17 (Norwegian Constitution Day), displaying a form of positive ethnocentrism.

people around the globe increasingly share a common popular culture. Others, of course, disagree with this perspective. Let's see what you think.

A Global Popular Culture?

Before taking this course, what was the first thing you thought about when you heard the term *culture*? In everyday life, culture is often used to describe the fine arts, literature, and classical music. When people say that a person is "cultured," they may mean that the individual has a highly developed sense of style or aesthetic appreciation of the "finer" things.

High Culture and Popular Culture

Some sociologists use the concepts of high culture and popular culture to distinguish between different cultural forms. *High culture* consists of classical music, opera, ballet, live theater, and other activities usually patronized by elite audiences, composed primarily of members of the upper-middle

and upper classes, who have the time, money, and knowledge assumed to be necessary for its appreciation. In the United States, high culture is often viewed as being international in scope, arriving in this country through the process of diffusion, because many art forms originated in European nations or other countries of the world.

By contrast, much of U.S. popular culture is often thought of as "homegrown" in this country. *Popular culture* consists of activities, products, and services that are assumed to appeal primarily to members of the middle and working classes. These include rock concerts, spectator sports, movies, and television soap operas and sitcoms. Although we will distinguish between "high" and "popular" culture in our discussion, it is important to note that some social analysts believe the rise of a consumer society in which luxury items have become more widely accessible to the masses has reduced the great divide between them and the activities and possessions associated with wealthy people or a social elite.

However, most sociological examinations of high culture and popular culture focus primarily on the link between culture and social class. French sociologist Pierre Bourdieu's (1984) *cultural capital theory* views high culture as a device used by the dominant class to exclude the subordinate classes. According to Bourdieu, people must be trained to appreciate and understand high culture. Individuals learn about high culture in upper-middle-class and upper-class families and in elite education systems, especially higher education. Once they acquire this trained capacity, they possess a form of cultural capital. Persons from poor and working-class backgrounds typically do not acquire this cultural capital. Because knowledge and appreciation of high culture are considered a prerequisite for access to the dominant class, its members can use their cultural capital to deny access to subordinate-group members and thus preserve and reproduce the existing class structure.

Forms of Popular Culture

Three prevalent forms of popular culture are fads, fashions, and leisure activities. A *fad* is a temporary but widely copied activity followed enthusiastically by large numbers of people. Most fads are short-lived novelties. Fads can be divided into four major categories. First, *object fads* are items that people purchase despite the fact that they have little use value, such as wristbands that make a statement or support a cause. Second, *activity fads* include pursuits such as body piercing or flash mobs. Third are *idea fads,* such as New Age ideologies. Fourth are *personality fads*—for example, those surrounding celebrities such as Taylor Swift, Ryan Seacrest, and Katy Perry (● Figure 3.15).

A *fashion* is a currently valued style of behavior, thinking, or appearance that is longer lasting and more widespread than a fad. Examples of fashion are found in many areas, including child rearing, education, arts, clothing, music, and sports. Soccer is an example of a fashion in sports. Until fairly recently, only schoolchildren played soccer in

Kevin Mazur/Getty Images

FIGURE 3.15 Television, the Internet and social media have provided celebrities like Katy Perry with a global platform from which they can spread popular-culture hits to their worldwide audience. What other personality fads can you identify?

the United States. Now it has become a popular sport, perhaps in part because of immigration from Latin America and other areas of the world where soccer is widely played.

Like soccer, other forms of popular culture move across nations. In fact, popular culture is one of the United States' largest exports to other nations. In turn, people in this country continue to be strongly influenced by popular culture from other nations. Will the spread of popular culture produce a homogeneous global culture? Critics argue that the world is not developing a global culture; rather, other cultures are becoming Westernized. Political and religious leaders in some nations oppose this process, which they view as *cultural imperialism*—the extensive infusion of one nation's culture into other nations. For example, some view the widespread infusion of the English language into countries that speak other languages as a form of cultural imperialism. On the other hand, the concept of cultural imperialism may fail to take into account various cross-cultural influences. For example, cultural diffusion of literature, music, clothing, and food has occurred on a global scale. A global culture, if it comes into existence, will most likely include components from many societies and cultures.

Sociological Analysis of Culture

Sociologists regard culture as a central ingredient in human behavior. Although all sociologists share a similar purpose, they typically see culture through somewhat different lenses as different theoretical perspectives in their research guide them. What do these perspectives tell us about culture?

Functionalist Perspectives

Functionalist perspectives are based on the assumption that society is a stable, orderly system with interrelated parts that serve specific functions. Anthropologist Bronislaw Malinowski (1922) suggested that culture helps people meet their *biological needs* (including food and procreation), *instrumental needs* (including law and education), and *integrative needs* (including religion and art). Societies in which people share a common language and core values are more likely to have consensus and harmony.

How might functionalist analysts view popular culture? According to many functionalist theorists, popular culture serves a significant function in society in that it may be the "glue" which holds society together. Regardless of race, class, sex, age, or other characteristics, many people are brought together (at least in spirit) to cheer teams competing in major sporting events such as Super Bowl Sunday and the Olympics. Television, the Internet, and social media help integrate recent immigrants into the mainstream culture, whereas longer-term residents may become more homogenized as a result of seeing the same images and being exposed to the same beliefs and values. However, functionalists acknowledge that all societies have dysfunctions, which produce a variety of societal problems. When a society contains numerous subcultures, discord may result from a lack of consensus about values and social norms. In fact, popular culture may undermine cultural values rather than reinforce them. For example, popular culture may glorify crime, rather than hard work, as the quickest way to get ahead. According to some analysts, excessive violence in music, video games, movies, and television shows may be harmful to children and young people. From this perspective, popular culture can be a factor in antisocial behavior such as hate crimes and fatal shootings.

A strength of the functionalist perspective on culture is its focus on the needs of society and the fact that stability

high culture
classical music, opera, ballet, live theater, and other activities usually patronized by elite audiences.

popular culture
activities, products, and services that are assumed to appeal primarily to members of the middle and working classes.

cultural imperialism
the extensive infusion of one nation's culture into other nations.

is essential for society's continued survival. A shortcoming is its overemphasis on harmony and cooperation. This approach also fails to fully account for factors embedded in the structure of society—such as class-based inequalities, racism, and sexism—that may contribute to conflict among people in the United States or to global strife.

Conflict Perspectives

Conflict perspectives are based on the assumption that social life is a continuous struggle in which members of powerful groups seek to control scarce resources. According to this approach, values and norms help create and sustain the privileged position of the powerful in society while excluding others. As early conflict theorist Karl Marx stressed, ideas are *cultural creations* of a society's most powerful members. Thus, it is possible for political, economic, and social leaders to use *ideology*—an integrated system of ideas that is external to, and coercive of, people—to maintain their positions of dominance in a society. As Marx stated,

> The ideas of the ruling class are in every epoch the ruling ideas, i.e., the class which is the ruling material force in society, is at the same time, its ruling intellectual force. The class, which has the means of material production at its disposal, has control at the same time over the means of mental production.... The ruling ideas are nothing more than the ideal expression of the dominant material relationships, the dominant material relationships grasped as ideas. (Marx and Engels, 1970/1845–1846: 64)

Many contemporary conflict theorists agree with Marx's assertion that ideas, a nonmaterial component of culture, are used by agents of the ruling class to affect the thoughts and actions of members of other classes. The role of the mass media in influencing people's thinking about the foods that they should—or should not—eat is an example of such ideological control.

How might conflict theorists view popular culture? Some conflict theorists believe that popular culture, which originated with everyday people, has been largely removed from their domain and has become nothing more than a part of the capitalist economy in the United States. From this approach, media conglomerates such as Time Warner and ABC/Disney create popular culture, such as films, television shows, and amusement parks, in the same way that they would produce any other product or service. Creating new popular culture also promotes consumption of *commodities*—objects outside ourselves that we purchase to satisfy our

human needs or wants. According to contemporary social analysts, consumption—even of things that we do not necessarily need—has become prevalent at all social levels, and some middle- and lower-income individuals and families now use as their frame of reference the lifestyles of the more affluent in their communities. As a result, many families live on credit in order to purchase the goods and services that they would like to have or that keep them on the competitive edge with their friends, neighbors, and coworkers.

A strength of the conflict perspective is that it stresses how cultural values and norms may perpetuate social inequalities. It also highlights the inevitability of change and the constant tension between those who want to maintain the status quo and those who desire change (● Figure 3.16). A limitation is its focus on societal discord and the divisiveness of culture.

Symbolic Interactionist Perspectives

Unlike functionalists and conflict theorists, who focus primarily on macrolevel concerns, symbolic interactionists engage in a microlevel analysis that views society as the sum of all people's interactions. From this perspective, people create,

FIGURE 3.16 Rapid changes in language and culture in the United States are reflected in this sign at a shopping center. How do functionalist and conflict theorists' views regarding language differ?

maintain, and modify culture as they go about their everyday activities. Symbols make communication with others possible because symbols provide us with shared meanings.

According to some symbolic interactionists, people continually negotiate their social realities. Values and norms are not independent realities that automatically determine our behavior. Instead, we reinterpret them in each social situation we encounter. However, the classical sociologist Georg Simmel warned that the larger cultural world—including both material culture and nonmaterial culture—eventually takes on a life of its own apart from the actors who daily re-create social life. As a result, individuals may be more controlled by culture than they realize.

Simmel (1990/1907) suggested that money is an example of how people may be controlled by their culture. According to Simmel, people initially create money as a means of exchange, but then money acquires a social meaning that extends beyond its purely economic function. Money becomes an end in itself, rather than a means to an end. Today, we are aware of the relative "worth" not only of objects but also of individuals. Many people revere wealthy entrepreneurs and highly paid celebrities, entertainers, and sports figures for the amount of money they make, not for their intrinsic qualities. According to Simmel (1990/1907), money makes it possible for us to *relativize* everything, including our relationships with other people. When social life can be reduced to money, people become cynical, believing that anything—including people, objects, beauty, and truth—can be bought if we can pay the price. Although Simmel acknowledged the positive functions of money, he believed that the social interpretations people give to money often produce individual feelings of cynicism and isolation.

A symbolic interactionist approach highlights how people maintain and change culture through their interactions with others. However, interactionism does not provide a systematic framework for analyzing how we shape culture and how it, in turn, shapes us. It also does not provide insight into how shared meanings are developed among people, and it does not take into account the many situations in which there is disagreement on meanings. Whereas the functional and conflict approaches tend to overemphasize the macrolevel workings of society, the interactionist viewpoint often fails to take these larger social structures into account.

Postmodernist Perspectives

Postmodernist theorists believe that much of what has been written about culture in the Western world is Eurocentric—that it is based on the uncritical assumption that European culture (including its dispersed versions in countries such as the United States, Australia, and South Africa) is the true, universal culture in which all the world's people ought to

believe (Lemert, 1997). By contrast, postmodernists believe that we should speak of *cultures,* rather than *culture.*

However, Jean Baudrillard, one of the best-known French social theorists and key figures in postmodern theory, believes that the world of culture today is based on *simulation,* not reality. According to Baudrillard, social life is much more a spectacle that simulates reality than it is reality itself. Many U.S. children, upon entering school for the first time, have already watched more hours of television than the total number of hours of classroom instruction they will encounter in their entire school careers. Add to this the number of hours that some children will have spent playing computer games or using the Internet, where they often find that it is more interesting to deal with imaginary heroes and villains than to interact with "real people" in real life. Baudrillard refers to this social creation as *hyperreality*—a situation in which the *simulation* of reality is more real than experiencing the event itself and having any actual connection with what is taking place. For Baudrillard, everyday life has been captured by the signs and symbols generated to represent it, and we ultimately relate to simulations and models as if they were reality.

Baudrillard (1983) uses Disneyland as an example of a simulation—one that conceals the reality that exists outside rather than inside the boundaries of the artificial perimeter (• Figure 3.17). According to Baudrillard, Disney-like theme parks constitute a form of seduction that substitutes symbolic (seductive) power for real power, particularly the ability to bring about social change. From this perspective, amusement park "guests" may feel like "survivors" after enduring the rapid speed and gravity-defying movements of the roller-coaster rides or see themselves as "winners" after

YOSHIKAZU TSUNO/Getty Images

FIGURE 3.17   Disneyland in Tokyo, Japan, illustrates the idea of postmodern social theorist Jean Baudrillard that theme parks provide visitors with a *simulation* of reality—their symbolic power as "guests" and "consumers" is a substitute for having real power to bring about change in the real world.

Analysis of Culture

Components of Culture	Symbol	Anything that meaningfully represents something else.
	Language	A set of symbols that expresses ideas and enables people to think and communicate with one another.
	Values	Collective ideas about what is right or wrong, good or bad, and desirable or undesirable in a particular culture.
	Norms	Established rules of behavior or standards of conduct.
Sociological Analysis of Culture	Functionalist Perspectives	Culture helps people meet their biological, instrumental, and expressive needs.
	Conflict Perspectives	Ideas are a cultural creation of society's most powerful members and can be used by the ruling class to affect the thoughts and actions of members of other classes.
	Symbolic Interactionist Perspectives	People create, maintain, and modify culture during their everyday activities; however, cultural creations can take on a life of their own and end up controlling people.
	Postmodern Perspectives	Much of culture today is based on simulation of reality (e.g., what we see on television) rather than reality itself.

surviving fights with hideous cartoon villains on the "dark rides." In reality, they have been made to *appear* to have power, but they do not actually possess any real power.

In their examination of culture, postmodernist social theorists thus make us aware of the fact that no single perspective can grasp the complexity and diversity of the social world. There is no one, single, universal culture. They also make us aware that reality may not be what it seems. According to the postmodernist view, no one authority can claim to know social reality, and we should deconstruct—take apart and subject to intense critical scrutiny—existing beliefs and theories about culture in hopes of gaining new insights.

Although postmodern theories of culture have been criticized on a number of grounds, we will mention only three. One criticism is postmodernism's lack of a clear conceptualization of ideas. Another is the tendency to critique other perspectives as being "grand narratives," whereas postmodernists offer their own varieties of such narratives. Finally, some analysts believe that postmodern analyses of culture lead to profound pessimism about the future.

This chapter's Concept Quick Review summarizes the components of culture as well as how the four major perspectives view it.

Looking Ahead: Culture, Social Change, and Your Future

Many changes are occurring in the United States. Increasing cultural diversity can either cause long-simmering racial and ethnic antagonisms to come closer to a boiling point or result in the creation of a truly "rainbow culture" in which diversity is respected and encouraged.

In the future the issue of cultural diversity will increase in importance, especially in schools (see "You Can Make a Difference"). Multicultural education that focuses on the contributions of a wide variety of people from different backgrounds will continue to be an issue of controversy from kindergarten through college. In the Los Angeles school district, for example, students speak more than 150 different languages and dialects (● Figure 3.18). Schools will face the challenge of embracing widespread cultural diversity while conveying a sense of community and national identity to students.

Technology will continue to have a profound effect on culture. Television and radio, films and videos, and digital communications will continue to accelerate the flow of information and expand cultural diffusion throughout the world. Global communication devices will move images of people's lives, behavior, and fashions instantaneously among almost all nations. Increasingly, computers and cyberspace will become people's window on the world and, in the process, promote greater integration or fragmentation among nations. Integration occurs when there is a widespread acceptance of ideas and items—such as democracy, hip-hop music, blue jeans, and hamburgers— among cultures. By contrast, fragmentation occurs when people in one culture disdain the beliefs and actions of other cultures. As a force for both cultural integration and fragmentation, technology will continue to revolutionize communications, but most of the world's population will not participate in this revolution.

Schools as Laboratories for Getting Along: Having Lunch Together

Where did you make many of your childhood friends? Where did you learn about their families and cultural backgrounds? Research has shown that schools and friendship groups can expose children and young people to cultures that are different from their own. Studies have also shown that it may be easier for children to set aside their differences and get to know one another than it is for adults to do so, particularly if they are able to spend time together at lunch or talking in other informal settings. Consider what happened among some children at the International Community School in Decatur, Georgia. Some students were born in the United States, but many were refugees from as many as forty war-torn countries. This school became a "laboratory for getting along," particularly as some children took the initiative to befriend and help others by doing such things as eating lunch with them (St. John, 2007). An excellent example is the friendship that developed between nine-year-old Dante Ramirez and Soung Oo Hlaing, an eleven-year-old Burmese refugee who spoke no English:

> The two boys met on the first day of school this year. Despite the language barrier, Dante managed to invite the newcomer to sit with him at lunch.
>
> "I didn't think he'd make friends at the beginning because he didn't speak that much English," Dante said. "So I thought I should be his friend."

In the next weeks, the boys had a sleepover. They trick-or-treated on Soung's first Halloween. Soung, a gifted artist, gave Dante pointers on how to draw. And Dante helped Soung with his English. "I use simple words that are easy to know and sometimes hand movements," Dante explained. "For 'huge,' I would make my hands bigger. And for 'big,' I would make my hands smaller than for huge." (St. John, 2007: A14)

Over time, as the boys got to know each other better, their mothers also developed a friendship and began to celebrate ethnic holidays together even though they largely relied on gestures (a form of nonverbal communication) to communicate with each other. Could "laboratories" such as this help more people from diverse cultures get along? Would you like to participate in a school or community effort, such as arranging lunches or other meals together, to help people get to know individuals from diverse cultural backgrounds?

REFLECT & ANALYZE

What examples can you provide that show how eating together or having other social gatherings may create new bonds across diverse cultural groups on your college campus or in the community where you reside?

Bill Aron/PhotoEdit

FIGURE 3.18 Although students often use similar digital equipment, their diverse backgrounds represent a challenge for teachers who wish to focus on everyone's cultural heritage but also provide a sense of community.

From a sociological perspective, the study of culture helps us not only understand our own "tool kit" of symbols, stories, rituals, and worldviews but also expands our insights to include those of other people of the world, who also seek strategies for enhancing their own lives. If we understand how culture is used by people, how cultural elements constrain or further certain patterns of action, what aspects of our cultural heritage have enduring effects on our actions, and what specific historical changes undermine the validity of some cultural patterns and give rise to others, we can apply our sociological imagination not only to our own society but to the entire world as well.

CHAPTER REVIEW Q & A

LO1 What is culture, and why is it important in helping people in their daily life?

Culture is the knowledge, language, values, customs, and material objects that are passed from person to person and from one generation to the next in a human group or society. Culture is essential for our individual survival and our communication with other people.

LO2 What is material culture, and what are the four nonmaterial components of culture that are common to all societies?

Material culture consists of physical and tangible creations that members of society make, use, and share. Changes in technology continue to shape the material culture of society. Nonmaterial components of culture are symbols, language, values, and norms.

LO3 What are cultural universals?

Cultural universals are customs and practices that exist in all societies and include activities and institutions such as storytelling, families, and laws. However, specific forms of these universals vary from one cultural group to another.

LO4 How do symbols and language reflect cultural values?

Symbols express shared meanings; through them, groups communicate cultural ideas and abstract concepts. Language is a set of symbols through which groups communicate. One of our most important human attributes is the ability to use language to share our experiences, feelings, and knowledge with others.

LO5 What are the differences among folkways, mores, and laws?

Folkways are norms that express the everyday customs of a group, whereas mores are norms with strong moral and ethical connotations and are essential to the stability of a culture. Laws are formal, standardized norms that are enforced by formal sanctions.

LO6 How have technological changes affected the culture of a nation and the world?

All parts of culture do not change at the same pace. However, technological change widens a cultural lag between a technical development in a society and the society's moral and legal institutions. The pace of technology modifies people's daily lives, as evident in the spread of information via smart technologies, such as handheld electronic devices and automated jobs.

LO7 Explain the importance of *culture shock, ethnocentrism*, and *cultural relativism*.

These terms are important for studying culture and for gaining a better understanding of how we relate to other people, whether we may realize it or not. Culture shock refers to the anxiety that people experience when they encounter cultures radically different from their own. Ethnocentrism is the assumption that one's own culture is superior to other cultures. Cultural relativism views and analyzes another culture in terms of that culture's own values and standards. Depending on which of these approaches we use in our views of others and communications with others, we may have quite different outcomes in our social interactions.

LO8 How do the major sociological perspectives view society and culture?

A functionalist analysis of culture assumes that a common language and shared values help produce consensus and harmony. According to some conflict theorists, certain groups may use culture to maintain their privilege and exclude others from society's benefits. Symbolic interactionists suggest that people create, maintain, and modify culture as they go about their everyday activities. Postmodern thinkers believe that there are many cultures within the United States alone. In order to grasp a better understanding of how popular culture may simulate reality rather than be reality, postmodernists believe that we need a new way of conceptualizing culture and society.

KEY TERMS

QUESTIONS for CRITICAL THINKING

1 Would it be possible today to live in a totally separate culture in the United States? Could you avoid all influences from the mainstream popular culture or from the values and norms of other cultures? How would you be able to avoid any change in your culture?

2 Consider a wide variety of fads and fashions: musical styles, computer and video games and other technologies, literature, and political, social, and religious ideas. Do fads and fashions reflect and reinforce or challenge and change the values and norms of a society?

3 You are doing a survey analysis of recent immigrants to the United States to determine the effects of popular culture on their views and behavior. What are some of the questions that you would use in your survey?

ANSWERS to the SOCIOLOGY QUIZ
ON GLOBAL FOOD AND CULTURE

1 **False** Although cheese is a popular food in many cultures, most of the people living in China find cheese distasteful and prefer delicacies such as duck's feet.

2 **True** In some cultures, round foods such as pears, grapes, and moon cakes are given to celebrate the birth of babies because the shape of the food is believed to symbolize family unity.

3 **False** Although wedding cakes are a tradition in virtually all nations and cultures, the ingredients of the cake—as well as other foods served at the celebration—vary widely. For example, the traditional wedding cake in Italy is made from biscuits.

4 **True** Many faiths, including Christianity, Judaism, Islam, Hinduism, and Buddhism, have dietary rules and rituals that involve food; however, these practices and beliefs vary widely among individuals and communities.

5 **True** Just as foods are divided into yin foods (e.g., bean sprouts and carrots) and yang foods (beef, chicken, and mushrooms), cooking methods are also referred to as having yin qualities (e.g., boiling and steaming) or yang qualities (roasting and stir-frying). For many Chinese Americans, yin and yang are complementary pairs that should be incorporated into all aspects of social life, including the ingredients and preparation of foods.

6 **False** Although more people now rely on fast foods, some cultural and religious communities—such as the Amish of Ohio, Pennsylvania, and Indiana—encourage families to prepare their food from scratch and to preserve their own fruits, vegetables, and meats. Rural families are more likely to grow their own food or prepare it from scratch than are urban families.

7 **False** Rice is a popular mainstay in the diets of people from diverse cultural backgrounds—from the Hmong and Vietnamese to Puerto Ricans and Mexican Americans—who have arrived in the United States over the past four decades. Among some in the younger generations, however, food choices have become increasingly Americanized, and items such as french fries and pizza have become popular.

8 **True** Cultural diversity is a major issue in eating, and people in some cultures, religions, and nations expect that even an "outsider" will have a basic familiarity with, and respect for, their traditions and practices. However, social analysts also suggest that we should not generalize or imply that certain characteristics apply to *all* people in a cultural group or nation.

© Jack Frog/Shutterstock.com

LEARNING OBJECTIVES

1   **Debate** the extent to which people would become human beings without adequate socialization.

2   **Discuss** the sociological perspective on human development, emphasizing the contributions of Charles Horton Cooley and George Herbert Mead.

3   **Contrast** functionalist and conflict theorists' perspectives on the roles that families play in the socialization process.

4   **Describe** how schools socialize children in both formal and informal ways.

5   **Explain** the role that peer groups and media play in socialization now, and predict the role that these agents will play in the future.

6   **Identify** ways in which gender socialization and racial–ethnic socialization occur.

7   **Discuss** the stages in the life course, and demonstrate why the process of socialization is important in each stage.

8   **Distinguish** between voluntary and involuntary resocialization, and give examples of each.

SOCIOLOGY & EVERYDAY LIFE

Class Attendance in Higher Education

Part of my job . . . is to give presentations to visiting middle and high school classes that come to see what a college campus is like. I nearly always start off my presentation by asking the students how they think college is different from high school. One day, a very bright high school student responded by saying, "You don't have to go to class if you don't want to." . . . I've worked on college campuses for the past nine years and one of the most commonly overheard conversations in the dining halls goes something like this:

Student A: Hey, did you go to class today?

Student B: Yeah, I was there, where were you?

Student A: I didn't feel like going. Did we do anything in class?

Student B: No, not really.

Student A: Oh cool, then I didn't miss anything.

Really? You think the professor just stared at the class blankly for the class period and didn't say a word?

—SETH MILLER (2011), admissions advisor at a U.S. university, tells this story to get us thinking about the important of class attendance.

Harvard University conducts an experiment to record students' attendance by setting up secret cameras in ten lecture halls. The cameras take a photo every one minute to show how many seats are empty

Tomas Rodriguez/Fancy/Corbis

You may wonder what class attendance in college has to do with socialization? Why are colleges and universities concerned about students attending classes? Simply stated, skipping classes may be one major factor associated with students' poor grades or lack of overall academic success. In fact, some analysts believe that class attendance is the best-known predictor of college grades, particularly in science, technology, engineering, and math (STEM) courses. Clearly, grades are also related to graduation rates. How do you learn in your classes? As you know, information is often communicated to students through formal instruction, such as in a professor's syllabus or class lecture. But we also learn informally as a result of our personal observations and interactions with others when we are in their physical presence or using digital communications.

How were you socialized to attend classes and to meet the requirements of your courses? As previously mentioned, some methods of socialization are more direct, such as what the professor states in the syllabus about attendance requirements. Some approaches are more covert. Are you aware of the "retention alert system" that many colleges are now using to promote attendance? This system involves the use of technology, such as digital chips

in students' ID cards, interactive software, or clickers, to record students' attendance. In a somewhat controversial move, some universities are now using video equipment to record images of everyone present in a lecture hall. As students, parents, professors, and administrators have become increasingly concerned about low graduation rates, particularly after four years, more emphasis has been placed on how to socialize students for success and ways to eliminate problems such as excessive absences. Concerns about the high cost of a college education, combined with rapidly increasing student debt, have made schools more aware that something must be done to ensure student success. Newer technologies will no doubt play an important part in twenty-first-century socialization, not only in higher education but in all areas of our social life. As you expand your knowledge of sociology, you will often find yourself using the term *socialization* to refer to the process by which people learn various social roles throughout their life and find out how to successfully participate in many groups and organizations.

In this chapter we examine the process of socialization and identify reasons why socialization is crucial to the well-being of individuals, groups, and societies. We discuss

or filled. Overall, about 2,000 undergraduate students are filmed to study classroom attendance. Professors whose classes were filmed are informed, but other faculty and students are unaware this is happening. A group of students and faculty claim that this action constitutes an invasion of privacy, and extensive debate about the issue takes place on campus and across the nation (*Harvard Magazine*, 2014).

How Much Do You Know About Socialization and the College Experience?

TRUE	FALSE		
T	F	1	Professors are the primary agents of socialization for college students.
T	F	2	Researchers have found that few students spend time studying with other students.
T	F	3	Many students find that college courses are stressful because the classes are an abrupt change from those found in high school.
T	F	4	Law and medical students often report high levels of academic pressure because they know that their classmates were top students during their undergraduate years.
T	F	5	Academic stress may be positive for students: It does not necessarily trigger psychological stress.
T	F	6	College students typically find the socialization process in higher education to be less stressful than the professional socialization process they experience when they enter an occupation or profession.
T	F	7	Students who hold jobs outside of school experience higher levels of stress than students who are not employed during their college years.
T	F	8	Getting good grades and completing schoolwork are the top sources of stress reported by college students.

Answers can be found at the end of the chapter.

both sociological and social psychological theories of human development. We look at the dynamics of socialization—how it occurs and what shapes it. We also focus on positive and negative aspects of the socialization process, including the daily stresses that may be involved in this process.

Before reading on, test your knowledge about socialization and the college experience by taking the "Sociology and Everyday Life" quiz.

 LO1 **Debate** the extent to which people would become human beings without adequate socialization.

Why Is Socialization Important Around the Globe?

Socialization is the lifelong process of social interaction through which individuals acquire a self-identity and the physical, mental, and social skills needed for survival in society (• Figure 4.1). It is the essential link between the individual and society because it helps us become aware of ourselves as members of the larger groups and organizations of which we are a part. Socialization also helps us to learn how to communicate with other people and to have knowledge of how other people expect us to behave in a variety of social settings. Briefly stated, socialization enables us to develop our human potential and to learn the ways of thinking, talking, and acting that are necessary for social living.

When do you think socialization is most important? Socialization is the most crucial during childhood because it is essential for the individual's survival and for human development. The many people who met the early material and social needs of each of us were central to our establishing our own identity. Can you identify some of the people in your own life who were the most influential in your earliest years of social development? During the first three years of our life, we begin to develop both a unique identity and the ability to manipulate things and to walk. We acquire

socialization
the lifelong process of social interaction through which individuals acquire a self-identity and the physical, mental, and social skills needed for survival in society.

FIGURE 4.1 The kind of person we become depends greatly on the people who surround us. How will this boy's life be shaped by his close and warm relationship with his mother?

sophisticated cognitive tools for thinking and for analyzing a wide variety of situations, and we learn effective communication skills. In the process we begin a socialization process that takes place throughout our lives and through which we also have an effect on other people who watch us.

What does socialization do for us beyond the individual level? Socialization is essential for the survival and stability of society. Members of a society must be socialized to support and maintain the existing social structure. From a functionalist perspective, individual conformity to existing norms is not taken for granted; rather, basic individual needs and desires must be balanced against the needs of the social structure. The socialization process is most effective when people conform to the norms of society because they believe that doing so is the best course of action. Socialization enables a society to "reproduce" itself by passing on its culture from one generation to the next.

How does socialization differ across cultures and ways of life? Although the techniques used to teach newcomers the beliefs, values, and rules of behavior are somewhat similar in many nations, the *content* of socialization differs greatly from society to society. How people walk, talk, eat, make love, and wage war are all functions of the culture in which they are raised. At the same time, we are also influenced by our exposure to subcultures of class, race, ethnicity, religion, and gender. In addition, each of us has unique experiences in our family and friendship groupings. The kind of human being that we become depends greatly on the particular society and social groups that surround us at birth and during early childhood. What we believe about ourselves, our society, and the world does not spring full-blown from inside ourselves; rather, we learn these things from our interactions with others. What examples can you think of from your own experiences with your family and other close associates?

Human Development: Biology and Society

What does it mean to be "human"? To be human includes being conscious of ourselves as individuals, with unique identities, personalities, and relationships with others. As humans, we have ideas, emotions, and values. We have the capacity to think and to make rational decisions. But what is the source of "humanness"? Are we born with these human characteristics, or do we develop them through our interactions with others?

Have you ever thought about what you were like when you were first born? When we are born, we are totally dependent on others for our survival. We cannot turn ourselves over, speak, reason, plan, or do many of the things that are associated with being human. Although we can nurse, wet, and cry, most small mammals can also do those things. As discussed in Chapter 3, we humans differ from nonhuman animals because we lack instincts and must rely on learning for our survival. Human infants have the potential to develop human characteristics if they are exposed to an adequate socialization process.

Do you think we are more the product of our biological inheritance or the people we are around? Every human being is a product of biology, society, and personal experiences—that is, of heredity and environment or, in even more basic terms, "nature" and "nurture." How much of our development can be explained by socialization? How much by our genetic heritage? Sociologists focus on how humans design their own culture and transmit it from generation to generation through socialization. By contrast, sociobiologists assert that nature, in the form of our genetic makeup, is a major factor in shaping human behavior. ***Sociobiology*** is the systematic study of "social behavior from a biological perspective" (Wilson and Wilson, 2007: 328). According to the zoologist Edward O. Wilson, who pioneered sociobiology, genetic inheritance underlies many forms of social behavior, such as war and peace, envy of and concern for others, and competition and cooperation. Most sociologists disagree with the notion that biological principles can be used to explain all human behavior. Obviously, however, some aspects of our physical makeup—such as eye color, hair color, height, and weight—are largely determined by our heredity.

How important is social influence ("nurture") in human development? There is hardly a single behavior that is not influenced socially. Except for simple reflexes, most human actions are social, either in their causes or in their consequences. Even solitary actions such as crying or brushing our teeth are ultimately social. We cry because someone has hurt us. We brush our teeth because our parents (or dentist) told us it was important. Social environment probably has a greater effect than heredity on the way we develop and the way we act. However, heredity does provide the basic material from which other people help to mold an individual's human characteristics.

How are our biological and emotional needs met, and how are they related? Children whose needs are met in settings characterized by affection, warmth, and closeness see the world as a safe and comfortable place and see other people as trustworthy and helpful. By contrast, infants and children who receive less-than-adequate care or who are emotionally rejected or abused often view the world as hostile and have feelings of suspicion and fear.

Martin Rogers/The Image Bank/Getty Images

FIGURE 4.2 As Harry and Margaret Harlow discovered, humans are not the only primates that need contact with others. Deprived of its mother, this infant monkey found a substitute.

instinctively clung to the cloth "mother" and would not abandon it until hunger drove them to the bottle attached to the wire "mother." As soon as they were full, they went back to the cloth "mother" seeking warmth, affection, and physical comfort.

The Harlows' experiments show the detrimental effects of isolation on nonhuman primates. When the young monkeys were later introduced to other members of their species, they cringed in the corner. Having been deprived of social contact during their first six months of life, they never learned how to relate to other monkeys or to become well-adjusted adults—they were fearful of or hostile toward other monkeys (Harlow and Harlow, 1962, 1977).

Because humans rely more heavily on social learning than do monkeys, the process of socialization is even more important for us.

Problems Associated with Social Isolation and Maltreatment

Social environment, then, is a crucial part of an individual's socialization. Even nonhuman primates such as monkeys and chimpanzees need social contact with others of their species in order to develop properly. As we will see, appropriate social contact is even more important for humans.

Isolation and Nonhuman Primates   Researchers have attempted to demonstrate the effects of social isolation on nonhuman primates raised without contact with others of their own species. In a series of laboratory experiments, the psychologists Harry and Margaret Harlow (1962, 1977) took infant rhesus monkeys from their mothers and isolated them in separate cages. Each cage contained two nonliving "mother substitutes" made of wire, one with a feeding bottle attached and the other covered with soft terry cloth but without a bottle (see ● Figure 4.2). The infant monkeys

Isolated Children   Of course, sociologists would never place children in isolated circumstances so that they could observe what happened to them. However, some cases have arisen in which parents or other caregivers failed to fulfill their responsibilities, leaving children alone or placing them in isolated circumstances. From analysis of these situations, social scientists have documented cases in which children were deliberately raised in isolation. A look at the lives of two children who suffered such emotional abuse provides important insights into the significance of a positive socialization process and the negative effects of social isolation.

Anna   Born in 1932 in Pennsylvania to an unmarried, mentally impaired woman, Anna was an unwanted child. She was kept in an attic-like room in her grandfather's house. Her mother, who worked on the farm all day and often went out at night, gave Anna just enough care to keep

sociobiology
the systematic study of "social behavior from a biological perspective."

her alive; she received no other care. Sociologist Kingsley Davis (1940) described Anna's condition when she was found in 1938:

> [Anna] had no glimmering of speech, absolutely no ability to walk, no sense of gesture, not the least capacity to feed herself even when the food was put in front of her, and no comprehension of cleanliness. She was so apathetic that it was hard to tell whether or not she could hear. And all of this at the age of nearly six years.

When she was placed in a special school and given the necessary care, Anna slowly learned to walk, talk, and care for herself. Just before her death at the age of ten, Anna reportedly could follow directions, talk in phrases, wash her hands, brush her teeth, and try to help other children (Davis, 1940).

Genie About three decades later, Genie was found in 1970 at the age of thirteen (• Figure 4.3). She had been locked in a bedroom alone, alternately strapped down to a child's potty chair or straitjacketed into a sleeping bag, since she was twenty months old. She had been fed baby food and beaten with a wooden paddle when she whimpered. She had not heard the sounds of human speech because no one talked to her and there was no television or radio in her room (Curtiss, 1977; Pines, 1981). Genie was placed in a pediatric hospital, where one of the psychologists described her condition:

> At the time of her admission she was virtually unsocialized. She could not stand erect, salivated continuously, had never been toilet-trained and had no control over her urinary or bowel functions. She was unable to chew solid food and had the weight, height and appearance of a child half her age. (Rigler, 1993: 35)

In addition to her physical condition, Genie showed psychological traits associated with neglect, as described by one of her psychiatrists:

> If you gave [Genie] a toy, she would reach out and touch it, hold it, caress it with her fingertips, as though she didn't trust her eyes. She would rub it against her cheek to feel it. So when I met her and she began to notice me standing beside her bed, I held my hand out and she reached out and took my hand and carefully felt my thumb and fingers individually, and then put my hand against her cheek. She was exactly like a blind child. (Rymer, 1993: 45)

Extensive therapy was used in an attempt to socialize Genie and develop her language abilities (Curtiss, 1977; Pines, 1981). These efforts met with limited success: In the 1990s, Genie was living in a board-and-care home for adults with intellectual disabilities (see Angier, 1993; Rigler, 1993; Rymer, 1993). From 2008, when the latest available reports on Genie were released by the news media, we know that she was 51 and living in a foster home where she had experienced further regression and was unable to speak (James, 2008). No further information about her is currently available.

Why do we discuss children who have been the victims of maltreatment when we are thinking about the socialization process? Because cases like this are important to our understanding of the socialization process and show the importance of the process. These cases also demonstrate how detrimental that social isolation and neglect can be to the well-being of people. Among other things, for children to experience proper grammatical development, they need linguistic stimulation from other people. If children do not hear language, they are unable to speak in sentences.

Child Maltreatment What do the terms *child maltreatment* and *child abuse* mean to you? When asked what constitutes child maltreatment, many people first think of cases that involve severe physical injuries or sexual abuse. However, neglect is the most frequent form of child maltreatment (Mattingly and Walsh, 2010). Child neglect

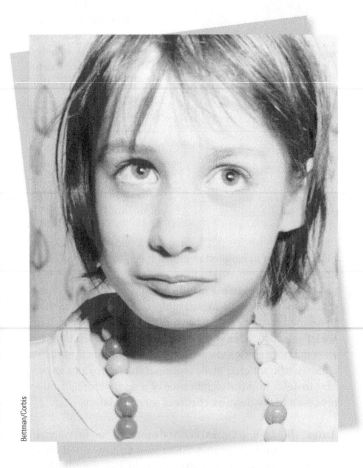

Bettman/Corbis

FIGURE 4.3 A victim of extreme child abuse, Genie was isolated from human contact and tortured until she was rescued at the age of thirteen. What are the consequences to children of isolation and physical abuse, as contrasted with social interaction and parental affection? Sociologists emphasize that the social environment is a crucial part of an individual's socialization.

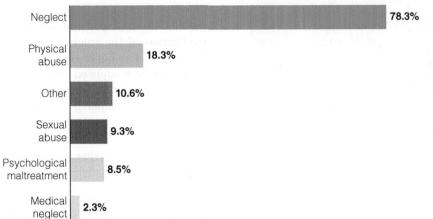

FIGURE 4.4   Types of Maltreatment Among Children Under Age 18*

Source: U.S. Department of Health and Human Services, Children's Bureau, 2013.

*Does not add up to 100 percent because a child may have suffered from multiple forms of maltreatment and was counted once for each maltreatment type.

occurs when children's basic needs—including emotional warmth and security, adequate shelter, food, health care, education, clothing, and protection—are not met, regardless of cause (Mattingly and Walsh, 2010). Neglect often involves acts of omission (where parents or caregivers fail to provide adequate physical or emotional care for children) rather than acts of commission (such as physical or sexual abuse). Neglect is the most common type of maltreatment among children under age eighteen (see • Figure 4.4). Of course, what constitutes child maltreatment differs from society to society.

Social Isolation and Loneliness   Up to this point, we have primarily looked at the effects of isolation on children in their formative years. However, social isolation and loneliness are central issues for persons across all age categories. In the twenty-first century, medical and social researchers continually produce new research documenting that lack of interaction and ongoing learning from others is problematic for everyone. Although we often think that we are more connected than people were in the past and that we have more "friends" than would have been possible for them (because of Facebook, Twitter, Instagram, Foursquare, Pinterest, and other social media sites), the reality is that many people are lonely and have few people to confide in. One study found that 20 percent of all individuals are, at any given time, unhappy because of social isolation (Cacioppo and Hawkley, 2003). According to one study, "People are so embarrassed about being lonely that no one admits it. Loneliness is stigmatized, even though everyone feels it at one time or another" (Seligman, 2009).

Living alone does not necessarily equal being lonely; people experience loneliness in different ways, and some people are more sensitive to social isolation than others. This is why the socialization process of learning how to interact with other people is important. Communicating with other people and learning from them links us to

a larger social world and is energizing for us. Gerontologists who study aging and the issues associated with this process are the first to tell us that older individuals are among the most likely to be socially isolated because of the structure of contemporary families and the greater likelihood that one spouse (typically the wife) will outlive the other partner by a good number of years. We will look into this issue in greater detail in Chapter 12, "Aging and Inequality Based on Age."

Social Psychological Theories of Human Development

Over the past hundred years, a variety of psychological and sociological theories have been developed not only to explain child abuse but also to describe how a positive process of socialization occurs. Although these are not sociological theories, it is important to be aware of the contributions of Freud, Piaget, Kohlberg, and Gilligan because knowing about them provides us with a framework for comparing various perspectives on human development.

Freud and the Psychoanalytic Perspective

The basic assumption in Sigmund Freud's (1924) psychoanalytic approach is that behavior and personality originate from unconscious forces within individuals. Freud (1856–1939), who is known as the founder of psychoanalytic theory, developed his major theories in the Victorian era, when biological explanations of human behavior were prevalent (• Figure 4.5). For example, Freud based his ideas on the belief that people have two basic tendencies: the urge to survive and the urge to procreate.

According to Freud (1924), human development occurs in three states that reflect different levels of the personality, which he referred to as the *id, ego,* and *superego.* The **id** is the component of personality that includes all of the individual's basic biological drives and needs that demand immediate gratification. For Freud, the newborn child's personality is all id, and from birth the child finds that urges for self-gratification—such as wanting to be held, fed, or changed—are not going to be satisfied immediately. However, id remains with people throughout their life in the form of *psychic energy,* the urges and desires that account for behavior.

Id

Sigmund Freud's term for the component of personality that includes all of the individual's basic biological drives and needs that demand immediate gratification.

FIGURE 4.5 Sigmund Freud, founder of the psychoanalytic perspective.

By contrast, the second level of personality—the *ego*—develops as infants discover that their most basic desires are not always going to be immediately met. The *ego* is the rational, reality-oriented component of personality that imposes restrictions on the innate pleasure-seeking drives of the id. The ego channels the desire of the id for immediate gratification into the most advantageous direction for the individual. The third level of personality—the superego—is in opposition to both the id and the ego. The *superego*, or conscience, consists of the moral and ethical aspects of personality. It is first expressed as the recognition of parental

control and eventually matures as the child learns that parental control is a reflection of the values and moral demands of the larger society. When a person is well adjusted, the ego successfully manages the opposing forces of the id and the superego. • Figure 4.6 illustrates Freud's theory of personality.

Piaget and Cognitive Development

Jean Piaget (1896–1980), a Swiss psychologist, was a pioneer in the field of cognitive (intellectual) development (• Figure 4.7). Cognitive theorists are interested in how people obtain, process, and use information—that is, in how we think. Cognitive development relates to changes over time in how we think.

Piaget (1954) believed that in each stage of development (from birth through adolescence), children's activities are governed by their perception of the world around them. His four stages of cognitive development are organized around specific tasks that, when mastered, lead to the acquisition of new mental capacities, which then serve as the basis for the next level of development. Piaget emphasized that all children must go through each stage in sequence

FIGURE 4.6 Freud's Theory of Personality

This illustration shows how Freud might picture a person's internal conflict over whether to commit an antisocial act such as stealing a candy bar. In addition to dividing personality into three components, Freud theorized that our personalities are largely unconscious—hidden from our normal awareness. To dramatize his point, Freud compared conscious awareness (portions of the ego and superego) to the visible tip of an iceberg. Most of personality—including the id, with its raw desires and impulses—lies submerged in our subconscious.

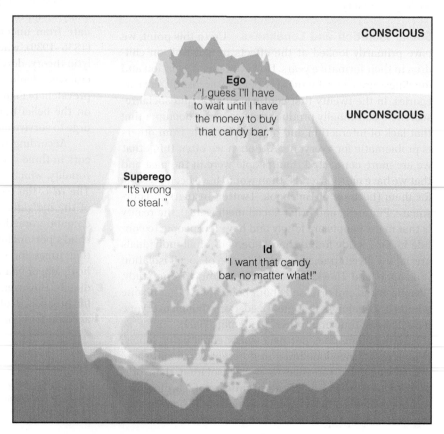

FIGURE 4.7 Jean Piaget, a pioneer in the field of cognitive development.

before moving on to the next one, although some children move through them faster than others.

1. *Sensorimotor stage* (birth to age two). During this period, children understand the world only through sensory contact and immediate action because they cannot engage in symbolic thought or use language. Toward the end of the second year, children comprehend *object permanence;* in other words, they start to realize that objects continue to exist even when the items are out of sight.

2. *Preoperational stage* (age two to seven). In this stage, children begin to use words as mental symbols and to form mental images. However, they are still limited in their ability to use logic to solve problems or to realize that physical objects may change in shape or appearance while still retaining their physical properties (see • Figure 4.8).

3. *Concrete operational stage* (age seven to eleven). During this stage, children think in terms of tangible objects and actual events. They can draw conclusions about the likely physical consequences of an action without always having to try the action out. Children

begin to take the role of others and start to empathize with the viewpoints of others.

4. *Formal operational stage* (age twelve through adolescence). By this stage, adolescents are able to engage in highly abstract thought and understand places, things, and events they have never seen. They can think about the future and evaluate different options or courses of action.

Kohlberg and the Stages of Moral Development

Lawrence Kohlberg (1927–1987) elaborated on Piaget's theories of cognitive reasoning by conducting a series of studies in which children, adolescents, and adults were presented with moral dilemmas that took the form of stories. Based on his findings, Kohlberg (1969, 1981) classified moral reasoning into three sequential levels:

1. *Preconventional level* (age seven to ten). Children's perceptions are based on punishment and obedience. Evil behavior is that which is likely to be punished; good conduct is based on obedience and avoidance of unwanted consequences.

2. *Conventional level* (age ten through adulthood). People are most concerned with how they are perceived by their peers and with how one conforms to rules.

3. *Postconventional level* (few adults reach this stage). People view morality in terms of individual rights; "moral conduct" is judged by principles based on human rights that transcend government and laws.

Gilligan's View on Gender and Moral Development

Psychologist Carol Gilligan (b. 1936) noted that both Piaget and Kohlberg did not take into account how gender affects the process of social and moral development. According to Gilligan (1982), Kohlberg's model was developed solely on the basis of research with male respondents, who often have different views from women on morality. Gilligan believes that men become more concerned with law and order but that women tend to analyze social relationships and the social consequences of behavior. Gilligan argues that men are more likely to use *abstract standards* of right and wrong when making moral decisions, whereas women are more likely to be concerned about the *consequences* of behavior. Does this constitute a "moral deficiency" on the part of either women or men? Not according to Gilligan, who believes that people make moral decisions according to both abstract principles of justice and principles of compassion and care.

ego
Sigmund Freud's term for the rational, reality-oriented component of personality that imposes restrictions on the innate pleasure-seeking drives of the id.

superego
Sigmund Freud's term for the conscience, consisting of the moral and ethical aspects of personality.

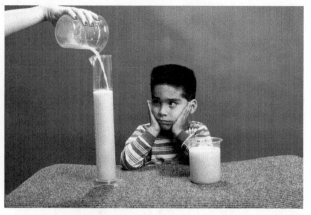

Tony Freeman/PhotoEdit

FIGURE 4.8 **The Preoperational Stage**
Psychologist Jean Piaget identified four stages of cognitive development, including the preoperational stage, in which children have limited ability to realize that physical objects may change in shape or appearance. Piaget showed children two identical beakers filled with the same amount of water. After the children agreed that both beakers held the same amount of water, Piaget poured the water from one beaker into a taller, narrower beaker and then asked them about the amounts of water in each beaker. Those still in the preoperational stage believed that the taller beaker held more water because the water line was higher than in the shorter, wider beaker.

 LO2 **Discuss** the sociological perspective on human development, emphasizing the contributions of Charles Horton Cooley and George Herbert Mead.

Sociological Theories of Human Development

Although social scientists acknowledge the contributions of social–psychological explanations of human development, sociologists believe that it is important to bring a sociological perspective to bear on how people develop an awareness of self and learn about the culture in which they live. Let's look at symbolic interactionist, functional, and conflict approaches to describing the socialization process and its outcomes.

Symbolic Interactionist Perspectives on Socialization

According to a symbolic interactionist approach to socialization, we cannot form a sense of self or personal identity without intense social contact with others. How do we develop ideas about who we are? How do we gain a sense of self? The self represents the sum total of perceptions and feelings that an individual has of being a distinct, unique person—a sense of who and what one is. When we speak of the "self," we typically use words such as *I, me, my, mine,* and *myself* (Cooley, 1998/1902). This sense of self (also referred to *self-concept*) is not present at birth; it arises in the process of social experience. **Self-concept** is the totality of our beliefs and feelings about ourselves. Four components

make up our self-concept: (1) the physical self ("I am tall"), (2) the active self ("I am good at soccer"), (3) the social self ("I am nice to others"), and (4) the psychological self ("I believe in world peace"). Between early and late childhood, a child's focus tends to shift from the physical and active dimensions of self toward the social and psychological aspects. Self-concept is the foundation for communication with others; it continues to develop and change throughout our lives.

Our *self-identity* is our perception about what kind of person we are and our awareness of our unique identity. Self-identity emerges when we ask the question "Who am I?" Factors such as individuality, uniqueness, and personal characteristics and personality are components of self-identity. As we have seen, socially isolated children do not have typical self-identities because they have had no experience of "humanness." According to symbolic interactionists, we do not know who we are until we see ourselves as we believe that others see us. The perspectives of symbolic interactionists Charles Horton Cooley and George Herbert Mead help us understand how our self-identity is developed through our interactions with others.

Cooley: Looking-Glass Self Charles Horton Cooley (1864–1929) was one of the first U.S. sociologists to describe how we learn about ourselves through social interaction with other people. Cooley used the concept of the *looking-glass self* to describe how the self emerges. The **looking-glass self** refers to the way in which a person's sense of self is derived from the perceptions of others. Our looking-glass self is based on our perception of *how* other people think of us (Cooley, 1998/1902). As

We imagine how we appear to other people.

We imagine how other people judge the appearance that we think we present.

If we think the evaluation is favorable, our self-concept is enhanced.

If we think the evaluation is unfavorable, our self-concept is diminished.

FIGURE 4.9   How the Looking-Glass Self Works

• Figure 4.9 shows, the looking-glass self is a self-concept derived from a three-step process:

1. We imagine how our personality and appearance will look to other people.

2. We imagine how other people judge the appearance and personality that we think we present.

3. We develop a self-concept. If we think the evaluation of others is favorable, our self-concept is enhanced. If we think the evaluation is unfavorable, our self-concept is diminished. (Cooley, 1998/1902)

Because the looking-glass self is based on how we *imagine* other people view us, we may develop self-concepts based on an inaccurate perception of what other individuals think about us. Consider, for example, the individual who believes that other people see him or her as "fat" when, in actuality, he or she is a person of an average height, weight, and build. The consequences of such a false perception may lead to excessive dieting or health problems such as anorexia, bulimia, and other eating disorders.

Mead: Role-Taking and Stages of the Self   George Herbert Mead (1863–1931) extended Cooley's insights by linking the idea of self-concept to *role-taking*—the process by which a person mentally assumes the role of another person or group in order to understand the world from that person's or group's point of view. Role-taking often occurs through play and games, as children try out different roles (such as being mommy, daddy, doctor, or teacher) and gain

an appreciation of them. First, people come to take the role of the other (role-taking). By taking the roles of others, the individual hopes to ascertain the intention or direction of the acts of others. Then the person begins to construct his or her own roles (role-making) and to anticipate other individuals' responses. Finally, the person plays at her or his particular role (role-playing).

According to Mead (1934), children in the early months of life do not realize that they are separate from others. However, they do begin early on to see a mirrored image of themselves in others. Shortly after birth, infants start to notice the faces of those around them, especially the significant others, whose faces start to have meaning because they are associated with experiences such as feeding and cuddling. ***Significant others*** are those persons whose care, affection, and approval are especially desired and who are most important in the development of the self. Gradually, we distinguish ourselves from our caregivers and begin to perceive ourselves in contrast to them. As we develop language skills and learn to understand symbols, we begin to develop a self-concept. When we can represent ourselves in our minds as objects distinct from everything else, our self has been formed.

As Mead (1934) points out, the self has two sides—the "me" and the "I." The "me" is what is learned by interaction with others in the larger social environment; it is the organized set of attitudes of others that an individual assumes. The "me" is the objective element of the self, which represents an internalization of the expectations and attitudes of others and the individual's awareness of those demands. By contrast, the "I" is the person's individuality—it is the response of the person to the attitudes of other individuals. We might think of the "me" as the social self and the "I" as the response to the "me." According to Mead, the "I" develops first, and the "me" takes form during the three stages of self-development (• Figure 4.10):

1. During the *preparatory stage,* up to about age three, interactions lack meaning, and children largely imitate the people around them, particularly parents and other family members. At this stage, children are preparing for role-taking.

self-concept
the totality of our beliefs and feelings about ourselves.

looking-glass self
Charles Horton Cooley's term for the way in which a person's sense of self is derived from the perceptions of others.

role-taking
the process by which a person mentally assumes the role of another person or group in order to understand the world from that person's or group's point of view.

significant others
those persons whose care, affection, and approval are especially desired and who are most important in the development of the self.

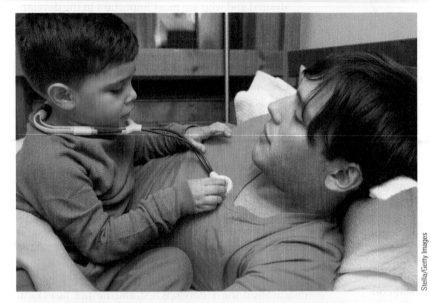

FIGURE 4.10 According to sociologist George Herbert Mead, the self develops through three stages. In the preparatory stage, children imitate others; in the play stay, children pretend to take the roles of specific people; and in the game stage, children become aware of the "rules of the game" and the expectations of others.

2. In the *play stage,* from about age three to five, children learn to use language and other symbols, thus enabling them to pretend to take the roles of specific people. At this stage, they begin to see themselves in relation to others, but they do not see role-taking as something they have to do.

3. During the *game stage,* which begins in the early school years, children understand not only their own social position but also the positions of others around them. In contrast to play, games are structured by rules, are often competitive, and involve a number of other "players." At this time, children become concerned about the demands and expectations of others and of the larger society.

Mead's concept of the **generalized other** refers to the child's awareness of the demands and expectations of the society as a whole or of the child's subculture. According to Mead, the generalized other is evident when a person takes into account other people and groups when he or she speaks or acts. In sum, both the "I" and the "me" are needed to form the social self. The unity of the two (the "generalized other") constitutes the full development of the individual and a more thorough understanding of the social world.

More-Recent Symbolic Interactionist Perspectives Symbolic interactionist approaches emphasize that socialization is a collective process in which children are active and creative agents, not just passive recipients of the socialization process. From this view, childhood is a *socially constructed* category. As children acquire language skills and interact with other people, they begin to construct their own shared meanings. Sociologist William A. Corsaro (2011) refers to this as the "orb web model," whereby the cultural knowledge that children possess consists not only of beliefs found in the adult world but also of unique interpretations from the children's own peer culture. According to Corsaro, children create and share their own *peer culture,* which is an established set of activities, routines, and beliefs that are in some ways different from adult culture. This peer culture emerges through interactions as children "borrow" from the adult culture but transform it so that it fits their own situation. In fact, according to Corsaro, peer culture is the most significant arena in which children and young people acquire cultural knowledge.

Functionalist Perspectives on Socialization

As discussed in Chapter 1, functionalist theorists such as Talcott Parsons and Robert Merton saw socialization as the process by which individuals internalize social norms and values. They believed that socialization is important to societies as well as to individuals because social institutions must be maintained and preserved for a nation to survive. For these institutions to be efficient, individuals must play their roles appropriately, or dysfunctions will occur. Simply stated, the socialization process plays an integral part in teaching the next generation, as well as new arrivals, about how to conform to the rules of the game, and this keeps the society functioning properly. As a result of adequate socialization, people come to support a society that is stable and orderly. Individuals learn to accept the values, beliefs, and behavioral expectations that keep society, and sometimes the larger global community, functioning effectively.

Some functionalist theorists identify three stages of socialization: **Primary socialization** refers to the process of learning that begins at birth and occurs in the home and family; by contrast, **secondary socialization** refers to the process of learning that takes place outside the home—in settings such as schools, religious organizations, and the workplace—and helps individuals learn how to act in appropriate ways in various situations. Secondary socialization often occurs when we are teenagers and young adults. **Tertiary socialization** takes place when adults move into new settings where they must accept certain ideas or engage in specific behaviors that are appropriate to that specific setting. (See • Figure 4.11.) For example, older persons entering a retirement community often have to internalize new social norms and values that are appropriate to the setting in which they now reside. From a functionalist approach, problems in the socialization process contribute not only to individual concerns but also to larger societal issues, such as high rates of crime and poverty, school dropouts and failures, and family discord.

© Shots Studio/Shutterstock.com

FIGURE 4.11 Some theorists identify three stages of socialization: primary, secondary, and tertiary. At what stage might socialization be occurring for the people working together in this photo?

Conflict Perspectives on Socialization

Based on an assumption that groups in society are engaged in a continuous power struggle for control of scarce resources, conflict theorists stress that socialization contributes to "false consciousness"—a lack of awareness and a distorted perception of the reality of class as it affects all aspects of social life. As a result, socialization reaffirms and reproduces the class structure in the next generation rather than challenging existing conditions. For example,

generalized other
George Herbert Mead's term for the child's awareness of the demands and expectations of the society as a whole or of the child's subculture.

primary socialization
the process of learning that begins at birth and occurs in the home and family.

secondary socialization
the process of learning that takes place outside the home—in settings such as schools, religious organizations, and the workplace—and helps individuals learn how to act in appropriate ways in various situations.

tertiary socialization
the process of learning that takes place when adults move into new settings where they must accept certain ideas or engage in specific behaviors that are appropriate to that specific setting.

Psychological and Sociological Theories of Human Development and Socialization

Social Psychological Theories	Freud's psychoanalytic perspective	
	Piaget's cognitive development	Children go through four stages of cognitive (intellectual) development, moving from understanding only through sensory contact to engaging in highly abstract thought.
	Kohlberg's stages of moral development	People go through three stages of moral development, from avoidance of unwanted consequences to viewing morality based on human rights.
	Gilligan: gender and moral development	Women go through stages of moral development from personal wants to the greatest good for themselves and others.
Symbolic Interactionist Theories	Cooley's looking-glass self	A person's sense of self is derived from his or her perception of how others view him or her.
	Mead's three stages of self-development	In the preparatory stage, children prepare for role-taking. In the play stage, they pretend to take the roles of specific people. In the game stage, they learn to take into account the demands and expectations of the larger society and to develop a generalized other.
Functionalist Theories	Parsons's and Merton's views	The socialization process serves a central function for both individuals and society by helping people learn the appropriate norms, values, and behaviors that support social institutions and the larger social group.
Conflict Theories	Based on Marx's work	Socialization contributes to false consciousness and reproduces inequalities in the class structure in the next generation as well as ignoring crucial differences based on gender, race/ethnicity, and other factors.

children in low-income families may be unintentionally socialized to believe that acquiring an education and aspiring to lofty ambitions are pointless because of existing economic conditions in the family. By contrast, middle- and upper-income families typically instill ideas of monetary and social success in children. As discussed later, schools may also provide different experiences to children depending on their gender, social class, racial–ethnic background, and other factors. This chapter's Concept Quick Review summarizes the major theories of human development and socialization.

Agents of Socialization

Agents of socialization are the persons, groups, or institutions that teach us what we need to know in order to participate in society. We are exposed to many agents of socialization throughout our lifetime; in turn, we have an influence on those socializing agents and organizations. In this section we look at the most pervasive agents of socialization in childhood—the family, the school, peer groups, and the mass media.

 LO3 **Contrast** functionalist and conflict theorists' perspectives on the roles that families play in the socialization process.

The Family

The family is the most important agent of socialization in all societies. From our infancy onward, our families transmit cultural and social values to us (● Figure 4.12). As discussed later in this book, families vary in size and structure. Some families consist of two parents and their biological children, whereas others consist of a single parent and one or more children. Still other families reflect changing patterns of divorce and remarriage, and an increasing number are made up of same-sex partners and their children. Over time, patterns have changed in some two-parent families so that fathers, rather than mothers, are the primary daytime agents of socialization for their young children.

Theorists using a functionalist perspective emphasize that families serve important functions in society because they are the basis for the procreation and socialization of children. Most of us form an emerging sense of self and

FIGURE 4.12 As this chess game attended by several generations of family members illustrates, socialization enables society to "reproduce" itself.

acquire most of our beliefs and values within the family context. We also learn about the larger dominant culture (including language, attitudes, beliefs, values, and norms) and the primary subcultures to which our parents and other relatives belong.

Families are also the primary source of emotional support. Ideally, people receive love, understanding, security, acceptance, intimacy, and companionship within families. The role of the family is especially significant because young children have little social experience beyond the family's boundaries; they have no basis for comparing or evaluating how they are treated by their own family.

To a large extent, the family is where we acquire our specific social position in society. From birth, we are a part of the specific racial, ethnic, class, religious, and regional subcultural grouping of our family. Many parents socialize their children somewhat differently based on race, ethnicity, and class. Some families instruct their children about the unique racial–ethnic and/or cultural backgrounds of their parents and grandparents so that they will have a better appreciation of their heritage. Other families teach their children primarily about the dominant, mainstream culture in hopes that this will help their children get ahead in life.

Some upper-class parents focus on teaching their children about the importance of wealth, power, and privilege; however, many downplay this aspect and want their children to make their own way in life, fearing that "spoiling them" will not be in their best interest. Middle-class parents have typically focused on academic achievement and the importance of hard work to achieve the American

Dream. However, with the global recession that hit the U.S. economy hard between December 2007 and June 2009, optimism that persons in the middle class had previously passed on to their children diminished as homes went into foreclosure, jobs were lost and not replaced, and people saw their standard of living slip. Even then, some research showed that middle-class families felt slightly more secure financially than families at working- and lower-income levels, where parents continually struggle to keep a roof overhead and food on the table. Parents in lower-income categories often felt that they had little time to help their children learn about important things that might help them succeed in school and life (Kendall, 2002, 2011). Problems such as these contribute to and reinforce social inequality, and this is one of many reasons why conflict theorists are concerned about the long-term effects of the socialization process.

However, we should note that socialization is a bidirectional process in which children and young people socialize their agents of socialization, including parents, teachers, and others, as well as receiving socialization from these important agents (• Figure 4.13). ***Reciprocal socialization*** is the process by which the feelings, thoughts, appearance, and behavior of individuals who are undergoing socialization also have a direct influence on those agents of socialization who are attempting to influence them. Examples of this process include parents whose preferences in music, hairstyles, and clothing are influenced by their children, and teachers whose choice of words ("cool," "you know," "LOL," and other slang terms) is similar to that of their students.

agents of socialization
the persons, groups, or institutions that teach us what we need to know in order to participate in society.

reciprocal socialization
the process by which the feelings, thoughts, appearance, and behavior of individuals who are undergoing socialization also have a direct influence on those agents of socialization who are attempting to influence them.

Jamie Grill/Getty Images

FIGURE 4.13 Students are sent to school to be educated. However, what else will they learn in school beyond the academic curriculum? Sociologists differ in their responses to this question.

LO4 **Describe** how schools socialize children in both formal and informal ways.

The School

As the amount of specialized technical and scientific knowledge has expanded rapidly and as the amount of time that children are in educational settings has increased, schools continue to play an enormous role in the socialization of young people. For many people, the formal education process is an undertaking that lasts up to twenty years.

As the number of one-parent families and families in which both parents work outside the home has increased dramatically, the number of children in day-care and preschool programs has also grown rapidly. Nearly 11 million children younger than age 5 whose mothers are working are in some type of child-care arrangement where they spend, on average, about 36 hours a week. Potentially, more than 15 million children under the age of 6 need child care (Child Care Aware of America, 2014).

Generally, studies have found that quality day-care and preschool programs have a positive effect on the overall socialization of children. These programs provide children with the opportunity to have frequent interactions with teachers and to learn how to build their language and literacy skills. High-quality programs also have a positive effect on the academic performance of children, particularly those from low-income families. Today, however, the cost of child-care programs has become a major concern for many families. For example, a year of center-based care for a four-year-old ranges from slightly more than $4,500 in Tennessee to more than $12,300 in Massachusetts (Child Care Aware of America, 2014).

In schools ranging from kindergarten through grade 12, students learn specific mandated knowledge and skills. However, schools also have a profound effect on children's self-image, beliefs, and values (Figure 4.13). As children enter school for the first time, they are evaluated and systematically compared with one another by the teacher. A permanent, official record is kept of each child's personal behavior and academic activities. From a functionalist perspective, schools are responsible for (1) socialization, or teaching students to be productive members of society; (2) transmission of culture; (3) social control and personal development; and (4) the selection, training, and placement of individuals on different rungs in the society (Ballantine and Hammack, 2012).

In contrast, conflict theorists assert that students have different experiences in the school system depending on their social class, their racial–ethnic background, the neighborhood in which they live, their gender, and other factors. For example, Langhout and Mitchell (2008), after investigating the "hidden curriculum" in a low-income elementary school, concluded that African American and Latino boys were disproportionately punished for violating the rules (e.g., raising your hand to speak) when compared to their white and female counterparts. Thus, schools do not socialize children for their own well-being but rather for their roles in school and the workforce, where it is important to be well-behaved and "know your place." Students who are destined for leadership or elite positions acquire different skills and knowledge than those who will enter working-class and middle-class occupations.

LO5 **Explain** the role that peer groups and media play in socialization now, and predict the role that these agents will play in the future.

Peer Groups

As soon as we are old enough to have acquaintances outside the home, most of us begin to rely heavily on peer groups as a source of information and approval about social behavior. A ***peer group*** is a group of people who are linked by common interests, equal social position, and (usually) similar age. In early childhood, peer groups are often composed of classmates in day care, preschool, and elementary school. Preadolescence—the latter part of the elementary school years—is an age period in which children's peer culture has an important effect on how children perceive themselves and how they internalize society's expectations (Robnett and Susskind, 2010). For example, boys who have

FIGURE 4.14 The pleasure of participating in activities with friends is one of the many attractions of adolescent peer groups. What groups have contributed the most to your sense of belonging and self-worth?

a large proportion of same-gender friends are more likely to reject "feminine" traits, which they associate with girls. As a result, this may play a part in socializing them to have negative attitudes toward femininity that they display later in life (Robnett and Susskind, 2010). In adolescence, peer groups are typically made up of people with similar interests and social activities. As adults, we continue to participate in peer groups of people with whom we share common interests and comparable occupations, income, and/or social position.

Peer groups function as agents of socialization by contributing to our sense of "belonging" and our feelings of self-worth (● Figure 4.14). As early as the preschool years, peer groups provide children with an opportunity for successful adaptation to situations such as gaining access to ongoing play, protecting shared activities from intruders, and building solidarity and mutual trust during ongoing activities (Corsaro, 2011). Unlike families and schools, peer groups provide children and adolescents with some degree of freedom from parents and other authority figures. They also teach cultural norms such as what constitutes "acceptable" behavior in a specific situation. Peer groups simultaneously reflect the larger culture and serve as a conduit for passing on culture to young people. As a result, the peer group is both a product of culture and one of its major transmitters.

Do you think there is such a thing as "peer pressure"? Most of us are acutely aware of such a social force. Individuals must earn their acceptance with their peers by conforming to a given group's norms, attitudes, speech patterns, and dress codes. When we conform to our peer group's expectations, we are rewarded; if we do not conform, we may be ridiculed or even expelled from the group. Conforming to the demands of peers frequently places children and adolescents at cross-purposes with their parents. For example, young people are frequently under pressure to obtain certain valued material possessions (such as toys, clothing, athletic shoes, or cell phones); they then pass the pressure on to their parents through emotional pleas to purchase the desired items.

Mass Media

An agent of socialization that has a profound impact on both children and adults is the ***mass media***, composed of large-scale organizations that use print or electronic means (such as radio, television, film, and the Internet) to communicate with large numbers of people. Today, the term *media* also includes the many forms of Web-based and social media such as Facebook, Twitter, and YouTube. For many years, the media have functioned as socializing agents in several ways: (1) they inform us about events; (2) they introduce us to a wide variety of people; (3) they provide an array of viewpoints on current issues; (4) they make us aware of products and services that, if we purchase them, will supposedly help us to be accepted by others; and (5) they entertain us by providing the opportunity to live vicariously (through other people's experiences). Although most of us take for granted that the media play an important part in contemporary socialization, we frequently underestimate the enormous influence that this agent of socialization may have on our attitudes and behavior.

As you are aware, the use of social media such as Facebook and Twitter has grown exponentially in recent years. Today, 95 percent of teens report that they use the Internet, and most indicate that they use it to interact with friends and watch video content online that they previously might have watched on television (Nielsen, 2013). Within

peer group
a group of people who are linked by common interests, equal social position, and (usually) similar age.

mass media
large-scale organizations that use print or electronic means (such as radio, television, film, and the Internet) to communicate with large numbers of people.

FIGURE 4.15 Texting, social networking, and using smartphones now provide us with instant access to friends, information, and entertainment around the clock. How does this compare to the socialization process when your parents or grandparents were children?

households where teens are present, smartphones and tablets are the fastest-growing devices. Social networking is a rapidly increasing layer on top of existing layers of other media use (• Figure 4.15).

Consider young people between the ages of 12 and 17, for example. Research has shown that the monthly time spent, on average, watching video on TV is 98 hours, 26 minutes, as compared to 7 hours, 48 minutes watching video on smartphones and 5 hours, 26 minutes watching video on the Internet (Nielsen, 2013). Smartphones, tablets, and other mobile media make it possible for young people to have access to media 24 hours per day, 7 days per week, with little time for other influences or activities in their life. Does this make a significant difference in childhood socialization? Future studies will no doubt continue to examine new media's effects on children and how increased use relates to grades, family interaction patterns, social networks, and other important issues in reaching maturity.

Parents, educators, social scientists, and public officials have widely debated the consequences of young people watching violence on television. In addition to concerns about violence in television programming, motion pictures, and electronic games, television shows have been criticized for projecting negative images of women and people of color. Although the mass media have changed some of the roles that they depict women as playing, some newer characters tend to reinforce existing stereotypes of women as sex objects even when they are in professional roles such as doctors or lawyers. What effect do you think this has on children and young people as they develop their own ideas about the "adult world"? Throughout this text, we will look at additional examples of how the media socialize us in ways that we may or may not realize.

LO6 **Identify** ways in which gender socialization and racial–ethnic socialization occur.

Gender Socialization

Gender socialization is the aspect of socialization that contains specific messages and practices concerning the nature of being female or male in a specific group or society. Through the process of gender socialization we learn about what attitudes and behaviors are considered to be appropriate for girls and boys, men and women, in a particular society. Different sets of gender norms are appropriate for females and males in the United States and most other nations. When do you first remember learning about gender-specific norms for your own appearance and behavior?

One of the primary agents of gender socialization is the family. In some families, this process begins even before the child's birth. Parents who learn the sex of the fetus through ultrasound or amniocentesis often purchase color-coded and gender-typed clothes, toys, and nursery decorations in anticipation of their daughter's or son's arrival. After birth, parents may respond differently toward male and female infants; they often play more roughly with boys and talk more lovingly to girls. Throughout childhood and adolescence, boys and girls are typically assigned different household chores and given different privileges such as boys being given more latitude to play farther away from home than girls and being allowed to stay out later at night (• Figure 4.16).

In regard to gender socialization practices among various racial–ethnic groups, some sociologists have found that children typically are not taught to think of gender strictly in "male–female" terms. Both daughters and sons

FIGURE 4.16 Do you believe that what this child is learning here will have an influence on his actions in the future? What other childhood experiences might offset early gender socialization?

are socialized toward autonomy, independence, self-confidence, and nurturance of children. Sociologist Patricia Hill Collins (2000) has suggested that "othermothers" (women other than a child's biological mother) play an important part in the gender socialization and motivation of African American children, especially girls. Othermothers often serve as gender-role models and encourage women to become activists on behalf of their children and community (Collins, 2000). In the past, by contrast, Korean American and Latino/a families typically engaged in more-traditional gender socialization, but evidence in the 2010s suggests that this pattern has continued to change as young women are spending more time away from older family members and are gaining greater freedom of expression at school and in the workplace.

Like the family, schools, peer groups, and the media also contribute to our gender socialization. From kindergarten through college, teachers and peers reward gender-appropriate attitudes and behavior. Sports reinforce traditional gender roles through a rigid division of events into male and female categories. The media are also a powerful source of gender socialization; starting very early in childhood, children's books, television programs, movies, and music provide subtle and not-so-subtle messages about how boys and girls should act (see Chapter 11, "Sex, Gender, and Sexuality").

Racial–Ethnic Socialization

In addition to gender-role socialization, we receive racial socialization throughout our lives. **Racial socialization** is the aspect of socialization that contains specific messages and practices concerning the nature of our racial or ethnic status as it relates to our identity, interpersonal relationships, and location in the social hierarchy. Racial socialization includes direct statements regarding race, modeling behavior (wherein a child imitates the behavior of a parent or other caregiver), and indirect activities such as exposure to an environment that conveys a specific message about a racial or ethnic group ("We are better than they are," for example).

The most important aspects of racial identity and attitudes toward other racial–ethnic groups are passed down in families from generation to generation. As the sociologist Martin Marger (1994: 97) notes, "Fear of, dislike for, and antipathy toward one group or another is learned in much the same way that people learn to eat with a knife or fork rather than with their bare hands or to respect others' privacy in personal matters." These beliefs can be transmitted in subtle and largely unconscious ways; they do not have to be taught directly or intentionally.

How early do you think racial socialization begins? Scholars have found that ethnic values and attitudes begin to crystallize among children as young as age four (Van Ausdale and Feagin, 2001). By this age, the society's ethnic hierarchy has become apparent to the child. Some minority parents feel that racial socialization is essential because it provides children with the skills and abilities that they will need to survive in the larger society.

Discuss the stages in the life course, and demonstrate why the process of socialization is important in each stage.

Socialization Through the Life Course

Why is socialization a lifelong process? Throughout our lives, we continue to learn. Each time we experience a change in status (such as becoming a college student or getting married), we learn a new set of rules, roles, and relationships. Even before we achieve a new status, we often participate in **anticipatory socialization**—the process by which knowledge and skills are learned for future roles. Many societies organize social activities according to age and gather data regarding the age composition of the people who live in that society. Some societies have distinct *rites of passage,* based on age or other factors that publicly dramatize and validate changes in a person's status. In the United States and other industrialized societies, the most common categories of age are childhood, adolescence, and adulthood (often subdivided into young adulthood, middle adulthood, and older adulthood).

Childhood

Some social scientists believe that a child's sense of self is formed at an early age and that it is difficult to change this self-perception later in life. Symbolic interactionists emphasize that during infancy and early childhood, family support and guidance are crucial to a child's developing self-concept. In some families, children are provided with emotional warmth, feelings of mutual trust, and a sense of security. These families come closer to our ideal cultural belief that childhood should be a time of carefree play, safety, and freedom from economic, political, and sexual responsibilities. However, other families reflect the discrepancy between cultural ideals and reality—children grow up in a setting characterized by fear, danger, and risks that are created by parental neglect, emotional maltreatment, or premature economic and sexual demands. Abused and neglected children often experience

gender socialization
the aspect of socialization that contains specific messages and practices concerning the nature of being female or male in a specific group or society.

racial socialization
the aspect of socialization that contains specific messages and practices concerning the nature of our racial or ethnic status as it relates to our identity, interpersonal relationships, and location in the social hierarchy.

anticipatory socialization
the process by which knowledge and skills are learned for future roles.

FIGURE 4.17 An important rite of passage for many Latinas is the *quinceañera*—a celebration of their fifteenth birthday and their passage into womanhood. Can you see how this occasion might also be a form of anticipatory socialization?

physical consequences, such as damage to their growing brains, which can lead to cognitive delays or emotional difficulties. Psychological problems can also occur that involve high-risk behavior such as smoking, alcohol or drug abuse, or similar activities. Other psychological problems manifest as low self-esteem, an inability to trust others, feelings of isolation and powerlessness, and denial of one's feelings.

Adolescence

Did you know that some societies have not had a period of time in the life of the individual known as "adolescence"? In contemporary societies, the adolescent (or teenage) years represent a buffer between childhood and adulthood. It is a time during which young people pursue their own routes to self-identity and adulthood. Anticipatory socialization is often associated with adolescence, with many young people spending time planning or being educated for future roles they hope to occupy. Although no specific rites of passage exist in the United States to mark *every* child's transition between childhood and adolescence or between adolescence and adulthood, some rites of passage are observed. For example, a celebration known as a Bar Mitzvah

is held for some Jewish boys on their thirteenth birthday, and a Bat Mitzvah is held for some Jewish girls on their twelfth birthday; these events mark the occasion upon which young people accept moral responsibility for their own actions and the fact that they are now old enough to own personal property. Similarly, some Latinas are honored with the *quinceañera*—a celebration of their fifteenth birthday that marks their passage into young womanhood (• Figure 4.17). Although it is not officially designated as a rite of passage, many of us think of the time when we get our first driver's license or graduate from high school as another way in which we mark the transition from one period of our life to the next.

Adolescence is often characterized by emotional and social unrest. In the process of developing their own identities, some young people come into conflict with parents, teachers, and other authority figures who attempt to restrict their freedom. Adolescents may also find themselves caught between the demands of adulthood and their own lack of financial independence and experience in the job market.

The experiences of individuals during adolescence vary according to race, class, and gender. Based on their family's economic situation and personal choices, some young people leave high school and move directly into the world of work, whereas others pursue a college education and may continue to receive advice and financial support from their parents. Others are involved in both the world of work and the world of higher education as they seek to support themselves and to acquire more years of formal education or vocational/career training. Whether or not a student works while in college may affect the process of adjusting to college life (see ● Figure 4.18). In the second decade of the twenty-first century, more college students are exploring international study programs as part of their adult socialization to help them gain new insights into divergent cultures and the larger world of which they are a part (see "Sociology in Global Perspective").

EARLY FALL ➤

- Adapting to new people and new situations
- Anticipation and excitement about studying in a new setting
- Insecurity about academic demands
- Homesickness
- If employed, trying to balance school and work life

MID FALL ➤

- Social pressures from others: What would my parents think?
- Anticipation (and dread) of midterm exams and major papers
- Time-management problems between school and social life
- Intense need for a break
- Concerns about role conflict between school and work

LATE FALL ➤

- Positive or negative assessment of grades so far
- Pre-final studying and jitters
- Making up for lost time and procrastination
- First college illnesses likely to occur because of late hours, poor eating habits, and proximity to others who become ill
- Potential problems with roommates or others who make excessive demands on one's time and/or personal space

END OF TERM ◆

- Final exams: late nights, extra effort, and stress
- Concerns about leaving new friends and college setting for winter break
- Anticipation (and tension) associated with going home for break for those who have been away
- Reassessment of college choice, major, and career options: Am I on the right track?
- Acknowledgment that growth has occurred and much has been learned, both academically and otherwise, during the first college term

FIGURE 4.18 **Time Line for First-Semester College Socialization**

Source: Based on the author's observations of student life and on Kansas State University, 2010.

SOCIOLOGY in GLOBAL PERSPECTIVE

Open Doors: Study Abroad and Global Socialization

[T]he first month or so of the study abroad experience feels like a vacation in that everything is exciting and new. After this "honeymoon" period, the experience becomes something other than merely a vacation or fleeting visit. You start to relate to the people, the culture, and life in that country not from the eyes of a tourist passing through, but progressively from the eyes of those around you—the citizens who were born and raised there. That is the perspective which is unattainable without actually living in another country, and a perspective which I have come to appreciate and understand more fully as I settle back into life here back at home.

—John R. R. Howie (2010), then a Boston College economics and Mandarin Chinese major, explaining what studying abroad at Peking University, in Beijing, meant to him. Howie has since graduated and is now employed as a financial analyst.

Studying abroad is an important part of the college socialization process for preparing to live and work in an interconnected world. Here are a few interesting facts about studying abroad (adapted from Institute of International Education, 2014):

- More than 289,000 U.S. students participated in study-abroad programs for credit in 2012–2013, and this number continues to increase each year.

- The United Kingdom, Italy, Spain, France, and China are the top destinations for study abroad; however, Latin America, Asia, Oceania—which includes Australia, New Zealand, and the South Pacific islands—and Africa are also popular destinations.

- The top fields of study for U.S. study-abroad students are STEM, social sciences, business, humanities, fine or applied arts, foreign languages, and education.

- More than 60 percent of study-abroad students remain in their host country for a short-term stay (summer or eight weeks or less during the academic year). About 37 percent spend one semester or one or two quarters in the host country, while 3 percent remain for an academic or calendar year.

Sociologists are interested in studying the profile of U.S. study-abroad students because the data provide interesting insights on differences in students' participation by classification, gender, race, and class. Based on the latest figures available (2012/2013), most students participating in study-abroad programs are classified as juniors (34.7 percent) or seniors (24.7 percent). Women make up 65.3 percent of all study-abroad students, and men make up 34.7 percent. White students make up the vast majority of study-abroad students (76.3 percent). Other groups include Hispanic or Latino(a) (7.6 percent), Asian or Pacific Islander (7.3 percent), and black or African American (5.3 percent) (Institute of International Education, 2014).

Socialization for life in the global community is necessary for all students because of the increasing significance of international understanding and the need to learn how to live and work in a diversified nation and world. Even more important may be the opportunity for each student to gain direction and meaning in his or her own life. Do you think that studying abroad might make an important contribution to your own socialization while in college? Why or why not?

REFLECT & ANALYZE

What are the positive aspects of study-abroad programs in the college socialization process? What are the limitations of such programs? If you are unable to participate in a study-abroad program, what other methods and resources might you use to gain "global socialization," which could be beneficial in helping you meet your goals for the future?

Adulthood

One of the major differences between child socialization and adult socialization is the degree of freedom of choice. If young adults are able to support themselves financially, they gain the ability to make more choices about their own lives. In early adulthood (usually until about age forty), people work toward their own goals of creating relationships with others, finding employment, and seeking personal fulfillment. Of course, young adults continue to be socialized by their parents, teachers, peers, and the media, but they also learn new attitudes and behaviors. For example, when we marry or have children, we learn new roles as partners or parents.

Workplace (occupational) socialization is one of the most important types of early adult socialization. This type of socialization tends to be most intense immediately after a person makes the transition from school to the workplace; however, many people experience continuous workplace socialization as a result of having more than one career in their lifetime.

In middle adulthood—between the ages of forty and sixty-five—people begin to compare their accomplishments

FIGURE 4.19 Throughout life, our self-concept is influenced by our interactions with others. How might the self-concept of these women be influenced by each other? By society at large?

with their earlier expectations. This is the point at which people either decide that they have reached their goals or recognize that they have attained as much as they are likely to achieve.

Some analysts divide late adulthood into three categories: (1) the "young-old" (ages sixty-five to seventy-four), (2) the "old-old" (ages seventy-five to eighty-five), and (3) the "oldest-old" (over age eighty-five). Others believe that these distinctions are arbitrary and that actual appearance and behavior are quite different based on people's health status, socioeconomic level, and numerous other factors. Although these are somewhat arbitrary divisions, the "young-old" are less likely to suffer from disabling illnesses, whereas some of the "old-old" are more likely to suffer such illnesses. Increasingly, studies in gerontology and the sociology of medicine have come to question these arbitrary categories and show that many persons defy the expectations of their age grouping based on their individual genetic makeup, lifestyle choices, and zest for living. Perhaps "old age" is what we make it!

Late Adulthood and Ageism

In older adulthood, some people are quite happy and content; others are not. Erik Erikson noted that difficult changes in adult attitudes and behavior occur in the last years of life, when people experience decreased physical ability,

lower prestige, and the prospect of death. Older adults in industrialized societies may experience *social devaluation*—wherein a person or group is considered to have less social value than other persons or groups. Social devaluation is especially acute when people are leaving roles that have defined their sense of social identity and provided them with meaningful activity (• Figure 4.19).

Negative images regarding older persons reinforce *ageism*—prejudice and discrimination against people on the basis of age, particularly against older persons. Ageism is reinforced by stereotypes, whereby people have narrow, fixed images of certain groups. Older persons are often stereotyped as thinking and moving slowly; as being bound to themselves and their past, unable to change and grow; as being unable to move forward and often moving backward.

Negative images also contribute to the view held by some that women are "old" ten or fifteen years sooner than men. In popular films, male characters increase in

social devaluation
a situation in which a person or group is considered to have less social value than other individuals or groups.

ageism
prejudice and discrimination against people on the basis of age, particularly against older persons.

leadership roles and powerful positions as they grow older; women are either moved into the background or are given stereotypical roles that disparage gender and aging. Similarly, the multibillion-dollar cosmetics industry helps perpetuate the myth that age reduces the "sexual value" of women but increases it for men. Men's sexual value is defined more in terms of personality, intelligence, and earning power than by physical appearance. For women, however, sexual attractiveness is based on youthful appearance. By idealizing this "youthful" image of women and playing up the fear of growing older, sponsors sell millions of products and services that claim to prevent or fix the "ravages" of aging.

Although not all people act on appearances alone, Patricia Moore, an industrial designer, found that many do. At age twenty-seven, Moore disguised herself as an eighty-five-year-old woman by donning age-appropriate clothing and placing baby oil in her eyes to create the appearance of cataracts. With the help of a makeup artist, Moore supplemented the "aging process" with latex wrinkles, stained teeth, and a gray wig. For three years, "Old Pat Moore" went to various locations, including a grocery store, to see how people responded to her:

> When I did my grocery shopping while in character, I learned quickly that the Old Pat Moore behaved— and was treated—differently from the Young Pat Moore. When I was 85, people were more likely to jockey ahead of me in the checkout line. And even more interesting, I found that when it happened, I didn't say anything to the offender, as I certainly would at age 27. It seemed somehow, even to me, that it was okay for them to do this to the Old Pat Moore, since they were undoubtedly busier than I was anyway. And further, they apparently thought it was okay, too! After all, little old ladies have plenty of time, don't they? And then when I did get to the checkout counter, the clerk might start yelling, assuming I was deaf, or becoming immediately testy, assuming I would take a long time to get my money out, or would ask to have the price repeated, or somehow become confused about the transaction. What it all added up to was that people feared I would be trouble, so they tried to have as little to do with me as possible. And the amazing thing is that I began almost to believe it myself. . . . I think perhaps the worst thing about aging may be the overwhelming sense that everything around you is letting you know that you are not terribly important any more. (Moore with Conn, 1985: 75–76)

Do you think we would find the same thing if we recreated Moore's study today? We might find out what many older persons already know—it is other people's *reactions* to their age, not their age itself, that place them at a disadvantage. Consider, for example, that researchers in one study searched on Facebook for groups that concentrate on older people and found 84 groups (with about 25,500 members) created by people between the ages of 20 and 29 years that were extremely derogatory, encouraged such things as banning older people from public activities such as shopping, infantilized them, or used negative terminology to describe them. Although Facebook policies on hate speech prohibit singling out people based on their sex, sexual orientation, gender, illness status, disability, race, ethnicity, national origin, or religion, no such policy exists in regard to age and the problem of ageism (Adler, 2013).

Many older people buffer themselves against ageism by continuing to view themselves as being in middle adulthood long after their actual chronological age would suggest otherwise. Other people begin a process of resocialization to redefine their own identity as mature adults.

L08 **Distinguish** between voluntary and involuntary resocialization, and give examples of each.

Resocialization

Resocialization is the process of learning a new and different set of attitudes, values, and behaviors from those in one's background and previous experience. Resocialization may be voluntary or involuntary. In either case, people undergo changes that are much more rapid and pervasive than the gradual adaptations that socialization usually involves.

Voluntary Resocialization

Resocialization is voluntary when we assume a new status (such as becoming a student, an employee, or a retiree) of our own free will. Sometimes, voluntary resocialization involves medical or psychological treatment or religious conversion, in which case the person's existing attitudes, beliefs, and behaviors must undergo strenuous modification to a new regime and a new way of life. For example, resocialization for adult survivors of emotional/physical child abuse includes extensive therapy in order to form new patterns of thinking and action, somewhat like Alcoholics Anonymous and its twelve-step program, which has become the basis for many other programs dealing with addictive behavior.

Involuntary Resocialization

Involuntary resocialization occurs against a person's wishes and generally takes place within a ***total institution***—a place where people are isolated from the rest of society for a set period of time and come under the control of the officials who run the institution (Goffman, 1961a). Military boot camps, jails and prisons, concentration camps, and some mental hospitals are considered total institutions. Involuntary resocialization is a two-step process. First, people are stripped of their former selves—or

FIGURE 4.20 New inmates are taught how to order their meals. Two fingers raised means two portions. There is no talking in line. Inmates must eat all their food. This "ceremony" suggests how much freedom and dignity an inmate loses when beginning the resocialization process.

depersonalized—through a degradation ceremony (Goffman, 1961a). For example, inmates entering prison are required to strip, shower, and wear assigned institutional clothing. In the process, they are searched, weighed, fingerprinted, photographed, and given no privacy even in showers and restrooms. Their official identification becomes not a name but a number. In this abrupt break from their former existence, they must leave behind their personal possessions and their family and friends. The depersonalization process continues as they are required to obey rigid rules and to conform to their new environment (• Figure 4.20).

The second step in the resocialization process occurs when the staff members at an institution attempt to build a more compliant person. A system of rewards and punishments (such as providing or withholding television or exercise privileges) encourages conformity to institutional norms.

Individuals respond to involuntary resocialization in different ways. Some people are rehabilitated; others become angry and hostile toward the system that has taken away their freedom. Although the assumed purpose of involuntary resocialization is to reform people so that they will conform to societal standards of conduct after their release, the ability of total institutions to modify offenders' behavior in a meaningful manner has been widely questioned. In many prisons, for example, inmates may conform to the norms of the prison or of other inmates but

have little respect for the norms and the laws of the larger society.

Looking Ahead: Socialization, Social Change, and Your Future

What do you think socialization will be like in the future? The family is likely to remain the institution that most fundamentally shapes and nurtures people's personal values and self-identity. However, other institutions, including education, religion, and the media, will continue to exert a profound influence on individuals of all ages. A central value-oriented issue facing parents and teachers as they attempt to socialize children is the dominance of television, the Internet, and social media, which make it possible for children and young people to experience many things

resocialization
the process of learning a new and different set of attitudes, values, and behaviors from those in one's background and previous experience.

total institution
Erving Goffman's term for a place where people are isolated from the rest of society for a set period of time and come under the control of the officials who run the institution.

outside their homes and schools and to communicate routinely with people around the world.

The socialization process in colleges and universities will become more diverse as students have an even wider array of options in higher education, including attending traditional classes in brick-and-mortar buildings, taking independent-study courses, enrolling in online courses and degree programs, participating in study-abroad programs, and facing options that are unknown at this time. However, it remains to be seen whether newer approaches to socialization in higher education will be more effective and less stressful than current methods.

A very important area of social change in regard to socialization has occurred with the distinction between "digital natives" and "digital immigrants" because people in each category supposedly see the world fundamentally differently. Also known as the Net Generation, Millennials, and Generation Y, individuals in the category of *digital natives*—which would include many of you reading this sentence—literally were born into the digital world, grew up with the Internet, and think absolutely nothing of the rapid changes that so quickly brought digital technology into all aspects of our lives. According to Marc Prensky (2001), who coined the terms *digital natives* and *digital immigrants,*

Today's students—K through college—represent the first generations to grow up with this new technology. They have spent their entire lives surrounded by and using computers, videogames, digital music players, video cams, cell phones, and all the other toys and tools of the digital age. Today's average college grads have spent less than 5,000 hours of their lives reading, but over 10,000 hours playing video games (not

to mention 20,000 hours watching TV). Computer games, email, the Internet, cell phones and instant messaging are integral parts of their lives. It is now clear that as a result of this ubiquitous environment and sheer volume of their interaction with it, today's students *think and process information fundamentally differently* from their predecessors.

By contrast, their predecessors, the "digital immigrants," are persons who have extensively used older technologies and were socialized differently from their children. Digital immigrants have to be resocialized to think and live in a world of digital immersion. For example, you might communicate by shooting a YouTube video while your parents would write a letter or an essay (*Economist*, 2010).

If there is validity to the distinction between digital natives and digital immigrants, then socialization will continue to change dramatically. Parents and teachers will seek to communicate in the language and style of their children and students. However, digital natives will need to be aware of, and tolerant toward, some of the more traditional ways of thinking and learning that may have unique merit for unraveling certain problems, learning specific forms of information, and completing specific projects.

Socialization in the future is linked to new technologies that are being developed now. Some people in the United States, and many people throughout the world, do not have access to the digital technology that many of us take for granted. These are important social, economic, and political issues for now and the future. One thing remains clear: The socialization process will continue to be a dynamic and important part of our life whether we are learning information from parents and teachers, from a smartphone, or from a robot. What kind of future would you like to see?

CHAPTER REVIEW Q & A

LO1 What is the extent to which people would become human beings without adequate socialization?

Socialization is the lifelong process through which individuals acquire their self-identity and learn the physical, mental, and social skills needed for survival in society. The kind of person we become depends greatly on what we learn during our formative years from our surrounding social groups and social environment. Social contact is essential in developing a self, or self-concept, which represents an individual's perceptions and feelings of being a distinct or separate person. Much of what we think about ourselves is gained from our interactions with others and from what we perceive that others think of us.

LO2 What is the sociological perspective on human development?

According to Charles Horton Cooley's concept of the looking-glass self, we develop a self-concept as we see ourselves through the perceptions of others. Our initial sense of self is typically based on how our families perceive and treat us. George Herbert Mead suggested that we develop a self-concept through role-taking and learning the rules of social interaction. According to Mead, the self is divided into the "I" and the "me." The "I" represents the spontaneous and unique traits of each person. The "me" represents the internalized attitudes and demands of other members of society.

LO3 How do the functionalist and conflict theorists' perspectives differ on the roles that families play in the socialization process?

Theorists using a functionalist perspective emphasize that families serve important functions in society because they are the primary locus for the socialization of children. The family influences an emerging sense of self and the acquisition of beliefs and values. Families are also the primary source of emotional support. Ideally, people receive love, understanding, security, acceptance, intimacy, and companionship within families. On the other hand, conflict theorists stress that socialization contributes to false consciousness—a lack of awareness and a distorted perception of the reality of class as it affects all aspects of social life. As a result, socialization reaffirms and reproduces the class structure in the next generation rather than challenging the conditions that presently exist.

LO4 How do schools socialize children in both formal and informal ways?

Schools continue to play an enormous role in the socialization of young people. Schools primarily teach knowledge and skills but also have a profound influence on the self-image, beliefs, and values of children.

LO5 What role do peer groups and media play in socialization now, and what role might these agents play in the future?

Peer groups contribute to our sense of belonging and self-worth, and are a key source of information about acceptable behavior. Peer groups simultaneously reflect the larger culture and serve as a conduit for passing on culture to young people. The media function as socializing agents by (1) informing us about world events; (2) introducing us to a wide variety of people; (3) providing an array of viewpoints on current issues; (4) making us aware of

products and services that, if we purchase them, will supposedly help us to be accepted by others; and (5) providing an opportunity to live vicariously through other people's experiences.

LO6 What are ways in which gender socialization and racial–ethnic socialization occur?

Through the process of gender socialization, we learn about what attitudes and behaviors are considered to be appropriate for girls and boys, men and women, in a particular society. One of the primary agents of gender socialization is the family. Racial socialization includes direct statements regarding race, modeling behavior, and indirect activities such as exposure to an environment that conveys a specific message about a racial or ethnic group.

LO7 What are the stages in the life course, and why is the process of socialization important in each stage?

Socialization is ongoing throughout the life course—from childhood to adolescence to adulthood and old age. We learn knowledge and skills for future roles through anticipatory socialization. Throughout childhood, we are socialized by our parents, schools, peers, and other groups. Workplace (occupational) socialization is one of the most important types of early adult socialization.

LO8 What is the difference between voluntary and involuntary resocialization?

Resocialization is the process of learning new attitudes, values, and behaviors, either voluntarily or involuntarily. Resocialization is voluntary when we assume a new status (such as becoming a student) of our own free will. Involuntary resocialization occurs against a person's wishes and generally takes place within a total institution, such as a jail or prison.

KEY TERMS

ageism 107
agents of socialization 98
anticipatory socialization 103
ego 92
gender socialization 102
generalized other 96
id 91
looking-glass self 94

mass media 101
peer group 100
primary socialization 97
racial socialization 103
reciprocal socialization 99
resocialization 108
role-taking 95
secondary socialization 97

self-concept 94
significant others 95
social devaluation 107
socialization 87
sociobiology 88
superego 92
tertiary socialization 97
total institution 108

QUESTIONS for CRITICAL THINKING

1 Consider the concept of the looking-glass self. How do you think others perceive you? Do you think most people perceive you correctly?

2 What are your "I" traits? What are your "me" traits? Which ones are stronger?

3 What are some different ways that you might study the effect of toys on the socialization of children? How could you isolate the toy variable from other variables that influence children's socialization?

4 How is socialization different in the digital age? Do you believe that a distinction can be made between "digital natives" and "digital immigrants"? Why or why not?

ANSWERS to the SOCIOLOGY QUIZ

ON SOCIALIZATION AND THE COLLEGE EXPERIENCE

1	False	Studies have concluded that although professors are important in helping students learn about the academic side of the college experience, our friends and acquaintances help us adapt to higher education.
2	False	Slightly more than 85 percent of first-year students at four-year colleges report that they have studied with other students. Similar data are not available for students at two-year schools. How might data for this group differ?
3	True	The college environment is stressful for many students, who find that it is an abrupt change from high school because workloads increase, students are expected to manage their time independently, and grades are increasingly important for a person's future endeavors.
4	True	The competitive nature of the admission process in law schools and medical schools virtually guarantees that new students will be surrounded by classmates who were exceptional students during their undergraduate years. However, this level of achievement may be a source of stimulation for some students rather than a source of stress.
5	True	Some amount of academic stress may be positive in helping students reach their academic and career goals; however, excessive academic stress may be detrimental if it results in high levels of psychological stress or problematic behaviors such as alcohol abuse.
6	False	Studies have found that stress levels among college students are higher than those of people entering a new occupation or profession. For this reason, students are encouraged to develop coping skills and build support networks of friends, family, and others.
7	False	Most research has not shown a significant relationship between the number of hours worked and levels of stress among students. Earning money for school and personal expenses appears to offset additional time and responsibility in the workplace.
8	True	The top stressors most frequently reported in college are getting good grades and completing schoolwork. However, first-year college students also report that changes in eating and sleeping habits, increased workloads and new responsibilities, and going home for holidays and other breaks are major sources of stress for them.

Sources: *Campus Times*, 2008; *Chronicle of Higher Education*, 2014; Messenger, 2009; and Reuters News Service, 2008.

SOCIAL STRUCTURE AND INTERACTION IN EVERYDAY LIFE

LEARNING OBJECTIVES

1 **Explain** why social structure is important in our interaction with others.

2 **Distinguish** among ascribed, achieved, and master statuses, and give examples of each.

3 **Explain** each of these concepts: role, role expectation, role performance, role conflict, role strain, and role exit.

4 **Compare** functionalist and conflict views on social institutions.

5 **Explain** how social change occurs in preindustrial, industrial, and postindustrial societies.

6 **Discuss** the symbolic interactionist views on the social construction of reality and the self-fulfilling prophecy.

7 **Compare** ethnomethodology and dramaturgical analysis as two research methods for observing how people deal with everyday life.

8 **State** three ways in which the sociology of emotions and the study of nonverbal communication add to our understanding of human behavior.

SOCIOLOGY & EVERYDAY LIFE

The Art of Diving for Dinner

I began Dumpster diving [scavenging in a large garbage bin] about a year before I became homeless. . . . The area I frequent is inhabited by many affluent college students. I am not here by chance; the Dumpsters in this area are very rich. Students throw out many good things, including food. In particular they tend to throw everything out when they move at the end of a semester, before and after breaks, and around midterm, when many of them despair of college. So I find it advantageous to keep an eye on the academic calendar. I learned to scavenge gradually, on my own. Since then I have initiated several companions into the trade. I have learned that there is a predictable series of stages a person goes through in learning to scavenge.

At first the new scavenger is filled with disgust and self-loathing. He is ashamed of being seen and may lurk around, trying to duck behind things, or he may dive at night. (In fact, most people instinctively look away from a scavenger. By skulking around, the novice calls attention to himself and arouses suspicion. Diving at night is ineffective and needlessly messy.) . . . That stage passes with experience. The scavenger finds a pair of running shoes that fit and look and

All activities in life—including scavenging in garbage bins and living "on the streets"—are social in nature.

smell brand-new. . . . He begins to understand: People throw away perfectly good stuff, a lot of perfectly good stuff.

Bob Collins/The Image Works

Are you familiar with Dumpster diving? College and university campuses are a frequent site for such activities, particularly around holidays and the ends of semesters, when many students vacate their dorm rooms or apartments. Often, students are able to take only the personal possessions they can put into their car or take on an airplane. Many leave valuable items, including food, clothing, and furniture, in the nearest dumpster. The behavior of your community dumpster divers, as well as Lars Eighner's activities, described above, reflects a specific pattern of social behavior. All activities in life—including scavenging in garbage bins or living "on the streets"—are social in nature. Homeless persons and domiciled persons (those with homes) participate in a social world that has predictable patterns. If you check YouTube or similar social media and social networking sites, you will find a variety of people who engage in dumpster diving not only for economic reasons but for the sport of it as well. Some "divers" report their daily "catch" to others through videos and blogs. In this chapter we look at the relationship between social structure and social interaction in everyday life. In the process, informal,

"on-the-street" activities, and homelessness are used as examples of how social problems may occur and how they may be either reduced or perpetuated within social structures and patterns of interaction in communities and nations.

Let's start by defining social interaction and social structure. Although we are frequently not aware of it, our daily interactions with others and the larger patterns found in the social world of which we are a part are important ingredients in the framework of our individual daily lives. **Social interaction** is the process by which people act toward or respond to other people and is the foundation for all relationships and groups in society. As discussed in Chapter 4, we learn virtually all of what we know from our interactions with other people.

Socialization is a small-scale process, whereas social structure is a much more encompassing framework. **Social structure** is the complex framework of societal institutions (such as the economy, politics, and religion) and the social practices (such as rules and social roles) that make up a society and that organize and establish limits on people's behavior. This structure is essential for the survival of society and for the well-being of individuals because it provides a social

At this stage, Dumpster shyness begins to dissipate. The diver, after all, has the last laugh. He is finding all manner of good things that are his for the taking. Those who disparage his profession are the fools, not he.

—Author LARS EIGHNER recalls his experiences as a dumpster diver while living under a shower curtain in a stand of bamboo in a public park. Eighner became homeless when he was evicted from his "shack" after being unemployed for about a year (Eighner, 1993: 111–119).

How Much Do You Know About Homeless People and the Social Structure of Homelessness?

TRUE	FALSE		
T	F	1	Local, state, and federal assistance to homeless people has shrunk in recent years.
T	F	2	A majority of people who are counted as homeless live on the streets or in cars, abandoned buildings, or other places not intended for human habitation.
T	F	3	Many homeless people have full-time employment.
T	F	4	Homelessness is affected by both income and the affordability of available housing.
T	F	5	Homeless people typically panhandle (beg for money) so that they can buy alcohol or drugs.
T	F	6	Shelters for the homeless consistently have clients who sleep on overflow cots, in chairs, in hallways, and in other nonstandard sleeping arrangements.
T	F	7	There have always been homeless people throughout the history of the United States.
T	F	8	"Doubled-up" populations (people who live with friends, family, or other nonrelatives for economic reasons) have decreased in recent years.

Answers can be found at the end of the chapter.

web of familial support and social relationships that connects each of us to the larger society. Many homeless people have lost this vital linkage. As a result, they often experience a loss of personal dignity and a sense of moral worth because of their "homeless" condition. Before reading on, learn more about homeless people and how the pressing national problem of homelessness is related to social structure and interaction by taking the "Sociology and Everyday Life" quiz.

 LO1 **Explain** why social structure is important in our interaction with others.

Social Structure: The Macrolevel Perspective

What does the term "social structure" mean to you? Social structure is very important in our social world because it provides the framework within which we interact

with others. This framework is an orderly, fixed arrangement of parts that together make up the whole group or society (see • Figure 5.1). You will notice that "Society" is at the top of this figure. As you will recall from Chapter 1, a *society* is a large social grouping that shares the same geographical territory and is subject to the same political authority and dominant cultural expectations. Note also that the social structure of a society has several essential elements: social institutions, statuses and roles, and social groups.

Functional theorists emphasize that social structure is essential because it creates order and predictability in

social interaction
the process by which people act toward or respond to other people: the foundation for all relationships and groups in society.

social structure
the complex framework of societal institutions (such as the economy, politics, and religion) and the social practices (such as rules and social roles) that make up a society and that organize and establish limits on people's behavior.

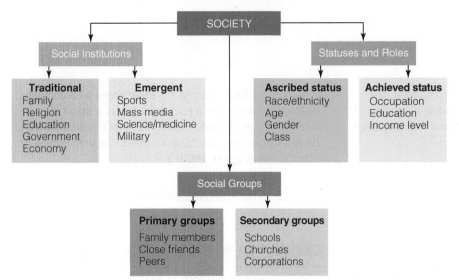

FIGURE 5.1 **Social Structure Framework**

will be the "outsiders." *Social marginality* is the state of being part insider and part outsider in the social structure. Sociologist Robert Park (1928) coined this term to refer to persons (such as immigrants) who simultaneously share the life and traditions of two distinct groups. Social marginality results in stigmatization. A *stigma* is any physical or social attribute or sign that so devalues a person's social identity that it disqualifies that person from full social acceptance (Goffman, 1963b). A convicted criminal wearing a prison uniform is an example of a person who has been stigmatized; the uniform says that the person has done something wrong and should not be allowed unsupervised outside the prison walls.

a society. Social structure is also important for our human development. As discussed in Chapter 3, we develop a self-concept as we learn the attitudes, values, and behaviors of the people around us. When these attitudes and values are part of a predictable structure, it is easier to develop that self-concept.

Social structure gives us the ability to interpret the social situations we encounter. How about the social structure in your own life? What about your family, your schools, and the police? For example, we expect our families to care for us, our schools to educate us, and our police to protect us. When our circumstances change dramatically, most of us feel an acute sense of anxiety because we do not know what to expect or what is expected of us. Newly homeless individuals are an example of this anxiety. Some of them may feel disoriented because they do not know how to function in their new setting. The person is likely to ask questions: "How will I survive on the streets?" "Where do I go to get help?" "Should I stay at a shelter?" "Where can I get a job?" Social structure helps people make sense out of their environment even when they find themselves on the streets.

By contrast, conflict theorists maintain that there is more to social structure than is readily visible and that we must explore the deeper, underlying structures that determine social relations in a society. For example, Karl Marx suggested that the way economic production is organized is the most important structural aspect of any society and that it puts some individuals at a distinct disadvantage. According to Marx, in capitalistic societies, where a few people control the labor of many, the social structure reflects a system of relationships of domination among categories of people (for example, owner–worker and employer–employee).

Social structure creates boundaries that define which persons or groups will be the "insiders" and which

LO2 **Distinguish** among ascribed, achieved, and master statuses, and give examples of each.

Components of Social Structure

The social structure of a society includes its social positions, the relationships among those positions, and the kinds of resources attached to each of the positions. Social structure also includes all the groups that make up society and the relationships among those groups (Smelser, 1988). We begin by examining the social positions that are closest to you, the individual.

Status

A *status* is a socially defined position in a group or society characterized by certain expectations, rights, and duties. Statuses exist independently of the specific people occupying them; the statuses of professional athlete, rock musician, professor, college student, and homeless person all exist exclusive of the specific individuals who occupy these social positions. Think about your own college experience. Although thousands of new students arrive on college campuses each year to occupy the status of first-year student, the status of college student and the expectations attached to that position have remained relatively stable for many years.

What do you think of when you hear the word "status"? Does the term refer only to high-level positions in society? No, not in the sociological sense we are using it. Although many people equate the term with high levels of prestige, sociologists use it to refer to all socially defined positions—high rank and low rank. For example, both the position of director of the Department of Health and Human

Services in Washington, D.C., and that of a homeless person who is paid about five dollars a week (plus bed and board) to clean up the dining room at a homeless shelter are social statuses.

Take a moment to ask yourself the question "Who am I?" To determine who you are, you must think about your social identity, which is derived from the statuses you occupy and is based on your status set. A *status set* comprises all the statuses that a person occupies at a given time. For example, Maria may be a psychologist, a professor, a wife, a mother, a Catholic, a school volunteer, a Texas resident, and a Mexican American. All of these socially defined positions constitute her status set.

Ascribed Status and Achieved Status Statuses are distinguished by the manner in which we acquire them. An *ascribed status* is a social position conferred at birth or received involuntarily later in life, based on attributes over which the individual has little or no control, such as race/ethnicity, age, and gender. For example, Maria is a female born to Mexican American parents; she was assigned these statuses at birth. She is an adult and—if she lives long enough—will someday become an "older adult," which is an ascribed status received involuntarily later in life.

An *achieved status* is a social position that a person assumes voluntarily as a result of personal choice, merit, or direct effort (• Figure 5.2). Achieved statuses (such as occupation, education, and income) are thought to be gained as a result of personal ability or successful competition. Most occupational positions in modern societies are achieved statuses. For instance, Maria voluntarily assumed the statuses of psychologist, professor, wife, mother, and school volunteer. However, not all achieved statuses are positions most people would want to attain; for example, being a criminal, a drug addict, or a homeless person is a negative achieved status.

Cheryl Gerber/Reuters

FIGURE 5.2 In the past, a person's status was primarily linked to his or her family background, education, occupation, and other sociological attributes. Today, some sociologists believe that celebrity status has overtaken the more traditional social indicators of status. Singer Taylor Swift, shown here, is an example of celebrity status.

Ascribed statuses have a significant influence on the achieved statuses that we occupy. Race/ethnicity, gender, and age affect each person's opportunity to acquire certain achieved statuses. Those who are privileged by their positive ascribed statuses are more likely to achieve the more-prestigious positions in a society. Those who are disadvantaged by their ascribed statuses may more easily acquire negative achieved statuses.

Master Status If we occupy many different statuses, how can we determine which is the most important? Sociologist Everett Hughes has stated that societies resolve this ambiguity by determining master statuses. A *master status* is the most important status that a person occupies; it dominates all the individual's other statuses and is the overriding ingredient in determining a person's general social position (Hughes, 1945). Being poor or rich is a master status that influences many other areas of life, including health, education, and life

status
a socially defined position in a group or society characterized by certain expectations, rights, and duties.

status set
all the statuses that a person occupies at a given time.

ascribed status
a social position conferred at birth or received involuntarily later in life, based on attributes over which the individual has little or no control, such as race/ethnicity, age, and gender.

achieved status
a social position that a person assumes voluntarily as a result of personal choice, merit, or direct effort.

master status
the most important status that a person occupies.

FIGURE 5.3 Sociologists believe that being rich or poor may be a master status in the United States. How do the lifestyles of these two men differ based on their master statuses?

opportunities (• Figure 5.3). For men, occupation has usually been the most important status, although occupation is increasingly a master status for many women as well. "What do you do?" is one of the first questions most people ask when meeting another. Occupation provides important clues to a person's educational level, income, and family background. An individual's race/ethnicity may also constitute a master status in a society in which dominant-group members single out members of other groups as "less worthy" than themselves on the basis of real or alleged physical, cultural, or nationality characteristics.

Master statuses confer high or low levels of personal worth and dignity on people. These are not characteristics that we inherently possess; they are derived from the statuses we occupy. For those who have no residence, being a homeless person readily becomes a master status regardless of the person's other attributes. Homelessness is a stigmatized master status that confers disrepute on its occupant because domiciled people often believe that a homeless person has a "character flaw." Sometimes this assumption is supported by how the media represent homeless people. For example, media analysts may focus on the problems of one homeless family during the holidays, describing how the parents and kids live in a car and eat meals from a soup kitchen. Journalists frequently highlight the fact that the adults have limited education, have spent time in jail or a mental facility, lack a stable work history, or possess other personal problems. What is not included many times is the macro view of the situation, including what structural factors (such as economic recessions, lack of affordable housing, and large-scale loss of jobs) might produce homelessness at national and global levels.

The circumstances under which someone becomes homeless may determine the extent to which that person is stigmatized (see • Figure 5.4). Snow and Anderson (1993: 199) observed the effects of homelessness as a master status:

It was late afternoon, and the homeless were congregated in front of [the Salvation Army shelter] for dinner. A school bus approached that was packed with Anglo junior high school students being bused from an eastside barrio school to their upper-middle and upper-class homes in the city's northwest neighborhoods. As the bus rolled by, a fusillade of coins came flying out the windows, as the students made obscene gestures and shouted, "Get a job." Some of the homeless gestured back, some scrambled for the scattered coins—mostly pennies—others angrily threw the coins at the bus, and a few seemed oblivious to the encounter. For the passing junior high schoolers, the

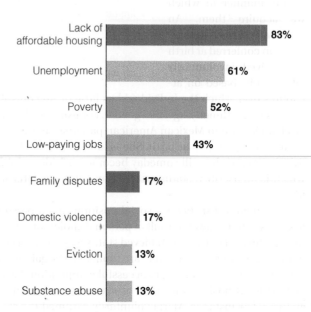

Note: Survey cities were asked to identify the three main causes of homelessness among families with children.

FIGURE 5.4 **Causes of Family Homelessness in 25 Cities**

Source: U.S. Conference of Mayors, 2014.

exchange was harmless fun, a way to work off the restless energy built up in school; but for the homeless it was a stark reminder of their stigmatized status and of the extent to which they are the objects of negative attention.

Status Symbols When people are proud of a particular social status that they occupy, they often choose to use visible means to let others know about their position. **Status symbols** are material signs that inform others of a person's specific status. For example, just as wearing a wedding ring proclaims that a person is married, for many people owning a Rolls-Royce announces that they have "made it." As we saw in Chapter 3, achievement and success are core U.S. values. For this reason, people who have "made it" tend to want to display symbols to inform others of their accomplishments.

Status symbols for the domiciled and for the homeless may have different meanings. Among affluent persons, a full shopping cart in the grocery store and bags of merchandise from expensive department stores indicate a lofty financial position. By contrast, among the homeless, bulging shopping bags and overloaded grocery carts suggest a completely different status. Carts and bags are essential to street life; there is no other place to keep things, as shown by this now-classic description of Darian, a homeless woman in New York City:

> The possessions in her grocery cart consist of a whole house full of things, from pots and pans to books, shoes, magazines, toilet articles, personal papers and clothing, most of which she made herself. . . .
>
> Because of its weight and size, Darian cannot get the cart up over the curb. She keeps it in the street near the cars. This means that as she pushes it slowly up and down the street all day long, she is living almost her entire life directly in traffic. She stops off along her route to sit or sleep for a while and to be both stared at as a spectacle and to stare back. Every aspect of her life including sleeping, eating, and going to the bathroom is constantly in public view. . . . [S]he has no space to call her own and she never has a moment's privacy. Her privacy, her home, is her cart with all its possessions. (Rousseau, 1981: 141)

Although this description is more than thirty years old, homeless persons today can still be spotted in large urban centers, such as New York, San Francisco, and Seattle, and smaller cities, such as Waco, Texas, where they live with all of their possessions in a shopping cart or other portable conveyance.

For homeless women and men, possessions are not status symbols as much as they are a link with the past, a hope for the future, and a potential source of immediate cash. As Snow and Anderson (1993: 147) note, selling personal possessions is not uncommon among most social classes; members of the working and middle classes hold garage sales, and those in the upper classes have estate sales. However, when homeless persons sell their personal possessions, they do so to meet their immediate needs, not because they want to "clean house."

 LO3 **Explain** each of these concepts: role, role expectation, role performance, role conflict, role strain, and role exit.

Role

A **role** is a set of behavioral expectations associated with a given status. For example, a carpenter (employee) hired to remodel a kitchen is not expected to sit down uninvited and join the family (employer) for dinner. A role is the dynamic aspect of a status. Whereas we occupy a status, we play a role.

Role expectation is a group's or society's definition of the way that a specific role ought to be played. By contrast, **role performance** is how a person actually plays the role. Role performance does not always match role expectation. Some statuses have role expectations that are highly specific, such as that of surgeon or college professor. Other statuses, such as friend or significant other, have less structured expectations. The role expectations tied to the status of student are more specific than those of being a friend. Role expectations are typically based on a range of acceptable behavior rather than on strictly defined standards.

Our roles are relational (or complementary); that is, they are defined in the context of roles performed by others. We can play the role of student because someone else fulfills the role of professor. Conversely, to perform the role of professor, the teacher must have one or more students.

Role ambiguity occurs when the expectations associated with a role are unclear. For example, it is not always clear when the provider–dependent aspect of the parent–child relationship ends. Should it end at age eighteen or twenty-one? When a person is no longer in school? Different people will answer these questions differently depending on their experiences and socialization, as well as on the parents' financial capability and psychological

---

status symbol
a material sign that informs others of a person's specific status.

role
a set of behavioral expectations associated with a given status.

role expectation
a group's or society's definition of the way that a specific role ought to be played.

role performance
how a person actually plays a role.

willingness to continue contributing to the welfare of their adult children.

Role Conflict and Role Strain

Most people occupy a number of statuses, each of which has numerous role expectations attached. For example, Charles is a student who attends morning classes at the university, and he is an employee at a fast-food restaurant, where he works from 3:00 to 10:00 P.M. He is also Stephanie's boyfriend, and she would like to see him more often. On December 7, Charles has a final exam at 7:00 P.M., when he is supposed to be working. Meanwhile, Stephanie is pressuring him to take her to a movie. To top it off, his mother calls, asking him to fly home because his father is going to have emergency surgery. How can Charles be in all these places at once? Such experiences of role conflict can be overwhelming.

Role conflict occurs when incompatible role demands are placed on a person by two or more statuses held at the same time. When role conflict occurs, we may feel pulled in different directions. To deal with this problem, we may prioritize our roles and first complete the one we consider to be most important. Or we may compartmentalize our lives and separate our roles from one another. That is, we may perform the activities linked to one role for part of the day and then engage in the activities associated with another role in some other time period or elsewhere. For example, under routine circumstances, Charles would fulfill his student role for part of the day and his employee role for another part of the day. In his current situation, however, he is unable to compartmentalize his roles.

Role conflict may occur as a result of changing statuses and roles in society. Research has found that women who engage in behavior that is gender-typed as "masculine" tend to have higher rates of role conflict than those who engage in traditional "feminine" behavior. Role conflict may sometimes be attributed not to the roles themselves but to the pressures that people feel when they do not fit into culturally prescribed roles. In a study of women athletes in college sports programs, female gymnasts and softball players faced a "female/athlete paradox," which led caused them to construct images that were based on femininity on some occasions and images that were based on athleticism on others. It appears that these young women had constructed their own approach for dealing with the conflict inherent in playing the female student–athlete role (Ross and Shinew, 2008).

Whereas role conflict occurs between two or more statuses, role strain takes place within one status. *Role strain* occurs when incompatible demands are built into a single status that a person occupies. For example, married or cohabiting women may experience more role strain than married or cohabiting men because many of them experience work overload, meaning that they work for wages outside the household but are also responsible for most of the parenting and household responsibilities within the family.

Recent social and economic changes in society may have increased role strain for men. In some families, men's traditional position of dominance has eroded as more women have entered the paid labor force and, in more cases, become the primary or sole breadwinner for the family. The concepts of role expectation, role performance, role conflict, and role strain are illustrated in • Figure 5.5.

Individuals frequently distance themselves from a role they find extremely stressful or otherwise problematic. *Role distancing* occurs when people consciously foster the impression of a lack of commitment or attachment to a particular role and merely go through the motions of role performance (Goffman, 1961b). People use distancing techniques when they do not want others to take them as the "self" implied in a particular role, especially if they think the role is "beneath them." While Charles is working in the fast-food restaurant, for example, he does not want people to think of him as a "loser in a dead-end job." He

Role Expectation: a group's or society's definition of the way a specific role *ought* to be played.

Role Performance: how a person *actually* plays a role.

Role Conflict: occurs when incompatible demands are placed on a person by two or more statuses held at the same time.

Role Strain: occurs when incompatible demands are built into a single status that a person holds.

FIGURE 5.5 **Role Expectation, Performance, Conflict, and Strain**
When playing the role of "student," do you sometimes personally encounter these concepts? How do you deal with such issues in your daily life?

Mario Anzuoni/Reuters/Landov

FIGURE 5.6 *Los Angeles Times* columnist Steve Lopez met a homeless man, Nathaniel Ayers (above), and learned that he had been a promising musician studying at The Juilliard School who had dropped out because of his struggle with mental illness. In his book *The Soloist*, Lopez chronicles the relationship that he developed with Ayers and how he eventually helped get Ayers off the street and treated for his schizophrenia. This story is an example of role exit, and you can see the movie version of *The Soloist* online.

wants them to view him as a college student who is working there just to "pick up a few bucks" until he graduates. When customers from the university come in, Charles talks to them about what courses they are taking, what they are majoring in, and what professors they have. He does not discuss whether the bacon cheeseburger is better than the chili burger. When Charles is really involved in role distancing, he tells his friends that he "works there but wouldn't eat there."

Role Exit   *Role exit* occurs when people disengage from social roles that have been central to their self-identity (• Figure 5.6). Sociologist Helen Rose Fuchs Ebaugh (1988) studied this process by interviewing ex-convicts, ex-nuns, retirees, divorced men and women, and others who had exited voluntarily from significant social roles. According to Ebaugh, role exit occurs in four stages. The first stage is doubt, in which people experience frustration or burnout when they reflect on their existing roles. The second stage involves a search for alternatives; here, people may take a leave of absence from their work or temporarily separate from their marriage partner. The third stage is the turning point at which people realize that they must take some final action, such as quitting their job or getting a divorce. The fourth and final stage involves the creation of a new identity. Consider, for example, attempting to exit the "homeless" role: This is a very difficult process because the longer an individual remains on the streets, the more that person's personal resources diminish, and his or her work experience and skills become outdated and unmarketable.

Groups

Groups are another important component of social structure. To sociologists, a *social group* consists of two or more people who interact frequently and share a common identity and a feeling of interdependence. Throughout our lives, most of us participate in groups, from our families and childhood friends, to our college classes, to our work and community organizations, and even to society.

Primary and secondary groups are the two basic types of social groups. A *primary group* is a small, less specialized group in which members engage in face-to-face, emotion-based interactions over an extended period of time. Primary groups include our family, close friends, and school- or work-related peer groups. By contrast, a *secondary group* is a larger, more specialized group in which members engage in more-impersonal, goal-oriented relationships for a limited period of time (• Figure 5.7). Schools, churches, and corporations are examples of secondary groups. In secondary groups, people have few, if any, emotional ties

role conflict
a situation in which incompatible role demands are placed on a person by two or more statuses held at the same time.

role strain
a condition that occurs when incompatible demands are built into a single status that a person occupies.

role exit
a situation in which people disengage from social roles that have been central to their self-identity.

social group
a group that consists of two or more people who interact frequently and share a common identity and a feeling of interdependence.

primary group
a small, less specialized group in which members engage in face-to-face, emotion-based interactions over an extended period of time.

secondary group
a larger, more specialized group in which members engage in more-impersonal, goal-oriented relationships for a limited period of time.

FIGURE 5.7 For many years, powerful "old-boy" social networks have dominated capitalism. Informal discussions, such as the one shown here, often close major business deals.

to one another. Instead, they come together for some specific, practical purpose, such as getting a degree or a paycheck. Secondary groups are more specialized than primary ones; individuals relate to one another in terms of specific roles (such as professor and student) and more-limited activities (such as course-related endeavors). Primary and secondary groups are further discussed in Chapter 6 ("Groups and Organizations").

Social solidarity, or cohesion, refers to a group's ability to maintain itself in the face of obstacles. Social solidarity exists when social bonds, attractions, or other forces hold members of a group in interaction over a period of time. For example, if a local church is destroyed by fire and congregation members still worship together in a makeshift setting, then they have a high degree of social solidarity.

Many of us build social networks that involve our personal friends in primary groups and our acquaintances in secondary groups. A *social network* is a series of social relationships that links an individual to others. Social networks work differently for men and women, for different races/ethnicities, and for members of different social classes. Traditionally, people of color and white women have been excluded from powerful "old-boy" social networks. At the middle- and upper-class levels, individuals tap social networks to find employment, make business deals, and win political elections. However, social networks typically do not work effectively for poor and homeless individuals. Snow and Anderson (1993) found that homeless men have fragile social networks that are plagued with instability. Most of the avenues for exiting the homeless role and acquiring housing are intertwined with the large-scale, secondary groups that sociologists refer to as *formal organizations.*

A ***formal organization*** is a highly structured group formed for the purpose of completing certain tasks or achieving specific goals. Many of us spend most of our time in formal organizations such as colleges, corporations, or the government. In Chapter 6 ("Groups and Organizations"), we analyze the characteristics of bureaucratic organizations; however, at this point we should note that these organizations are a very important component of social structure in all industrialized societies. We expect such organizations to educate us, solve our social

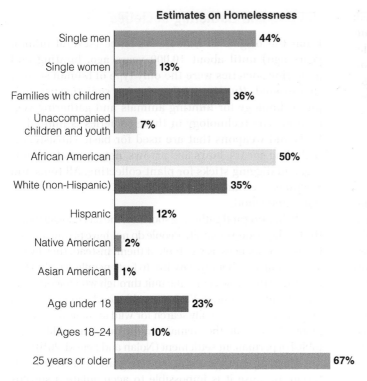

Estimates on Homelessness

Category	Percent
Single men	44%
Single women	13%
Families with children	36%
Unaccompanied children and youth	7%
African American	50%
White (non-Hispanic)	35%
Hispanic	12%
Native American	2%
Asian American	1%
Age under 18	23%
Ages 18–24	10%
25 years or older	67%

FIGURE 5.8 Who Are the Homeless?
Source: U.S. Conference of Mayors, 2014.

problems (such as crime and homelessness), and provide us work opportunities.

Today, formal organizations such as the U.S. Conference of Mayors and the National Law Center on Homelessness and Poverty work with groups around the country to make people aware that homelessness must be viewed within the larger context of poverty and to educate the public on the nature and extent of homelessness among various categories of people in the United States (see ● Figure 5.8 for statistics on homelessness).

 LO4 Compare functionalist and conflict views on social institutions.

Social Institutions

At the macrolevel of all societies, certain basic activities routinely occur—children are born and socialized, goods and services are produced and distributed, order is preserved, and a sense of purpose is maintained. Social institutions are the means by which these basic needs are met. A *social institution* is a set of organized beliefs and rules that establishes how a society will attempt to meet its basic social needs. In the past, these needs centered around five basic social institutions: the family, religion, education, the economy, and the government or politics. Today, mass media, sports, science and medicine, and the military are also considered to be social institutions.

What do you think is the difference between a group and a social institution? A group is composed of specific, identifiable people; an institution is a standardized way of doing something. The concept of "family" helps to distinguish between the two. When we talk about "your family" or "my family," we are referring to a specific family. When we refer to the family as a social institution, we are talking about ideologies and standardized patterns of behavior that organize family life. For example, the family as a social institution contains certain statuses organized into well-defined relationships, such as husband–wife, parent–child, and brother–sister. However, specific families do not always conform to these ideologies and behavior patterns.

Functionalist Views on Social Institutions Functional theorists emphasize that social institutions exist because they perform five essential tasks:

1. *Replacing members.* Societies and groups must have socially approved ways of replacing members who move away or die.

2. *Teaching new members.* People who are born into a society or move into it must learn the group's values and customs.

3. *Producing, distributing, and consuming goods and services.* All societies must provide and distribute goods and services for their members.

4. *Preserving order.* Every group or society must preserve order within its boundaries and protect itself from attack by outsiders.

5. *Providing and maintaining a sense of purpose.* In order to motivate people to cooperate with one another, a sense of purpose is needed.

Although this list of functional prerequisites is shared by all societies, the institutions in each society perform these tasks in somewhat different ways depending on their specific cultural values and norms.

Conflict Views on Social Institutions Conflict theorists agree with functionalists that social institutions are originally organized to meet basic social needs. However, they do not believe that social institutions work for the common good of everyone. For example, some conflict theorists might point out that families may be a source of problems (rather than solutions) for

formal organization
a highly structured group formed for the purpose of completing certain tasks or achieving specific goals.

social institution
a set of organized beliefs and rules that establishes how a society will attempt to meet its basic social needs.

young people. Some children are abused or neglected; others have arguments with their parents or other authority figures in the household that contribute to a decision to run away from home and try their luck living on the streets. Traumatic incidents in families may trigger fear, anxiety, and dread that contribute to homelessness among young people.

Societies, Technology, and Sociocultural Change

As you and I think about homeless people today, it is difficult to realize that for people in some societies, being without a place of residence is a way of life. Where people live and the mode(s) of production they use to generate a food supply are related to *subsistence technology*—the methods and tools that are available for acquiring the basic needs of daily life. Social scientists have identified five types of societies based on various levels of subsistence technology: hunting and gathering, horticultural and pastoral, agrarian, industrial, and postindustrial societies. These types of societies are said to have different *technoeconomic bases* related to the technology that is available and the economic structure of the society. (The features of the different types of societies, distinguished by technoeconomic base, are summarized in • Table 5.1.) The first three of these societies—hunting and gathering, horticultural and pastoral, and agrarian—are also referred to as preindustrial societies. According to social scientist Gerhard Lenski, societies change over time through the process of *sociocultural evolution,* the changes that occur as a society gains new technology (see Nolan and Lenski, 2010).

Hunting and Gathering Societies

From the origins of human existence (several million years ago) until about 10,000 years ago, hunting and gathering societies were the only type of human society that existed. ***Hunting and gathering societies*** use simple technology for hunting animals and gathering vegetation. The technology in these societies is limited to tools and weapons that are used for basic subsistence, including spears, bows and arrows, nets, traps for hunting, and digging sticks for plant collecting. All tools and weapons are made of natural materials such as stone, bone, and wood.

In hunting and gathering societies, the basic social unit is the kinship group or family. People do not have private households or residences as we think of them. Instead, they live in small groups of about twenty-five to forty people. Kinship ties constitute the basic economic unit through which food is acquired and distributed. With no stable food supply, hunters and gatherers continually search for wild animals and edible plants. As a result, they remain on the move and seldom establish a permanent settlement (Nolan and Lenski, 2010).

Hunting and gathering societies are relatively egalitarian. Because it is impossible to accumulate a surplus of food, there are few resources upon which individuals or groups can build a power base. Some specialization (*division of labor*) occurs, primarily based on age and sex. Young children and older people are expected to contribute what they can to securing the food supply, but healthy adults of both sexes are expected to obtain most of the food (• Figure 5.9). In some societies, men hunt for animals, and women gather plants; in others, both women and men

TABLE 5.1

Technoeconomic Bases of Society					
	Hunting and Gathering	**Horticultural and Pastoral**	**Agrarian**	**Industrial**	**Postindustrial**
Change from Prior Society	—	Use of hand tools, such as digging stick and hoe	Use of animal-drawn plows and equipment	Invention of steam engine	Invention of computer and development of "high-tech" society
Economic Characteristics	Hunting game, gathering roots and berries	Planting crops, domestication of animals for food	Labor-intensive farming	Mechanized production of goods	Information and service economy
Control of Surplus	None	Men begin to control societies	Men own land or herds	Men own means of production	Corporate shareholders and high-tech entrepreneurs
Inheritance	None	Shared—patrilineal and matrilineal	Patrilineal	Bilateral	Bilateral
Control Over Procreation	None	Increasingly controlled by men	Men—to ensure legitimacy of heirs	Men—but less so in later stages	Mixed
Women's Status	Relative equality	Decreasing in move to pastoralism	Low	Low	Varies by class, race, and age

Source: Adapted from Lorber, 1994: 140.

FIGURE 5.9 In contemporary hunting and gathering societies, women contribute to the food supply by gathering plants and sometimes hunting small animals. These women of the Kalahari in Botswana gather and share edible roots.

gather plants and hunt for wild game, with women more actively participating when smaller animals are nearby.

Contemporary hunting and gathering societies are located in relatively isolated geographical areas. However, some analysts predict that these groups will soon cease to exist, as food producers with more-dominating technologies usurp the geographic areas from which these groups have derived their food supply (Nolan and Lenski, 2010).

Horticultural and Pastoral Societies

The period between 13,000 and 7,000 B.C.E. marks the beginning of horticultural and pastoral societies. During this period, there was a gradual shift from *collecting* food to *producing* food, a change that has been attributed to three factors: (1) the depletion of the supply of large game animals as a source of food, (2) an increase in the size of the human population to feed, and (3) dramatic weather and environmental changes that probably occurred by the end of the Ice Age.

Why did some societies become horticultural while others became pastoral? Water supply, terrain, and soils are three critical factors in whether horticultural activities or pastoral activities became a society's primary mode of food production. *Pastoral societies* are based on technology that supports the domestication of large animals to provide food and emerged in mountainous regions and areas with low amounts of annual rainfall. Pastoralists—people in pastoral societies—typically remain nomadic as they seek new grazing lands and water sources for their animals. *Horticultural societies* are based on technology that supports the cultivation of plants to provide food. These societies emerged in more-fertile areas that were better suited for growing plants through the use of hand tools.

The family is the basic unit in horticultural and pastoral societies. Because they typically do not move as often as hunter-gatherers or pastoralists, horticulturalists establish more-permanent family ties and create complex systems for tracing family lineage. Some social analysts believe that the invention of a hoe with a metal blade was a contributing factor to the less nomadic lifestyle of the horticulturalists because this made planting more efficient and productive. As a result, people become more *sedentary,* remaining settled for longer periods in the same location.

Unless there are fires, floods, droughts, or environmental problems, herding animals and farming are more reliable sources of food than hunting and gathering. When food is no longer in short supply, more infants are born, and children have a greater likelihood of surviving. When people are no longer nomadic, children are viewed as an economic asset: They can cultivate crops, tend flocks, or care for younger siblings.

Division of labor increases in horticultural and pastoral societies. As the food supply grows, not everyone needs to be engaged in food production. Some pursue activities such as weaving cloth or carpets, crafting jewelry, serving as priests, or creating the tools needed for building the society's structure. Horticultural and pastoral societies are less egalitarian than hunter-gatherers, and the idea of property rights emerges as people establish more-permanent settlements. At this stage, families with the largest surpluses have an economic advantage and gain prestige and power.

hunting and gathering societies
societies that use simple technology for hunting animals and gathering vegetation.

pastoral societies
societies based on technology that supports the domestication of large animals to provide food.

horticultural societies
societies based on technology that supports the cultivation of plants to provide food.

In contemporary horticultural societies, women do most of the farming while men hunt game, clear land, work with arts and crafts, make tools, participate in religious and ceremonial activities, and engage in war. Gender inequality is greater in pastoral societies because men herd the large animals and women contribute relatively little to subsistence production. In some herding societies, women's primary value is their ability to produce male offspring so the family lineage is preserved and a sufficient number of males are available to protect against enemy attack.

Agrarian Societies

About five to six thousand years ago, agrarian (or agricultural) societies emerged, first in Mesopotamia and Egypt and slightly later in China (• Figure 5.10). *Agrarian societies* use the technology of large-scale farming, including animal-drawn or energy-powered plows and equipment, to produce their food supply. Farming made it possible for people to spend their entire lives in the same location, and food surpluses made it possible for people to live in cities, where they were not directly involved in food production. The use of animals to pull plows made it possible for people to generate a large surplus of food. The land can be used more or less continuously because the plow turns the topsoil, thus returning more nutrients to the soil. In some cases, farmers reap several harvests each year from the same plot of land.

In agrarian societies, social inequality is the highest of all preindustrial societies in terms of both class and gender. The two major classes are the landlords and the peasants. The landlords own the fields and the harvests produced by the peasants. Inheritance becomes important as families of wealthy landlords own the same land for generations. By contrast, the landless peasants enter into an agreement with the landowners to live on and cultivate a parcel of land in exchange for part of the harvest or other economic incentives. Over time, the landlords grow increasingly wealthy and powerful as they extract labor, rent, and taxation from the landless workers. Politics is based on a feudal system controlled by a political–economic elite made up of the ruler, his royal family, and members of the landowning class. Peasants have no political power and may be suppressed through the use of force or military power.

Gender-based inequality grows dramatically in agrarian societies. Men gain control over both the disposition of the food surplus and the kinship system. Because agrarian tasks require more labor and greater physical strength than horticultural ones, men become more involved in food production. Women may be excluded from these tasks because they are seen as too weak for the work or because it is believed that their child-care responsibilities are incompatible with the full-time labor that the tasks require. Today, gender inequality continues in agrarian societies; the division of labor between women and men is very distinct in areas such as parts of the Middle East. Here, women's work takes place in the private sphere (inside the home), and men's work occurs in the public sphere, providing men with more recognition and greater formal status.

Industrial Societies

Industrial societies are based on technology that mechanizes production. Originating in England during the Industrial Revolution, this mode of production dramatically transformed predominantly rural and agrarian societies into urban and industrial societies. Chapter 1 describes how the revolution first began in Britain and then spread to other countries, including the United States.

Industrialism involves the application of scientific knowledge to the technology of production, thus making it

FIGURE 5.10 In the twenty-first century, many people around the globe still reside in agrarian societies that are in various stages of industrialization. Open-air markets such as this one in Bali, where people barter or buy their food from one another, are a common sight in agrarian societies.

possible for machines to do the work previously done by people or animals. New technologies, such as the invention of the steam engine and fuel-powered machinery, stimulated many changes. Previously, machines were run by natural power sources (such as wind or water mills) or harnessed power (either human or animal power). The steam engine made it possible to produce goods by machines powered by fuels rather than undependable natural sources or physical labor.

As inventions and discoveries build upon one another, the rate of social and technological change increases. For example, the invention of the steam engine brought about new types of transportation, including trains and steamships. Inventions such as electric lights made it possible for people to work around the clock without regard to whether it was daylight or dark outside. Industrialism changes the nature of subsistence production. In countries such as the United States, large-scale agribusinesses have practically replaced small, family-owned farms and ranches. However, large-scale agriculture has produced many environmental problems while providing solutions to the problem of food supply.

In industrial societies a large proportion of the population lives in or near cities. Large corporations and government bureaucracies grow in size and complexity. The nature of social life changes as people come to know one another more as statuses than as individuals. In fact, a person's occupation becomes a key defining characteristic in industrial societies, whereas his or her kinship ties are most important in preindustrial societies.

Social institutions are transformed by industrialism. The family diminishes in significance as the economy, education, and political institutions grow in size and complexity. The family is now a consumption unit, not a production unit. Although the influence of traditional religion is diminished in industrial societies, religion remains a powerful institution. Religious organizations are important in determining what moral issues will be brought to the forefront (e.g., unapproved drugs, abortion, and violence and sex in the media) and in trying to influence lawmakers to pass laws regulating people's conduct. Politics in industrial societies is usually based on a democratic form of government. As nations such as South Korea, the People's Republic of China, and Mexico have become more industrialized, many people in these nations have intensified their demands for political participation.

Although the standard of living rises in industrial societies, social inequality remains a pressing problem. As societies industrialize, the status of women tends to decline further. For example, after industrialization occurred in the United States, the division of labor between men and women in the middle and upper classes became much more distinct: Men were responsible for being "breadwinners"; women were seen as "homemakers." This gendered division of labor increased the economic and political subordination of women. In short, industrial societies have brought about some of the greatest innovations in all of human history, but they have also maintained and perpetuated some of the greatest problems.

Postindustrial Societies

A *postindustrial society* is one in which technology supports a service- and information-based economy. As discussed in Chapter 1, postmodern (or "postindustrial") societies are characterized by an *information explosion* and an economy in which large numbers of people either provide or apply information (IT specialists, for example) (● Figure 5.11) or are employed in service jobs (such as fast-food servers or health care workers). For example, banking, law, and the travel industry are characteristic forms of employment in postindustrial societies, whereas producing steel or automobiles is representative of employment in industrial societies. In fact, some analysts refer to postindustrial societies as "service economies" because many workers provide services for others. However, most of the new service occupations pay relatively low wages and offer limited opportunities for advancement.

Postindustrial societies produce knowledge that becomes a commodity. This knowledge can be leased or sold to others, or it can be used to generate goods, services, or more knowledge. In the previous types of societies we have examined, machinery or raw materials are crucial to how the economy operates. In postindustrial societies, the economy is based on involvement with people and communications technologies such as the mass media, computers, and the Web.

Previous forms of production, including agriculture and manufacturing, do not disappear in postindustrial societies. Instead, they become more efficient through computerization and other technological innovations. Work that relies on manual labor is often shifted to less technologically advanced societies, where workers are paid low wages to produce profits for corporations based in industrial and postindustrial societies.

Knowledge is viewed as the basic source of innovation and policy formulation in postindustrial societies. As a result, formal education and other sources of information become crucial to the success of individuals and organizations. Scientific research becomes institutionalized, and newer industries—such as computer manufacturing and software development—come into existence that would

agrarian societies
societies that use the technology of large-scale farming, including animal-drawn or energy-powered plows and equipment, to produce their food supply.

industrial societies
societies based on technology that mechanizes production.

postindustrial societies
societies in which technology supports a service- and information-based economy.

Ansgar Photography/Fancy/Corbis

FIGURE 5.11 In postindustrial economies, the focus is often on service- and information-based jobs. Technology is the key to many occupations and professions.

not have been possible without the new knowledge and technological strategies.

LO5 Explain how social change occurs in preindustrial, industrial, and postindustrial societies.

Sociological Perspectives on Stability and Change in Society

How do you think changes in social structure affect individuals, groups, and societies? These changes have a dramatic impact at all levels. Social arrangements in contemporary societies have grown more complex with the introduction of new digital technologies, changes in values and norms, and the rapidly shrinking "global village." How do societies maintain some degree of social solidarity in the face of such changes? Sociologists Emile Durkheim and Ferdinand Tönnies developed typologies to explain the processes of stability and change in the social structure of societies. A *typology* is a classification scheme containing two or more mutually exclusive categories that are used to compare different kinds of behavior or types of societies.

Durkheim: Mechanical and Organic Solidarity

Emile Durkheim (1933/1893) was concerned with the question "How do societies manage to hold together?" According to Durkheim, social solidarity derives from a society's social structure, which, in turn, is based on the society's division of labor. *Division of labor* refers to how the various tasks of a society are divided up and performed. People in diverse societies (or in the same society at different points in time) divide their tasks somewhat differently, based on their own history, physical environment, and level of technological development. Durkheim claimed that preindustrial societies are held together by strong traditions and by the members' shared moral beliefs and values. As societies industrialized and developed more-specialized economic activities, social solidarity came to be rooted in the members' shared dependence on one another.

To explain social change, Durkheim categorized societies as having either mechanical or organic solidarity. *Mechanical solidarity* refers to the social cohesion of preindustrial societies, in which there is minimal division of labor and people feel united by shared values and common social bonds. Durkheim used the term *mechanical solidarity* because he believed that people in such preindustrial societies feel a more or less automatic sense of belonging.

Social cohesion comes from the similarity of individuals who feel connected because they engage in the same kind of work and have similar education, religious beliefs, and lifestyles. Social interaction is characterized by face-to-face, intimate, primary-group relationships. Because everyone is engaged in similar work, little specialization is found in the division of labor. In societies of this kind, the focus is on the group, not the individual, and social interaction is much more personal.

Organic solidarity refers to the social cohesion found in industrial (and perhaps postindustrial) societies, in which people perform very specialized tasks and feel united by their mutual dependence. Durkheim chose the term *organic solidarity* because he believed that individuals in industrial societies come to rely on one another in much the same way that the organs of the human body function interdependently. Social interaction is less personal, more status oriented, and more focused on specific goals and objectives. People no longer rely on morality or shared values for social solidarity; instead, they are bound together by interdependence and practical considerations.

Tönnies: *Gemeinschaft* and *Gesellschaft*

German sociologist Ferdinand Tönnies (1855–1936) used the terms *Gemeinschaft* and *Gesellschaft* to characterize the degree of social solidarity and social control found in societies. He was especially concerned about what happens to social solidarity in a society when a "loss of community" occurs.

The **Gemeinschaft (guh-MINE-shoft)** is a traditional society in which social relationships are based on personal bonds of friendship and kinship and on intergenerational stability. Tönnies (1963/1887) used the German term *Gemeinschaft* because it means "commune" or "community"; social solidarity and social control are maintained by the community. In this kind of society, relationships are based on ascribed (from birth) status rather than achieved (acquired) status. For example, the child of a farmer is likely to become, and remain, a farmer as well. In the *Gemeinschaft,* people have a commitment to the entire group and feel a sense of togetherness: They tend to focus more on the needs and interests of the group rather than their own self-interest. Members have a strong sense of belonging, but they also have very limited privacy. External social control is seldom needed because control is maintained through informal means such as persuasion or gossip.

By contrast, the **Gesellschaft (guh-ZELL-shoft)** is a large, urban society in which social bonds are based on impersonal and specialized relationships, with little long-term commitment to the group or consensus on values. Tönnies (1963/1887) selected the German term *Gesellschaft* because it means "association"; relationships are based on achieved statuses, and interactions among people are both rational and calculated. For example, achieved status might be based on education level or the kind of work that people do rather than the family into which they were born. In such societies, most people are "strangers" who perceive that they have very little in common with most other people.

TABLE 5.2

Comparing *Gemeinschaft* and *Gesellschaft* Societies

Gemeinschaft	*Gesellschaft*
Characterized by rural life	Characterized by urban life
Sense of community based on similarity	Lack of feeling of community
Intimate, face-to-face social interactions	Impersonal and task-oriented relationships
Primary focus on personal relationships	Primary focus on tasks or goals to be accomplished
Social control on an informal basis	Formal social control
Ascribed statuses most important	Achieved statuses most important
Limited social change	Social change more prevalent

Consequently, self-interest dominates, and little consensus exists regarding values. • Table 5.2 compares the characteristics of *Gemeinschaft* and *Gesellschaft* societies.

Social Structure and Homelessness

In *Gesellschaft* societies such as the United States, a prevailing value is that people should be able to take care of themselves. Consequently, some politicians and everyday people argue that social agencies and institutions have little or no responsibility to assist homeless persons. Others believe that the social structure of societies should provide a safety net for everyone regardless of their economic status. Clearly, there is no simple answer to questions about what should be done to help homeless persons. Nor, as you will notice in "Sociology and Social Policy," is there any

division of labor
how the various tasks of a society are divided up and performed.

mechanical solidarity
Emile Durkheim's term for the social cohesion of preindustrial societies, in which there is minimal division of labor and people feel united by shared values and common social bonds.

organic solidarity
Emile Durkheim's term for the social cohesion found in industrial (and perhaps postindustrial) societies, in which people perform very specialized tasks and feel united by their mutual dependence.

Gemeinschaft (guh-MINE-shoft)
a traditional society in which social relationships are based on personal bonds of friendship and kinship and on intergenerational stability.

Gesellschaft (guh-ZELL-shoft)
a large, urban society in which social bonds are based on impersonal and specialized relationships, with little long-term commitment to the group or consensus on values.

What's Going on in "Paradise?"—Homeless Rights Versus Public Space

Police officers came up to me and said it looks much better. Residents who live there came up to me and said, "Mayor, what are you doing? It looks so much better."

—Mayor Kirk Caldwell of Honolulu, Hawaii, comments on efforts to remove homeless persons and their belongings from Waikiki, a beachfront neighborhood that is a favorite destination of tourists worldwide. (Grube, 2014)

It's a totally different experience. It's a totally different life. It's hard to be excluded from society so fast, and the people that work for the city and state that used to work for us now work against us. ... My daughter is three years old today and if this bill passes, you're going to label her as a criminal because she sleeps on the sidewalk? You know we have no place to go. . . . Each sweep that happens actually paralyzes us, pushes us back ten steps out of the five steps we move forward.

—Tabatha Martin explains what life had been like on the streets since her husband lost his job after a heart attack and stated her concern about a new homeless ordinance (that was not passed). After this news story was aired, Kathryn Xian, an activist and soap maker in Honolulu, helped Tabatha's family find housing and a job. (KITV.com, 2014; Cave, 2015)

In Honolulu, like many other cities, a pitched battle has continued for years between persons without permanent residences and law enforcement officials and community

Sidewalk clearance and public space protection are controversial topics in cities where law enforcement officials have been instructed to remove homeless individuals and their possessions from public spaces. What are the central issues in this social policy debate? Why should this problem be of concern to each of us?

AP images/Cathy Bussewitz

consensus on what rights the homeless have to occupy public spaces such as parks and city sidewalks. The answers we derive as a society and as individuals are often based on our social construction of the reality of life for the homeless.

a macrosociological overview because they concentrate on large-scale events and broad social features. By contrast, the symbolic interactionist perspective takes a microsociological approach, asking how social institutions affect our daily lives.

LO6 **Discuss** the symbolic interactionist views on the social construction of reality and the self-fulfilling prophecy.

Social Interaction: The Microlevel Perspective

So far in this chapter, we have focused on society and social structure from a macrolevel perspective, seeing how the structure of society affects the statuses we occupy, the roles we play, and the groups and organizations to which we belong. Functionalist and conflict perspectives provide

Social Interaction and Meaning

When you are with other people, do you often wonder what they think of you? If so, you are not alone! Because most of us are concerned about the meanings that others ascribe to our behavior, we try to interpret their words and actions so that we can plan how we will react toward them (Blumer, 1969). We know that others have expectations of us. We also have certain expectations about them. For example, if we enter an elevator that has only one other person in it, we do not expect that individual to confront us and stare into our eyes. As a matter of fact, we would be quite upset if the person did so.

Social interaction within a given society has certain shared meanings across situations (• Figure 5.12).

leaders who believe they must "take back the streets" to save their community and the local tourism industry.

"Public space protection" has increasingly become an issue as record numbers of homeless individuals and families seek refuge in public places because they have nowhere else to go or do not wish to sleep in homeless shelters, if they are available. However, the seemingly individualistic problem of not having a home is actually linked to larger social concerns, including long-term unemployment, lack of education and affordable housing, cutbacks in government and social service budgets, and the physical and mental conditions of military veterans, such as those who suffer from PTSD (posttraumatic stress disorder).

The problem of homelessness raises significant social policy issues, such as the extent to which cities can make it illegal for people to remain for extended periods of time in public spaces, use public restrooms at all hours, or sleep in motor vehicles. Should homeless people be allowed to sleep on sidewalks, in parks, and in other public areas? As cities have sought to improve their downtown areas and public spaces, they have taken measures to enforce city ordinances controlling loitering (standing around or sleeping in public spaces), "aggressive panhandling," and disorderly conduct. Advocates for the homeless and civil liberties groups have filed lawsuits claiming that the rights of the homeless are being violated by the enforcement of these laws. The lawsuits assert that the homeless have a right to sleep in parks because no affordable housing is available for them. Advocates also argue that panhandling is a legitimate means of livelihood for some of the homeless and is protected speech under the U.S. Constitution's First Amendment. In addition,

they accuse public and law enforcement officials of seeking to punish the homeless on the basis of their "status," a cruel and unusual punishment prohibited by the Eighth Amendment.

The "homeless problem" is not a new one for city governments. Of the limited public funding that is designated for the homeless, most has been spent on shelters that are frequently overcrowded and otherwise inadequate. Officials in some cities, including Honolulu, have given homeless people a one-way ticket to another city or back to the mainland. Still others have routinely run them out of public spaces or tried to relocate them to marginalized areas of the city.

What responsibility do you think society should have for the homeless? Are laws restricting the hours that public areas or parks are open to the public unfair to homeless persons? These questions highlight pressing social policy concerns because affordable housing and job opportunities are not available for many people, often leaving homeless persons with nowhere to go.

Sources: Based on Grube, 2014; National Coalition for the Homeless, 2015.

REFLECT & ANALYZE

Do you think it is possible for communities to maintain public safety for everyone and not criminalize homelessness? How does your city deal with the issue of homeless rights and public space? What do you think should be done?

For instance, our reaction would be the same regardless of *which* elevator we rode in *which* building. Sociologist Erving Goffman (1961b) described these shared meanings in his observation about two pedestrians approaching each other on a public sidewalk. He noted that each will tend to look at the other just long enough to acknowledge the other's presence. By the time they are about eight feet away from each other, both individuals will tend to look downward. Goffman referred to this behavior as *civil inattention*—the ways in which an individual shows an awareness that another is present without making this person the object of particular attention. The fact that people engage in civil inattention demonstrates that interaction does have a pattern, or *interaction order,* which regulates the form and processes (but not the content) of social interaction.

Does everyone interpret social interaction rituals in the same way? No. Race/ethnicity, gender, and social class play a part in the meanings we give to our interactions with others, including chance encounters on elevators or

the street. Our perceptions about the meaning of a situation vary widely based on the statuses we occupy and our unique personal experiences. For example, women often do not perceive street encounters to be "routine" rituals. They fear for their personal safety and try to avoid comments and propositions that are sexual in nature. African Americans may also feel uncomfortable in street encounters. A middle-class African American college student described his experiences walking home at night from a campus job:

So, even if you wanted to, it's difficult just to live a life where you don't come into conflict with others. . . . Every day that you live as a black person you're reminded how you're perceived in society. You walk the streets at night; white people cross the streets. I've seen white couples and individuals dart in front of cars to not be on the same side of the street. Just the other day, I was walking down the street, and this white female with a child, I saw her pass a young white male about

FIGURE 5.12 How do our expectations of social interaction affect how we behave in a crowded bus or subway?

20 yards ahead. When she saw me, she quickly dragged the child and herself across the busy street.... [When I pass,] white men tighten their grip on their women. I've seen people turn around and seem like they're going to take blows from me.... So, every day you realize [you're black]. Even though you're not doing anything wrong; you're just existing. You're just a person. But you're a black person perceived in an unblack world. (qtd. in Feagin, 1991: 111–112)

Although this statement was made more than twenty years ago, some current students of color reading it can still relate to the experiences described here. As this passage indicates, social encounters have different meanings for men and women, whites and people of color, individuals from different social classes, and sometimes people from different areas of the country. Members of the dominant classes regard the poor, unemployed, and working class as less worthy of attention, frequently subjecting them to subtle yet systematic "attention deprivation" (Derber, 1983). The same can certainly be said about how members of the dominant classes "interact" with the homeless. There are some who say that all of this has changed since the time that some of this research was

originally conducted, but, unfortunately, violent incidents are periodically reported in the media which renew the concerns of people that we have a long way to go in protecting the safety and security of all individuals in our nation.

The Social Construction of Reality

If we interpret other people's actions so subjectively, can we have a shared social reality? Some symbolic interaction theorists believe that there is very little shared reality beyond that which is socially created. Symbolic interactionists refer to this as the *social construction of reality*—the process by which our perception of reality is largely shaped by the subjective meaning that we give to an experience (Berger and Luckmann, 1967). This meaning strongly influences what we "see" and how we respond to situations.

As we discussed previously, our perceptions and behavior are influenced by how we initially define situations: We act on reality as we see it. Sociologists describe this process as the *definition of the situation,* meaning that we analyze a social context in which we find ourselves, determine what is in our best interest, and adjust our attitudes and actions accordingly. This process can result in a *self-fulfilling prophecy*—a false belief or prediction that produces behavior that makes the originally false belief come true (Merton, 1968). An example would be a person who has been told repeatedly that she or he is not a good student; eventually, this person might come to believe it to be true, stop studying, and receive failing grades. Dominant-group members with prestigious statuses may have the ability to establish how other people define "reality" (Berger and Luckmann, 1967: 109).

An example of the self-fulfilling prophecy is a study of homeless persons in the United Kingdom where extensive interviews were conducted with eight homeless individuals to learn about their experiences with health-related social services. When the study was conducted early in the twenty-first century, approximately 400,000 people were considered to be "hidden homeless" in the United Kingdom, or approximately more than two in every thousand people in that nation. According to the researchers, homeless people in the UK were trapped between the *homed* system—the social service system, which offered formal help but was quick to label and stigmatize them, while enforcing extensive rules on them—and the *homeless* system of informal help, where other homeless individuals were supportive but often intensified their problems by offering alcohol, drugs, or other "solutions" that did not help their overall situation. The researchers concluded that a self-fulfilling prophecy occurs when homeless persons who have been labeled as "sofa surfers," "homeless," or similar terms come to view the label as central to how they see themselves and then decide to embed themselves within the homeless system, where they can have the support of other homeless people who also live without establishment-ordered rules and routines (Ogden and Avades, 2011). From the choices these individuals made, the self-fulfilling prophecy became a reality because their homeless situation was unchanged, and social organizations and institutions within the community were unable to meet their needs.

LO7 **Compare** ethnomethodology and dramaturgical analysis as two research methods for observing how people deal with everyday life.

Ethnomethodology

How do we know how to interact in a given situation? What rules do we follow? Ethnomethodologists are interested in the answers to these questions. *Ethnomethodology* is the study of the commonsense knowledge that people use to understand the situations in which they find themselves (Heritage, 1984: 4). Sociologist Harold Garfinkel (1967) initiated this approach and coined the term: *ethno* for "people" or "folk" and *methodology* for a "system of methods." Garfinkel was critical of mainstream sociology for not recognizing the ongoing ways in which people create reality and produce their own world. Consequently, ethnomethodologists examine existing patterns of conventional behavior in order to uncover people's background expectancies—that is, their shared interpretation of objects and events, as well as their resulting

actions. According to ethnomethodologists, interaction is based on assumptions of shared expectancies. For example, when you are talking with someone, what expectations do you have that you will take turns? Based on your background expectancies, would you be surprised if the other person talked for an hour and never gave you a chance to speak?

To uncover people's background expectancies, ethnomethodologists frequently break "rules" or act as though they do not understand some basic rule of social life so that they can observe other people's responses. In a series of *breaching experiments,* Garfinkel assigned different activities to his students to see how breaking the unspoken rules of behavior created confusion.

The ethnomethodological approach contributes to our knowledge of social interaction by making us aware of subconscious social realities in our daily lives. However, a number of sociologists regard ethnomethodology as a frivolous approach to studying human behavior because it does not examine the impact of macrolevel social institutions—such as the economy and education—on people's expectancies. Some scholars suggest that ethnomethodologists fail to do what they claim to do: look at how social realities are created. Rather, they take ascribed statuses (such as race, class, gender, and age) as "givens," not as *socially created* realities.

Dramaturgical Analysis

Can you compare everyday life to watching a dramatic presentation? Erving Goffman suggested that day-to-day interactions have much in common with being on stage or in a dramatic production. *Dramaturgical analysis* is the study of social interaction that compares everyday life to a theatrical presentation. In this presentation, there is a stage, actors, and an audience to observe and analyze the social interactions of the actors. The actors have a *social script*—a playbook that the actors use to guide their verbal replies and overall performance to achieve the desired goal of the conversation or fulfill the role they are playing. Although most of us do not have scripted conversations, we have a good idea how a social exchange will occur.

---

social construction of reality
the process by which our perception of reality is shaped largely by the subjective meaning that we give to an experience.

self-fulfilling prophecy
a situation in which a false belief or prediction produces behavior that makes the originally false belief come true.

ethnomethodology
the study of the commonsense knowledge that people use to understand the situations in which they find themselves.

dramaturgical analysis
Erving Goffman's term for the study of social interaction that compares everyday life to a theatrical presentation.

social script
a "playbook" that "actors" use to guide their verbal replies and overall performance to achieve the desired goal of the conversation or fulfill the role they are playing.

FIGURE 5.13 Erving Goffman believed that people spend a great amount of time and effort managing the impression that they present. What kinds of impressions are these men presenting to others?

For example, when someone asks us how we are doing, we expect to reply, "Fine, thanks." If we take our vehicle to the drive-up order station at a fast-food restaurant, we expect to hear a voice say, "May I take your order, please?" We do not expect to reply, "Do you know what the high temperature will be today?"

Because we are familiar with most scripts in our daily lives, we know what to expect; however, there is often more than one way to interpret a script, leading to confusion and sometimes to conflict. According to Goffman (1959, 1963a), members of our "audience" judge our performance and are aware that we may slip and reveal our true character. Consequently, most of us attempt to play our role as well as possible and to create and sustain favorable impressions. *Impression management* (*presentation of self*) refers to people's efforts to present themselves to others in ways that are most favorable to their own interests or image (● Figure 5.13). For example, suppose that a professor has returned graded exams to your class. Will you discuss the exam and your grade with others in the class? If you are like most people, you probably play your student role differently depending on whom you are talking to and what grade you received on the exam. Your "presentation" may vary depending on the grade earned by the other person (your "audience"). In one study, students who all received high grades ("Ace–Ace encounters") willingly talked with one another about their grades and sometimes engaged in a little bragging about how they had "aced" the test. However, encounters between students who had received high grades and those who had received low or failing grades ("Ace–Bomber encounters") were uncomfortable. The Aces felt as if they had to minimize their own grade. Consequently, they tended to attribute their success to "luck" and were quick to offer the Bombers words of encouragement. On the other hand, the Bombers believed that they had to praise the Aces and hide their own feelings of frustration and disappointment. Students who received low or failing grades ("Bomber–Bomber encounters") were more comfortable when they talked with one another because they could share their negative emotions. They often indulged in self-pity and relied on face-saving excuses (such as an illness or an unfair exam) for their poor performances (Albas and Albas, 1988, 2011).

In Goffman's terminology, *face-saving behavior* refers to the strategies we use to rescue our performance when we experience a potential or actual loss of face. When the

Bombers made excuses for their low scores, they were engaged in face-saving; the Aces attempted to help them save face by asserting that the test was unfair or that it was only a small part of the final grade. Why would the Aces and Bombers both participate in face-saving behavior? In most social interactions, all role players have an interest in keeping the "play" going so that they can maintain their overall definition of the situation in which they perform their roles.

Goffman noted that people consciously participate in *studied nonobservance,* a face-saving technique in which one role player ignores the flaws in another's performance to avoid embarrassment for everyone involved. Most of us remember times when we have failed in our role and know that it is likely to happen again; thus, we may be more forgiving of the role failures of others.

Social interaction, like a theater, has a front stage and a back stage. The *front stage* is the area where a player performs a specific role before an audience. The *back stage* is the area where a player is not required to perform a specific role because it is out of view of a given audience. For example, when the Aces and Bombers were talking with one another at school, they were on the "front stage." When they were in the privacy of their own residences, they were in "back stage" settings—they no longer had to perform the Ace and Bomber roles and could be themselves.

The need for impression management is most intense when role players have widely divergent or devalued statuses. As we have seen with the Aces and Bombers, the participants often play different roles under different circumstances and keep their various audiences separated from one another. If one audience becomes aware of other roles that a person plays, the impression being given at that time may be ruined. For example, people facing or experiencing homelessness are not only stigmatized but may also find that they lose the opportunity to get a job if their homelessness becomes known. However, many homeless individuals do not passively accept the roles into which they are cast. For the most part, they attempt—as we all do—to engage in impression management in their everyday lives.

The dramaturgical approach helps us think about the roles we play and the audiences who judge our presentation of self; however, this perspective has also been criticized for focusing on appearances and not the underlying substance. This approach may not place enough emphasis on the ways in which our everyday interactions with other people are influenced by occurrences within the larger society. For example, if some political leaders or social elites in a community deride homeless people by saying they are "lazy" or "unwilling to work," it may become easier for everyday people walking down a street to treat homeless individuals poorly. Similarly, in the 2012 presidential campaign, one candidate repeatedly claimed that poor children have no work habits or concept of earning money and will likely remain poor and perhaps homeless throughout their lives (Dover, 2011). Overall, however, Goffman's dramaturgical analysis has been highly influential in the development of the sociology of emotions, an important area of contemporary theory and research.

LO8 **State** three ways in which the sociology of emotions and the study of nonverbal communication add to our understanding of human behavior.

The Sociology of Emotions

Why do we laugh, cry, or become angry? Are these emotional expressions biological or social in nature? To some extent, emotions are a biologically given sense (like hearing, smell, and touch), but they are also social in origin. We are socialized to feel certain emotions, and we learn how and when to express (or not express) those emotions (Hochschild, 1983).

How do we know which emotions are appropriate for a given role? Sociologist Arlie Hochschild (1983) suggests that we acquire a set of *feeling rules* that shapes the appropriate emotions for a given role or specific situation (• Figure 5.14). These rules include how, where, when, and with whom an emotion should be expressed. For example, for the role of a mourner at a funeral, feeling rules tell us which emotions are required (sadness and grief, for example), which are acceptable (a sense of relief that the deceased no longer has to suffer), and which are unacceptable (enjoyment of the occasion expressed by laughing out loud) (see Hochschild, 1983: 63–68).

Feeling rules also apply to our occupational roles. For example, the truck driver who handles explosive cargos must be able to suppress fear. Although all jobs place some burden on our feelings, *emotional labor* occurs only in jobs that require personal contact with the public or the production of a state of mind (such as hope, desire, or fear) in others (Hochschild, 1983). With emotional labor, employees must display only certain carefully selected emotions. For example, flight attendants are required to act friendly toward passengers, to be helpful and open to requests, and to maintain an "omnipresent smile" in order to enhance the customers' status. By contrast, bill collectors are encouraged to show anger and make threats to customers, thereby supposedly deflating the customers' status and wearing down their presumed resistance to paying past-due bills. In both jobs, the employees are expected to show feelings that are often not their true ones (Hochschild, 1983).

Social class and race are determinants in managed expression and emotion management. Emotional labor is emphasized in middle- and upper-class families. Because middle- and upper-class parents often work with people, they are more likely to teach their children the importance of emotional labor in their own careers than are working-class parents, who tend to work with things, not people (Hochschild, 1983). Race is also an important factor in emotional labor. People of color spend much of their life engaged in

impression management (presentation of self)
Erving Goffman's term for people's efforts to present themselves to others in ways that are most favorable to their own interests or image.

face-saving behavior
Erving Goffman's term for the strategies we use to rescue our performance when we experience a potential or actual loss of face.

Tom Prettyman/PhotoEdit

Nonverbal Communication

In a typical stage drama, the players not only speak their lines but also use nonverbal communication to convey information. ***Nonverbal communication*** is the transfer of information between persons without the use of words (• Figure 5.15). It includes not only visual cues (gestures, appearances) but also vocal features (inflection, volume, pitch) and environmental factors (use of space, position) that affect meanings. Facial expressions, head movements, body positions, and other gestures carry as much of the total meaning of our communication with others as our spoken words do.

Functions of Nonverbal Communication   We obtain first impressions of others from various kinds of nonverbal communication, such as the clothing they wear and their body positions. Head and facial movements may provide us with information about other people's emotional states, and others receive similar information from us. Through our body posture and eye contact, we signal that we do or do not wish to speak to someone. For example, we may look down at the sidewalk or off into the distance when we pass homeless persons who look as if they are going to ask for money.

© Maxisport/Shutterstock.com

FIGURE 5.14   Are there different gender-based expectations in the United States about the kinds of emotions that men, as compared with women, are supposed to show? What feeling rules shape the emotions of the man in this photo?

emotional labor because racist attitudes and discrimination make it continually necessary to manage one's feelings.

Emotional labor may produce feelings of estrangement from one's "true" self. C. Wright Mills (1956) suggested that when we "sell our personality" in the course of selling goods or services, we engage in a seriously self-alienating process. In other words, the "commercialization" of our feelings may dehumanize our work-role performance and create alienation and contempt that spill over into other aspects of our life.

Hochschild (2012) also conducted research to demonstrate how many middle- and upper-income individuals outsource emotional labor in the more-intimate aspects of their lives, such as having other people professionally plan their family birthday parties and weddings, selecting names for their children, overseeing the daily lives of their children, and assuming major caregiving responsibilities for their elderly parents. Hochschild (2012) uses the term the "outsourced self" to refer to what happens when individuals defer most of their emotional labor to others, particularly as the market continues to invade private life.

FIGURE 5.15   Nonverbal communication may be thought of as an international language. What message do you receive from the facial expression, body position, and gestures of each of these people? Is it possible to misinterpret these messages?

Nonverbal communication establishes the relationship among people in terms of their responsiveness to and power over one another. For example, we show that we are responsive toward or like another person by maintaining eye contact and attentive body posture and perhaps by touching and standing close. Studies of communications in the doctor–patient relationship confirm that trust is vital for quality health care outcomes and that the best communications are established through actions more than words. Nonverbal communication, such as good eye contact and attentive listening posture, show that the doctor is paying attention to what the patient is saying, and this helps to build rapport and establish the agenda of the professional in relation to that of the patient (Brown et al., 2011).

Goffman (1956) suggested that *demeanor* (how we behave or conduct ourselves) is relative to social power. People in positions of dominance are allowed a wider range of permissible actions than are their subordinates, who are expected to show deference. *Deference* is the symbolic means by which subordinates give a required permissive response to those in power; it confirms the existence of inequality and reaffirms each person's relationship to the other.

Facial Expression, Eye Contact, and Touching   Deference behavior is important in regard to facial expression, eye contact, and touching. This type of nonverbal communication is symbolic of our relationships with others. Who smiles? Who stares? Who makes and sustains eye contact? Who touches whom? All these questions relate to demeanor and deference; the key issue is the status of the person who is doing the smiling, staring, or touching relative to the status of the recipient (Goffman, 1967).

Facial expressions, especially smiles and eye contact, also reflect gender-based patterns of dominance and subordination in society. Typically, women have been socialized to smile and frequently do so even when they are not actually happy (LaFrance and Hecht, 2000). Jobs held predominantly by women (including flight attendant, secretary and administrative assistant, elementary schoolteacher, and nurse) are more closely associated with being pleasant and smiling than are "men's jobs." By contrast, men tend to display less emotion through smiles or other facial expressions and instead seek to show that they are reserved and in control. Even as women have entered previously male-dominated professions, such as medicine, law, and college teaching and administration, some expectation that they will be more personable, more understanding, and more nurturing than their male counterparts has remained in the minds of some patients, clients, customers, colleagues, students, supervisors, and others who come into daily contact with the woman professional.

Women are also more likely to sustain eye contact during conversations (but not otherwise) as a means of showing their interest in and involvement with others. By contrast, men are less likely to maintain prolonged eye contact during conversations but are more likely to stare at other people (especially men) in order to challenge them and assert their own status (Hall, Carter, and Horgan, 2000).

Eye contact can be a sign of domination or deference. For example, in a classic participant observation study of domestic (household) workers and their employers, the sociologist Judith Rollins (1985) found that household workers were supposed to show deference by averting their eyes when they talked to their employers. Deference also required that they present an "exaggeratedly subservient demeanor" by standing less erect and walking tentatively. This kind of behavior is depicted in the best-selling book and movie *The Help.* More-recent relationships of domination and deference in household relationships among immigrant workers and their affluent employers in the United States are discussed in the research of the sociologist Pierrette

nonverbal communication
the transfer of information between persons without the use of words.

Hondagneu-Sotelo (2007). Other contemporary examples of using eye contact as a means to express domination and deference are when a teacher or principal is correcting a student and when a detainee is in a juvenile facility or prison.

Touching is another form of nonverbal behavior that has many different shades of meaning. Although touching is a universal aspect of people's communication with one another, it varies greatly by gender, age, and culture. Gender and power differences are evident in tactile communication from birth: Boys are touched more roughly and playfully, whereas girls are handled more gently and protectively. This pattern continues into adulthood, with women touched more frequently than men. Clearly, touching has a different meaning to women than to men. Women may hug and touch others to indicate affection and emotional support, but men are more likely to touch others to give directions, assert power, and express sexual interest.

Age is also a factor in touching: Different patterns of touching have been identified in people under thirty years of age as compared with those of older adults. Younger men behave more possessively and women more submissively in regard to touching behavior. Touching behavior also involves large cultural differences in personal space, such as the handshake as a preferred means of personal greeting in the United States as compared to extensive hugs and kisses in some other countries. Let's look more closely at personal space.

Personal Space *Personal space* is the immediate area surrounding a person that the person claims as private. Our personal space is contained within an invisible boundary surrounding our body, much like a snail's shell or an invisible bubble or zone around a person. This space is fluctuating, and it is part of a communication style. When others invade our space, we may retreat, stand our ground, or even lash out, depending on our cultural background.

Anthropologist Edward Hall (1959, 1966) first described the concept of personal space and identified four dimensions of this space among people in the United States: *Intimate distance* (or *intimate zone*) involves a high level of intimacy between two persons, including touching; *personal distance* (or *personal zone*) is the distance between two persons who know each other with a relative intimacy, such as friends, brothers, sisters, or other relatives; and *social distance* (or *social zone*) is the more impersonal form of communication or business interaction that takes place among individuals at a social gathering or employees in a business environment. The fourth dimension is *public distance* (or *public zone*), in which no intimacy exists between a speaker on a platform and an audience where the distance is greater than thirteen feet and may be as much as twenty-seven or more feet in a formal setting.

Age, gender, and cultural differences are important factors in the allocation of personal space. With regard to age, adults generally do not hesitate to enter the personal space of a child. Women in the United States tend to interact at closer distances than men, and this appears to remain relatively consistent across age categories. Cross-cultural

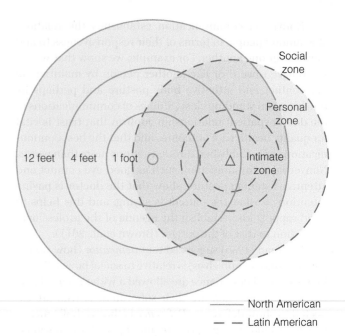

FIGURE 5.16 **North American and Latin American Social Distance Rules**

Source: From *Cultural Anthropology* by Paul Hiebert. Copyright © 1983 by Baker Academic, a division of Baker Publishing Group. Used by permission.

studies have confirmed that people in the United States are more comfortable with larger zones of personal space than are individuals from Latin American countries. Less personal space is also required among South Americans, Southern and Eastern Europeans, and Arabs, while larger amounts of personal space are required among Asians and Northern Europeans (Beaulieu, 2004). • Figure 5.16 illustrates differences in social distance rules between two contrasting cultures, North America and Latin America, which are known for having different social distance rules.

In sum, all forms of nonverbal communication are influenced by gender, race, social class, and the personal contexts in which they occur. Although it is difficult to generalize about people's nonverbal behavior, we still need to think about our own nonverbal communication patterns. Recognizing that differences in social interaction exist is important. Learning to understand and respect alternative styles of social interaction enhances our personal effectiveness by increasing the range of options we have for communicating with different people in diverse contexts and for varied reasons. (The Concept Quick Review summarizes the microlevel approach to social interaction.)

Looking Ahead: Social Change, Social Structure, and Interaction in the Future

The social structure in the United States has been changing rapidly in recent decades, and we can no longer think of our future as separate from the larger world of which

YOU CAN MAKE A DIFFERENCE

Offering a Helping Hand to Homeless People

When you pull up at an intersection and see a person holding a piece of cardboard with a handwritten sign on it, how do you react? Many of us shy away from encounters such as this because we know, without actually looking, that the sign says something like "Homeless, please help." In an attempt to avoid eye contact with the person, we suddenly look with newfound interest at something lying on our car seat, or we check our appearance in the rearview mirror, or we adjust the radio. In fact, we do just about whatever it takes to divert our attention, making eye contact with this person impossible until the traffic light changes and we can be on our way.

Does this scenario sound familiar? Many of us see homeless individuals on street corners and elsewhere as we go about our daily routine. We are uncomfortable in their presence because we don't know what we can do to help them, or even if we should. Frequently, we hear media reports stating that some allegedly homeless people abuse the practice of asking for money on the streets and that many are faking injury or poverty so that they can take advantage of generous individuals. Stereotypes such as this are commonplace, but they are far from the entire picture: Many homeless people are in need of assistance, and many of the homeless are children, persons with disabilities, and people with other problems that make it difficult, if not impossible, for them to earn enough money to pay for housing.

Do all of these "big-picture" problems in our society mean that we have no individual responsibility to help homeless people? We do not necessarily have to hand money over to the person on the street to help individuals who are homeless. There are other, and perhaps even better, ways in which we can provide help to the homeless through our small acts of generosity and kindness. Here are a few ways in which you and others at your school might help homeless individuals and families in your community:

- *Understand who the homeless are* so that you can help dispel the stereotypes often associated with homeless people. Learn what causes homelessness, and remember that each person's story is unique.

- *Buy a street newspaper sold by homeless people if you live in an urban area where these newspapers are sold.* Homeless people receive a small amount for every paper they sell.

An example of paying-it-forward for homeless persons is Rosa's Fresh Pizza in Center City Philadelphia, where customers can pay extra to buy slices for homeless persons who come into the store throughout the day. Owner Mason Wartman (shown here) keeps track of these purchases, and people place notes of encouragement or thanks on the walls.

- *Give to organizations that aid the homeless.* In addition to money and clothing, recyclable cans and bottles are helpful because they can help pay for living expenses.

- *Volunteer at a shelter, soup kitchen, or battered women's shelter* where you can help meet the needs of homeless people as well as women and children who need assistance in getting away from abusive relationships.

- *Look for campus organizations that work with the homeless,* or create your own and enlist friends and existing organizations (such as your service organization, sorority, or fraternity) to engage in community service projects.

For additional ways you can help the homeless, check with shelters in your area. You may also want to visit the websites of organizations such as the following:

- Just Give

- The Doe Fund

- U.S. Department of Housing and Urban Development

---

we are a part. Currently, there are more possible statuses for persons to occupy and roles to play than at any other time in history. Although achieved statuses are considered very important, ascribed statuses still have a significant effect on people's options and opportunities. National issues are now fused with global concerns regarding economic crises, actual and potential war and terrorism, more-frequent occurrences of natural disasters, and other problems that affect people around the world.

personal space
the immediate area surrounding a person that the person claims as private.

Social Interaction: The Microlevel Perspective

Social Interaction and Meaning	In a given society, forms of social interaction have shared meanings, although these may vary to some extent based on race/ethnicity, gender, and social class.
Social Construction of Reality	The process by which our perception of reality is largely shaped by the subjective meaning that we give to an experience.
Ethnomethodology	Studying the commonsense knowledge that people use to understand the situations in which they find themselves makes us aware of subconscious social realities in daily life.
Dramaturgical Analysis	The study of social interaction that compares everyday life to a theatrical presentation. This approach includes impression management (people's efforts to present themselves favorably to others).
Sociology of Emotions	We are socialized to feel certain emotions, and we learn how and when to express (or not express) them.
Nonverbal Communication	The transfer of information between persons without the use of words, such as by facial expressions, head movements, and gestures.

Ironically, at a time when we have more ability to communicate than ever before, more technological capability, more leisure activities and types of entertainment, and more quantities of material goods available for consumption, many people experience problems that are beyond their individual ability to resolve, such as chronic unemployment, hunger, and homelessness (see "You Can Make a Difference").

As the sociological imagination suggests, these individuals in need are dealing not only with their own personal troubles but also with broader public issues that affect large numbers of people and require solutions at community, societal, and global levels. The future of our country rests on our collective ability to deal with major social problems at both the macrolevel and the microlevel of the social world.

CHAPTER REVIEW Q & A

LO1 Why is social structure important in our interaction with others?

The stable patterns of social relationships within a particular society make up its social structure. Social structure is a macrolevel influence because it shapes and determines the overall patterns in which social interaction occurs. Social structure provides an ordered framework for society and for our interactions with others. Social structure comprises statuses, roles, groups, and social institutions.

LO2 What are the differences among ascribed, achieved, and master statuses?

A status is a specific position in a group or society and is characterized by certain expectations, rights, and duties. Ascribed statuses, such as gender, class, and race/ethnicity, are acquired at birth or involuntarily later in life. Achieved statuses, such as education and occupation, are assumed voluntarily as a result of personal choice, merit, or direct effort. A master status is the most important status a person occupies. For some, occupation is the chief indicator of their status. Occupation provides important

clues to a person's educational level, income, and family background. Master statuses confer high or low levels of personal worth and dignity on people.

LO3 How are role, role expectation, role performance, role conflict, role strain, and role exit alike or different?

A role is a set of behavioral expectations associated with a given status. Role expectation is a group's definition of the way that a specific role ought to be played, whereas role performance is how a person actually plays the role. When role conflict occurs, we may feel pulled in different directions. To deal with this problem, we may prioritize our roles and first complete the one we consider to be most important. Role conflict may occur as a result of changing statuses and roles in society. Role strain occurs when incompatible demands are built into a single status that a person occupies. For example, married women might feel role strain when they have the majority of responsibility to work full time, manage household duties, and take care of the family. Role exit occurs when people disengage from social roles that have been central to their self-identity. Role exit is a four-stage process, ending with the creation of a new identity.

LO4 What are the functionalist and conflict perspectives on social institutions?

According to functionalist theorists, social institutions perform several prerequisites of all societies: replace members; teach new members; produce, distribute, and consume goods and services; preserve order; and provide and maintain a sense of purpose. Conflict theorists suggest that social institutions do not work for the common good of all individuals: Institutions may enhance and uphold the power of some groups but exclude others, such as the homeless.

LO5 How does social change occur in preindustrial, industrial, and postindustrial societies?

According to Emile Durkheim, although changes in social structure may dramatically affect individuals and groups, societies manage to maintain some degree of stability. According to Durkheim, social solidarity derives from a society's social structure, which, in turn, is based on the society's division of labor. People in preindustrial societies are united by mechanical solidarity because they have shared values and common social bonds. As societies industrialized and developed more-specialized economic activities, social solidarity came to be rooted in the members' shared dependence on one another. Industrial societies are characterized by organic solidarity, which refers to the cohesion that results when people perform specialized tasks and are united by mutual dependence.

LO6 What are the symbolic interactionist views on the social construction of reality and the self-fulfilling prophecy?

Symbolic interactionists refer to this as the social construction of reality—the process by which our perception of reality is largely shaped by the subjective meaning that we give to an experience. We analyze a social context in which we find ourselves, determine what is in our best interest, and adjust our attitudes and actions accordingly. This process can result in a self-fulfilling prophecy—a false belief or prediction that produces behavior that makes the originally false belief come true.

LO7 What are dramaturgical analysis and ethnomethodology?

According to Erving Goffman's dramaturgical analysis, our daily interactions are similar to dramatic productions. Presentation of self refers to efforts to present our own self to others in ways that are most favorable to our interests or self-image. Ethnomethodology is the study of the commonsense knowledge that people use to understand the situations in which they find themselves. Ethnomethodologists frequently break "rules" or act as though they do not understand some basic rule of social life so that they can observe other people's responses.

LO8 How do the sociology of emotions and the study of nonverbal communication add to our understanding of human behavior?

Our emotions are not always private, and specific emotions may be demanded of us on certain occasions. Feeling rules shape the appropriate emotions for a given role or specific situation. Nonverbal communication is the transfer of information between persons without the use of words. It establishes the relationship among people in terms of their responsiveness to and power over one another.

KEY TERMS

achieved status 119
agrarian society 128
ascribed status 119
division of labor 130
dramaturgical analysis 135
ethnomethodology 135
face-saving behavior 136
formal organization 124
Gemeinschaft 131
Gesellschaft 131
horticultural society 127
hunting and gathering society 126
impression management (presentation of self) 136

industrial society 128
master status 119
mechanical solidarity 130
nonverbal communication 138
organic solidarity 131
pastoral society 127
personal space 140
postindustrial society 129
primary group 123
role 121
role conflict 122
role exit 123
role expectation 121
role performance 121

role strain 122
secondary group 123
self-fulfilling prophecy 134
social construction of reality 134
social group 123
social institution 125
social interaction 116
social script 135
social structure 116
status 118
status set 119
status symbol 121

QUESTIONS for CRITICAL THINKING

1 Think of a person you know well who often irritates you or whose behavior grates on your nerves (it could be a parent, friend, relative, or teacher, among others). First, list that person's statuses and roles. Then analyze the person's possible role expectations, role performance, role conflicts, and role strains. Does anything you find in your analysis help to explain the irritating behavior? How helpful are the concepts of social structure in analyzing individual behavior?

2 You are conducting field research on gender differences in nonverbal communication styles. How are you going to account for variations among age, race, and social class?

3 When communicating with other genders, races, and ages, is it better to express and acknowledge different styles or to develop a common, uniform style?

ANSWERS to the SOCIOLOGY QUIZ

ON HOMELESS PEOPLE AND THE SOCIAL STRUCTURE
OF HOMELESSNESS

1 **True** Debt and deficit reduction at the federal level, combined with fiscal crises at the local and state levels, has reduced funds for assistance available to homeless people and the organizations that assist them.

2 **False** A majority of people who are counted as homeless are found in emergency shelters or transitional housing programs, but these organizations offer only a temporary break from the larger problem of homelessness.

3 **True** Many homeless people do have full-time employment, but they are among the working poor. The minimum-wage jobs they hold do not pay enough for them to support their families and pay the high rents that are typical in many cities.

4 **True** Although many people think of homelessness as being based solely on lack of income to pay for housing, another significant factor is the cost of housing. In some cities and regions, the available housing simply is not affordable to people who have limited financial resources.

5 **False** Many homeless people panhandle to pay for food, a bed at a shelter, or other survival needs.

6 **True** Overcrowded shelters throughout the nation often attempt to accommodate as many homeless people as possible on a given night, particularly when the weather is bad. As a result, any available spaces—including offices, closets, and hallways—are used as sleeping areas until the individuals can find another location or weather conditions improve.

7 **True** Scholars have found that homelessness has always existed in the United States. However, the number of homeless people has increased or decreased with fluctuations in the national economy.

8 **False** The "doubled-up" population in the United States increased by more than 50 percent from 2005 to 2010 as many people faced difficult economic times during the Great Recession. Many people lost their homes to foreclosure, others were unable to pay rent because of losing their jobs, and still others found themselves in a downward economic spiral.

Source: U.S. Conference of Mayors, 2014.

GROUPS AND ORGANIZATIONS

Disaster
Relief

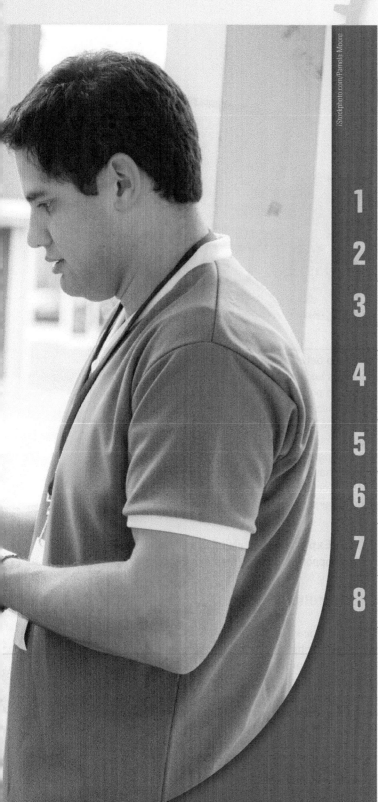

LEARNING OBJECTIVES

1 **Explain** what constitutes a social group as opposed to an aggregate or a category.

2 **Distinguish** among ingroups, outgroups, and reference groups, and give an example of each.

3 **Discuss** how a group's size shapes members' communication, leadership styles, and pressures to conform.

4 **Applying** the concept of groupthink, describe how people often respond differently in a group context than they might if they were alone.

5 **Identify** the three categories of formal organizations and state how they differ in membership.

6 **Debate** the strengths and weaknesses of bureaucracies in contemporary nations such as the United States.

7 **Define** the iron law of oligarchy and apply the concept to a brief analysis of the U.S. government.

8 **Identify** alternative forms of organization that exist today in nations such as Japan.

SOCIOLOGY & EVERYDAY LIFE

Social Media and the Classroom

At my university, professors are divided about whether they should meddle [with students who bring smartphones, iPads or other tablets, and computers to class]. Our students, some say, are grown-ups. It is not for us to dictate how they take notes or to get involved if they let their attention wander from class-related materials. But when I stand in back of our Wi-Fi enabled lecture halls, students are on Facebook and YouTube, and they are shopping, mostly for music. I want to engage my students in conversation. I don't think they should use class time for any other purpose. One year, I raised the topic for general discussion and suggested using notebooks (the paper kind) for note taking. Some of my students claimed to be relieved. "Now I won't be tempted by Facebook messages," said one sophomore. Others were annoyed, almost surly. . . . I maintained my resolve, but the following year, I bowed to common practice and allowed students to do what they wished.

—SHERRY TURKLE (2011), a professor at MIT, describing her feelings about students' use of digital technology in classrooms where professors are attempting to form groups and build community among students

Blend Images/Ariel Skelley/the Agency Collection/Getty Images

Although books are still an integral part of higher education, computers and other digital technology are rapidly changing the social and learning environments of today's colleges and universities.

According to sociologists, we need groups and organizations—just as we need culture and socialization—to live and participate in a society. Historically, the basic premise of groups and organizations was that individuals engage in face-to-face interactions in order to be part of such a group; however, millions of people today communicate with others through the Internet, cell phones, and other forms of information technology that make it possible for them to "talk" with individuals they have never met and who may live thousands of miles away. A variety of networking websites, including Facebook, LinkedIn, Twitter, and Google Plus, now compete with or, in some cases, replace live, person-to-person communications. For many college students, Facebook has become a fun way to get to know other people, to join online groups with similar interests or activities, and to plan "real-life" encounters. Despite the wealth of information and opportunities for new social connections that such websites offer, many of our daily activities require that we participate in social groups and formal organizations where *face time*—time spent interacting with others on a face-to-face basis, rather than via Internet or cell phone—is necessary.

What do social groups and formal organizations mean to us in an age of rapid telecommunications? What is the relationship between information and social organizations in societies such as ours? How can we balance the information that we provide to other people about us with our own right to privacy and need for security? These questions are of interest to sociologists who seek to apply the sociological imagination to their studies of social groups, bureaucratic organizations, social networking, and virtual communities. Before we have a closer look at groups and organizations, take the "Sociology and Everyday Life" quiz on issues pertaining to personal privacy in groups and organizations.

 LO1   **Explain** what constitutes a social group as opposed to an aggregate or a category.

Social Groups

If you see three strangers standing at a street corner waiting for a traffic light to change, do they constitute a group? Five hundred women and men are first-year graduate students at a university. Do they constitute a group? In everyday usage, we use the word *group* to mean any collection of people. According to sociologists, however, the answer to

How Much Do You Know About Privacy in Groups and Organizations?

TRUE	FALSE		
T	F	1	A college student's privacy is protected when using a school-owned computer as long as he or she deletes from the computer all e-mails or other documents he or she has worked on and thus prevents anyone else from examining those documents.
T	F	2	Parents of students at all U.S. colleges and universities are entitled to obtain a transcript of their children's college grades, regardless of the student's age.
T	F	3	If you work for a business that monitors phone calls with a pen register (an electronic device that records information about calls to or from a particular phone extension), your employer has the right to maintain and examine a list of phone numbers dialed by your extension and how long each call lasted.
T	F	4	Members of a high school football team can be required to submit to periodic, unannounced drug testing.
T	F	5	A company has the right to keep its employees under video surveillance anywhere at the company's place of business—even in the restrooms.
T	F	6	A professor can legally post students' grades in public, using the student's Social Security number as an identifier, as long as the student's name does not appear with the number.
T	F	7	Students at a church youth-group meeting who hear one member of the group confess to an illegal act can be required to divulge what that member said.
T	F	8	If you apply for a job at a company that has more than 25 employees, your employer can require that you provide a history of your medical background or take a physical examination prior to offering you a job.

Answers can be found at the end of the chapter.

these questions is no; individuals who happen to share a common feature or to be in the same place at the same time do not constitute social groups.

Groups, Aggregates, and Categories

As you will recall from Chapter 5, a *social group* is a collection of two or more people who interact frequently with one another, share a sense of belonging, and have a feeling of interdependence. Several people waiting for a traffic light to change constitute an ***aggregate***—a collection of people who happen to be in the same place at the same time but share little else in common. Shoppers in a department store and passengers on an airplane flight are also examples of aggregates. People in aggregates share a common purpose (such as purchasing items or arriving at their destination) but generally do not interact with one another, except perhaps briefly. The first-year graduate students, at least initially, constitute a ***category***—a number of people who may never have met one another but share a similar characteristic, such as education level, age, race, or gender. Men and women make up categories, as do Native Americans and Latinos/as, and victims of sexual or racial

harassment. Categories are not social groups because the people in them do not usually create a social structure or have anything in common other than a particular trait.

Occasionally, people in aggregates and categories form social groups. For instance, people within the category known as "graduate students" may become an aggregate when they get together for an orientation to graduate school. Some of them may form social groups as they interact with one another in classes and seminars, find that they have mutual interests and concerns, and develop a sense of belonging to the group. Information technology raises new and interesting questions about what constitutes a group.

Where do our social media "friends" fit into these categories? Some social scientists believe that virtual communities established on the Internet constitute true

aggregate
a collection of people who happen to be in the same place at the same time but share little else in common.

category
a number of people who may never have met one another but share a similar characteristic, such as education level, age, race, or gender.

communities (see Wellman, 2001), but others do not. According to sociologists Robyn Bateman Driskell and Larry Lyon, although the Internet provides us with the opportunity to share interests with others whom we have not met and to communicate with people whom we already know, the original concept of community, which "emphasized local place, common ties, and social interaction that is intimate, holistic, and all-encompassing," is lacking in social media (Driskell and Lyon, 2002: 6). Why? Because virtual communities on the Internet do not have geographic and social boundaries, are limited in their scope to specific areas of interest, are psychologically detached from close interpersonal ties, and have only limited concern for their members. In fact, people who spend hours in isolation on social media may reduce community rather than enhance it, or there is a chance that they will create a weak replacement for people based on specialized ties they develop through extended, remote interaction with others (Driskell and Lyon, 2002). What do you think?

LO2 **Distinguish** among ingroups, outgroups, and reference groups, and give an example of each.

Types of Groups

As you will recall from Chapter 5, groups have varying degrees of social solidarity and structure. This structure is flexible in some groups and more rigid in others. Some groups are small and personal; others are large and impersonal. We more closely identify with the members of some groups than we do with others.

Cooley's Primary and Secondary Groups

Sociologist Charles H. Cooley (1963/1909) used the term *primary group* to describe a small, less specialized group in which members engage in face-to-face, emotion-based interactions over an extended period of time. We have primary relationships with other individuals in our primary groups—that is, with our *significant others,* who frequently serve as role models.

In contrast, as you will recall, a *secondary group* is a larger, more specialized group in which the members engage in more-impersonal, goal-oriented relationships for a limited period of time. The size of a secondary group may vary. Twelve students in a graduate seminar may start out as a secondary group but eventually become a primary group as they get to know one another and communicate on a more personal basis.

Formal organizations are secondary groups, but they also contain many primary groups within them. For example, how many primary groups do you think there are within the secondary-group setting of your college?

Sumner's Ingroups and Outgroups All groups set boundaries by distinguishing between insiders who are members and outsiders who are not members (● Figure 6.1). Sociologist William Graham Sumner (1959/1906) coined the terms *ingroup* and *outgroup* to describe people's feelings toward members of their own and other groups. An **ingroup** is a group to which a person belongs and with which the person feels a sense of identity. An **outgroup** is a group to which a person does not belong and toward which the person may feel a sense of competitiveness or hostility. Distinguishing between our ingroups and our outgroups

FIGURE 6.1 Sometimes, the distinction between what constitutes an ingroup and an outgroup is subtle. Other times, it is not subtle at all. Would you feel comfortable entering or joining this club?

helps us establish our individual identity and self-worth. Likewise, groups are solidified by ingroup and outgroup distinctions; the presence of an enemy or a hostile group binds members more closely together (Coser, 1956).

Group boundaries may be formal, with clearly defined criteria for membership. For example, a private city club or country club that requires an applicant for membership to be recommended by four current members and to pay a $100,000 initiation fee has clearly set requirements for its members and established "ingroup" and "outgroup" distinctions. Formal group boundaries are reinforced by "invitation-only" policies for membership and the privacy and exclusivity of life inside the club. Many clubs have "Members Only" signs to indicate that the organization does not welcome outsiders within club walls. As a result, club members develop a *consciousness of kind*—the awareness that individuals have when they believe that they share important commonalities with certain other people. Consciousness of kind is strengthened by membership in clubs ranging from country clubs to college sororities, fraternities, and other by-invitation-only college or university social clubs (Kendall, 2008).

In our own lives, most of us are aware that our ingroups provide us with a unique sense of identity. But sometimes we are less aware that they also give us the ability to exclude individuals whom we do not want to be in our inner circle of friends. The early sociologist Max Weber captured this idea in his description of the *closed relationship*—a setting in which the "participation of certain persons is excluded, limited, or subjected to conditions" (Gerth and Mills, 1946: 139). Ingroup and outgroup distinctions may encourage social cohesion among members, but they may also promote classism, racism, sexism, and ageism. Ingroup members typically view themselves positively and members of outgroups negatively. These feelings of group superiority, or *ethnocentrism,* are somewhat inevitable. Some group members may never act on these beliefs of superiority and inferiority because the larger organization of which they are a part actively discourages ethnocentric beliefs and discriminatory actions. However, other organizations may covertly foster ethnocentrism and negative ingroup/outgroup distinctions by denying that these beliefs exist among group members or by failing to take action when misconduct occurs that is rooted in racism, sexism, and/or ageism. An example is a college Greek letter organization in which the fraternity's or sorority's national leadership strongly opposes theme parties with racist or sexist overtones sponsored on local campuses, but its affiliates continue to hold social gatherings with decorations, clothing, music, and slogans that ridicule subordinate-group members such as persons of color, older individuals, persons with a disability, or women who have been turned into sex objects. Although campus social organizations often promote social cohesion among members by making them feel like they are the "in group" and everyone else is the "out group," such beliefs and practices may also promote classism, racism, sexism, and/or ageism.

Reference Groups   Ingroups provide us not only with a source of identity but also with a point of reference. A *reference group* is a group that strongly influences a person's behavior and social attitudes, regardless of whether that individual is an actual member. When we attempt to evaluate our appearance, ideas, or goals, we automatically refer to the standards of some group. Sometimes, we will refer to our membership groups, such as family or friends. Other times, we will rely on groups to which we do not currently belong but that we might wish to join in the future, such as a social club or a profession.

Reference groups help explain why our behavior and attitudes sometimes differ from those of our membership groups. We may accept the values and norms of a group with which we identify rather than one to which we belong. We may also act more like members of a group that we want to join than members of groups to which we already belong. In this case, reference groups are a source of anticipatory socialization. For most of us, our reference-group attachments change many times during our life course, especially when we acquire a new status in a formal organization.

Networks   A *network* is a web of social relationships that links one person with other people and, through them, with other people they know. Frequently, networks connect people who share common interests but who otherwise might not identify and interact with one another. For example, if A is tied to B, and B is tied to C, then a network is more likely to be formed among individuals A, B, and C. Today, the term *networking* is widely used to describe the contacts that people make to find jobs or other opportunities; however, sociologists have studied social networks for many years in an effort to learn more about the linkages between individuals and their group memberships.

What are your networks? For a start, your networks consist of all the people linked to you by primary ties, including your relatives and close friends. Your networks also include your secondary ties, such as acquaintances, classmates, professors, and—if you are employed—your supervisor and coworkers. However, your networks actually extend far beyond these ties to include not only the people that you *know* but also the people that you *know of*—and who know of you—through your primary and secondary ties. In fact, your networks potentially include a pool of

ingroup
a group to which a person belongs and with which the person feels a sense of identity.

outgroup
a group to which a person does not belong and toward which the person may feel a sense of competitiveness or hostility.

reference group
a group that strongly influences a person's behavior and social attitudes, regardless of whether that individual is an actual member.

network
a web of social relationships that links one person with other people and, through them, with other people they know.

FIGURE 6.2 According to the sociologist Georg Simmel, interaction patterns change when a third person joins a dyad—a group composed of two members. How might the conversation between these two women change when another person arrives to talk to them?

between 500 and 2,500 acquaintances if you count the connections of everyone in your networks (Milgram, 1967).

The Purpose of Groups: Multiple Perspectives

What purpose do groups serve? Why are individuals willing to relinquish some of their freedom to participate in groups? According to functionalists, people form groups to meet instrumental and expressive needs. *Instrumental,* or task-oriented, needs cannot always be met by one person, so the group works cooperatively to fulfill a specific goal. Groups help members do jobs that are impossible to do alone or that would be very difficult and time-consuming at best. For example, think of how hard it would be to function as a one-person football team or to single-handedly build a skyscraper. In addition to instrumental needs, groups also help people meet their *expressive,* or emotional, needs, especially those involving self-expression and support from family, friends, and peers.

Although not disputing that groups ideally perform such functions, conflict theorists suggest that groups also involve a series of power relationships whereby the needs of individual members may not be equally served. Symbolic interactionists focus on how the size of a group influences the kind of interaction that takes place among members. To many postmodernists, groups and organizations—like other aspects of postmodern societies—are generally characterized by superficiality and depthlessness in social relationships. For example, fast-food restaurant employees and customers interact in extremely superficial ways that are

largely scripted: The employees follow scripts in taking and filling customers' orders ("Would you like fries and a drink with that?"), and the customers respond with their own "recipied" action. According to the sociologist George Ritzer (1997: 226), "[C]ustomers are mindlessly following what they consider tried-and-true social recipes, either learned or created by them previously, on how to deal with restaurant employees and, more generally, how to work their way through the system associated with the fast-food restaurant." What examples can you think of that fit this description?

 Discuss how a group's size shapes members' communication, leadership styles, and pressures to conform.

Group Characteristics and Dynamics

We will now look at certain characteristics of groups, such as how size affects group dynamics.

Group Size

The size of a group is one of its most important features. Interactions are more personal and intense in a ***small group***, a collectivity small enough for all members to be acquainted with one another and to interact simultaneously.

Sociologist Georg Simmel (1950/1902–1917) suggested that small groups have distinctive interaction patterns

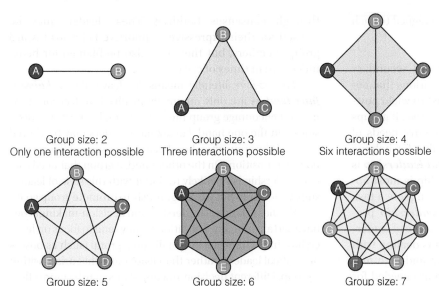

Group size: 2
Only one interaction possible

Group size: 3
Three interactions possible

Group size: 4
Six interactions possible

Group size: 5
Ten interactions possible

Group size: 6
Fifteen interactions possible

Group size: 7
Twenty-one interactions possible

FIGURE 6.3 **Growth of Possible Social Interaction Based on Group Size**

that do not exist in larger groups. According to Simmel, in a *dyad*—a group composed of two members—the active participation of both members is crucial to the group's survival. If one member withdraws from interaction or "quits," the group ceases to exist. Examples of dyads include two people who are best friends, married couples, and domestic partnerships. Dyads provide members with an intense bond and a sense of unity not found in most larger groups.

When a third person is added to a dyad, a *triad*, a group composed of three members, is formed. The nature of the relationship and interaction patterns changes with the addition of the third person (• Figure 6.2). In a triad, even if one member ignores another or declines to participate, the group can still function. In addition, two members may unite to create a coalition that can subject the third member to pressure to conform. A *coalition* is an alliance created in an attempt to reach a shared objective or goal. If two members form a coalition, the other member may be seen as an outsider or intruder.

As the size of a group increases beyond three people, members tend to specialize in different tasks, and everyday communication patterns change. For instance, in groups of more than six or seven people, it becomes increasingly difficult for everyone to take part in the same conversation; therefore, several conversations will probably take place simultaneously. Members are also likely to take sides on issues and form a number of coalitions. In groups of more than ten or twelve people, it becomes virtually impossible for all members to participate in a single conversation unless one person serves as moderator and guides the discussion. As shown in • Figure 6.3, when the size of the group increases, the number of possible social interactions also increases.

Although large groups typically have less social solidarity than small ones, they may have more power. However, the relationship between size and power is more complicated

than it might initially seem. The power relationship depends on both a group's *absolute* size and its *relative* size (Simmel, 1950/1902–1917; Merton, 1968). The absolute size is the number of members the group actually has; the relative size is the number of potential members. For example, suppose that 300 people band together to "march on Washington" and demand enactment of a law on some issue that they strongly believe to be important. Although 300 people is a large number in some contexts, opponents of this group would argue that the low turnout (compared with the number of people in this country) demonstrates that most people don't believe the issue is important. At the same time, the power of a small group to demand change may be based on a "strength in numbers" factor if the group is seen as speaking on behalf of a large number of other people (who are also voters).

Larger groups typically have more-formalized leadership structures, and their leaders are expected to perform a variety of roles, some related to the internal workings of the group and others related to external relationships with other groups.

Group Leadership

What role do leaders play in groups? *Leadership* refers to the ability to influence what goes on in a group or social system. Leaders are responsible for directing plans and activities so that the group completes its task or fulfills its goals. Primary groups generally have informal leadership. For example, most of us do not elect or appoint leaders in our own families. Various family members may assume a leadership role at various times or act as leaders for specific tasks. In traditional families, the father or eldest male is usually the leader. However, in today's more-diverse families, leadership and power are frequently in question, and power relationships may be quite different, as discussed later in this text. By comparison, larger groups typically have more-formalized leadership structures. For example, leadership in secondary groups (such as colleges, governmental agencies, and corporations) involves a clearly defined chain

small group
a collectivity small enough for all members to be acquainted with one another and to interact simultaneously.

dyad
a group composed of two members.

triad
a group composed of three members.

leadership
the ability to influence what goes on in a group or social system.

of command, with written responsibilities assigned to each position in the organizational structure.

Leadership Functions Both primary and secondary groups have some type of leadership or positions that enable certain people to be leaders, or at least to wield power over others. From a functionalist perspective, if groups exist to meet the instrumental and expressive needs of their members, then leaders are responsible for helping the group meet those needs. *Instrumental leadership* is goal or task oriented; this type of leadership is most appropriate when the group's purpose is to complete a task or reach a particular goal. *Expressive leadership* provides emotional support for members; this type of leadership is most appropriate when the group is dealing with emotional issues, and harmony, solidarity, and high morale are needed. Both kinds of leadership are needed for groups to work effectively.

Leadership Styles Three major styles of leadership exist in groups: authoritarian, democratic, and laissez-faire. *Authoritarian leaders* make all major group decisions and assign tasks to members. These leaders focus on the instrumental tasks of the group and demand compliance from others (• Figure 6.4). In times of crisis, such as a war or natural disaster, authoritarian leaders may be commended for their decisive actions. In other situations, however, they may be criticized for being dictatorial and for fostering intergroup hostility. By contrast, *democratic leaders* encourage group discussion and decision making through consensus building. These leaders may be praised for their expressive, supportive behavior toward group members, but they may also be blamed for being indecisive in times of crisis.

Laissez-faire literally means "to leave alone." *Laissez-faire leaders* are only minimally involved in decision making and encourage group members to make their own decisions. On the one hand, laissez-faire leaders may be viewed positively by group members because they do not flaunt their power or position. On the other hand, a group that needs active leadership is not likely to find it with this style of leadership, which does not work vigorously to promote group goals.

Studies of kinds of leadership and decision-making styles have certain inherent limitations. They tend to focus on leadership that is imposed externally on a group (such as bosses or political leaders) rather than leadership that arises within a group. Different decision-making styles may be more effective in one setting than another. For example, imagine attending a college class in which the professor asked the students to determine what should be covered in the course, what the course requirements should be, and how students should be graded. It would be a difficult and cumbersome way to start the semester; students might spend the entire term negotiating these matters and never actually learn anything.

Group Conformity

To what extent do groups exert a powerful influence on our lives? Groups have a significant amount of influence on our values, attitudes, and behavior. In order to gain and then retain our membership in groups, most of us are willing to exhibit a high level of conformity to the wishes of other group members. *Conformity* is the process of maintaining or changing behavior to comply with the norms established by a society, subculture, or other group. We often experience powerful pressure from other group members to conform. In some situations, this pressure may be almost overwhelming.

Researchers have found that the pressure to conform may cause group members to say they see something that is contradictory to what they are actually seeing or to do something that they would otherwise be unwilling to do. Conforming to group pressure begins as early as preschool

FIGURE 6.4 Organizations have different leadership styles based on the purpose of the group. How do leadership styles in the military differ from those on college and university campuses and in office workplaces, for example?

Robert Nickelsberg/Getty Images

age (Haun and Tomasello, 2011). As we look at two classic studies on group conformity (which would be impossible to conduct today for ethical reasons), ask yourself what you might have done if you had been involved in this research.

Asch's Research Pressure to conform is especially strong in small groups in which members want to fit in with the group. In a series of experiments conducted by Solomon Asch (1955, 1956), the pressure toward group conformity was so great that participants were willing to contradict their own best judgment if the rest of the group disagreed with them.

One of Asch's experiments involved groups of undergraduate men (seven in each group) who were allegedly recruited for a study of visual perception. All the men were seated in chairs. However, the person in the sixth chair did not know that he was the only actual subject; all the others were assisting the researcher. The participants were first shown a large card with a vertical line on it and then a second card with three vertical lines (see ● Figure 6.5). Each of the seven participants was asked to indicate which of the three lines on the second card was identical in length to the "standard line" on the first card.

In the first test with each group, all seven men selected the correct matching line. In the second trial, all seven still answered correctly. In the third trial, however, the actual subject became very uncomfortable when all the others selected the incorrect line. The subject could not understand what was happening and became even more confused as the others continued to give incorrect responses on eleven out of the next fifteen trials.

Asch (1955) found that about one-third of all subjects chose to conform by giving the same (incorrect) responses as Asch's assistants. In discussing the experiment afterward, most of the subjects who gave incorrect responses indicated that they had known the answers were wrong but decided to go along with the group in order to avoid ridicule or ostracism.

Asch concluded that the size of the group and the degree of social cohesion felt by participants were important influences on the extent to which individuals respond to group pressure. If you had been in the position of the subject, how would you have responded? Would you have continued to give the correct answer, or would you have been swayed by the others?

Milgram's Research How willing are we to do something because someone in a position of authority has told us to do it? How far are we willing to go to follow the demands of that individual? Stanley Milgram (1963, 1974) conducted a series of controversial experiments to find answers to these questions about people's obedience to authority. *Obedience* is a form of compliance in which people follow direct orders from someone in a position of authority.

Milgram's subjects were men who had responded to an advertisement seeking individuals to participate in an experiment. When the first (actual) subject arrived, he was told that the study concerned the effects of punishment on learning. After the second subject (an assistant of Milgram's) arrived, the two men were instructed to draw slips of paper from a hat to get their assignments as either the "teacher" or the "learner." Because the drawing was rigged, the actual subject always became the teacher, and the assistant the learner. Next, the learner was strapped into a chair with protruding electrodes that looked something like an electric chair. The teacher was placed in an adjoining room and given a realistic-looking but nonoperative shock generator. The "generator's" control panel showed levels that went from "Slight shock" (15 volts) on the left, to "Intense shock" (255 volts) in the middle, to "Danger: severe shock" (375 volts), and finally "XXX" (450 volts) on the right.

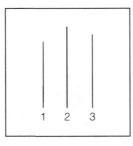

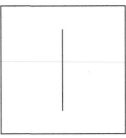

FIGURE 6.5 Asch's Cards
Although Line 2 is clearly the same length as the line in the lower card, Solomon Asch's research assistants tried to influence "actual" participants by deliberately picking Line 1 or Line 3 as the correct match. Many of the participants went along rather than risking the opposition of the "group."
Source: Asch, 1955.

instrumental leadership
goal- or task-oriented leadership.

expressive leadership
leadership that provides emotional support for members.

authoritarian leaders
leaders who make all major group decisions and assign tasks to members.

democratic leaders
leaders who encourage group discussion and decision making through consensus building.

laissez-faire leaders
leaders who are only minimally involved in decision making and who encourage group members to make their own decisions.

conformity
the process of maintaining or changing behavior to comply with the norms established by a society, subculture, or other group.

The teacher was instructed to read aloud a pair of words and then repeat the first of the two words. At that time, the learner was supposed to respond with the second of the two words. If the learner could not provide the second word, the teacher was instructed to press the lever on the shock generator so that the learner would be punished for forgetting the word. Each time the learner gave an incorrect response, the teacher was supposed to increase the shock level by 15 volts. The alleged purpose of the shock was to determine whether punishment improves a person's memory.

What was the maximum level of shock that a "teacher" was willing to inflict on a "learner"? The learner had been instructed (in advance) to beat on the wall between him and the teacher as the experiment continued, pretending that he was in intense pain. The teacher was told that the shocks might be "extremely painful" but that they would cause no permanent damage. At about 300 volts, when the learner quit responding at all to questions, the teacher often turned to the experimenter to see what he should do next. When the experimenter indicated that the teacher should give increasingly painful shocks, 65 percent of the teachers administered shocks all the way up to the "XXX" (450-volt) level (see • Figure 6.6). By this point in the process, the teachers were frequently sweating, stuttering, or biting on their lip. According to Milgram, the teachers (who were free to leave whenever they wanted to) continued in the experiment because they were being given directions by a person in a position of authority (a university scientist wearing a white coat).

What can we learn from Milgram's study? The study provides evidence that obedience to authority may be more common than most of us would like to believe. None of the "teachers" challenged the process before they had applied 300 volts. Almost two-thirds went all the way to what could have been a deadly jolt of electricity if the shock generator had been real. For many years, Milgram's findings were found to be consistent in a number of different settings and with variations in the research design (Miller, 1986).

This research once again raises some questions concerning research ethics. As was true of Asch's research, Milgram's subjects were deceived about the nature of the study in which they were asked to participate. Many of them found the experiment extremely stressful. Such conditions cannot be ignored by social scientists because subjects may receive lasting emotional scars from this kind of research. Today, it would be virtually impossible to obtain permission to replicate this experiment in a university setting.

 LO4 **Applying** the concept of groupthink, describe how people often respond differently in a group context than they might if they were alone.

Groupthink

As we have seen, individuals often respond differently in a group context than they might if they were alone. Social psychologist Irving Janis (1972, 1989) examined group decision making among political experts and found that major blunders in U.S. history can be attributed to pressure toward group conformity. To describe this phenomenon, he coined the term *groupthink*—the process by which members of a cohesive group arrive at a decision that many individual members privately believe is unwise. Why not speak up at the time? Members usually want to be "team players." They may

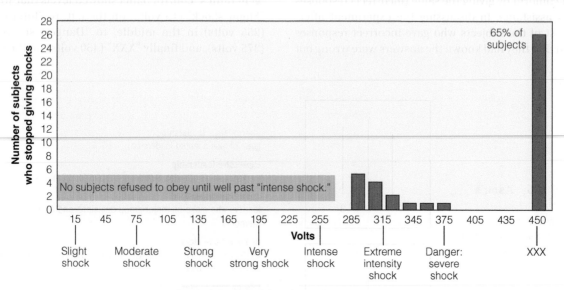

FIGURE 6.6 **Results of Milgram's Obedience Experiment**
Even Milgram was surprised by subjects' willingness to administer what they thought were severely painful and even dangerous shocks to a helpless "learner."
Source: Milgram, 1963.

Process of Groupthink	Example: Deepwater Horizon Explosion
PRIOR CONDITIONS Isolated, cohesive, homogeneous decision-making group Lack of impartial leadership High stress	Millions of dollars had been spent on production of mobile offshore drilling unit; BP was running behind schedule and was under pressure to complete work despite reports of a leak in the rig's blowout preventer.
SYMPTOMS OF GROUPTHINK Closed-mindedness Rationalization Squelching of dissent "Mindguards" Feelings of righteousness and invulnerability	Although rig workers reported pieces of rubber seal coming loose, superiors stated this happened often. Superiors closed off debate by saying blowout-preventer problem would be resolved if something went wrong.
DEFECTIVE DECISION MAKING Incomplete examination of alternatives Failure to examine risks and contingencies Incomplete search for information	In March 2010, when rig workers informed superiors of actual leaks in gasket on blowout preventer on rig, no decision was made to repair the rubber seal or to stop work. Superiors began to hide bad news, and decisions were made without a clear sense of what risks were involved.
CONSEQUENCES Poor decisions	No one stopped production. The BP oil rig exploded, killing 11 and injuring 17. The spill was the largest of its kind in the history of the petroleum industry. BP, Transocean, and others were blamed for making a series of bad decisions based on money, time pressures, and too many people thinking alike.

FIGURE 6.7 Janis's Description of Groupthink

In Janis's model, prior conditions such as a highly homogeneous group with committed leadership can lead to potentially disastrous "groupthink," which short-circuits careful and impartial deliberation. Events leading up to the tragic 2010 explosion of the BP oil rig have been cited as an example of this process.

Source: Mackin, 2010.

not want to be the ones who undermine the group's consensus or who challenge the group's leaders. Consequently, members often limit or withhold their opinions and focus on consensus rather than on exploring all of the options and determining the best course of action. • Figure 6.7 summarizes the dynamics and results of groupthink.

The tragic 2010 explosion of the BP Deepwater Horizon oil rig, owned by British Petroleum and located in the Gulf of Mexico, is an example of this process. Errors in decision making contributed to one of the worst oil spills and marine and wildlife disasters in U.S. history. Eleven people were killed and seventeen were injured in the rig explosion, and it is impossible to estimate the full extent of the damage done to the Gulf Coast and the fishing and tourism industries because of this massive accident. Why is this disaster an example of groupthink? Because officials for BP, Transocean, and Halliburton, the major transnational corporations responsible for this error in decision making, closed off their discussions about safety and hid bad news from one another and public officials; because they began to think alike in their assumption about safety, namely that a blow-out preventer would keep such a massive disaster from occurring; and because their companies were already behind schedule, had put millions of dollars into production, and did not want to stop to check out reports that a rubber safety seal was broken.

groupthink
the process by which members of a cohesive group arrive at a decision that many individual members privately believe is unwise.

AP Images/Anonymous/US Coast Guard

Identify the three categories of formal organizations and state how they differ in membership.

Formal Organizations in Global Perspective

Over the past century, the number of formal organizations has increased dramatically in the United States and other industrialized nations. Previously, everyday life was centered in small, informal, primary groups, such as the family and the village. With the advent of industrialization and urbanization (as discussed in Chapter 1), people's lives became increasingly dominated by large, formal, secondary organizations. A *formal organization,* you will recall, is a highly structured secondary group formed for the purpose of achieving specific goals in the most efficient manner. Formal organizations (such as corporations, schools, and government agencies) usually keep their basic structure for many years in order to meet their specific goals.

Types of Formal Organizations

We join some organizations voluntarily and others out of necessity. Sociologist Amitai Etzioni (1975) classified formal organizations into three categories—normative, coercive, and utilitarian—based on the nature of membership in each.

Normative Organizations We voluntarily join *normative organizations* when we want to pursue some common interest or gain personal satisfaction or prestige from being a member (• Figure 6.8). Political parties, ecological activist groups, religious organizations, parent–teacher associations, and college sororities and fraternities are examples of normative, or voluntary, associations.

Class, gender, and race are important determinants of a person's participation in a normative association. Class (socioeconomic status based on a person's education, occupation, and income) is the most significant predictor of whether a person will participate in mainstream normative organizations; membership costs may exclude some from joining. Those with higher socioeconomic status are more likely to be not only members but also active participants in these groups. Gender is also an important determinant. Historically, all-male voluntary organizations have had a higher level of prestige than many women's organizations. In the twenty-first century, some of these patterns have changed.

Throughout history, people of all racial–ethnic categories have participated in voluntary organizations to bring about racial equality and social justice. Women have often taken leadership roles in these movements. African American women were actively involved in antislavery societies in the nineteenth century and in the civil rights movement

FIGURE 6.8 Normative organizations such as the Red Cross rely on volunteers to fulfill their goals.

Characteristics of Groups and Organizations

Types of Social Groups	Primary group	Small, less specialized group in which members engage in face-to-face, emotion-based interaction over an extended period of time
	Secondary group	Larger, more specialized group in which members engage in more-impersonal, goal-oriented relationships for a limited period of time
	Ingroup	A group to which a person belongs and with which the person feels a sense of identity
	Outgroup	A group to which a person does not belong and toward which the person may feel a sense of competitiveness or hostility
	Reference group	A group that strongly influences a person's behavior and social attitudes, regardless of whether the person is actually a member
Group Size	Dyad	A group composed of two members
	Triad	A group composed of three members
	Formal organization	A highly structured secondary group formed for the purpose of achieving specific goals
Types of Formal Organizations	Normative	Organizations that we join voluntarily to pursue some common interest or gain personal satisfaction or prestige by joining
	Coercive	Associations that people are forced to join (total institutions such as boot camps and prisons are examples)
	Utilitarian	Organizations that we join voluntarily when they can provide us with a material reward that we seek

in the twentieth century. Similarly, Native American women participated in the American Indian Movement, a group organized to fight problems ranging from police brutality to housing and employment discrimination. Mexican American women have held a wide range of leadership positions in La Raza Unida Party and the League of United Latin American Citizens, organizations oriented toward civic activities and protests against injustices.

Coercive Organizations   People do not voluntarily become members of *coercive organizations*—associations that people are forced to join. Total institutions, such as boot camps, prisons, and some mental hospitals, are examples of coercive organizations. As discussed in Chapter 4, the assumed goal of total institutions is to resocialize people through incarceration. These environments are characterized by restrictive barriers (such as locks, bars, and security guards) that make it impossible for people to leave freely. When people leave without being officially dismissed, their exit is referred to as an "escape."

Utilitarian Organizations   We voluntarily join *utilitarian organizations* when they can provide us with a material reward that we seek. To make a living or earn a college degree, we must participate in organizations that can provide us these opportunities. Although we have some choice regarding where we work or attend school, utilitarian organizations are not always completely voluntary. For example,

most people must continue to work even if the conditions of their employment are less than ideal. (This chapter's Concept Quick Review summarizes the types of groups, sizes of groups, and types of formal organizations.)

 Debate the strengths and weaknesses of bureaucracies in contemporary nations such as the United States.

Bureaucracies

The bureaucratic model of organization remains the most universal organizational form in government, business, education, and religion. A **bureaucracy** is an organizational model characterized by a hierarchy of authority, a clear division of labor, explicit rules and procedures, and impersonality in personnel matters.

Sociologist Max Weber (1968/1922) was interested in the historical trend toward bureaucratization that accelerated during the Industrial Revolution. To Weber, bureaucracy was the most "rational" and efficient means of attaining organizational goals because it contributed to coordination and control. According to Weber,

bureaucracy
an organizational model characterized by a hierarchy of authority, a clear division of labor, explicit rules and procedures, and impersonality in personnel matters.

Characteristics

Effects

Characteristics	Effects
• Division of labor	• Inefficiency and rigidity
• Hierarchy of authority	• Resistance to change
• Rules and regulations	• Perpetuation of race, class, and
• Qualification-based employment	gender inequalities
• Impersonality	

FIGURE 6.9 **Characteristics and Effects of Bureaucracy**
The very characteristics that define Weber's idealized bureaucracy can create or exacerbate the problems that many people associate with this type of organization. Can you apply this model to an organization with which you are familiar?

rationality is the process by which traditional methods of social organization, characterized by informality and spontaneity, are gradually replaced by efficiently administered formal rules and procedures. Bureaucracy can be seen in all aspects of our lives, from small colleges with perhaps a thousand students to multinational corporations employing many thousands of workers worldwide.

In his study of bureaucracies, Weber relied on an ideal-type analysis, which he adapted from the field of economics. An *ideal type* is an abstract model that describes the recurring characteristics of some phenomenon (such as bureaucracy). To develop this ideal type, Weber abstracted the most characteristic bureaucratic aspects of religious, educational, political, and business organizations. Weber acknowledged that no existing organization would exactly fit his ideal type of bureaucracy.

Ideal Characteristics of Bureaucracy Weber set forth several ideal-type characteristics of bureaucratic organizations. His model (see • Figure 6.9) highlights the organizational efficiency and productivity that bureaucracies strive for in these five central elements of the ideal organization:

• *Division of labor.* Bureaucratic organizations are characterized by specialization, and each member has highly specialized tasks to fulfill.

• *Hierarchy of authority.* In a bureaucracy, each lower office is under the control and supervision of a higher one. Those few individuals at the top of the hierarchy have more power and exercise more control than do the many at the lower levels. Those who are lower in the hierarchy report to (and often take orders from) those above them in the organizational pyramid. Persons at the upper levels are responsible not only for their own actions but also for those of the individuals they supervise.

• *Rules and regulations.* Rules and regulations establish authority within an organization. These rules are typically standardized and provided to members

in a written format. In theory, written rules and regulations offer clear-cut standards for determining satisfactory performance so that each new member does not have to reinvent the rules (• Figure 6.10).

• *Qualification-based employment.* Bureaucracies require competence and hire staff members and professional employees based on specific qualifications. Individual performance is evaluated against specific standards, and promotions are based on merit as spelled out in personnel policies.

• *Impersonality.* Bureaucracies require that everyone must play by the same rules and be treated the same.

FIGURE 6.10 Colleges and universities rely a great deal on the use of standardized tests to evaluate students. How do such tests relate to Weber's model of bureaucracy?

Personal feelings should not interfere with organizational decisions.

Contemporary Applications of Weber's Theory

How well do Weber's theory of rationality and his ideal-type characteristics of bureaucracy withstand the test of time? More than 100 years later, many organizational theorists still apply Weber's perspective. For example, the sociologist George Ritzer used Weber's theories to examine fast-food restaurants such as McDonald's. According to Ritzer, the process of "McDonaldization" has become a global phenomenon that can be seen in fast-food restaurants and other "speedy" or "jiffy" businesses (such as Sir Speedy Printing and Jiffy Lube). *McDonaldization* is the term coined by Ritzer to describe the process of rationalization, which takes a task and breaks it down into smaller tasks. This process is repeated until all tasks have been broken down into the smallest possible level. The resulting tasks are then rationalized to find the single most efficient method for completing each task. The result is an efficient, logical sequence of methods that can be completed the same way every time to produce the desired outcome. Ritzer (2014) identifies four dimensions of formal rationality (McDonaldization) found in fast-food restaurants:

- *Efficiency* means the search for the best means to the end; the drive-through window is a good example of heightening the efficiency of obtaining a meal.

- *Predictability* means a world of no surprises; the Big Mac in Los Angeles is indistinguishable from the one in New York. Similarly, the one we consume tomorrow or next year will be just like the one we eat today.

- *Emphasis on quantity rather than quality*. The Big Mac is a good example of this emphasis on quantity rather than quality.

- *Control through nonhuman technologies* such as unskilled cooks following detailed directions and assembly-line methods applied to the cooking and serving of food.

Finally, such a formally rational system brings with it various irrationalities, most notably the dehumanization of the dining experience. For example, people in a McDonaldized world may

FIGURE 6.11   How do people use this informal "grapevine" to spread information? Is it faster than the organization's official channels of communication? Is it more or less accurate than official channels?

become more enthusiastic about quickly purchasing extremely large portions of relatively inexpensive foods than with having a "slow" dining experience where people bond over the experience of cooking and eating a more nutritious (not mass-produced) meal with their friends or family. How applicable are some of Weber's ideas today? While still useful, Weber's ideal type largely failed to take into account the informal side of bureaucracy.

The Informal Side of Bureaucracy   When we look at an organizational chart, the official, formal structure of a bureaucracy is readily apparent. In practice, however, a bureaucracy has patterns of activities and interactions that cannot be accounted for by its organizational chart. These have been referred to as *bureaucracy's other face* (Page, 1946).

The *informal side of a bureaucracy* is composed of those aspects of participants' day-to-day activities and interactions that ignore, bypass, or do not correspond with the official rules and procedures of the bureaucracy. An example is an informal "grapevine" that spreads information (with varying degrees of accuracy) much faster than do official channels of communication, which tend to be slow and unresponsive (• Figure 6.11). The informal structure has also been referred to as *work culture* because it includes the ideology and practices of workers on the job. Workers create this work culture in order to confront, resist, or adapt to the constraints of their jobs, as well as to guide and interpret social relations on the job. Today, computers, smartphones,

rationality
the process by which traditional methods of social organization, characterized by informality and spontaneity, are gradually replaced by efficiently administered formal rules and procedures.

ideal type
an abstract model that describes the recurring characteristics of some phenomenon.

informal side of a bureaucracy
those aspects of participants' day-to-day activities and interactions that ignore, bypass, or do not correspond with the official rules and procedures of the bureaucracy.

and tablets offer additional opportunities for workers to enhance or degrade their work culture. Some organizations have sought to control offensive communications so that workers will not be exposed to a hostile work environment brought about by colleagues, but such control has raised significant privacy issues (see "Sociology and Social Policy").

Is the informal side of bureaucracy good or bad? Should it be controlled or encouraged? Two schools of thought have emerged with regard to these questions. One approach emphasizes control of informal groups in order to ensure greater worker productivity. By contrast, the other school of thought asserts that informal groups should be nurtured because such networks may serve as a means of communication and cohesion among individuals. Large organizations would be unable to function without strong informal norms and relations among participants.

Informal networks thrive in contemporary organizations because people can communicate with one another continually without ever having to engage in face-to-face interaction. The need to meet at the water fountain or the copy machine in order to exchange information is long gone: Workers now have an opportunity to tell one another—and higher-ups, as well—what they think.

Problems of Bureaucracies

The characteristics that make up Weber's "rational" model of bureaucracy have a dark side that has frequently given this type of organization a bad name. Three of the major problems of bureaucracies are (1) inefficiency and rigidity, (2) resistance to change, and (3) perpetuation of race, class, and gender inequalities.

Inefficiency and Rigidity

Bureaucracies experience inefficiency and rigidity at both the upper and lower levels of the organization. The self-protective behavior of officials at the top may render the organization inefficient. One type of self-protective behavior is the monopolization of information in order to maintain control over subordinates and outsiders. Information is a valuable commodity in organizations, and those persons in positions of authority guard information because it is a source of power for them— others cannot "second-guess" their decisions without access to relevant (and often "confidential") information.

When those at the top tend to use their power and authority to monopolize information, they also fail to communicate with workers at the lower levels. As a result, they are often unaware of potential problems facing the organization and of high levels of worker frustration. Bureaucratic regulations are written in far greater detail than is necessary in order to ensure that almost all conceivable situations are covered. *Goal displacement* occurs when the rules become an end in themselves rather than a means to an end, and organizational survival becomes more important than achievement of goals (Merton, 1968).

Inefficiency and rigidity occur at the lower levels of the organization as well. Workers often engage in *ritualism*; that is, they become most concerned with "going through the

motions" and "following the rules." According to Robert Merton (1968), the term *bureaucratic personality* describes those workers who are more concerned with following correct procedures than they are with getting the job done correctly. Such workers are usually able to handle routine situations effectively but are frequently incapable of handling a unique problem or an emergency. Thorstein Veblen (1967/1899) used the term *trained incapacity* to characterize situations in which workers have become so highly specialized or have been given such fragmented jobs to do that they are unable to come up with creative solutions to problems. Workers who have reached this point also tend to experience bureaucratic alienation— they really do not care what is happening around them.

Resistance to Change

Once bureaucratic organizations are created, they tend to resist change. This resistance not only makes bureaucracies virtually impossible to eliminate but also contributes to bureaucratic enlargement. Because of the assumed relationship between size and importance, officials tend to press for larger budgets and more staff and office space. To justify growth, administrators and managers must come up with more tasks for workers to perform.

Resistance to change may also lead to incompetence. Based on organizational policy, bureaucracies tend to promote people from within the organization. As a consequence, a person who performs satisfactorily in one position is promoted to a higher level in the organization. Eventually, people reach a level that is beyond their knowledge, experience, and capabilities.

Perpetuation of Race, Class, and Gender Inequalities

Some bureaucracies perpetuate inequalities of race, class, and gender because this form of organizational structure creates a specific type of work or learning environment. This structure was typically created for middle-class and upper-middle-class white men, who for many years were the predominant organizational participants.

For people of color, *entry* into a dominant white bureaucratic organization does not equal actual *integration*. Instead, many have experienced an internal conflict between the bureaucratic ideals of equal opportunity and fairness and the prevailing norms of discrimination and hostility that exist in some organizations. Research has found that people of color are more adversely affected than dominant-group members by hierarchical bureaucratic structures and exclusion from informal networks.

Like racial inequality, social-class divisions may be perpetuated in bureaucracies. The theory of a "dual labor market" has been developed to explain how social-class distinctions are perpetuated through different types of employment. Middle- and upper-middle-class employees are more likely to have careers characterized by higher wages, more job security, and opportunities for advancement. By contrast, poor and working-class employees work in occupations characterized by low wages, lack of job security, and few opportunities for promotion. The "dual economy" not only reflects but may also perpetuate people's current class position. Conflict theorists point out that persons in the lowest-wage and

SOCIOLOGY & SOCIAL POLICY

Technological and Social Change in the Workplace: BYOD?

What if I take my own smartphone or tablet to the office so that I can do both company and personal work on it? Can my employer demand to see what's on my mobile device? Can the IT people wipe it clean or take the device from me if I am no longer employed there?

—frequently asked questions by persons who are employed at companies with BYOD ("bring your own device") policies

Do employers really have the right to monitor everything that their employees do at work? Are company-owned computers different from worker-owned digital devices that employees bring to work and use for business and personal activities? Generally speaking, the majority of U.S. companies monitor employee use of company-owned computers and other electronic devices. These employers assert that they have the right to engage in surveillance because it may be necessary for their own protection. Employers state that they own the computer network and the monitors, pay for the Internet service, and pay the employee to spend time on company business. Unchecked Internet activity can expose a company's network and systems to malware and other intrusions that the company otherwise might not encounter. As a result, many employers take the position that First Amendment (privacy) rights are left at the office door when a person agrees to work for a private employer. In most instances, courts have upheld monitoring of employees by employers (see, for example, *Bourke v. Nissan, Smyth v. Pillsbury,* and *Shoars v. Epson).*

But what about situations in which employees bring their own mobile devices so that they can work virtually anywhere?

Do you think that employers should have the right to monitor everything that their employees do on company-owned computers?

What if employees use these devices for both personal and work purposes? Loss of the devices or loss of information stored on the devices may pose a significant security risk for the organization, and more companies have established policies that provide them with access to employees' mobile devices if they are used at work. New laws are emerging as more employees use their own mobile devices; however, laws typically vary from state to state and are not necessarily the same for all employers. Rules for government employees may differ from those for private companies, and highly regulated industries such as health care and finance may have more-stringent rules. As well, employees are encouraged to think about whether they actually want to use their own personal electronic devices at work if doing so would grant other people access to their personal lives through their e-mail, photos, social media sites, and other "private" information posted online.

Sociologically speaking, what is the bottom line here? New technologies necessitate change in social policy and law to address issues such as the meaning of privacy in the workplace. There are valid arguments for surveillance to create adequate security, but there are also valid arguments against invasion of privacy. Employees should have a reasonable expectation of privacy—a reasonable belief that neither fellow workers nor employers are prying into their private lives. Organizations should make employees and others aware of surveillance policies, and this is what *endpoint security* in businesses suggests: Be forthright with people, and let everyone know what is being tracked and why. Use the employee handbook or an orientation session to inform employees that their behavior may be monitored under certain circumstances, and let them know what those circumstances are. (For additional information, visit the Privacy Rights Clearinghouse website.) What do you think about this? Would you prefer to keep your own mobile devices separate from the workplace, or do you think it would be more convenient to have everything available on one portable device?

REFLECT & ANALYZE

Are you concerned about privacy in your own life? Should businesses and colleges have the right to monitor our digital communications? If so, how should they go about this process?

---

goal displacement
a process that occurs in organizations when the rules become an end in themselves rather than a means to an end, and organizational survival becomes more important than achievement of goals.

bureaucratic personality
a psychological construct that describes those workers who are more concerned with following correct procedures than they are with getting the job done correctly.

Jack Kurtz/The Image Works

FIGURE 6.12 According to conflict theorists, members of the capitalist class benefit from the work of laborers such as the people shown here, who are harvesting onions on a farm in the Texas Rio Grande Valley. How do low wages and lack of job security contribute to class-based inequalities in the United States?

highest-potential-for-injury jobs, such as agricultural harvesters and other seasonal laborers, are among the workers most harmed by the presence of a dual economy and its role in perpetuating race-, gender-, and class-based inequalities in the United States and other nations (● Figure 6.12).

Gender inequalities are also perpetuated in bureaucracies. Women in traditionally male organizations may feel more visible and experience greater performance pressure. They may also find it harder to gain credibility in management positions.

Inequality in organizations has many consequences. People who lack opportunities for integration and advancement tend to be pessimistic and to have lower self-esteem. Believing that they have few opportunities, they may resign themselves to staying put and surviving at that level. By contrast, those who enjoy full access to organizational opportunities tend to have high aspirations and high self-esteem. They often feel loyalty to the organization and typically see their job as a means for mobility and growth.

LO7 **Define** the iron law of oligarchy and apply the concept to a brief analysis of the U.S. government.

Bureaucracy and Oligarchy

Why do a small number of leaders at the top make all the important organizational decisions? According to the German political sociologist Robert Michels (1949/1911), all organizations encounter the *iron law of oligarchy*—the tendency to

become a bureaucracy ruled by the few. His central idea was that those who control bureaucracies not only wield power but also have an interest in retaining their power. For example, formal and informal political party leaders often do not want to relinquish their control over the party because they are able to influence who runs for public office and how campaigns are conducted. Officials elected to Congress frequently choose to serve multiple terms in office because it provides them with the opportunity to become more involved not only in service to their country but also in bureaucratic power. Some members of Congress have served more than half a century as elected officials (Manning, 2011).

Michels found that the hierarchical structures of bureaucracies and oligarchies go hand in hand. On the one hand, power may be concentrated in a few people because rank-and-file members must inevitably delegate a certain amount of decision-making authority to their leaders. Leaders have access to information that other members do not have, and they have "clout," which they may use to protect their own interests. On the other hand, oligarchy may result when individuals have certain outstanding qualities that make it possible for them to manage, if not control, others. The members choose to look to their leaders for direction; the leaders are strongly motivated to maintain the power and privileges that go with their leadership positions.

Are there limits to the iron law of oligarchy? The leaders in most organizations do not have unlimited power. Divergent groups within a large-scale organization often compete for power, and informal networks can be used to "go behind the backs" of leaders. In addition, members routinely challenge leaders' decisions, and sometimes they (or the organization's governing board) can remove leaders when they are not pleased with the leaders' actions.

LO8 **Identify** alternative forms of organization that exist today in nations such as Japan.

Alternative Forms of Organization

Many organizations have sought new and innovative ways to organize work more efficiently than the traditional hierarchical model.

Humanizing Bureaucracy

In the early 1980s there was a movement in the United States to *humanize bureaucracy*—to establish an organizational environment that develops rather than impedes human resources. More-humane bureaucracies are characterized by (1) less rigid hierarchical structures and greater sharing of power and responsibility by all participants, (2) encouragement of participants to share their ideas and try new approaches to problem solving, and (3) efforts to reduce the number of people in dead-end jobs, train people in needed skills and competencies, and help people meet outside family responsibilities while still receiving equal treatment inside the organization (Kanter, 1983, 1985, 1993/1977). However, this movement has been overshadowed by globalization and the perceived strengths of systems of organizing work in other nations, such as Japan.

Organizational Structure in Japan, Russia, and India

For several decades the Japanese model of organization was widely praised for its innovative structure because it focused on lifetime employment and company loyalty. Although the practice of lifetime employment has largely been replaced by the concept of long-term employment, many workers in Japan have higher levels of job security than do U.S. workers. According to advocates of the Japanese system, this model encourages worker loyalty and a high level of productivity. Managers move through various parts of the organization and acquire technical knowledge about the workings of many aspects of the corporation, unlike their U.S. counterparts, who tend to become highly specialized. Unlike top managers in the United States who have given themselves pay raises and bonuses even when their companies were financially strapped, many Japanese managers have taken pay cuts under similar circumstances. Japanese management is characterized as being people oriented, taking a long-term view, and having a culture that focuses on *how* work gets done rather than on the result alone.

In the twenty-first century the Japanese organization is often based on a management style where information flows from the bottom to the top. As a result, senior managers serve in a supervisory capacity, rather than taking a "hands-on" approach, and policies usually originate at middle organizational levels and then are passed upward for senior managers' approval. According to analysts, this approach is beneficial because the same persons responsible for implementing policies are the ones who have an active role in initially shaping the rules, policies, and procedures (Bizshifts-Trends, 2011).

In the Japanese organization, managers are expected to be "father figures" and create an environment in which groups can succeed and goals can be met (• Figure 6.13). Effective leadership is not based on individual personality or a dictatorial manner, and there is disapproval for those who appear to be overly ambitious.

Unlike Japanese organizational structure and management style, organizations in Russia and India are more likely to be hierarchical, centralized, and highly directive. Most organizations also have a "top-down" approach in which chief executives or the highest leaders issue orders for subordinates to follow, and very little consultation takes place with persons in the lower sectors. Leaders who allow too much participation in organizational decision making are often viewed as weak and indecisive. However, middle managers who have privileged access to top elites often become more powerful managers than managers who lack such access. Looking specifically at India, many organizations are family-owned businesses that are tightly controlled across generations; however, there are indications that Western management styles have become more prevalent in that nation as the children and grandchildren of company founders increasingly have been educated in universities in the United States or other high-income nations.

What can we learn by examining alternative organizational structures in other countries? We can see that all organizations are not established on the same premises about how leadership should operate and how decisions should be made. We can also see that different types of leadership affect how organizations will go about their tasks. Some leadership styles are more democratic (managers delegate authority to subordinates in the decision-making process), others are more autocratic (decisions are made solely by those at the top of the hierarchy), and others are more participatory. Finally, we can see that cultural differences do have an important effect on how organizations operate and how leaders think and act (Bizshift-Trends, 2011).

Looking Ahead: Social Change and Organizations in the Future

What is the best organizational structure for the future? Of course, this question is difficult to answer because it requires the ability to predict economic, political, and social conditions. Nevertheless, we can make several observations.

Socially Sustainable Organizations

First, organizations have been affected by growing social inequality in the United States and other nations because of heightening differences between high- and low-income segments of populations. Having *socially sustainable organizations* is of increasing importance because television, the Internet, and international travel have made people more aware of the wide disparities in the resources and power of "haves" and "have-nots" both within a single country and across nations.

iron law of oligarchy
according to Robert Michels, the tendency of bureaucracies to be ruled by a few people.

FIGURE 6.13 The Japanese model of organization—including planned group-exercise sessions for employees—has become a part of the workplace in many nations. Would it be a positive change if more workplace settings, such as the one shown here, were viewed as an extension of the family? Why or why not?

The term *socially sustainable organizations* is used here to refer to those organizations that take into account the social effects of organizational activities on workers and other persons in the community, the nation, and sometimes the world. Researchers have shown how organizations interact with their physical environments and may produce problems such as pollution and environmental degradation. But the focus of the socially sustainable organizational approach is more on the human and social environment and what organizations can do to sustain and sometimes enhance those aspects of the environment that are not strictly physical or biological. As a result of emphasizing the social sustainability factor, organizations must be developed that are both economically efficient and as equitable as possible.

Some organizational and management analysts suggest that more attention must be paid to the "stakeholders" of an organization. Stakeholder theory is based on the assumption that organizations and their managers must focus on morals and values in goal-setting and decision-making processes. For example, at a college or university, stakeholders would include (but not be limited to)

students, faculty and staff, administrators, alumni, major contributors, boards of regents, suppliers, the community where the school is located, and the society as a whole. The management structure and the morals and values of the institution should reflect the interests of those constituent groups. The goals of the organization should be based on taking into account the interests of these various stakeholders and working toward organizational goals and outcomes that will not only ensure organizational success but also provide the greatest good for the greatest number of stakeholders. Although academic success, winning sports teams, and college financial stability are important in higher education, other criteria should also be used in assessing the effectiveness and overall output of the college community. In other types of organizations, similar stakeholders can be identified and goals established to meet the needs of various constituencies.

Globalization, Technology, and "Smart Working"

Second, *globalization* is the key word for management and change in many organizations, and the use of technology

YOU CAN MAKE A DIFFERENCE

Can Facebook, Twitter, and Other Social Media Make You a Better, More Helpful Person?

"We are the service generation!!!" @BEXwithanX tweeted. And @sjtetreault picked this quote from the first lady to share: "'You didn't think I'd show up here without another challenge, did you? Be yourself, just take it global.' Michelle Obama."

—reactions from two students who (along with about 25,000 other people) listened to First Lady Michelle Obama's commencement address at George Washington University (Johnson, 2010)

Although it is not unusual for political leaders and their spouses to be keynote speakers at university commencements, Michelle Obama's address at George Washington University (GWU) was unique in that it was her payoff in a bet in which she challenged students to do 100,000 hours of community service in exchange for a graduation speech. GWU students easily met the deadline because of social networking and students, such as Christine French, who were highly motivated not only to reach but also to surpass the goal. According to VolunteerMatch.org (2010), a website that links volunteers with community service opportunities, "You can't really major in volunteering, but if you could your schedule might look a lot like Christine French's." In addition to leading the charge to complete the 100,000 service hours by graduation day, Christine was president of the Human Service Student Organization and the Teach for America chapter at GWU.

What unique factors contribute to the success of college volunteers as they make a difference in people's lives? Christine French believes that this is the secret: "I think it's that I listen to people. Often, all people really need is someone to listen to them and validate their feelings. We all just want human connection and to know that we are loved and valuable. This is what I can do for others, and it's more important than the fact that I am a hard worker or a critical thinker."

Given this model for making a difference, how might we connect with individuals and organizations that are in need of our assistance? Online social networks connect people together: people with similar interests, people who may come to know one another. Volunteer organizations use online networks as one way to find a new generation of supporters and activists.

Can Facebook, Twitter, and other social networking sites successfully inspire us to get active in the real world? It seems that the answer is a resounding "Yes!" Worldwide, a new generation of volunteers is being recruited through the power of the media and social networking. Why not explore your favorite social networking site and your school's volunteer information system to learn more about available opportunities where you might share your time and resources with other people in your community and around the world?

---

is intricately linked with performing flexible, mobile work anywhere in the world. Based on the assumption that organizations must respond to a rapidly changing environment or they will not thrive, several twenty-first-century organizational models are based on the need to relegate traditional organizational structure to dinosaur status and to move ahead with structures that fully use technology and focus on the need to communicate more effectively. As the pace of communication has increased dramatically and information overload has become prevalent, the leaders of organizations are seeking new ways in which to more efficiently manage their organizations and to be ahead of change, rather than merely adapting to change after it occurs.

One recent approach is referred to as "smart working," which is based on the assumption that innovation is crucial and that organizational leaders must be able to use the talents and energies of the people who work with them. At one level, "smart working" refers to "anytime, anywhere" ways of work that have become prevalent because of communications technologies such as smartphones and computers. However, another level focuses on the ways in which smart working makes it possible for people to have flexibility and autonomy in where, when, and how they work (chiefexecutive.com, 2010). According to one management specialist,

> It turns out that the sort of collaborative, challenging work with potential for learning and personal development that people find satisfying is exactly the sort of work needed to adapt to current turbulent global operating conditions. Smart working is an outcome of designing organizational systems that are good for business and good for people. (chiefexecutive.com, 2010)

From this perspective, organizations must adapt to change; empower all organizational participants to become involved in collaboration, problem solving, and innovation; and create a work environment that people find engaging and that inspires them to give their best to the organization (•Figure 6.14).

Exactly how these organizations might look is not fully clear, although some analysts suggest that corporations

FIGURE 6.14 "Smart working" is based on the assumption that innovation is crucial and that people should have flexibility in where, when, and how they work. How do nontraditional office spaces such as this one reflect the idea of "smart working"?

such as Google, Microsoft, and other high-tech companies have actively sought to redefine organizational culture and environment by being responsive to employees, customers, and other stakeholders. Although management continues to exist, the distinction between managers and the managed becomes less prevalent, and the idea that management knowledge will be everyone's responsibility becomes more predominant. Emphasis is also placed on the importance of improving communication and on acquiring the latest technologies to make this process even faster, more secure, and more efficient. Overall, there is a focus on change and the assumption that people in an organization should be change agents, not individuals who merely respond to change after it occurs.

Ultimately, everyone has a stake in seeing that organizations operate in an effective, humane manner and that opportunities are widely available to all people regardless of race, gender, class, or age. Workers and students alike can benefit from organizational environments that make it possible for people to explore their joint interests without fear of losing their privacy or being pitted against one another in a competitive struggle for advantage. (For an example of students working together on a meaningful activity that benefits others, see "You Can Make a Difference.")

CHAPTER REVIEW Q & A

LO 1 How do sociologists distinguish among social groups, aggregates, and categories?

Sociologists define a social group as a collection of two or more people who interact frequently, share a sense of belonging, and depend on one another. People who happen to be in the same place at the same time are considered an aggregate. Those who share a similar characteristic are considered a category. Neither aggregates nor categories are considered social groups.

LO 2 How do sociologists distinguish among ingroups, outgroups, and reference groups?

Sociologists distinguish between primary groups and secondary groups. Primary groups are small and personal, and members engage in emotion-based interactions over an extended period. Secondary groups are larger and more specialized, and members have less personal and more-formal, goal-oriented relationships. Sociologists also divide groups into ingroups, outgroups, and reference groups.

Ingroups are groups to which we belong and with which we identify. Outgroups are groups we do not belong to or perhaps feel hostile toward. Reference groups are groups that strongly influence people's behavior whether or not they are actually members.

LO 3 How does the size of a group shape its members' communication, leadership styles, and pressures to conform?

In small groups, all members know one another and interact simultaneously. In groups with more than three members, communication dynamics change, and members tend to assume specialized tasks. Leadership may be authoritarian, democratic, or laissez-faire. Authoritarian leaders make major decisions and assign tasks to individual members. Democratic leaders encourage discussion and collaborative decision making. Laissez-faire leaders are minimally involved and encourage members to make their own decisions. Groups may have significant influence on members' values, attitudes, and behaviors.

LO4 Applying the concept of groupthink, how do people often respond differently in a group context than they might if they were alone?

Groupthink is the process by which members of a cohesive group arrive at a decision that many individual members privately believe is unwise. In order to maintain ties with a group, many members are willing to conform to norms established and reinforced by group members.

LO5 What are the three types of formal organizations, and how do they differ in membership?

Normative, coercive, and utilitarian organizations are formal organizations. We voluntarily join normative organizations when we want to pursue some common interest or gain personal satisfaction or prestige from being a member. People do not voluntarily become members of coercive organizations—associations that people are forced to join. We voluntarily join utilitarian organizations when they can provide us with a material reward that we seek.

LO6 What are the strengths and weaknesses of bureaucracies in contemporary nations such as the United States?

A bureaucracy is a formal organization characterized by hierarchical authority, division of labor, explicit procedures, and impersonality. According to Max Weber, bureaucracy supplies a rational means of attaining organizational goals because it contributes to coordination and control. A bureaucracy also has an informal structure, which includes the daily activities and interactions that bypass the official rules and procedures. The informal structure may enhance productivity or may be counterproductive to the organization. A bureaucracy may be inefficient, resistant to change, and a vehicle for perpetuating class, race, and gender inequalities.

LO7 What is the iron law of oligarchy, and how does the concept apply to the U.S. government?

The iron law of oligarchy is the tendency to become a bureaucracy ruled by the few. Those who control bureaucracies not only wield power but also have an interest in retaining their power. For example, officials elected to the U.S. Congress frequently choose to serve multiple terms in office because it provides them with the opportunity to become more involved not only in service to their country but also in bureaucratic power.

LO8 What alternative forms of organization exist today in nations such as Japan?

Some organizations have adopted Japanese management techniques based on long-term employment and company loyalty as alternative forms of bureaucratic structures. Unlike Japanese organizational structure and management style, organizations in Russia and India are more likely to be hierarchical, centralized, and highly directive. More recently, having socially sustainable organizations is becoming increasingly important.

KEY TERMS

aggregate 149
authoritarian leaders 154
bureaucracy 159
bureaucratic personality 162
category 149
conformity 154
democratic leaders 154
dyad 153

expressive leadership 154
goal displacement 162
groupthink 156
ideal type 160
informal side of a bureaucracy 161
ingroup 150
instrumental leadership 154
iron law of oligarchy 164

laissez-faire leaders 154
leadership 153
network 151
outgroup 150
rationality 160
reference group 151
small group 152
triad 153

QUESTIONS for CRITICAL THINKING

1 Who might be more likely to conform in a bureaucracy, those with power or those wanting more power?

....

2 Do the insights gained from Milgram's research on obedience outweigh the elements of deception and stress that were forced on its subjects?

....

3 How would you organize a large-scale organization or company for the second decade of the twenty-first century?

....

ANSWERS to the SOCIOLOGY QUIZ

ON PRIVACY IN GROUPS AND ORGANIZATIONS

| 1 | **False** | Deleting an e-mail or other document from a computer does not actually remove it from the computer's memory. Until other files are entered that write over the space where the document was located, experts can retrieve the document that was deleted. |

| 2 | **False** | The Family Educational Right to Privacy Act, which allows parents of a student under age 18 to obtain their child's grades, requires the student's consent once he or she has attained age 18. However, that law applies only to institutions that receive federal educational funds. |

| 3 | **True** | Telephone numbers called from a company's phone extensions can be recorded on a pen register, and this information can be used by the employer in evaluating the amount of time employees have spent talking with clients—or with other people. However, personal cell phones now provide employees with a way to talk to friends, family, and others without being detected by their employer. |

| 4 | **True** | The U.S. Supreme Court has ruled that schools may require students to submit to random drug testing as a condition of participating in extracurricular activities. |

| 5 | **False** | An employer may not engage in video surveillance of its employees in situations where they have a reasonable right of privacy. At least in the absence of a sign warning of such surveillance, employees have this right in company restrooms. |

| 6 | **False** | The Federal Educational Rights and Privacy Act states that Social Security numbers are "personally identifiable information" that may not be released without written consent from the student. Posting grades by Social Security number violates this provision unless the student has consented to the number being disclosed. |

| 7 | **True** | Confidential communications made privately to a minister, priest, rabbi, or other religious leader (or to an individual the person reasonably believes to hold such a position) generally cannot be divulged without the consent of the person making the communication. This does not apply when other people are present who are likely to hear the statement. |

| 8 | **False** | The Americans with Disabilities Act prohibits employers in companies with more than 25 employees from asking job applicants about medical information or requiring a physical examination prior to employment. |

DEVIANCE AND CRIME

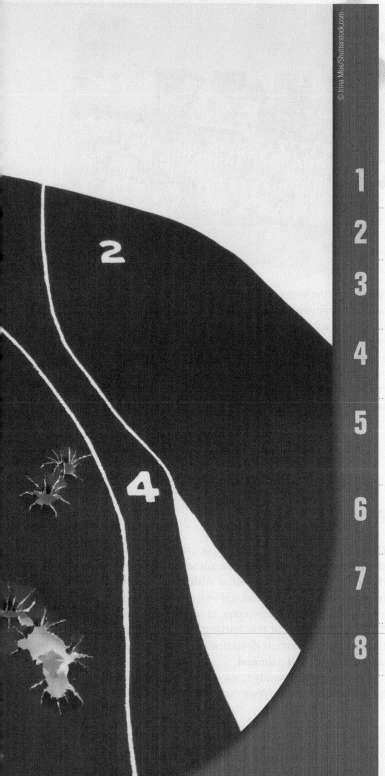

LEARNING OBJECTIVES

1   **Define** deviance and explain when deviant behavior is considered a crime.

2   **Identify** and compare the key functionalist perspectives on deviance.

3   **State** the key ideas of conflict explanations of deviance and crime that focus on power relations, capitalism, feminism, and the intersection of race, class, and gender.

4   **Explain** these symbolic interactionist perspectives on deviance: differential association theory, social bond theory, and labeling theory.

5   **Describe** how postmodern perspectives on deviance differ from other theoretical approaches, specifically identifying Michel Foucault's contributions to the study of deviance and social control.

6   **Define** the following types of crime: violent crime, property crime, public order crime, occupational crime, corporate crime, organized crime, and political crime.

7   **Explain** why official statistics may not be a good indicator of how many crimes are committed, particularly in regard to factors such as age, race, gender, and class.

8   **Identify** the components of the criminal justice system and list the goals of punishment.

SOCIOLOGY & EVERYDAY LIFE

When the Unspeakable Happens

STATE OF CONNECTICUT
DEPARTMENT OF EMERGENCY SERVICES
    AND PUBLIC PROTECTION
OFFICE OF THE COMMISSIONER
December 27, 2013
    (Cover Letter, Final Report on the Sandy Hook
    Elementary School Killings)

First and foremost, my thoughts and prayers . . .
continue to be with the families of the victims, the survivors,
and the entire Newtown community. On December 14, 2012,
your lives were changed forever in a matter of moments. The
horror and unspeakable sadness of that day and the lives
that were lost will never be forgotten.

In the midst of the darkness of that day, we also saw
remarkable heroism and glimpses of grace. We saw Sandy
Hook Elementary School faculty and staff doing everything
in their power to protect their charges. We saw law enforcement officers from dozens of local, state, and federal
agencies running as one into harm's way at enormous
personal risk. We saw dispatchers calmly and efficiently
coordinating resources and directing personnel while at
the very same time compassionately reassuring victims.

We saw fire and EMS personnel from all over the region
offering their services, state and municipal employees and
contractors working countless hours in innumerable ways
to support the first responders and the community at large,
and thousands of volunteers and donors from across the
globe reaching out in any way they could. For everything

© Gina Jacobs/Shutterstock.com

Mass shootings have become an all-too-common occurrence in the United States. Some of the most
traumatic shootings have been in elementary or secondary schools and at colleges or universities. Violence at
Columbine High School in Littleton, Colorado, and at Virginia
Tech in Blacksburg, Virginia, were well documented, and
then came the tragic occurrence at Sandy Hook Elementary
School in Newtown, Connecticut, where an armed shooter
killed twenty young children, six teachers, and his own
mother (in another location) prior to committing suicide.

People worldwide were left reeling by the Newtown
deaths and turned to multiple explanations, none satisfactory,
of why such a terrible event had occurred. Blame was placed
on lax gun-control policies, mental illness, divorced parents,
violent video games, media glorification of violence and killing,
and a myriad of other possible "causes" of this horrific event.
The final report, issued one year later, did not provide a definitive answer to pressing questions about what triggered the
Newtown murder/suicide rampage. Although this is often the
case, we continue to look for, and try to prevent, the causes
of violent crime if we have any hope of reducing these attacks
and other forms of criminal conduct in the future.

For this reason, crimes such as mass shootings are of
great interest and concern to all who teach and study sociology because of the harm that such incidents inflict not only on
victims and their families but also on entire communities, the

nation, and the larger global population linked through instantaneous communications and social media networks.

In this chapter we look at deviance and crime to learn
more about sociological perspectives on deviance, the nature
and extent of crime in the United States, and how the criminal justice system operates. Before reading on, take the
"Sociology and Everyday Life" quiz on violence and guns in the
United States to see how much you know about the subject.

LO1 **Define** deviance and explain when deviant
behavior is considered a crime.

What Is Deviance?

Deviance is any behavior, belief, or condition that violates
significant social norms in the society or group in which
it occurs. We are most familiar with *behavioral* deviance,
based on a person's intentional or inadvertent actions. For
example, a person may engage in intentional deviance by
drinking too much or comimtting a bank robbery, or participate in inadvertent deviance by losing money in a casino or laughing at a funeral.

Although we usually think of deviance as a type of behavior, people may be regarded as deviant if they express a

that was done by all of these and more, I am profoundly grateful. . . .

I hope that the release of this [final] report, though painful, will allow those who have been affected by it to continue their personal process of healing, and will provide helpful information that can be put to use to prevent such tragedies in the future.

Sincerely,
Reuben F. Bradford
Commissioner

How Much Do You Know About Violence and Guns in the United States?

TRUE	FALSE		
T	F	1	The total number of reported violent crimes increases each year in the United States.
T	F	2	Aggravated assaults account for the highest number of violent crimes in this country.
T	F	3	Firearms are used in more than half of all murders in the United States.
T	F	4	Despite extensive media coverage about mass killings, shooting sprees are rare in the United States.
T	F	5	The West is the most violent region in the United States.
T	F	6	The presence of more guns tends to indicate a likelihood of more homicides.
T	F	7	Residents in urban areas experience the highest rates of violence.
T	F	8	Each year about 50 percent of violent victimizations are reported to the police.

Answers can be found at the end of the chapter.

radical or unusual *belief system*. Members of far-right-wing or far-left-wing political groups may be considered deviant when their religious or political beliefs become known to people with more-conventional cultural beliefs. However, individuals who are considered to be "deviant" by one category of people may be seen as conformists by another group. For example, when people who believe in Bigfoot (also known as Sasquatch) are surrounded by other like-minded individuals, they think of their beliefs as normal and gain a personal sense of belonging to the group (• Figure 7.1). Sociologist Carson Mencken observed a group of Bigfoot hunters and concluded that they did not see themselves as deviant at all but actually treated one another with the kind of respect that they did not receive from nonbelievers. Instead of seeing themselves as deviant, searching for Bigfoot was akin to a religion for them.

In addition to behaving in a specific way and holding certain beliefs, individuals may be regarded as deviant if they possess a specific *condition* or *characteristic*. A wide range of conditions and characteristics have been identified by others as "deviant," including being obese, having excessive tattoos or body piercings, or being diagnosed with certain kinds of illnesses or diseases (• Figure 7.2). A stigma is often attached to a condition, such as obesity, in which blame may be placed on the patient because some people believe that the problem was caused by the individual's behavior. Chapter 5 defines a *stigma* as any physical or social attribute or sign that so devalues a person's social identity that it disqualifies the person from full social acceptance (Goffman, 1963b). Based on this definition, the stigmatized person has a "spoiled identity" as a result of being negatively evaluated by others (Goffman, 1963b). To avoid or reduce stigma, many people seek to conceal the characteristic or condition that might lead to stigmatization.

Who Defines Deviance?

Are some behaviors, beliefs, and conditions inherently deviant? In commonsense thinking, deviance is often viewed as inherent in certain kinds of behavior or people. For sociologists, however, deviance is related to social situations and social structures rather than to the behavior of individual actors. As the sociologist Kai T. Erikson (1964: 11) explains,

> Deviance is not a property inherent in certain forms of behavior; it is a property conferred upon these forms by the audiences which directly or indirectly witness them. *The critical variable in the study of deviance, then, is the social audience rather than the individual actor,*

deviance
any behavior, belief, or condition that violates significant social norms in the society or group in which it occurs.

FIGURE 7.1 Searching for Bigfoot—allegedly a large, hairy, bipedal humanoid—is a hobby for some people. Others believe that these hunters are deviant persons deluded by a myth. Some sociological studies have found that Bigfoot seekers do not perceive themselves as deviant at all. What do you think?

since it is the audience which eventually determines whether or not any episode of behavior or any class of episodes is labeled deviant. [emphasis added]

Based on this statement, we can conclude that deviance is *relative*—that is, an act becomes deviant when it is socially defined as such. Definitions of deviance vary widely from place to place, from time to time, and from group to group, as we have seen in the case of Bigfoot hunters. Clothing styles are another example: What some people wear in public today might have landed them in jail in their grandparents' or great-grandparents' day.

Deviant behavior also varies in *degree of seriousness*, ranging from mild transgressions of folkways to more-serious infringements of mores to quite serious violations of the law. Have you skipped class or pretended that you were sick so that you would have more time to complete a homework assignment or study for an exam? If so, you have violated a folkway. Others probably view your infraction as relatively minor; at most, you might receive a lower grade. Violations of mores—such as falsifying a college application or cheating on an examination—are viewed as more-serious infractions and are punishable by stronger sanctions, such as academic probation or expulsion. Some forms of deviant behavior violate the criminal law, which defines the behaviors that society labels as criminal. A *crime* is behavior that violates criminal law and is punishable with fines, jail terms, and/or other negative sanctions. Crimes range from minor offenses (such as traffic violations) to major offenses (such as murder). A subcategory, *juvenile delinquency*, refers to a violation of law or the commission of a status offense by

young people. Note that the legal concept of juvenile delinquency includes not only crimes but also *status offenses,* which are illegal only when committed by younger people (such as cutting school or running away from home).

What Is Social Control?

Societies not only have norms and laws that govern acceptable behavior; they also have various mechanisms to control people's behavior. *Social control* refers to the systematic practices that social groups develop in order to encourage conformity to norms, rules, and laws and to discourage deviance. Social control mechanisms may be either internal or

FIGURE 7.2 Do you consider this man's appearance to be deviant? In what types of groups might he be considered a conformist?

external. *Internal social control* takes place through the socialization process: Individuals *internalize* societal norms and values that prescribe how people should behave and then follow those norms and values in their everyday lives. By contrast, *external social control* involves the use of negative sanctions that proscribe certain behaviors and set forth the punishments for rule breakers and nonconformists. In contemporary societies the criminal justice system, which includes the police, the courts, and the prisons, is the primary mechanism of external social control.

If most actions deemed deviant do little or no direct harm to society or its members, why is social control so important to groups and societies? Why is the same belief or action punished in one group or society and not in another? These questions pose interesting theoretical concerns and research topics for sociologists and criminologists who examine issues pertaining to law, social control, and the criminal justice system. **Criminology** is the systematic study of crime and the criminal justice system, including the police, courts, and prisons.

The primary interest of sociologists and criminologists is not questions of how crime and criminals can best be controlled but rather social control as a social product. Sociologists do not judge certain kinds of behavior or people as being "good" or "bad." Instead, they attempt to determine what types of behavior are defined as deviant, who does the defining, how and why people become deviants, and how society deals with deviants. Although sociologists have developed a number of theories to explain deviance and crime, no one perspective is a comprehensive explanation of all deviance. Each theory provides a different lens through which we can examine aspects of deviant behavior.

 LO2 **Identify** and compare the key functionalist perspectives on deviance.

Functionalist Perspectives on Deviance

As we have seen in previous chapters, functionalists focus on societal stability and the ways in which various parts of society contribute to the whole. According to functionalists, a certain amount of deviance contributes to the smooth functioning of society.

What Causes Deviance, and Why Is It Functional for Society?

Sociologist Emile Durkheim believed that deviance is rooted in societal factors such as rapid social change and lack of social integration among people. As you will recall, Durkheim attributed the social upheaval he saw at the end of the nineteenth century to the shift from mechanical to organic solidarity, which was brought about by rapid industrialization and urbanization. Although many people continued to follow the dominant morals (norms, values, and

laws) as best they could, rapid social change contributed to *anomie*—a social condition in which people experience a sense of futility because social norms are weak, absent, or conflicting. According to Durkheim, as social integration (bonding and community involvement) decreased, deviance and crime increased. However, from his perspective, this was not altogether bad because he believed that deviance has positive social functions in terms of its consequences. For Durkheim (1964a/1895), deviance is a natural and inevitable part of all societies. Likewise, contemporary functionalist theorists suggest that deviance is universal because it serves three important functions:

1. *Deviance clarifies rules.* By punishing deviant behavior, society reaffirms its commitment to the rules and clarifies their meaning.

2. *Deviance unites a group.* When deviant behavior is seen as a threat to group solidarity and people unite in opposition to that behavior, their loyalties to society are reinforced.

3. *Deviance promotes social change.* Deviants may violate norms in order to get them changed. For example, acts of *civil disobedience*—including lunch counter sit-ins and bus boycotts—were used to protest and eventually correct injustices such as segregated buses and lunch counters in the South. More recently, this is what organizers of groups such as Occupy Wall Street hoped to accomplish, but their objective of redistribution of some wealth from the richest 1 percent to the bottom 99 percent constituted a very complex issue.

Functionalists acknowledge that deviance may also be dysfunctional for society. If too many people violate the norms, everyday existence may become unpredictable, chaotic, and even violent. If even a few people commit acts that are so violent that they threaten the survival of a society, then deviant acts move into the realm of the criminal and even the unthinkable. One example that stands out in everyone's mind is terrorist attacks around the world and the fear that remains constantly present as a result.

Although there is a wide array of contemporary functionalist theories regarding deviance and crime, many of these theories focus on social structure. For this reason, the first theory we will discuss is referred to as a structural functionalist approach. It describes the relationship between the

crime
behavior that violates criminal law and is punishable with fines, jail terms, and/or other negative sanctions.

juvenile delinquency
a violation of law or the commission of a status offense by young people.

social control
systematic practices that social groups develop in order to encourage conformity to norms, rules, and laws and to discourage deviance.

criminology
the systematic study of crime and the criminal justice system, including the police, courts, and prisons.

society's economic structure and why people might engage in various forms of deviant behavior.

Strain Theory: Goals and Means to Achieve Them

Modifying Durkheim's (1964a/1895) concept of *anomie,* the sociologist Robert Merton (1938, 1968) developed strain theory. According to **strain theory**, people feel strain when they are exposed to cultural goals that they are unable to obtain because they do not have access to culturally approved means of achieving those goals. The goals may be material possessions and money; the approved means may include an education and jobs. When denied legitimate access to these goals, some people seek access through deviant means.

Merton identified five ways in which people adapt to cultural goals and approved ways of achieving them: conformity, innovation, ritualism, retreatism, and rebellion (see • Table 7.1). According to Merton, *conformity* occurs when people accept culturally approved goals and pursue them through approved means. Persons who want to achieve success through conformity work hard, save their money, and so on. Even people who find that they are blocked from achieving a high level of education or a lucrative career may take a lower-paying job and attend school part time, join the military, or seek alternative (but legal) avenues, such as playing the lottery to "strike it rich."

Merton classified the remaining four types of adaptation as deviance:

- *Innovation* occurs when people accept society's goals but adopt disapproved means of achieving them. Innovations for acquiring material possessions or money cover a wide variety of illegal activities, including theft and drug dealing.
- *Ritualism* occurs when people give up on societal goals but still adhere to the socially approved means of achieving them. Ritualism is the opposite of innovation; persons

who cannot obtain expensive material possessions or wealth may nevertheless seek to maintain the respect of others by being a "hard worker" or "good citizen."

- *Retreatism* occurs when people abandon both the approved goals and the approved means of achieving them. Merton included persons such as skid-row alcoholics and drug addicts in this category; however, not all retreatists are destitute. Some may be middle- or upper-income individuals who see themselves as rejecting the conventional trappings of success or the means necessary to acquire them.
- *Rebellion* occurs when people challenge both the approved goals and the approved means for achieving them and advocate an alternative set of goals or means. To achieve their alternative goals, rebels may use violence (such as rioting) or may register their displeasure with society through acts of vandalism or graffiti.

Opportunity Theory: Access to Illegitimate Opportunities

Expanding on Merton's strain theory, sociologists Richard Cloward and Lloyd Ohlin (1960) suggested that for deviance to occur, people must have access to **illegitimate opportunity structures**—circumstances that provide an opportunity for people to acquire through illegitimate activities what they cannot achieve through legitimate channels. For example, in studies of juvenile gangs, researchers have found that gang members may have insufficient legitimate means to achieve conventional goals of status and wealth but have illegitimate opportunity structures—such as theft, drug dealing, or robbery—through which they can achieve these goals (• Figure 7.3). In his classic sociological study of the "Diamonds," a Chicago street gang whose members are second-generation Puerto Rican youths, sociologist Felix M. Padilla (1993) found that gang membership was linked to the members' belief that they might reach their aspirations by transforming the gang into a business

TABLE 7.1

Merton's Strain Theory of Deviance			
Mode of Adaptation	**Method of Adaptation**	**Seeks Culture's Goals**	**Follows Culture's Approved Ways**
Conformity	Accepts culturally approved goals; pursues them through culturally approved means	Yes	Yes
Innovation	Accepts culturally approved goals; adopts disapproved means of achieving them	Yes	No
Ritualism	Abandons society's goals but continues to conform to approved means	No	Yes
Retreatism	Abandons both approved goals and the approved means to achieve them	No	No
Rebellion	Challenges both the approved goals and the approved means to achieve them	No—seeks to replace	No—seeks to replace

FIGURE 7.3 Members of the California group known as the Culver City Boyz typify how gang members use items of clothing and gang signs made with their hands to assert their membership in the group and solidarity with one another. Researchers have found that some gang members may have insufficient legitimate means to achieve conventional goals of status and wealth but have illegitimate opportunity structures through which they can achieve these goals.

as drug trafficking, weapon smuggling, and extortion (Egley and Howell, 2012).

Other early research by Cloward and Ohlin (1960) identified three basic gang types—criminal, conflict, and retreatist—that emerge on the basis of what type of illegitimate opportunity structure is available in a specific area. The *criminal gang* is devoted to theft, extortion, and other illegal means of securing an income. For young men who grow up in a criminal gang, running drug houses and selling drugs on street corners make it possible for them to support themselves and their families as well as purchase material possessions to impress others. By contrast, *conflict gangs* emerge in communities that do not provide either legitimate or illegitimate opportunities. Members of conflict gangs seek to acquire a "rep" (reputation) by fighting over "turf" (territory) and adopting a value system of toughness, courage, and similar qualities. On some Native American reservations, for example, homegrown gangs routinely fight their rivals, often over a minor incident or slight, and engage in thefts, assaults, and property crimes in some of the nation's poorest, most neglected places, including the Pine Ridge Indian Reservation (Eckholm, 2009). Unlike criminal and conflict gangs, members of *retreatist gangs* are unable to gain success through legitimate means and are unwilling to do so through illegal ones. As a result, the consumption of drugs is stressed, and addiction is prevalent.

How useful are social structural approaches such as opportunity theory and strain theory in explaining deviant behavior? Although there are weaknesses in these approaches, they focus our attention on one crucial issue: the close association between certain forms of deviance and social class position. If we view gangs as a microcosm of

enterprise. Coco, one of the Diamonds, explains the importance of sticking together in the gang's income-generating business organization:

> We are a group, a community, a family—we have to learn to live together. If we separate, we will never have a chance. We need each other even to make sure that we have a spot for selling our supply [of drugs]. You know, there is people around here, like some opposition, that want to take over your *negocio* [business]. And they think that they can do this very easy. So we stick together, and that makes other people think twice about trying to take over what is yours. In our case, the opposition has never tried messing with our hood, and that's because they know it's protected real good by us fellas. (qtd. in Padilla, 1993: 104)

Although Padilla's study is more than two decades old, his findings continue to be supported by research, popular culture, and media accounts of how gangs stick together in income-generating business organizations. In the latest available study of national youth gangs, law enforcement agencies estimate that 29,400 gangs operate across the United States and that about 756,000 individuals are gang members. Some gangs are believed to be involved in criminal activities such

strain theory
the proposition that people feel strain when they are exposed to cultural goals that they are unable to obtain because they do not have access to culturally approved means of achieving those goals.

illegitimate opportunity structures
circumstances that provide an opportunity for people to acquire through illegitimate activities what they cannot achieve through legitimate channels.

SOCIOLOGY in GLOBAL PERSPECTIVE

A Wider Perspective on Gangs: Look and Listen Around the World!

- What are gangs like throughout the world?

- What messages are sent through gang music?

Why is an understanding of the music of gangs important for addressing the "gang problem"? For many of us, gangs are most closely associated with urban slums and drug-dealing outlaws in U.S. border towns. However, this perspective provides too limited a view of what actually constitutes gangs today, and it provides little or no explanation of why they exist. According to research by criminologist John M. Hagedorn (2009), gangs are a universal feature of daily life in cities throughout the world. In his study of gang formation in Chicago, Illinois; Rio de Janeiro, Brazil; and Cape Town, South Africa, Hagedorn concluded that gang formation is a strategy employed by people who believe that they have no other way to deal with poverty, injustice, and racial and ethnic oppression. In sum, demoralized people come to view gangs as a replacement for the government in providing security, needed services, and economic viability. When some people feel demoralized and believe that official channels are doing little or nothing for them, they may create alternative organizational structures—in this case, gangs—to help them adapt to their environment.

What does music have to do with this? Hagedorn argues that rap and hip-hop music provide an outlet for gang members to express their anger and show defiance toward political leaders and their country's oppressive system. With

music, gang members develop a culture of rebellion and gain a resistance identity based on their street experiences. Using the "power of negativity" as expressed in music, gang members describe problematic community conditions that are unlikely to change and try to find a way to adapt to the conditions in which they live. In the process, gang members hope to gain a sense of power and identity apart from standardized, acceptable means. Based on this idea, Hagedorn suggests that policies should be implemented that incorporate gang members into social movements and help them work for the betterment of their community and their own self-empowerment.

Studies like this are important for our understanding of deviance and crime because they point out global commonalities across a variety of urban areas around the world. They also point out ways in which we could build off of earlier theories to understand contemporary issues such as gang behavior worldwide.

REFLECT & ANALYZE

Which theoretical perspectives do you find most useful for explaining why people create or join gangs? What theory can you develop about the importance of music in encouraging conformity or deviance in a society?

---

the larger society, we can see connections between poverty and inequality and larger patterns of crime not only in the United States but throughout the world (see "Sociology in Global Perspective").

 State the key ideas of conflict explanations of deviance and crime that focus on power relations, capitalism, feminism, and the intersection of race, class, and gender.

Conflict Perspectives on Deviance

Who determines what kinds of behavior are deviant or criminal? Different branches of conflict theory offer somewhat divergent answers to this question. One branch emphasizes power as the central factor in defining deviance and crime: People in positions of power maintain their advantage by using the law to protect their interests. Another branch emphasizes the relationship between deviance and capitalism, whereas a third focuses on feminist

perspectives and the confluence of race, class, and gender issues in regard to deviance and crime.

Deviance and Power Relations

Conflict theorists who focus on power relations in society suggest that the lifestyles considered deviant by political and economic elites are often defined as illegal. According to this approach, norms and laws are established for the benefit of those in power and do not reflect any absolute standard of right and wrong. As a result, the activities of poor and lower-income individuals are more likely to be defined as criminal than those of persons from middle- and upper-income backgrounds. The media often contribute to this perception by frequent reporting of African American perpetrators that portrays them as criminal through both the context of the story and through the social structural context in which the news stories are reported (Bjornstrom et al., 2010).

Moreover, the criminal justice system is more focused on, and is less forgiving of, deviant and criminal behavior engaged in by people in specific categories. For example, research shows that young, single, urban males are more likely to be perceived as criminals and receive stricter sentences in courts (Rehavi and Starr, 2012). One study of racial disparity in federal criminal sentencing found that African Americans receive almost 10-percent longer sentences than comparable white Americans arrested for the same offenses (Rehavi and Starr, 2012). This finding may partly be attributed to the fact that prosecutors are almost twice as likely to file charges against African Americans that carry a mandatory minimum sentence. This is especially true in drug-offense cases. By contrast, another study found that black–white disparity in sentencing of offenders is less frequent in courts in those counties where more African American lawyers practice, suggesting that having representation by a person from one's own racial or ethnic category may bring greater scrutiny to criminal proceedings (King, Johnson, and McGeever, 2010).

Deviance and Capitalism

A second branch of conflict theory—Marxist/critical theory—views deviance and crime as a function of the capitalist economic system. Although the early economist and social thinker Karl Marx wrote very little about deviance and crime, many of his ideas are found in a critical approach that has emerged from earlier Marxist and radical perspectives on criminology. The critical approach is based on the assumption that the laws and the criminal justice system protect the power and privilege of the capitalist class. As you may recall from Chapter 1, Marx based his critique of capitalism on the inherent conflict that he believed existed between the capitalists (bourgeoisie) and the working class (proletariat). In a capitalist society, social institutions (such as law, politics, and education, which make up the superstructure) legitimize existing class inequalities and maintain the capitalists' superior position in the class structure. According to Marx, capitalism produces haves and have-nots, who engage in different forms of deviance and crime.

According to sociologist Richard Quinney (2001/1974), people with economic and political power define as criminal any behavior that threatens their own interests. The powerful use laws to control those who are without power. For example, drug laws enacted early in the twentieth century were actively enforced in an effort to control immigrant workers, especially the Chinese, who were being exploited by the railroads and other industries (Tracy, 1980). By contrast, antitrust legislation passed at about the same time was seldom enforced against large corporations owned by prominent families such as the Rockefellers, Carnegies, and Mellons. Having antitrust laws on the books merely shored up the government's legitimacy by making it appear responsive to public concerns about big business.

In sum, the Marxist/critical approach argues that criminal law protects the interests of the affluent and powerful. The way that laws are written and enforced benefits the capitalist class by ensuring that individuals at the bottom of the social class structure do not infringe on the property or threaten the safety of those at the top (Reiman and Leighton, 2010). However, others assert that critical theorists have not shown that powerful economic and political elites actually manipulate lawmaking and law enforcement for their own benefit. Rather, people of all classes share a consensus about the criminality of certain acts. For example, laws that prohibit murder, rape, and armed robbery protect not only middle- and upper-income people but also low-income people, who are frequently the victims of such violent crimes.

Feminist Approaches

Do you think that theories developed to explain male behavior can be used to understand female deviance and crime? According to feminist scholars, the answer is no. An interest in women and deviance developed in 1975 when two books—Freda Adler's *Sisters in Crime* and Rita James Simons's *Women and Crime*—declared that women's crime rates were going to increase significantly as a result of the women's liberation movement. Although this so-called *emancipation theory* of female crime has been refuted by subsequent analysts, Adler's and Simons's works encouraged feminist scholars (both women and men) to examine more closely the relationship among gender, deviance, and crime. More recently, feminist scholars have developed theories and conducted research to fill the void in our knowledge about gender and crime. For example, in a study by Janet Davidson and Meda Chesney-Lind (2009), the authors conclude that sociobiographical variables such as a history of physical and sexual abuse are more predictive of female criminality than of male criminality. Although there is no single feminist perspective on deviance and crime, three schools of thought have emerged.

Why do women engage in deviant behavior and commit crimes? According to the *liberal feminist approach,* women's deviance and crime are a rational response to the gender discrimination that women experience in families and the workplace. From this view, lower-income and minority women typically have fewer opportunities not only for education and good jobs but also for "high-end" criminal endeavors.

By contrast, the *radical feminist approach* views the cause of women's crime as originating in patriarchy (male domination over females). From this view, arrests and prosecution for crimes such as prostitution reflect our society's sexual double standard whereby it is acceptable for a man to pay for sex but unacceptable for a woman to accept money for such services. Although state laws usually view both the female prostitute and the male customer as violating the law, in most states the woman is far more likely than the man to be arrested, brought to trial, convicted, and sentenced.

The third school of feminist thought, the *Marxist (socialist) feminist approach,* is based on the assumption that women are exploited by both capitalism and patriarchy. Because many females have relatively low-wage jobs (if any) and few economic resources, crimes such as prostitution and shoplifting become a means to earn money or acquire consumer goods. However, instead of freeing women from

FIGURE 7.4   This young woman is being arrested after a local prostitution ring was broken up by Dallas police officers. Which feminist theory of women's crime might best explain the offenses of women like the one pictured here?

their problems, prostitution institutionalizes women's dependence on men and makes women more vulnerable to criminal prosecution. Lower-income women are further victimized by the fact that they are often the targets of violent acts by lower-class males, who perceive themselves as being powerless in the capitalist economic system.

These approaches make a contribution to our understanding of deviance and crime by focusing on gender as a central concern; however, some theories neglect the centrality of race, ethnicity, class, and sexual orientation in deviance, crime, and the criminal justice system (● Figure 7.4).

Approaches Focusing on the Interaction of Race, Class, and Gender

Some studies have focused on the simultaneous effects of race, class, and gender on deviant behavior and crime. Multiracial feminist approaches have examined how the intersecting systems of race, class, and gender act as "structuring forces" that affect how people act, what opportunities they have available, and how their behavior is socially defined. Examples include how the legal system responds to individual offenders based on their social location in hierarchies based on race, class, and/or gender. Some studies have found, for instance, that young unemployed African

American (black) and Hispanic men are treated differentially in sentencing decisions by judges and juries based on their location at the margins of race, age, and gender systems (see Burgess Proctor, 2006). Other work has also been conducted using an intersectional approach to learn more about domestic violence and other crimes in society.

LO4   **Explain** these symbolic interactionist perspectives on deviance: differential association theory, social bond theory, and labeling theory.

Symbolic Interactionist Perspectives on Deviance

Symbolic interactionists focus on *social processes,* such as how people develop a self-concept and learn conforming behavior through socialization. According to this approach, deviance is learned in the same way as conformity—through interaction with others. Although there are a number of symbolic interactionist perspectives on deviance, we will examine three major approaches—differential association and differential reinforcement theories, control theory, and labeling theory.

Richard Wong/Alamy

FIGURE 7.5 Is this example of graffiti likely to be the work of isolated artists or of gang members? In what ways do gangs reinforce such behavior?

Differential Association Theory and Differential Reinforcement Theory

How do people learn deviant behavior through their interactions with others? According to the sociologist Edwin Sutherland (1939), people learn the necessary techniques and the motives, drives, rationalizations, and attitudes of deviant behavior from people with whom they associate. *Differential association theory* states that people have a greater tendency to deviate from societal norms when they frequently associate with individuals who are more favorable toward deviance than conformity. From this approach, criminal behavior is learned within intimate personal groups such as one's family and peer groups.

Differential association theory contributes to our knowledge of how deviant behavior reflects the individual's learned techniques, values, attitudes, motives, and rationalizations. It calls attention to the fact that criminal activity is more likely to occur when a person has frequent, intense, and long-lasting interactions with others who violate the law. However, it does not explain why many individuals who have been heavily exposed to people who violate the law still engage in conventional behavior most of the time.

Criminologist Ronald Akers (1998) has combined differential association theory with elements of psychological learning theory to create *differential reinforcement theory,* which suggests that both deviant behavior and conventional behavior are learned through the same social processes. Akers starts with the fact that people learn to evaluate their own behavior through interactions with significant others. If the persons and groups that a particular individual considers most significant in his or her life

define deviant behavior as being "right," the individual is more likely to engage in deviant behavior; likewise, if the person's most significant friends and groups define deviant behavior as "wrong," the person is less likely to engage in that behavior. This approach helps explain not only juvenile gang behavior but also how peer cliques on high school campuses have such a powerful influence on people's behavior.

Rational Choice Theory

Another approach to studying deviance is rational choice theory, which suggests that people weigh the rewards and risks involved in certain types of behavior and then decide which course of action to follow. Rational choice theory is based on the assumption that when people are faced with several courses of action, they will usually do what they believe is likely to have the best overall outcome (Elster, 1989). The *rational choice theory of deviance* states that deviant behavior occurs when a person weighs the costs and benefits of nonconventional or criminal behavior and determines that the benefits will outweigh the risks involved in such actions. Rational choice approaches suggest that most people who commit crimes do not engage in random acts of antisocial behavior. Instead, they make careful decisions based on weighing the available information regarding *situational factors,* such as the place of the crime, suitable targets, and the availability of people to deter the behavior, and *personal factors,* such as what rewards they may gain from their criminal behavior (● Figure 7.5).

How useful is rational choice theory in explaining deviance and crime? A major strength of this theory is that it explains why high-risk youths do not constantly engage in delinquent acts: They have learned to balance risk against the potential for criminal gain in each situation. Moreover, rational choice theory is not limited by the underlying assumption of most social structural theories,

differential association theory
the proposition that people have a greater tendency to deviate from societal norms when they frequently associate with persons who are more favorable toward deviance than conformity.

rational choice theory of deviance
the proposition that deviant behavior occurs when a person weighs the costs and benefits of nonconventional or criminal behavior and determines that the benefits will outweigh the risks involved in such actions.

which is that the primary participants in deviant and criminal behaviors are people in the lower classes. Rational choice theory also has important policy implications regarding crime reduction or prevention, suggesting that people must be taught that the risks of engaging in criminal behavior far outweigh any benefits they may gain from their actions. Thus, people should be taught *not* to engage in crime.

Control Theory: Social Bonding

Another approach to studying deviance is control theory, which suggests that conformity is often associated with a person's bonds to other people. According to the sociologist Walter Reckless (1967), society produces pushes and pulls that move people toward criminal behavior; however, some people "insulate" themselves from such pressures by having positive self-esteem and good group cohesion. Reckless suggests that many people do not resort to deviance because of *inner containments*—such as self-control, a sense of responsibility, and resistance to diversions—and *outer containments*—such as supportive family and friends, reasonable social expectations, and supervision by others (• Figure 7.6). Those with the strongest containment mechanisms are able to withstand external pressures that might cause them to participate in deviant behavior. As you can see, control/social bonding theories have elements of functionalist and symbolic interactionist perspectives embedded within them because they focus on both social control and on the bonds that tie people together.

Extending Reckless's containment theory, sociologist Travis Hirschi's (1969) social control theory is based on the assumption that deviant behavior is minimized when people have strong bonds that bind them to families, schools, peers, churches, and other social institutions. **Social bond theory** holds that the probability of deviant behavior increases when a person's ties to society are weakened or broken. According to Hirschi, social bonding consists of (1) *attachment* to other people, (2) *commitment* to conformity, (3) *involvement* in conventional activities, and (4) *belief* in the legitimacy of conventional values and norms. Later, Michael R. Gottfredson and Hirschi (1990) modified the earlier theory that strong social bonds minimize criminal conduct and focused instead on the importance of self-control as a determinant of who will be likely to commit crime. According to Gottfredson and Hirschi (1990), high self-control is related to an individual's likelihood of conforming to norms and laws; low self-control can help explain a person's propensity to commit or refrain from committing crimes. From this perspective, young children who are adequately socialized and have behavioral problems are more likely to grow into delinquents and then into adult offenders. Social bond and

FIGURE 7.6 According to control theory, strong bonds—including close family ties—are a factor in explaining why many people do not engage in deviant behavior. Why do some sociologists believe that quality family time is more important in discouraging delinquent behavior than is time spent with other young people?

social control theory are both rooted in a functionalist assumption about the division of labor in families between men and women, with women being primarily responsible for how children are socialized and whether they become conformists or deviants in their behavior.

Labeling Theory

Labeling theory states that deviance is a socially constructed process in which social control agencies designate certain people as deviants and they, in turn, come to accept the label placed upon them and begin to act accordingly. Based on the symbolic interactionist theory of Charles H. Cooley and George H. Mead, labeling theory focuses on the variety of symbolic labels that people are given in their interactions with others.

How does the process of labeling occur? The act of fixing a person with a negative identity, such as "criminal" or "mentally ill," is directly related to the power and status of

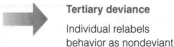

FIGURE 7.7   A Closer Look at Labeling Theory

those persons who *do* the labeling and those who are *being labeled*. Behavior, then, is not deviant in and of itself; it is defined as such by a social audience (Erikson, 1962). According to the sociologist Howard Becker (1963), *moral entrepreneurs* are often the ones who create the rules about what constitutes deviant or conventional behavior. Becker believes that moral entrepreneurs use their own perspectives on "right" and "wrong" to establish the rules by which they expect other people to live. They also label others as deviant. Often these rules are enforced on persons with less power than the moral entrepreneurs. Becker (1963: 9) concludes that the deviant is "one to whom the label has successfully been applied; deviant behavior is behavior that people so label."

As the definition of labeling theory suggests, several stages may occur in the labeling process (see • Figure 7.7). ***Primary deviance*** refers to the initial act of rule breaking (Lemert, 1951). However, if individuals accept the negative label that has been applied to them as a result of the primary deviance, they are more likely to continue to participate in the type of behavior that the label was initially meant to control. ***Secondary deviance*** occurs when a person who has been labeled a deviant accepts that new identity and continues the deviant behavior. For example, a person may shoplift an item of clothing from a department store but not be apprehended or labeled as a deviant. The person may subsequently decide to forgo such behavior in the future. However, if the person shoplifts the item, is apprehended, is labeled as a "thief," and subsequently accepts that label, then the person may shoplift items from stores on numerous occasions. A few people engage in ***tertiary deviance***, which occurs when a person who has been labeled a deviant seeks to normalize the behavior by relabeling it as nondeviant (Kitsuse, 1980). An example would be drug users who believe that using marijuana or other illegal drugs is no more deviant than drinking alcoholic beverages and therefore should not be stigmatized.

How can labeling theory be used in research on deviance? In a now-classic study that continues to show how labeling theory works, the sociologist William Chambliss (1973) studied two groups of adolescent boys in a high school: the "Saints" and the "Roughnecks." Members of both groups were constantly involved in acts of truancy, drinking, wild parties, petty theft, and vandalism. Although the Saints committed more offenses than the Roughnecks, the Roughnecks were the ones who were labeled as "troublemakers" and arrested by law enforcement officials. By contrast, the Saints were described as being the "most likely to succeed," and none of the Saints were ever arrested. According to Chambliss (1973), the Roughnecks were more likely to be labeled as deviants because they came from lower-income

families, did poorly in school, and were generally viewed negatively, whereas the Saints came from "good families," did well in school, and were generally viewed positively. Although both groups engaged in similar behavior, only the Roughnecks were stigmatized by a deviant label. Findings by William Chambliss about the significance of labeling theory in explaining deviance, particularly in regard to juvenile offenses, have been reaffirmed by numerous other studies over the past four decades (see, for example, Ascani, 2012; Bernburg, Krohn, and Rivera, 2006).

What are the specific contributions of labeling theory to explaining deviance and social control? One contribution of labeling theory is that it calls attention to the way in which social control and personal identity are intertwined: Labeling may contribute to the acceptance of deviant roles and self-images. What are the limitations of labeling theory? Although it has a number of shortcomings, the most obvious weaknesses of labeling theory is that it does not explain what caused the original acts that constituted primary deviance. Labeling theory also does not provide insight into why some people accept deviant labels that are put upon them but other individuals do not.

Describe how postmodern perspectives on deviance differ from other theoretical approaches, specifically identifying Michel Foucault's contributions to the study of deviance and social control.

Postmodernist Perspectives on Deviance

Departing from other theoretical perspectives on deviance, some postmodern theorists emphasize that the study of deviance reveals how the powerful exert control over the

social bond theory
the proposition that the probability of deviant behavior increases when a person's ties to society are weakened or broken.

labeling theory
the proposition that deviance is a socially constructed process in which social control agencies designate certain people as deviants and they, in turn, come to accept the label placed upon them and begin to act accordingly.

primary deviance
the initial act of rule breaking.

secondary deviance
the process that occurs when a person who has been labeled a deviant accepts that new identity and continues the deviant behavior.

tertiary deviance
deviance that occurs when a person who has been labeled a deviant seeks to normalize the behavior by relabeling it as nondeviant.

FIGURE 7.8 Michael Foucault contended that new means of surveillance would make it possible for prison officials to use their knowledge of prisoners' activities as a form of power over the inmates. These guards are able to monitor the activities of many prisoners without ever leaving their station.

AP Images/Brett Coomer

they were under constant scrutiny by officials in the observation post. If we think of this in contemporary terms, we can see how cameras, computers, and other devices have made continual surveillance quite easy in virtually all institutions. In such cases, social control and discipline are based on the use of knowledge, power, and technology.

Foucault's view on deviance and social control has influenced other social analysts, including researchers who have looked at a variety of social issues, ranging from workplace surveillance to squelching music pirating on the Internet. These analyses typically see the world as a modern panopticon that gives supervisors and law enforcement officials virtually unlimited capabilities for surveillance. Today, smartphones and social media outlets provide new opportunities for surveillance by government officials, corporate supervisors, and others who are not visible to the individuals being watched.

We have examined functionalist, conflict, symbolic interactionist, and postmodernist perspectives on social control, deviance, and crime (see the Concept Quick Review). All of these explanations contribute to our understanding of the causes and consequences of deviant behavior; however, we now turn to the subject of crime itself.

powerless by taking away their free will to think and act as they might choose. From this approach, institutions such as schools, prisons, and mental hospitals use knowledge, norms, and values to categorize people into "deviant" subgroups such as slow learners, convicted felons, or criminally insane individuals, and then to control them through specific patterns of discipline.

An example of this idea is found in social theorist Michel Foucault's *Discipline and Punish* (1979), in which Foucault examines the intertwining nature of power, knowledge, and social control. In this study of prisons from the mid-1800s to the early 1900s, Foucault found that many penal institutions ceased torturing prisoners who disobeyed the rules and began using new surveillance techniques to maintain social control. Although the prisons appeared to be more humane in the post-torture era, Foucault contends that the new means of surveillance impinged more on prisoners and brought greater power to prison officials. To explain, he described the *panopticon*—a structure that gives prison officials the possibility of complete observation of criminals at all times. Typically, the panopticon was a tower located in the center of a circular prison from which guards could see all the cells (• Figure 7.8). Although the prisoners knew they could be observed at any time, they did not actually know when their behavior was being scrutinized. As a result, prison officials were able to use their knowledge as a form of power over the inmates. Eventually, the guards did not even have to be present all the time because prisoners believed that

Define the following types of crime: violent crime, property crime, public order crime, occupational crime, corporate crime, organized crime, and political crime.

Crime Classifications and Statistics

Crime in the United States can be divided into different categories. We will look first at the legal classifications of crime and then at categories typically used by sociologists and criminologists.

How the Law Classifies Crime

Crimes are divided into felonies and misdemeanors. The distinction between the two is based on the seriousness of the crime. A *felony* is a serious crime such as rape,

Theoretical Perspectives on Deviance

	Theory	Key Elements
Functionalist Perspectives		
Robert Merton	Strain theory	Deviance occurs when access to the approved means of reaching culturally approved goals is blocked. Innovation, ritualism, retreatism, or rebellion may result.
Richard Cloward/Lloyd Ohlin	Opportunity theory	Lower-class delinquents subscribe to middle-class values but cannot attain them. As a result, they form gangs to gain social status and may achieve their goals through illegitimate means.
Conflict Perspectives		
Karl Marx and Richard Quinney	Critical approach	The powerful use law and the criminal justice system to protect their own class interests.
Kathleen Daly and Meda Chesney-Lind	Feminist approach	Historically, women have been ignored in research on crime. Liberal feminism views women's deviance as arising from gender discrimination, radical feminism focuses on patriarchy, and socialist feminism emphasizes the effects of capitalism and patriarchy on women's deviance.
Symbolic Interactionist Perspectives		
Edwin Sutherland	Differential association	Deviant behavior is learned in interaction with others. A person becomes delinquent when exposure to law-breaking attitudes is more extensive than exposure to law-abiding attitudes.
Travis Hirschi	Social control/social bonding	Social bonds keep people from becoming criminals. When ties to family, friends, and others become weak, an individual is most likely to engage in criminal behavior.
Howard Becker	Labeling theory	Acts are deviant or criminal because they have been labeled as such. Powerful groups often label less powerful individuals.
Edwin Lemert	Primary/secondary deviance	Primary deviance is the initial act. Secondary deviance occurs when a person accepts the label of "deviant" and continues to engage in the behavior that initially produced the label.
Postmodernist Perspective		
Michel Foucault	Knowledge as power	Power, knowledge, and social control are intertwined. In prisons, for example, new means of surveillance that make prisoners think they are being watched all the time give officials knowledge that inmates do not have. Thus, the officials have a form of power over the inmates.

homicide, or aggravated assault, for which punishment typically ranges from more than a year's imprisonment to death. A *misdemeanor* is a minor crime that is typically punished by less than one year in jail. In either event, a fine may be part of the sanction as well. Actions that constitute felonies and misdemeanors are determined by the legislatures in the various states; thus, their definitions vary from jurisdiction to jurisdiction.

Other Crime Categories

The *Uniform Crime Report (UCR)* is the major source of information on crimes reported in the United States. The UCR has been compiled since 1930 by the Federal Bureau of Investigation based on information filed by law enforcement agencies throughout the country. This report is available online at the Federal Bureau of Investigation website and is listed as "Crime in the United States" (showing the latest year available). The UCR focuses on violent crime and property crime (which, prior to 2004, were jointly referred to in that report as "index crimes") but also contains data on other types of crime (see • Figure 7.9). In 2013 an estimated 11,302,102 arrests were made for all criminal infractions (excluding traffic violations) in the United States. This number was down slightly from 2012, when an estimated 12,196,959 arrests were made. Although the UCR gives some indication of crime, the figures do not reflect the *actual* number and kinds of crimes, as will be discussed later.

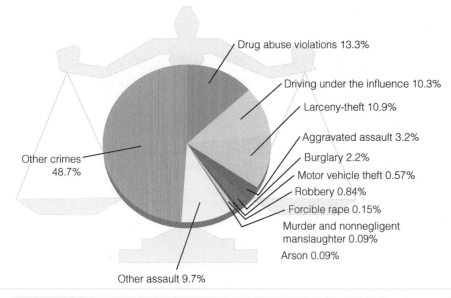

Drug abuse violations 13.3%

Driving under the influence 10.3%

Larceny-theft 10.9%

Aggravated assault 3.2%

Burglary 2.2%

Motor vehicle theft 0.57%

Robbery 0.84%

Forcible rape 0.15%

Murder and nonnegligent manslaughter 0.09%

Arson 0.09%

Other crimes 48.7%

Other assault 9.7%

FIGURE 7.9 Distribution of Arrests by Type of Offense, 2014

Note: Percentages may not equal 100% because of rounding.

Source: FBI, 2014.

Violent Crime *Violent crime* consists of actions—murder, rape, robbery, and aggravated assault—involving force or the threat of force against others. Although only 4.3 percent of all arrests in the United States in 2013 were for violent crimes, this category is probably the most anxiety-provoking of all criminal behavior: Most of us know someone who has been a victim of violent crime, or we have been so ourselves. Victims are often physically injured or even lose their lives; the psychological trauma can last for years after the event.

Violent crime also receives the most sustained attention from the media. In the aftermath of every U.S. mass shooting in recent years, communities have been over-run with worldwide media outlets bringing in broadcasting trucks, bright lights, cameras, and hundreds of reporters and commentators to continually describe, analyze, and reanalyze what might have happened to cause such a tragic event. Some information provided by the media later proves to be accurate; however, other statements are found to be flawed or completely inaccurate. On the one hand, people learn details about crime and violence that they otherwise might not know about or experience firsthand. This information may help them form an opinion on issues such as school safety, gun control, and the nature and extent of violence in the United States. On the other hand, continual media reports about a tragedy provide us with endless details about the crime and the fear that it has produced in survivors, families, neighbors, and others in the community. Many media presentations of crime prey on fear, disbelief, and distrust. For people who are already fearful of violent crime, media representations confirm these concerns and make individuals more afraid to leave home and interact with others. Violence leaves millions worrying about their safety and that of their children and other family members, even if the statistical odds of a similar occurrence in their own community are small.

Property Crime *Property crimes* include burglary (breaking into private property to commit a serious crime), motor vehicle theft, larceny-theft (theft of property worth $50 or more), and arson. In the United States a property crime occurs, on average, once every 3.7 seconds; a violent crime occurs, on average, once every 27 seconds (see • Figure 7.10). In most property crimes the primary motive is to obtain money or some other desired valuable.

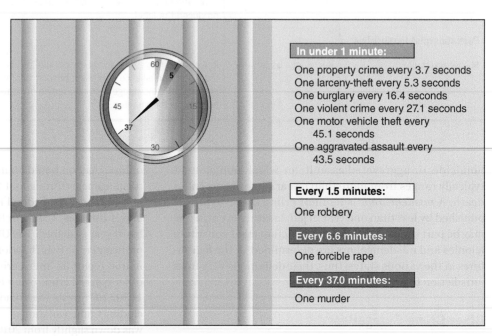

In under 1 minute:

One property crime every 3.7 seconds
One larceny-theft every 5.3 seconds
One burglary every 16.4 seconds
One violent crime every 27.1 seconds
One motor vehicle theft every
    45.1 seconds
One aggravated assault every
    43.5 seconds

Every 1.5 minutes:

One robbery

Every 6.6 minutes:

One forcible rape

Every 37.0 minutes:

One murder

FIGURE 7.10 The FBI Crime Clock, 2014

Source: FBI, 2014.

FIGURE 7.11 Persons who are accused of occupational and corporate crimes may be treated differently by law enforcement officials and the criminal justice system than individuals accused of certain violent crimes. What reasons can you give for possible disparities in treatment and sentencing based on the types of crime committed?

Public Order Crime *Public order crimes* (sometimes referred to as "morals" crimes) involve an illegal action voluntarily engaged in by the participants, such as prostitution, illegal gambling, the private use of illegal drugs, and illegal pornography. Many people assert that such conduct should not be labeled as a crime; these offenses are often referred to as *victimless crimes* because they involve a willing exchange of illegal goods or services among adults. However, public order crimes can include children and adolescents as well as adults. Young children and adolescents may unwillingly become child pornography "stars" or prostitutes.

Occupational and Corporate Crime Although the sociologist Edwin Sutherland (1949) developed the theory of white-collar crime more than sixty years ago, it was not until the 1980s that the public became fully aware of its nature. *Occupational (white-collar) crime* comprises illegal activities committed by people in the course of their employment or financial affairs (• Figure 7.11).

In addition to acting for their own financial benefit, some white-collar offenders become involved in criminal conspiracies designed to improve the market share or profitability of their companies. This is known as *corporate crime*—illegal acts committed by corporate employees on behalf of the corporation and with its support. Examples include antitrust violations; tax evasion; misrepresentations in advertising; infringements on patents, copyrights,

and trademarks; price fixing; and financial fraud. These crimes are a result of deliberate decisions made by corporate personnel to enhance resources or profits at the expense of competitors, consumers, and the general public.

Although people who commit occupational and corporate crimes can be arrested, fined, and sent to prison, some people are less likely to regard such behavior as "criminal." In many cases, punishment for such offenses has been a fine and a relatively brief prison sentence in settings sometimes referred to as "country club prisons" because of their amenities.

violent crime
actions—murder, rape, robbery, and aggravated assault—involving force or the threat of force against others.

property crimes
burglary (breaking into private property to commit a serious crime), motor vehicle theft, larceny-theft (theft of property worth $50 or more), and arson.

victimless crimes
crimes involving a willing exchange of illegal goods or services among adults.

occupational (white-collar) crime
illegal activities committed by people in the course of their employment or financial affairs.

corporate crime
illegal acts committed by corporate employees on behalf of the corporation and with its support.

For many years, public concern and media attention focused primarily on the street crimes disproportionately committed by persons who were poor, powerless, and nonwhite. Today, however, part of our focus has shifted to crimes committed by top banking officials in corporate suites, such as fraud, tax evasion, and insider trading by executives at large and well-known corporations. For example, in 2014 Credit Suisse, a global banking giant headquartered in Zurich, Switzerland, pleaded guilty to conspiring to help U.S. citizens hide their wealth from federal tax authorities in secret offshore accounts. Credit Suisse agreed to pay $2.6 billion in fines. Although this is not an isolated example of this kind of crime, it is one of the few cases in which banks or other corporations actually pled guilty to criminal wrongdoing and received a substantial penalty for doing so. According to U.S. media reports, Credit Suisse will survive with very limited damage to its reputation and financial worth. Like other major transnational corporations, international banks are often considered "too big to jail," and some of their illegal practices have been going on for at least a century (Protess and Silver-Greenberg, 2014).

How damaging to the public are corporate crimes? Some corporate crimes are more costly in terms of money and lives lost than street crimes. Thousands of jobs and billions of dollars have been lost annually as a result of corporate crime (Reiman and Leighton, 2010). Deaths resulting from corporate crimes such as polluting the air and water, manufacturing defective products, and selling unsafe foods and drugs far exceed the number of deaths caused by homicide each year. Other costs include the effect on the moral climate of society (Simon, 2012). Throughout the United States the confidence of everyday people in the nation's economy has been shaken badly by the greedy and illegal behavior of corporate insiders.

Internet Crime *Internet crime* consists of FBI-related scams, identity theft, advance fee fraud, nonauction/nondelivery of merchandise, and overpayment fraud. The proliferation of computers and Internet access worldwide has contributed to the growth of lucrative crimes in which the victim never meets the perpetrator. Auto fraud is one of the most frequently reported Internet crimes, in which perpetrators attempt to sell vehicles they do not own. Typically, they advertise vehicles for sale on the Internet at prices well below market value and claim that they must sell the vehicles quickly because of personal issues, such as relocation for work, deployment by the military, or a family crisis. Victims are instructed to wire full or partial payment to a third-party agent without meeting the seller or inspecting the vehicle prior to purchase. Because victims

think they are getting a good deal, they contact their bank for a wire transfer of the money. After the third-party agent receives the victim's money, the perpetrator pockets the cash but does not deliver the vehicle.

Another frequently reported Internet crime is the FBI-related scam, in which a perpetrator poses as a high-ranking government official or as the FBI to defraud victims (see • Figure 7.12). The perpetrator sends an e-mail that looks legitimate because it contains an "official" government letterhead or the FBI seal, and the message contains information about money or property that the recipient allegedly inherited, bogus lottery-winning notifications, or occasionally extortion threats. However, the recipient must contribute a specific amount of money to gain the inheritance, lottery, or other prize. Although no monetary losses are incurred in some cases, public trust is still undermined, and these scams pose a viable threat to national security (Internet Crime Complaint Center, 2014).

Real estate fraud occurs when perpetrators find websites that list homes for sale and take legitimate information and misuse it with their own e-mail addresses on Craigslist (without consent) under the housing rental category. Victims are instructed to send money for the first and last month's rent via a wire transfer to an overseas account. By the time they find out that this is a scam, their money is long gone, and they have no legitimate claim on the house that they hoped to occupy for a below-market rental rate. Other forms of real estate fraud include time-share marketing schemes and loan-modification scams in which a bogus loan company contacts a homeowner who is having financial problems and offers a loan-modification plan in which the homeowner stops making mortgage payments and starts sending the money, along with additional fees, to the bogus loan company instead.

Other frequently reported Internet crimes include the intimidation/extortion scam, such as payday loans or grandparent scams, in which older individuals are targeted

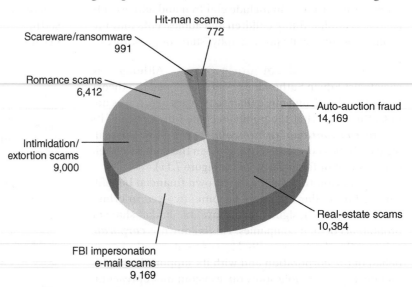

FIGURE 7.12 Top Reported Internet Crime Types
Source: Internet Crime Complaint Center, 2015.

by fraudsters who claim to be a relative in a legal or financial crisis and make an urgent plea for money to be sent to help them. Using another twist, romance scams are perpetrated by individuals who promise love and romance to online victims they meet in chat rooms, dating sites, and social media networks. Romance scams have grown in popularity because it is relatively easy to target individuals who seek companionship or romance online. The victim believes that he or she is becoming acquainted with, or dating, a real person online because scammers sometimes write poetry or send flowers and gifts to victims until they persuade them to part with their money. Romance scams alone accounted for $65.6 million lost in 2013.

Scareware and ransomware schemes are hoaxes pulled on computer users through pop-up messages which alert individuals that their computers are infected with viruses and bait them into purchasing nonexistent or useless software that allegedly will remove the viruses from their computer. One of the most frightening of the Internet schemes is the Hit Man Scam, in which scammers send an e-mail claiming to be a hit man hired to kill the victim. Recipients must pay money to make sure that the hit man does not carry out the death contract. More recently, scammers using this approach have gained personal information about the alleged victim based on what he or she has posted on social media sites, giving more credibility to their threat.

Internet crime continues to grow because perpetrators can so easily change their tactics. According to the FBI, the total loss involving known Internet crimes is about $782 million annually (Internet Crime Complaint Center, 2014).

Organized Crime
Organized crime is a business operation that supplies illegal goods and services for profit. Premeditated, continuous illegal activities of organized crime include drug trafficking, prostitution, loan-sharking, money laundering, and large-scale theft such as truck hijackings (Simon, 2012). No single organization controls all organized crime; rather, many groups operate at all levels of society. In recent decades, organized crime in the United States has become increasingly transnational in nature. Globalization of the economy and the introduction of better communications technology have made it possible for groups around the world to operate in the United States and other nations. The FBI has identified a number of major categories of organized crime threats in the United States, as shown in • Figure 7.13.

Organized crime thrives because there is great demand for illegal goods and services. Criminal organizations initially gain control of illegal activities by combining threats and promises. For example, small-time operators running drug or prostitution rings may be threatened with violence if they compete with organized crime or fail to make required payoffs.

Apart from their illegal enterprises, organized crime groups have infiltrated the world of legitimate business. Known linkages between legitimate businesses and organized crime exist in banking, hotels and motels, real estate, garbage collection, vending machines, construction,

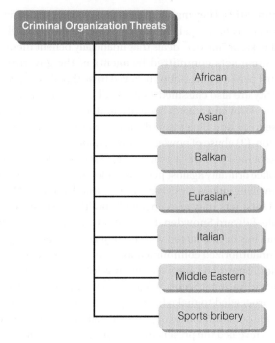

FIGURE 7.13 **Organized Crime Threats in the United States**

*Includes groups born in or with families from the former Soviet Union or Central Europe.

Source: http://www.fbi.gov/about-us/investigate/organizedcrime, 2015.

delivery and long-distance hauling, garment manufacturing insurance, stocks and bonds, vacation resorts, and funeral homes. In addition, some law enforcement and government officials are corrupted through bribery, campaign contributions, and favors intended to buy them off.

In the future, organized crime groups may further intensify their illegal activities in global drug and sex trafficking, identity theft, Internet scams, and other forms of cybercrime that are even more difficult to detect and can produce millions of dollars in a short period of time.

Political Crime
The term *political crime* refers to illegal or unethical acts involving the usurpation of power by government officials or illegal/unethical acts perpetrated against the government by outsiders seeking to make a political statement, undermine the government, or overthrow it. Government officials may use their authority unethically or illegally for the purpose of material gain or political power

Internet crime
illegal acts committed by criminals on the Internet, including FBI-related scams, identity theft, advance fee fraud, nonauction/nondelivery of merchandise, and overpayment fraud.

organized crime
a business operation that supplies illegal goods and services for profit.

political crime
illegal or unethical acts involving the usurpation of power by government officials or illegal/unethical acts perpetrated against the government by outsiders seeking to make a political statement, undermine the government, or overthrow it.

(Simon, 2012). They may engage in graft (taking advantage of political position to gain money or property) through bribery, kickbacks, or "insider" deals that financially benefit them.

Some acts committed by agents of the government against persons and groups believed to be threats to national security are also classified as political crimes. Four types of political deviance have been attributed to some officials: (1) secrecy and deception designed to manipulate public opinion, (2) abuse of power, (3) prosecution of individuals because of their political activities, and (4) official violence, such as police brutality against people of color or the use of citizens as unwilling guinea pigs in scientific research (Simon, 2012).

Outsiders may also engage in political crime. For example, Edward Snowden, formerly a U.S. computer professional, has been charged with theft of government property and unauthorized communication of national defense information. Snowden's alleged crime was leaking classified data from the National Security Agency to various media outlets that proceeded to make formerly confidential government information available to a worldwide audience. Some documents revealed the presence of global surveillance programs implemented by the U.S. government and raised concerns about the balance between national security and individual privacy. At the time of this writing, Snowden remains in Russia while seeking asylum in Switzerland. He continues to be a fugitive being sought by the U.S. government.

This type of political crime was possible because newer technologies made it easy for information to be gathered and stored beyond the reach of any one government and for large amounts of data to be released instantaneously and globally without going through processes such as the "declassification" of information, which typically takes months or years to accomplish.

Explain why official statistics may not be a good indicator of how many crimes are committed, particularly in regard to factors such as age, race, gender, and class.

Crime Statistics

How useful are crime statistics as a source of information about crime? As mentioned previously, official crime statistics provide important information on crime; however, the data reflect only those crimes that have been reported to the police.

Why are some crimes not reported? People are more likely to report crime when they believe that something can be done about it (apprehension of the perpetrator or retrieval of their property, for example). About half of all assault and robbery victims do not report the crime because they may be embarrassed or fear reprisal by the perpetrator. Thus, the number of crimes reported to police represents only the proverbial "tip of the iceberg" when compared with all offenses actually committed. Official statistics are problematic in social science research because of these limitations.

The *National Crime Victimization Survey* was developed by the Bureau of Justice Statistics as an alternative means of collecting crime statistics. Known as the NCVS, this annual survey collects information on nonfatal crimes reported and not reported to the police against persons age 12 or older from a nationally representative sample of U.S. households. In 2013 members of 90,630 households were interviewed in person to determine if they had been the victims of crimes even if they had not reported the crime to law enforcement officials. Between 2012 and 2013, the number of violent crimes, known and unknown to the police, reported in the NCVS decreased slightly from 26.1 per 1,000 persons in 2012 to 23.2 per 1,000 persons in 2013. Similarly, the rate of property crime reported in the NCVS decreased from 155.8 per 1,000 persons in 2012 to 131.4 per 1,000 in 2013.

Studies based on anonymous self-reports of criminal behavior often reveal much higher rates of crime than those found in official statistics. For example,

Sunshinepress/Getty Images

FIGURE 7.14   Edward Snowden, referred to by some media analysts as an NSA whistleblower, is shown here being interviewed about allegations of U.S. spying in Germany. How have newer technologies changed the nature of political crime?

FIGURE 7.15   Terrorism remains a major worldwide concern in the twenty-first century. The people shown leaving flowers are visiting the site of the Paris, France, offices of the satirical magazine, *Charlie Hebdo*. Four hostages and three suspects were killed when police ended two separate sieges—one at the magazine's office and the other at a kosher supermarket—by alleged terrorists.

self-reports tend to indicate that adolescents of all classes violate criminal laws. However, not all children who commit juvenile offenses are apprehended and referred to court. Some children from white, affluent families have their cases handled outside the juvenile justice system (for example, a youth may be sent to a private school or hospital rather than to a juvenile court and a public correctional facility).

Some crimes committed by persons of higher socioeconomic status in the course of business are handled by an administrative or quasi-judicial body, such as the Securities and Exchange Commission or the Federal Trade Commission, or by civil courts. As a result, many elite crimes are never classified as "crimes," nor are the businesspeople who commit them labeled as "criminals."

Terrorism and Crime

In the twenty-first century, the United States and other nations are confronted with a difficult prospect: how to deal with terrorism. *Terrorism* is the calculated, unlawful use of physical force or threats of violence against persons or property in order to intimidate or coerce a government, organization, or individual for the purpose of gaining some political, religious, economic, or social objective. At the time that I am writing this, the most recent act of massive terrorism took place in January 2015, when seventeen people were killed in the Paris office of the French satirical magazine *Charlie Hebdo* and at a kosher grocery store (● Figure 7.15). This magazine published satirical pieces on religion,

politics, and other topics not viewed favorably by certain radical groups. A cartoon depiction of the Prophet Muhammad was thought to have set off this particular attack.

The United States has not been immune to terrorism either. The 2013 bombings at the Boston Marathon targeted innocent civilians who were participants in, and spectators at, a 26-mile runners' marathon. Of course, the most devastating act of terrorism in the United States took place on September 11, 2001 (the "9/11 attacks"), when nearly 3,000 people, including 19 airplane hijackers, lost their lives in attacks on the World Trade Center in New York and the Pentagon in Washington, D.C. In the aftermath of this tragic event, national security in the United States was revamped and intensified in an effort to prevent a recurrence of these tragedies.

How are sociologists and criminologists to explain world terrorism, which may have its origins in more than one nation and include diverse "cells" of terrorists who operate in a somewhat gang-like manner but are believed to be following directives from leaders elsewhere? In order to deal with the aftermath of terrorist attacks, government officials typically focus on "known enemies" such as Osama bin Laden, who was killed by U.S. military forces in 2011. The nebulous nature of the "enemy" and the problems faced by any one

terrorism
the calculated, unlawful use of physical force or threats of violence against persons or property in order to intimidate or coerce a government, organization, or individual for the purpose of gaining some political, religious, economic, or social objective.

government trying to identify and apprehend the perpetrators of acts of terrorism have resulted in a global "war on terror."

Social scientists who use a rational choice approach suggest that terrorists are rational actors who constantly calculate the gains and losses of participation in violent—and sometimes suicidal—acts against others. For example, terrorists created a climate of fear prior to the 2014 Winter Olympic Games, hosted by Russia, through a series of deadly terror attacks by suicide bombers prior to the games. Major sporting events, including the Boston Marathon, have been the targets of terrorists who act when the whole world is watching and seek to put their beliefs or causes out for everyone to see. Unfortunately, the bombings of buses, schools, and other public accommodations become frequent occurrences and provide a venue for persons to kill or threaten to kill many innocent civilians and to destroy millions of dollars in property. *Rational choice* seems an odd term for behavior such as this.

Street Crimes and Criminals

Given the limitations of official statistics, is it possible to determine who commits crimes? We have much more information available about conventional (street) crime than elite crime; therefore, statistics concerning street crime do not show who commits all types of crime. Gender, age, class, and race are important factors in official statistics pertaining to street crime.

Gender and Crime   There is a gender gap in crime statistics: Males are arrested for significantly more crimes than females (• Figure 7.16). In 2013 more than 73 percent of all persons arrested nationwide were male. Males made up slightly less than 80 percent of persons arrested for violent crimes and about 62 percent of all persons arrested for property crimes (FBI, 2014). Females have higher arrests rates than males only in the category of prostitution and commer-

Sean Cayton/The Image Works

FIGURE 7.16   Most of the crimes that women commit are nonviolent ones. Nevertheless, many women are incarcerated. What effects might a mother's imprisonment have on the lives of her children?

cial vice. In all other categories, males have higher arrest rates.

Before further consideration of differences in crime rates by males and females, three similarities should be noted. First, the most common arrest categories for both men and women are driving under the influence of alcohol, drug abuse violations, larceny, and minor or criminal mischief types of offenses. Second, liquor-law violations (such as underage drinking), simple assault, and disorderly conduct are middle-range offenses for both men and women. Third, the rate of arrests for murder, arson, and embezzlement is relatively low for both men and women.

The most important gender differences in arrest rates are reflected in the proportionately greater involvement of men in major property crimes (such as robbery, fraud, and larceny-theft) and violent crime, as shown in • Figure 7.17. In 2013 men accounted for about 88 percent of murders, 86 percent of robberies, and almost 57 percent of all larceny-theft arrests in the United States. Property crimes for which women are most frequently arrested are nonviolent in nature, including shoplifting, theft of services, passing bad checks, credit card fraud, and employee pilferage. In the past, when women were arrested for serious violent and property crimes, they were seen as accomplices of the men who planned the crime and instigated its commission; however, this assumption frequently does not prove true today. Some women play an active role in planning and carrying out robberies and other major crimes.

Age, Class, and Crime   Of all factors associated with crime, the age of the offender is one of the most significant. Arrest rates for violent crime and property crime are highest for people between the ages of 13 and 25, with the peak being between ages 16 and 17. In 2013 individuals under age 18 accounted for about 11 percent of all arrests for violent crime and 16 percent of all arrests for property crime. Persons under age 18 were arrested for 23 percent of all vandalism and 20 percent of all disorderly conduct charges (FBI, 2014).

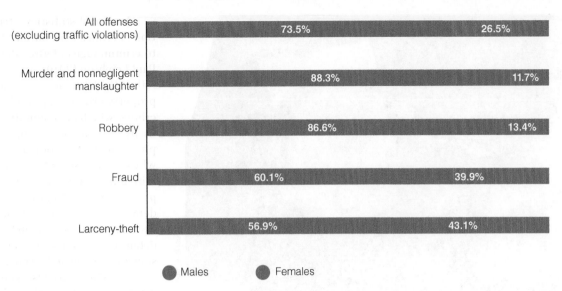

FIGURE 7.17   Arrest Rates by Gender, 2013 (Selected Offenses)
Source: www/fbi.gov/hq/cid/orgcrime/ocshome.htm.

The median age of those arrested for aggravated assault and homicide is somewhat older, generally in the late twenties. Typically, white-collar criminals are even older because it takes time to acquire both a high-ranking position and the skills needed to commit this particular type of crime. At every age and for nearly all offenses, rates of arrest remain higher for males than females. This female-to-male ratio remains fairly constant across all age categories.

Individuals from all classes commit crimes; they simply commit different kinds of crimes. Persons from lower socioeconomic backgrounds are more likely to be arrested for violent and property crimes. By contrast, persons from the upper part of the class structure generally commit white-collar or elite crimes, although only a

small proportion of these individuals will be arrested or convicted of a crime.

Race and Crime   Are people from some racial–ethnic categories more likely to be arrested for committing a crime? In 2013 whites (including Hispanics or Latinos/as) accounted for nearly 69 percent of *all arrests*. When these data were gathered, it was impossible to separate out arrest rates for white (non-Hispanic) offenders and Hispanic (Latino/a) offenders. However, the FBI started collecting nationwide data on ethnicity in 2014 to bring UCR data closer in line with U.S. Census Bureau statistics. In 2013 the total estimated arrest rate for African Americans was slightly over 28 percent, as shown in • Figure 7.18. The total

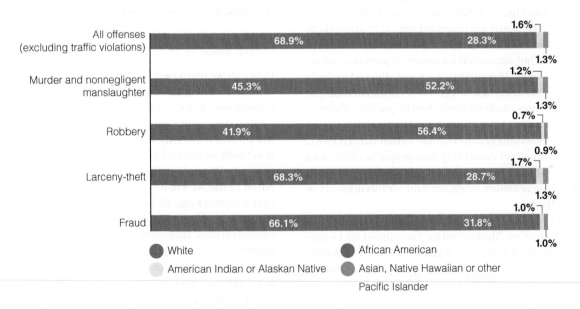

FIGURE 7.18   Arrest by Race, 2013 (Selected Offenses)
Note: Classifications as used in the Uniform Crime Report.
Source: FBI, 2014.

Nick Koudis/Photodisc/Getty Images

FIGURE 7.19 According to Karl Marx, capitalism produces haves and have-nots, and each group engages in different types of crime. Statistically, the man being arrested here is much more likely to be suspected of a financial crime than a violent crime.

Are arrest statistics a true reflection of crimes committed? According to criminologists Jeffrey Reiman and Paul Leighton (2010), arrest statistics reflect the UCR's focus on violent and property crimes, especially property crimes, which are committed primarily by low-income people. This emphasis draws attention away from the white-collar and elite crimes committed by middle- and upper-income people (• Figure 7.19). Police may also demonstrate bias and racism in their decisions regarding whom to question, detain, or arrest under certain circumstances (Reiman and Leighton, 2010).

Statistics may also show that a disproportionate number of people of color are arrested because law enforcement officials may focus on certain types of crime and direct their attention to specific "high-crime" neighborhoods in which deviance is believed to be more prevalent. As discussed previously, many poor, young, central-city males turn to criminal activity because of lack of education, job skills, and employment opportunities, giving them little or no chance of earning a living wage through legitimate employment. Law enforcement has focused on drug-related offenses, particularly among young people of color, and arrest rates have been high for this type of crime.

Finally, arrests should not be equated with guilt: Being arrested does not necessarily mean that a person is guilty of the crime with which he or she has been charged. In the United States, individuals accused of crimes are, at least theoretically, "innocent until proven guilty."

arrest rate for American Indians or Alaskan Natives was 1.6 percent, slightly above the 1.3 percent for Asian, Native Hawaiian, and other Pacific Islander. African Americans made up more than 52 percent of all arrests for murder and nonnegligent manslaughter. By contrast, arrest rates for whites were higher for nonviolent property crimes, including more than 68 percent of all arrests for larceny-theft and slightly more than 66 percent for fraud (FBI, 2014).

Although official arrest records reveal certain trends, these data tell us very little about the actual dynamics of crime by racial–ethnic category. According to official statistics, African Americans are overrepresented in arrest data. African Americans made up over 13 percent of the U.S. population in 2013 but accounted for nearly 28 percent of all arrests. In some areas, African Americans were the vast majority of those arrested for certain categories of crime. For example, African Americans made up more than 66 percent of arrests for gambling, 56 percent of arrests for robbery, and 41 percent of arrests for prostitution and commercialized vice. Likewise, African Americans are more likely than people in other racial or ethnic classifications to become crime victims (FBI, 2014). What sociological factors might account for this disparity in crime statistics?

In 2013 Native Americans (designated in the UCR as "American Indian" or "Alaskan Native") accounted for 1.6 percent of all arrests, although the highest-ranking crimes were drunkenness, liquor-law violations, vagrancy, disorderly conduct, and offenses against the family (FBI, 2014). In that same year, 1.3 percent of all arrests were of Asian, Native Hawaiian, or other Pacific Islander. Among the higher percentages of arrests for members of this category were prostitution and commercialized vice, gambling, and embezzlement (FBI, 2014).

Crime Victims

How can we learn more about crime victims? The National Crime Victimization Survey (NCVS), compiled by the U.S. Census Bureau for the Bureau of Justice Statistics, provides annual data about crimes (reported and not reported to the police) from which we can find out more about who is actually victimized by crime. In 2013 about 160,040 persons age 12 or older from 90,630 U.S. households participated in the NCVS. Based on this data, researchers found that residents age 12 or older experienced an estimated 6.1 million violent victimizations and 16.8 million property victimizations. However, the overall victimization rate for violent crime declined slightly, from 26.1 victimizations per 1,000 persons age 12 or older in 2012 to 23.2 victimizations per 1,000 persons age 12 or older in 2013 (Truman and Langton, 2014). According to the study, a slight decrease in simple assault accounted for most of the decline in total violence between 2012 and 2013. Property crimes also decreased, from 155.8 victimizations per 1,000 households in

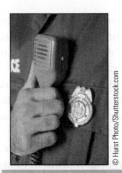
Police

Prosecutors

Judges or Magistrates

• Enforce specific laws • Investigate specific crimes • Search people, vicinities, buildings • Arrest or detain people	• File charges or petitions for judicial decision • Seek indictments • Drop cases • Reduce charges • Recommend sentences	• Set bail or conditions for release • Accept pleas • Determine delinquency • Dismiss charges • Impose sentences • Revoke probation

FIGURE 7.20 **Discretionary Powers in Law Enforcement**

2012 to 131.4 victimizations per 1,000 in 2013. According to the report, most of the decline in property crime can be attributed to a decline in reported theft.

Victimization studies look at the sex and race of persons being victimized. In 2013, there was no significant difference in male (23.7 per 1,000) and female (22.7 per 1,000) violent victimization rates. In 2013, 1.6 million males age 12 or older (1.2 percent) reported that that they had experienced one or more violent victimizations. Similarly, 1.5 million females (1.1 percent) in the same age category reported violent victimizations. By definition, violent victimization includes rape/sexual assault, robbery, total assault, aggravated assault, and simple assault. As might be expected, robbery victimization was higher for males; victimization for rape or sexual assault was higher for females. Race is also an important factors in studying victimization. Unlike previous years, in 2013 the rates of violent victimization for blacks (25.1 per 1,000) were similar to those of whites (22.2 per 1,000) and Hispanics (24.8 per 1,000). After several years of increases in victimization reporting, the overall violent and property crime rates reported for 2013 decreased. Was this a temporary fluke? Does it represent a trend? Check online for the U.S. Department of Justice's "Criminal Victimization" report for the latest year available, and compare these statistics (Truman and Langton, 2014).

What can we learn from victimization studies such as this? Because not all crimes are reported, we can find out more about the nature and extent to which people in specific regions of the country, income categories, racial or ethnic groupings, ages, and other demographic characteristics are victimized by violent and property crimes in the United States. These data can be compared with official crime statistics to see if arrest rates and convictions are an accurate reflection of crime in the United States. The NCVS particularly helps us to learn about the types and number of offenses that go unreported to police or other law enforcement officials who are part of the criminal justice system.

Identify the components of the criminal justice system and list the goals of punishment.

The Criminal Justice System

Of all the agencies of social control (including families, schools, and churches) in contemporary societies, only the criminal justice system has the power to control crime and punish those who are convicted of criminal conduct. The **criminal justice system** refers to the local, state, and federal agencies that enforce laws, adjudicate crimes, and treat and rehabilitate criminals. The system includes the police, the courts, the correctional facilities, and the people in myriads of police agencies, courts, prosecutorial agencies, correctional institutions, and probation and parole departments.

The term *criminal justice system* is somewhat misleading because it implies that law enforcement agencies, courts, and correctional facilities constitute one large, integrated system, when, in reality, the criminal justice system is made up of many bureaucracies that have considerable discretion in how decisions are made. *Discretion* refers to the use of personal judgment by police officers, prosecutors, judges, and other criminal justice system officials regarding whether and how to proceed in a given situation (see • Figure 7.20). The police are a prime example of discretionary processes because they have the power to selectively enforce the law and have on many occasions been accused of being too harsh or too lenient on alleged offenders.

criminal justice system
the local, state, and federal agencies that enforce laws, adjudicate crimes, and treat and rehabilitate criminals.

The Police

The role of the police in the criminal justice system continues to expand. The police are responsible for crime control and maintenance of order, but local police departments now serve numerous other human-service functions, including improving community relations, resolving family disputes, and helping people during emergencies. Not all "police officers" are employed by local police departments; they are employed in governmental agencies ranging from local jurisdictions to federal levels. However, we will focus primarily on metropolitan police departments because they constitute the vast majority of the law enforcement community.

Metropolitan police departments are made up of a chain of command (similar to the military), with ranks such as officer, sergeant, lieutenant, and captain, and each rank must follow specific rules and procedures. However, individual officers maintain a degree of discretion in the decisions they make as they respond to calls and try to apprehend fleeing or violent offenders. The problem of police discretion is most acute when decisions are made to use force (such as grabbing, pushing, or hitting a suspect) or deadly force (shooting and killing a suspect). Generally, deadly force is allowed only in situations in which a suspect is engaged in a felony, is fleeing the scene of a felony, or is resisting arrest and has endangered someone's life.

Although many police departments have worked to improve their public image in recent years, the practice of *racial profiling*—the use of ethnic or racial background as a means of identifying criminal suspects—remains a highly charged issue. Officers in some police departments have singled out for discriminatory treatment African Americans, Latinos/as, and other people of color, treating them more harshly than white (Euro-American) individuals. Racial profiling was recently in the headlines and a hot topic in social media after the police shooting of Michael Brown, an unarmed teenager in Ferguson, Missouri. According to records maintained by Missouri officials, 86 percent of the stops and 92 percent of the searches made by the Ferguson Police Department in 2013 were of African Americans, while only 67 percent of the Ferguson population is black. After extensive protests and numerous riots in Ferguson and other cities throughout the nation, federal authorities called for a full investigation. No civil rights violations were brought against the police officer who fired the fatal shot, but as of March 2015 the Justice Department investigation into allegations against the Ferguson Police Department regarding discriminatory traffic stops and use of excessive force against persons of color remains open (Apuzzo and Schmidt, 2015).

The belief that differential treatment takes place on the basis of race contributes to a negative image of police among many people of color who believe that they have been hassled by police officers, and this assumption is intensified by the fact that police departments have typically been made up of white male personnel at all levels. As many social analysts and media commentators point out, however, this is not a problem unique to Ferguson, Missouri: Nationwide, a massive problem of racial injustice exists in the criminal justice system, and Ferguson is only one of many symbols of this reality.

When situations such as this arise, police-department officials typically contend that race is only one factor in determining why individuals are questioned or detained as they go about everyday activities such as driving a car or walking down the street. By contrast, equal-justice advocacy groups argue that differential treatment of minority-group members amounts to a race-based double standard, which they believe exists not only in police work but also throughout the criminal justice system.

In recent years, this situation has slowly begun to change. Slightly less than 25 percent of all *sworn officers*—those who have taken an oath and been given the powers to make arrests and use necessary force in accordance with their duties—are women and minorities. Women accounted for about 13 percent of all sworn officers in local police departments in 2013. The largest percentage of women and minority police officers are located in cities with a population of 250,000 or more. African Americans make up a larger percentage of the police department in cities with a larger proportion of African American residents (such as Detroit), but Latinos/as constitute a larger percentage in cities such as San Antonio and El Paso, Texas, where Latinos/as make up a larger proportion of the population. Women officers of all races are more likely to be employed in departments in cities of more than 250,000 as compared with smaller communities (cities of fewer than 50,000), where women officers constitute a small percentage of the force.

Police departments now place greater emphasis on *community-oriented policing*—an approach to law enforcement that focuses on police officers building ties to the community by working closely with other community members. To accomplish this, officers maintain a presence in the community by walking up and down the streets or riding bicycles, getting to know people, and holding public service meetings at schools, churches, and other neighborhood settings. Community-oriented policing is often limited by budget constraints and the lack of available personnel to conduct this type of "hands-on" community involvement. In many jurisdictions, police officers believe that they have only enough time to keep up with reports of serious crime and life-threatening occurrences and that the level of available personnel and resources does not allow officers to take on a greatly expanded role in the community.

The Courts

Criminal courts determine the guilt or innocence of those persons accused of committing a crime. In theory, justice is determined in an adversarial process in which the prosecutor (an attorney who represents the state) argues that the accused is guilty, and the defense attorney asserts that the accused is innocent. In reality, judges have considerable discretion but are still constrained in some cases by *structured sentencing* guidelines. Structured sentencing is also referred to as *determinate sentencing* or *mandatory sentencing.* A determinate sentence sets the term of imprisonment at a fixed period of time (such as three years) for a specific offense. Mandatory-sentencing guidelines are established by law and require that a person convicted of a specific offense or series of offenses be given a penalty within a fixed range. Although these practices limit judicial discretion in sentencing, many critics are concerned about other negative effects that these sentencing approaches may have on the accused and the criminal justice system.

Prosecuting attorneys also have considerable leeway in deciding which cases to prosecute and when to negotiate a plea bargain with a defense attorney. As cases are sorted through the legal machinery, a steady attrition occurs. At each stage, various officials determine what alternatives will be available for those cases still remaining in the system.

About 90 percent of criminal cases are never tried in court; instead, they are resolved by plea bargaining, a process in which the prosecution negotiates a reduced sentence for the accused in exchange for a guilty plea. Defendants (especially those who are poor and cannot afford to pay an attorney) may be urged to plead guilty to a lesser crime in return for not being tried for the more serious crime for which they were arrested. Prison sentences given in plea bargains vary widely from one region to another and even from judge to judge within one state.

Those who advocate the practice of plea bargaining believe that it allows for individualized justice for alleged offenders because judges, prosecutors, and defense attorneys can agree to a plea and to a punishment that best fits the offense and the offender. They also believe that this process helps reduce the backlog of criminal cases in the court system as well as the lengthy process often involved in a criminal trial. However, those who seek to abolish plea bargaining believe that this practice leads to innocent people pleading guilty to crimes they have not committed or pleading guilty to a crime other than the one they actually committed because they are offered a lesser sentence. More-serious crimes, such as murder, felonious assault, and rape, are more likely to proceed to trial than other forms of criminal conduct; however, many of these cases do not reach the trial stage.

One of the most important activities of the court system is establishing the sentence of the accused after he or she has been found guilty or has pleaded guilty. Typically, sentencing involves the following kinds of sentences or dispositions: fines, probation, alternative or intermediate sanctions (such as house arrest or electronic monitoring), incarceration, and capital punishment. However, adult courts operate differently from those established for juvenile offenders (● Figure 7.21).

Juvenile Courts   Juvenile courts were established under a different premise than courts for adults. Under the doctrine of *parens patriae* (the state as parent), the official purpose of juvenile courts has been to care for, rather than punish, youthful offenders. In theory, less weight is given to offenses and more weight to the youth's physical, mental, or social condition. The juvenile court seeks to change or resocialize offenders through treatment or therapy, not to punish them. Consequently, judges in juvenile courts are given relatively wide latitude, or discretion, in the decisions they mete out regarding young offenders.

Unlike adult offenders, juveniles are not always represented by legal counsel. A juvenile hearing is not a trial but rather an informal private hearing before a judge or probation officer, with only the young person and a parent or guardian present. No jury is convened, and the juvenile offender does not cross-examine his or her accusers. In addition, the offender is not "sentenced"; rather, the case is "adjudicated" or "disposed of." Finally, the offender is not "punished" but instead may be "remanded to the custody" of a youth authority in order to receive training, treatment, or care.

Image Source/Getty Images

FIGURE 7.21   Although TV and movie crime dramas often prominently feature a judge and jury in a courtroom, about 90 percent of criminal cases are never tried in court. Nevertheless, jury duty is considered to be an important civic responsibility for citizens to perform, so that those who are accused have their case heard by a jury of their peers.

Because of judicial discretion, courts may treat juveniles differently based on gender. Considerable disparity exists in the disposition of juvenile cases, with much of the variation thought to result from judges' beliefs rather than objective facts in the case. Female offenders are more likely than males to be institutionalized for committing status offenses such as truancy, running away from home, and other offenses that serve as "buffer charges" for suspected sexual misconduct.

Disparity also exists on the basis of race and class. Judges tend to see youths from white, middle- or upper-class families as being very much like their children and to believe that the families will take care of the problem on their own. They may view juveniles from lower-income families or other racial–ethnic groups as delinquents in need of attention from authorities. Furthermore, some judges view gang members from impoverished central cities as "guilty by association" because of their companions.

The political climate may have an effect on how judges dispose of juvenile cases. In the process of dealing with the public perception that the juvenile justice system is too lenient, some judges may have inadvertently contributed to other problems. Many more youths have been remanded to overcrowded juvenile detention facilities that are unable to provide necessary educational, health, and social services. Based on a judge's discretion, many juvenile offenders are incarcerated under indeterminate sentences and placed in a detention facility that may serve merely as a school for adult criminality.

Punishment and Corrections

Although the United States makes up less than 5 percent of the world's population, our nation accounts for almost 25 percent of the world's prison population. Some analysts suggest that our laws prescribe greater punishment for some offenses than those in other nations, resulting in Americans being locked up for crimes that would rarely result in prison sentences in other countries. In 2013 approximately 1.57 million persons were incarcerated in state and federal prisons. When local jail populations of about 721,654 are included, the number rises to about 2.3 million people who are incarcerated at any given time.

It should be noted that jails differ from prisons. Most jails are run by local governments or a sheriff's department. They are designed to hold people before they make bail, when they are awaiting trial, or when they are serving short sentences for committing a misdemeanor. By contrast, prisons are operated by state governments and the Federal Bureau of Prisons, and are designed to hold individuals convicted of felonies. Some prisons are operated by private contractors that build and control the facilities while receiving public monies for their operation. Both jails and prisons are based on the assumption that punishment and/or corrections are necessary to protect the public good and to effectively deal with those who violate laws.

Punishment is any action designed to deprive a person of things of value (including liberty) because of some offense the person is thought to have committed. Historically, punishment has had four major goals:

1. *Retribution* is punishment that a person receives for infringing on the rights of others. Retribution imposes a penalty on the offender and is based on the premise that the punishment should fit the crime: The greater the degree of social harm, the more the offender should be punished. For example, an individual who murders should be punished more severely than one who shoplifts.

2. *General deterrence* seeks to reduce criminal activity by instilling a fear of punishment in the general public. However, we most often focus on *specific deterrence,* which inflicts punishment on individual criminals to discourage them from committing future crimes. Recently, criminologists have debated whether imprisonment has a deterrent effect, given the fact that high rates (between 30 and 50 percent) of those who are released from prison become recidivists (previous offenders who commit new crimes).

3. *Incapacitation* is based on the assumption that offenders who are detained in prison or are executed will be unable to commit additional crimes. This approach is often expressed as "lock 'em up and throw away the key!" In recent years, more emphasis has been placed on *selective incapacitation,* which means that offenders who repeat certain kinds of crimes are sentenced to long prison terms.

4. *Rehabilitation* seeks to return offenders to the community as law-abiding citizens by providing therapy or vocational or educational training. Based on this approach, offenders are treated, not punished, so that they will not continue their criminal activity. However, many correctional facilities are seriously understaffed and underfunded in the rehabilitation programs that exist. The job skills (such as agricultural work) that many offenders learn in prison do not transfer to the outside world, nor are offenders given any assistance in finding work that fits their skills once they are released.

Other approaches have also been advocated for dealing with criminal behavior. Key among these is the idea of *restoration,* which is designed to repair the damage done to the victim and the community by an offender's criminal act. This approach is based on the *restorative justice perspective,* which states that the criminal justice system should promote a peaceful and just society; therefore, the system should focus on peacemaking rather than on punishing offenders. Advocates of this approach believe that punishment of offenders actually encourages crime rather than deterring it and are in favor of approaches such as probation with treatment. Opponents of this approach

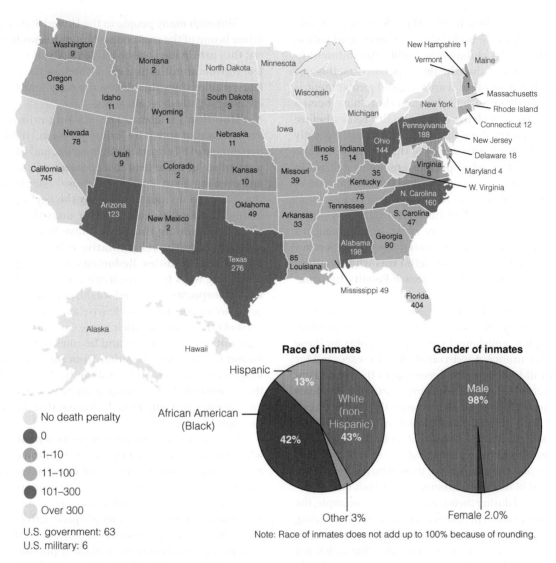

Washington 9
Oregon 36
Montana 2
North Dakota
Minnesota
New Hampshire 1
Vermont
Maine 1
Idaho 11
Wyoming 1
South Dakota 3
Wisconsin
Michigan
Massachusetts
New York
Rhode Island
Nevada 78
Utah 9
Colorado 2
Nebraska 11
Iowa
Illinois 15
Indiana 14
Ohio 144
Pennsylvania 188
Connecticut 12
New Jersey
Delaware 18
Maryland 4
California 745
Kansas 10
Missouri 39
Kentucky 35
Virginia 8
W. Virginia
Arizona 123
New Mexico 2
Oklahoma 49
Arkansas 33
Tennessee 75
N. Carolina 160
S. Carolina 47
Texas 276
Louisiana 85
Alabama 198
Georgia 90
Mississippi 49
Florida 404
Alaska
Hawaii

Race of inmates

Hispanic 13%
White (non-Hispanic) 43%
African American (Black) 42%
Other 3%

Gender of inmates

Male 98%
Female 2.0%

No death penalty
0
1–10
11–100
101–300
Over 300

U.S. government: 63
U.S. military: 6

Note: Race of inmates does not add up to 100% because of rounding.

FIGURE 7.22   Death Row Census, October 1, 2014
Source: Death Penalty Information Center, 2015.

suggest that increased punishment of offenders leads to lower crime rates and that the restorative justice approach amounts to "coddling criminals." However, numerous restorative justice programs are now in operation, and many have been found to reduce recidivism rates below the ones associated with more-conventional criminal justice sanctions (Barkan, 2012).

Instead of the term *punishment,* the term *corrections* is often used. It refers not only to prisons and jails but also to a number of programs and organizations that manage individuals who have been either accused or convicted of crimes. Included in the field of corrections are halfway houses, probation, work release and education programs, parole supervision, counseling, and community service.

The Death Penalty

Historically, removal from the group has been considered one of the ultimate forms of punishment. For many years,

capital punishment, or the death penalty, has been used in the United States as an appropriate and justifiable response to very serious crimes. In 2014, 43 inmates were executed (as contrasted with 98 in 1999), and 3,035 people awaited execution, having received the death penalty under federal law or the laws of one of the states that have the death penalty (Death Penalty Information Center, 2015). By far, the largest numbers of death row inmates are in states such as California, Florida, Texas, Alabama, and Pennsylvania (see • Figure 7.22).

Because of the finality of the death penalty, it has been a subject of much controversy and numerous Supreme Court debates about the decision-making process

punishment
any action designed to deprive a person of things of value (including liberty) because of some offense the person is thought to have committed.

involved in capital cases. In 1972 the U.S. Supreme Court ruled (in *Furman v. Georgia*) that *arbitrary* application of the death penalty violates the Eighth Amendment to the Constitution but that the death penalty itself is not unconstitutional. In other words, capital punishment is legal if it is fairly imposed. Although there have been a number of cases involving death penalty issues before the Supreme Court since that time, the Court typically has upheld the constitutionality of this practice. Yet the fact remains that racial disparities are highly evident in the death row census and executions. Over 75 percent of the murder victims in cases that resulted in an execution were white despite the fact that nationally only about 50 percent of murder victims are white. In Louisiana, for example, the odds of a death sentence were 97 percent higher for perpetrators whose victim was white than for those whose victim was black (Death Penalty Information Center, 2015).

People who have lost relatives and friends as a result of criminal activity often see the death penalty as justified. However, capital punishment raises many doubts for those who fear that innocent individuals may be executed for crimes they did not commit. According to the Death Penalty Information Center (2015), more than 140 individuals have been released from death row since 1973 based on evidence that was later presented to demonstrate their innocence. For still others, the problem of racial discrimination in the sentencing process poses troubling questions. Other questions involve the execution of those who are believed to be insane and of those defendants who did not have effective legal counsel during their trial. In 2002, for example, the Supreme Court ruled (in *Atkins v. Virginia*) that executing the mentally retarded is unconstitutional. In another landmark case (*Ring v. Arizona*), the Court ruled that juries, not judges, must decide whether a convicted murderer should receive the death penalty.

Executions resumed in 2008 after a *de facto* moratorium was lifted by the Supreme Court when it decided to uphold lethal injection as not constituting "cruel and unusual punishment" despite the fact that some scientists disagree. Although only 72 new death sentences were handed down in 2014 (after a high of 328 in 1994), the issue of the death penalty is far from resolved; the debate, which has taken place for more than two centuries, will no doubt continue for many years in the future.

Looking Ahead: Deviance and Crime in the Future

Three pressing questions pertaining to deviance and crime will continue to face us in the future: Is the solution to our "crime problem" more law and order? Is equal justice under the law possible? Is more-stringent gun control the way to reduce violent crime in the United States? (See "Sociology and Social Policy" for this discussion.)

Although many people in the United States agree that crime is one of the most important problems in this country, they are divided over what to do about it. Some of the frustration about crime might be based on unfounded fears; studies show that the *overall* crime rate has been decreasing slightly in recent years.

One thing is clear: The existing criminal justice system cannot solve the "crime problem." If only a small percentage of all crimes result in arrest, and even a smaller percentage of those lead to a conviction in serious cases, and even less than 5 percent of those result in a jail term, the "lock 'em up and throw away the key" approach has little chance of succeeding. Nor does the high rate of recidivism among those who have been incarcerated speak well for the rehabilitative efforts of our existing correctional facilities. Reducing street crime may hinge on finding ways to short-circuit criminal behavior. Likewise, if corporate elites and wealthy offenders continue to get away with crimes that leave their employees and millions of others without their homes, pensions, and life savings, not only individuals and families but also the entire nation will continue to suffer from the consequences of misplaced priorities and a system that benefits the few at the expense of the law-abiding many who attempt to follow the rules and adhere to the system that they have been socialized to accept.

One of the greatest challenges is juvenile offenders, who may become the adult criminals of tomorrow. However, instead of military-style boot camps or other stopgap measures, *structural solutions*—such as more and better education and jobs, affordable housing, more equality and less discrimination, and socially productive activities— are needed to reduce street crime. In the past, structural solutions such as these have made it possible for youths who initially committed street crimes to leave the streets, get jobs, and lead productive lives. Ultimately, the best approach for reducing delinquency and crime would be prevention: to work with young people *before* they become juvenile offenders to help them establish family relationships, build self-esteem, choose a career, and get an education that will help them pursue that career. Criminologist Steven E. Barkan (2012) proposes seven strategies for reducing crime and delinquency that are more structural in nature:

1. Create decent jobs that pay a living wage.
2. Provide economic aid for people who are unemployed or are barely making it.
3. End racial segregation in housing.
4. Strengthen social integration and social institutions in urban neighborhoods.
5. Reduce housing and population density.
6. Change male socialization.
7. Reduce economic inequality.

SOCIOLOGY & SOCIAL POLICY

The Long War Over Gun Control

The good news is there's already a growing consensus for us to build from. A majority of Americans support banning the sale of military-style assault weapons. A majority of Americans support banning the sale of high-capacity ammunition clips. A majority of Americans support laws requiring background checks before all gun purchases, so that criminals can't take advantage of legal loopholes to buy a gun from somebody who won't take the responsibility of doing a background check at all.

—President Barack Obama's comments on the urgency of enacting social policies that will deter mass shootings and other gun-related violence (whitehouse.gov, 2012)

The only thing that stops a bad guy with a gun is a good guy with a gun.

—National Rifle Association Chief Executive Wayne LaPierre, responding to demands for more-effective gun control in the aftermath of the Sandy Hook Elementary School shootings in Newtown, Connecticut (qtd. in Gold and Mason, 2012)

Pro-gun-control advocates believe that we need social policies that regulate the gun industry and gun ownership. However, opponents of gun-control measures argue that regulation will not curb random violence perpetrated by a few disturbed or frustrated individuals and that more, rather than fewer, people need to be armed to protect others against future mass violence.

Some people view social policy as a means of "declaring war" on a problem such as school violence (Best, 1999). Social policy discussions on gun-related violence typically focus on how to win the "war" on guns and how to keep violent incidents from happening again. However, a lack of consensus exists on both the causes of gun violence and what should be done about it. Some favor regulation of the gun industry and gun ownership; others believe that additional regulations will not curb random violence perpetrated by individuals.

Underlying the arguments for and against gun control are these words from the Second Amendment to the U.S. Constitution: "A well regulated Militia, being necessary to the security of a free State, the right of the people to keep and bear Arms, shall not be infringed." Those in favor of legislation to regulate the gun industry and gun ownership argue that the Second Amendment does not guarantee an individual's right to own guns: The right to keep and bear arms applies only to those citizens who do so as part of an official state militia. However, in 2008 the U.S. Supreme Court ruled in *District of Columbia v. Heller* that the Second Amendment protects an individual's right to own a gun for personal use. From this perspective, individuals possess a constitutionally protected right to have a loaded gun at home for self-defense. This argument has been made for many years by the National Rifle Association (NRA), one of the most powerful interest groups in Washington and in legislatures across the nation. The NRA has stated that gun-control regulations violate the individual's constitutional right to own a gun. The NRA's response to recent school violence has been to call for armed guards in all schools.

Those who call for more-stringent gun-control laws in the United States point out, for comparison, that in the horrific attack on twenty-two children and one adult that occurred in the Henan Province of China near the date of the Newtown mass killings in 2012, everyone under attack survived the violence because the perpetrator wielded only a knife, not a semiautomatic assault rifle (Associated Press, 2012; CNN.com, 2012b). The fast and extremely lethal nature of gun attacks became all the more poignant in the contrast between these two events.

What solutions exist for the quandary over gun regulations? Declaring war on gun-related violence is difficult because it is virtually impossible to rally our political leaders and the general public behind a single policy. Instead, time is spent arguing over how to proceed and how to identify the real enemy, and gun violence remains a chronic concern in the United States. Clearly, this is an issue in which we have yet to successfully address a pressing problem that is continuing to take precious lives in this nation.

REFLECT & ANALYZE

Do you believe that arming more people would reduce mass shootings in the United States? Should armed guards be placed in all schools? Should college students be allowed to carry guns on campus for their own safety? What do you think?

Many people still ask if equal justice under the law is possible. As long as racism, sexism, classism, and ageism exist in our society, many individuals will see deviant and criminal behavior through a selective lens. To solve the problems addressed in this chapter, we must ask ourselves what we can do to ensure the rights of everyone, including the poor, people of color, and women and men alike. Many of us can counter classism, racism, sexism, and ageism where they occur. Perhaps the only way that the United States can have equal justice under the law (and, perhaps, less crime as a result) in the future is to promote social justice for individuals regardless of their race, class, gender, or age.

The Future of Transnational Crime and the Global Criminal Economy

Transnational crime occurs across multiple national borders. This type of crime not only involves crossing borders between countries but also crimes in which crossing national borders is essential to the criminal activity. Much of transnational crime is conducted by organized criminal groups that use systematic violence and corruption to achieve their goals. Often these criminal networks are able to prey on less-powerful governments that do not have the resources to oppose them. According to the National Institute of Justice (2011), transnational crime should be of concern to people living in the United States because this type of criminal activity has a detrimental effect on everyone, not just the nations that are destabilized by these activities, often through the use of bribery, violence, or terror.

Transnational crime is fueled by globalization, which has brought increased travel, expanded international trade, and advances in telecommunications and computer technology. This type of criminal activity cannot be controlled by one nation alone. How much money and other resources change hands in the global criminal economy? Although the exact amount of profits and financial flows originating in the global criminal economy is impossible to determine, the United Nations Conference on Global Organized Crime estimated that more than $600 billion (in U.S. currency) per year is accrued in the global trade in drugs alone. Profits from all kinds of global criminal activities are estimated to be as high as $5 trillion per year (United Nations

Development Programme, 2011). Some analysts believe that even this figure may underestimate the true nature and extent of the global criminal economy. The highest-income-producing activities of global criminal organizations include trafficking in drugs, weapons, and nuclear material; smuggling of things and people (including migrants); trafficking in women and children for the sex industry; and trafficking in body parts such as corneas and major organs for the medical industry. Undergirding the entire criminal system are money laundering and various complex financial schemes and international trade networks that make it possible for people to use the resources they obtain through illegal activity for the purposes of consumption and investment in the formal ("legitimate") economy. Theft of critical U.S. intellectual property, including intrusions into corporate computer networks, makes this country more vulnerable to significant business losses. Cybercrime poses a threat to banking, stock markets, and credit card services, just as penetration of intelligence services makes nations more vulnerable to terrorism. (• Figure 7.23).

Can anything be done about transnational crime? The United Nations Office of Drugs and Crime (2014) suggests that specific steps must be taken: (1) coordination at the international level for identifying, investigating, and prosecuting the people and groups behind the crimes; (2) education and raising awareness about these kinds of crime and how they affect individuals' everyday lives; (3) intelligence and technology to help law enforcement officials combat powerful criminal networks; and (4) assistance for developing countries to help them build their capacity for countering terrorist threats.

In the United States, the White House has also developed a strategy to combat transnational organized crime that focuses on stemming the southward flow of guns and money that contributes to an increase in drug violence. From this perspective, the demand for illegal drugs in the United States fuels the global drug trade, which is a key source of funding for transnational organized crime. If this source can be reduced or cut off, this will be a beginning toward reducing problems of terrorism and insurgent networks. Among the policy objectives set forth in the U.S. National Security Strategy are the following:

Protect Americans and our partners from the harm, violence, and exploitation of transnational criminal networks. . . . We will target the

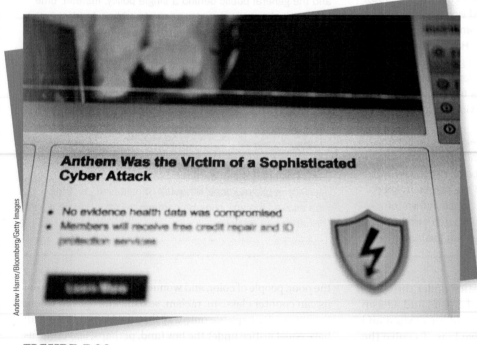

Andrew Harrer/Bloomberg/Getty Images

Anthem Was the Victim of a Sophisticated Cyber Attack

- No evidence health data was compromised
- Members will receive free credit repair and ID protection services

FIGURE 7.23   One of a growing number of cyberattacks was reported to the FBI by Anthem, Inc., the second largest U.S. health insurer, when hackers obtained data on tens of millions of current and former customers and employees from Anthem's IT system.

networks that pose the gravest threat to citizen safety and security, including those that traffic illicit drugs, arms, and people—especially women and children; sell and distribute substandard, tainted and counterfeit goods; rob Americans of their prosperity; carry out kidnappings for ransom and extortion; and seek to terrorize and intimidate through acts of torture and murder. (whitehouse.gov, 2011)

Will the United States and other nations be able to curb transnational organized crime? This remains to be seen; however, public safety, public health, democratic institutions, and economic stability rely on the ability of the U.S. government and others to combat networks that pose a strategic threat to Americans and U.S. interests. Crime around the world, as well as at home, can be a destabilizing influence on the social order.

CHAPTER REVIEW Q & A

LO1 What is deviance, and when is deviant behavior considered a crime?

Deviance is any behavior, belief, or condition that violates significant social norms in the society or group in which it occurs. Some forms of deviant behavior violate the criminal law, which defines the behaviors that society labels as criminal. Sociologists are interested in what types of behavior are defined by societies as "deviant," who does that defining, how individuals become deviant, and how those individuals are dealt with by society.

LO2 What are the key functionalist perspectives on deviance?

Functionalist perspectives on deviance include strain theory and opportunity theory. Strain theory focuses on the idea that when people are denied legitimate access to cultural goals, such as a good job or a nice home, they may engage in illegal behavior to obtain them. Opportunity theory suggests that for deviance to occur, people must have access to illegitimate means to acquire what they want but cannot obtain through legitimate means.

LO3 What are the key ideas of conflict explanations of deviance and crime that focus on power relations, capitalism, feminism, and the intersection of race, class, and gender?

Conflict theorists who focus on power relations in society suggest that the lifestyles considered deviant by political and economic elites are often defined as illegal. Marxist conflict theorists link deviance and crime to the capitalist society, which divides people into haves and have-nots, leaving crime as the only source of support for those at the bottom of the economic ladder. Feminist approaches to deviance focus on the relationship between gender and deviance. Multiracial feminist approaches have examined how the intersecting systems of race, class, and gender act as "structuring forces" that affect how people act, what opportunities they have available, and how their behavior is socially defined.

LO4 What are some key symbolic interactionist perspectives on deviance, including differential association theory, social bond theory, and labeling theory?

According to symbolic interactionists, deviance is learned through interaction with others. Differential association theory states that individuals have a greater tendency to deviate from societal norms when they frequently associate with persons who tend toward deviance instead of conformity. According to social control theories, everyone is capable of committing crimes, but social bonding (attachments to family and to other social institutions) keeps many from doing so. According to labeling theory, deviant behavior is that which is labeled deviant by those in powerful positions.

LO5 How do postmodern perspectives on deviance differ from other theoretical approaches?

Postmodernist views on deviance focus on how the powerful control others through discipline and surveillance. This control may be maintained through largely invisible forces such as the Panoptican, as described by Michel Foucault, or by newer technologies that place everyone—not just "deviants"—under constant surveillance by authorities, who use their knowledge as power over others.

LO6 How do sociologists define the following types of crime: violent crime, property crime, public order crime, occupational crime, corporate crime, organized crime, and political crime?

Violent crime consists of actions involving force or the threat of force against others. Property crimes include burglary, motor vehicle theft, larceny-theft, and arson. Public order crimes involve an illegal action voluntarily engaged in by the participants, such as prostitution. Occupational crime comprises illegal activities committed by people in the course of their employment or financial affairs. Corporate crime consists of illegal acts committed by corporate employees on behalf of the corporation and with its support. Organized crime is a business operation that supplies illegal goods and services for profit. Political crime refers to illegal or unethical acts involving the usurpation of power by government officials or illegal/unethical acts perpetrated against the government by outsiders seeking to make a political statement, undermine the government, or overthrow it.

LO7 Why may official statistics not be a good indicator of how many crimes are committed, particularly in regard to factors such as age, race, gender, and class?

Official crime statistics are taken from the Uniform Crime Report, which lists crimes reported to the police, and the National Crime Victimization Survey, which interviews households to determine the incidence of crimes, including those not reported to police. Studies show that many more crimes are committed than are officially reported. Age is the key factor in crime statistics. Younger people are more likely to have the highest criminal victimization rates. The elderly tend to be fearful of crime but are the least likely to be victimized. Males are arrested for significantly more crimes than females. Persons from lower socioeconomic backgrounds are more likely to be arrested for violent and property crimes; white-collar crime is more likely to occur among the upper socioeconomic classes.

LO8 What are the components of the criminal justice system and the goals of punishment?

The criminal justice system refers to the local, state, and federal agencies that enforce laws, adjudicate crimes, and treat and rehabilitate criminals. The system includes the police, the courts, the correctional facilities, and the people in myriads of police agencies, courts, prosecutorial agencies, correctional institutions, and probation and parole departments. Historically, punishment has had four major goals: retribution, general deterrence, incapacitation, and rehabilitation.

KEY TERMS

corporate crime 189

crime 176

criminal justice system 197

criminology 177

deviance 174

differential association theory 183

illegitimate opportunity structures 178

Internet crime 190

juvenile delinquency 176

labeling theory 184

occupational (white-collar) crime 189

organized crime 191

political crime 191

primary deviance 185

property crimes 188

punishment 200

rational choice theory of deviance 183

secondary deviance 185

social bond theory 184

social control 176

strain theory 178

terrorism 193

tertiary deviance 185

victimless crimes 189

violent crime 188

QUESTIONS for CRITICAL THINKING

1 Why does the crime rate fluctuate in the United States? How might we use sociology to explain the fact that the crime rate has not increased dramatically even in the recession and the slow economic recovery?

2 Should so-called victimless crimes, such as prostitution and recreational drug use, be decriminalized? Do these crimes harm society?

3 As a sociologist armed with a sociological imagination, how would you propose to deal with the problem of crime in the United States and around the world?

ANSWERS to the SOCIOLOGY QUIZ

ON VIOLENCE AND GUNS IN THE UNITED STATES

1	**False**	Looking at recent trends, the total number of reported violent crimes decreased by about 15 percent between 2007 and 2011.
2	**True**	Aggravated assaults in recent years accounted for the highest number (62.4 percent) of violent crimes reported to law enforcement officials.
3	**True**	FBI data show that firearms are used in 67.7 percent of the nation's murders.
4	**False**	Over the past three decades, at least 61 mass murders have been carried out with firearms in the United States. These killings have taken place in 30 different states, ranging from Massachusetts to Hawaii.
5	**False**	The South is the most violent region in the United States.
6	**True**	The Harvard Injury Control Research Center found evidence that links higher rates of gun ownership with more murders in various U.S. states as well as in different countries.
7	**True**	Residents in urban areas experience the highest rates of violent crime when compared to suburban and rural residents.
8	**True**	About 50 percent of violent victimizations are reported to the police annually. Crimes such as domestic violence, serious violent crimes involving weapons, and serious violent crimes involving injury are the most likely to be reported.

Sources: Based on FBI, 2014; and Truman and Langton, 2014.

CLASS AND STRATIFICATION IN THE UNITED STATES

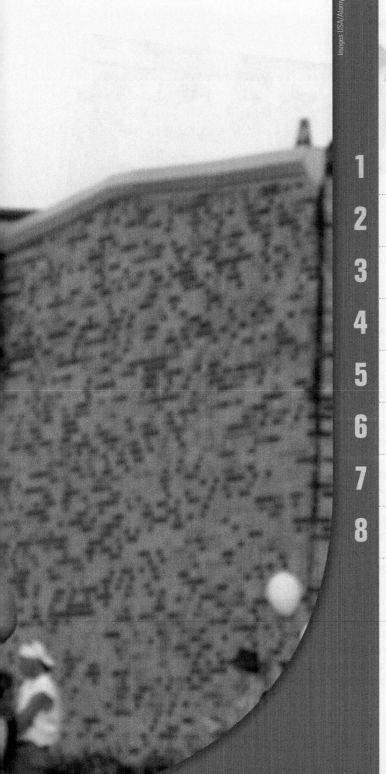

LEARNING OBJECTIVES

1 **Define** *social stratification* and distinguish among three major systems of stratification.

2 **Explain** Karl Marx's views on social class and stratification.

3 **Discuss** Max Weber's multidimensional approach to social stratification.

4 **List** and note the key characteristics of U.S. social classes.

5 **Distinguish** between income inequality and wealth inequality.

6 **Describe** three important consequences of inequality in the United States.

7 **Identify** the characteristics of the U.S. poor based on age, gender, and race/ethnicity.

8 **Compare** and contrast functionalist, conflict, and symbolic interactionist perspectives on social inequality.

SOCIOLOGY & EVERYDAY LIFE

The Power of Class

When they were ten years old, the middle-class youth [in my study] seemed worldly, blasé, and hard to impress. For them, pizza parties were very common and thus no special treat. Spring concerts drew shrugs. . . . Although the working-class and poor children were the same age as the middle-class children, they seemed younger, bouncier, and more childlike. They smiled broadly while on stage for the spring concert, were ecstatic over a pizza party, and entertained themselves for hours on weekends and evenings. Ten years later, the pattern had reversed: it was the middle-class youth who seemed younger. Now college students, they were excited about the way the world was opening up for them. . . . By contrast, the working-class and poor youth were generally working full-time in jobs they did not like, and they had various pressing responsibilities such as raising children, paying for food and board, and making monthly car payments. . . . [A]s the children moved from fourth grade into adulthood, the power of class pushed their lives in such different directions that I could not pose the same interview questions to the group as a whole. Middle-class youths' interviews were filled with questions about their college preparation classes, college searches, college choice, and college

adjustment. . . . Working-class and poor youths' interviews were filled with discussions of their difficulties in high school, challenges and work, and uncertain future goals. . . . The follow-up study suggests that over time the gap that existed between the families when the children were ten widened rather than narrowed. . . . [I]t is important to recognize that in American society, people who are blessed with class advantages tend to be unaware of these benefits

How does class affect young people?

People living in the United States want to achieve the American Dream. Throughout the history of this country, people have aspired to have more than their parents and grandparents. Young people from low-income and poverty-level families are often no exception. Upward mobility in the U.S. class system has been linked to opportunities in education and employment; however, recent national and global economic crises have made it more difficult for many people to achieve their dream. Some sociological studies, such as the one conducted by Annette Lareau (2011), also emphasize that the social-class origins of children have powerful and long-term effects on young people because more-affluent parents typically are better able to orchestrate their children's lives in institutional settings such as schools.

From a sociological perspective, this issue of class-based inequalities experienced by individuals brings us to a larger, macrolevel question in studying class and stratification in the United States: What is the American Dream? Simply stated, the American Dream is the belief that if people work hard and play by the rules, they will have a chance to get ahead. Moreover, each generation will be able to have a higher standard of living than that of its parents. The American Dream is based on the assumption that people in the United States have equality of opportunity regardless of their race, creed, color, national origin, gender, or religion.

For some people the American Dream means that each subsequent generation will be able to acquire more material possessions and wealth than family members in the preceding generation. For other people the American Dream means having an opportunity to secure a job, own a home, and get a good education for their children. For still others it is the chance to rise from poverty by working hard enough.

When we talk about the American Dream, it is important to realize that not all people will achieve success. The way a society is stratified (arranged from top to bottom) has a major influence on a person's position in the class structure. In fact, the growing income divide between the rich and the rest of the people in the United States has truly dampened the belief of many individuals that the American Dream still exists in this nation. According to recent surveys, the number of people in the United States who actually believe that there is plenty of opportunity to get ahead through hard work has declined by 16 percent since 2000 (Kohut, 2015). In this chapter we examine systems of social stratification and describe how the U.S. class system may make it easier for some individuals to attain (or maintain) top positions in society while others have great difficulty moving up from low-income origins or have problems in getting into, or staying in, the middle class. But before we explore class and stratification, test your knowledge of wealth, poverty, and the American Dream by taking the "Sociology and Everyday Life" quiz.

and privileges. . . . They downplay, or do not even notice, the social class benefits bestowed upon them.

—sociologist ANNETTE LAREAU (2011: 309–310) describing how class influences white and African American children in

poor, working-, and middle-class families as she revisits young people from her earlier study who had moved from fourth grade to adulthood

How Much Do You Know About Wealth, Poverty, and the American Dream?

TRUE	FALSE		
T	F	1	Most U.S. adults say that they have achieved the American Dream.
T	F	2	Individuals over age 65 have the highest rate of poverty.
T	F	3	Compared with other industrialized countries, the United States has the highest rate of childhood poverty.
T	F	4	In the United States a family of four is considered to be "poor" if the household earns less than $40,000.
T	F	5	According to media reports, the wealthiest person in the world lives in the United States.
T	F	6	Slavery still exists in the United States.
T	F	7	Some states have a higher minimum wage than the federal minimum wage.
T	F	8	Since the 1970s, the gap between the rich and the poor in the United States has decreased significantly.

Answers can be found at the end of the chapter.

 LO1 **Define** *social stratification* and distinguish among three major systems of stratification.

What Is Social Stratification?

Social stratification is the hierarchical arrangement of large social groups based on their control of basic resources. Stratification involves patterns of structural inequality that are associated with membership in each of these groups, as well as the ideologies that support inequality. Sociologists examine the social groups that make up the hierarchy in a society and seek to determine how inequalities are structured and persist over time.

Max Weber's term *life chances* refers to the extent to which individuals have access to important societal resources such as food, clothing, shelter, education, and health care. According to sociologists, more-affluent people typically have better life chances than the less affluent; they have greater access to quality education, safe neighborhoods, high-quality nutrition and health care, police and private security protection, and an extensive array of other goods and services. In contrast, persons with low and poverty-level incomes tend to have limited access to these resources. *Resources* are anything valued in a society, ranging from money

and property to medical care and education; resources are considered to be scarce because of their unequal distribution among social categories. If we think about the valued resources available in the United States, for example, the differences in life chances are readily apparent. As one analyst suggested, "Poverty narrows and closes life chances. The victims of poverty experience a kind of arteriosclerosis of opportunity. Being poor not only means economic insecurity, it also wreaks havoc on one's mental and physical health" (Ropers, 1991: 25). Our life chances are intertwined with our class, race, gender, and age.

All societies distinguish among people by age. Young children typically have less authority and responsibility than older persons. Older persons, especially those without wealth or power, may find themselves at the bottom of the social hierarchy. Similarly, all societies differentiate

social stratification
the hierarchical arrangement of large social groups based on their control of basic resources.

life chances
Max Weber's term for the extent to which individuals have access to important societal resources such as food, clothing, shelter, education, and health care.

between females and males: Women are often treated as subordinate to men. From society to society, people are also treated differently as a result of their religion, race/ethnicity, appearance, physical strength, disabilities, or other distinguishing characteristics. All of these differentiations result in inequality. However, systems of stratification are also linked to the specific economic and social structure of a society and to a nation's position in the system of *global* stratification, which is so significant for understanding social inequality that we will devote the next chapter to this topic.

Systems of Stratification

Around the globe, one of the most important characteristics of systems of stratification is their degree of flexibility. Sociologists distinguish among such systems based on the extent to which they are open or closed. In an *open system,* the boundaries between levels in the hierarchies are more flexible and may be influenced (positively or negatively) by people's achieved statuses. Open systems are assumed to have some degree of social mobility. *Social mobility* is the movement of individuals or groups from one level in a stratification system to another. This movement can be either upward or downward. *Intergenerational mobility* is the social movement experienced by family members from one generation to the next. For example, Sarah's father is a carpenter who makes good wages in good economic times but is often unemployed when the construction industry slows to a standstill. Sarah becomes a neurologist, earning $350,000 a year, and moves from the working class to the upper-middle class. Between her father's generation and her own, Sarah has experienced upward social mobility.

By contrast, *intragenerational mobility* is the social movement of individuals within their own lifetime. Consider, for example, RaShandra, who began her career as a high-tech factory worker and through increased experience and taking specialized courses in her field became an entrepreneur, starting her own highly successful Internet-based business. RaShandra's advancement is an example of upward intragenerational social mobility. However, note that both intragenerational mobility and intergenerational mobility can be downward as well as upward.

In a *closed system*, the boundaries between levels in the hierarchies of social stratification are rigid, and people's positions are set by ascribed status (• Figure 8.1). Open and closed systems are ideal-type constructs; no actual stratification system is completely open or closed. The systems of stratification that we will examine—slavery, caste, and class—are characterized by different hierarchical structures and varying degrees of mobility. Let's examine these three systems of stratification to determine how people acquire their positions in each and what potential for social movement they have.

James Brunker/Alamy

FIGURE 8.1 Max Weber's term *life chances* refers to the extent to which people have access to resources such as food, clothing, shelter, education, and health care. How might the life chances of the people living in these buildings be different?

Slavery

Slavery is an extreme form of stratification in which some people are owned or controlled by others for the purpose of economic or sexual exploitation. It is a closed system in which people designated as "slaves" are treated as property and have little or no control over their lives. Those of us living in the United States are aware of the legacy of slavery in our own country, beginning in the 1600s, when slaves were forcibly imported to the United States as a source of cheap labor. Slavery was defined in law and custom by the 1750s, making it possible for one person to own another person. In fact, early U.S. Presidents including George Washington, James Madison, and Thomas Jefferson owned slaves.

As practiced in the United States, slavery had four primary characteristics: (1) it was lifelong and was inherited (children of slaves were considered to be slaves); (2) slaves were considered property, not human beings; (3) slaves were denied rights; and (4) coercion was used to keep slaves "in their place" (Noel, 1972). Although most slaves were

Hulton Archive/Staff/Getty Images

Amanda Ahn/dbimages/Alamy

John Lund/Drew Kelly/Blend Images/Getty Images

FIGURE 8.2 Systems of stratification include slavery, caste, and class. As shown in these photos, the life chances of people living in each of these systems differ widely.

powerless to bring about change, some were able to challenge slavery—or at least their position in the system—by engaging in activities such as sabotage, intentional carelessness, work slowdowns, or running away from owners and working for the abolition of slavery (Healey, 2002). Despite the fact that slavery in this country officially ended about 150 years ago, sociologists such as Patricia Hill Collins (1990) believe that its legacy is deeply embedded in current patterns of prejudice and discrimination against African Americans.

Slavery is not simply an unfortunate historical legacy in American society. The U.S. State Department (2015) estimates that as many as 25,000 foreigners are brought to the United States every year and are enslaved. The three main areas of "trafficking" people into the United States for the purpose of enslavement are agricultural slavery, domestic servitude, and sexual exploitation. The problem of global slavery and human trafficking is so prevalent in the second decade of the twenty-first century that a period called National Slavery and Human Trafficking Prevention Month is set aside to call attention to the problem. "Sociology in Global Perspective" shows how modern slavery may affect your daily life without your ever knowing it.

The Caste System

Like slavery, caste is a closed system of social stratification. A *caste system* is a system of social inequality in which people's status is permanently determined at birth based on their parents' ascribed characteristics. Vestiges of caste systems exist in contemporary India and South Africa (• Figure 8.2).

In India, caste is based in part on occupation; thus, families typically perform the same type of work from generation to generation. By contrast, the caste system of South Africa was based on racial classifications and the belief of white South Africans (Afrikaners) that they were morally superior to the black majority. Until the

social mobility
the movement of individuals or groups from one level in a stratification system to another.

intergenerational mobility
the social movement experienced by family members from one generation to the next.

intragenerational mobility
the social movement of individuals within their own lifetime.

slavery
an extreme form of stratification in which some people are owned or controlled by others for the purpose of economic or sexual exploitation.

caste system
a system of social inequality in which people's status is permanently determined at birth based on their parents' ascribed characteristics.

SOCIOLOGY in GLOBAL PERSPECTIVE

A Day in Your Life: How Are You Touched by Modern Slavery?

According to the U.S. State Department (2014), many of the things that we enjoy on a daily basis may have been produced or touched by those held in involuntary servitude. Consider this typical day in the life of a person:

6:00 A.M.: Wake Up and Get Ready for the Day

The clothes on your back could have been produced by a man, woman, or child in a garment factory in Asia, the Middle East, or Latin America who is subjected to forced labor, including withholding of passports, no pay, long working hours to meet quotas, and physical and sexual abuse. The jewelry you wear may include gold mined by trafficked children in Africa, Asia, and Latin America.

8:00 A.M.: Sit Down at Your Desk

The electronics you use may be dependent on minerals that are produced in conflict-affected areas of Africa where children and adults are forced to work in mines under conditions of forced labor and sexual servitude.

10:00 A.M.: Take a Caffeine Break

The coffee you drink may have been touched by modern slaves who work under conditions of forced labor on coffee plantations in Latin America and Africa. The sugar in your coffee may have come from plantations where children and men in Latin America, Asia, and Africa are subjected to conditions of forced labor and debt bondage.

12:00 P.M.: Eat Lunch

The fish you eat may have been caught by men in Southeast Asia and by children as young as four years old in West Africa, who are subjected to conditions of forced labor where they have been deprived of wages, food, water, and shelter while working long hours and suffering abuse.

2:00 P.M.: Afternoon Snack

The chocolate snack that you eat may have come from plantations in Africa where the children who produce the cocoa—the key ingredient in chocolate—are subjected to conditions

of forced labor. As many as 300,000 children may work in cocoa production worldwide.

4:00 P.M.: Drive to Meeting

The tires on your car are made of rubber, which may have been produced on rubber plantations in Asia and Africa where entire families are forced to work excessive hours on the plantations for little or no pay and in hazardous working conditions.

6:00 P.M.: Arrive at Home

Bricks in the walls of your house may have been produced in brick kilns in Asia or Latin America by men, women, and children who are forced to work in hazardous conditions as victims of bonded labor.

8:00 P.M.: Enjoy Dinner

The food you eat for dinner may have been touched by men and children subjected to forced labor—with little or no pay and extensive physical and emotional abuse—on cattle ranches and farms in the United States, Latin America, and Africa.

11:00 P.M.: Go to Bed

The cotton in your sheets may have been picked by men, women, and children—some as young as three years old—in cotton fields in Central Asia and Africa.

According to the U.S. Department of State, this scenario accounts for only one day. If you want to see how modern slavery may touch the rest of your life, check out the Slavery Footprint website.

REFLECT & ANALYZE

Why is it important for us to be aware of human slavery? Is this concern relevant to our daily lives? Is there any relationship between the American Dream and the exploitation of people on a global basis?

Source: U.S. State Department, 2014, 2015.

1990s, the Afrikaners controlled the government, the police, and the military by enforcing *apartheid*—the separation of the races. Blacks were denied full citizenship and restricted to segregated hospitals, schools, residential neighborhoods, and other facilities. Whites held almost all of the desirable jobs; blacks worked as manual laborers and servants.

In a caste system, marriage is endogamous, meaning that people are allowed to marry only within their own group. In India, parents traditionally have selected marriage partners for their children. In South Africa, interracial marriage was illegal until 1985.

Cultural beliefs and values sustain caste systems. Hinduism, the primary religion of India, reinforces the caste system by teaching that people should accept their fate in life and work hard as a moral duty. Caste systems grow weaker as societies industrialize; the values reinforcing the system break down, and people start to focus on the types of skills needed for industrialization.

As we have seen, in closed systems of stratification, group membership is hereditary, and it is almost impossible to move up within the structure. Custom and law frequently perpetuate privilege and ensure that higher-level positions are reserved for the children of the advantaged.

The Class System

The *class system* is a type of stratification based on the ownership and control of resources and on the type of work that people do. At least theoretically, a class system is more open than a caste system because the boundaries between classes are less distinct than the boundaries between castes. In a class system, status comes at least partly through achievement rather than entirely by ascription.

In class systems, people may become members of a class other than that of their parents through both intergenerational and intragenerational mobility, either upward or downward. *Horizontal mobility* occurs when people experience a gain or loss in position and/or income that does not produce a change in their place in the class structure. For example, a person may get a pay increase and a more prestigious title but still not move from one class to another.

By contrast, movement up or down the class structure is *vertical mobility*. For example, Bruce is a physician, but he was the first person in his family to attend college, much less to graduate from medical school. His father was a day laborer picked up each day by contractors outside the local lumberyard to work on various building sites. His mother was a stay-at-home-mom who took care of Bruce and his four younger siblings. Bruce's parents did not complete high school and had little support from home because their families had limited economic means. In high school, Bruce became involved in the Upward Bound program, which encourages young people to remain in school and provides mentoring, tutoring, and enrichment activities that ultimately helped him achieve his dream.

Bruce's situation reflects upward mobility; however, people may also experience downward mobility, caused by any number of reasons, including a lack of jobs, low wages and employment instability, marriage to someone with fewer resources and less power than oneself, and changing social conditions.

Classical Perspectives on Social Class

Early sociologists grappled with the definition of class and the criteria for determining people's location in the class structure. Both Karl Marx and Max Weber viewed class as an important determinant of social inequality and social change, and their works have had a profound influence on how we view the U.S. class system today.

 LO2 **Explain** Karl Marx's views on social class and stratification.

Karl Marx: Relationship to the Means of Production

According to Karl Marx, class position and the extent of our income and wealth are determined by our work situation, or our relationship to the means of production. As we

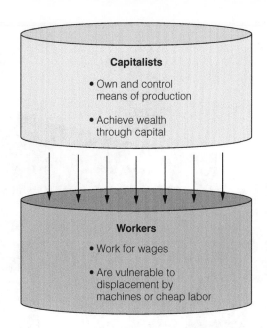

FIGURE 8.3 Marx's View of Stratification

have previously seen, Marx stated that capitalistic societies consist of two classes—the capitalists and the workers. The *capitalist class (bourgeoisie)* consists of those who own the means of production—the land and capital necessary for factories and mines, for example. The *working class (proletariat)* consists of those who must sell their labor to the owners in order to earn enough money to survive (see • Figure 8.3).

According to Marx, class relationships involve inequality and exploitation. Capitalists maximize their profits by exploiting workers, paying them less than the resale value of what they produce but do not own. This exploitation results in workers' *alienation*—a feeling of powerlessness and estrangement from other people and from oneself. In Marx's view, alienation develops as workers manufacture goods that embody their creative talents but the goods do not belong to them. Workers are also alienated from the work itself because they are forced to perform it in order to live. Because the workers' activities are not their own, they feel self-estrangement. Moreover, workers are separated from others in the factory because they individually sell their labor power to the capitalists as a commodity.

class system
a type of stratification based on the ownership and control of resources and on the type of work that people do.

capitalist class (bourgeoisie)
Karl Marx's term for those who own and control the means of production.

working class (proletariat)
Karl Marx's term for those who must sell their labor to the owners in order to earn enough money to survive.

alienation
a feeling of powerlessness and estrangement from other people and from oneself.

FIGURE 8.4   Karl Marx believed that class relationships involve inequality and exploitation. How is his idea represented in this photo?

prediction does not mean that his analysis of capitalism and his theoretical contributions to sociology are without validity.

Marx had a number of important insights into capitalist societies. First, he recognized the economic basis of class systems (Gilbert, 2014). Second, he noted the relationship between people's social location in the class structure and their values, beliefs, and behavior. Finally, he acknowledged that classes may have opposing (rather than complementary) interests. For example, capitalists' best interests are served by a decrease in labor costs and other expenses and a corresponding increase in profits; workers' best interests are served by well-paid jobs, safe working conditions, and job security (● Figure 8.4).

In Marx's view, the capitalist class maintains its position at the top of the class structure by control of the society's *superstructure,* which is composed of the government, schools, churches, and other social institutions that produce and disseminate ideas perpetuating the existing system of exploitation. Marx predicted that the exploitation of workers by the capitalist class would ultimately lead to *class conflict*—the struggle between the capitalist class and the working class. According to Marx, when the workers realized that capitalists were the source of their oppression, they would overthrow the capitalists and their agents of social control, leading to the end of capitalism. The workers would then take over the government and create a more egalitarian society.

Why has no workers' revolution occurred? Capitalism may have persisted because it has changed significantly since Marx's time. Individual capitalists no longer own and control factories and other means of production; today, ownership and control have largely been separated. For example, contemporary transnational corporations are owned by a multitude of stockholders but are run by paid officers and managers. Similarly, many (but by no means all) workers have experienced a rising standard of living, which may have contributed to a feeling of complacency. During the twentieth century, workers pressed for salary increases and improvements in the workplace through their activism and labor union membership. They also gained more legal protection in the form of workers' rights and benefits such as workers' compensation insurance for job-related injuries and disabilities. For these reasons, and because of a myriad of other complex factors, the workers' revolution predicted by Marx has not come to pass. However, the failure of his

 LO3   **Discuss** Max Weber's multidimensional approach to social stratification.

Max Weber: Wealth, Prestige, and Power

Max Weber's analysis of class builds upon earlier theories of capitalism (particularly those by Marx) and of money (particularly those by Georg Simmel, as discussed in Chapter 1). Living in the late nineteenth and early twentieth centuries, Weber was in a unique position to see the transformation that occurred as individual, competitive, entrepreneurial capitalism went through the process of shifting to bureaucratic, industrial, corporate capitalism. As a result, Weber had more opportunity than Marx to see how capitalism changed over time.

Weber agreed with Marx's assertion that economic factors are important in understanding individual and group behavior. However, Weber emphasized that no single factor (such as economic divisions between capitalists and workers) was sufficient for defining the location of categories of people within the class structure. According to Weber, the access that people have to important societal resources (such as economic, social, and political power) is crucial in determining people's life chances. To highlight the importance of life chances for categories of people, Weber developed a multidimensional approach to social stratification that reflects the interplay among wealth, prestige, and power. In his analysis of these dimensions of

FIGURE 8.5   What level of wealth, power, and prestige do you believe these women have?

class structure, Weber viewed the concept of "class" as an *ideal type* (one that can be used to compare and contrast various societies) rather than as a specific social category of "real" people (Bourdieu, 1984).

Wealth is the value of all of a person's or family's economic assets, including income, personal property, and income-producing property. Weber placed categories of people who have a similar level of wealth and income in the same class. For example, he identified a privileged commercial class of *entrepreneurs*—wealthy bankers, ship owners, professionals, and merchants who possess similar financial resources. He also described a class of *rentiers*—wealthy individuals who live off their investments and do not have to work. According to Weber, entrepreneurs and rentiers have much in common. Both are able to purchase expensive consumer goods, control other people's opportunities to acquire wealth and property, and monopolize costly status privileges (such as education) that provide contacts and skills for their children.

Weber divided those who work for wages into two classes: the middle class and the working class. Traditionally, according to this type of classification, the middle class consists of white-collar workers, public officials, managers, and professionals. The working class consists of skilled, semiskilled, and unskilled workers. As we shall see

later, these categories have shifted with the introduction of new technologies and media in the contemporary world economic structure.

The second dimension of Weber's system of stratification is *prestige*—the respect or regard that a person or status position is given by others. Fame, respect, honor, and esteem are the most common forms of prestige. A person who has a high level of prestige is assumed to receive deferential and respectful treatment from others. Weber suggested that individuals who share a common level of social prestige belong to the same status group regardless of their level of wealth (● Figure 8.5). They tend to socialize with one another, marry within their own group of social equals, spend their leisure time together, and safeguard their status by restricting outsiders' opportunities to join their ranks.

class conflict
Karl Marx's term for the struggle between the capitalist class and the working class.

wealth
the value of all of a person's or family's economic assets, including income, personal property, and income-producing property.

prestige
the respect or regard that a person or status position is given by others.

The other dimension of Weber's system is *power*—the ability of people or groups to achieve their goals despite opposition from others. The powerful can shape society in accordance with their own interests and direct the actions of others. According to Weber, bureaucracies hold social power in modern societies; individual power depends on a person's position within the bureaucracy. Weber suggested that the power of modern bureaucracies was so strong that even a workers' revolution (as predicted by Marx) would not lessen social inequality.

Weber stated that wealth, prestige, and power are separate continuums on which people can be ranked from high to low, as shown in • Figure 8.6. Individuals may be high on one dimension while being low on another. For example, people may be very wealthy but have little political power (for example, a recluse who has inherited a large sum of money). They may also have prestige but not wealth (for instance, a college professor who receives teaching excellence awards but lives on a relatively low income). In Weber's multidimensional approach, people are ranked on all three dimensions. Sociologists often use the term *socioeconomic status (SES)* to refer to a combined measure that attempts to classify individuals, families, or households in terms of factors such as income, occupation, and education to determine class location.

What important contribution does Weber make to our understanding of social stratification and class? Weber's analysis of social stratification contributes to our understanding by emphasizing that people behave according to both their economic interests and their values. He also added to Marx's insights by developing a multidimensional explanation of the class structure and by identifying additional classes. Both Marx and Weber emphasized that capitalists and workers are the primary players in a class society, and both noted the importance of class to people's life chances. However, they saw different futures for capitalism and the social system. Marx saw these structures being overthrown; Weber saw the increasing bureaucratization of life even without capitalism.

LO4 **List** and note the key characteristics of U.S. social classes.

Contemporary Sociological Models of the U.S. Class Structure

How many social classes exist in the United States? What criteria are used for determining class membership? No broad consensus exists about how to characterize the class structure in this country. In fact, many people deny that class distinctions exist. Most people like to think of themselves as middle class; it puts them in a comfortable middle position—neither rich nor poor. Sociologists have developed two models of the class structure: One is based on a Weberian approach, the other on a Marxian approach. We will examine both models briefly.

The Weberian Model of the U.S. Class Structure

Expanding on Weber's analysis of class structure, sociologist Dennis Gilbert (2014) uses a model of social classes based on three elements: (1) education, (2) occupation of family head, and (3) family income (see • Figure 8.7).

The Upper (Capitalist) Class

The upper class is the wealthiest and most powerful class in the United States. About 1 percent of the population is included in this class, whose members own substantial income-producing assets and operate on both the national and international levels. According to Gilbert (2014), people in this class have an influence on the economy and society far beyond their numbers. He estimates their annual income to be in the $2 million range.

Some models further divide the upper class into upper-upper ("old money") and lower-upper ("new money") categories (Warner and Lunt, 1941; Coleman and Rainwater, 1978; Kendall, 2002). Members of the

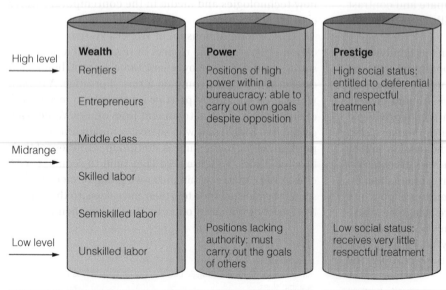

FIGURE 8.6 **Weber's Multidimensional Approach to Social Stratification**
According to Max Weber, wealth, power, and prestige are separate continuums. Individuals may rank high in one dimension and low in another, or they may rank high or low in more than one dimension. Also, individuals may use their high rank in one dimension to achieve a comparable rank in another. How does Weber's model compare with Marx's approach, as shown in Figure 8.3?

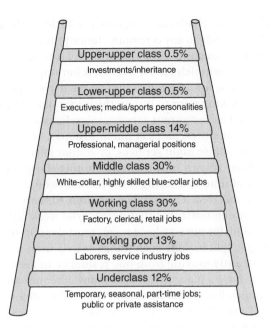

FIGURE 8.7 Stratification Based on Education, Occupation, and Income

upper-upper class come from prominent families which possess great wealth that they have held for several generations. In the past, family names—such as Rockefeller, Mellon, Du Pont, and Kennedy—were well-known and often held in high esteem. Today, some upper-class family names are well-known, but many of the individuals and families involved have often made their fortunes much more recently. Persons in the upper-upper class tend to have strong feelings of in-group solidarity. They belong to the same exclusive clubs and support high culture (such as the opera, symphony orchestras, ballet, and art museums). Children are educated in prestigious private schools and Ivy League universities; many acquire strong feelings of privilege from birth, as upper-class author Lewis H. Lapham (1988: 14) states:

> Together with my classmates and peers, I was given to understand that it was sufficient accomplishment merely to have been born. Not that anybody ever said precisely that in so many words, but the assumption was plain enough, and I could confirm it by observing the mechanics of the local society. A man might become a drunkard, a concert pianist or an owner of companies, but none of these occupations would have an important bearing on his social rank.

Children of the upper class are socialized to view themselves as different from others; they also learn that they are expected to marry within their own class (Warner and Lunt, 1941; Mills, 1959a; Kendall, 2002).

Members of the lower-upper class may be extremely wealthy but not have attained as much prestige as members of the upper-upper class. The "newer rich" have earned most of their money in their own lifetime as entrepreneurs,

presidents of major corporations, sports or entertainment celebrities, or top-level professionals. Gilbert (2014) refers to people in this category as the "working rich."

The Upper-Middle Class Persons in the upper-middle class are often highly educated professionals who have built careers as physicians, attorneys, stockbrokers, or corporate managers. Others derive their income from family-owned businesses. According to Gilbert (2014), about 14 percent of the U.S. population is in this category, and they have an annual income in the $150,000 range. A combination of three factors qualifies people for the upper-middle class: university degrees, authority and independence on the job, and high income. Of all the class categories, the upper-middle class is the one that is most shaped by formal education.

Over the past fifty years, Asian Americans, Latinos/as, and African Americans have placed great importance on education as a means of attaining the American Dream. Many people of color have moved into the upper-middle class by acquiring higher levels of education. However, higher levels of educational attainment are not always a guarantee of wealth and success for persons in any racial–ethnic category.

The Middle Class In past decades a high school diploma was necessary to qualify for most middle-class jobs. Today, two-year or four-year college degrees have replaced the high school diploma as an entry-level requirement for employment in many middle-class occupations, including medical technicians, nurses, legal and medical assistants, lower-level managers, semiprofessionals, and nonretail sales workers. An estimated 30 percent of the U.S. population is in the middle class, even though most people in this country think of themselves as middle class. Annual income in the middle class is in the $70,000 range (Gilbert, 2014).

Traditionally, most middle-class occupations have been relatively secure and have provided more opportunities for advancement (especially with increasing levels of education and experience) than working-class positions. Recently, however, several factors have eroded the hope for upward mobility and achievement of the American Dream for this class: (1) the housing crisis in the first decade of the twenty-first century, beginning with an overpriced housing market and too-easy credit, followed by high rates of foreclosures and a banking meltdown; (2) job loss and instability in some economic sectors,

power
the ability of people or groups to achieve their goals despite opposition from others.

socioeconomic status (SES)
a combined measure that attempts to classify individuals, families, or households in terms of factors such as income, occupation, and education to determine class location.

with long-term unemployment and blocked mobility on the job; and (3) national economic sluggishness, starting with the 2007 U.S. recession that eventually brought some economic recovery, but primarily for persons in the top of the class structure.

The Working Class

An estimated 30 percent of the U.S. population is in the working class, which is characterized by an annual income in the $40,000 range (Gilbert, 2014). In the past, this class was typically made up of people in occupational categories such as semiskilled machine operators in factories. Today, many in the working class are employed in the service sector as clerks, salespeople, and fast-food workers whose job responsibilities involve routine, mechanized tasks requiring little skill beyond basic literacy and a brief period of on-the-job training (Gilbert, 2014). Some people in the working class are employed in *pink-collar occupations*—relatively low-paying, nonmanual, semiskilled positions primarily held by women, such as day-care workers, checkout clerks, cashiers, and restaurant servers.

How does life in working-class families compare with that of individuals in middle-class families? According to sociologists, working-class families not only earn less than middle-class families, but they also have less financial security, particularly because of high rates of layoffs and plant closings in some regions of the country. Few people in the working class have more than a high school diploma, and many have less, which makes job opportunities scarce for them in a "high-tech" society (Gilbert, 2014). Others find themselves in low-paying jobs in the service sector of the economy, particularly fast-food restaurants, a condition that often places them among the working poor (• Figure 8.8).

The Working Poor

The working poor account for about 13 percent of the U.S. population. Members of the working-poor class have an annual income in the $25,000 range, living from just above to just below the poverty line. They typically hold unskilled jobs, seasonal migrant jobs in agriculture, lower-paid factory jobs, and service jobs (such as counter help at restaurants). Employed single mothers often belong to this class; consequently, children are overrepresented in this category. African Americans and other people of color are also overrepresented among the working poor. Living from paycheck to paycheck makes it impossible for the working poor to save sufficient money for emer-gencies like a vehicle breaking down or job loss, which are constant threats to any economic stability they may have.

Over a decade ago, social critic and journalist Barbara Ehrenreich (2001, 2011) left her upper-middle-class lifestyle for a period of time to see if it was possible for the working poor to live on the wages that they were being paid as restaurant servers, salesclerks at discount department stores, aides at nursing homes, housecleaners for franchise maid services, or similar jobs. She conducted her research by actually holding those jobs for periods of time and seeing if she could live on the wages that she received. Through her research, Ehrenreich persuasively demonstrated that people who work full time, year round, for poverty-level wages must develop survival strategies that include such things as help from relatives or constantly moving from one residence to another in order to have a place to live. Like many other researchers, Ehrenreich found that minimum-wage jobs cannot cover the full cost of living, such as rent, food, and the rest of an adult's monthly needs, even without taking into consideration the needs of children or other family members. If Ehrenreich's study were to be replicated now, we would probably find that the tasks she sought to accomplish are even more difficult today because jobs have become more scarce, more people are working part time who desire full-time employment, and the cost of living continues to increase in many regions.

At some point in our lives, most of us have held a job paying the minimum wage, and we know the limitations of trying to survive on such low earnings. The federal *minimum wage* is the hourly rate that (with certain exceptions) is the lowest amount an employer can legally pay

Craig Warga/Bloomberg/Getty Images

FIGURE 8.8 In which segment of the class structure would sociologists place this worker? What are the key elements of that social class?

its employees (each state may adopt a higher minimum wage but not a lower one). In 2015 the federal minimum wage was $7.25 per hour, where it had been since 2009. A person earning minimum wage and working forty hours every week, fifty-two weeks per year (in other words, no time off, no vacation), would still earn an amount slightly above the *official poverty line* (and slightly below that line for a person with two children). Several states have a minimum wage above the federal requirement: Washington, D.C.'s minimum wage was increased to $10.50 an hour in 2015—making it the first to cross the $10.00 per hour threshold. Numerous discussions have taken place about increasing the federal minimum wage, as well as the wage in various states, but political leaders are often extremely divided on such issues, and little change actually occurs in the laws. Increasing social and economic inequality in the United States has been partly attributed to the vast divide between the low wages paid to workers, based on a low federal minimum wage, and the astronomically high salaries and compensation packages given to some major corporate CEOs and others at the top of the socioeconomic ladder.

The Underclass   People in the **underclass** are poor, seldom employed, and caught in long-term deprivation that results from low levels of education and income and high rates of unemployment. About 12 percent of the U.S. population is in the underclass, where the annual income is in the $15,000 range (Gilbert, 2014). Some persons in this category are unable to work because of age or disability; others experience discrimination based on race/ethnicity. Single mothers are overrepresented in this class because of the lack of jobs, lack of affordable child care, and many other impediments to the mother's future and that of her children. People without a "living wage" often must rely on public- or private-assistance programs for their survival.

Gaining work-related skills and having employment opportunities are two critical issues for people on the lowest rungs of the class ladder. Many of the jobs that exist today require specialized knowledge or skills that are inaccessible to people in the underclass. Skills and jobs are essential for people to have the opportunity to earn a decent wage; have medical coverage; live meaningful, productive lives; and raise their children in a safe environment. These issues are closely tied to upward mobility and the American Dream that we have been discussing in this chapter.

The Marxian Model of the U.S. Class Structure

The earliest Marxian model of class structure identified ownership or nonownership of the means of production as the distinguishing feature of classes. From this perspective, classes are social groups organized around property ownership, and social stratification is created and maintained by one group in order to protect and enhance its own economic interests. Moreover, societies are organized around classes in conflict over scarce resources. From this perspective, inequality results from the most powerful exploiting the less powerful.

Contemporary Marxian (or conflict) models examine class in terms of people's relationship to others in the production process. For example, conflict theorists attempt to determine the degree of control that workers have over the decision-making process and the extent to which they are able to plan and implement their own work. They also analyze the type of supervisory authority, if any, that a worker has over other workers. According to this approach, most employees are a part of the working class because they do not control either their own labor or that of others.

Erik Olin Wright (1979, 1985, 1997, 2010), one of the leading stratification theorists to examine social class from a Marxian perspective, has concluded that Marx's definition of "workers" does not fit the occupations found in advanced capitalist societies. For example, many top executives, managers, and supervisors who do not own the means of production (and thus would be "workers" in Marx's model) act like capitalists in their zeal to control workers and maximize profits. Likewise, some experts hold positions in which they have control over money and the use of their own time even though they are not owners. Wright views Marx's category of "capitalist" as being too broad as well. For instance, small-business owners might be viewed as capitalists because they own their own tools and have a few people working for them, but they have little in common with large-scale capitalists and do not share the interests of factory workers. • Figure 8.9 compares Marx's and Wright's models.

Wright (1979) also argues that classes in modern capitalism cannot be defined simply in terms of different levels of wealth, power, and prestige, as in the Weberian model. Consequently, he outlines four criteria for placement in the class structure: (1) ownership of the means of production, (2) purchase of the labor of others (employing others), (3) control of the labor of others (supervising others on the job), and (4) sale of one's own labor (being employed by someone else). Wright (1978) assumes that these criteria can be used to determine the class placement of all workers, regardless of race/ethnicity, in a capitalist society. Let's take a brief look at Wright's (1979, 1985) four classes—(1) the capitalist class, (2) the managerial class, (3) the small-business class, and (4) the working class—so that you can compare them to those found in the Weberian model.

The Capitalist Class   According to Wright, this class holds most of the wealth and power in society through ownership of capital—for example, banks, corporations, factories, mines, news and entertainment industries, and agribusiness

pink-collar occupations
relatively low-paying, nonmanual, semiskilled positions primarily held by women.

underclass
those who are poor, seldom employed, and caught in long-term deprivation that results from low levels of education and income and high rates of unemployment.

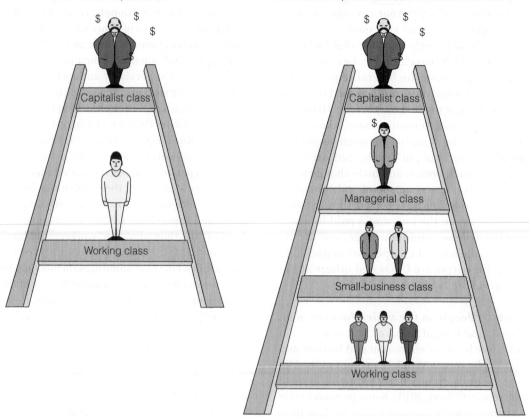

Marx's Model	Wright's Model
Based on relationship to the means of production:	Takes into account both ownership and control of the means of production and the labor of others:
Capitalist class	Capitalist class
	Managerial class
	Small-business class
Working class	Working class

FIGURE 8.9 Comparison of Marx's and Wright's Models of Class Structure

firms. The "ruling elites," or "ruling class," within the capitalist class hold political power and are often elected or appointed to influential political and regulatory positions.

This class is composed of individuals who have inherited fortunes, own major corporations, or are top corporate executives with extensive stock holdings or control of company investments. Even though many top executives have only limited *legal ownership* of their corporations, they have substantial economic ownership and exert extensive control over investments, distribution of profits, and management of resources. The major sources of income for the capitalist class are profits, interest, and very high salaries. Members of this class make important decisions about the workplace, including which products and services to make available to consumers and how many workers to hire or fire.

Forbes magazine's 2015 list of the richest people in the world identified Bill Gates (cofounder of Microsoft Corporation) as the wealthiest capitalist, with a net worth of $76 billion. Mexican telecom entrepreneur Carlos Slim Helu was the second-wealthiest person, with a net worth of $72 billion (*Forbes,* 2015). Although many men who made the *Forbes* wealthiest list gained their fortunes through entrepreneurship or being CEOs of large corporations,

most women who made the list acquired their wealth through inheritance, marriage, or both (● Figure 8.10). The wealthiest women in the world in 2015 were Christy Walton, U.S. heir to the Walmart empire ($36.7 billion); Liliane Bettencourt, French heir to the L'Oreal cosmetics fortune ($34.5 billion); and Alice Walton ($34.3 billion), also of the U.S. Walmart family (*Forbes,* 2015). To wrap your head around these figures, consider, for example, that it would take the wealth of 52.5 million American families to equal the net worth held by six members of the Walton family (the Walmart heirs) in the United States (Bivens, 2014).

The Managerial Class People in the managerial class have substantial control over the means of production and over workers. However, these upper-level managers, supervisors, and professionals usually do not participate in key corporate decisions such as how to invest profits. Lower-level managers may have some control over employment practices, including the hiring and firing of some workers.

Top professionals such as physicians, attorneys, accountants, and engineers may control the structure of their own work; however, they typically do not own the means of production and may not have supervisory authority over

FIGURE 8.10 Men who make the *Forbes* wealthiest list usually gain their fortunes through entrepreneurship while women are more likely to acquire their wealth through inheritance, marriage, or both. Carlos Slim Helu is an entrepreneur with a net worth of more than $65 billion that he made in telecommunications and real estate. He is seen here speaking at the Clinton Foundation Future of the Americas summit. By contrast, Alice Walton's more than $34 billion net worth came primarily from an inheritance from her father, Sam Walton, the founder of Walmart.

craftspeople—who may hire employees but also do some of their own work (● Figure 8.11). Some own businesses such as "mom-and-pop" grocery stores, retail clothing stores, and jewelry stores. Others have businesses in landscaping and groundskeeping, business and household cleaning, home health and personal care, and construction. Still others are professionals such as lawyers, management analysts, and accountants and auditors who receive relatively high incomes from selling their knowledge and services. Some of these professionals share attributes with members of the capitalist class because they have formed corporations that hire and control employees who produce profits for the professionals.

Of the approximately 10 million self-employed workers in the United States, nearly two-thirds are men, and more than 30 percent are 55 years of age or older, particularly workers who retire from wage and salary jobs and become self-employed to supplement their retirement income.

The Working Class The working class is made up of a number of subgroups, one of which is blue-collar workers, some of whom are highly skilled and well paid and others of whom are unskilled and poorly paid. Skilled blue-collar workers include electricians, plumbers, and carpenters; unskilled blue-collar workers include janitors and gardeners.

White-collar workers are another subgroup of the working class. Referred to by some as a "new middle class," these workers are actually members of the working class because they do not own the means of production, do not control the work of others, and are relatively powerless in the workplace. Secretaries, administrative assistants and other clerical workers, and salesworkers are members of the white-collar faction of the working class. They take orders from others and tend to work under constant supervision.

more than a few people. Even so, they may influence the organization of work and the treatment of other workers. Members of the capitalist class often depend on these professionals for their specialized knowledge.

The Small-Business Class This class consists of people who are self-employed—small-business owners and

Thus, these workers are at the bottom of the class structure in terms of domination and control in the workplace. The working class contains about half of all employees in the United States.

Although Marxian and Weberian models of the U.S. class structure show differences in people's occupations and access to valued resources, neither fully reflects the nature

Bo Zaunders/Encyclopedia/Corbis

FIGURE 8.11 Many immigrants believe that they can achieve the American Dream by starting a small business, such as a store or restaurant. Other immigrants are employed in major retail concessions, such as the man shown here who works at an airport newsstand.

and extent of inequality in the United States. In the next section we will take a closer look at the unequal distribution of income and wealth in the United States and the effects of inequality on people's opportunities and life chances.

 LO5 **Distinguish** between income inequality and wealth inequality.

Inequality in the United States

Throughout human history, people have argued about the distribution of scarce resources in society. Disagreements often center on whether the share we get is a fair reward for our effort and hard work. Recently, social analysts have pointed out that (except during temporary economic downturns) the old maxim "the rich get richer" continues to be valid in the United States. To understand how this happens, we must take a closer look at the distribution of income and wealth in this country.

Distribution of Income and Wealth

Money is essential for acquiring goods and services. People without money cannot purchase food, shelter, clothing, medical care, legal aid, education, and the other things they need or desire. Money—in the form of both income and wealth—is very unevenly distributed in the United States. Among high-income nations, the United States remains number one in inequality of income distribution.

Income Inequality *Income* is the economic gain derived from wages, salaries, income transfers (governmental aid), and ownership of property. Or, to put it another way, income comes from the money, wages, and payments that people receive from their occupation or investments. Data from the U.S. Census Bureau typically provide income estimates that are based solely on money income before taxes and do not include the value of noncash benefits such as health care coverage or retirement benefits.

Sociologist Dennis Gilbert (2014) compares the distribution of income to a national pie that has been sliced into portions, ranging in size from stingy to generous, for distribution among segments of the population. As shown in • Figure 8.12, in 2013 the wealthiest 20 percent of U.S. households received more than half of the total income "pie," while the poorest 20 percent of households received slightly more than 3 percent of all income. The top 5 percent *alone* received 22 percent of all income—an amount greater than that received by the bottom 40 percent of all households.

The gap between the rich and the poor has widened significantly in recent decades. Between 1981 and 2013, the median income of the top one-fifth of U.S. households increased by more than four times—about twice the rate of increase of the bottom one-fifth of families (U.S. Census Bureau Historical Income Tables, 2015) (see • Figure 8.13). Although the recent recession may have reduced the annual income of some

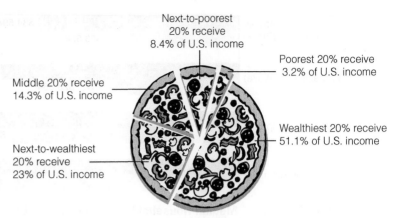

FIGURE 8.12   Distribution of Pretax Income in the United States, 2013
Thinking of personal income in the United States (before taxes) as a large pizza helps us to see which segments of the population receive the largest and smallest portions.
Source: DeNavas-Walt and Proctor, 2014.

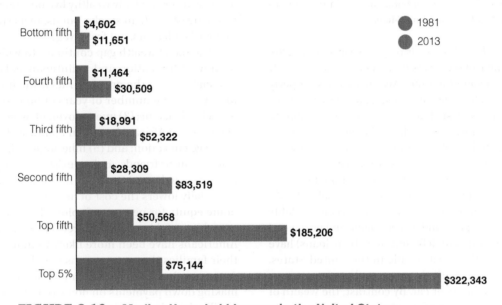

FIGURE 8.13   Median Household Income in the United States
This chart shows the distribution of before-tax income in the United States. Notice the dramatic increase in income for the top 5 percent of households. During the past three decades the difference in income between the richest and poorest has become much more pronounced.
Source: U.S. Census Bureau Historical Income Tables, 2015.

rich Americans, this does not mean that they have become poor. Their income may have dropped by several million dollars, causing them to slightly modify their affluent lifestyles, but their overall standard of living is protected by other wealth they possess. For example, the net worth of top U.S. billionaires such as Bill Gates and Warren Buffett may have dropped several billion dollars, but they are far from living in poverty. However, the picture is different for people situated in the middle and bottom sectors of the income pie who are faced with high mortgage rates or rent payments and increasing costs for food, fuel, transportation, and other necessities.

Income distribution varies by race/ethnicity as well as class. • Figure 8.14 compares median household income by race/ethnicity, showing not only the disparity among groups but also the consistency of that disparity over almost two decades. Although households across all racial–ethnic categories have experienced some decline in real annual median income, the income gap between African American (black) households and white (non-Hispanic)

income
the economic gain derived from wages, salaries, income transfers (governmental aid), and ownership of property.

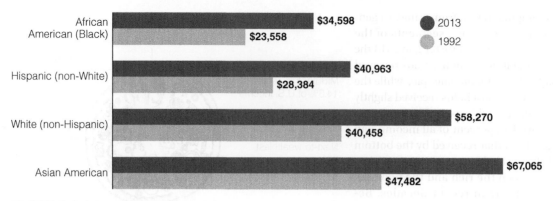

FIGURE 8.14 Median Household Income by Race/Ethnicity in the United States
Sources: DeNavas-Walt, Cleveland, and Webster, 2003; DeNavas-Walt and Proctor, 2014.

and Asian or Pacific Islanders is striking. In 2013 African American households had the lowest median income, $34,598, as compared with Asian households, which had the highest median, $67,065. Non-Hispanic white households had a median income of $58,270, as compared with Hispanic households, which had $40,963.

Wealth Inequality Income is only one aspect of wealth. Wealth is the value of the assets that an individual or family holds at a specific point in time. Wealth includes property such as buildings, land, farms, houses, factories, and cars, as well as other assets such as bank accounts, corporate stocks, bonds, and insurance policies. Wealth is computed by subtracting all debt obligations and converting the remaining assets into cash. Wealth is also referred to as "net worth," and it is a very important component in determining the standard of living for families and households. *Net worth* is the total household wealth after secured debts (such as home mortgages and vehicle loans) and unsecured debts (such as credit card debt and some bank loans) have been subtracted. For most people in the United States, wealth is invested primarily in property that generates no income, such as a home or car. By contrast, the wealth of elites is often in the form of income-producing property or investments that make money for them. Wealth is especially important because it can be transferred across generations and thus increases inequality over time in a nation.

Wealth is even more unevenly distributed than income. In 2013 the top 10 percent of U.S. households received 47.3 percent of U.S. income but owned 75.3 percent of U.S. wealth (Federal Reserve Bulletin, 2014). The top 3 percent of U.S. households owned more than half (54.4 percent) of all U.S. wealth, followed by the next 7 percent at 20.9 percent. This means that the bottom 90 percent of households held slightly less than one-quarter (24.7 percent) of all wealth in 2013, down from nearly one-third (33.2 percent) in 1989 (Federal Reserve Bulletin, 2014).

Since the 1970s, the wealth of the richest 1 percent in the United States has increased dramatically. In fact, the net worth of each household in the wealthiest 1 percent is nearly 300 times greater than the net worth of the median U.S. household. (The "median U.S. household" refers to families that are in the exact middle of the distribution for their group. Half of the people in the group will have more than the median, and half will have less.) Many people in less-affluent income categories live on wages from their jobs; however, the truly wealthy live on, and become richer from, investments in stocks, bonds, real estate, and other financial endeavors, not from wages.

The racial wealth gap continues to widen in terms of both wealth creation and maintenance. Factors closely associated with the growing racial wealth divide are as follows: (1) the number of years of homeownership; (2), household income; (3) unemployment, which is more prevalent among African American families; (4) possession of a college education; and (5) inheritance, financial support from family or friends, and preexisting family wealth (Shapiro, Meschede, and Osoro, 2013). Residential segregation artificially lowers the cost of housing and the building of home equity for persons of color who own residences in nonwhite neighborhoods. Moreover, for many years white Americans have been more likely to inherit money from their families than are persons of color. Typically, white Americans have also received more parental assistance in making down payments on home purchases because their families have had greater financial reserves. White Americans have also had easier access to credit or have been able to acquire credit with lower interest rates and lending costs (Shapiro, Meschede, and Osoro, 2013).

Given these factors, the median wealth of white households is about thirteen times higher than the median wealth of black (African American) households and more than ten times that of Hispanic (Latino/a) households. • Figure 8.15 shows the racial divide in net worth in 2013 dollars. White (non-Hispanic) Americans, at $141,900, have much higher net worth than Hispanic Americans (Latinos/as), at $13,700, and African Americans (blacks), at $11,000 (Pew Research Center, 2014). So if we go back to our example of the Walton family ("Walmart heirs"), it would take the net worth of the bottom 67.4 percent of nonwhite families to match the net worth of the Waltons (Bivens, 2014). The current gap between African Americans and whites is higher than it has been since the late 1980s. We are unable to calculate where Asian

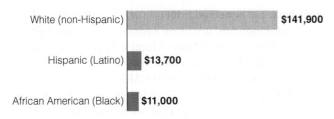

White (non-Hispanic) $141,900

Hispanic (Latino) $13,700

African American (Black) $11,000

FIGURE 8.15 **Racial Divide in Net Worth, 2013**
Source: Federal Reserve Bulletin, 2014.

Americans specifically fit in the wealth mix because in recent data Asian Americans and other racial–ethnic groups are not identified separately by the Federal Reserve.

LO6 **Describe** three important consequences of inequality in the United States.

Consequences of Inequality

Income and wealth are not simply statistics; they are intricately related to the American Dream and our individual life chances. Persons with a high income or substantial wealth have more control over their own lives. They have greater access to goods and services; they can afford better housing, more education, and a wider range of medical services. Persons with less income, especially those living in poverty, must spend their limited resources to acquire the basic necessities of life.

Physical Health, Mental Health, and Nutrition People who are wealthy, well educated, and have high-paying jobs are much more likely to be healthy than are poor people. As people's economic status increases, so does their health status. The poor have shorter life expectancies and are at greater risk for chronic illnesses such as type 2 diabetes, heart disease, and cancer, as well as infectious diseases such as tuberculosis. Compared with adults living in households with at least four times the poverty level, adults living below the poverty line are five times more likely to report their health as being fair or poor and more than eight times more likely to report serious psychological distress (National Center for Health Statistics, 2014).

Children born into poor families are at much greater risk of dying during their first year of life. Some die from disease, accidents, or violence. Others are unable to survive because they are born with low birth weight, a condition linked to birth defects and increased probability of infant mortality. Low birth weight in infants is attributed, at least in part, to the inadequate nutrition received by many low-income pregnant women. Most of the poor do not receive preventive medical and dental checkups; many do not receive adequate medical care after they experience illness or injury (Children's Defense Fund, 2015).

Many high-poverty areas lack an adequate supply of doctors and medical facilities. Even in areas where such services are available, the inability to pay often prevents people from seeking medical care when it is needed. Some "charity" clinics and hospitals may provide indigent patients (those who cannot pay) with minimal emergency care but make them feel stigmatized in the process. For some poor individuals living in states that did not expand Medicaid coverage under the Affordable Care Act, the lack of adequate health insurance remains a pressing problem. The Census Bureau classifies health insurance coverage as private coverage or government coverage. Private health insurance is a plan provided through an employer or a union, or is a plan purchased by an individual from a private company. By contrast, government health insurance includes such programs as Medicare, Medicaid, military health care, the Children's Health Insurance Program (CHIP), and individual state health plans.

As provisions of the Affordable Care Act were being implemented in 2013–2014, approximately 42 million adults in the United States—13.4 percent of the U.S. population—had no health insurance coverage for the entire year (Smith and Medalia, 2014). • Figure 8.16 shows that the rate of uninsured people increases as household income decreases. About 20 percent of all households with income less than $25,000 were uninsured in 2013, as compared with only 5.3 percent of households uninsured when income was $150,000 and over (Smith and Medalia, 2014). Many people rely on their employers for health coverage, but some employers are cutting back on health coverage, particularly for employees' family members. However, the overall number and percentage of people without health insurance decreased between 2011 and 2013 as a result of an increase in the number of people enrolled in plans established by the Affordable Care Act.

In 2013 the majority of people (64.2 percent) with health insurance were covered by private plans, primarily employment-based health insurance. The percentage of people covered by government health programs increased from 32.2 percent in 2011 to 34.3 percent in 2013. Medicaid accounted for 54 million people, slightly over 17 percent of all individuals

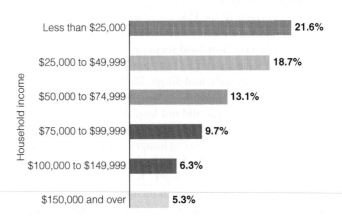

Less than $25,000 — 21.6%

$25,000 to $49,999 — 18.7%

$50,000 to $74,999 — 13.1%

$75,000 to $99,999 — 9.7%

$100,000 to $149,999 — 6.3%

$150,000 and over — 5.3%

FIGURE 8.16 **Rate of Uninsurance by Household Income, 2013**
Source: Smith and Medalia, 2014.

covered by government health insurance plans. As discussed in more detail in Chapter 18, Medicaid is government-funded health insurance coverage for persons who meet specific lower-income guidelines and other criteria. The percentage of people covered by Medicare in 2013 was 15.6 percent (or 49 million individuals) (Smith and Medalia, 2014).

When the effects of the Affordable Care Act are seen in the future, further reductions in the number and percentage of uninsured persons will probably occur because the law includes a requirement that individuals must purchase coverage under various circumstances. Those who do not enroll in a health care plan will be assessed increasing penalties (estimated to range from $95 to $695 per adult and $47.50 to $347.50 per child up to a specified dollar figure) in the years 2014–2016 and beyond.

Although the precise relationship between class and health is not known, analysts suggest that people with higher income and wealth tend to smoke less, exercise more, maintain a healthy body weight, and eat nutritious meals. In addition, many lower-paying jobs are often the most dangerous and have the greatest health hazards. Black lung disease, cancers caused by asbestos, and other environmental hazards found in the workplace are more likely to affect manual laborers and low-income workers, as are job-related accidents.

Good health is basic to good life chances; in turn, adequate nutrition is essential for good health. Hunger is related to class position and income inequality. Studying the problem of hunger has become more complex in recent years because the Department of Agriculture stopped using the word *hunger* in its reports in 2006. *Food insecure* is now used to identify people whose access to adequate food is limited by lack of money and other resources (Coleman-Jensen, Gregory, and Singh, 2014). People living with food insecurity include those who are unable to afford the basics, those who are unable to get to the grocery store, and those who are unable to find fresh, nutritious produce and other foods to eat because they are surrounded by fast-food stores that do not provide foods with proper nutrition. In 2013, 17.5 million Americans (or 14.3 percent of the total U.S. population) lived in households that were "food insecure" (Coleman-Jensen, Gregory, and Singh, 2014). Of the 14.3 percent of the U.S. population that reported that they were in food-insecure households, 8.7 percent were in households with "low food security," and 5.6 percent were in those households with "very low food security" (Coleman-Jensen, Gregory, and Singh, 2014). Indicators of "low food security" are factors such as "worrying food would run out," "food bought did not last," and "could not afford balanced meals." "Very low food security" is characterized by these factors, plus "being hungry but not eating," "losing weight," "did not eat whole day," and "did not eat whole day for three months or longer." Rates of food insecurity are highest in households with incomes near or below the poverty line, households with children headed by single women or single men, and black and Hispanic households.

Sixty-two percent of individuals in food-insecure households reported in one survey that they, or other family members, had benefited from one or more of the three largest federal food and nutrition assistance programs: SNAP, WIC, and the National School Lunch Program. The number of people who received benefits from the Supplemental Nutrition Assistance Program (SNAP)—formerly the "Food Stamp Program"—was 46.5 million in 2014. The average benefit received was $125 per person per month. WIC refers to the Special Supplemental Nutrition Program for Women, Infants, and Children. This federally funded program serves about 8.7 million participants who receive nutrition education and vouchers that they can use to acquire supplemental food packages at authorized food stores. Still other food assistance comes from the National School Lunch Program, which annually provides 30.7 million children with food in more than 100,000 public and private schools and residential child-care facilities. Approximately 62 percent of all meals in the National School Lunch Program are served free, another 8 percent are provided at reduced prices, and the remaining one-third are paid for by either students or schools (Coleman-Jensen, Gregory, and Singh, 2014).

Housing As discussed in Chapter 5, homelessness is a major problem in the United States. It is estimated that about 1.2 million children are homeless each year, and 40 percent of all homeless children are 5 years of age or younger (Children's Defense Fund, 2015). The lack of affordable housing is a pressing concern for many low-income individuals and families. Housing is considered affordable when a household spends no more than 30 percent of its income on rent or mortgage payments. But about 12 million renter and homeowner households pay more than half of their annual income for housing (U.S. Department of Housing and Urban Development, 2015).

Lack of *affordable* housing is one central problem brought about by economic inequality. A family with one full-time worker earning the minimum wage cannot afford the local fair-market rent for a two-bedroom apartment anywhere in the United States (U.S. Department of Housing and Urban Development, 2015). Another concern is *substandard* housing, which refers to facilities that have inadequate heating, air conditioning, plumbing, electricity, or structural durability. Structural problems—caused by faulty construction or lack of adequate maintenance—exacerbate the potential for other problems such as damage from fire, falling objects, or floors and stairways collapsing. A final issue is the extent to which a family lives in a neighborhood with a high poverty rate. Studies have shown that high school graduation rates are as much as 20 percentage points lower in neighborhoods with a high poverty rate than graduation rates in higher-income areas (Children's Defense Fund, 2015).

Education Educational opportunities and life chances are directly linked. Some functionalist theorists view education as the "elevator" to social mobility. Improvements in the educational achievement levels (measured in number of years of schooling completed) of the poor, people of color, and white women have been cited as evidence that students'

FIGURE 8.17 Conflict theorists see schools as agents of the capitalist class system that perpetuate social inequality: Upper-class students are educated in well-appointed environments such as the one shown here, whereas children of the poor tend to go to antiquated schools with limited facilities.

abilities are now more important than their class, race, or gender. From this perspective, inequality in education is declining, and students have an opportunity to achieve upward mobility through achievements at school. Functionalists generally see the education system as flexible, allowing most students the opportunity to attend college if they apply themselves (Ballantine and Hammack, 2012).

In contrast, most conflict theorists stress that schools are agencies for reproducing the capitalist class system and perpetuating inequality in society. From this perspective, education perpetuates poverty. Parents with low educational attainment and limited income are often not able to provide the same educational opportunities for their children as families where at least one parent has more formal education.

Today, great disparities exist in the distribution of educational resources. Because funding for education comes primarily from local property taxes, school districts in wealthy suburban areas generally pay higher teachers' salaries, have newer buildings, and provide state-of-the-art equipment (• Figure 8.17). By contrast, schools in poorer areas have a limited funding base. Students in central-city schools and poverty-stricken rural areas often attend dilapidated schools that lack essential equipment and teaching materials. As far back as the early 1990s, author Jonathan Kozol (1991, qtd. in Feagin and Feagin, 1994: 191) documented the effect of a two-tiered system on students, which remains in many of today's schools:

> Kindergartners are so full of hope, cheerfulness, high expectations. By the time they get into fourth grade, many begin to lose heart. They see the score, understanding

they're not getting what others are getting. . . . They see suburban schools on television. . . . They begin to get the point that they are not valued much in our society. By the time they are in junior high, they understand it. "We have eyes and we can see; we have hearts and we can feel. . . . We know the difference."

As previously noted, poverty extracts such a toll that many young people will not have the opportunity to finish high school, much less enter college.

 LO7 **Identify** the characteristics of the U.S. poor based on age, gender, and race/ethnicity.

Poverty in the United States

So far, we have examined various forms of inequality in the United States and their effects. Let's now focus more closely on the problem of poverty in this country.

The United States has the highest rate of poverty among wealthy countries. When many people think about poverty, they think of people who are unemployed or on welfare. However, many hardworking people with full-time jobs live in poverty. The U.S. Social Security Administration has established an *official poverty line*, which is based on

official poverty line
the income standard that is based on what the federal government considers to be the minimum amount of money required for living at a subsistence level.

what the federal government considers to be the minimum amount of money required for living at a subsistence level. The poverty level (or poverty line) is computed by determining the cost of a minimally nutritious diet (a low-cost food budget on which a family could survive nutritionally on a short-term, emergency basis) and multiplying this figure by three to allow for nonfood costs. In 2014 about 14.5 percent of the U.S. population had income below the official government poverty level for a family of four with two adults and two children under the age of eighteen. In 2015 the federal government increased the official poverty level for a family of four to $24,250.

Poverty rates vary widely across the United States. The percentage of people living below the poverty line is higher in some states and regions than in others (see • Figure 8.18). The highest poverty rates by state are in Mississippi (23.9 percent) and New Mexico (21.4 percent). The lowest poverty rates are in New Hampshire (9.0 percent), Alaska (10.1 percent), and Maryland (10.2 percent). Based on region, the highest rates of poverty are in the South and West, with the lowest being in the Northeast and Midwest.

When sociologists define poverty, they distinguish between absolute and relative poverty. *Absolute poverty* exists when people do not have the means to secure the most-basic necessities of life. This definition comes closest to that used by the federal government. Absolute poverty often has life-threatening consequences, such as when a homeless person freezes to death on a park bench. By comparison, *relative poverty* exists when people may be able

to afford basic necessities but are still unable to maintain an average standard of living. A family must have income substantially above the official poverty line in order to afford the basic necessities, even when these are purchased at the lowest possible cost. But many families do not earn enough money to afford living comfortably and must survive on an economy budget, as described below:

Members of families existing on the economy budget never go out to eat, for it is not included in the food budget; they never go out to a movie, concert, or ball game or indeed to any public or private establishment that charges admission, for there is no entertainment budget; they have no cable television, for the same reason; they never purchase alcohol or cigarettes; never take a vacation or holiday that involves any motel or hotel or, again, any meals out; never hire a baby-sitter or have any other paid child care; never give an allowance or other spending money to the children; never purchase any lessons or home-learning tools for the children; never buy books or records for the adults or children, or any toys, except in the small amounts available for birthday or Christmas presents ($50 per person over the year); never pay for a haircut; never buy a magazine; have no money for the feeding or veterinary care of any pets; and, never spend any money for preschool for the children, or educational trips for them away from home, or any summer camp or other activity with a fee.

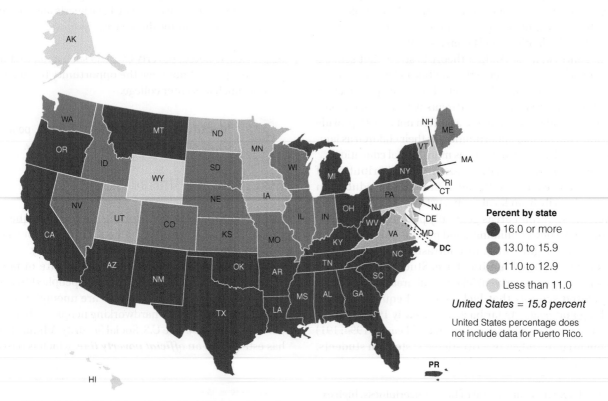

FIGURE 8.18 Percentage of People in Poverty in the Past 12 Months by State, 2013

Source: U.S Census Bureau, 2012 American Community Survey, 2012 Puerto Rico Community Survey.

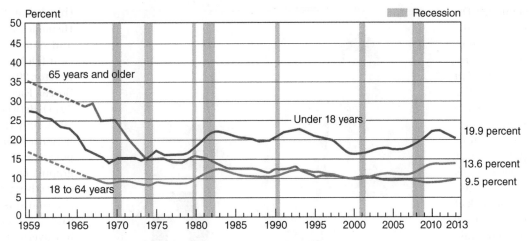

FIGURE 8.19 **U.S. Poverty Rates by Age, 1959–2013**

Note: The data points are placed at the midpoints of the respective years. Data for people 18 to 64 and 65 and older are not available from 1960 to 1965.

Source: DeNavas-Walt and Proctor, 2014, based on U.S. Census Bureau, Current Population Survey, 1960 to 2014 Annual Social and Economic Supplements.

Take a guess: When do you think this statement was written? Recently? No, this statement was written more than twenty years ago (1992) by social scientists John Schwarz and Thomas Volgy to describe the limited resources of people at or on the edge of poverty. Today, many people live in similar or worse conditions. Some participate in the occasional protest, such as the brief Occupy movement; others hope that they will not sink to the bottom rungs of poverty.

Who Are the Poor?

Poverty in the United States is not randomly distributed but rather is highly concentrated according to age, gender, and race/ethnicity.

Age In the past, persons over age 65 were at the greatest risk of being poor; however, older individuals today have the lowest poverty rate of all age categories, largely because of Social Security. Actually titled "Old Age, Survivors, Disability, and Health Insurance," Social Security is a federal insurance program established in 1935 that protects against loss of income caused by retirement, disability, or death. When Social Security was established in 1935, it was intended to supplement other savings and assets. But for more than half of Americans over age 65, Social Security provides more than fifty percent of their income, and without Social Security income, nearly half of all U.S. seniors would be living in poverty (Edwards, Turner, and Hertel-Fernandez, 2011). Instead, fewer than one in ten senior citizens lives below the poverty line, making Social Security the most successful antipoverty program in the United States. However, because Social Security benefits are based on the number of years of paid employment and preretirement earnings, women and minorities, who often earn less during their employment years, receive less in Social Security benefits and are less protected against poverty in old age (AARP.com, 2012).

The child poverty rate in the United States is higher than in other industrialized countries. In the United States today, children under age 18 have the highest rate of poverty, followed by people age 18 to 64 (see • Figure 8.19). In 2013 the number of U.S. children in poverty fell from 16.1 million to 14.7 million (a percentage change from 21.8 percent to 19.9 percent). However, children under age 18 represent nearly 24 percent of the total population but almost one-third (32.3 percent) of individuals living in poverty (DeNavas-Walt and Proctor, 2014). The precarious position of African American and Latino/a children is even more striking. Nearly two in five African American (black) children, one in three Hispanic children, and more than one in three American Indian/Native Alaskan children are from families living below the official poverty line, as compared with nearly one in seven white (non-Hispanic) children.

Gender In 2013 the poverty rate was 13.1 percent for males of all ages, as compared with 15.8 percent for females of all ages. However, these figures do not tell the entire story. Gender differences in poverty rates are more pronounced for people age 65 and older and among younger women who head single-parent families. The poverty rate for women age 65 and older was 11.6 percent, while it was 6.8 percent for men. Single-parent families headed by women in 2013 had a 30.6 percent poverty rate as compared with a 15.9 percent rate for male-householder-with-no-wife-present families and a 5.8 percent rate for married-couple two-parent families (DeNavas-Walt and Proctor, 2014). In her now-classical

absolute poverty
a level of economic deprivation that exists when people do not have the means to secure the most basic necessities of life.

relative poverty
a level of economic deprivation that exists when people may be able to afford basic necessities but are still unable to maintain an average standard of living.

study, the sociologist Diana Pearce (1978) coined a term to describe this problem of gender-specific poverty: The *feminization of poverty* refers to the trend in which women are disproportionately represented among individuals living in poverty. Over the decades since Pearce's study, women have continued to face a higher risk of being poor because they bear the major economic and emotional burdens of raising children when they are single heads of households. This problem is compounded by the fact that the female-to-male earnings ratio was 78 percent in 2013, which means that for every dollar that a male worker earns, a female worker earns 78 cents. This constitutes a so-called gender wage gap of 22 percent. This issue is further discussed in Chapter 11, "Sex, Gender, and Sexuality," and Chapter 13, "The Economy and Work in Global Perspective."

Does the feminization of poverty explain poverty in the United States today? Clearly, this thesis highlights a genuine problem—the link between gender and poverty (• Figure 8.20). However, all women are not equally vulnerable to poverty: Many in the upper and upper-middle classes have the financial resources, education, and skills to support themselves regardless of the presence of a man in the household. Moreover, poverty is everyone's problem, not just women's. When women are impoverished, so are their children. Likewise, many of the poor in our society are men, especially those who are chronically unemployed, older men, the homeless, men with disabilities, and men of color.

Race/Ethnicity In 2013 whites (non-Hispanic) accounted for 62.4 percent of the U.S. population but 41.5 percent of people in poverty. Whites (non-Hispanic) had the lowest rate of poverty—9.6 percent—of any racial–ethnic group. The highest rate of poverty was among American Indian (Native Americans) and Alaskan Native populations at 29.1 percent, followed by African Americans

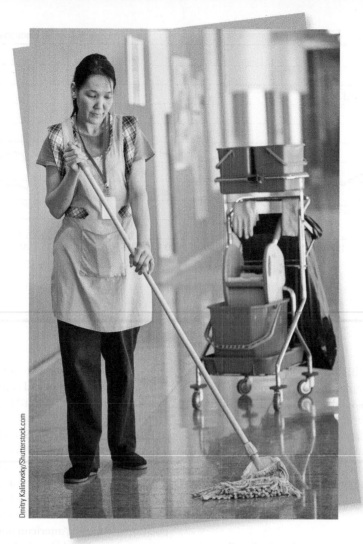

Dmitry Kalinovsky/Shutterstock.com

FIGURE 8.20 Many women are among the "working poor," who, although employed full time, have jobs in service occupations that are typically lower paying and less secure than jobs in other sections of the labor market. Does the nature of women's work contribute to the feminization of poverty in the United States?

(blacks) at 27.2 percent, Hispanics at 23.5 percent, and Asians at 10.5 percent (DeNavas-Walt and Proctor, 2014; U.S. Census Bureau Facts for Features, 2014). Between 2010 and 2013, only Hispanics experienced a decline in the poverty rate. Some demographic factors no doubt contributed to slight fluctuations in poverty rates by race/ethnicity, but overall rates remain persistently stubborn to change.

Economic and Structural Sources of Poverty

Social inequality and poverty have both economic and structural sources. Unemployment is a major cause of contemporary poverty that involves both economic and structural components. Tough economic times provide fewer opportunities for individuals to get a position that may help them gain a toehold in U.S. society. In January 2015, 9 million persons were unemployed in the United States, making the unemployment rate 5.7 percent. Although the unemployment rate for adult women was slightly lower (at 5.6 percent) than the average, much higher rates were reported for teenagers (18.8 percent) and African Americans (10.3 percent). About 32 percent of the unemployed (2.8 million) had been jobless for 27 weeks or more in early 2015 (U.S. Bureau of Labor Statistics, 2015). Even as unemployment rates declined some in 2014–2015, more people did not necessarily find work. Persons are no longer counted as unemployed if they drop out of the labor market and are no longer actively looking for a job.

In addition to unemployment, low wages paid for many jobs are another major cause of poverty: Although most working families are not officially poor, many are "near-poor" or "low-income," struggling to pay for basic needs such as housing, health care, food, child care, and transportation. Nearly one in three working families in the United States is a "low-income" family, earning less than 200 percent of the U.S. poverty threshold.

Structural problems contribute to both unemployment and underemployment. Corporations have been disinvesting in the United States, displacing millions of people from their jobs. Economists refer to this displacement as the *deindustrialization of America.* Even as they have closed their U.S. factories and plants, many corporations have opened new facilities in other countries where lower-wage labor exists because people will, of necessity, work for lower wages. Many analysts have documented how the relocation of domestic manufacturing offshore has drained millions of manufacturing jobs from the U.S. economy.

Job deskilling—a reduction in the proficiency needed to perform a specific job—leads to a corresponding reduction in the wages for that job or in the use of nonhuman technologies to perform the work. This kind of deskilling has resulted from the computerization and automation of the workplace. A significant step in job deskilling occurs when the primary responsibility of human operators is to monitor automated systems. The eventual outcome of such deskilling is that human operators either do not know what to do when the system fails or they are replaced entirely by automated technology. Other structural problems include the overall shift from manufacturing to service occupations in the United States, which has resulted in the loss of higher-paying positions and their replacement with lower-paying and less-secure positions that do not offer the wages, job stability, or advancement potential of the disappearing jobs.

Technological advances and changing patterns of consumerism have also contributed to unemployment in the United States and other high-income nations. For example, the introduction of smartphones, tablets, and other electronic devices means that fewer people now own a watch, camera, calculator, printed calendar, or numerous other features (or apps) found on the typical smartphone or tablet. As a result, fewer people are needed to design, make, and repair numerous items that created jobs for millions in the past. The Internet has also contributed to a decline in hundreds of thousands of jobs in the postal, publishing, and printing industries as people now e-mail or text one another or self-publish their ideas rather than going the traditional route. For example, thousands of postal jobs have disappeared because of a dramatic reduction in the number of items to be delivered, and automated systems have reduced the number of workers employed at the post office itself.

FIGURE 8.21   This California electronic benefit transfer (EBT) card represents a modern approach to helping people of limited income purchase groceries. Data-encoded cards such as this one were developed to prevent the trading or selling of traditional food stamps. However, one drawback of this technology is that many of California's popular farmers' markets are not able to process EBT cards.

Solving the Poverty Problem

The United States has attempted to solve the poverty problem in several ways. One of the most enduring is referred to as social welfare. When most people think of "welfare," they think of food stamps (currently called SNAP) and programs such as Temporary Assistance for Needy Families (TANF) or the earlier program it replaced, Aid to Families with Dependent Children (AFDC). Some who receive benefits from welfare programs tend to be stigmatized, even when our nation describes itself as having compassion for the less fortunate (● Figure 8.21).

 LO8   **Compare** and contrast functionalist, conflict, and symbolic interactionist perspectives on social inequality.

Sociological Explanations of Social Inequality in the United States

Obviously, some people are disadvantaged as a result of social inequality. In this section we examine some sociological explanations of social inequality. In doing so we will see

feminization of poverty
the trend in which women are disproportionately represented among individuals living in poverty.

job deskilling
a reduction in the proficiency needed to perform a specific job that leads to a corresponding reduction in the wages for that job or in the use of nonhuman technologies to perform the work.

how different sociologists answer this question: Is inequality always harmful to society?

Functionalist Perspectives

According to the well-known classical sociologists Kingsley Davis and Wilbert Moore (1945), inequality is not only inevitable but also necessary for the smooth functioning of society. The Davis–Moore thesis, which has become the definitive functionalist explanation for social inequality, can be summarized as follows:

1. All societies have important tasks that must be accomplished and certain positions that must be filled.

2. Some positions are more important for the survival of society than others.

3. The most important positions must be filled by the most qualified people (• Figure 8.22).

4. The positions that are the most important for society and that require scarce talent, extensive training, or both must be the most highly rewarded.

5. The most highly rewarded positions should be those that are functionally unique (no other position can perform the same function) and on which other positions rely for expertise, direction, or financing.

Davis and Moore use the physician as an example of a functionally unique position. Doctors are very important to society and require extensive training, but individuals would not be motivated to go through years of costly and stressful medical training without incentives to do so. The Davis–Moore thesis assumes that social stratification results in *meritocracy*—a hierarchy in which all positions are rewarded based on people's ability and credentials.

A key problem with the Davis–Moore thesis is that it ignores inequalities based on inherited wealth and intergenerational family status. The thesis assumes that

FIGURE 8.22 According to a functionalist perspective, people such as these Harvard Law School graduates attain high positions in society because they are the most qualified and they work the hardest. Is our society a meritocracy? How would conflict theorists answer this question?

economic rewards and prestige are the only effective motivators for people and fails to take into account other intrinsic aspects of work, such as self-fulfillment. It also does not adequately explain how such a reward system guarantees that the most-qualified people will gain access to the most highly rewarded positions.

Conflict Perspectives

From a conflict perspective, people with economic and political power are able to shape and distribute the rewards, resources, privileges, and opportunities in society for their own benefit. Conflict theorists do not believe that inequality serves as a motivating force for people; they argue that powerful individuals and groups use ideology to maintain their favored positions at the expense of others. Core values in the United States emphasize the importance of material possessions, hard work, individual initiative to get ahead, and behavior that supports the existing social structure. These same values support the prevailing resource-distribution system and contribute to social inequality.

Are wealthy people smarter than others? According to conflict theorists, certain stereotypes suggest that this is the case; however, the wealthy may actually be "smarter" than others only in the sense of having "chosen" to be born to wealthy parents from whom they could inherit assets. Conflict theorists also note that laws and informal social norms support inequality in the United States. For the first half of the twentieth century, both legalized and institutionalized segregation and discrimination reinforced employment discrimination and produced higher levels of economic inequality. Although laws have been passed to make these overt acts of discrimination illegal, many forms of discrimination still exist in educational and employment opportunities.

Symbolic Interactionist Perspectives

Symbolic interactionists focus on microlevel concerns and usually do not analyze larger structural factors that contribute to inequality and poverty. However, many significant insights on the effects of wealth and poverty on people's lives and social interactions can be derived from applying a symbolic interactionist approach. Using qualitative research methods and influenced by a symbolic interactionist approach, researchers have collected the personal narratives of people across all social classes, ranging from the wealthiest to the poorest people in the United States.

A few studies provide rare insights into the social interactions between people from vastly divergent class locations. In sociologist Judith Rollins's (1985) classic study of the relationship between household workers and their employers, she examined rituals of deference that were often demanded by elite white women of their domestic workers, who were frequently women of color. According to the sociologist Erving Goffman (1967), *deference* is a type of ceremonial activity that functions as a symbolic means whereby appreciation is regularly conveyed to

CONCEPT QUICK REVIEW

Sociological Explanations of Social Inequality in the United States

Functionalist Perspectives	Some degree of social inequality is necessary for the smooth functioning of society (in order to fill the most important positions) and thus is inevitable.
Conflict Perspectives	Powerful individuals and groups use ideology to maintain their favored positions in society at the expense of others.
Symbolic Interactionist Perspectives	The beliefs and actions of people reflect their class location in society.

a recipient. In fact, deferential behavior between nonequals (such as employers and employees) confirms the inequality of the relationship and each party's position in the relationship relative to the other. Rollins identified three types of linguistic deference between domestic workers and their employers: use of the first names of the workers, contrasted with titles and last names (Mrs. Adams, for example) of the employers; use of the term *girls* to refer to female household workers regardless of their age; and deferential references to employers, such as "Yes, ma'am." Spatial demeanor, including touching and how close one person stands to another, is an additional factor in deference rituals across class lines. Rollins (1985: 232) concludes that

> The employer, in her more powerful position, sets the essential tone of the relationship; and that tone ... is one that functions to reinforce the inequality of the relationship, to strengthen the employer's belief in the rightness of her advantaged class and racial position, and to provide her with justification for the inegalitarian social system.

Many concepts introduced by the sociologist Erving Goffman (1959, 1967) could be used as springboards for examining microlevel relationships between inequality and people's everyday interactions. What could you learn about class-based inequality in the United States by using a symbolic interactionist approach to examine a setting with which you are familiar?

The Concept Quick Review summarizes the three major perspectives on social inequality in the United States.

meritocracy
a hierarchy in which all positions are rewarded based on people's ability and credentials.

Students Helping Others Through Campus Kitchen

Since 2001:
Pounds of food recovered: 4.16 million
Meals prepared: 2.34 million

The Campus Kitchen Project

My life has gotten to the point where if I'm not in class I'm sleeping or doing something for Campus Kitchen. . . . So, I'll let y'all in on a little secret: Campus Kitchen is worth being excited about. For the uninitiated, Campus Kitchen is a student-led organization. We rescue and cook food to distribute to those in need. Five days a week students go to the dining halls and pick up pans of food and take them to the Salvation Army. On Tuesday afternoons, students are busy in the Family and Consumer Sciences kitchen creating healthy snacks to be given to children at local schools. On Thursdays, the kitchen crew cooks a meal that is delivered to the women and children at the Family Abuse Center.

Excited yet?

—Amy Heard (2011), a Baylor University student, describing how much she enjoys volunteering for Campus Kitchen

What is Campus Kitchen? Currently, more than 42 college campuses are involved in Campus Kitchen. This on-campus student service program is part of the Campus Kitchen Project, begun by Robert Egger, director of the nonprofit D.C. Central Kitchen in Washington, D.C. As Amy Heard explains, students go to dining halls, cafeterias, and local food banks or restaurants to pick up un-served, usable food and to make sure that meals gets to a local organization that feeds persons in need. Each college provides on-campus space for the "Campus Kitchen," such as a dining hall at off-hours or a classroom/kitchen, where students can prepare meals and snacks using the donated food. Students then deliver the meals to individuals and families in need of food assistance and to organizations such as homeless shelters and soup kitchens in their community. In addition to food recovery, delivery, and cooking, some campus groups, including the one at Baylor University, have community gardens in which people water, prune, harvest, or deliver fresh produce. This is most important for

people living in low-income areas who have a difficult time getting fresh, affordable fruits and vegetables for a balanced diet and proper nutrition.

In addition to helping start Campus Kitchen, Robert Egger has been an inspiration for many others who make a difference by helping feed the hungry. Egger is one of the people responsible for an innovative chef-training program that feeds hope as well as hunger. At the Central Kitchen, located in the nation's capital, staff and guest chefs annually train homeless persons in three-month-long kitchen-arts courses. While the trainees are learning about food preparation, which will help them get starting jobs in the restaurant industry, they are also helping feed homeless persons each day. Much of the food is prepared using donated goods such as turkeys that people have received as gifts at office parties and given to the kitchen, and leftover food from grocery stores, restaurants, hotels, and college cafeterias.

Can you think of ways that leftover food could be recovered from your college or university or other places where you eat so the food could be redistributed to persons in need? For more information, check the websites of the following organizations:

- Campus Kitchen Project
- D.C. Central Kitchen

Here is an example of Campus Kitchen at work. These Gonzaga University undergraduates are using leftovers from the dining hall to put together meals for the needy. Does your college have a similar program?

Looking Ahead: U.S. Stratification in the Future

The United States continues to face one of the greatest economic challenges it has experienced since the Great Depression of the 1930s. Although we have strong hopes that the American Dream will remain alive and well, many people are concerned about the lack of upward mobility for many Americans and a decline for others. The nationwide slump in housing and jobs has distressed people across all income levels, and continued high rates of unemployment and a shifting stock market bring about weekly predictions that things are either getting only slightly better or are becoming worse. Perhaps one of the most critical factors contributing to a lack of optimism about future mobility in the United States is the vast wealth gap between the rich and everyone else, and the depth of poverty in this country.

Given the current economic situation, it is difficult to predict the future of the U.S. system of stratification. What will happen with the great economic imbalance in the United States? Economist and former Secretary of Labor Robert Reich (2010: 146) has summed up the problem as follows:

> None of us can thrive in a nation divided between a small number of people receiving an ever larger share of the nation's income and wealth, and everyone else receiving a declining share. The lopsidedness not only diminishes economic growth but also tears at the fabric of our society. America cannot succeed if the basic bargain at the heart of our economy remains broken. The most fortunate among us who have reached the pinnacles of power and success depend on a stable economic and political system. That stability rests on the public's trust that the system operates in the interest of us all. Any loss of such trust threatens the well-being of everyone.

Given this assessment, politicians, business leaders, and ordinary people must do all they can to reinvigorate the American Dream, or everyone's future—young and old alike—will look much dimmer as we continue on through the century.

In the middle of the second decade of the twenty-first century, it is alarming to see headlines such as "Middle Class Shrinks Further as More Fall Out Instead of Climbing Up" (Searcey and Gebeloff, 2015). If more than half of U.S. households were in the middle class in the late 1960s, what has happened? Why have more people fallen to the bottom of the class structure? Middle-class couples with children are among those having the hardest time holding their place in the structure, at least partly because of the Great Recession, when many middle-income jobs were lost, only to be replaced by lower-wage positions. The main route to the middle class has been through higher education, and as no one needs to tell you as a college student that this is an often expensive and very time-consuming pursuit.

Until median incomes improve and more middle-class jobs are available, people have to find innovative ways to increase their income and improve their lifestyle. Some work several jobs; others create a niche for themselves in social media or other newer technologies that did not even exist a few years ago. Overall, the bottom-line question becomes how to handle the rapidly growing gap between the highest income and wealth group and everyone else. A Pew Research Center (2014) study found that the gap between the upper-income and middle-income families in the United States has reached its highest level on record. More than that, upper-income families have a median net worth that is nearly 70 times that of lower-income families. For any meaningful change to occur, there must be consistent improvements in the job market, the unemployment rate, the stock market, and housing and oil and gas prices. Because many middle-class families had their homes as their primary source of net worth, the Great Recession damaged their overall economic well-being, and they have not fully recovered.

Are we sabotaging our future if we do not work constructively to eliminate vast income inequalities and high rates of poverty? It has been said that a chain is no stronger than its weakest link. If we apply this idea to the problem of vast income inequality and high rates of poverty, then it is to our advantage to see that those who cannot find work or do not have a job that provides a living wage receive adequate training and employment. Innovative programs can combine job training with producing something useful to meet the immediate needs of people living in poverty. Children of today—the adults of tomorrow—need nutrition, education, health care, and safety as they grow up (see "You Can Make a Difference").

CHAPTER REVIEW Q & A

LO 1 What is social stratification, and how do the three major systems of stratification compare?

Social stratification is the hierarchical arrangement of large social groups based on their control over basic resources. People are treated differently based on where they are positioned within the social hierarchies of class, race, gender, and age. Stratification systems include slavery, caste, and class. Slavery, an extreme form of stratification in which people are owned or controlled by others, is a closed system. The caste system is also a closed one in which people's status is determined at birth based on their parents' position in society. The class system, which exists in the United States, is a type of stratification based on ownership of resources and on the type of work that people do.

LO2 How did Karl Marx view social class and stratification?

Marx viewed social class as a key determinant of social inequality and social change. For Marx, class position and the extent of our income and wealth are determined by our work situation, or our relationship to the means of production. Marx stated that capitalistic societies consist of two classes—the capitalists and the workers—and class relationships involve inequality and exploitation.

LO3 What is Max Weber's multidimensional approach to social stratification?

Weber emphasized that no single factor (such as economic divisions between capitalists and workers) was sufficient for defining the location of categories of people within the class structure. Weber developed a multidimensional concept of stratification that focuses on the interplay of wealth, prestige, and power.

LO4 What are the key characteristics of social classes in the United States?

No broad consensus exists about how to characterize the class structure in this country. Sociologists have developed two models of the class structure: One is based on a Weberian approach, the other on a Marxian approach. In the Weberian-based approach, social classes are based on three elements: (1) education, (2) occupation of family head, and (3) family income. This approach to class structure consists of the upper class, the upper-middle class, the middle class, the working class, the working poor, and the underclass. Contemporary Marxian models examine class in terms of people's relationship to others in the production process.

LO5 What is the difference between income inequality and wealth inequality?

Income is the economic gain derived from wages, salaries, income transfers (governmental aid), and ownership of property. In 2013 the wealthiest 20 percent of U.S. households received more than half of the total income "pie," while the poorest 20 percent of households received slightly more than 3 percent of all income. Wealth includes property such as buildings, land, farms, houses, factories, and cars, as well as other assets such as bank accounts, corporate stocks, bonds, and insurance policies. Wealth is even more unevenly distributed than income.

LO6 What are three important consequences of inequality in the United States?

The stratification of society into different social groups results in wide discrepancies in income and wealth and in variable access to available goods and services. People with high income or wealth have greater opportunity to control their own lives. They can afford better housing, more education, and a wider range of medical services. People with less income have fewer life chances and must spend their limited resources to acquire basic necessities.

LO7 What are the characteristics of the U.S. poor based on age, gender, and race/ethnicity?

Age, gender, and race tend to be factors in poverty. Children have a greater risk of being poor than do the elderly, and women have a higher rate of poverty than do men. Although whites account for approximately two-thirds of those below the poverty line, people of color account for a disproportionate share of the impoverished in the United States.

LO8 How do functionalist, conflict, and symbolic interactionist perspectives on social inequality compare?

Functionalist perspectives view classes as broad groupings of people who share similar levels of privilege on the basis of their roles in the occupational structure. According to the Davis–Moore thesis, stratification exists in all societies, and some inequality is not only inevitable but also necessary for the ongoing functioning of society. The positions that are most important within society and that require the most talent and training must be highly rewarded. Conflict perspectives on class are based on the assumption that social stratification is created and maintained by one group (typically the capitalist class) in order to enhance and protect its own economic interests. Conflict theorists measure class according to people's relationships with others in the production process. Unlike functionalist and conflict perspectives that focus on macrolevel inequalities in societies, symbolic interactionist views focus on microlevel inequalities such as how class location may positively or negatively influence one's identity and everyday social interactions. Symbolic interactionists use terms such as *social cohesion* and *deference* to explain how class binds some individuals together while categorically separating out others.

KEY TERMS

absolute poverty 230	caste system 213	feminization of poverty 232
alienation 215	class conflict 216	income 224
capitalist class (bourgeoisie) 215	class system 215	intergenerational mobility 212

QUESTIONS for CRITICAL THINKING

1 Based on Max Weber's multidimensional class model of wealth, power, and prestige, how do entertainers and reality TV personalities, such as Lady Gaga and the Kardashians, become wealthy and well-known worldwide? Where do individuals such as these fit in Weber's system of stratification?

2 Should employment be based on meritocracy, need, or affirmative-action-type policies designed to bring about greater diversity?

3 What might happen in the United States if the gap between rich and poor continues to widen?

ANSWERS to the SOCIOLOGY QUIZ
ON WEALTH, POVERTY, AND THE AMERICAN DREAM

1 **False** In a national poll, less than a third (31 percent) of U.S. adults said that they had achieved the American Dream.

2 **False** The age group that includes people ages 65 and over actually has the lowest rate of poverty of any age group in the United States. Several decades ago the poverty rate of the elderly was much higher than it is today, but largely because of Social Security, rates of poverty among the elderly have dropped significantly.

3 **True** The United States has the highest rate of childhood poverty of any industrialized country. In 2015 more than one in five U.S. children lived in poverty.

4 **False** In 2015 a family of four in the 48 contiguous U.S. states would have to earn less than $24,250 to be considered poor. By contrast, the poverty threshold for a family of four was $27,890 in Hawaii and $30,320 in Alaska.

5 **True** *Forbes* magazine's 2015 list of the richest people in the world identified Microsoft founder, Bill Gates, as the wealthiest person in the world, with a net worth of $76 billion.

6 **True** The U.S. State Department estimates that at least 20,000 people are brought to the United States each year and are enslaved in such jobs as agriculture, domestic work, and sex work.

7 **True** As of July 2015, twenty-nine states, plus the District of Columbia, and nearly two dozen cities and counties had set higher minimum wages than the federal minimum wage.

8 **False** The gap between the rich and the poor has widened significantly in recent decades. In 1979 the richest 1 percent of Americans took in less than 9 percent of the country's total income. In 2014 the top 1 percent took more than one-fifth of all income earned by people in the United States. The top 1 percent also own about 40 percent of the nation's wealth.

Sources: Based on DeNavas-Walt and Proctor, 2014; Federal Reserve Bulletin, 2014; *Forbes*, 2015.

GLOBAL STRATIFICATION

LEARNING OBJECTIVES

1   **Define** *global stratification* and explain how it contributes to economic inequality within and between nations.

2   **Describe** the levels of development approach for studying global inequality.

3   **Identify** the World Bank classifications of nations into four economic categories, and explain why organizations such as this have problems measuring wealth and poverty on a global basis.

4   **Discuss** the relationship between global poverty and key human development issues such as life expectancy, health, education, and literacy.

5   **Define** modernization theory and list the four stages of economic development identified by Walt Rostow.

6   **Define** dependency theory and explain why this theory is often applied to newly industrializing countries.

7   **Describe** the key components of world systems theory and identify the three major types of nations set forth in this theory.

8   **Discuss** the new international division of labor theory and explain how it might be useful in the twenty-first century.

SOCIOLOGY & EVERYDAY LIFE

Leaving the Snare of Poverty

Angeline Mugwendere's parents were impoverished farmers in Zimbabwe, and she was mocked by classmates when she went to school barefoot and in a torn dress with nothing underneath. Teachers would sternly send her home to collect school fees that were overdue, even though everyone knew there was no way her family could pay them. Yet Angeline suffered the humiliations and teasing and pleaded to be allowed to remain in school. Unable to buy school supplies, she cadged [asked for] what she could.

"At break time, I would go to a teacher's house and say, 'Can I wash your dishes?'" she remembers. "And in return, they would sometimes give me a pen."

—In their best-selling book *Half the Sky*, the journalists Nicholas D. Kristof and Sheryl WuDunn (2009: 179) describe how some

How does poverty affect the educational opportunities of children in poor countries such as Zimbabwe?

impoverished girls have benefited from efforts to eradicate global poverty. Through help from organizations such as Campaign for Female Education (Camfed), Angeline acquired an education, moved out of poverty, and eventually became Camfed's executive director.

Global poverty affects people in a variety of ways: For some, it means absolute poverty because they do not have even the basic necessities to survive. For others, poverty is relative: Their standard of living remains below that of other people who reside in their nation and around the world. Without the intervention of individuals or organizations such as the one that helped Angeline attain more education and find productive employment, young people may be mired in hardscrabble poverty from which there is little possibility of escape. Consequently, their children are also born into poverty, and the cycle continues across generations.

Regardless of where people live in the world, social and economic inequalities are pressing daily concerns. Poverty and inequality know no political boundaries or national borders. Even within countries that are designated as "high income," many people live in poverty. Likewise, some wealthy people live in "low-income nations." Disparities between the rich and the poor within one high-income country may be greater than inequalities based on wealth and income that exist among some people who live in nations identified as middle income or low income. In this chapter we examine global stratification and inequality, and discuss sociological perspectives that have been developed to explain the nature and extent of this problem. Before reading on, test your knowledge of global wealth and poverty by taking the "Sociology and Everyday Life" quiz.

 Define *global stratification* and explain how it contributes to economic inequality within and between nations.

Wealth and Poverty in Global Perspective

What do we mean by global stratification? *Global stratification* refers to the unequal distribution of wealth, power, and prestige on a global basis, resulting in people having vastly different lifestyles and life chances both within and among the nations of the world. Just as the United States is divided into classes, the world is divided into unequal segments characterized by extreme differences in wealth and poverty. For example, the income gap between the richest and the poorest percentage of the world population continues to widen (see • Figure 9.1).

As previously defined, *high-income countries* have highly industrialized economies; technologically advanced industrial, administrative, and service occupations; and relatively high levels of national and per capita (per person) income. In contrast, *middle-income countries* have industrializing economies, particularly in urban areas, and moderate levels of national and personal income. *Low-income countries* have little industrialization and low levels of national and personal income.

Although some progress has been made in reducing extreme poverty and child mortality rates while improving

How Much Do You Know About Global Wealth and Poverty?

TRUE	FALSE		
T	F	1	Poverty has been increasing in the United States but decreasing in other nations because of globalization.
T	F	2	The assets of the world's 500 richest people are more than the combined income of over 50 percent of the world's population.
T	F	3	More than one billion people worldwide live below the international poverty line, earning less than $1.25 each day.
T	F	4	Although poverty is a problem in most areas of the world, relatively few people die of causes arising from poverty.
T	F	5	In low-income countries, the problem of poverty is unequally shared between men and women.
T	F	6	The majority of people with incomes below the poverty line live in urban areas of the world.
T	F	7	Two-thirds of adults (15 years and older) worldwide who are not able to read and write are men.
T	F	8	Poor people in low-income countries meet most of their energy needs by burning wood, dung, and agricultural wastes, which increases health hazards and environmental degradation.

Answers can be found at the end of the chapter.

health and literacy rates in some lower-income countries, the overall picture remains bleak. Many people have sought to address the issue of world poverty and to determine ways in which resources can be used to meet the urgent challenge of poverty. However, not much progress has been made on this front despite a great deal of talk and billions of dollars in "foreign aid" flowing from high-income nations to low-income nations. The idea of "development" has become one of the primary means used in attempts to reduce social and economic inequalities and alleviate the worst effects of poverty in the less industrialized nations of the world.

As we take a closer look at global stratification, there are a number of problems inherent in studying this issue, one of which is what terminology should be used to describe various nations. As we shall now see, a lack of consensus exists among political, economic, and social leaders on this topic.

Problems in Studying Global Inequality

One of the primary problems encountered by social scientists studying global stratification and social and economic inequality is what terminology should be used to refer to the distribution of resources in various nations. During the past sixty years, major changes have occurred in the way that inequality is addressed by organizations such as the United Nations and the World Bank. Most definitions of inequality are based on comparisons of levels of income or economic development, whereby countries are identified in terms of the "three worlds" or upon their levels of economic development.

The "Three Worlds" Approach

After World War II, the terms "First World," "Second World," and "Third World" were introduced by social analysts to distinguish among nations on the basis of their levels of economic development and the standard of living of their citizens. *First World* nations were said to consist of the rich, industrialized nations that primarily had capitalist economic systems and democratic political systems. The most frequently noted First World nations were the United States, Canada, Japan, Great Britain, Australia, and New Zealand. *Second World* nations were said to be countries with at least a moderate level of economic development and a moderate standard of living. These nations include China, Vietnam, Cuba, and portions of the former Soviet Union. According to social analysts, although the quality of life in Second World nations was not comparable to that of life in the First World, it was far greater than that of people living in the *Third World*—the poorest countries, with little

global stratification
the unequal distribution of wealth, power, and prestige on a global basis, resulting in people having vastly different lifestyles and life chances both within and among the nations of the world.

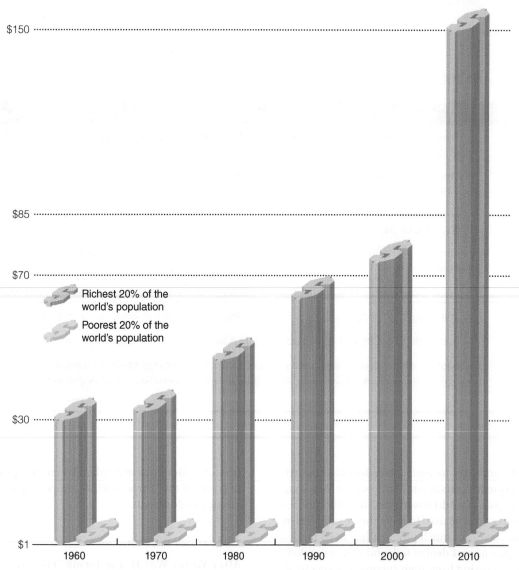

FIGURE 9.1 **Income Gap Between the World's Richest and Poorest People**
The income gap between the richest and poorest people in the world continued to grow between 1960 and 2010 (the last year for which information is available). As this figure shows, in 1960 the highest-income percentage of the world's population received $30 for each dollar received by the lowest-income percentage. By 2010, the disparity had increased: $150 to $1.

Source: International Monetary Fund, 2012; World Bank, 2012.

or no industrialization and the lowest standards of living, shortest life expectancies, and highest rates of mortality. Examples of these nations include Democratic Republic of the Congo, Liberia, Niger, and Sierra Leone.

 Describe the levels of development approach for studying global inequality.

The Levels of Development Approach

Among the most controversial terminology used for describing world poverty and global stratification has been the language of development. Terminology based on levels of development includes concepts such as developed nations, developing nations, less-developed nations, and underdevelopment. Let's look first at the contemporary origins of the idea of "underdevelopment" and "underdeveloped nations."

Following World War II, the concepts of *underdevelopment* and *underdeveloped nations* emerged out of the Marshall Plan (named after U.S. Secretary of State George C. Marshall), which provided massive sums of money in direct aid and loans to rebuild the European economic base destroyed during World War II. Given the Marshall Plan's success in rebuilding much of Europe, U.S. political leaders decided that the Southern Hemisphere nations that had recently been released from European colonialism could also benefit from a massive financial infusion and rapid economic development. Leaders of the developed nations argued that urgent problems such as poverty, disease, and famine could be reduced through the transfer of finance, technology, and experience from the developed nations to lesser-developed countries. From this viewpoint, economic development is the primary way to solve the poverty problem: Hadn't economic growth brought the developed nations to their own high standard of living?

Ideas regarding *underdevelopment* were popularized by President Harry S. Truman in his 1949 inaugural address. According to Truman, the nations in the Southern Hemisphere were "underdeveloped areas" because of their low gross national product, which today is referred to as *gross national income* (GNI)—a term that refers to all the goods and services produced in a country in a given year, plus the net income earned outside the country by individuals or corporations. If nations could increase their GNI, then social and economic inequality among the citizens within the country could also be reduced. Accordingly, Truman believed that it was necessary to assist the people of economically underdeveloped areas to raise their *standard of living,* by which he meant material well-being that can be measured by the quality of

goods and services that may be purchased by the per capita national income. Thus, an increase in the standard of living meant that a nation was moving toward economic development, which typically included the exploitation of natural resources by industrial development.

What has happened to the issue of development since the post–World War II era? After several decades of economic development fostered by organizations such as the United Nations and the World Bank, it became apparent by the 1970s that improving a country's GNI did not tend to reduce the poverty of the poorest people in that country. In fact, global poverty and inequality were increasing, and the initial optimism of a speedy end to underdevelopment faded.

Why did inequality increase even with greater economic development? Some analysts in the developed nations began to link growing social and economic inequality on a global basis to relatively high rates of population growth taking place in the underdeveloped nations (•Figure 9.2). Organizations such as the United Nations and the World Health Organization stepped up their efforts to provide family planning services to the populations so that they could control their own fertility. However, population researchers are now aware that issues such as population

FIGURE 9.2   Some analysts believe that growing global social and economic inequality is related to high rates of population growth taking place in underdeveloped nations. Why might this be so?

growth, economic development, and environmental problems must be seen as interdependent concerns. After the U.N. Conference on Environment and Development in Rio de Janeiro, Brazil (the "Earth Summit"), in 1992, terms such as *underdevelopment* were dropped by many analysts.

L03 **Identify** the World Bank classifications of nations into four economic categories, and explain why organizations such as this have problems measuring wealth and poverty on a global basis.

Classification of Economies by Income

The World Bank classifies nations into four economic categories and establishes the upper and lower limits for the gross national income (GNI) in each category. *Low-income economies* had a GNI per capita of less than $1,045 in 2015, *lower-middle-income economies* had a GNI per capita between $1,046 and $4,125, *upper-middle-income economies* had a GNI per capita between $4,126 and $12,745, and *high-income economies* had a GNI per capita of $12,746 or more (World Bank, 2015).

Low-Income Economies

Currently, about thirty-four nations are classified by the World Bank (2015) as low-income economies. In these economies, many people engage in agricultural pursuits, reside in nonurban areas, and are impoverished. As shown in • Figure 9.3, low-income economies are primarily found in countries in Asia and Africa, where half of the world's population resides.

Among those most affected by poverty in low-income economies are women and children. Why is this true? Fertility rates remain high in low-income economies. In all nations the poor have higher fertility rates than the wealthy residing within the same country. Other factors that contribute to the poverty of women and children are lack of educational opportunity, disadvantage in control over resources and assets in the household, gender disparities in work, and lower overall pay than men. To learn more about gender inequality worldwide, go online to the World Bank's *World Development Report.*

Middle-Income Economies

About one-third of the world's population resides in a middle-income economy. As previously stated, the World Bank has subdivided the middle-income economies into two categories—the lower-middle income ($1,046 to $4,125) and the upper-middle income ($4,126 to $12,745). Countries classified as lower-middle income include Armenia, Ghana, Guatemala, Honduras, Lesotho, Nigeria, Ukraine, and Vietnam. In recent years, millions of people have migrated from the world's poorest nations in hopes of finding better economic conditions elsewhere.

As compared with lower-middle-income economies, nations having upper-middle-income economies typically

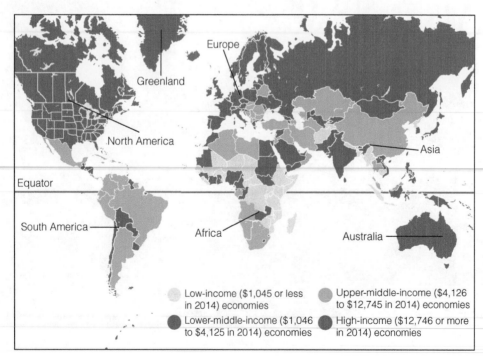

FIGURE 9.3 **High-, Middle-, and Low-Income Economies in Global Perspective**
Source: International Monetary Fund, 2012; World Bank, 2012.

have a somewhat higher standard of living. Nations with upper-middle-income economies include Angola, Argentina, Brazil, China, Colombia, Costa Rica, Cuba, Iraq, Jordan, Mexico, and Turkey. Some of these nations export a diverse variety of goods and services, ranging from manufactured goods to raw materials and fuels. For example, Kazakhstan, in Central Asia, is well-known for tobacco growing and harvesting; however, this country has been accused of hazardous child labor practices by the United Nations and the international media.

Bloomberg/Getty Images

FIGURE 9.4 Although capital flight and deindustrialization have produced problems in the U.S. economy, many job sectors continue to offer opportunities for workers. These Tesla employees on the assembly line in Fremont, California, are adding seat belts to the company's Model S electric sedans.

High-Income Economies

High-income economies (a gross national income per capita of $12,746 or more in 2015) are found in nations such as the United States, Canada, Japan, Australia, Portugal, Ireland, Israel, Italy, Norway, and Germany. According to the World Bank, people in high-income economies typically have a higher standard of living than those in low- and middle-income economies, but income is only one indicator of overall human development.

Nations with high-income economies continue to dominate the world economy, despite the fact that shifts in the global marketplace have affected workers who have a mismatch between their schooling and workplace skills and the availability of job opportunities. Another problem is *capital flight*—the movement of jobs and economic resources from one nation to another—because transnational corporations have found a ready-and-willing pool of workers worldwide to perform jobs for lower wages. In the United States and other industrialized

nations, the process of *deindustrialization*—the closing of plants and factories—because of their obsolescence or the movement of work to other regions of the country or to other nations has contributed to shifts in the global marketplace. Meanwhile, the 2007 U.S. financial crisis and the Great Recession that followed produced higher rates of unemployment worldwide. However, in the second decade of the twenty-first century, there is some hope for continued job recovery as companies such as Tesla Motors, the designer and manufacturer of electric vehicles, have opened plants in the United States and returned thousands of jobs to areas where other industries were previously located (• Figure 9.4). For example, Tesla's Fremont, California, plant is in the former New United Motor Manufacturing plant, a failed joint venture between General Motors and Toyota that closed in 2010. (The Concept Quick Review describes economies classified by income.)

CONCEPT QUICK REVIEW

Classification of Economies by Income

	Low-Income Economies	Middle-Income Economies	High-Income Economies
Previous Categorization	Third World, underdeveloped	Second World, developing	First World, developed
Per Capita Income (GNI)	$1,045 or less	Lower middle: $1,046 to $4,125 Upper middle: $4,126 to $12,745	High income: $12,746 or more
Type of Economy	Largely agricultural	Diverse, from agricultural to manufacturing	Information-based and postindustrial

Source: World Bank, 2015.

Measuring Global Wealth and Poverty

On a global basis, measuring wealth and poverty is a difficult task because of conceptual problems and problems in acquiring comparable data from various nations. As well, over time, some indicators, such as the literacy rate, become less useful in helping analysts determine what progress is being made in reducing poverty.

Absolute, Relative, and Subjective Poverty

How is poverty defined on a global basis? Isn't it more a matter of comparison than an absolute standard? According to social scientists, defining poverty involves more than comparisons of personal or household income; it also involves social judgments made by researchers. From this point of view, *absolute poverty*—previously defined as a condition in which people do not have the means to secure the most basic necessities of life—would be measured by comparing personal or household income or expenses with the cost of buying a given quantity of goods and services. As noted earlier, the World Bank has defined absolute poverty as living on less than $1.25 per day (as measured in 2005 international prices). Similarly, *relative poverty*—which exists when people may be able to afford basic necessities but are still unable to maintain an average standard of living—would be measured by comparing one person's income with the incomes of others. Finally, *subjective poverty* would be measured by comparing the actual income against the income earner's expectations and perceptions.

However, for low-income nations in a state of economic transition, data on income and levels of consumption are typically difficult to obtain and are often ambiguous when they are available. Defining levels of poverty involves several dimensions: (1) how many people are poor, (2) how far below the poverty line people's incomes fall, and (3) how long they have been poor (is the poverty temporary or long term?). • Figure 9.5 provides a unique portrayal of human poverty in which the territory size shows the proportion of the world population living in poverty in each region.

The Gini Coefficient and Global Quality-of-Life Issues

One measure of income inequality is the *Gini coefficient,* which measures the degree of inequality in the distribution of family income in a country. In technical terms the Gini Index measures the extent to which the distribution of income (or consumption expenditures) among individuals or households within an economy deviates from a perfectly equal distribution. The lower a country's score on the Gini coefficient, the more equal the income distribution. The index ranges from zero (meaning that everyone has the same income) to 100 (one person receives all the income). According to World Bank data, income inequality tends to be lower in Northern Europe, with countries such as Sweden, Norway, and Finland showing some of the world's lowest GINI coefficients. It is also surprisingly low in much-less-affluent countries such as Afghanistan and Ethiopia. The highest levels of income inequality are found in countries such as the Central African Republic, Honduras, Angola, Haiti, South Africa, and Namibia. However, these data are not always strictly comparable because of differing methods and types of data collection in various countries.

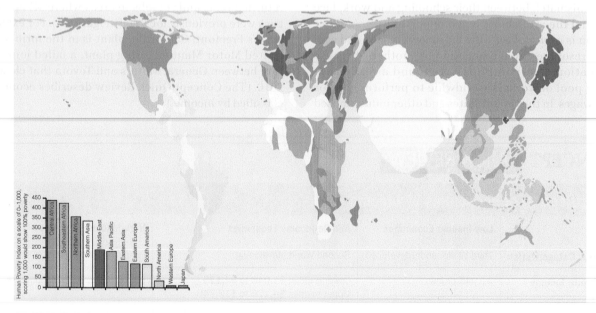

FIGURE 9.5 **Proportion of World's Population Living in Poverty (by Region)**

Source: Copyright 2006 SASI Group (University of Sheffield) and Mark Newman (University of Virginia). Map courtesy of worldmapper.org.

Discuss the relationship between global poverty and key human development issues such as life expectancy, health, education, and literacy.

Global Poverty and Human Development Issues

Income disparities are not the only factor that defines poverty and its effect on people. Although the average income per person in lower-income countries has doubled in the past thirty years and for many years economic growth has been seen as the primary way to achieve development in low-income economies, the United Nations since the 1970s has more actively focused on human development as a crucial factor in fighting poverty. In 1990 the United Nations Development Program introduced the *Human Development Index (HDI)*, establishing three new criteria—in addition to GNI—for measuring the level of development in a country: life expectancy, education, and living standards. According to the United Nations, human development is the process of increasing the number of choices that people have so that they can lead life to its fullest and be able to take action for themselves to improve their lives (United Nations Development Programme, 2013). (• Figure 9.6 compares indicators such as life expectancy and per capita gross national income of various regions around the world.) The United Nations continues to monitor the progress of nations in regard to life expectancy, educational attainment, and other factors that are related to length and quality of life, as discussed in "Sociology and Social Policy."

The top level of development category used by the United Nations is "Very High Human Development." According to the *Human Development Report,* on average, people who live in countries in the highest-human-development categories on average are better educated,

will live longer, and will earn more. In a nation with very high human development, for example, the gross national income per capita averages about $40,046, as compared to $2,904 in countries in the low human development category. The top three countries in the HDI are Norway, Australia, and Switzerland. Recently, the United States moved from third to fifth place. The Central African Republic, the Democratic Republic of the Congo, and Niger are the three bottom countries in the 2014 HDI. In terms of human development, a child born in the Democratic Republic of the Congo in 2014, for example, has a life expectancy of 50 years—nearly 32 years less than a child born in the same year in Norway, who will have a life expectancy of 81.5 years (United Nations Development Programme, 2014).

Life Expectancy

The good news is that people everywhere are living longer (World Health Organization, 2014). According to WHO's annual statistics, low-income nations have made the most progress in increases in life expectancy. Global life expectancy has improved primarily because fewer children are dying before their fifth birthday. However, we should not become too optimistic: People in high-income, highly developed nations still have a much greater chance of living longer than people in low-income countries.

Although some advances have been made in middle- and low-HDI countries regarding increasing life expectancy, major problems still exist. The average life expectancy at birth of people in medium- or middle-HDI countries remains about 12 years less (67.9 years) than that of people in very-high-HDI countries (80.2 years). Moreover, the life expectancy of people in low-HDI nations is about 20 years fewer than that of people in very high-HDI nations. Consider these figures: A child born in a low-HDI country has a life expectancy at birth of just 59.4 years,

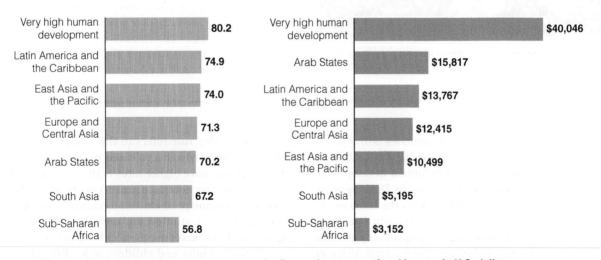

a. **Life expectancy in years for persons born in 2013**

b. **Per capita gross national income in U.S. dollars**

FIGURE 9.6 Indicators of Human Development

Source: United Nations Development Programme, 2014.

SOCIOLOGY &

Fighting Extreme Poverty One Social Policy at a Time

The primary aim of the United Nations Millennium Development Goals "should be to eradicate extreme poverty. That should be our rallying cry. We can do it in our generation."

—At a UN conference, UK Prime Minister David Cameron, chairman of the G8 organization (Canada, France, Germany, Italy, Japan, Russia, the United Kingdom, and the United States), made this call for a rapid reduction in extreme poverty worldwide (qtd. in Wintour, 2013).

Organizations such as the United Nations establish social policy initiatives to reduce extreme poverty (defined as living on under $1.25 per day) and to promote education, health, and well-being. Developing effective policies for dealing with global inequality and extreme poverty is crucial for bringing about social change. Social policy is an area of research and action that looks at the social relations necessary for people's well-being and works to build systems that promote well-being. The United Nations established eight Millennium Development Goals to be achieved and continues to monitor

efforts in each of these areas: end poverty and hunger; ensure universal education, gender equality, child health, and maternal health; combat HIV/AIDs; work toward environmental sustainability; and establish global partnerships.

All of the G8 leaders signed on to help implement this policy, and in 2012 the World Bank announced that the first goal, cutting extreme poverty in half, was ahead of the 2015 deadline. Although it is an outstanding accomplishment that an estimated 663 million people moved above the extreme poverty line (of $1.25 per day), it is important to recall that they still remain very poor, even at $2 per day, based on the standards of middle- and higher-income nations. Looking at differences in urban and rural areas, the bad news is that extreme poverty engulfs almost 29 percent of the urban population and 34 percent of the rural population. However, the good news is that urban poverty is down about 7 percent and rural poverty has decreased by more than 9 percent since 2005.

Will it be possible to eliminate poverty through social policy? According to David Cameron, it will be possible to eradicate poverty only if people see the issue as "not only about money" (qtd. in Gill, 2012). He suggests that these goals will be impossible to attain if there is widespread corruption, lack of justice, and limited access to law. As a result, Cameron believes that people must stand up for these governance issues in order to eliminate extreme poverty and help reduce overall inequality.

The Sydney Morning Herald/Getty Images

Malnutrition is a widespread problem in many low-income nations. What other kinds of health problems are related to global poverty?

REFLECT & ANALYZE

Do you believe that social policy is an effective tool for reducing poverty? Are global policies established by international organizations such as the United Nations effective in reducing poverty? Should each nation have its own policies to reduce inequality?

---

while the average in medium-HDI countries is 67.9 years. In high-human-development countries, life expectancy rises to 74.5 years, and in very-high-human-development countries it is 80.2 years (United Nations Development Programme, 2014). Regions also vary on the basis of life expectancies: An infant born in 2013 in Latin American or the Caribbean had an average life expectancy at birth of 74.9 years, while a child born in sub-Saharan Africa in the same year had a life expectancy at birth of 56.8 years (United Nations Development Programme, 2014).

One major cause of shorter life expectancy in low-income nations is the high rate of infant mortality. The *infant mortality rate* refers to the number of deaths per thousand live births in a calendar year. Low-income countries typically have higher rates of illness and disease, and they do not have adequate health care facilities. Malnutrition is a common problem among children, many of whom are underweight, stunted, and anemic—a nutritional deficiency with serious consequences for child mortality.

Among adults and children alike, life expectancies are strongly affected by hunger and malnutrition. It is estimated that people in the United States spend more than $66 billion each year on diet soft drinks, health club memberships, and other weight-loss products (Wyatt, 2014), whereas the world's poorest people suffer from chronic malnutrition, and many die each year from hunger-related diseases. Inadequate nutrition affects people's ability to work and to earn the income necessary for a minimum standard of living. Although some gains have been made in reducing the rate of malnourishment in some lower-income nations, about one billion people around the world are malnourished, and 63 percent of these are in Asia and the Pacific, 26 percent in sub-Saharan Africa, and 1 percent in developed countries (United Nations Development Programme, 2013).

On the plus side of the life-expectancy problem, some nations have made positive gains, seeing average life expectancy increase in the past four decades. For example, life expectancy has improved in sub-Saharan Africa, mainly because of reductions in deaths from diarrhea, lower respiratory tract infections, and neonatal disorders. In countries with high- and very-high human development, life-expectancy gains are mainly driven by reductions in cardiovascular disease, some cancers, transport injuries (motor vehicle accidents), and chronic respiratory conditions. Of course, problems of illness and mortality remain grave in regions such as sub-Saharan Africa, where overall longevity and quality of life remain highly problematic.

Health

Health refers to a condition of physical, mental, and social well-being. In other words, it is more than the absence of illness or disease. Many people in low-income nations are far from having physical, mental, and social well-being. In fact, more than 25 million people die each year from AIDS, malaria, tuberculosis, pneumonia, diarrheal diseases, measles, and other infectious and parasitic illnesses (World Health Organization, 2014). According to the World Health Organization, infectious diseases are far from under control in many nations: Infectious and parasitic diseases are the leading killers of children and young adults, and these diseases have a direct link to environmental conditions and poverty, especially to unsanitary and overcrowded living conditions.

Some middle-income countries are experiencing rapid growth in degenerative diseases such as cancer and coronary heart disease, and many more deaths are expected from smoking-related illnesses. Despite the decrease in tobacco smoking in high-income countries, globally there has been an increase in per capita consumption of tobacco products. Today, nearly 80 percent of the world's one billion smokers live in low- or middle-income nations (WHO Tobacco Facts, 2014).

Education and Literacy

Education is fundamental to improving life chances and reducing both individual and national poverty (● Figure 9.7). People with more years of formal education tend to earn higher wages and have better jobs. Progress in education has been made in many nations, and people around the world have higher levels of education than in the past. For this reason the United Nations *Human Development Report* uses "mean years of schooling"—completed years of educational attainment—and "expected years of schooling"—the years of schooling that a child can expect to receive given current enrollment rates—to measure progress in education.

In nations with very high human development, the mean years of schooling received by people ages 25 and older is 11.7 years; in low-HDI nations, the average number of years is only 4.2 years. Medium-HDI countries have 5.5 mean years of schooling, compared with 8.1 years for high-HDI countries. However, the United Nations also calculates *expected* years of schooling because the agency believes that these figures better reflect changing education opportunities in developing countries. For example, the average incoming elementary school

FIGURE 9.7 In functional adult literacy programs, such as this one sponsored by the Community Action Fund for Women in Africa, women hope to gain educational skills that will lift them and their families out of poverty.

student in a low-HDI country is expected to complete 9.0 years of schooling, as compared to estimates of 16.3 years in very-high-HDI countries. Some progress has been made in recent years, as most low-HDI countries have achieved or are advancing toward full enrollment in elementary school.

What is the relationship between education and literacy? The United Nations Educational, Scientific and Cultural Organization (UNESCO) defines a literate person as "someone who can, with understanding, both read and write a short, simple statement on their everyday life" (United Nations, 1997: 89). Based on this definition, people who can write only with figures, their name, or a memorized phrase are not considered literate. Literacy was previously used by the United Nations as a measure to determine education in relation to levels of human development. As improvements have occurred in global literacy rates, researchers found that other measures of educational attainment were more useful in assessing the knowledge dimension of human development.

Literacy rates continue to rise among adults and youths, and gender gaps are narrowing. It is estimated that 84 percent of the global adult population (ages 15 and above) is able to read and write. Although 89 percent of young people around the world have basic literacy and numeracy skills, there are still more than 123 million young people who are unable to read or write (United Nations, 2013). The largest increases in youth literacy rates were identified in Northern Africa and Southern Asia. The literacy rate among young women is growing faster than that of young men, particularly in these areas of the world. Although the literacy rate among adult women has continued to rise, women still represent two-thirds of those who are illiterate worldwide (United Nations, 2013). Literacy is crucial for everyone, but it is especially important for women because it has been closely linked to decreases in fertility, improved child health, and increased earnings potential.

A Multidimensional Measure of Poverty

Can we solve the problem of poverty? This is an easy question to ask, but the answer is very difficult. In part, it is important to note that there are various kinds of poverty. Sometimes global poverty is defined as earning $1.25 per day or less, but this approach overlooks other kinds of poverty indicators. Many people living in poverty daily face overlapping disadvantages, including poor health and nutrition, low education and usable skills, inadequate livelihoods, bad housing conditions, and social exclusion and lack of participation. So thinking critically about human development and poverty requires that we examine the issue of human deprivation. As a result, the United Nations developed a global *Multidimensional Poverty Index (MPI)* to help identify overlapping deprivations that are suffered by households regarding health, education, and living standards. The three dimensions of the MPI—health,

education, and living standards—are subdivided into ten indicators:

- Health—nutrition and child mortality
- Education—years of schooling (deprived if no household member has completed five years of school) and children enrolled
- Living standards—cooking fuel, toilet, water, electricity, floor (deprived if the household has a dirt, sand, or dung floor), and assets (deprived if the household does not own more than one of the following: radio, television, telephone, bike, or motorbike)

To be considered multidimensionally poor, households must be deprived in at least six standard-of-living indicators or in three standard-of-living indicators and one health or education indicator.

How many people are considered to be poor by these measures? Almost 1.5 billion people in the 91 countries covered by the MPI experience multidimensional poverty. This number exceeds the estimated 1.2 billion people who are defined as poor based on income because they live on $1.25 per day or less. It is estimated that another 800 million people are vulnerable to falling to poverty if they have financial setbacks (United Nations Development Programme, 2014).

A major contribution of the Multidimensional Poverty Index is that it focuses on many aspects of poverty and calls our attention to the idea that human development involves much more than money: It includes life chances and opportunities that contribute to human well-being. Overall, countries with less human development have more multidimensional inequality and poverty. As a result, the MPI is most useful in analyzing poverty in the less developed countries of South Asia and sub-Saharan Africa and in the poorest Latin American countries.

Even with a better understanding of how to identify poverty, much remains to be done. The good news is that multidimensional poverty is on the decline for some people in some nations; however, the bad news is that a great deal of poverty still remains throughout the world.

Persistent Gaps in Human Development

Some middle- and lower-income countries have made progress in certain indicators of human development. The gap between some richer and middle- or lower-income nations has narrowed significantly for life expectancy, health, education, and income. Some of the countries in Africa that have seen notable improvements in school attendance, life expectancy, and per capita income growth have recently emerged from lengthy periods of armed conflict within their borders. For example, the Democratic Republic of the Congo has sought to reduce the negative effects of genocide, war, and periodic volcanic eruptions on people's everyday lives while also working to provide them with better educational and health care facilities (• Figure 9.8). However, the overall picture for the world's poorest people remains dismal. The

FIGURE 9.8 The Heal Africa hospital (shown here) is an example of the efforts that are being made in some middle- and lower-income countries to improve quality of life for people who have experienced violence and who live in poverty.

gap between the richest and poorest *countries* has widened to a gulf, and the gap between the richest and poorest *people* within individual countries has also widened to a gulf. As previously stated, the countries with the highest percentages of "poor" on the Multidimensional Poverty Index are all in Africa: Ethiopia (87 percent), Liberia (84 percent), and Mozambique (79 percent). Overall, however, South Asia has the largest absolute number of multidimensionally poor people, and about 612 million of them live in India alone (United Nations Development Programme, 2013).

Poverty, food shortages, hunger, and rapidly growing populations are pressing problems for at least two billion people, most of them women and children living in a state of absolute poverty. Although more women around the globe have paid employment than in the past, more and more women are still finding themselves in poverty because of increases in single-person and single-parent households headed by women and the fact that low-wage work is often the only source of livelihood available to them.

Human development research has reached a surprising conclusion: Economic growth and higher incomes in low- and medium-development nations are not always necessary to bring about improvements in health and education. According to the *Human Development Report,* technological improvements and changes in societal structure allow even poorer countries to bring about significant changes in health and education even without significant gains in income.

Theories of Global Inequality

Social scientists have developed a variety of theories that view the causes and consequences of global inequality. We will examine modernization theory, dependency theory, world systems theory, and the new international division of labor theory.

 Define modernization theory and list the four stages of economic development identified by Walt Rostow.

Development and Modernization Theory

According to some social scientists, global wealth and poverty are linked to the level of industrialization and economic development in a given society. Although the process by which a nation industrializes may vary somewhat, industrialization almost inevitably brings with it a higher standard of living and some degree of social mobility for individual participants in the society. Specifically, the traditional caste system becomes obsolete as industrialization progresses. Family status, race/ethnicity, and gender are said to become less significant in industrialized nations than in agrarian-based societies. As societies industrialize, they also urbanize as workers locate their residences near factories, offices, and other places of work. Consequently, urban values and folkways overshadow the beliefs and practices of the rural areas. Analysts using a development framework typically view industrialization and economic development as essential steps that nations must go through in order to reduce poverty and increase life chances for their citizens.

The most widely known development theory is **modernization theory**—a perspective that links global inequality to different levels of economic development and suggests that low-income economies can move to middle- and high-income economies by achieving self-sustained economic growth. According to modernization theory, the low-income, less-developed nations can improve their standard of living only with a period of intensive economic growth and accompanying changes in people's beliefs, values, and attitudes toward work. As a result of modernization, the values of people in developing countries supposedly become more similar to those of people in high-income nations. The number of hours that people work at their jobs each week is one measure of the extent to which individuals subscribe to the *work ethic,* a core value widely believed to be of great significance in the modernization process. Of course, this assumption may be false because much research indicates that people in developing nations typically work longer hours per day than individuals in industrialized nations.

Perhaps the best-known modernization theory is that of Walt W. Rostow (1971, 1978), who, as an economic advisor to U.S. President John F. Kennedy, was highly instrumental in shaping U.S. foreign policy toward Latin America in the 1960s. Rostow suggested that all countries go through four stages of economic development, with identical content, regardless of when these nations started the process of industrialization. He compared the stages of economic development to an airplane ride. The first stage is the *traditional stage,* in which very little social change takes place, and people do not think much about changing their current circumstances. According to Rostow, societies in this stage are slow to change because the people hold a fatalistic value system, do not subscribe to the work ethic, and save very little money. The second stage is the *take-off stage*—a period of economic growth accompanied by a growing belief in individualism, competition, and achievement. During this stage people start to look toward the future, to save and invest money, and to discard traditional values. According

to Rostow's modernization theory, the development of capitalism is essential for the transformation from a traditional, simple society to a modern, complex one. With the financial help and advice of the high-income countries, low-income countries will eventually be able to "fly" and enter the third stage of economic development (● Figure 9.9). In the third stage the country moves toward *technological maturity.* At this point, the country will improve its technology, reinvest in new industries, and embrace the beliefs, values, and social institutions of the high-income, developed nations. In the fourth and final stage the country reaches the phase of *high mass consumption* and a correspondingly high standard of living.

Modernization theory has had both its advocates and its critics. According to proponents of this approach, studies have supported the assertion that economic development occurs more rapidly in a capitalist economy. In fact, the countries that have been most successful in moving from low- to middle-income status have typically been those that are most centrally involved in the global capitalist economy. For example, the nations of East Asia have successfully made the transition from low-income to higher-income economies through factors such as a high rate of savings and the fostering of a market economy.

Critics of modernization theory point out that it tends to be Eurocentric in its analysis of low-income countries, which it implicitly labels as backward. In particular, modernization theory does not take into account the possibility that all nations do not industrialize in the same manner. In contrast, some analysts have suggested that

FIGURE 9.9 Although Iraq is no longer categorized as a low-income country, displaced Iraqis who fled their homeland following an Islamic State (IS) offensive must rely on supplies donated by the World Food Program to provide meals for their families. This example shows how many factors affect economic development.

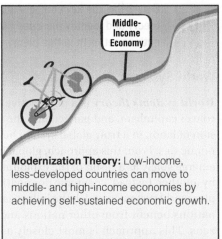

Modernization Theory: Low-income, less-developed countries can move to middle- and high-income economies by achieving self-sustained economic growth.

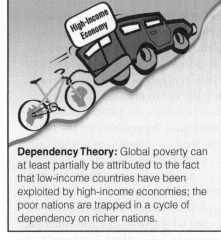

Dependency Theory: Global poverty can at least partially be attributed to the fact that low-income countries have been exploited by high-income economies; the poor nations are trapped in a cycle of dependency on richer nations.

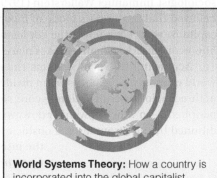

World Systems Theory: How a country is incorporated into the global capitalist economy (e.g., a core, semiperipheral, or peripheral nation) is the key feature in determining how economic development takes place in that nation.

The New International Division of Labor Theory: Commodity production is split into fragments, each of which can be moved (e.g., by a transnational corporation) to whichever part of the world can provide the best combination of capital and labor.

FIGURE 9.10   Approaches to Studying Global Inequality
What causes global inequality? Social scientists have developed a variety of explanations, including the four theories shown here.

modernization of low-income nations today will require novel policies, sequences, and ideologies that are not accounted for by Rostow's approach.

Which sociological perspective is most closely associated with the development approach? Modernization theory is based on a market-oriented perspective which assumes that "pure" capitalism is good and that the best economic outcomes occur when governments follow the policy of laissez-faire (or hands-off) business, giving capitalists the opportunity to make the "best" economic decisions, unfettered by government restraints or cumbersome rules and regulations. In today's global economy, however, many analysts believe that national governments are no longer central corporate decision makers and that transnational corporations determine global economic expansion and contraction. Therefore, corporate decisions to relocate manufacturing processes around the world make the rules and regulations of any one nation irrelevant and national boundaries obsolete. Just as modernization theory most closely approximates a functionalist approach to explaining inequality, dependency theory, world systems theory, and the new international division of labor theory

are perspectives rooted in the conflict approach. All four of these approaches are depicted in ● Figure 9.10.

LO6 **Define** dependency theory and explain why this theory is often applied to newly industrializing countries.

Dependency Theory

Dependency theory states that global poverty can at least partially be attributed to the fact that the low-income countries have been exploited by the high-income countries.

modernization theory
a perspective that links global inequality to different levels of economic development and suggests that low-income economies can move to middle- and high-income economies by achieving self-sustained economic growth.

dependency theory
the belief that global poverty can at least partially be attributed to the fact that the low-income countries have been exploited by the high-income countries.

Analyzing events as part of a particular historical process—the expansion of global capitalism—dependency theorists see the greed of the rich countries as a source of increasing impoverishment of the poorer nations and their people. Dependency theory disputes the notion of the development approach, and modernization theory specifically, that economic growth is the key to meeting important human needs in societies. In contrast, the poorer nations are trapped in a cycle of structural dependency on the richer nations because of their need for infusions of foreign capital and external markets for their raw materials, making it impossible for the poorer nations to pursue their own economic and human development agendas.

Dependency theory has been most often applied to the newly industrializing countries (NICs) of Latin America, whereas scholars examining the NICs of East Asia found that dependency theory had little or no relevance to economic growth and development in that part of the world. Therefore, dependency theory had to be expanded to encompass transnational economic linkages that affect developing countries, including foreign aid, foreign trade, foreign direct investment, and foreign loans. On the one hand, in Latin America and sub-Saharan Africa, transnational linkages such as foreign aid, investments by transnational corporations, foreign debt, and export trade have been significant impediments to development within a country. On the other hand, East Asian countries such as Taiwan, South Korea, and Singapore have historically also had high rates of dependency on foreign aid, foreign trade, and interdependence with transnational corporations but have still experienced high rates of economic growth despite dependency (• Figure 9.11).

Dependency theory makes a positive contribution to our understanding of global poverty by noting that "underdevelopment" is not necessarily the cause of inequality. Rather, it points out that exploitation not only of one country by another but also of countries by transnational corporations may limit or retard economic growth and human development in some nations.

Picture Alliance/Daniel Kalker/Newscom

FIGURE 9.11   A variety of factors—such as foreign investment and the presence of transnational corporations—have contributed to the economic growth of nations such as South Korea.

Describe the key components of world systems theory and identify the three major types of nations set forth in this theory.

World Systems Theory

World systems theory is a perspective that examines the role of capitalism, and particularly the transnational division of labor, in a truly global system held together by economic ties. From this approach, global inequality does not emerge solely as a result of the exploitation of one country by another. Instead, economic domination involves a complex world system in which the industrialized, high-income nations benefit from other nations and exploit their citizens. This approach is most closely associated with the sociologist Immanuel Wallerstein (1979, 1984, 2011), who believed that a country's mode of incorporation into the capitalist work economy is the key feature in determining how economic development takes place in that nation.

According to world systems theory, the capitalist world economy is a global system divided into a hierarchy of three major types of nations—core, semiperipheral, and peripheral—in which upward or downward mobility is conditioned by the resources and obstacles that characterize the international system. *Core nations* are dominant capitalist centers characterized by high levels of industrialization and urbanization. Core nations such as the United States, Japan, and Germany possess most of the world's capital and technology. Even more importantly for their position of domination, they exert massive control over world trade and economic agreements across national boundaries. Some cities in core nations are referred to as *global cities* because they serve as international centers for political, economic, and cultural concerns. New York, Tokyo, and London are the largest global cities, and they are often referred to as the "command posts" of the world economy.

Semiperipheral nations are more developed than peripheral nations but less developed than core nations. Nations in this category typically provide labor and raw materials to core nations within the world system. These nations constitute a midpoint between the core and peripheral nations that promotes the stability and legitimacy of the three-tiered world economy. These nations include South Korea and Taiwan in East Asia, Mexico and Brazil in Latin America, India in South Asia, and Nigeria and South Africa in Africa. Only two global cities are located in semiperipheral nations: São Paulo, Brazil, which is the center of the Brazilian economy, and Singapore, which is the economic center of a multicountry region in Southeast Asia. According to Wallerstein, semiperipheral nations exploit peripheral nations, just as the core nations exploit both the semiperipheral and the peripheral nations.

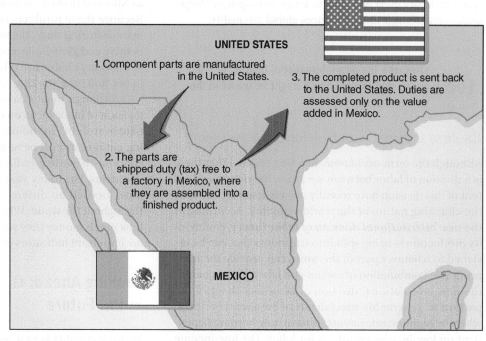

FIGURE 9.12 *Maquiladora* **Plants**
Here is the process by which transnational corporations establish plants in Mexico so that profits can be increased by using low-wage workers there to assemble products that are then brought into the United States for sale.

Most low-income countries in Africa, South America, and the Caribbean are ***peripheral nations***—nations that are dependent on core nations for capital, have little or no industrialization (other than what may be brought in by core nations), and have uneven patterns of urbanization. According to Wallerstein (1979, 1984, 2011), the wealthy in peripheral nations benefit from the labor of poor workers and from their own economic relations with core-nation capitalists, whom they uphold in order to maintain their own wealth and position. At a global level, uneven economic growth results from capital investment by core nations; disparity between the rich and the poor within the major cities in these nations is increased in the process.

The U.S.–Mexico border is an example of disparity and urban growth: Transnational corporations have built *maquiladora* plants so that goods can be assembled by low-wage workers to keep production costs down. As compared to a federal minimum wage of $7.25 per hour in the United States, the minimum wage in Mexico is about 70 pesos per *day*, which is about $4.51 in U.S. dollars based on the early 2015 exchange rate. Many people are paid more than this wage, but often not enough to be a living wage. For example, a person working in a factory that builds pickups and tractor-trailer cabs might earn $7.50 to $8.00 each day. • Figure 9.12 describes this process. Today, more than 5,000 *maquiladora* plants located in Mexico employ about 2 million people, in factories manufacturing apparel; electronic accessories such as computers, televisions, and small appliances; and motor vehicles and parts.

As Wallerstein's world systems theory might suggest, one threat to the Mexican *maquiladora* plants is *offshoring*—the movement of work by some transnational corporations to China and to countries in Central America, where wages are lower. However, in 2015 many high-tech businesses appear to be reversing the trend of offshoring. Instead, these corporations are *nearshoring*—moving their production to nearby countries such as Mexico and Canada where they can take advantage of cost efficiencies associated with closer proximity to the United States.

Not all social analysts agree with Wallerstein's perspective on the hierarchical position of nations in the global economy. However, nations throughout the world are influenced by a relatively small number of cities and transnational corporations that have prompted a shift from an international to a more global economy. World

world systems theory
a perspective that examines the role of capitalism, and particularly the transnational division of labor, in a truly global system held together by economic ties.

core nations
according to world systems theory, nations that are dominant capitalist centers characterized by high levels of industrialization and urbanization.

semiperipheral nations
according to world systems theory, nations that are more developed than peripheral nations but less developed than core nations.

peripheral nations
according to world systems theory, nations that are dependent on core nations for capital, have little or no industrialization (other than what may be brought in by core nations), and have uneven patterns of urbanization.

systems theory must continue to adapt to long-term, large-scale social change that influences global inequality.

LO8 **Discuss** the new international division of labor theory and explain how it might be useful in the twenty-first century.

The New International Division of Labor Theory

Although the term *world trade* has long implied that there is a division of labor between societies, the nature and extent of this division have recently been reassessed based on the changing nature of the world economy. According to the **new international division of labor theory**, commodity production is being split into fragments that can be assigned to whatever part of the world can provide the most profitable combination of capital and labor. Consequently, the new international division of labor has changed the pattern of geographic specialization between countries, whereby high-income countries have now become dependent on low-income countries for labor. The low-income countries provide transnational corporations with a situation in which they can pay lower wages and taxes and face fewer regulations regarding workplace conditions and environmental protection. Overall, a global manufacturing system that includes the offshoring of jobs has emerged in which transnational corporations establish labor-intensive, assembly-oriented export production, ranging from textiles and clothing to technologically sophisticated exports such as computers, in middle- and lower-income nations.

At the same time, manufacturing technologies are shifting from the large-scale, mass-production assembly lines of the past toward a more flexible production process involving microelectronic technologies. Even service industries—such as processing insurance claims forms, reading MRI and CT scans, and preparing tax forms—that were formerly thought to be less mobile have become exportable through electronic transmission and the Internet. The global nature of these activities has been referred to as *global commodity chains,* a complex pattern of international labor and production processes that results in a finished commodity ready for sale in the marketplace.

Some commodity chains are producer-driven, whereas others are buyer-driven. *Producer-driven commodity chains* is the term used to describe industries in which transnational corporations play a central part in controlling the production process. Industries that produce automobiles, computers, and other capital- and technology-intensive products are typically producer-driven. In contrast, *buyer-driven commodity chains* is the term used to refer to industries in which large retailers (such as Walmart-like corporations), brand-name merchandisers, and trading companies set up decentralized production networks in various middle- and low-income countries. This type of chain is most common in labor-intensive, consumer-goods industries such as toys, garments, and footwear. Athletic footwear companies such as Nike and Reebok are examples of the buyer-driven model. Because these products tend to be labor intensive at the manufacturing stage, the typical factory system is very competitive and globally decentralized. Workers in buyer-driven commodity chains are often exploited by low wages, long hours, and poor working conditions.

Although most discussions of the new international division of labor focus on changes occurring in the lives of people residing in industrialized urban areas of developing nations, it must be remembered that millions of people continue to live in grinding poverty in rural regions of these countries. For many years, sociologists studying poverty have focused on differences in rural and urban poverty throughout the world. Where people live strongly influences how much money they will make, and income inequalities are important indicators of the life chances of entire families.

Looking Ahead: Global Inequality in the Future

Social inequality is vast both within and among the countries of the world. In 2015 it was estimated that 80 of the wealthiest individuals in the world control as much wealth as the poorest 50 percent of the world's population (3.5 billion people). No, you did not read that incorrectly: It is 80 persons as compared with 3.5 billion. Even in high-income nations where wealth is highly concentrated, many poor people coexist with the affluent. In middle- and low-income countries, there are small pockets of wealth in the midst of poverty and despair. Although some political and business elites in local economies benefit greatly from partnerships with transnational corporations, everyday people residing in these nations have often continued to be exploited in both industrial and agricultural work. In China, for example, some people have accumulated vast wealth in urban areas while poverty has increased dramatically in some regions and particularly in rural areas.

What are the future prospects for greater equality across and within nations? Not all social scientists agree on the answer to this question. Depending on the theoretical framework that they apply in studying global inequality, social analysts may describe either an optimistic or a pessimistic scenario for the future. Moreover, some analysts highlight the human rights issues embedded in global inequality, whereas others focus primarily on an economic framework.

In some regions, high rates of unemployment and persistent poverty undermine human development and future possibilities for socioeconomic change. In the second decade of the twenty-first century, high unemployment rates in the United States and the Euro zone (nations in the European Union such as Italy, France, and Germany) continue to threaten global economic growth and limit opportunities for unemployed and underemployed workers and young people seeking to get into the labor market. Worldwide, millions of young people are unemployed, and many are destined to experience long-term unemployment. Estimates suggest that

YOU CAN MAKE A DIFFERENCE

Global Networking to Reduce World Hunger and Poverty

When many of us think about problems such as world poverty, we tend to see ourselves as powerless to bring about change in so vast an issue. However, a recurring message from social activists and religious leaders is that each person can contribute something to the betterment of other people and sometimes the entire world.

An initial way for each of us to be involved is to become more informed about global issues and to learn how we can contribute time and resources to organizations seeking to address social issues such as illiteracy and hunger. We can also find out about meetings and activities of organizations and participate in online discussion forums where we can express our opinions, ask questions, share information, and interact with other people interested in topics such as international relief and development. It may not feel like you are doing much to address global problems; however, information and education are the first steps in promoting greater understanding of social problems and of the world's people. Likewise, it is important to help our own nation's children understand that they can make a difference in ending hunger in the United States and other nations.

Would you like to function as a catalyst for change? You can learn how to proceed by gathering information from organizations that seek to reduce problems such as poverty and to provide forums for interacting with other people. Here are a few starting points for your search:

- CARE International is a confederation of fourteen global national members in North America, Europe, Japan, and Australia. CARE assists the world's poor in their efforts to achieve social and economic well-being. Programs include emergency relief, education,

health and population, children's health, reproductive health, water and sanitation, small economic activity development, agriculture, community development, and environment.

Other organizations fighting world hunger and health problems include the following; check out their websites:

- WhyHunger

- "Kids Can Make a Difference," an innovative program developed by the International Education and Resource Network

- World Health Organization

United Nations Relief and Works Agency volunteers provide a Palestinian woman with food supplied by the European Union and the World Food Programme. What other goods and services might volunteers be able to provide around the world?

as many as 200 million people are without a job and that about 39 million people have dropped out of the labor market altogether (International Labor Organization, 2013).

Problems such as long-term unemployment contribute to gross inequality, which has high financial and quality-of-life costs to people, even for those who are not the poorest of the poor. In the future, continued population growth, urbanization, environmental degradation, and violent conflict threaten even the meager living conditions of those residing in low-income nations. However, from this approach the future looks dim not only for people in low-income and middle-income countries but also for those in high-income countries, who will see their quality of life diminish as natural resources are depleted, the environment is polluted, and high rates of immigration and global political unrest threaten the high standard of living that many people have come to enjoy. According to some social analysts, transnational corporations and financial institutions such as the

World Bank and the International Monetary Fund will further solidify and control a globalized economy, which will transfer the power to make significant choices to these organizations and away from the people and their governments. Further loss of resources and means of livelihood will affect people and countries around the globe.

As a result of global corporate domination, there could be a leveling out of average income around the world, with wages falling in high-income countries and wages increasing significantly in low- and middle-income countries. If this pessimistic scenario occurs, there is likely to be greater polarization of the rich and the poor and more potential for ethnic and national conflicts over such issues

new international division of labor theory
the perspective that views commodity production as being split into fragments that can be assigned to whatever part of the world can provide the most profitable combination of capital and labor.

as worsening environmental degradation and who has the right to natural resources. For example, pulp-and-paper companies in Indonesia, along with palm-oil plantation owners, have continued clearing land for crops by burning off vast tracts of jungle, producing high levels of smog and pollution across seven Southeast Asian nations and creating havoc for millions of people.

On the other hand, a more optimistic scenario is also possible. With modern technology and worldwide economic growth, it might be possible to reduce absolute poverty and to increase people's opportunities. Trends that have the potential to bring about more-sustainable patterns of development are the socioeconomic progress made in many low- and middle-income countries over the past thirty years as technological, social, and environmental improvements have occurred. For example, technological innovation continues to improve living standards for some people.

Fertility rates are declining in some regions, but remain high in others, where there remains grave cause for concern about the availability of adequate natural resources for the future.

Finally, health and education may continue to improve in lower-income countries. Healthy, educated populations are crucial for the future in order to reduce global poverty. The education of women is of primary importance in the future if global inequality is to be reduced. All aspects of schooling and training are crucial for the future, including agricultural extension services in rural areas to help women farmers in regions such as western Kenya produce more crops to feed their families. From this viewpoint, we can enjoy prosperity only by ensuring that other people have the opportunity to survive and thrive in their own surroundings (see "You Can Make a Difference"). The problems associated with global poverty are therefore of interest to a wide-ranging set of countries and people.

CHAPTER REVIEW

LO1 What is global stratification, and how does it contribute to economic inequality?

Global stratification refers to the unequal distribution of wealth, power, and prestige on a global basis, which results in people having vastly different lifestyles and life chances both within and among the nations of the world. Today, the income gap between the richest and the poorest percent of the world population continues to widen, and within some nations the poorest one-fifth of the population has an income that is only a slight fraction of the overall average per capita income for that country.

LO2 What is the levels of development approach for studying global inequality?

One of the primary problems encountered by social scientists studying global stratification and social and economic inequality is what terminology should be used to refer to the distribution of resources in various nations. Most definitions of inequality are based on comparisons of levels of income or economic development, whereby countries are identified in terms of the "three worlds" or upon their levels of economic development. Terminology based on levels of development includes concepts such as developed nations, developing nations, less-developed nations, and underdevelopment.

LO3 How does the World Bank classify nations into four economic categories, and why do organizations such as this have problems measuring wealth and poverty on a global basis?

The World Bank classifies nations into four economic categories and establishes the upper and lower limits for the

gross national income (GNI) in each category. *Low-income economies* had a GNI per capita of less than $1,045 in 2015, *lower-middle-income economies* had a GNI per capita between $1,046 and $4,125, *upper-middle-income economies* had a GNI per capita between $4,126 and $12,745, and *high-income economies* had a GNI per capita of $12,746 or more (World Bank, 2015). Defining poverty is more than just personal and household income; it also involves social judgments made by researchers. Absolute poverty is a condition in which people do not have the means to secure the most basic necessities of life. It would be measured by comparing personal or household income or expenses with the cost of buying a given quantity of goods and services. Relative poverty exists when people may be able to afford basic necessities but are still unable to maintain an average standard of living. This would be measured by comparing one person's income with the incomes of others.

LO4 What is the relationship between global poverty and key human development issues such as life expectancy, health, education, and literacy?

Income disparities are not the only factor that defines poverty and its effect on people. The United Nations' Human Development Index measures the level of development in a country through indicators such as life expectancy, infant mortality rate, proportion of underweight children under age five, and adult literacy rate for low-income, middle-income, and high-income countries. People who live in countries in the highest-human-development categories on average are better educated, will live longer, and will earn more. The adult literacy rate in the low-income countries is significantly lower than that of high-income countries, and for women the rate is even lower.

LO5 What is modernization theory, and what are the four stages of economic development identified by Walt Rostow?

Modernization theory is a perspective that links global inequality to different levels of economic development and suggests that low-income economies can move to middle- and high-income economies by achieving self-sustained economic growth. Walt Rostow suggested that all countries go through four stages of economic development, with identical content, regardless of when these nations started the process of industrialization. The stages of economic development are as follows: the traditional stage, in which very little social change takes place and people do not think much about changing their current circumstances. The second stage is the take-off stage—a period of economic growth accompanied by a growing belief in individualism, competition, and achievement. In the third stage the country moves toward technological maturity. In the fourth and final stage the country reaches the phase of high mass consumption and a correspondingly high standard of living.

LO6 What is dependency theory, and why is this theory often applied to newly industrializing countries?

Dependency theory states that global poverty can at least partially be attributed to the fact that the low-income countries have been exploited by the high-income countries. Whereas modernization theory focuses on how societies can reduce inequality through industrialization and economic development, dependency theorists see the greed of the rich countries as a source of increasing impoverishment of the poorer nations and their people.

LO7 What is world systems theory, and what are the three major types of nations set forth in this theory?

According to world systems theory, the capitalist world economy is a global system divided into a hierarchy of three major types of nations: Core nations are dominant capitalist centers characterized by high levels of industrialization and urbanization, semiperipheral nations are more developed than peripheral nations but less developed than core nations, and peripheral nations are those countries that are dependent on core nations for capital, have little or no industrialization (other than what may be brought in by core nations), and have uneven patterns of urbanization.

LO8 What is the new international division of labor theory, and how might it be useful in the twenty-first century?

The new international division of labor theory is based on the assumption that commodity production is split into fragments that can be assigned to whichever part of the world can provide the most profitable combination of capital and labor. This division of labor has changed the pattern of geographic specialization between countries, whereby high-income countries have become dependent on low-income countries for labor. The low-income countries provide transnational corporations with a situation in which they can pay lower wages and taxes and face fewer regulations regarding workplace conditions and environmental protection.

KEY TERMS

core nations 256
dependency theory 255
global stratification 242

modernization theory 254
new international division
  of labor theory 258

peripheral nations 257
semiperipheral nations 256
world systems theory 256

QUESTIONS for CRITICAL THINKING

1 You have decided to study global wealth and poverty. How would you approach your study? What research methods would provide the best data for analysis? What might you find if you compared your research data with popular presentations—such as films and advertising—of everyday life in low- and middle-income countries?

2 How would you compare the lives of poor people living in the low-income nations of the world with those in central cities and rural areas of the United States? In what ways are their lives similar? In what ways are they different?

3 Should U.S. foreign policy include provisions for reducing poverty in other nations of the world? Should U.S. domestic policy include provisions for reducing poverty in the United States? How are these issues similar? How are they different?

4 Using the theories discussed in this chapter, devise a plan to alleviate global poverty. Assume that you have the necessary wealth, political power, and other resources necessary to reduce the problem. Share your plan with others in your class, and create a consolidated plan that represents the best ideas and suggestions presented.

ANSWERS to the SOCIOLOGY QUIZ
ON GLOBAL WEALTH AND POVERTY

1	**False**	In the twenty-first century, the income gap between the richest and poorest has increased on a global basis as well as in the United States.
2	**False**	The wealth of the world's 80 richest people is more than the combined wealth of 50 percent of the world's population (3.5 billion people).
3	**True**	The World Bank estimates that nearly 1.3 billion people (or 22 percent of the developing world's population) live below the international poverty line, which is defined as earning less than $1.25 each day.
4	**False**	One of the consequences of extreme poverty is hunger, and millions of people—including six million children under the age of five—die of hunger-related diseases or chronic malnutrition each year.
5	**True**	In almost all low-income countries (as well as middle- and high-income countries), poverty is a more chronic problem for women because of sexual discrimination, resulting in a lack of educational and employment opportunities.
6	**False**	Although the number of poor people residing in urban areas is growing rapidly, the majority of people with incomes below the poverty line live in rural areas of the world.
7	**False**	Two-thirds of adults worldwide who are unable to read and write are women. Gender disparity is greatest in southern Asia, where slightly more than half of all women are able to read and write.
8	**True**	Although these fuels are inefficient and harmful to health, many low-income people cannot afford appliances, connection charges, and so forth. In some areas, electric hookups are not available.

RACE AND ETHNICITY

LEARNING OBJECTIVES

1   **Distinguish** between the terms *race* and *ethnicity*.

2   **Explain** how racial and ethnic classifications continue to change in the United States.

3   **Define** *prejudice*, *stereotypes*, *racism*, *scapegoat*, and *discrimination*.

4   **Compare** the major sociological perspectives on race and ethnic relations.

5   **Discuss** the unique historical experiences of Native Americans and WASPs in the United States.

6   **Describe** how slavery, segregation, lynching, and persistent discrimination have uniquely affected the African American experience in this country.

7   **Identify** the major categories of Asian Americans and describe their historical and contemporary experiences.

8   **Describe** the unique experiences of Latinos/as (Hispanics) and Middle Eastern Americans in the United States.

SOCIOLOGY & EVERYDAY LIFE

Race and Moral Imagination: From Selma to Ferguson and Back

[T]here are places and moments in America where this nation's destiny has been decided. . . . Selma is such a place. . . . What they did here will reverberate through the ages. Not because the change they won was preordained, not because their victory was complete, but because they proved that nonviolent change is possible; that love and hope can conquer hate. . . .

Just this week, I was asked whether I thought the Department of Justice's Ferguson report [documenting explicit racism and abusive policing directed at African Americans in Ferguson, Missouri, where a white police officer was acquitted in the shooting of an unarmed black man] shows that, with respect to race, little has changed in this country. . . . But I rejected the notion that nothing's changed. What happened in Ferguson may not be unique, but it's no longer endemic. It's no longer sanctioned by law or by custom. And before the Civil Rights Movement, it most surely was. . . .

From 1965 when Amelia Boynton Robinson (shown here with President Barack Obama) marched across the Edmund Pettus Bridge as part of the civil rights march from Selma to Montgomery to the second decade of the twenty-first century, when grassroots activists have continued to proclaim "Black Lives Matter," racial inequality has been a pressing social issue in the United States. What progress do you believe this nation has made? What remains to be done to promote greater racial equality?

We know the march is not yet over. . . . There's nothing America can't handle if we actually look squarely at the problem. And this is work for all Americans, not just some. Not just whites. Not just blacks. If we want to honor the courage of those who marched [in Selma] that day, then all of us are called to possess their moral imagination. All of us need to feel as they did the fierce urgency of now. . . . And that's what the young people . . . must take away from this day. You are America. . . . For everywhere in this country, there are first steps to be taken, there's new ground to cover, there are more bridges to be crossed. And it is you, the young and

SAUL LOEB/AFP/Getty Images

D id you see the film *Selma*? This movie depicts the 1965 conflict that ensued in Selma, Alabama, when 600 demonstrators embarked on a fifty-mile march to Montgomery, the state capital, to demand the right to vote. However, as the civil rights protesters crossed the bridge out of town, they were confronted by state troopers who beat them with clubs and fired tear gas in an effort to disperse them. The fiftieth anniversary of this historic event, at which President Obama spoke, coincided with the release of a 2015 U.S. Justice Department report on the death of Michael Brown, an unarmed eighteen-year-old black man fatally shot by a white police officer in Ferguson, Missouri. Brown's death resulted in nationwide protests and chants of "Hands up, don't shoot!" According to the Justice Department report, Ferguson public officials and law enforcement officers were found to have routinely demonstrated explicit racism, discrimination, and abuse toward persons of color in that community. When the report was released, many concerned citizens, civil rights leaders, and journalists emphatically stated that Ferguson was not an isolated case. They pointed out many other occurrences that demonstrated that Ferguson was just the tip of the iceberg: Racial injustice and police brutality directed toward African

Americans and other persons of color are nationwide problems (Robertson, Dewan, and Apuzzo, 2015).

Ironically, shortly after President Obama's speech at the fiftieth anniversary of "Bloody Sunday" in Selma, Alabama, and the release of the Ferguson, Missouri, report by the Justice Department, a video came to light that shook the entire nation because of its racist overtones in the second decade of the twenty-first century. An article in *The Chronicle of Higher Education* described the situation as follows: "The hand-held video showed members of the Sigma Alpha Epsilon fraternity [at the University of Oklahoma, Norman] aboard a bus. A white student, in a bow tie, was leading the passengers in a song. To the tune of 'If You're Happy and You Know It,' they vowed to never have blacks students in their fraternity, using a racist slur and invoking imagery of lynching" (Berrett, 2015). The persistence of racism, as evidenced in Selma, Ferguson, Norman, and elsewhere, is not only a structural issue deeply embedded in U.S. history but also a pressing social problem that remains today.

In this chapter we examine prejudice, discrimination, sociological perspectives on race and ethnicity, and commonalities and differences in the experiences of racial and ethnic groups in

fearless at heart, the most diverse and educated generation in our history, who the nation is waiting to follow. . . .

—In President Barack Obama's speech honoring the fiftieth anniversary of the civil rights march from Selma to Birmingham (Obama, 2015), the president highlighted the role that he believed everyone should play in bringing about greater racial justice.

How Much Do You Know About Race, Ethnicity, and Sports?

TRUE	FALSE		
T	F	1	Some deeply held ideas about race and racial differences are often expressed in our beliefs about sports and athletic ability.
T	F	2	Most elite sprinters are of West African descent because they possess a greater percentage of fast-twitch muscles that can be attributed to a specific racial gene.
T	F	3	The chances (statistical odds) of becoming a professional athlete are better than the chance of getting struck by lightning or writing a *New York Times* best-seller.
T	F	4	In the twenty-first century the number of African Americans in head coaching positions in the 120 Football Bowl Subdivision (FBS), formerly known as NCAA Division I-A, has increased significantly.
T	F	5	Today, all university presidents of the 120 FBS (NCAA Division I) colleges are white.
T	F	6	African Americans account for more than half of all FBS football student-athletes.
T	F	7	All of the conference commissioners of the Football Bowl Subdivision (FBS) are white men.
T	F	8	Visible racial–ethnic diversity on college sports teams has little effect on how younger athletes perceive their future educational and sporting opportunities in college and beyond.

Answers can be found at the end of the chapter.

the United States. In the process we will examine how people, past and present, have been singled out for negative treatment on the basis of their perceived race or ethnicity. We will also show how they have sought to overcome prejudice and discrimination through endeavors such as sports, where both problems and progress in racial and ethnic relations are evident. Before reading on, test your knowledge about race, ethnicity, and sports by taking the "Sociology and Everyday Life" quiz.

Race and Ethnicity

What is race? Some people think it refers to skin color (the Caucasian "race"); others use it to refer to a religion (the Jewish "race"), nationality (the British "race"), or the entire human species (the human "race") (Marger, 2015). Popular usages of the word have been based on the assumption that a race is a grouping or classification based on *genetic* variations in physical appearance, particularly skin color. However, social scientists, biologists, and genetic anthropologists dispute the idea that biological race is a meaningful concept. Researchers with the Human Genome Project, which was commissioned to map all of the genes on the 23 pairs of human chromosomes and to sequence the 3.1 billion DNA base pairs that make up the chromosomes, made this statement about genes and race:

> DNA studies do not indicate that separate classifiable subspecies (races) exist within modern humans. While different genes for physical traits such as skin and hair color can be identified between individuals, no consistent patterns of genes across the human genome exist to distinguish one race from another. There also is no genetic basis for divisions of human ethnicity. People who have lived in the same geographic region for many generations may have some alleles in common [an allele is one member of a pair or series of genes that occupy a specific position on a specific chromosome], but no allele will be found in all members of one population and in no members of any other. (genomics.energy.gov, 2007)

The idea of race has little meaning in a biological sense because of the enormous amount of interbreeding that has taken place within the human population. For these reasons, sociologists sometimes place "race" in quotation marks to show that categorizing individuals and population groups on biological characteristics is neither accurate nor based on valid distinctions between the genetic makeup of differently identified "races."

Today, sociologists emphasize that race is a *socially constructed reality,* not a biological one. Race as a *social construct* means that races as such do not actually exist but that some groups are still racially defined because the *idea* persists in many people's minds that races are distinct biological categories with physically distinguishable characteristics and a shared common cultural heritage. The process of creating a socially constructed reality involves three key activities: *collective agreement, imposition,* and *acceptance of a specific construction* (Lusca, 2008).

Collective agreement means that people jointly agree on the idea of race and that they accept that it exists as an important component in how we describe or explain the individual's experiences in everyday life. Examples of collective agreement include a widely held acceptance of the view that "racial differences affect people's athletic ability" or of the assumption that "physical differences based on race cause cultural differences among various distinct categories of people."

Imposition refers to the fact that throughout much of human history, the notion of race has been defined by members of dominant groups who have the power to establish a system that hierarchically organizes racial categories (as superior or inferior, for example) to establish and maintain permanent status differentials among individuals and groups. These differences are demonstrated by the level of access that dominant- and subordinate-group members have to necessary and desired goods and services, such as education, housing, employment, health care, and legal services.

Finally, *acceptance of a specific construction* means that ideas pertaining to race become so widely accepted that they become embedded in law and social customs in a society and become much more difficult to change or eliminate. When a significant *number* of people, or a number of *significant* people, accept a social construction as absolute and real, the prevailing group typically imposes its beliefs and practices upon others through tradition and law. Over time, ideas about race, inadequate or false though they may be, are passed on from generation to generation.

In sum, the social significance that people accord to race is more important than any biological differences that might exist among people who are placed in arbitrary categories. Although race does not exist in an objective way, it does have *real* consequences and effects in the social world.

 LO1 **Distinguish** between the terms *race* and *ethnicity.*

Comparing Race and Ethnicity

A *race* is a category of people who have been singled out as inferior or superior, often on the basis of real or alleged physical characteristics such as skin color, hair texture, eye shape, or other subjectively selected attributes (Feagin and Feagin, 2012). Racial categories identified by the U.S. Census Bureau include white, black, Asian or Pacific Islander, and American Indian or Alaska Native.

As compared with race, *ethnicity* refers to one's cultural background or national origin. An ***ethnic group*** is a collection of people distinguished, by others or by themselves, primarily on the basis of cultural or nationality characteristics (Feagin and Feagin, 2012) (● Figure 10.1). Ethnic groups share five main characteristics: (1) *unique cultural traits,* such as language, clothing, holidays, or religious practices; (2) *a sense of community;* (3) *a feeling of ethnocentrism;* (4) *ascribed membership from birth;* and (5) *territoriality,* or the tendency to occupy a distinct geographic area (such as Little Italy or Little Moscow) by choice and/or for self-protection. Examples of ethnic groups include Jewish Americans, Irish Americans, Italian Americans, and

Steven Widoff/Alamy

FIGURE 10.1 New York City's Chinatown is an ethnic enclave where people participate in social interaction with other individuals in their ethnic group and feel a sense of shared identity. Ethnic enclaves provide economic and psychological support for recent immigrants as well as for those who were born in the United States.

Russian Americans. Many people mistakenly believe that the classification "Hispanic" or "Latino/a" is a "race." According to the U.S. Census Bureau, Hispanic or Latino/a is an ethnicity that refers to a person of Cuban, Mexican, Puerto Rican, South or Central American, or other Spanish culture or origin (Ennis, Rios-Vargas, and Albert, 2011). Persons who are Hispanic or Latino/a can be of any race.

Although some people do not identify with any ethnic group, others participate in social interaction with individuals in their ethnic group and feel a sense of common identity based on cultural characteristics such as language, religion, or politics. However, ethnic groups are not only influenced by their own history but also by patterns of ethnic domination and subordination in societies. It is important to note that terminology pertaining to racial–ethnic groups is continually in flux and that people within the category as well as outsiders often contest these changes. Examples include the use of *African American,* as compared to *black,* and *Hispanic,* as compared to *Latino/a.*

The Social Significance of Race and Ethnicity

Race and ethnicity take on great social significance because how people act in regard to these terms drastically affects other people's lives, including what opportunities they have, how they are treated, and even how long they live. According to the now-classic works on the effects of race by the sociologists Michael Omi and Howard Winant (1994: 158), race "permeates every institution, every relationship, and every individual" in the United States:

> As we . . . compare real estate prices in different neighborhoods, select a radio channel to enjoy while we drive to work, size up a potential client, customer, neighbor, or teacher, stand in line at the unemployment office, or carry out a thousand other normal tasks, we are compelled to think racially, to use the racial categories and meaning systems into which we have been socialized. (Omi and Winant, 1994: 158)

Historically, stratification based on race and ethnicity has pervaded all aspects of political, economic, and social life. Consider sports as an example. Throughout the early history of the game of baseball, many African Americans had outstanding skills as players but were categorically excluded from Major League teams because of their skin color. Even in 1947, after Jackie Robinson broke the "color line" to become the first African American in the Major Leagues, his experience was marred by racial slurs, hate letters, death threats against his infant son, and assaults on his wife (Ashe, 1988; Peterson, 1992/1970). With some professional athletes from diverse racial–ethnic categories having multimillion-dollar contracts and lucrative endorsement deals, it is easy to assume that racism in sports—as well as in the larger society—is a thing of the past. However, this *commercialization* of sports does not mean that racial prejudice and discrimination no longer exist (Coakley, 2009).

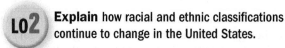

LO2 **Explain** how racial and ethnic classifications continue to change in the United States.

Racial Classifications and the Meaning of Race

If we examine racial classifications throughout history, we find that in ancient Greece and Rome, a person's race was the group to which she or he belonged, associated with an ancestral place and culture. From the Middle Ages until about the eighteenth century, a person's race was based on family and ancestral ties, in the sense of a *line,* or ties to a national group. During the eighteenth century, physical differences such as the darker skin hues of Africans became associated with race, but racial divisions were typically based on differences in religion and cultural tradition rather than on human biology. With the intense (though misguided) efforts that surrounded the attempt to justify black slavery and white dominance in all areas of life during the second half of the nineteenth century, *races* came to be defined as distinct biological categories of people who were not all members of the same family but who shared inherited physical and cultural traits that were alleged to be different from those traits shared by people in other races. Hierarchies of races were established, placing the "white race" at the top, the "black race" at the bottom, and others in between.

However, racial classifications in the United States have changed over the past century. If we look at U.S. Census Bureau classifications, for example, we can see how the meaning of race continues to change. First, race is defined by perceived skin color: white or nonwhite. Whereas one category exists for "whites" (who vary considerably in actual skin color and physical appearance), all of the remaining categories are considered "nonwhite."

Second, categories of official racial classifications may (over time) create a sense of group membership or "consciousness of kind" for people within a somewhat arbitrary classification. When people of European descent were classified as "white," some began to see themselves as different from "nonwhite." Consequently, Jewish, Italian, and Irish immigrants may have felt more a part of the Northern European white mainstream in the late eighteenth and early nineteenth centuries. Whether Chinese Americans, Japanese Americans, Korean Americans, and Filipino Americans come to think of themselves collectively as "Asian Americans" because of official classifications remains to be seen.

Third, racial purity is assumed to exist. Prior to the 2000 census, for example, the true diversity of the U.S. population was not revealed in census data because

race
a category of people who have been singled out as inferior or superior, often on the basis of real or alleged physical characteristics such as skin color, hair texture, eye shape, or other subjectively selected attributes.

ethnic group
a collection of people distinguished, by others or by themselves, primarily on the basis of cultural or nationality characteristics.

multiracial individuals were forced to either select a single race as being their "race" or to select the vague category of "other." Census 2000 made it possible—for the first time—for individuals to classify themselves as being of more than one race. In the 2010 census, nine million people in the United States—about 3 percent of the total population—identified themselves as multiracial (Humes, Jones, and Ramirez, 2011). Between 2000 and 2010, the percentage of Americans identifying as more than one race increased by 32 percent. Among U.S. children, the mixed-race population increased by nearly 50 percent, making mixed-race children the fastest-growing youth group in the United States (Saulny, 2011a). With one in seven new marriages in the United States involving spouses of different races or ethnicities, the multiracial population is likely to continue to increase (Passel, Wang, and Taylor, 2010).

Multiracial individuals do not always identify as such. Although President Barack Obama is the product of an interracial couple (his mother was white, and his father was black), for the 2010 census Obama checked only one box: black (Saulny, 2011b). For many multiracial individuals, choosing a racial–ethnic identity is not simple. Consider the case of Michelle López-Mullins, a mixed-race college student whose father is both Asian and Latino, and whose mother, with her long blonde hair, is mostly European in ancestry but is mixed with some Cherokee and Shawnee Native American. In grade school, Ms. López-Mullins was frequently asked "What are you?" and "Where are you from?" As she explains,

> I hadn't even learned the word "Hispanic" until I came home from school one day and asked my dad what I should refer to him as, to express what I am. . . . Growing up with my parents, I never thought we were different from any other family. . . . I was always having to explain where my parents are from because just saying "I'm from Takoma Park, Maryland," was not enough. . . . Saying "I'm an American" wasn't enough. . . . Now when people ask what I am, I say, "How much time do you have?" . . . Race will not automatically tell you my story. (qtd. in Saulny, 2011b)

When asked what box she checks on forms such as the census, López-Mullins replied, "Hispanic, white, Asian American, Native American. . . . I'm pretty much checking everything" (qtd. in Saulny, 2011b).

As noted earlier, the way that people are classified remains important because such classifications affect their access to employment, education, housing, social services, federal aid, and other public and private goods and services that might be available to them.

Dominant and Subordinate Groups

The terms *majority group* and *minority group* are widely used, but their meanings are less clear as the composition of the U.S. population continues to change. Accordingly, many sociologists prefer the terms *dominant* and

subordinate to identify power relationships that are based on perceived racial, ethnic, or other attributes and identities. To sociologists, a **dominant group** is a racial or ethnic group that has the greatest power and resources in a society (Feagin and Feagin, 2012). In the United States, whites with Northern European ancestry (often referred to as Euro-Americans, white Anglo-Saxon Protestants, or WASPs) have been considered to be the dominant group for many years. A **subordinate group** is one whose members, because of physical or cultural characteristics, are disadvantaged and subjected to unequal treatment and discrimination by the dominant group. Historically, African Americans and other persons of color have been considered to be subordinate-group members, particularly when they are from lower-income categories.

It is important to note that, in the sociological sense, the word *group* as used in these two terms is misleading because people who merely share ascribed racial or ethnic characteristics do not constitute a group. However, the terms *dominant group* and *subordinate group* do give us a way to describe relationships of advantage/disadvantage and power/exploitation that exist in contemporary nations.

 L03 **Define** *prejudice, stereotypes, racism, scapegoat,* and *discrimination.*

Prejudice

Although there are various meanings of the word part *dice,* sociologists define **prejudice** as a negative attitude based on faulty generalizations about members of specific racial, ethnic, or other groups. The term *prejudice* is from the Latin words *prae* ("before") and *judicium* ("judgment"), which means that people may be biased either for or against members of other groups even before they have had any contact with them. Although prejudice can be either *positive* (bias in favor of a group—often our own) or *negative* (bias against a group—one we deem less worthy than our own), it most often refers to the negative attitudes that people may have about members of other racial or ethnic groups (● Figure 10.2).

Stereotypes

Prejudice is rooted in ethnocentrism and stereotypes. When used in the context of racial and ethnic relations, *ethnocentrism* refers to the tendency to regard one's own culture and group as the standard—and thus superior—whereas all other groups are seen as inferior. Ethnocentrism is maintained and perpetuated by **stereotypes**—overgeneralizations about the appearance, behavior, or other characteristics of members of particular categories.

Although stereotypes can be either positive or negative, examples of negative stereotyping abound in sports. Think about the Native American names, images, and mascots used by sports teams such as the Atlanta Braves, Cleveland

FIGURE 10.2 Contemporary prejudice and discrimination cannot be understood without taking into account the historical background. School integration in the 1950s was accomplished despite white resistance. Today, integration in education, housing, and many other areas of social life remains a pressing social issue.

Indians, and Washington Redskins. Members of Native American groups have been actively working to eliminate the use of stereotypic mascots (with feathers, buckskins, beads, spears, and "warpaint"), "Indian chants," and gestures (such as the "tomahawk chop"), which they claim trivialize and exploit Native American culture. According to sociologist Jay Coakley (2009), the use of stereotypes and words such as *redskin* symbolizes a lack of understanding of the culture and heritage of native peoples and is offensive to many Native Americans. Although some people see these names and activities as "innocent fun," others view them as a form of racism.

Racism

What is racism? *Racism* is a set of attitudes, beliefs, and practices that is used to justify the superior treatment of one racial or ethnic group and the inferior treatment of another racial or ethnic group. The world has seen a long history of racism: It can be traced from the earliest

civilizations. At various times throughout U.S. history, various categories of people, including Irish Americans, Italian

dominant group
a racial or ethnic group that has the greatest power and resources in a society.

subordinate group
a group whose members, because of physical or cultural characteristics, are disadvantaged and subjected to unequal treatment and discrimination by the dominant group.

prejudice
a negative attitude based on faulty generalizations about members of specific racial, ethnic, or other groups.

stereotypes
overgeneralizations about the appearance, behavior, or other characteristics of members of particular categories.

racism
a set of attitudes, beliefs, and practices that is used to justify the superior treatment of one racial or ethnic group and the inferior treatment of another racial or ethnic group.

SOCIOLOGY & SOCIAL POLICY

Racist Hate Speech on Campus Versus First Amendment Right to Freedom of Speech

From news reports and social media:

- Racist graffiti scrawled on the walls of college dorms and other buildings

- Derogatory comments from current students directed toward persons of color visiting a campus on recruitment weekend

- Fraternity members singing a disparaging chant that targets African Americans

All of these forms of racist hate speech have been reported in recent years on college and university campuses throughout the United States. What is "hate speech" anyway? Hate speech refers to speech that offends, threatens, or insults groups, based on race, color, religion, national origin, sexual orientation, disability, or other traits. Hate speech is often directed at historically oppressed racial or religious minorities, or persons in other subordinate-power groups, with the intent to insult and demean them.

In our government and political science courses, most of us learned that the First Amendment to the United States Constitution protects the right to speech even when others disagree with the speech or find it contemptible. The First Amendment, in part, states, "Congress shall make no law . . . abridging the freedom of speech. . . ." And public colleges and universities are not exempt from this provision. U.S. Supreme Court decisions, such as *Healy v. James* (1972), have reaffirmed that free speech rights extend to college campuses because they are not "enclaves immune from the sweep of the First Amendment."

Although racist speech is constitutionally protected, many institutions of higher education have established speech codes to ban offensive expression on campus, to specify what speech and behaviors are prohibited, and to foster a productive learning environment for all students. A speech code is a set of rules or regulations that limit, restrict, or ban speech beyond strict legal limitations upon freedom of speech or press found in the legal definitions of

Like all other areas of social life, college campuses are not immune to racist hate speech and blatant acts of racism that target persons of color. Shown here, protesting students chant "No Diversity, No University" after a noose was found hanging on an African American professor's door at Columbia University's Teachers College in New York City. Are such actions a reflection of a climate of racism in our society? Why or why not?

Americans, Jewish Americans, African Americans, and Latinos/as, have been the objects of racist ideology.

Racism may be overt or subtle. *Overt racism* is more blatant and may take the form of public statements about the "inferiority" of members of a racial or ethnic group. An example of overt racism would be the 2015 situation in which some University of Oklahoma fraternity members were videorecorded singing a song that was extremely derogatory toward African Americans and which indicated that a black student would never be a member of their organization. It was later revealed that these students had learned the chant at a nationwide leadership retreat hosted by the national organization of Sigma Alpha Epsilon (SAE). However, hateful words such as this are often protected by the First Amendment's guarantee of freedom of speech, creating a quandary for those seeking to end overt racism and discrimination (see "Sociology and Social Policy"). Similarly, in other organizations,

including sports teams and their fan bases, researchers have documented that racist behavior continues to occur across generations, despite some people's belief that racism is a thing of the past. Examples of recurring patterns of racist behavior by sports team members or their fans include calling a player of color by a derogatory name, participating in racist chants, and writing racist graffiti on game-day signs, locker rooms, and school buildings. These actions are blatant and highly visible, but many subtle forms of racism also exist.

Subtle racism is hidden from sight and more difficult to recognize. Examples of *subtle racism* in sports include descriptions of African American athletes that suggest that they have "natural" athletic abilities and are better suited for those team positions that require speed and agility rather than the ability to think or process information quickly. By contrast, white athletes are depicted as being more intelligent, dependable, and possessing the right

harassment, slander, libel, and fighting words. Advocates for speech codes in higher education argue that even though hateful, racist speech is protected by the First Amendment, there still must be protection against speech that might constitute a direct threat to an individual or might provoke an immediate violent response. One university's code, for example, prohibits "conduct that is sufficiently severe and pervasive that it alters the conditions of education or employment and creates an environment that a reasonable person would find intimidating, harassing or humiliating" (*Dallas Morning News,* 2015). In lawsuits involving public universities, the courts have typically ruled that the rights bestowed by the Constitution take precedent over any speech codes the institutions might devise (with specific exceptions that are beyond the scope of our discussion).

However, other legal scholars argue that college speech codes are unconstitutional because they limit students' freedom of speech and send the wrong idea about what values should govern a free society. This approach is taken by the American Civil Liberties Union (ACLU, 2015):

> Free speech rights are indivisible. Restricting the speech of one group or individual jeopardizes everyone's rights because the same laws or regulations used to silence bigots can be used to silence you. Conversely, laws that defend free speech for bigots can be used to defend the rights of civil rights workers, anti-war protesters, lesbian and gay activists and others fighting for justice.

Therefore, the fundamental right to free speech should not be restricted even for bigots because this might mean that rights become restricted for other persons who are fighting for tolerance of diversity and justice. The assumption is that aggrieved individuals will engage in *counterspeech* that off-sets the negative, racist, or sexist speech. *Counterspeech* refers to the process of using more speech to contradict the negative and add new thoughts and values to the marketplace of ideas. But as Boston College Law School professor Kent Greenfield (2015) has stated, "Those not targeted by the [hate] speech can sit back and recite how distasteful such racism or sexism is, and isn't it too bad so little can be done. Meanwhile, those targeted by the speech are forced to speak out, yet again, to reassert their right to be treated equally, to be free to learn or work in an environment that does not threaten them with violence." According to Professor Greenfield, counterspeech is both "exhausting" and "distracting" because individuals continually have to be speaking up and standing up for their rights, emphasizing why they should not be oppressed by other people. From this perspective, individuals who are underrepresented—those with less power—do not have equal access to freedom of speech. In the case of the Oklahoma fraternity members' racist speech, for example, countering with a protest of their own does not offset the negative effects of the racist chant.

The debate over hate speech versus First Amendment rights has gone on for decades and no doubt will continue for many years to come.

REFLECT & ANALYZE

How do laws and court interpretations affect how we perceive race and racism in this country? Is there a possibility that counterspeech might produce new ideas about race and how to get along with each other? Why or why not?

leadership and decision-making skills needed in positions with higher levels of responsibility and control on the team.

Theories of Prejudice

Are some people more prejudiced than others? To answer this question, some theories focus on how individuals may transfer their internal psychological problem onto an external object or person. Others look at factors such as social learning and personality types.

The *frustration–aggression hypothesis* states that people who are frustrated in their efforts to achieve a highly desired goal will respond with a pattern of aggression toward others (Dollard et al., 1939). The object of their aggression becomes the **scapegoat**—a person or group that is incapable of offering resistance to the hostility or aggression of others (Marger, 2015). Scapegoats are often used as substitutes for the actual source of the frustration. For example, members of subordinate racial and ethnic groups are often blamed for local problems (such as the home team losing a football, basketball, or soccer game) or societal problems (such as large-scale unemployment or an economic recession) over which they believe they have little or no control (● Figure 10.3).

According to some symbolic interactionists, prejudice results from social learning; in other words, it is learned from observing and imitating significant others, such as parents and peers. Initially, children do not have a frame of reference from which to question the prejudices of their relatives and friends. When they are rewarded with smiles or laughs for telling derogatory jokes or making negative comments about outgroup members, children's prejudiced attitudes may be reinforced.

scapegoat
a person or group that is incapable of offering resistance to the hostility or aggression of others.

Michael Greenlar/The Image Works

FIGURE 10.3 According to the frustration–aggression hypothesis, members of white supremacy groups such as the Ku Klux Klan often use members of subordinate racial and ethnic groups as scapegoats for societal problems over which they have no control.

Psychologist Theodor W. Adorno and his colleagues (1950) concluded that highly prejudiced individuals tend to have an **authoritarian personality**, which is characterized by excessive conformity, submissiveness to authority, intolerance, insecurity, a high level of superstition, and rigid, stereotypic thinking. This type of personality is most likely to develop in a family environment in which dominating parents who are anxious about status use physical discipline but show very little love in raising their children (Adorno et al., 1950). Other scholars have linked prejudiced attitudes to traits such as submissiveness to authority, extreme anger toward outgroups, and conservative religious and political beliefs (Altemeyer, 1981, 1988; Weigel and Howes, 1985).

Whereas prejudice is an attitude, **discrimination** involves actions or practices of dominant-group members (or their representatives) that have a harmful effect on members of a subordinate group. Prejudiced attitudes do not always lead to discriminatory behavior. As shown in ● Figure 10.4, the sociologist Robert K. Merton (1949) identified four combinations of attitudes and responses. *Unprejudiced nondiscriminators* are not personally prejudiced and do not discriminate against others. For example, two players on a professional sports team may be best friends although they are of different races. *Unprejudiced discriminators* may have no personal prejudice but still engage in discriminatory behavior because of peer-group pressure or economic, political, or social interests. For example, on some sports teams, players may hold no genuine prejudice toward players from diverse racial or ethnic origins but believe that they have to impress their "friends" by making disparaging remarks about persons of color so

that they can get into, or remain in, a peer group. By contrast, *prejudiced nondiscriminators* hold personal prejudices but do not discriminate because of peer pressure, legal demands, or a desire for profits. For example, professional sports teams' owners and coaches who hold prejudiced beliefs may hire a player of color to enhance the team's ability to win. Finally, *prejudiced discriminators* hold personal prejudices and actively discriminate against others. For example, a baseball umpire who is personally prejudiced against persons of color may intentionally call a play incorrectly based on that prejudice. Of course, we hope that such an umpire does not exist or that his or her actions would be quickly sanctioned if such an event occurred. But the purpose of Merton's typology is to show that prejudice and discrimination do not always coexist as directly and specifically as many of us might imagine.

Discriminatory actions vary in severity, from the use of derogatory labels to violence against individuals and groups. The ultimate form of discrimination occurs when

	Prejudiced attitude?	Discriminatory behavior?
Unprejudiced nondiscriminator	No	No
Unprejudiced discriminator	No	Yes
Prejudiced nondiscriminator	Yes	No
Prejudiced discriminator	Yes	Yes

FIGURE 10.4 Merton's Typology of Prejudice and Discrimination

Merton's typology shows that some people may be prejudiced but not discriminate against others. Do you think that it is possible for a person to discriminate against some people without holding a prejudiced attitude toward them? Why or why not?

people are considered to be unworthy to live because of their race or ethnicity. ***Genocide*** is the deliberate, systematic killing of an entire people or nation. Examples of genocide include the killing of thousands of Native Americans by white settlers in North America and the extermination of six million European Jews by Nazi Germany. A lack of consensus exists as to whether genocide has occurred in the twenty-first century. Some analysts believe that the mass slaughter and rape of Darfuri men, women, and children in Western Sudan that began in 2003 should be classified as genocide. However, international governing bodies typically have ruled that this situation does not fit the description because for something to be identified as genocide, the perpetrators must have the intent to destroy an entire group. By contrast, inflicting damage on a group or removing the population from a location does not qualify. More recently, the term *ethnic cleansing* has been used to define a policy of "cleansing" geographic areas by forcing persons of other races or religions to flee—or die.

Discrimination varies in how it is carried out. Individuals may act on their own, or they may operate within the context of large-scale organizations and institutions, such as schools, churches, corporations, and governmental agencies. How does individual discrimination differ from institutional discrimination? ***Individual discrimination*** consists of one-on-one acts by members of the dominant group that harm members of the subordinate group or their property. Individual discrimination is often considered to be based on the prejudicial beliefs of bigoted individuals who overtly express those beliefs through discriminatory actions. For example, a college student may write racist graffiti on the dorm door of another student because the perpetrator possesses bigoted attitudes about the superiority or inferiority of others based on their race or ethnicity.

However, sociologists emphasize that individual discrimination is not purely *individual*. As sociologists in the past moved beyond studying individual racial discrimination, they found that a close relationship exists between individual and institutional discrimination because they are two aspects of the same phenomenon. Simply stated, when individuals engage in racial discrimination, their actions are *shaped* by structural racial inequalities in the existing society or social system, and, in turn, their actions *reinforce* existing large-scale patterns of discrimination, which we refer to as institutional discrimination. ***Institutional discrimination*** consists of the day-to-day practices of organizations and institutions that have a harmful effect on members of subordinate groups. For example, a bank might consistently deny loans to people of a certain race; a university might not accept additional Asian American students in its first-year class or medical school because of an institutional assumption that persons in this racial–ethnic category are already overrepresented at the school. However, it is important to note that institutional discrimination is carried out by the *individuals* who implement the policies and procedures of organizations.

Sociologist Joe R. Feagin has identified four major types of discrimination:

1. *Isolate discrimination* is harmful action intentionally taken by a dominant-group member against a member of a subordinate group. This type of discrimination occurs without the support of other members of the dominant group in the immediate social or community context. For example, a prejudiced judge may give harsher sentences to African American defendants but may not be supported by the judicial system in that action.

2. *Small-group discrimination* is harmful action intentionally taken by a limited number of dominant-group members against members of subordinate groups. This type of discrimination is not supported by existing norms or other dominant-group members in the immediate social or community context. For example, a small group of white students may hang nooses (that signify the practice of racial lynching in the past) on the door of an African American professor's office without the support of other students or faculty members.

3. *Direct institutionalized discrimination* is organizationally prescribed or community-prescribed action that intentionally has a differential and negative impact on members of subordinate groups. These actions are routinely carried out by a number of dominant-group members based on the norms of the immediate organization or community (Feagin and Feagin, 2012). Intentional exclusion of people of color from public accommodations in the past is an example of this type of discrimination.

4. *Indirect institutionalized discrimination* refers to practices that have a harmful effect on subordinate-group members even though the organizationally or community-prescribed norms or regulations guiding these actions were initially established with no intent to harm. For example, special education classes were originally intended to provide extra educational opportunities for children with various types of disabilities. However, critics claim that these programs have amounted to racial segregation in many school districts.

Various types of racial and ethnic discrimination call for divergent remedies if we are to reduce discriminatory

authoritarian personality
a personality type characterized by excessive conformity, submissiveness to authority, intolerance, insecurity, a high level of superstition, and rigid, stereotypic thinking.

discrimination
actions or practices of dominant-group members (or their representatives) that have a harmful effect on members of a subordinate group.

genocide
the deliberate, systematic killing of an entire people or nation.

individual discrimination
behavior consisting of one-on-one acts by members of the dominant group that harm members of the subordinate group or their property.

institutional discrimination
the day-to-day practices of organizations and institutions that have a harmful effect on members of subordinate groups.

actions and practices in contemporary social life. Since the 1950s and 1960s, many U.S. sociologists have analyzed the complex relationship between prejudice and discrimination. Some have reached the conclusion that prejudice is difficult, if not seemingly impossible, to eradicate because of the deeply held racist beliefs and attitudes that are often passed on from person to person and from one generation to the next. However, the persistence of prejudicial attitudes and beliefs does not mean that racial and ethnic discrimination should be allowed to flourish until such a time as prejudice is effectively eliminated. From this approach, discrimination must be aggressively tackled through demands for change and through policies that specifically target patterns of discrimination.

LO4 **Compare** the major sociological perspectives on race and ethnic relations.

Sociological Perspectives on Race and Ethnic Relations

Symbolic interactionist, functionalist, and conflict analysts examine race and ethnic relations in different ways. Symbolic interactionists examine how microlevel contacts between people may produce either greater racial tolerance or increased levels of hostility. Functionalists focus on the macrolevel intergroup processes that occur between members of dominant and subordinate groups in society. Conflict theorists analyze power and economic differentials between the dominant group and subordinate groups.

Symbolic Interactionist Perspectives

What happens when people from different racial and ethnic groups come into contact with one another? Symbolic interactionists claim that intergroup contact may either intensify or reduce racial and ethnic stereotyping and prejudice, depending on the context. In the *contact hypothesis*, symbolic interactionists point out that contact between people from divergent groups should lead to favorable attitudes and behavior *when certain factors are present:* Members of each group must (1) have equal status, (2) pursue the same goals, (3) cooperate with one another to achieve their goals, and (4) receive positive feedback when they interact with one another in positive, nondiscriminatory ways (• Figure 10.5). However, if these factors are not present, intergroup contact may lead to increased stereotyping and prejudice.

Of course, intergroup contact does not always include the four factors described above. What then happens when individuals meet someone who does not conform to their existing stereotype? According to symbolic interactionists, they frequently ignore anything that contradicts the stereotype, or they interpret the situation to support their prejudices. For example, a person who does not fit the stereotype may be seen as an exception—"You're not like other [persons of a particular race]." Conversely, when a person is seen as conforming to a stereotype, he or she may be treated simply as one of "you people."

Symbolic interactionist perspectives make us aware of the importance of intergroup contact and the fact that it may either intensify or reduce racial and ethnic stereotyping and prejudice.

Functionalist Perspectives

How do members of subordinate racial and ethnic groups become a part of the dominant group? To answer this question, early functionalists studied immigration and patterns of dominant- and subordinate-group interactions.

FIGURE 10.5 Symbolic interactionists believe that intergroup contact can reduce stereotyping and prejudice if group members have equal status, pursue the same goals and cooperate to achieve them, and receive positive feedback when they interact with one another in positive ways. How do sports teams enable such interaction?

Jeff Gross/Getty Images

Assimilation *Assimilation* is a process by which members of subordinate

Hulton Archive/Getty Images

FIGURE 10.6 Segregation exists when specific ethnic groups are set apart from the dominant group and have unequal access to power and privilege. What examples of segregation do you see today?

racial and ethnic groups become absorbed into the dominant culture. To some analysts, assimilation is functional because it contributes to the stability of society by minimizing group differences that might otherwise result in hostility and violence.

Assimilation occurs at several distinct levels, including the cultural, structural, biological, and psychological stages. *Cultural assimilation,* or *acculturation,* occurs when members of an ethnic group adopt dominant-group traits, such as language, dress, values, religion, and food preferences. Cultural assimilation in this country initially followed an "Anglo conformity" model; members of subordinate ethnic groups were expected to conform to the culture of the dominant white Anglo-Saxon population. However, members of some groups refused to be assimilated and sought to maintain their unique cultural identity.

Structural assimilation, or *integration,* occurs when members of subordinate racial or ethnic groups gain acceptance in everyday social interaction with members of the dominant group. This type of assimilation typically starts in large, impersonal settings such as schools and workplaces, and only later (if at all) results in close friendships and intermarriage. *Biological assimilation,* or *amalgamation,* occurs when members of one group marry those of other social or ethnic groups. Biological assimilation has been more complete in some other countries, such as Mexico and Brazil, than in the United States.

Psychological assimilation involves a change in racial or ethnic self-identification on the part of an individual. Rejection by the dominant group may prevent psychological assimilation by members of some subordinate racial and ethnic groups, especially those with visible characteristics such as skin color or facial features that differ from those of the dominant group.

Ethnic Pluralism Instead of complete assimilation, many groups share elements of the mainstream culture while remaining culturally distinct from both the dominant group and other social and ethnic groups. *Ethnic pluralism* is the coexistence of a variety of distinct racial and ethnic groups within one society.

Equalitarian pluralism, or *accommodation,* is a situation in which ethnic groups coexist in equality with one another. Switzerland has been described as a model of equalitarian pluralism; more than six million people with French, German, and Italian cultural heritages peacefully coexist there. *Inequalitarian pluralism,* or *segregation,* exists when specific ethnic groups are set apart from the dominant group and have unequal access to power and privilege. *Segregation* is the spatial and social separation of categories of people by race, ethnicity, class, gender, and/or religion (● Figure 10.6). Segregation may be enforced by law. *De jure segregation* refers to laws that systematically enforced the physical and social separation of African Americans in all areas of public life. For example, Jim Crow laws legalized the separation of the races in public accommodations (such as hotels, restaurants, transportation, hospitals, jails, schools, churches, and cemeteries) in the southern United States after the Civil War (Feagin and Feagin, 2012).

Segregation may also be enforced by custom. *De facto segregation*—racial separation and inequality enforced by custom—is more difficult to document than de jure segregation. For example, residential segregation is still prevalent in many U.S. cities; owners, landlords, real estate agents, and apartment managers often use informal mechanisms to maintain their properties for "whites only."

assimilation
a process by which members of subordinate racial and ethnic groups become absorbed into the dominant culture.

ethnic pluralism
the coexistence of a variety of distinct racial and ethnic groups within one society.

segregation
the spatial and social separation of categories of people by race, ethnicity, class, gender, and/or religion.

Even middle-class people of color find that racial polarization is fundamental to the residential layout of many cities.

Although functionalist explanations provide a description of how some early white ethnic immigrants assimilated into the cultural mainstream, they do not adequately account for the persistent racial segregation and economic inequality experienced by people of color.

Conflict Perspectives

Conflict theorists focus on economic stratification and access to power in their analyses of race and ethnic relations. Some emphasize the caste-like nature of racial stratification, others analyze class-based discrimination, and still others examine internal colonialism and gendered racism.

The Caste Perspective

The caste perspective views racial and ethnic inequality as a permanent feature of U.S. society. According to this approach, the African American experience must be viewed as different from that of other racial or ethnic groups. African Americans were the only group to be subjected to slavery; when slavery was abolished, a caste system was instituted to maintain economic and social inequality between whites and African Americans (Feagin and Feagin, 2012).

The caste system was strengthened by *antimiscegenation laws,* which prohibited sexual intercourse or marriage between persons of different races. Most states had such laws, which were later expanded to include relationships between whites and Chinese, Japanese, and Filipinos.

Class Perspectives

Although the caste perspective points out that racial stratification may be permanent because of structural elements such as the law, it has been criticized for not examining the role of class in perpetuating racial inequality. Class perspectives emphasize the role of the capitalist class in racial exploitation. Based on early theories of race relations by the African American scholar W. E. B. Du Bois, the sociologist Oliver C. Cox (1948) suggested that African Americans were enslaved because they were the cheapest and best workers the owners could find for heavy labor in mines and on plantations. Thus, the profit motive of capitalists, not skin color or racial prejudice, accounts for slavery.

Sociologists have also debated the relative importance of class and race in explaining the unequal life chances of African Americans. Sociologists William Julius Wilson and Richard P. Taub (2007) have suggested that race, cultural factors, social psychological variables, and social class must *all* be taken into account in examining the life chances of "inner-city residents." Their analysis focuses on how race, ethnicity, and class tensions are all important in assessing how residents live their lives in four low-income Chicago neighborhoods and why this finding is important for the rest of America as well.

How do conflict theorists view the relationship among race, class, and sports? Simply stated, sports reflects the interests of the wealthy and powerful. At all levels, sports exploits athletes (even highly paid ones) in order to gain high levels of profit and prestige for coaches, managers, and owners. In particular, African American athletes and central-city youths are exploited by the message of rampant consumerism. Many are given the unrealistic expectation that sports can be a ticket out of the ghetto or barrio. If they try hard enough (and wear the right athletic gear), they too can become wealthy and famous.

Internal Colonialism

Why do some racial and ethnic groups continue to experience subjugation after many years? According to the sociologist Robert Blauner (1972), groups that have been subjected to internal colonialism remain in subordinate positions longer than groups that voluntarily migrated to the United States. *Internal colonialism* occurs when members of a racial or ethnic group are conquered or colonized and forcibly placed under the economic and political control of the dominant group. This idea has been so widely received that it is often referred to as the "Blauner hypothesis" and is still used in research.

In the United States, indigenous populations (including groups known today as Native Americans and Mexican Americans) were colonized by Euro-Americans and others who invaded their lands and conquered them. In the process, indigenous groups lost property, political rights, aspects of their culture, and often their lives. The capitalist class acquired cheap labor and land through this government-sanctioned racial exploitation. The effects of past internal colonialism are reflected today in the number of Native Americans who live on government reservations and in the poverty of Mexican Americans who lost their land and had no right to vote (• Figure 10.7).

The internal colonialism perspective is rooted in the historical foundations of racial and ethnic inequality in the United States. However, it tends to view all voluntary immigrants as having many more opportunities than do members of colonized groups. Thus, this model does not explain the continued exploitation of some immigrant groups, such as the Chinese, Filipinos, Cubans, Vietnamese, and Haitians, and the greater acceptance of others, primarily those from Northern Europe.

The Split-Labor-Market Theory

Who benefits from the exploitation of people of color? Dual- or split-labor-market theory states that white workers and members of the capitalist class both benefit from the exploitation of people of color. *Split labor market* refers to the division of the economy into two areas of employment, a primary sector or upper tier, composed of higher-paid (usually dominant-group) workers in more-secure jobs, and a secondary sector or lower tier, composed of lower-paid (often subordinate-group) workers in jobs with little security and hazardous working conditions (Bonacich, 1972, 1976). According to this perspective, white workers in the upper tier may use racial discrimination against

FIGURE 10.7 Grinding poverty is a pressing problem for families living along the border between the United States and Mexico. Economic development has been limited in areas where *colonias* such as this one are located, and the wealthy have derived far more benefit than others from recent changes in the global economy. How might the concept of internal colonialism be used to explain the impoverished conditions shown in this photo?

nonwhites to protect their positions. These actions most often occur when upper-tier workers feel threatened by lower-tier workers hired by capitalists to reduce labor costs and maximize corporate profits. In the past, immigrants were a source of cheap labor that employers could use to break strikes and keep wages down. Throughout U.S. history, higher-paid workers have responded with racial hostility and joined movements to curtail immigration and thus do away with the source of cheap labor (Marger, 2015).

Proponents of the split-labor-market theory suggest that white workers benefit from racial and ethnic antagonisms. However, these analysts typically do not examine the interactive effects of race, class, and gender in the workplace.

Perspectives on Race and Gender The term *gendered racism* refers to the interactive effect of racism and sexism on the exploitation of women of color. According to the social psychologist Philomena Essed (1991), women's particular position must be explored within each racial or ethnic group because their experiences will not have been the same as men's in each grouping.

Capitalists do not equally exploit all workers. Gender and race or ethnicity are important in this exploitation. Historically, white men have monopolized the high-paying primary labor market. Many people of color

Internal colonialism
according to conflict theorists, a practice that occurs when members of a racial or ethnic group are conquered or colonized and forcibly placed under the economic and political control of the dominant group.

split labor market
the division of the economy into two areas of employment, a primary sector or upper tier, composed of higher-paid (usually dominant-group) workers in more-secure jobs, and a secondary sector or lower tier, composed of lower-paid (often subordinate-group) workers in jobs with little security and hazardous working conditions.

gendered racism
the interactive effect of racism and sexism on the exploitation of women of color.

Sociological Perspectives on Race and Ethnic Relations

	Focus	Theory/Hypothesis
Symbolic Interactionist	Microlevel contacts between individuals	Contact hypothesis

Functionalist	Macrolevel intergroup processes	1. Assimilation
		   a. cultural
		   b. biological
		   c. structural
		   d. psychological
		2. Ethnic pluralism
		   a. equalitarian pluralism
		   b. inequalitarian pluralism (segregation)

Conflict	Power/economic differentials between dominant and subordinate groups	1. Caste perspective 2. Class perspective 3. Internal colonialism 4. Split labor market 5. Gendered racism 6. Racial formation
Critical Race Theory	Racism as an ingrained feature of society that affects everyone's daily life	Laws may remedy overt discrimination but have little effect on subtle racism. Interest convergence is required for social change.

and white women hold lower-tier jobs. Below that tier is the underground sector of the economy, characterized by illegal or quasi-legal activities such as drug trafficking, prostitution, and working in sweatshops that do not meet minimum wage and safety standards. Many undocumented workers and some white women and people of color attempt to earn a living in this sector, as further described in Chapter 13, "The Economy and Work in Global Perspective."

Racial Formation The **theory of racial formation** states that actions of the government substantially define racial and ethnic relations in the United States. Government actions range from race-related legislation to imprisonment of members of groups believed to be a threat to society. Sociologists Michael Omi and Howard Winant (2013) suggest that the U.S. government has shaped the politics of race through actions and policies that cause people to be treated differently because of their race. For example, immigration legislation reflects racial biases. The Naturalization Law of 1790 permitted only white immigrants to qualify for naturalization; the Immigration Act of 1924 favored Northern Europeans and excluded Asians and Southern and Eastern Europeans.

Social protest movements of various racial and ethnic groups periodically challenge the government's definition of racial realities. When this social rearticulation occurs, people's understanding about race may be restructured somewhat. For example, the African American protest movements of the 1950s and 1960s helped redefine the rights of people of color in the United States.

An Alternative Perspective: Critical Race Theory

Emerging out of scholarly law studies on racial and ethnic inequality, *critical race theory* derives its foundation from the U.S. civil rights tradition. Critical race theory has several major premises, including the belief that racism is such an ingrained feature of U.S. society that it appears to be ordinary and natural to many people (Delgado, 1995). As a result, civil rights legislation and affirmative action laws (formal equality) may remedy some of the more overt, blatant forms of racial injustice but have little effect on subtle, business-as-usual forms of racism that people of color experience as they go about their everyday lives. Although many minority-group members participate in collegiate and professional sports, studies of sports and media show that overt and covert forms of racism persist in the twenty-first century.

According to this approach, the best way to document racism and ongoing inequality in society is to listen to the lived experiences of people who have encountered such discrimination. In this way we can learn what actually happens in regard to racial oppression and the many effects it has on people, including alienation, depression, and certain physical illnesses. Central to this argument is the belief that *interest convergence* is a crucial factor in bringing about social change. According to the legal scholar Derrick Bell, white elites tolerate or encourage racial advances for

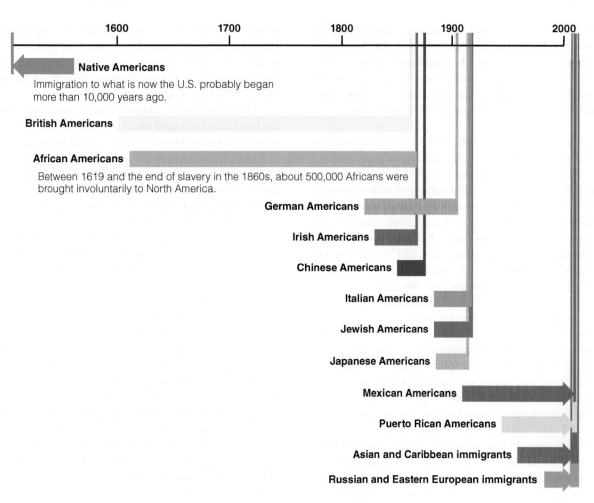

FIGURE 10.8 Time Line of Racial and Ethnic Groups in the United States

people of color *only* if the dominant-group members believe that their own self-interest will be served in so doing (cited in Delgado, 1995). From this approach, civil rights laws have typically benefited white Americans as much (or more) as people of color because these laws have been used as mechanisms to ensure that "racial progress occurs at just the right pace: change that is too rapid would be unsettling to society at large; change that is too slow could prove destabilizing" (Delgado, 1995: xiv). The Concept Quick Review outlines the key aspects of each sociological perspective on race and ethnic relations.

Racial and Ethnic Groups in the United States

How do racial and ethnic groups come into contact with one another? How do they adjust to one another and to the dominant group over time? Sociologists have explored these questions extensively; however, a detailed historical account of the unique experiences of each group is beyond the scope of this chapter. Instead, we will look briefly at intergroup contacts. In the process, sports will be used as an example of how members of some

groups have attempted to gain upward mobility and become integrated into society.

 Discuss the unique historical experiences of Native Americans and WASPs in the United States.

Native Americans and Alaska Natives

Native Americans and Alaska Natives are believed to have migrated to North America from Asia thousands of years ago, as shown on the time line in • Figure 10.8. One of the most widely accepted beliefs about this migration is that the first groups of Mongolians made their way across a natural bridge of land called Beringia into present-day Alaska. From there, they moved to what is now Canada and the northern United States, eventually making their way as far south as the tip of South America.

As schoolchildren are taught, Spanish explorer Christopher Columbus first encountered the native inhabitants

theory of racial formation
the idea that actions of the government substantially define racial and ethnic relations in the United States.

in 1492 and referred to them as "Indians." When European settlers (or invaders) arrived on this continent, the native inhabitants' way of life was changed forever. Experts estimate that approximately two million native inhabitants lived in North America at that time; however, their numbers had been reduced to fewer than 240,000 by 1900.

Genocide, Forced Migration, and Forced Assimilation

Native Americans have been the victims of genocide and forced migration. Although the United States never had an official policy that set in motion a pattern of deliberate extermination, many Native Americans were either massacred or died from European diseases (such as typhoid, smallpox, and measles) and starvation. In battle, Native Americans were often no match for the Europeans, who had "modern" weaponry. Europeans justified their aggression by stereotyping the Native Americans as "savages" and "heathens."

After the Revolutionary War, the federal government offered treaties to the Native Americans so that more of their land could be acquired for the growing white population. Scholars note that the government broke treaty after treaty as it engaged in a policy of wholesale removal of indigenous nations in order to clear the land for settlement by Anglo-Saxon "pioneers." Entire nations were forced to move in order to accommodate the white settlers. The "Trail of Tears" was one of the most disastrous of the forced migrations. In the coldest part of the winter of 1832, over half of the members of the Cherokee Nation died during or as a result of their forced relocation from the southeastern United States to the Indian Territory in Oklahoma.

Native Americans were subjected to forced assimilation on the reservations after 1871. Native American children were placed in boarding schools operated by the Bureau of Indian Affairs to hasten their assimilation into the dominant culture. About 98 percent of native lands had been expropriated by 1920. This process was aided by the Dawes Act (1877), which allowed the federal government to usurp Native American lands for the benefit of corporations and other non-native settlers who sought to turn a profit from oil and gas exploration and grazing.

Native Americans and Alaska Natives Today

Currently, about 5.2 million Native Americans and Alaska Natives, including those of more than one race, live in the United States, including Aleuts, Inuit (Eskimos), Cherokee, Navajo, Choctaw, Chippewa, Sioux, and more than 500 other nations of varying sizes and different locales. There is a wide diversity among the people in this category: Each nation has its own culture, history, and unique identity, and more than 250 Native American languages are spoken today. Slightly more than 20 percent of Native Americans and Alaska Natives ages five and older have reported that they spoke a language other than English at home.

Although Native Americans live in a number of states, they are concentrated in specific regions of the country. About 22 percent of American Indians and Alaska Natives reside in federal American Indian reservations and/or off-reservation trust lands or other tribal-designated areas. There are 325 federally recognized American Indian reservations in this country and a total of 630 legal and statistical areas (U.S. Census Bureau, "American Indian and Alaska Native Heritage Month," 2014).

Data continue to indicate that Native Americans are the most disadvantaged racial or ethnic group in the United States in terms of income, employment, housing, nutrition, and health. As compared to a median household income of $52,176 for the nation as a whole in 2013, for example, the median household income of American Indian and Alaska Native households was $36,252. In the same year, nearly 30 percent of American Indians and Alaska Natives lived in poverty at a time when the national poverty rate was slightly less than 16 percent. American Indians and Alaska Natives have higher rates of infant mortality than white American (non-Hispanic) infants, and American Indian and Alaska Native infants are four times more likely to die from pneumonia and influenza. American Indian and Alaska Native suicide rates are nearly 50 percent higher than those of white Americans (non-Hispanic). Suicide is particularly a concern among American Indian and Alaska Native males and among persons under age 25 (cdc.gov, 2014; U.S. Department of Health and Human Services, 2014).

Historically, Native Americans have had very limited educational opportunities and very high rates of unemployment. Educational opportunities have largely been tied to community colleges. Since the introduction of six tribally controlled community colleges in the 1970s, a growing network of tribal colleges and universities now serves over 30,000 students from more than 250 tribal nations (see • Figure 10.9). This network has been successful in providing some Native Americans with the necessary education to move into the ranks of the skilled working class and beyond (• Figure 10.10). Across the nation, Native Americans own and operate many types of enterprises, such as construction companies, computer-graphic-design firms, grocery stores, and management consulting businesses. Casino gambling operations and cigarette shops on Native American reservations—resulting from a reinterpretation of federal law in the 1990s—have brought more income to some of the tribal nations. However, this change has not been without its critics, who believe that such businesses bring new problems for Native Americans.

In 2009 Native Americans received a $3.4 billion settlement from the federal government after the conclusion of *Cobell v. Salazar,* a thirteen-year-old lawsuit that accused the government of mishandling revenues generated by the extraction of natural resources from American Indian land trusts as a result of the Dawes Act. Although the federal government was responsible for leasing tribal lands for use by mining, lumber, oil, and gas industries and passing on royalty payments to the Native Americans to whom the lands belonged, Native Americans derived little benefit because of the government's massive abuse of the trust funds.

Native Americans are currently in a transition from a history marked by prejudice and discrimination to a

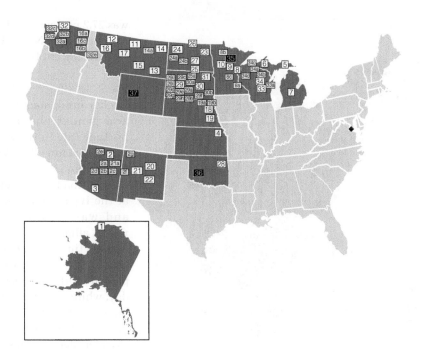

Member Tribal Colleges

Alaska
1 Ilisagvik College
Barrow, AK

Arizona
2 Diné College
Tsaile, AZ

2a Chinle, AZ

2b Ganado, AZ

2c Window Rock, AZ

2d Tuba City, AZ

2e Kayenta, AZ

2f Crownpoint, NM

2g Shiprock, NM

3 Tohono O'odham Community College
Sells, AZ

Kansas
4 Haskell Indian Nations University
Lawrence, KS

Michigan
5 Bay Mills Community College
Brimley, MI

6 Keweenaw Bay Ojibwa Community College
Baraga, MI

7 Saginaw Chippewa Tribal College
Mount Pleasant, MI

Minnesota
8 Fond du Lac Tribal and Community College
Cloquet, MN

8a Minneapolis, MN

8b Red Lake, MN

8c Onamia, MN

9 Leech Lake Tribal College
Cass Lake, MN

10 White Earth Tribal and Community College
Mahnomen, MN

Montana
11 Aaniiih Nakoda College
Harlem, MT

12 Blackfeet Community College
Browning, MT

13 Chief Dull Knife College
Lame Deer, MT

14 Fort Peck Community College
Poplar, MT

14a Wolf Point, MT

15 Little Big Horn College
Crow Agency, MT

16 Salish Kootenai College
Pablo, MT

16a Colville, WA

16b Spokane, WA

16c Wellpinit, WA

17 Stone Child College
Box Elder, MT

Nebraska
18 Little Priest Tribal College
Winnebago, NE

19 Nebraska Indian Community College
Macy, NE

19a Niobrara, NE

19b South Sioux City, NE

New Mexico
20 Institute of American Indian Arts
Santa Fe, NM

21 Navajo Technical University
Crownpoint, NM

21a Chinle, AZ

22 Southwestern Indian Polytechnic Institute
Albuquerque, NM

North Dakota
23 Cankdeska Cikana Community College
Fort Totten, ND

24 Fort Berthold Community College
New Town, ND

24a Mandaree, ND

24b White Shield, ND

25 Sitting Bull College
Fort Yates, ND

25a McLaughlin, SD

26 Turtle Mountain Community College
Belcourt, ND

27 United Tribes Technical College
Bismarck, ND

Oklahoma
28 College of the Muscogee Nation
Okmulgee, OK

South Dakota
29 Oglala Lakota College
Kyle, SD

29a Allen, SD

29b East Wakpamni, SD

29c Eagle Butte, SD

29d Manderson, SD

29e Porcupine, SD

29f Martin, SD

29g Oglala, SD

29h Pine Ridge, SD

29i Wambli, SD

29j Rapid City, SD

30 Sinte Gleska University
Mission, SD

30a Lower Brule, SD

30b Marty, SD

31 Sisseton Wahpeton College
Sisseton, SD

Virginia
◆ American Indian Higher Education Consortium
Alexandria, VA

Washington
32 Northwest Indian College
Bellingham, WA

32a Auburn, WA

32b Tulalip, WA

32c La Conner, WA

32d Kingston, WA

32e Lapwai, ID

Wisconsin
33 College of Menominee Nation
Keshena, WI

33a Green Bay-Oneida Campus

34 Lac Courte Oreilles Ojibwa Community College
Hayward, WI

34a Odanah, WI

34b Lac du Flambeau, WI

34c Bayfield, WI

34d Hertel, WI

AIHEC Associate Members

Minnesota
35 Red Lake Tribal College
Red Lake, MN

Oklahoma
36 Comanche Nation College
Lawton, OK

Wyoming
37 Wind River Tribal College
Ethete, WY

FIGURE 10.9 U.S. Tribal Colleges and Universities
Source: American Indian College Fund, 2012.

national visibility as athletes in football, baseball, and track and field. Teams at boarding schools such as the Carlisle Indian Industrial School in Pennsylvania and the Haskell Institute in Kansas were well-known. However, after the first three decades of the twentieth century, Native Americans became less prominent in sports. Native American scholar Joseph B. Oxendine (2003) attributes the lack of athletic participation to these factors: (1) a reduction in opportunities for developing sports skills, (2) restricted opportunities for participation, and (3) a lessening of Native Americans' interest in competing with and against non–Native Americans. However, in the twenty-first century, Native Americans slowly began making their mark in a few professional sports. Sam Bradford (Cherokee Nation) has made inroads in professional football. Other notables are in golf (Notah Begay III), lacrosse (Brett Bucktooth), bowling (Mike Edwards), rodeo (Clint Harry), and baseball (Kyle Lohse). More Native American college athletes are also being recognized in the twenty-first century (visit the website for NDNSPORTS.com).

White Anglo-Saxon Protestants (British Americans)

Whereas Native Americans have been among the most disadvantaged peoples in this country, white Anglo-Saxon Protestants (WASPs) have been the most privileged group. Although many English settlers initially came to North America as indentured servants or as prisoners, they quickly emerged as the dominant group, creating a core culture (including language, laws, and holidays) to which all other groups were expected to adapt. Most of the WASP immigrants arriving from Northern Europe were advantaged over later immigrants because they were highly skilled and did not experience high levels of prejudice and discrimination.

contemporary life in which they may find new opportunities. Many see the challenge for Native Americans today as erasing negative stereotypes while maintaining their heritage and obtaining recognition for their contributions to this nation's development and growth. For the poorest of poor, however, access to opportunities is very limited.

Native Americans and Sports Early in the twentieth century, Native Americans such as Jim Thorpe gained

Class, Gender, and WASPs Like members of other racial and ethnic groups, not all WASPs are alike. Social class

FIGURE 10.10   Historically, Native Americans have had a low rate of college attendance. However, the development of a network of tribal colleges has provided them with a local source for upward mobility.

was 27.2 percent for African Americans (DeNavas-Walt and Proctor, 2014).

WASPs and Sports   Family background, social class, and gender play an important role in the sports participation of WASPs. Contemporary North American football was invented at the Ivy League colleges and was dominated by young, affluent WASPs who had the time and money to attend college and participate in sports activities. Today, whites are more likely than any other racial or ethnic group to become professional athletes in all sports except football and basketball. Although current data are not available to document differences among racial and ethnic categories by types of sports, we know that the probability of competing in athletics beyond the high school interscholastic level is extremely low. For example, only .03 percent of high school men's basketball players will become professional athletes, as will only .02 percent of women's basketball players. For football, the percentage of high school players who will become professional athletes is .08 percent; for baseball, .4 percent; for men's ice hockey, .4 percent; and for men's soccer, .08 percent. Even the odds of advancing from high school athletics to NCAA college sports remain low: 3.2 percent for men's basketball, 3.6 percent for women's basketball, 6.1 percent for football, 6.6 percent for baseball, 10.7 percent for men's ice hockey, and 5.7 percent for men's soccer (National Collegiate Athletic Association, 2012).

Affluent WASP women participated in intercollegiate women's basketball in the late 1800s, and various other sporting events were used as a means to break free of restrictive codes of femininity. Until fairly recently, however, most women have had little chance for any involvement in college and professional sports.

and gender affect their life chances and opportunities. For example, members of the working class and the poor do not have political and economic power; men in the capitalist class do. WASPs constitute the majority of the upper class and maintain cohesion through listings such as the *Social Register* and interactions with one another in elite settings such as private schools and country clubs (Kendall, 2002).

Today, the U.S. Census Bureau uses the term "white" to refer to a person having origins in any of the original peoples of Europe, the Middle East, or North Africa. In the latest (2010) census, people who indicated that they were Caucasian, white, Irish, German, Polish, Arab, Lebanese, Palestinian, Algerian, Moroccan, and Egyptian, among others, were included in the white racial category (Hixson, Hepler, and Kim, 2011). Seventy-two percent of all persons (at the time nearly 224 million people) included in the census identified as white alone. An additional 7.5 million people (2 percent) reported white in combination with one or more other races. Today, the majority of people in these population categories reside in the South (in states such as Alabama, Arkansas, Florida, Georgia, Texas, and Virginia) and the Midwest (in states such as Illinois, Indiana, Michigan, Nebraska, Ohio, and Wisconsin). The fastest growth in white population occurred in states in the South and in the West (Hixson, Hepler, and Kim, 2011).

As noted in Chapter 8, the median household income of whites (non-Hispanic) of $58,270 in 2013 is second only to that of Asian Americans ($67,065) and significantly above that of Hispanic or Latino/a and African American (black) residents of the United States. Likewise, the poverty rate for whites (non-Hispanic) was 9.6 percent while it

 Describe how slavery, segregation, lynching, and persistent discrimination have uniquely affected the African American experience in this country.

African Americans

The African American (black) experience has been one uniquely marked by slavery, segregation, and persistent discrimination. There is a lack of consensus about whether

African American or *black* is the most appropriate term to refer to the 45 million Americans of African descent who live in the United States today. Those who prefer the term *black* point out that it incorporates many African-descent groups living in this country that do not use *African American* as a racial or ethnic self-description. For example, many people who trace their origins to Haiti, Puerto Rico, or Jamaica typically identify themselves as "black" but not as "African American." Although African Americans reside throughout the United States, eighteen states have an estimated black population of at least one million. About 3.7 million African Americans lived in New York State in 2013, but the District of Columbia had the highest percentage of blacks (51 percent), followed by Mississippi (18 percent), in the total population (see • Figure 10.11).

Although the earliest African Americans probably arrived in North America with the Spanish conquerors in the fifteenth century, most historians trace their arrival to about 1619, when the first groups of indentured servants were brought to the colony of Virginia. However, by the 1660s, indentured servanthood had turned into full-fledged slavery because of the enactment of laws that sanctioned the enslavement of African Americans. Although the initial status of persons of African descent in this country may not have been too different from that of the English indentured servants, all of that changed with the passage of laws turning human beings into property and making slavery a status from which neither individuals nor their children could escape (Franklin, 1980).

Between 1619 and the 1860s, about 500,000 Africans were forcibly brought to North America, primarily to work on southern plantations, and these actions were justified by the devaluation and stereotyping of African Americans. Some analysts believe that the central factor associated with the development of slavery in this country was the plantation system, which was heavily reliant on cheap and dependable manual labor. Slavery was primarily beneficial to the wealthy southern plantation owners, but many of the stereotypes used to justify slavery were eventually institutionalized in southern custom and practice (Wilson, 1978). However, some slaves and whites engaged in active resistance against slavery and its barbaric practices, which eventually resulted in slavery being outlawed in the northern states by the late 1700s. Slavery continued in the South until 1863, when it was abolished by the Emancipation Proclamation (Takaki, 1993).

Segregation and Lynching   Gaining freedom did not give African Americans equality with whites. African Americans were subjected to many indignities because of race. Through informal practices in the North and *Jim Crow laws* in the South, African Americans experienced segregation in housing, employment, education, and all public accommodations. African Americans who did not stay in their "place" were often the victims of violent attacks and lynch mobs (Franklin, 1980). *Lynching* is a killing carried out by a group of vigilantes seeking revenge for an actual or imagined crime by the victim. The practice of lynching was used

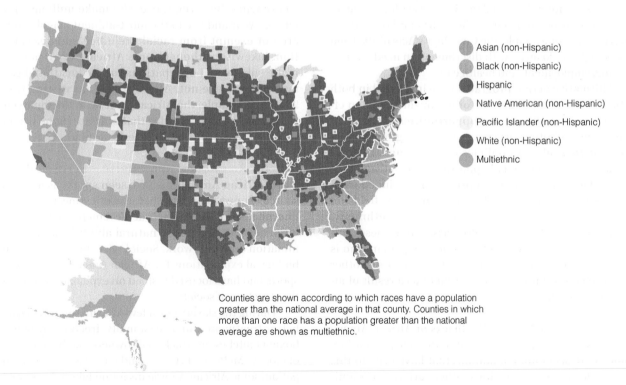

Asian (non-Hispanic)

Black (non-Hispanic)

Hispanic

Native American (non-Hispanic)

Pacific Islander (non-Hispanic)

White (non-Hispanic)

Multiethnic

Counties are shown according to which races have a population greater than the national average in that county. Counties in which more than one race has a population greater than the national average are shown as multiethnic.

FIGURE 10.11   U.S. Racial and Ethnic Distribution
While minority populations do continue to grow, regional differences in racial makeup are still quite pronounced, as this map shows.
Source: U.S. Census Bureau, 2011f.

by whites to intimidate African Americans into staying "in their place." It is estimated that as many as 6,000 lynchings occurred from the end of the Civil War to the present, at least half of which have gone unrecorded (Feagin and Feagin, 2012). In spite of all odds, many African American women and men resisted oppression and did not give up in their struggle for equality.

Discrimination   In the twentieth century the lives of many African Americans were changed by industrialization and two world wars. When factories were built in the northern United States, many African American families left the rural South in hopes of finding jobs and a better life.

During World Wars I and II, African Americans were a vital source of labor in war production industries; however, racial discrimination continued both on and off the job. In World War II, many African Americans fought for their country in segregated units in the military; after the war, they sought—and were denied—equal opportunities in the country for which they had risked their lives.

African Americans began to demand sweeping societal changes in the 1950s. Initially, the Reverend Dr. Martin Luther King Jr., and the civil rights movement used *civil disobedience*—nonviolent action seeking to change a policy or law by refusing to comply with it—to call attention to racial inequality and to demand greater inclusion of African Americans in all areas of public life. Subsequently, leaders of the Black Power movement, including Malcolm X and Marcus Garvey, advocated black pride and racial awareness among African Americans. Gradually, racial segregation was outlawed by the courts and the federal government. For example, the Civil Rights Acts of 1964 and 1965 sought to do away with discrimination in education, housing, employment, and health care.

Affirmative action programs were instituted in both public-sector and private-sector organizations in an effort to bring about greater opportunities for African Americans and other previously excluded groups. *Affirmative action* refers to policies or procedures that are intended to promote equal opportunity for categories of people deemed to have been previously excluded from equality in education, employment, and other fields on the basis of characteristics such as race or ethnicity. Critics of affirmative action often assert that these policies amount to *reverse discrimination*—a person who is better qualified being denied a position because another person received preferential treatment as a result of affirmative action.

African Americans Today   Blacks or African Americans make up about 13.2 percent of the U.S. population. Some are descendants of families that have been in this country for many generations; others are recent immigrants from Africa and the Caribbean. Black Haitians make up the largest group of recent Caribbean immigrants; others come from Jamaica and Trinidad and Tobago. Recent African immigrants are primarily from Nigeria, Ethiopia, Ghana, and Kenya. They have been simultaneously "pushed" out of their countries of origin by severe economic and political turmoil and "pulled" by perceived opportunities for a better life in the United States. Recent immigrants are often victimized by the same racism that has plagued African Americans as a people for centuries.

Since the 1960s, many African Americans have made significant gains in politics, education, employment, and income. Between 1964 and 2010, the number of African Americans elected to political office (on local, state, and federal levels) increased from about 100 to more than 10,000 nationwide (Hayes, 2011). At the local level, African Americans have won mayoral elections in many major cities that have large African American populations, such as Atlanta, Houston, New Orleans, Philadelphia, and Washington, D.C. At the national level, the top accomplishment of an African American in politics was the election of Barack Obama to the presidency in 2008 and his reelection to a second term of office in 2012 (● Figure 10.12).

In 2015 there were 46 African Americans serving in the 114th U.S. Congress: 44 were in the House of Representatives and 2 in the U.S. Senate. Over the past five decades, more African Americans have made impressive occupational gains and joined the top socioeconomic classes in terms of earnings. Some of these individuals have become professionals while others have achieved great wealth and fame as entertainers, athletes, and entrepreneurs. But even those who make millions of dollars per year and live in the most-affluent neighborhoods are not exempt from racial prejudice and discrimination. Likewise, although some African Americans have made substantial occupational and educational gains, many more have not. At the time I am writing this in 2015, for example, the African American unemployment rate of 10.4 percent is twice that of white (non-Hispanic) Americans (4.7) and Asian Americans (4.0).

African Americans and Sports   In recent decades many African Americans have seen sports as a possible source of upward mobility because other means have been unavailable. However, their achievements in sports have often been attributed to "natural ability" and not determination and hard work. Sociologists have rejected such biological explanations for African Americans' success in sports and have focused instead on explanations rooted in the structure of society.

During the slavery era, a few African Americans gained better treatment and, occasionally, freedom by winning boxing matches on which their owners had bet large sums of money (McPherson, Curtis, and Loy, 1989). After emancipation, some African Americans found jobs in horse racing and baseball. For example, fourteen of the fifteen jockeys in the first Kentucky Derby (in 1875) were African Americans. A number of African Americans played on baseball teams; a few played in the Major Leagues until the Jim Crow laws

FIGURE 10.12 In August 2008, Barack Obama made history by becoming the first African American to receive the presidential nomination of a major political party, and on Election Day he was voted in as the first African American president of the United States. In 2012 Obama was reelected to his second term and is shown here taking the oath of office from Chief Justice John Roberts on January 21, 2013.

White Ethnic Americans

The term *white ethnic Americans* is applied to a wide diversity of immigrants who trace their origins to Ireland and to Eastern and Southern European countries such as Poland, Italy, Greece, Germany, Yugoslavia, and Russia and other former Soviet republics. Unlike the WASPs, who migrated primarily from Northern Europe and assumed a dominant cultural position in society, white ethnic Americans arrived late in the nineteenth century and early in the twentieth century to find relatively high levels of prejudice and discrimination directed at them by nativist organizations that hoped to curb the entry of non-WASP European immigrants. Because many of the people in white ethnic American categories were not Protestant, they experienced discrimination because they were Catholic, Jewish, or members of other religious bodies, such as the Eastern Orthodox churches.

Discrimination Against White Ethnics Many white ethnic immigrants entered the United States between 1830 and 1924. Irish Catholics were among the first to arrive, driven out of Ireland by English oppression and famine and seeking jobs in the United States (Feagin and Feagin, 2012). When they arrived, they found that British Americans controlled the major institutions of society. The next arrivals were Italians, who had been recruited for low-wage industrial and construction jobs. British Americans viewed Irish and Italian immigrants as "foreigners": The Irish were stereotyped as ape-like, filthy, bad-tempered, and heavy drinkers; the Italians were depicted as lawless, knife-wielding thugs looking for a fight, "dagos," and "wops" (short for "without papers") (Feagin and Feagin, 2012).

Both Irish Americans and Italian Americans were subjected to institutionalized discrimination in employment. Employment ads read "Help Wanted—No Irish Need Apply" and listed daily wages at $1.30–$1.50 for "whites" and $1.15–$1.25 for "Italians" (Gambino, 1975: 77). In spite of discrimination, white ethnics worked hard to establish themselves in the United States, often founding mutual self-help organizations and becoming politically active (Mangione and Morreale, 1992).

forced them out. Then they formed their own "Negro" baseball and basketball leagues (Peterson, 1992/1970).

Since Jackie Robinson broke baseball's "color line," in 1947, many African American athletes have played collegiate and professional sports. Even now, however, persistent class inequalities between whites and African Americans are reflected in the fact that, until recently, African Americans have primarily excelled in sports (such as basketball or football) that do not require much expensive equipment and specialized facilities in order to develop athletic skills. According to one sports analyst, African Americans typically participate in certain sports and not others because of the *sports opportunity structure*—the availability of facilities, coaching, and competition in the schools and community recreation programs in their area (Phillips, 1993).

Regardless of the sport in which they participate, African American men athletes continue to experience inequalities in coaching, management, and ownership opportunities in professional sports. In recent years, only five of the thirty-two National Football League teams and fourteen of the teams in the NCAA elite "Football Bowl Subdivision"—postseason bowl-eligible competitors—had African American head coaches. By contrast, 60 percent of the players on the top 25 FBS teams are black. Today, African Americans remain significantly underrepresented in many other sports, including hockey, skiing, figure skating, golf, volleyball, softball, swimming, gymnastics, sailing, soccer, bowling, cycling, and tennis.

Between 1880 and 1920, a wave of Eastern European Jewish immigrants arrived in the United States and settled in the Northeast. Jewish Americans differ from other white ethnic groups in that some focus their identity primarily on their religion whereas others define their Jewishness in terms of ethnic group membership (Feagin and Feagin, 2012). In any case, Jews continued to be the victims of *anti-Semitism*—prejudice, hostile attitudes, and discriminatory behavior targeted at Jews. For example, signs in hotels read "No Jews Allowed," and some "help wanted" ads stated "Christians Only" (Levine, 1992: 55). In spite of persistent discrimination, Jewish Americans achieved substantial success in many areas, including business, education, the arts and sciences, law, and medicine.

However, old biases remain deeply embedded in the fabric of American life and are passed on from one generation to the next. An example of this kind of lingering prejudice surfaced in 2015 when a Jewish student at UCLA who was being considered for the student council's judicial board was asked the following by a another student: "Given that you are a Jewish student and very active in the Jewish community, how do you see yourself being able to maintain an unbiased view?" (Nagourney, 2015). Although the vast majority of Jewish American students on college campuses nationwide are not asked questions such as this, the discussion about possible bias in this students' decision-making skills because of affiliation with Jewish organizations raises anew the fears and concerns of eras, apparently not all gone, when prejudice and discrimination were directed toward people because they were identified as Jews in this country and worldwide.

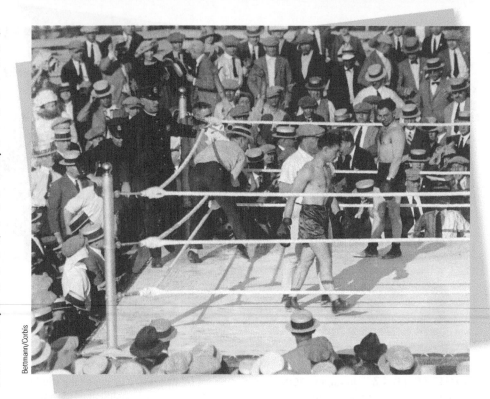

Bettmann/Corbis

FIGURE 10.13 Early-twentieth-century Jewish American and Italian American boxers not only produced intragroup ethnic pride but also earned a livelihood through boxing matches.

White Ethnics and Sports   Sports provided a pathway to assimilation for many white ethnics. The earliest collegiate football players who were not white Anglo-Saxon Protestants were of Irish, Italian, and Jewish ancestry. Sports participation provided educational opportunities that some white ethnics would not have had otherwise.

Boxing became a way to make a living for white ethnics who did not participate in collegiate sports (• Figure 10.13). Boxing promoters encouraged ethnic rivalries to increase their profits, pitting Italians against Irish or Jews, and whites against African Americans (Levine, 1992; Mangione and Morreale, 1992). Eventually, Italian Americans graduated from boxing into baseball and football. Jewish Americans found that sports lessened the shock of assimilation and gave them an opportunity to refute stereotypes about their physical weaknesses and to counter anti-Semitic charges that they were "unfit to become Americans" (Levine, 1992: 272). Today, assimilation is so complete that little attention is paid to the origins of white ethnic athletes.

Identify the major categories of Asian Americans and describe their historical and contemporary experiences.

Asian Americans

Recent research has found that Asian Americans have the highest income and the most formal education of any racial group in the United States. They are also the fastest-growing racial group in the nation (Pew Research Center Social and Demographic Trends, 2013).

The U.S. Census Bureau uses the term *Asian Americans* to designate the many diverse groups with roots in Asia. Chinese and Japanese immigrants were among the earliest Asian Americans. Many Filipinos, Asian Indians, Koreans, Vietnamese, Cambodians, Pakistani, and Indonesians have arrived more recently. Asian Americans who reported only one race constituted about 5.3 percent of the U.S. population in 2013. From 2000 to 2013, there was a nearly 50 percent growth in the population of Asian Americans (reported alone or in combination with other racial–ethnic categories) in the United States. In 2013 about 18.9 million people in the United States identified themselves as residents of Asian descent or Asian in combination with one or more other races (U.S. Census Bureau Asian/Pacific American Heritage Month, 2014). About three-quarters (74 percent) of all Asian American adults were born in other countries.

Chinese Americans

Chinese Americans are the largest Asian American group, at 4.2 million (U.S. Census Bureau Asian/Pacific American Heritage Month, 2014). The initial wave of Chinese immigration occurred between 1850 and 1880, when more than 200,000 Chinese men were "pushed" from China by harsh economic conditions and "pulled" to the United States by the promise of gold in California and employment opportunities in the construction of transcontinental railroads. Far fewer Chinese women immigrated; however, many were brought to the United States against their will, and some were forced into prostitution.

Chinese Americans were subjected to extreme prejudice and stereotyped as "coolies," "heathens," and "Chinks." Some Asians were attacked and even lynched by working-class whites who feared that they would lose their jobs to these immigrants. Passage of the Chinese Exclusion Act of 1882 brought Chinese immigration to a halt. The Exclusion Act was not repealed until World War II, when Chinese Americans who were contributing to the war effort by working in defense plants pushed for its repeal. After immigration laws were further relaxed in the 1960s, the second and largest wave of Chinese immigration occurred, with immigrants coming primarily from Hong Kong and Taiwan. These recent immigrants have had more education and workplace skills than earlier arrivals, and they brought families and capital with them to pursue the American Dream.

Today, many Asians of Chinese descent reside in large urban enclaves in California, Texas, New York, and

Joe Sohm/VisionsofAmerica/Photodisc/Getty Images

FIGURE 10.14 During World War II, nearly 120,000 Japanese Americans—some of whom are still alive today—were interned in camps such as the Manzanar Relocation Center in California, where this statue memorializes their ordeal.

Hawaii. As a group, Asian Americans have enjoyed considerable upward mobility, and Chinese Americans are no exception. Many have become highly successful professionals and business entrepreneurs. However, other Chinese Americans remain in the lower tier of the working class—providing low-wage labor in personal services, repair, and maintenance (Pew Research Center, 2012c).

Japanese Americans

Most of the early Japanese immigrants were men who worked on sugar plantations in the Hawaiian Islands in the 1860s. Like Chinese immigrants, the Japanese American workers were viewed as a threat by white workers, and immigration of Japanese men was curbed in 1908. However, Japanese women were permitted to enter the United States for several years thereafter because of the shortage of women on the West Coast. Although some Japanese women married white men, laws prohibiting interracial marriage stopped this practice.

With the exception of the forced migration and genocide experienced by Native Americans and the enslavement of African Americans, Japanese Americans experienced one of the most vicious forms of discrimination ever sanctioned by U.S. laws. During World War II, when the United States was at war with Japan, nearly 120,000 Japanese Americans were placed in internment camps, where they remained for more than two years despite the total lack of evidence that they posed a security threat to this country (● Figure 10.14). This action was a direct violation of the citizenship rights of many *Nisei* (second-generation Japanese Americans), who were born in the United States. Only Japanese Americans were singled out for such harsh treatment; German Americans avoided this fate even though the United States was also at war with Germany. Four decades later, the U.S. government issued an apology for its actions and eventually paid $20,000 each to some of those who had been placed in internment camps.

Since World War II, many Japanese Americans have been very successful. The annual household income of Japanese Americans is between $66,000 and $68,000, as contrasted with approximately $49,800 for the total U.S. population. But, many Japanese Americans (and other Asian Americans as well) reside in states with higher than average incomes and higher costs of living than the national average.

Korean Americans Male workers primarily made up the first wave of Korean immigrants who arrived in Hawaii between 1903 and 1910. The second wave came to the U.S. mainland following the Korean War in 1954. This cohort was made up primarily of the wives of servicemen and Korean children who had lost their parents during the war. The third wave arrived after the Immigration Act of 1965 permitted well-educated professionals to migrate to the United States. Korean Americans have helped one another open small businesses by pooling money through the *kye*— an association that grants members money on a rotating basis to gain access to more capital.

Today, an estimated 1.8 million Korean Americans reside in the United States, constituting the fifth-largest category of Asian Americans and about 10 percent of the total adult Asian population in the nation. Many Korean Americans live in California and New York, where there is a concentration of Korean-owned businesses. The median annual household income for Korean Americans is slightly above $50,000, which is lower than for all Asians Americans but slightly higher than for the U.S. population as a whole.

Filipino Americans Today, Filipino Americans constitute the second-largest category of Asian Americans, with about 3.6 million U.S. residents reporting that they are Filipino alone or in combination with one or more additional racial–ethnic categories. To understand the status of Filipino Americans, it is important to look at the complex relationship between the Philippine Islands and the U.S. government. After Spain lost the Spanish-American War, the United States established colonial rule over the islands, a rule that lasted from 1898 to 1946. Despite control by the United States, Filipinos were not granted U.S. citizenship, but male Filipinos were allowed to migrate to Hawaii and the U.S. mainland to work in agriculture and in fish canneries in Seattle and Alaska. Like other Asian Americans, Filipino Americans were accused of taking jobs away from white workers and suppressing wages, and Congress restricted Filipino immigration to fifty people per year between the Great Depression and the aftermath of World War II.

The second wave of Filipino immigrants came following the Immigration Act of 1965, when large numbers of physicians, nurses, technical workers, and other professionals moved to the U.S. mainland. Most Filipinos have not had the start-up capital necessary to open their own businesses, and many have been employed in the low-wage

sector of the service economy. However, the average household income of Filipino American families is relatively high, at nearly $77,000, because, among other reasons, Filipinos have among the highest level of educational attainment among Asian Americans.

Indochinese Americans Indochinese Americans include people from Vietnam, Cambodia, Thailand, and Laos. Vietnamese refugees who had the resources to flee at the beginning of the Vietnam War were the first to arrive. The next to arrive were Cambodians and lowland Laotians, referred to as "boat people" by the media. Many who tried to immigrate did not survive at sea; others were turned back when they reached this country or were kept in refugee camps for long periods of time. When they arrived in the United States, inflation was high, the country was in a recession, and many native-born citizens feared that they would lose their jobs to these new refugees, who were willing to work very hard for low wages.

In 2013 it was estimated that about 1.9 million adult Vietnamese Americans resided in the United States, constituting the fourth-largest group of Asian Americans. About 84 percent of Vietnamese Americans were foreign born, but nearly 80 percent possess U.S. citizenship. The median household income of Vietnamese Americans is $55,132.

Like Vietnamese Americans, other Indochinese Americans from Cambodia, Thailand, and Laos are often first- or second-generation residents of the United States; about half live in the western states, especially California. Even though most first-generation Indochinese immigrants spoke no English when they arrived in this country, their children and grandchildren have done very well in school and have been stereotyped as "brains."

Asian Indian Americans Asian Indian Americans (also known as Indian Americans or Indo Americans) trace their origins to India and make up about 1 percent of the U.S. population. Slightly more than 2.8 million people count themselves as "Asian Indian alone" in U.S. Census Bureau surveys; however, when counted in combination with one or more races, they account for nearly 3.2 million people. Some earlier Asian Indian immigrants arrived on the West Coast in the 1900s to work in agriculture, but it was not until the 1960s that their population increased significantly.

Initially, Asian Indians were classified as Caucasian and allowed to become citizens, but they were later barred from citizenship. It was not until the 1950s that legislation was passed to lift this restriction, bringing several waves of immigration. Among the first to arrive were well-educated professionals and managers and their families. Later groups were less well educated and found jobs in the service industry, such as driving taxis, working in fast food, or opening small family-owned businesses such as restaurants.

Since the 1980s, many Asian Indian Americans have been in top positions in the high-tech Silicon Valley of California,

FIGURE 10.15   Asian American workers, such as these software engineers, now make up a larger percentage of the high-tech workforce than white Americans and persons in other racial or ethnic categories. This change constitutes a dramatic shift in technology-related jobs and the corresponding distribution of higher wages and benefits provided by this employment sector.

particularly in companies such as Google and Microsoft. However, slightly less than 25 percent of all adult Asian Indian Americans live in the West, as compared with nearly half (47 percent) of adult Asian Americans overall. The largest populations of Asian Indian Americans are found in New Jersey, New York City, Atlanta, Raleigh-Durham, Baltimore-Washington, Boston, Chicago, Dallas-Fort Worth, Houston, Los Angeles, Philadelphia, and the San Francisco Bay Area.

The median household income of Asian Indian Americans ($88,000) is higher than that of Asian Americans as a whole ($66,000) and of the U.S. population as a whole ($49,800). Asian Indian Americans have a higher level of educational attainment than other groups in the United States. Among Asian Indian Americans, 32 percent of adults age 25 and older have a bachelor's degree. Of Asian Americans as a whole, 29 percent hold a bachelor's degree, as compared to 18 percent of the U.S. population. Similarly, 38 percent of the Asian Indian Americans hold advanced degrees, as compared to 20 percent of Asian Americans as a whole and 10 percent of the U.S. population as a whole (Pew Research Center, 2015) (● Figure 10.15).

Asian Indian Americans have experienced hostility and discrimination in some areas of the country, at least partly because of their perceived success and the fear that they are taking opportunities away from native-born Americans. In the 1980s, Asian Indian Americans were targeted by the "Dotbusters" in New Jersey because some

wore a distinctive dot on their forehead. Others were discriminated against in the workplace because U.S. workers believed that they were losing their jobs to outsourcing in countries such as India. Some Asian Indian American students have taken legal action against a number of Ivy League universities, claiming that they were the victims of discrimination because the schools did not want an overrepresentation of Asian Americans in their student population. One of the most recent tragedies occurred when a white supremacist killed four people and injured others at a Sikh gurdwara, a place of worship, in Wisconsin.

Asian Americans and Sports

Asian American athletes have begun to receive recognition in a variety of sports, in the past winning acclaim in the Olympics and in other major sports: Kyla Ross (gymnastics), Nathan Adrian (swimming), Jeremy Lin (basketball), Nonito Donaire (boxing), Julie Chu (ice hockey), Ed Wang (football), and Ichiro Suzuki and Tim Lincecum (baseball). These and a number of other Asian Americans continue to be recognized as top athletes. Sports analysts have pointed out the importance of having outstanding Asian American athletes because they provide role models for all young people, but especially for their own communities, exemplifying the integrity, discipline, and hard work that are necessary to become a success in sports and in life.

 LO8   **Describe** the unique experiences of Latinos/as (Hispanics) and Middle Eastern Americans in the United States.

Latinos/as (Hispanic Americans)

The terms *Latino* (for males), *Latina* (for females), and *Hispanic* are used interchangeably to refer to people who trace their origins to Spanish-speaking Latin America and the Iberian peninsula. However, as racial–ethnic scholars have pointed out, the label *Hispanic* was first used by the U.S. government to designate people of Latin American and Spanish descent living in the United States, and it has not been fully accepted as a source of identity by the more than 54 million Latinos/as who live in the United States today (U.S. Census Bureau Hispanic

Heritage Month, 2014). Instead, many of the people who trace their roots to Spanish-speaking countries think of themselves as Mexican Americans, Chicanos/as, Puerto Ricans, Cuban Americans, Salvadorans, Guatemalans, Nicaraguans, Costa Ricans, Argentines, Hondurans, Dominicans, or members of other categories. Many also think of themselves as having a combination of Spanish, African, and Native American ancestry.

Across all Hispanic categories, more than 38 million persons ages 5 and older in the United States speak Spanish as the first language at home. This is a 121 percent increase between 1990 (17.3 million people) and 2012. However, more than half (58 percent) of all Hispanics who speak Spanish indicate in U.S. Census Bureau surveys that they also speak English "very well." As discussed in Chapter 8, Hispanic households have lower median household incomes and higher rates of poverty (about 25 percent) than white (non-Hispanic) Americans.

Mexican Americans or Chicanos/as

Mexican Americans—including both native-born and foreign-born people of Mexican origin—are the largest segment (64 percent) of the Latino/a population in the United States. Most Mexican Americans live in the southwestern region of the United States, although more have moved throughout the United States in recent years.

Immigration from Mexico is the primary vehicle by which the Mexican American population grew in this country. Initially, Mexican-origin workers came to work in agriculture, where they were viewed as a readily available cheap and seasonal labor force. Many initially entered the United States as undocumented workers ("illegal aliens"); however, they were more vulnerable to deportation than other illegal immigrants because of their visibility and the proximity of their country of origin. For more than a century, there has been a "revolving door" between the United States and Mexico that has been open when workers were needed and closed during periods of economic recession and high rates of U.S. unemployment.

Mexican Americans have long been seen as a source of cheap labor, while at the same time they have been stereotyped as lazy and unwilling to work. As has been true of other groups, when white workers viewed Mexican Americans as a threat to their jobs, they demanded that the "illegal aliens" be sent back to Mexico. Consequently, U.S. citizens who happen to be Mexican American have been asked for proof of their citizenship, especially when anti-immigration sentiments are running high. Many Mexican American families have lived in the United States for five or six generations—they have fought in wars, made educational and political gains, and consider themselves to be solid U.S. citizens. Thus, it is a great source of frustration for them to be viewed as illegal immigrants or to be asked "How long have you been in this country?"

The U.S. recession that began in 2007 and the gradual economic recovery of the second decade of the twenty-first century considerably reduced the flow of immigration from Mexico to the United States. The collapse of the U.S. housing market reduced the number of jobs in the construction industry, and other employment opportunities were also lost as the financial crisis took away positions in manufacturing, personal service, leisure, and other sectors. However, it is clear that Mexican Americans will continue to make a major contribution to the U.S. population because the Mexican-origin population increased by more than 50 percent (from 20.6 million to 32.9 million), with the largest numerical increase of any racial or ethnic category, between 2000 and 2010 (Pew Research Center, 2012a).

Puerto Ricans

Today, the nearly 5 million Puerto Rican Americans residing in the United States make up 9.4 percent of Hispanic-origin people in this country. When Puerto Rico became a territory of the United States in 1917, Puerto Ricans acquired U.S. citizenship and the right to move freely to and from the mainland. In the 1950s, many migrated to the mainland when the Puerto Rican sugar industry collapsed, settling primarily in New York and New Jersey. Today, more than half of all Puerto Rican Americans reside in the Northeast, followed by the South, primarily Florida (Pew Research Center, 2012a).

Although living conditions have improved substantially for some Puerto Ricans, life has been difficult for the many living in poverty in Spanish Harlem and other barrios. Nevertheless, in recent years Puerto Ricans have made dramatic advances in education, the arts, and politics. Puerto Rican Americans have higher levels of educational attainment than the Hispanic population overall: Among Puerto Ricans ages 25 and older, 16 percent have obtained at least a bachelor's degree, as compared to 13 percent of all U.S. Hispanics (Pew Research Center, 2012a). However, the annual median household income ($36,000) of Puerto Rican Americans is considerably less than that of the U.S. population as a whole.

Cuban Americans

Cuban Americans live primarily in the Southeast, especially Florida. As a group, they have fared somewhat better than other Latinos/as because many Cuban immigrants were affluent professionals and businesspeople who fled Cuba after Fidel Castro's 1959 Marxist revolution. This early wave of Cuban immigrants has median incomes well above those of other Latinos/as; however, this group is still below the national average. The second wave of Cuban Americans, arriving in the 1970s, has fared worse. Many had been released from prisons and mental hospitals in Cuba, and their arrival fueled an upsurge in prejudice against all Cuban Americans. The more-recent arrivals have developed their own ethnic and economic enclaves in Miami's Little Havana, and many of the earlier immigrants have become mainstream professionals and entrepreneurs.

Latinos/as and Sports

For more than a century, Latinos have played Major League Baseball in the United

Middle Eastern Americans

Since 1970, many immigrants have arrived in the United States from countries located in the "Middle East," which is the geographic region from Afghanistan to Libya and including Arabia, Cyprus, and Asiatic Turkey. Placing people in the "Middle Eastern" American category is somewhat like placing wide diversities of people in the categories of Asian American or Latino/a; some U.S. residents trace their origins to countries such as Bahrain, Egypt, Iran, Iraq, Kuwait, Lebanon, Oman, Qatar, Saudi Arabia, Syria, UAE (United Arab Emirates), and Yemen. Middle Eastern Americans speak a variety of languages and have diverse religious backgrounds: Some are Muslim, some are Coptic Christian, and others are Melkite Catholic. Although some are from working-class families, Lebanese Americans, Syrian Americans, Iranian Americans, and Kuwaiti Americans primarily come from middle- and upper-income family backgrounds. For example, numerous Iranian Americans are scientists, professionals, and entrepreneurs.

Arab Americans In the twenty-first century, about 3.5 million people in the United States identify their family's country of origin as being an Arab country. The primary countries of origin are Lebanon, Syria, Palestine, Egypt, and Iraq. Although Arab Americans live throughout the United States, nearly half live in California, Michigan, New York, Florida, and New Jersey. One-third of all Arab Americans reside in one of three major metropolitan areas—Detroit, Los Angeles, and New York. Most Arab Americans were born in the United States, and over 80 percent are U.S. citizens.

Since the 2010 U.S. Census, the population of Arab Americans in the United States has become better known because of campaigns promoting the slogan "check it right, you ain't White" that were launched by various Arab American groups to encourage Arab Americans to check the "Other" box when they filled out the 2010 Census form and then to identify themselves as "Arab" or to indicate their specific country of origin.

Iranian (Persian) Americans About 1.5 million Iranian Americans live in the United States in the 2010s. However, no official statistics are available because these data are not collected by the Census Bureau. Instead, the annual American Community Survey, a sample survey, asks questions of ancestry that provide this information.

The terms *Iranian American* and *Persian American* are used interchangeably because Iran was called Persia prior

FIGURE 10.16 Professional sports, particularly baseball, increasingly reflects the growing racial–ethnic and national diversity of the U.S. population.

States (• Figure 10.16). Originally, Cubans, Puerto Ricans, and Venezuelans were selected for their light skin as well as for their skill as players. Baseball became a major means of assimilation for earlier Latinos in the United States. By 2012, Latinos represented more than 25 percent of Major League Baseball players, and growing numbers also participated in football, hockey, and basketball at all levels of competition. In women's sports, golf, soccer, and basketball also had rising numbers of Latina athletes.

In addition to baseball, Latinos have made impressive gains in Major League Soccer in the United States and Canada; however, many people in the United States are less enthusiastic fans of soccer as compared to football. Consequently, salaries of professional soccer players are lower than in the high-profile sports, fan bases are smaller, and revenues from sales of team clothing are not as lucrative. However, the sixteen U.S. and three Canadian soccer teams provide opportunities for Latino athletes to be highly visible in professional sports in regions with large Latino/a populations.

FIGURE 10.17 Muslims in the United States who wear traditional attire may face prejudice and/or discrimination as they go about their daily lives.

States has the highest number of Iranian residents outside of Iran. More than 80 percent of Iranian Americans are U.S. citizens. Many Iranian Americans have high levels of educational attainment and are employed in professional positions in business, academia, and science.

Discrimination Despite such high level of achievement, Iranian Americans, like Arab Americans, have experienced persistent discrimination, particularly if they are Muslim (• Figure 10.17) or if there has been a recent terrorist scare in the United States. Following the September 11, 2001, attacks on the United States by terrorists whose origins were traced to the Middle East, there was an escalation in the number of hate crimes and other types of discrimination against persons assumed to be Arabs, Arab Americans, Iranian Americans, or Muslims. In the aftermath of this terrorist attack, the U.S. Patriot Act was passed. This law gave the federal government greater authority to engage in searches and surveillance of persons suspected of terrorist activity than in the past. The Patriot Act caused heightened concern among many individuals and groups because it was believed that this law might be used to target individuals who appear to be of Middle Eastern origins.

What about the Muslim experience in the United States? In cities across this country, Muslims have established social, economic, and ethnic enclaves for social stability and personal safety. Islamic schools and centers often bring together people from a diversity of countries such as Egypt and Pakistan. Many Muslim leaders and parents focus on how to raise children to be good Muslims and good U.S. citizens. In the second decade of the twenty-first century, some Middle Eastern Americans experience discrimination based on their speech patterns, appearance, and clothing (such as the *hijabs,* or "head-to-toe covering" that leaves only the face exposed, which many girls and women wear). The idea that Middle Easterners are somehow associated with terrorism has also been difficult to remove from media representations and some people's thinking, producing ongoing hardship for many upstanding citizens of this nation.

Middle Eastern Americans and Sports Although an increasing number of Islamic schools now focus on sports for teenage boys, overall there has been less emphasis placed on competitive athletics among Middle Eastern Americans when compared to other groups. Based on popular sporting events in their countries of origin,

to 1935. Many Iranian Americans refer to themselves as "Persian" rather than "Iranian" because of the perceived negativity associated with the political history of the country of Iran and its relationship to the United States. It should be noted that Persian Americans are not considered to be Arab because they speak Farsi and have a different culture.

The most extensive immigration of Iranians to the United States began in the late 1970s and early 1980s, when early immigrants, particularly college students, left Iran as the Iranian revolution was taking place. When the Islamic Republic was established after the revolution, many Iranian students decided to remain in the United States, and other Iranians also left their country and established a new life in this nation. Today, the United

some Middle Eastern Americans play golf or soccer. As well, some Iranian Americans follow the soccer careers of professional players from Iran who now play for German, Austrian, Belgian, and Greek clubs. Over time, sports participation will probably continue to increase among Middle Eastern American males, particularly in soccer and golf; however, girls and women in more-traditional Muslim families typically have not participated in athletic activities unless they are conducted privately.

Looking Ahead: The Future of Global Racial and Ethnic Inequality

Throughout the world, many racial and ethnic groups seek *self-determination*—the right to choose their own way of life. As many nations are currently structured, however, self-determination is impossible.

Worldwide Racial and Ethnic Struggles

The cost of self-determination is the loss of life and property in ethnic warfare. Ethnic violence has persisted in Mali, Myanmar, Bangladesh, India, China, South Sudan, and many other regions where hundreds of thousands have died from warfare, disease, and refugee migration. Ethnic wars have a high price even for survivors, whose life chances can become bleaker even after the violence subsides.

In the twenty-first century, the struggle between the Israeli government and various Palestinian factions over the future and borders of Palestine continues to make headlines. Discord in this region has heightened tensions among people not only in Israel and Palestine but also in the United States and around the world as deadly clashes continue and political leaders are apparently unable to reach a lasting solution to the decades-long strife.

Growing Racial and Ethnic Diversity in the United States

Racial and ethnic diversity is increasing in the United States. African Americans, Latinos/as, Asian Americans, Native Americans, and mixed-race individuals constitute more than a third (37 percent) of the U.S. population—up from 30.9 percent in 2000 (Humes, Jones, and Ramirez, 2011). As shown in • Figure 10.18, states vary in their percentage of population that are minorities. Today, non-Hispanic white Americans make up 63 percent of the population, in contrast to 80 percent in 1980. It is predicted that by 2056, the roots of the average U.S. resident will be in Africa, Asia, Hispanic countries, the Pacific islands, and the Middle East—not white Europe.

What effect will these changes have on racial and ethnic relations? Several possibilities exist. On the one hand, conflicts may become more overt and confrontational as people continue to use *sincere fictions*—personal beliefs that reflect larger societal mythologies,

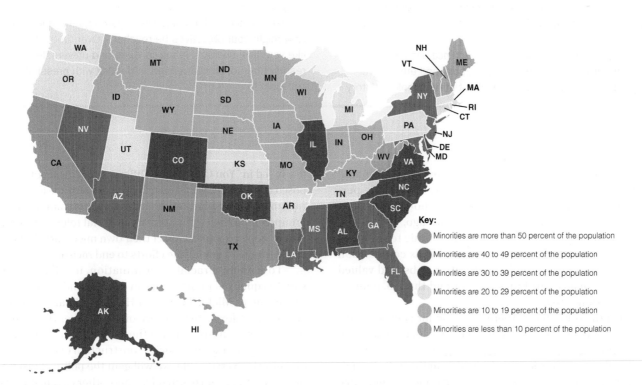

FIGURE 10.18   Minority Populations and Percentages by State, 2010*

*Percentages are rounded to nearest whole number.
Source: Humes, Jones, and Ramirez, 2011.

Working for Racial and Gender Harmony on College Campuses

How can you promote racial and gender harmony on your college campus? One student, Morgane Richardson, decided to establish a group called "Refuse the Silence." The organization collected stories of women of color who were either attending or had graduated from an "elite liberal arts college in the United States" and had experienced problems such as sexual assault. Organizations such as Refuse the Silence seek to identify the unique problems of students of color, particularly women, in regard to racism and sexual assault on elite campuses. However, concerns about racial inequality, discrimination, and assault exist on all campuses. You can help by establishing a similar organization or website at your institution to determine how to best address this pressing problem and bring greater racial and gender harmony on campus.

If you are interested in starting your own organization or developing a blog to look at racism, sexism, or similar issues, consider how the following factors contribute to the problem: (1) divisiveness between different cultural and ethnic communities; (2) persistent lack of trust; (3) the fact that many people never really communicate with one another, despite the omnipresence of social media; (4) the need to bring different voices into the curriculum and college life generally; and (5) the need to learn respect for people from different backgrounds. Your group could also develop a set of questions to be answered. Consider these topics for developing questions on campus racism:

1. *Encouraging inclusion and acceptance.* Do members of our group reflect the college's racial and ethnic diversity? How much do I know about other people's history and culture? How can I become more tolerant—or accepting—of people who are different from me?

2. *Raising consciousness.* What is racism? What causes it? Can people participate in racist language and behavior without realizing what they are doing? What is our college or university doing to reduce racism?

3. *Becoming more self-aware.* How much do I know about my own family roots and ethnic background? How do the families and communities where we grow up affect our perceptions of racial and ethnic relations?

4. *Using available resources.* What resources are available for learning more about working to reduce racism? Here are some agencies to contact:

   - The American Civil Liberties Union (ACLU)

   - The Anti-Defamation League (ADL)

   - The National Association for the Advancement of Colored People (NAACP)

   - The National Council of La Raza

What additional items would you add to the list of problem areas on your campus? Would you like to also address Morgane Richardson's concern about woman of color and sexual violence? What goals would your organization have? How might your objectives be reached? Over time, students like you have changed many colleges and universities as a result of personal involvement in dealing with pressing social issues!

---

such as "I am not a racist" or "I have never discriminated against anyone"—even when these are inaccurate perceptions (Feagin and Vera, 1995). Although the term *sincere fictions* was coined two decades ago, we face the real possibility in the future that interethnic tensions, as well as many other forms, may increase as competition for scarce resources such as education, jobs, and valued goods in society continues to grow and the U.S. population continues to age.

On the other hand, there is reason for cautious optimism. Throughout U.S. history, members of diverse racial and ethnic groups have struggled to gain the freedom and rights that were previously withheld from them. Today, minority grassroots organizations are pressing for affordable housing, job training, and educational opportunities. As discussed in "You Can Make a Difference," movements composed of both whites and people of color continue to oppose racism in everyday life, to seek to heal divisions among racial groups, and to teach children about racial tolerance. Many groups hope not only to affect their own microcosm but also to contribute to worldwide efforts to end racism.

To eliminate racial discrimination, it will be necessary to equalize opportunities in schools and workplaces. According to Michael Omi and Howard Winant (2013), it is important for us to be aware of race, rather than ignoring it, if we wish to challenge the problem of racism. If we are aware that race as a social construction exists and has meaning in everyday life, we will gain the political insights necessary to mobilize ourselves and others against injustice and inequality in our society.

CHAPTER REVIEW Q & A

LO1 How do race and ethnicity differ?

A race is a category of people who have been singled out as inferior or superior, often on the basis of physical characteristics such as skin color, hair texture, or eye shape. An ethnic group is a collection of people distinguished primarily by cultural or national characteristics, including unique cultural traits, a sense of community, a feeling of ethnocentrism, ascribed membership, and territoriality.

LO2 How do racial and ethnic classifications continue to change in the United States?

Racial classifications in the United States have changed over the past century. If we look at U.S. Census Bureau classifications, for example, we can see how the meaning of race continues to change. First, race is defined by perceived skin color: white or nonwhite. Census 2000 made it possible—for the first time—for individuals to classify themselves as being of more than one race.

LO3 What are prejudice, stereotypes, racism, scapegoat, and discrimination?

Prejudice is a negative attitude often based on stereotypes, which are overgeneralizations about the appearance, behavior, or other characteristics of all members of a group. Stereotypes are overgeneralizations about the appearance, behavior, or other characteristics of members of particular categories. Racism is a set of attitudes, beliefs, and practices that is used to justify the superior treatment of one racial or ethnic group and the inferior treatment of another racial or ethnic group. A scapegoat is a person or group that is incapable of offering resistance to the hostility or aggression of others Discrimination involves actions or practices of dominant-group members that have a harmful effect on members of a subordinate group.

LO4 How do sociologists view racial and ethnic group relations?

Symbolic interactionists claim that intergroup contact may either intensify or reduce racial and ethnic stereotyping and prejudice, depending on the context. In the *contact hypothesis,* symbolic interactionists point out that contact between people from divergent groups should lead to favorable attitudes and behavior when certain factors are present. Functionalists stress that members of subordinate groups become a part of the mainstream through assimilation, the process by which members of subordinate groups become absorbed into the dominant culture. Conflict theorists focus on economic stratifica-tion and access to power in race and ethnic relations. The caste perspective views inequality as a permanent feature of society, whereas class perspectives focus on the link between capitalism and racial exploitation. According to racial formation theory, the actions of the U.S. government substantially define racial and ethnic relations.

LO5 What are the unique historical experiences of Native Americans and WASPs in the United States?

Experts estimate that approximately two million native inhabitants lived in North America in 1492; their numbers had been reduced to fewer than 240,000 by 1900. Native Americans have been the victims of genocide and forced migration. After the Revolutionary War, the federal government broke treaty after treaty as it engaged in a policy of wholesale removal of indigenous nations in order to clear the land for settlement by Anglo-Saxon "pioneers." Data continue to show that Native Americans are the most disadvantaged racial or ethnic group in the United States in terms of income, employment, housing, nutrition, and health. Whereas Native Americans have been among the most disadvantaged peoples, white Anglo-Saxon Protestants (WASPs) have been the most privileged group in this country. Although many English settlers initially came to North America as indentured servants or as prisoners, they quickly emerged as the dominant group, creating a core culture (including language, laws, and holidays) to which all other groups were expected to adapt.

LO6 How have slavery, segregation, lynching, and persistent discrimination uniquely affected the African American experience in this country?

The African American (black) experience has been one uniquely marked by slavery, segregation, and persistent discrimination. Between 1619 and the 1860s, about 500,000 Africans were forcibly brought to North America, primarily to work on southern plantations, and these actions were justified by the devaluation and stereotyping of African Americans. Following the abolishment of slavery in 1863, African Americans were still subjected to segregation, discrimination, and lynchings. Despite civil rights legislation and economic and political gains by many African Americans, racial prejudice and discrimination continue to exist.

LO7 What are the major categories of Asian Americans, and what are their historical and contemporary experiences?

The term *Asian Americans* designates the many diverse groups with roots in Asia. Chinese and Japanese immigrants were among the earliest Asian Americans. Many Filipinos, Asian Indians, Koreans, Vietnamese, Cambodians, Pakistani, and Indonesians have arrived more recently. The subgroups are listed as Chinese Americans (the largest Asian American group), Japanese Americans, Korean Americans, Filipino Americans (the second-largest category of Asian Americans), and Indochinese Americans (which include people from Vietnam, Cambodia, Thailand, and Laos). Asian American immigrants as a group have enjoyed considerable upward mobility in U.S. society in recent decades, but many Asian Americans still struggle to survive by working at low-paying jobs and living in urban ethnic enclaves.

LO8 What have been the unique experiences of Latinos/as (Hispanics) and Middle Eastern Americans in the United States?

Mexican Americans—including both native-born and foreign-born people of Mexican origin—are the largest segment (approximately two-thirds) of the Latino/a population in the United States. Today, Puerto Rican Americans make up 9 percent of Hispanic-origin people in the United States. Although some Latinos/as have made substantial political, economic, and professional gains in U.S. society, as a group they are nevertheless subjected to anti-immigration sentiments. Since 1970, many immigrants have arrived in the United States from countries located in the "Middle East," which is the geographic region from Afghanistan to Libya and includes Arabia, Cyprus, and Asiatic Turkey. Middle Eastern immigrants to the United States speak a variety of languages and have diverse religious backgrounds. Because they generally come from middle-class backgrounds, they have made inroads into mainstream U.S. society. However, some Middle Eastern Americans experience discrimination based on their speech patterns, appearance, and clothing. The idea that Middle Easterners are somehow associated with terrorism has also been difficult to remove from media representations and some people's thinking, which produces ongoing hardship for many upstanding citizens of this nation.

KEY TERMS

assimilation 276
authoritarian personality 274
discrimination 274
dominant group 270
ethnic group 268
ethnic pluralism 277
gendered racism 279

genocide 275
individual discrimination 275
institutional discrimination 275
internal colonialism 278
prejudice 270
race 268
racism 271

scapegoat 273
segregation 277
split labor market 278
stereotypes 270
subordinate group 270
theory of racial formation 280

QUESTIONS for CRITICAL THINKING

1 Do you consider yourself defined more strongly by your race or by your ethnicity? How so?

....

2 Given that subordinate groups have some common experiences, why is there such deep conflict between some of these groups?

....

3 What would need to happen in the United States, both individually and institutionally, for a positive form of ethnic pluralism to flourish in the twenty-first century?

....

ANSWERS to the SOCIOLOGY QUIZ

ON RACE, ETHNICITY, AND SPORTS

1 **True** Many people continue to believe that differences in athletic abilities can largely be attributed to racial physiology (such as in the term "natural athlete") rather than personal attributes such as "hard worker" or "good decision maker."

2 **False** This is an example of the "white men can't jump" assumption that links racial differences to athletic ability; however, no "racial" gene has been found to account for differences in muscle fiber in black runners and white runners. (Human muscles contain a genetically determined mixture of both slow and fast fiber types. Fast-twitch fibers produce force at a higher rate for short bursts of speed and can be an asset to a short-distance sprinter. Slow-twitch fibers produce a lower rate of force that lasts longer and can be an asset to a distance runner.)

3 **False** A person has a better chance of getting struck by lightning or writing a *New York Times* best-seller (or marrying a millionaire) than becoming a professional athlete, where the current odds are about 24,550 to 1.

4 **True** Research has found improvements in the number of African Americans in football head–coaching positions. At the beginning of the 2014 football season, there were 14 African American coaches in the Football Bowl Subdivision (FBS) Division I. Nearly 89 percent of the FBS coaches were white.

5 **False** Although 88 percent of all FBS university presidents in 2014 were white, there were seven African American, six Asian American, and two Latino presidents. There were no Native American university presidents. At one time, all presidents of NCAA Division I schools were white (non-Hispanic).

6 **True** In the 2014 season the percentage of African American football student-athletes was approximately 53 percent, as compared to about 42 percent for white players, 2 percent each for Latino and Asian American players, and 0.1 percent for Native American players.

7 **True** All FBS conference commissioners, who make major decisions about how their conferences are run, are white men, despite the relatively large number of student-athletes and coaches who are persons of color and/or women.

8 **False** Many children and young adolescents look upon college athletes as role models. When younger people see players who look like them, they may be more disposed to believe that they too could attend college and participate in sports.

Source: Based on Lapchick, 2014.

SEX, GENDER, AND SEXUALITY

CandyBox Images/Shutterstock.com

LEARNING OBJECTIVES

1   **Distinguish** between sex and gender.

2   **Discuss** prejudice and discrimination based on sexual orientation.

3   **Define** *gender role, gender identity, body consciousness,* and *sexism.*

4   **Describe** how the division of labor between women and men differs in various kinds of societies.

5   **Identify** the primary agents of gender socialization and note their role in socializing people throughout life.

6   **Discuss** ways in which the contemporary workplace reflects gender stratification.

7   **Compare** functionalist and conflict perspectives on gender inequality.

8   **Describe** four feminist perspectives on gender inequality.

SOCIOLOGY & EVERYDAY LIFE

When Gender, Sexual Orientation, and Weight Bias Collide

"SeaWorld, Shamu Got Out!" (words yelled out a car window by a male passenger to Whitney Thore, the 380-pound TLC star of the series *My Big Fat, Fabulous Life,* as she was walking down the street). In the words of Thore,

There's PC terms for everything these days but fat people are fair game. Someone has to fight for us. . . . I'm Whitney Thore. I'm 30 years old, and I live at home with my parents. I'm a fat dancer. . . . I hate people thinking that I'm lazy. I have polycystic ovarian syndrome (that's PCOS for short). It's an endocrine disorder that does a lot to your body. Two-thirds of women with PCOS are overweight or obese. . . . It makes it really easy for me to gain weight and really difficult for me to lose weight. . . . When I put my fat dancing woman video on YouTube and it went viral, suddenly people were coming out of the woodwork thanking me for being myself and for showing them that they can do the same thing. Living with PCOS isn't easy but I still have a choice in how I let it affect my attitude. I may be fat, but I am also fabulous. I have one life to live . . . and it sure better count!

—WHITNEY WAY THORE (TLC Channel, 2015), who uploaded "A Fat Girl Dancing," a video that went viral on YouTube, has become a sensation among some

Activists such as Whitney Way Thore (shown here) and Louis Peitzman (not pictured) call our attention to the fact that gender and sexual orientation do make a difference in how individuals labeled as overweight or obese are treated by others in their everyday lives. What examples of weight bias targeting specific categories of people have you observed in daily life?

Alberto E. Rodriguez/Getty Images

Although the lived experiences of Whitney Thore and Louis Peitzman differ in many ways, one common theme emerges from their comments: One's gender and sexual orientation do matter in how people treat individuals whom they believe are extremely overweight or obese. People—particularly women and gay men—who are weight challenged are fair game for the ridicule of those who perceive of themselves as being of average body weight. Some of these individuals believe that they have the right to set the standards for everyone else. The intersectionality of gender and weight is not a new topic in the social sciences. Feminist analysts in the 1970s and 1980s called attention to the gendered nature of weight preoccupation, disordered eating, and discrimination against persons based on weight and physical appearance. Many early studies focused on appearance pressures and harassment experienced by heterosexual women. However, researchers have found that overweight and obese gay men have also had have similar concerns because they live in a dominant culture that practices weight stigma and bias plus being part of a gay male culture that is not "gay fat" friendly. If you Google "gay fat," you will come across this definition in Urbandictionary.com (2015): "A gay man who does not have a gym-perfect body, but rather carries a body fat percentage in the 12%–20% range. A man who is considered gay fat within the community would likely be considered athletic, physically fit, and in-shape within the greater cultural context."

mainstream and social media fans. She has her own TLC show and has created the "#NoBodyShame" campaign. However, her critics claim that she is promoting obesity.

I was once told that coming out as a gay man was like being welcomed into the best club in the world. It was maybe an overstatement, but I understand the sentiment: When you first come out, you're automatically granted inclusion—if not by friends and family, then by the gay community as a whole. They get it. They get you. And they're eager to let you know that you're not alone, and that you have a seat at the table. Unless, of course, you're also fat, in which case, no, you can't sit with us.

—Louis Peitzman (2013), an editor for BuzzFeed News, calls attention to the issue of weight and gay men in his response to the slogan for the "It Gets Better" Project. This campaign communicates to lesbian, gay, bisexual, and transgender youths that things get better and encourages other people to help make things better.

How Much Do You Know About Gender, Sexual Orientation, and Weight Bias?

TRUE	FALSE		
T	F	1	Both men and women are equally vulnerable to weight bias and discrimination in employment, education, health care, and interpersonal relationships.
T	F	2	Gender differences in weight bias and discrimination often differ by ethnicity.
T	F	3	Many young girls and women believe that being even slightly "overweight" makes them less "feminine."
T	F	4	Physical attractiveness is a more central part of self-concept for women than for men.
T	F	5	Regardless of their sexual orientation, men have less concern than women about their weight and body image.
T	F	6	Thinness has always been the "ideal" body image for women.
T	F	7	The topic of eating disorders is more "taboo" among men than women because it is perceived to be a weakness associated with femininity.
T	F	8	The media play a significant role in shaping societal perceptions about the ideal weight for men and women and about body image.

Answers can be found at the end of the chapter.

As gender studies scholar Jason Whitesel (2014: 2) concludes in his book *Fat Gay Men,* "Big gay men incur social wounds produced by the stigmas of their size and sexuality combined. As looks are one of the organizing features of the gay world, gay big men have an added exclusion that has not been fully explored." In an attempt to overcome this stigma, some of the men join organizations such as Girth and Mirth to help them gain dignity and respect in spite of the shaming, desexualization, exclusion, and marginalization they experience from both mainstream and gay society (Whitesel, 2014).

Similarly, Whitney Thore of *My Big Fat, Fabulous Life* seeks to overcome the negative views of people who call her "lazy" by having an upbeat attitude and doing her "fat girl"

dancing on YouTube and television shows. But the "gay big men" and Whitney's "fat girl" face a culture in which many people objectify other individuals by seeing them primarily in terms of their outward appearance and their perceived sexual attractiveness. This process is referred to as objectification.

What specifically is objectification? *Objectification* is the process whereby some people treat other individuals as if they were objects or things, not human beings. We objectify other individuals when we judge them strictly on the basis of their physical appearance, rather than on their individual qualities, attributes, or actions (Hatton and Trautner, 2011). Objectification of girls and women is common in the United States and many other nations, and this problem

TABLE 11.1

The Objectification of Women

General Aspects of Objectification	Objectification Based on Cultural Preoccupation with "Looks"
Women are responded to primarily as "females," whereas their personal qualities and accomplishments are of secondary importance.	Women are often seen as the objects of sexual attraction, not full human beings—for example, when they are stared at because of their physical characteristics.
Women are seen as "all alike."	Women are seen by some as depersonalized body parts—for example, "a piece of ass."
Women are seen as being subordinate and passive, so things can easily be "done to a woman"—for example, discrimination, harassment, and violence.	Depersonalized female sexuality is used for cultural and economic purposes—such as in the media, advertising, the fashion and cosmetics industries, and pornography.
Women are seen as easily ignored or trivialized.	Women are seen as being "decorative" and status-conferring objects to be bought (sometimes collected) and displayed by men and sometimes by other women.
Women are judged on appearance more than men are.	Women are evaluated according to prevailing, narrow "beauty" standards and often feel pressure to conform to appearance norms.

Source: Adapted from Schur, 1983.

is intensified when weight and appearance are also issues (see ● Table 11.1). Compared to heterosexual men, studies have found more-extensive objectification of gay males, including negative evaluations of their physical appearance, weight, and sexuality. Gay men like Louis are scrutinized and objectified for being "gay fat" and thus are deemed less sexually desirable. In sum, all women and men are objectified and sexualized to some degree in society, but the process takes on numerous additional dimensions when sexual orientation and/or weight are also involved.

What does it mean to be "sexualized" or to go through the process of "sexualization"? According to the American Psychological Association (2010), **sexualization** is the act or processes whereby an individual or group is seen as sexual in nature or persons become aware of their sexuality. Both cognitive (mental processes of perception, judgment, and reason) and emotional consequences occur when persons are sexualized: (1) a person's value comes only from his or her sexual appeal or sexual behavior, to the exclusion of other characteristics; (2) a person is held to a standard that equates physical attractiveness (narrowly defined) with being sexy; (3) a person is sexually objectified; and/or (4) sexuality is inappropriately imposed upon a person.

Why is this important for our study of sociology? The way that people think about themselves and others in regard to personal appearance and body image provides us with important information about the larger society's cultural norms, expectations, and values in regard to females and males. This social construction of reality involves what we consider to be appropriate, or inappropriate, behavior for men and women. Obviously, some differences between men and women are biological in nature. However, many differences between the sexes are socially constructed. Studying sociology makes us aware of differences that relate to gender (a social concept) as well as differences that are based on a person's biological

makeup, or sex. In this chapter we examine the issue of gender: what it is and how it affects us. Before reading on, test your knowledge about gender, sexual orientation, and weight bias by taking the "Sociology and Everyday" Life quiz.

 LO1 **Distinguish** between sex and gender.

Sex: The Biological Dimension

Whereas the word *gender* is often used to refer to the distinctive qualities of men and women (masculinity and femininity) that are *culturally* created, **sex** refers to the biological and anatomical differences between females and males. At the core of these biological and anatomical differences is the chromosomal information transmitted at the moment a child is conceived. The mother contributes an X chromosome and the father either an X (which produces a female embryo) or a Y (which produces a male embryo). At birth, male and female infants are distinguished by **primary sex characteristics**: the genitalia used in the reproductive process. At puberty, an increased production of hormones results in the development of **secondary sex characteristics**: the physical traits (other than reproductive organs) that identify an individual's sex. For women, these include larger breasts, wider hips, and narrower shoulders; a layer of fatty tissue throughout the body; and menstruation. For men, they include development of enlarged genitals, a deeper voice, greater height, a more muscular build, and more body and facial hair.

Intersex and Transgender Persons

Sex is not always clear-cut. An **intersex person** is an individual who is born with a reproductive or sexual anatomy that does not correspond to the typical definitions of male or female; in other words, the person's sexual differentiation is

autopsy is performed at death. It is possible for some intersex people to live and die with intersexed anatomy but never know that the condition exists. According to the Intersex Society of North America (2015),

> Intersex is a socially constructed category that reflects real biological variation. Nature presents us with sex anatomy spectrums [, but] nature doesn't decide where the category of "male" ends and the category of "intersex" begins, or where the category of "intersex" ends and the category of "female" begins. *Humans decide.* Humans (today, typically doctors) decide how small a penis has to be, or how unusual a combination of parts has to be, before it counts as intersex. Humans decide whether a person with XXY chromosomes and XY chromosomes and androgen insensitivity will count as intersex.

Some people may be genetically of one sex but have a gender identity of the other. That is true for a ***transgender person***—an individual whose gender identity (self-identification as woman, man, neither, or both) does not match the person's assigned sex (identification by others as male, female, or intersex based on physical/genetic sex). Consequently, transgender persons may believe that they have the opposite gender identity from that of their sex organs and may be aware of this conflict between gender identity and physical sex as early as the preschool years. Some transgender individuals choose to take hormone treatments or have a sex change operation to alter their genitalia so that they can have a body congruent with their sense of gender identity (• Figure 11.1). Many then go on to lead lives that they view as being compatible with their true gender identity. But the issue of hormonal and surgical sex reassignment remains highly politicized. The "Standards of Care," a set of guidelines set up by the Harry Benjamin International Gender Dysphoria Association, establishes standards by which transgender persons may

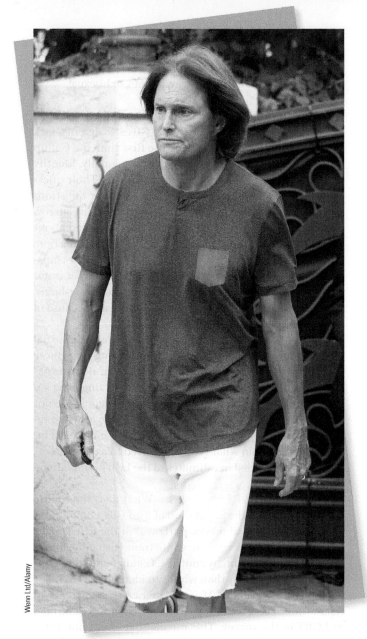

FIGURE 11.1 Caitlyn Jenner (formerly known as Bruce Jenner) went through a very public gender transformation from male to female and became a spokesperson for transgender persons. What influence do you think high-profile people like Caitlyn Jenner have on the attitudes and actions of other people in regard to the LGBTQ community?

ambiguous. Formerly referred to as *hermaphrodites* by some in the medical community, intersex persons may appear to be female on the outside at birth but have mostly male-type anatomy on the inside, or they may be born with genitals that appear to be in between the usual male and female types. For example, a chromosomally normal (XY) male may be born with a penis just one centimeter long and a urinary opening similar to that of a female. However, although intersexuality is considered to be an inborn condition, intersex anatomy is not always known or visible at birth. In fact, intersex anatomy sometimes does not become apparent until puberty, when an adult is found to be infertile, or when an

sexualization
the act or processes whereby an individual or group is seen as sexual in nature or persons become aware of their sexuality.

sex
the biological and anatomical differences between females and males.

primary sex characteristics
the genitalia used in the reproductive process.

secondary sex characteristics
the physical traits (other than reproductive organs) that identify an individual's sex.

intersex person
an individual who is born with a reproductive or sexual anatomy that does not correspond to typical definitions of male or female; in other words, the person's sexual differentiation is ambiguous.

transgender person
an individual whose gender identity (self-identification as woman, man, neither, or both) does not match the person's assigned sex (identification by others as male, female, or intersex based on physical/genetic sex).

obtain hormonal and surgical sex reassignment to help ensure that people choosing such options are informed about what is involved in a gender transition.

Western societies acknowledge the existence of only two sexes; some other societies recognize three—men, women, and *berdaches* (or *hijras* or *xaniths*): biological males who behave, dress, work, and are treated in most respects as women. The closest approximation of a third sex in Western societies is a **crossdresser** (formerly known as a *transvestite*), a male who dresses as a woman or a female who dresses as a man but does not alter his or her genitalia. Although crossdressers are not treated as a third sex, they often "pass" for members of that sex because their appearance and mannerisms fall within the range of what is expected from members of the other sex. Most crossdressers are heterosexual men, many of whom are married, but gay men, lesbians, and straight women may also be crossdressers. Crossdressing can occur in conjunction with homosexuality, but this is frequently not the case. Researchers and analysts continue to engage in dialogue about the correct terminology to use when referring to persons in the diverse groups that now make up this segment of the population.

Sexual Orientation

Sexual orientation refers to an individual's preference for emotional–sexual relationships with members of the different sex (heterosexuality), the same sex (homosexuality), or both (bisexuality). In referring to homosexuality, many organizations representing lesbian, gay, bisexual, transgender, and queer persons have adopted the acronym *LGBTQ*. The term *lesbian* refers to females who prefer same-sex relationships; *gay* refers to males who prefer same-sex relationships. As noted above, *bisexual* is the term used to describe a person's physical or romantic attraction to both males and females, whereas *transgender* is a term applied to persons whose appearance, behavior, and/or gender identity does not match that individual's assigned sex. The "Q" in LGBTQ variously means "questioning" or "queer," and sometimes the acronym is written LGBTQQ to include both "questioning" and "queer." When the "Q" stands for "questioning," it refers to a person who is uncertain about his or her sexual orientation. When the "Q" stands for "queer," it is an umbrella term for the Queer Movement to indicate pride in one's sexual orientation and a rejection of the older, derogatory use of the word *queer* to disparage a nonheterosexual person's orientation.

What criteria have social scientists used to study sexual orientation? A definitive study of sexuality conducted by researchers at the University of Chicago established three criteria for identifying people as homosexual or bisexual: (1) *sexual attraction* to persons of one's own gender, (2) *sexual involvement* with one or more persons of one's own gender, and (3) *self-identification* as a gay, lesbian, or bisexual (Laumann et al., 1994). According to these criteria, then, having engaged in a homosexual act does not necessarily classify a person as homosexual. In fact, many respondents in the

University of Chicago study indicated that although they had at least one homosexual encounter when they were younger, they were no longer involved in homosexual conduct and never identified themselves as gay, lesbian, or bisexual.

Measuring Sexual Orientation It is difficult to determine how many people identify as LGBT because of a lack of official statistics. In 2012, for the first time the Gallup survey asked this question: "Do you, personally, identify as lesbian, gay, bisexual, or transgender?" (Their questionnaire did not include the "Q" for queer or questioning.) More than 120,000 people responded to this survey, making it the largest study of its kind to date, and about 3.4 percent of U.S. adults answered "yes" to the question, thereby self-identifying as lesbian, gay, bisexual, or transgender (Gates and Newport, 2012). Unfortunately, it is not possible to separately consider differences among lesbians, gay men, bisexuals, or transgender individuals because of the way the data were collected.

Gallup researchers emphasize that measuring sexual orientation and gender identity is "challenging since these concepts involve complex social and cultural patterns" (Gates and Newport, 2012: 2). Because of a lingering social stigma attached to the LGBT identity, people are not always forthcoming about their identity when asked to respond to a survey. As a result, an unknown number of individuals remain in what is often referred to as the "closet" and are not included in estimates of the LGBT population.

LGBT Population Estimates What does the Gallup study tell us about the population of the LGBT community? Although the typical media portrayal of the LGBT community is of persons who are white (non-Hispanic), highly educated, and affluent, by contrast, Gallup found that nonwhites are more likely than whites to identify as LGBT. For example, 4.6 percent of African Americans, 4.3 percent of Asian Americans, and 4.0 percent of Hispanics identified as LGBT in the survey. This means that about one-third (33 percent) of LGBT identifiers in the Gallup study are nonwhite, as compared with 27 percent of non-LGBT individuals. Persons with lower levels of education were also more likely to identify as LGBT: 3.5 percent of those who identified as LGBT had a high school education or less, 4.0 percent had some college education but not a college degree, 2.8 percent had a college degree, and 3.2 percent had postgraduate education (Gates and Newport, 2012). A higher proportion of people with lower incomes identify as LGBT: More than 5 percent of those with incomes of less than $24,000 per year self-identified, as compared with 2.8 percent of those making $60,000 per year or more. However, among Gallup participants who reported their income, about 16 percent of LGBT-identified individuals had incomes above $90,000 per year (Gates and Newport, 2012: 5).

Women accounted for more than 53 percent of LGBT individuals who self-identified in the Gallup study, and young adults between the ages of 18 and 29 were more than three times as likely as persons over the age of 65 to identify as LGBT. Younger women between the ages of 18 and 29 were more likely to self-identify (8.3 percent) than men the same age (4.6 percent). These figures may reflect continuing societal opposition among some political leaders and persons in the general public to equal rights and opportunities for persons in the LGBT community. (To see the full report, go to the Gallup website and search for "Special Report: 3.4 % of U.S. Adults Identify as LGBT.")

How valid are these estimates of the LGBT population in the United States? The figures from Gallup are relatively consistent with a previous study by the Williams Institute at the University of California at Los Angeles School of Law, in which researchers estimated that approximately nine million people (about 3.8 percent of all Americans) identify as gay, lesbian, bisexual, or transgender. According to this report, bisexuals make up 1.8 percent of the U.S. population, with more women than men typically identifying as bisexual. Only a slightly smaller proportion (1.7 percent) identify as being gay or lesbian. Transgender adults make up 0.3 percent of the population (Gates, 2011).

Gallup and other organizations continue to engage in research on the LGBT population. One of the Gallup Organization's most-recent surveys focused on where LGBTQ Americans live. Based on surveys of more than 374,000 people, the study identified ten metropolitan areas with the largest shares of LGBT people as residents. The highest percentage (6.2 percent) was found in San Francisco–Oakland–Hayward, California, followed by Portland–Vancouver–Hillsboro, Oregon–Washington (5.4 percent), and then Austin–Round Rock, Texas (5.3 percent). Other areas included in the top ten were New Orleans, Seattle, Boston, Salt Lake City, Los Angeles, Denver, and Hartford. The lowest percentages of LGBT populations were found in Birmingham–Hoover, Alabama (2.6 percent), Pittsburgh, Pennsylvania (3.0 percent), Memphis, San Jose, Raleigh, Cincinnati, Houston, Oklahoma City, Richmond, Nashville, and Milwaukee. However, the study concluded that respondents did not see the issue of openness to and acceptance of the gay population as a major concern when they chose where to live. According to the Gallup researchers, this fact might be an indication that more people in the LGBT community now perceive of this country as being more tolerant of diversity regardless of where you live. Complete results of this study are available on the Gallup Organization website at "San Francisco Metro Area Ranks Highest in LGBT Percentage."

LO2 **Discuss** prejudice and discrimination based on sexual orientation.

Discrimination Based on Sexual Orientation

The United States has numerous forms of discrimination based on sexual orientation. One of the most obvious issues was the fact that, throughout most of U.S. history, LGBTQ couples could not enter into legally recognized marital relationships. Many states passed constitutional amendments that limited marriage to a union between a man and a woman, and in other states, legislators had passed statutes with similar language. Prior to the 2015 U.S. Supreme Court ruling in *Obergefell v. Hodges,* which legalized same-sex marriage across the United States, thirty-seven states had legalized same-sex marriage as a result of court decisions, state laws passed by legislatures, or popular vote. In 2015 thirteen states still banned same-sex marriage through constitutional amendment and/or state law. Now that the U.S. Supreme Court has determined that the Constitution guarantees a right to same-sex marriage, many other issues pertaining to inequalities based on sexual orientation remain to be resolved. Among these are marital property rights, the ability to adopt children, and equal access to benefits that previously have been provided only to persons in legal heterosexual marriages. Consider, for example, parental rights.

Parental rights remain an issue of grave concern to LGBT couples in a number of states. Among the ways in which persons in the LGBT community become parents are by adoption, foster parenting, donor insemination, surrogacy, and having children from previous heterosexual relationships. Laws governing family relationships vary significantly from state to state. In some states, same-sex partners who want to adopt a child or are raising children together (typically from a previous heterosexual marriage) learn that only one partner is legally recognized as the child's parent or guardian. The LGBT community has struggled to gain the same parental rights in regard to legal and physical custody of children as heterosexual couples, including the right to physical access or visitation with a child, and various other rights pertaining to the property and well-being of a child. If gay and lesbian couples are denied parental rights by law and in the courts of the land, they have little or no legal recourse and are unable to exert authority over their children's lives, health care, or property.

Another pressing issue is housing discrimination. Housing discrimination is a problem in the LGBT community because the Fair Housing Act, which affords some redress for some other minority groups, does not apply. According to HUD.gov (U.S. Department of Housing and Urban Development, 2015), "The Fair Housing Act does not specifically include sexual orientation and gender identity as prohibited bases. However, a lesbian, gay, bisexual, or transgender (LGBT) person's experience with sexual orientation or gender identity housing discrimination may still be covered by the Fair Housing Act." Sometimes, HUD (Housing and Urban Development) guidelines come into play in ensuring equal access to housing for LGBT persons. Examples of housing discrimination include LGBT

crossdresser
a male who dresses as a woman or a female who dresses as a man but does not alter his or her genitalia.

sexual orientation
a person's preference for emotional–sexual relationships with members of the different sex (heterosexuality), the same sex (homosexuality), or both (bisexuality).

persons who have been discriminated against by real estate agents who refuse to show them houses in "family-oriented" apartments, condo buildings, or neighborhoods. Some finance and insurance companies have treated same-sex couples differently from other prospective homebuyers or lessees. Transgender persons have been particularly harmed by discriminatory practices in housing. One study found that transgender respondents were four times more likely to live in extreme poverty, and one in five respondents stated that they had experienced homelessness at some time in the past because of their gender identity (thetaskforce.org, 2011).

Health care is another area of discrimination based on sexual orientation. Although improvements have occurred in some areas of health care delivery, LGBT people are still are turned away from some medical facilities or face overt discrimination when they are seeking medical treatment. Prior to the Affordable Care Act, many LGBT people were unable to afford the high cost of health insurance coverage, and some were unable to acquire employer-provided health insurance because they were not allowed by their partner's employer to be counted as a dependent under the partner's insurance plan. This remains true in some areas, but changes have been made in others as state laws and the political climate in some areas have changed. Prior to the Affordable Care Act, LGBT individuals were denied insurance on the basis of preexisting conditions such as HIV/AIDS. The health care problem remains especially pronounced among transgender people, some of whom report that they have been refused care because of bias. A recent study found that 42 percent of female-to-male transgender adults reported verbal harassment, physical assault, or denial of equal treatment in a doctor's office or hospital. It is difficult, if not impossible, for many transgender people to identify themselves on medical forms as anything other than male or female (Seaman, 2015). Transgender respondents also have over four times the national average of HIV infection, which contributes to some health care professionals' lack of desire to provide medical treatment. Further confounding the problem of discrimination is the race, ethnicity, and/or class of LGBT persons.

Occupational discrimination remains a pressing problem for people in the LGBT community. Despite laws prohibiting discrimination in employment on the basis of sexual orientation, openly LGBT people have often experienced bias in hiring, retention, and promotion in public-sector and private-sector employment. However, in recent years, greater inclusion has occurred as there has been greater acceptance in society at large. In the twenty-first century, more Fortune 500 companies have included gender identity in their employee nondiscrimination policies, and other corporations have done likewise. Of course, it remains to be seen the extent to which actual compliance with these policies occurs and the workplace becomes truly more diversified and accepting of the LGBT community.

Historically, one of the most widely publicized forms of discrimination against gays and lesbians has been in the military. The "Don't Ask, Don't Tell" policy implemented in 1993 by the Clinton Administration required that commanders not ask a serviceperson about his or her sexual orientation. Gays and lesbians were allowed to serve in the military as long as they did not reveal their orientation. However, various studies showed that this policy led to differential treatment of many gays and lesbians in the military. As many as 13,000 military personnel may have been discharged under this law, and gay rights organizations advocated for its repeal, arguing that the rules were discriminatory and that they kept gay troops from seeking medical care or reporting domestic abuse for fear of being exposed and expelled from their military branch. In 2010 President Barack Obama signed the repeal of the policy, thus allowing gay and lesbian Americans to serve openly in the armed forces.

Various organizations of gays, lesbians, and transgender persons have been unified in their desire to reduce discrimination and other forms of **homophobia**—extreme prejudice and sometimes discriminatory actions directed at gays, lesbians, bisexuals, transgender persons, and others who are perceived as not being heterosexual (•Figure 11.2). Homophobia involves an aversion to LGBT people or their lifestyle or culture, and it sometimes includes behavior or an act, such as a hate crime, based on this aversion. Because of violence against LGBT individuals in the past, laws have been passed such as the Matthew Shepard and James Byrd Jr. Hate Crimes Prevention Act that attempt to prevent such crimes or to bring to justice those individuals who perpetrate such violent acts in the future.

Some of the more recently publicized forms of potential discrimination against the LGBT community are the "religious freedom" or "religious liberty" bills that twenty-one states have passed as of mid-year 2015 and sixteen other states have introduced as new legislation. For example, supporters of the 2015 Religious Freedom Restoration Act (RFRA) in Indiana and Arkansas claim that the laws are merely for protection of religious freedom in for-profit corporations. Critics of RFRA view this type of law as a possible vehicle to promote discrimination against members of the LGBT community by allowing conservative Christian vendors to decline to provide various wedding-related services (such as flowers, wedding cakes, ceremony planning, and venues) for same-sex partners. The laws apply religious rights to businesses and corporations, so it is possible that these vendors and service companies could use the laws to refuse to serve partners who are planning same-sex weddings. Because of the increase in the number of states allowing same-sex marriage, the intent of such laws may be to keep businesspeople from having to participate in any way if their religious convictions dictate otherwise. It is unclear what, if any, effect these laws will have on the LGBT community or whether additional states will pass similar legislation in the future. Despite changes in marriage laws in more states in recent years, RFRA laws are an indication that battles among diverse ideological viewpoints and constituencies, as well as the struggle for equal rights for the LGBT community, are far from over. Some analysts believe that until a federal law and/or laws in all fifty states are passed protecting the various classes of sexual orientation and gender identity, LGBT people will not achieve greater equality in the United States (Ford, 2015).

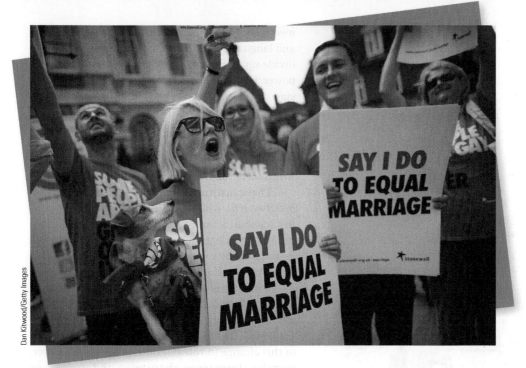

FIGURE 11.2 For many years, lesbian, gay, bisexual, and transgender persons, and others who support their cause, have participated in rallies to highlight the problem of homophobia and demand changes in laws that discriminate against LGBT persons. Public opinion and law eventually changed on the issue of same-sex marriage through the efforts of persons in social movements such as the one shown here.

or should exist between them. In a now-classic statement, the sociologist Judith Lorber (1994: 6) summarized the importance of gender:

> Gender is a human invention, like language, kinship, religion, and technology; like them, gender organizes human social life in culturally patterned ways. Gender organizes social relations in everyday life as well as in the major social structures, such as social class and the hierarchies of bureaucratic organizations.

Virtually everything social in our lives is *gendered:* People continually distinguish between males and females and evaluate them differentially. Gender is an integral part of the daily experiences of both women and men (Kimmell and Messner, 2012).

A *microlevel* analysis of gender focuses on how individuals learn gender roles and acquire a gender identity. **Gender role** refers to the attitudes, behavior, and activities that are socially defined as appropriate for each sex and that are learned through the socialization process. For example, in U.S. society males are traditionally expected to demonstrate aggressiveness and toughness, whereas females are expected to be passive and nurturing (• Figure 11.3). **Gender identity** is a person's perception of the self as female or male. Typically established between eighteen months and three years of age, gender identity is a powerful aspect of our self-concept. Although this identity is

How might we describe the type of prejudice and discrimination experienced by the LGBT community? Some social scientists use the term *heterosexism* to describe an ideological system that denies, denigrates, and stigmatizes any nonheterosexual form of behavior, identity, relationship, or community. This term is used as a parallel to other forms of prejudice and discrimination, including racism, sexism, ageism, and anti-Semitism. Clearly, from this perspective, issues pertaining to homosexuality and heterosexism are not just biological issues but also social constructions that involve societal customs and institutions. Let's turn to the cultural dimension of gender to see how socially constructed differences between females and males are crucial in determining how we identify ourselves as girls or boys, women or men.

LO3 **Define** *gender role, gender identity, body consciousness,* and *sexism.*

Gender: The Cultural Dimension

Gender refers to the culturally and socially constructed differences between females and males found in the meanings, beliefs, and practices associated with "femininity" and "masculinity." Although biological differences between women and men are very important, in reality most "sex differences" are socially constructed "gender differences." According to sociologists, social and cultural processes, not biological "givens," are most important in defining what females and males are, what they should do, and what sorts of relations do

homophobia
extreme prejudice and sometimes discriminatory actions directed at gays, lesbians, bisexuals, transgender persons, and others who are perceived as not being heterosexual.

gender
the culturally and socially constructed differences between females and males found in the meanings, beliefs, and practices associated with "femininity" and "masculinity."

gender role
the attitudes, behavior, and activities that are socially defined as appropriate for each sex and that are learned through the socialization process.

gender identity
a person's perception of the self as female or male.

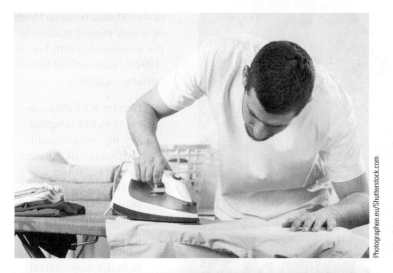

FIGURE 11.3 Which of these pictures contradicts our society's traditional gender roles for men? Do you see this trend as a positive one? Why or why not?

an individual perception, it is developed through interaction with others. As a result, most people form a gender identity that matches their biological sex: Most biological females think of themselves as female, and most biological males think of themselves as male. However, some people think of gender as a *continuum* (a continuous succession or whole) in which biological females perceive of themselves as more female than male, and biological males perceive of themselves as more male than female. Of course, this is a matter for individual consideration, as is the issue of body consciousness, which is also a part of gender identity. **Body consciousness** is how a person perceives and feels about his or her body; it also includes an awareness of social conditions in society that contribute to this self-knowledge. As we grow up, we become aware that the physical shape of our bodies subjects us to the approval or disapproval of others. Being small and weak may be considered positive attributes for women, but they are considered negative characteristics for "true men."

A *macrolevel* analysis of gender examines structural features, external to the individual, that perpetuate gender

inequality. Gender is embedded in the images, ideas, and language of a society and is used as a means to divide up work, allocate resources, and distribute power. For example, every society uses gender to assign certain tasks—ranging from child rearing to warfare—to females and to males, and differentially rewards those who perform these duties. These structures have been referred to as *gendered institutions,* meaning that gender is one of the major ways by which social life is organized in all sectors of society.

These institutions are reinforced by a *gender belief system,* which includes all the ideas regarding masculine and feminine attributes that are held to be valid in a society. This belief system is legitimated by religion, science, law, and other societal values. For example, gender belief systems may change over time as gender roles change. Many fathers take care of young children today while women are the primary income earners in the family, and there is a much greater acceptance of this change in roles by both partners. However, popular stereotypes about men and women, as well as cultural norms about gender-appropriate appearance and behavior, still linger and sometimes reinforce gendered institutions in society.

The Social Significance of Gender

Gender is a social construction with important consequences in everyday life. Just as stereotypes regarding race/ethnicity have built-in notions of superiority and inferiority, gender stereotypes hold that men and women are inherently different in attributes, behavior, and aspirations. Stereotypes define men as strong, rational, dominant, independent, and less concerned with their appearance. Women are stereotyped as more emotional, nurturing, dependent, and anxious about their appearance.

The social significance of gender stereotypes is illustrated by eating disorders. The three most common eating problems are anorexia, bulimia, and obesity. Some studies estimate that as many as 65 percent of women between the ages of 25 and 45 have disordered eating behaviors (ScienceDaily, 2008). With *anorexia,* a person has an overriding obsession with food and thinness that constantly controls his or her activities and eating patterns, resulting in a body weight of less than 85 percent of the average weight for a person of that individual's age and height group. With *bulimia,* a person binges by consuming large quantities of food and then purges the food by induced vomiting, excessive exercise, laxatives, or subsequent fasting. In the past, *obesity* was defined as being 20 percent or more above a person's desirable weight, as established by the medical profession. Today, however, medical professionals use the BMI (body mass index) to define obesity. To determine this index, a person's weight in kilograms is divided by his or her height in meters and squared to yield the BMI. Obesity is defined as a BMI of 30 and above (about

30 pounds overweight for the average person). In the past it was assumed that the individuals most likely to have eating disorders were white, middle-class, heterosexual women; however, such problems also exist among women of color, working-class women, lesbians, and gay men.

Bodybuilding is another gendered experience. *Bodybuilding* is the process of deliberately cultivating an increase in the mass and strength of the skeletal muscles by means of lifting and pushing weights. In the past, bodybuilding was predominantly a male activity; musculature connoted power, domination, and virility. Today, however, an increasing number of women engage in this activity. As gendered experiences, eating problems and bodybuilding have more in common than we might think. As some women's studies scholars have pointed out, the anorexic body and the muscled body are not opposites: Both are united against the common enemy of soft, flabby flesh (• Figure 11.4). In other words, the *body* may be objectified both through compulsive dieting and compulsive bodybuilding.

In *Muscle Boys: Gay Gym Culture*, writer and personal trainer Erick Alvarez (2008) describes a globalized subculture of bodybuilding and physical fitness training among gay men that focuses on a "built" muscular body. Drawing from his own experience as a personal trainer in a San Francisco gay gym club, he identifies categories of gay men—including the Muscle Bear, Muscle Boy, Circuit-Boy, and Older Male—that emerge from this subculture, with its distinctive experiences in physical training and bodybuilding. He concludes that many of the men who go to the gym are primarily concerned with body image and the need to look muscular and attractive and to be part of a distinct community. They are extremely vigilant about their workouts, training regimens, and diet schedules because they need to compete with other gay men in the LGBT social marketplace as well as in the world at large.

Sexism

Sexism is the subordination of one sex, usually female, based on the assumed superiority of the other sex. Sexism directed at women has three components: (1) negative attitudes toward women; (2) stereotypical beliefs that reinforce, complement, or justify the prejudice; and (3) discrimination—acts that exclude, distance, or keep women separate.

Can men be victims of sexism? Although women are more often the target of sexist remarks and practices, men can be victims of sexist assumptions. Examples of sexism directed against men are the assumption that men should not be employed in certain female-dominated occupations, such as nurse or elementary school teacher, and the belief that it is somehow more harmful for families when female soldiers are killed in battle than male soldiers.

Like racism, sexism is used to justify discriminatory treatment. Obvious manifestations of sexism are found in the undervaluing of women's work and in hiring and promotion practices that effectively exclude women from an organization or confine them to the bottom of the organizational

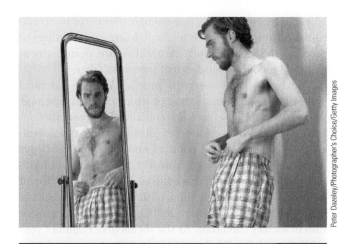

FIGURE 11.4 Not all anorexics are women, and not all bodybuilders are men. However, some analysts suggest that these two issues are manifestations of the same desire: to avoid having soft, flabby flesh.

body consciousness
how a person perceives and feels about his or her body.

sexism
the subordination of one sex, usually female, based on the assumed superiority of the other sex.

hierarchy. Even today, some women who enter nontraditional occupations (such as firefighting and welding) or professions (such as dentistry, architecture, or investment banking) encounter hurdles that men do not face.

Sexism is interwoven with **patriarchy**—a hierarchical system of social organization in which cultural, political, and economic structures are controlled by men. By contrast, **matriarchy** is a hierarchical system of social organization in which cultural, political, and economic structures are controlled by women; however, few (if any) societies have been organized in this manner. Patriarchy is reflected in the way that men may think of their position as men as a given, whereas women may deliberate on what their position in society should be. As the sociologist Virginia Cyrus (1993: 6) explains, "Under patriarchy, men are seen as 'natural' heads of households, Presidential candidates, corporate executives, college presidents, etc. Women, on the other hand, are men's subordinates, playing such supportive roles as housewife, mother, nurse, and secretary." Gender inequality and a division of labor based on male dominance are nearly universal, as we will see in the following discussion on the origins of gender-based stratification.

LO4 **Describe** how the division of labor between women and men differs in various kinds of societies.

Gender Stratification in Historical and Contemporary Perspective

How do tasks in a society come to be defined as "men's work" or "women's work"? Three factors are important in determining the gendered division of labor in a society: (1) the type of subsistence base, (2) the supply of and demand for labor, and (3) the extent to which women's child-rearing activities are compatible with certain types of work. *Subsistence* refers to the means by which a society gains the basic necessities of life, including food, shelter, and clothing. The three factors vary according to a society's *technoeconomic base*—the level of technology and the organization of the economy in a given society. Five such bases have been identified: hunting and gathering societies, horticultural and pastoral societies, agrarian societies, industrial societies, and postindustrial societies, as shown in • Table 11.2.

Hunting and Gathering Societies

The earliest known division of labor between women and men is in hunting and gathering societies. While the men hunt for wild game, women gather roots and berries. A relatively equitable relationship exists because neither sex has the ability to provide all the food necessary for survival. When wild game is nearby, both men and women may hunt. When it is far away, hunting becomes incompatible with child rearing (which women tend to do because they breast-feed their young), and women are placed at a disadvantage in terms of contributing to the food supply (Lorber, 1994). In most hunting and gathering societies, women are full economic partners with men; relations between them tend to be cooperative and relatively egalitarian (Bonvillain, 2001). Little social stratification of any kind is found because people do not acquire a food surplus.

Horticultural and Pastoral Societies

In horticultural societies, which first developed ten to twelve thousand years ago, a steady source of food becomes available. People are able to grow their own food because of hand tools, such as the hoe. Women make an important contribution to food production because hoe cultivation is

TABLE 11.2

Technoeconomic Bases of Society					
	Hunting and Gathering	Horticultural and Pastoral	Agrarian	Industrial	Postindustrial
Change from Prior Society	—	Use of hand tools, such as digging stick and hoe	Use of animal-drawn plows and equipment	Invention of steam engine	Invention of computer and development of "high-tech" society
Economic Characteristics	Hunting game, gathering roots and berries	Planting crops, domestication of animals for food	Labor-intensive farming	Mechanized production of goods	Information and service economy
Control of Surplus	None	Men begin to control societies	Men who own land or herds	Men who own means of production	Corporate shareholders and high-tech entrepreneurs
Women's Status	Relative equality	Decreasing in move to pastoralism	Low	Low	Varies by class, race, and age

Source: Adapted from Lorber, 1994: 140.

compatible with child care. A fairly high degree of gender equality exists because neither sex controls the food supply.

When inadequate moisture in an area makes planting crops impossible, *pastoralism*—the domestication of large animals to provide food—develops. Herding is primarily done by men, and women contribute relatively little to subsistence production in such societies. In some herding societies, women have relatively low status; their primary value is their ability to produce male offspring so that the family lineage can be preserved and enough males will exist to protect the group against attack.

In contemporary horticultural societies, women do most of the farming while men hunt game, clear land, work with arts and crafts, make tools, participate in religious and ceremonial activities, and engage in war. A combination of horticultural and pastoral activities is found in some contemporary societies in Asia, Africa, the Middle East, and South America. These societies are characterized by more gender inequality than in hunting and gathering societies but less gender inequality than in agrarian societies (Bonvillain, 2001).

Agrarian Societies

In agrarian societies, which first developed about eight to ten thousand years ago, gender inequality and male dominance become institutionalized. The most extreme form of gender inequality developed about five thousand years ago in societies in the Fertile Crescent around the Mediterranean Sea. Agrarian societies rely on agriculture—farming done by animal-drawn or mechanically powered plows and equipment. Because agrarian tasks require more labor and greater physical strength than horticultural ones, men become more involved in food production. It has been suggested that women are excluded from these tasks because they are viewed as too weak for the work and because child-care responsibilities are considered incompatible with the full-time labor that the tasks require.

Why does gender inequality increase in agrarian societies? Scholars cannot agree on an answer; some suggest that it results from private ownership of property. When people no longer have to move continually in search of food, they can acquire a surplus. Men gain control over the disposition of the surplus and the kinship system, and this control serves men's interests. The importance of producing "legitimate" heirs to inherit the surplus increases significantly, and women's lives become more secluded and restricted as men attempt to ensure the legitimacy of their children. Premarital virginity and marital fidelity are required; indiscretions are punished. However, some scholars argue that male dominance existed before the private ownership of property (Firestone, 1970; Lerner, 1986).

Male dominance is very strong in agrarian societies. Women are secluded, subordinated, and mutilated as a means of regulating their sexuality and protecting paternity. Most of the world's population currently lives in agrarian societies in various stages of industrialization.

Industrial Societies

An *industrial society* is one in which factory or mechanized production has replaced agriculture as the major form of economic activity. As societies industrialize, the status of women tends to decline further. Industrialization in the United States created a gap between the unpaid work performed by middle- and upper-class women at home and the paid work that was increasingly performed by men and unmarried girls. Husbands were responsible for being "breadwinners"; wives were seen as "homemakers."

This gendered division of labor increased the economic and political subordination of women. It also became a source of discrimination against women of color based on both their race and the fact that many of them had to work in order to survive. In the late 1800s and into the 1900s, many African American women were employed as domestic servants in affluent white households.

As people moved from a rural, agricultural lifestyle to an urban existence, body consciousness increased. People who worked in offices often became sedentary and exhibited physical deterioration from their lack of activity. As gymnasiums were built to fight this lack of physical fitness, images of masculinity shifted from the physique of the farmer or factory workman to the middle-class office man who exercised and lifted weights. As industrialization progressed and food became more plentiful, the social symbolism of women's body weight and size also changed, and middle-class women became more preoccupied with body fitness.

Postindustrial Societies

As previously defined, *postindustrial societies* are ones in which technology supports a service- and information-based economy. In such societies the division of labor in paid employment is increasingly based on whether people provide or apply information or are employed in service jobs such as fast-food counter help or health care workers. For both women and men in the labor force, formal education is increasingly crucial for economic and social success. However, although some women have moved into entrepreneurial, managerial, and professional occupations, many others have remained in the low-paying service sector, which affords few opportunities for upward advancement (• Figure 11.5).

How do new technologies influence gender relations in the workplace? Although some analysts presumed that technological developments would reduce the boundaries between women's and men's work, researchers have found that the gender stereotyping associated with specific jobs

patriarchy
a hierarchical system of social organization in which cultural, political, and economic structures are controlled by men.

matriarchy
a hierarchical system of social organization in which cultural, political, and economic structures are controlled by women.

FIGURE 11.5 In contemporary societies, women do a wide variety of work and are responsible for many diverse tasks. The women shown here are employed in the industrial, factory sector and the postindustrial, biotechnology sector of the U.S. economy. Do you think issues of gender inequality might be different for these two women? Why or why not?"

has remained remarkably stable even when the nature of work and the skills required to perform it have been radically transformed. Today, men and women continue to be segregated into different occupations, and this segregation is particularly visible within individual workplaces (as discussed later in the chapter).

How does the division of labor change in families in postindustrial societies? For a variety of reasons, more households are headed by women with no adult male present. In 2014 nearly 10 million U.S. children lived with their mother only (as contrasted with just 1.9 million who resided with their father only). Among African American children, 50 percent lived with their mother only (Child Trends Data Bank, 2015). This means that women in these households truly have a double burden, both from family responsibilities and from the necessity of holding gainful employment in the labor force.

In postindustrial societies such as the United States, approximately 60 percent of adult women are in the labor force, meaning that finding time to care for children, help aging parents, and meet the demands of the workplace will continue to place a heavy burden on women, despite living in an information- and service-oriented economy.

How people accept new technologies and the effect that these technologies have on gender stratification are related to how people are socialized into gender roles. However, gender-based stratification remains rooted in the larger social structures of society, which individuals have little ability to control.

 Identify the primary agents of gender socialization and note their role in socializing people throughout life.

Gender and Socialization

We learn gender-appropriate behavior through the socialization process. Our parents, teachers, friends, and the media all serve as gendered institutions that communicate to us our earliest, and often most-lasting, beliefs about the social meanings of being male or female and thinking and behaving in masculine or feminine ways. Some gender roles have changed dramatically in recent years; others have remained largely unchanged over time.

Some parents prefer boys to girls because of stereotypical ideas about the relative importance of males and females to the future of the family and society. Research suggests that social expectations play a major role in this preference. We are socialized to believe that it is important to have a son, especially for a first or only child. For many years it was assumed that only a male child could support his parents in their later years and carry on the family name.

Across cultures, boys are preferred to girls, especially when the number of children that parents can have is limited by law or economic conditions. In China and India, fewer girls are born each year than boys because a disproportionate number of female fetuses are aborted. Starting in the 1970s, China had a one-child-per-family law that favored males over females. However, in 2013 the policy was revised so that couples would be allowed to have two children if one parent was an only child. What effect this will have on the birth of female children remains to be seen. In India a strong cultural belief exists that a boy is an asset to his family while a girl is liability. Beliefs such as this contribute to the selective abortion of female fetuses. As a result of these past practices, nations such as China and India are faced with a shortage of marriageable young women and many other problems that result from an imbalance in the sex ratio. Perhaps seeing the consequences of favoring one sex over the other will produce new ideas among parents regarding sex and gender socialization.

Parents and Gender Socialization

From birth, parents act differently toward children on the basis of the child's sex. Baby boys are perceived to be less fragile than girls and tend to be treated more roughly by their parents. Girl babies are thought to be "cute, sweet, and cuddly" and receive more-gentle treatment. Parents strongly influence the gender-role development of children by passing on—both overtly and covertly—their own beliefs about gender. Although contemporary parents tend to play more similarly with their male and female children than their own parents or grandparents might have played with them as they were growing up, there remains a difference in how they respond toward their children based on gender even when "roughhousing" with them or engaging in sports events or other activities.

Children's toys reflect their parents' gender expectations (• Figure 11.6). Gender-appropriate toys for boys

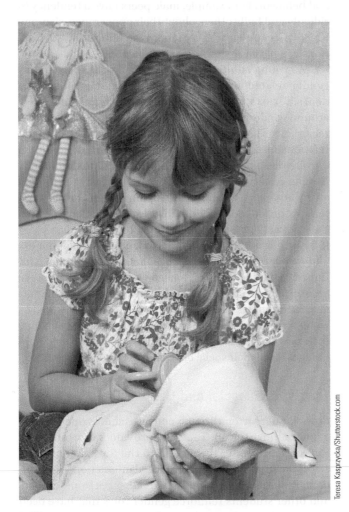

FIGURE 11.6 Are children's toys a reflection of their own preferences and choices? How do toys reflect gender socialization by parents and other adults?

include video games, trucks and other vehicles, sports equipment, and war toys such as guns and soldiers. Girls' toys include stuffed animals and dolls, makeup and dress-up clothing, and homemaking items. Ads for children's toys appeal to boys and girls differently. Most girl and boy characters are shown in gender-specific toy commercials that target either females or males. These commercials typically show boys playing outdoors and engaging in competitive activities. Girls are more often engaged in cooperative play in the ads, and this is in keeping with gender expectations about their behavior (Kaklenberg and Hein, 2010).

When children are old enough to help with household chores, they are often assigned different tasks. Girls often spend more time doing housework than boys (Belkin, 2009). Boys are more likely to be paid for doing chores at home than girls (University of Michigan, 2007). Parents are more likely to assign maintenance chores (such as mowing the lawn) to boys, whereas domestic chores (such as shopping, cooking, clearing the table, and taking care of young siblings) are assigned to girls.

In the past, most studies of gender socialization focused on white, middle-class families and paid little attention to ethnic differences. According to earlier studies, children from middle- and upper-income families are less likely to be assigned gender-linked chores than children from lower-income backgrounds. In addition, gender-linked chore assignments occur less frequently in African American families, where both sons and daughters tend to be socialized toward independence, employment, and child care (McHale et al., 2006). In contrast, gender socialization in Hispanic (Latino/a) families suggests that adolescent females often receive different gender socialization by their parents than do their male siblings. Many Latinas are allowed less interaction with members of the opposite sex than are the adolescent males in their families. Rules for dating, school activities, and part-time jobs are more stringent for the girls because many parents want to protect their daughters and keep them closer to home. Moreover, studies continue to show that many Latinas are primarily socialized by their families to become wives and mothers, while less emphasis is placed on educational attainment and careers (Landale and Oropesa, 2007). Some contemporary Latinas find that they must struggle with both cultural and structural barriers to achieving their academic and professional goals.

Across classes and racial–ethnic categories, mothers typically play a stronger role in gender socialization of daughters, whereas fathers do more to socialize sons than daughters, particularly when it comes to racial and gender socialization (McHale et al., 2006). However, many parents are aware of the effect that gender socialization has on their children and make a conscientious effort to provide gender-neutral experiences for them.

Peers and Gender Socialization

Peers help children learn prevailing gender-role stereotypes, as well as gender-appropriate and gender-inappropriate behavior. During the preschool years, same-sex peers have a powerful effect on how children see their gender roles. Children are more socially acceptable to their peers when they conform to implicit societal norms governing the "appropriate" ways that girls and boys should act in social situations and what prohibitions exist in such cases.

Male peer groups place more pressure on boys to do "masculine" things than female peer groups place on girls to do "feminine" things. For example, girls wear jeans and other "boy" clothes, play soccer and softball, and engage in other activities traditionally associated with males. By contrast, if a boy wears a dress, plays hopscotch with girls, and engages in other activities associated with being female, he will be ridiculed by his peers. This distinction between the relative value of boys' and girls' behaviors strengthens the cultural message that masculine activities and behavior are more important and more acceptable.

During adolescence, peers are often stronger and more-effective agents of gender socialization than adults. Peers are thought to be especially important in boys' development of gender identity. Male bonding that occurs during adolescence is believed to reinforce masculine identity and to encourage gender-stereotypical attitudes and behavior. For example, male peers have a tendency to ridicule and bully others about their appearance, size, and weight. Because peer acceptance is so important, such actions can have very harmful consequences.

As young adults, men and women still receive many gender-related messages from peers. Among college students, for example, peer groups are organized largely around gender relations and play an important role in career choices and the establishment of long-term intimate relationships. In a study of women college students at two universities (one primarily white, the other predominantly African American), anthropologists Dorothy C. Holland and Margaret A. Eisenhart (1990) found that the peer system propelled women into a world of romance in which their attractiveness to men counted most. Although peers initially did not influence the women's choices of majors and careers, they did influence whether the women continued to pursue their original goals, changed their course of action, or were "derailed." Subsequent research has also found that some African American women, as well as women from other racial–ethnic categories, may change their occupational aspirations partly based on peer-group influence and their social environment (Frome et al., 2006).

Teachers, Schools, and Gender Socialization

From kindergarten through college, schools operate as a gendered institution. Teachers provide important messages about gender through both the formal content of classroom assignments and informal interactions with students. Sometimes, gender-related messages from teachers and other students reinforce gender roles that have been taught at home; however, teachers may also contradict parental socialization. During the early years of a child's

FIGURE 11.7 Teachers often use competition between boys and girls because they hope to make a learning activity more interesting. Here, a middle school girl leads other girls against boys in a Spanish translation contest. What are the advantages and disadvantages of gender-based competition in classroom settings?

schooling, teachers' influence is very powerful; many children spend more hours per day with their teachers than they do with their own parents.

According to some researchers, the quantity and quality of teacher–student interactions often vary between the education of girls and that of boys (Sadker and Zittleman, 2009). One of the messages that teachers may communicate to students is that boys are more important than girls. Research spanning the past thirty years shows that unintentional gender bias occurs in virtually all educational settings. **_Gender bias_** consists of showing favoritism toward one gender over the other. Researchers consistently find that teachers devote more time, effort, and attention to boys than to girls (Sadker and Zittleman, 2009). Males receive more praise for their contributions and are called on more frequently in class, even when they do not volunteer.

Teacher–student interactions influence not only students' learning but also their self-esteem (Sadker and Zittleman, 2009). A comprehensive study of gender bias in schools suggested that girls' self-esteem is undermined in school through such experiences as (1) a relative lack of attention from teachers; (2) sexual harassment by male peers; (3) the stereotyping and invisibility of females in textbooks, especially in science and math texts; and (4) test bias based on assumptions about the relative importance of quantitative and visual–spatial ability, as compared with verbal ability, that restricts some girls' chances of being admitted to the most-prestigious colleges and being awarded scholarships.

Teachers also influence how students treat one another during school hours. Many teachers use sex segregation as a way to organize students, resulting in unnecessary competition between females and males (● Figure 11.7). In addition, teachers may take a "boys will be boys" attitude when girls complain of sexual harassment. Even though sexual harassment is prohibited by law and teachers and administrators are obligated to investigate such incidents, the complaints may be dealt with superficially. If that

gender bias
behavior that shows favoritism toward one gender over the other.

FIGURE 11.8   NCAA women's sports such as basketball are popularizing athletics for young women and making it easier for girls to become actively involved in sports at a young age.

happens, the school setting can become a hostile environment rather than a site for learning.

Sports and Gender Socialization

Children spend more than half of their nonschool time in play and games, but the type of games played differs with the child's sex. Studies indicate that boys are socialized to participate in highly competitive, rule-oriented games with a larger number of participants than games played by girls. Young girls have been socialized to play exclusively with others of their own age, in groups of two or three, in activities such as hopscotch and jump rope that involve a minimum of competitiveness. Other research shows that boys express much more favorable attitudes toward physical exertion and exercise than girls do. Some analysts believe this difference in attitude is linked to ideas about what is gender-appropriate behavior for boys and girls. For males, competitive sport becomes a means of "constructing a masculine identity, a legitimated outlet for violence and aggression, and an avenue for upward mobility" (Lorber, 1994: 43). Now more girls play soccer and softball and participate in sports formerly regarded as exclusively "male"

activities (• Figure 11.8). Girls who go against the grain and participate in masculine play as children are more likely to participate in sports as young women and adults (Giuliano, Popp, and Knight, 2000).

Many women athletes believe that they have to manage the contradictory statuses of being both "women" and "athletes." An earlier study found that women college basketball players dealt with this contradiction by dividing their lives into segments. On the basketball court, the women "did athlete": They pushed, shoved, fouled, ran hard, sweated, and cursed. Off the court, they "did woman": After the game, they showered, dressed, applied makeup, and styled their hair, even if they were only getting in a van for a long ride home (Watson, 1987). A more recent study found that female athletes who played softball, soccer, or basketball engaged in "apologetic behavior" after the game through their efforts to look feminine, their apologies for their aggression during the game, and the ways in which they marked themselves as heterosexual (Davis-Delano, Pollock, and Vose, 2009). According to some social analysts, being able to identify the paradox between "female" and "athlete" and the problems that women in sports experience is the beginning of confronting socially constructed gender norms and polarized views of masculinity and femininity in Western culture (Paloian, 2015).

Mass Media and Gender Socialization

The media—including newspapers, magazines, television, movies, and social media—are powerful sources of gender stereotyping. Although some critics argue that the media simply reflect existing gender roles in society, others point out that the media have a unique ability to shape ideas. Think of the impact that television might have on children if they spend one-third of their waking time watching it, as has been estimated. From children's cartoons to adult shows, television programs are sex typed, and many are male oriented. More male than female roles are portrayed, and male characters are typically more aggressive and direct. By contrast, females are depicted as either acting deferential toward other people and being manipulated by them or as being overly aggressive, overbearing, and even downright "bitchy."

In prime-time television, a number of significant changes in the past three decades have reduced gender stereotyping; however, men still outnumber women as leading characters, and they are often "in charge" in any setting where both men's and women's roles are portrayed. Recently, retro series on network and cable television have brought back an earlier era when men were dominant in public and family life and women played a subordinate role to them. Having recently concluded its final season and now available on DVD and Netflix, the award-winning series *Mad Men* (on AMC) is set in a 1960s New York advertising agency, where secretaries were expected to wear tight sweaters and skirts and bring men hot coffee throughout the day, while the men's wives were supposed to be the perfect companions and hostesses at home. Although many other TV series, such as *Modern Family,* have changed traditional norms, offering a wide diversity of families, including gay dads with a child, the shift to retro gender roles in some television programming and films in the second decade of the twenty-first century has raised questions about the extent to which change actually occurs in the portrayal of women and men in the media.

Advertising—whether on television and billboards or in magazines and newspapers—can be very persuasive. The intended message is clear to many people: If they embrace traditional notions of masculinity and femininity, their personal and social success is assured; if they purchase the right products and services, they can enhance their appearance and gain power over other people. A study by the sociologist Anthony J. Cortese (2004) found that women—regardless of what they were doing in a particular ad—were frequently shown in advertising as being young, beautiful, and seductive. Other research shows that TV ads such as the ones shown on Super Bowl Sunday are created to sell products but that they also contribute to the sexual objectification of women. For example, chocolate commercials often objectify women, turning them into

sexual objects whose seductive behavior is caused by the chocolate being advertised. Although such depictions may sell products, they may also have the effect of influencing how we perceive ourselves and others with regard to issues of power and subordination.

As we all know, social media can be used in both positive and negative ways. Facebook, Twitter, Instagram, Pinterest, Tumblr, Flickr, Vine, and other social networking sites are very effective tools for communicating with others, but they also offer prime venues in which to bully others and spread derogatory comments and photos relating to race, gender, and/or sexual orientation (Pew Research Center Internet, Science and Tech, 2015). Today, 91 percent of teens use a mobile device to go online, so they are no longer under the supervision of parents or other adults who might oversee their television-viewing habits or supervise a phone conversation. According to the Pew Research Center Internet, Science and Tech (2015), a typical teenager sends and receives thirty text messages per day. Although this research does not include questions about the content of these texts, data from other sources suggest that the texts often relate to the physical appearance of the sender and others, particularly in regard to sexual appeal, appearance, and behavior that identifies individuals by sexual orientation. Extensive research will be necessary to learn how social networking sites function as agents of socialization in regard to sexuality, weight, and body image, but these sites present a new and relatively unchallenged arena in which one's own beliefs and biases can be not only projected but also amplified to tens of thousands of other people.

Adult Gender Socialization

Gender socialization continues as women and men complete their training or education and join the workforce. Men and women are taught the "appropriate" type of conduct for persons of their sex in a particular job or occupation—both by employers and by coworkers. However, men's socialization usually does not include a measure of whether their work can be successfully combined with having a family; it is often assumed that men can and will do both. Even today, the reason given for women not entering some careers and professions is that this kind of work is not suitable for women because of their physical capabilities or assumed child-care responsibilities.

Different gender socialization may occur as people reach their forties and enter "middle age." A double standard of aging exists that affects women more than men (• Figure 11.9). Often, men are considered to be at the height of their success as their hair turns gray and their face gains a few wrinkles. By contrast, not only do other people in society make middle-age women feel as if they are "over the hill," but multimillion-dollar advertising campaigns continually call attention to women's every weakness, every pound gained, and every bit of flabby flesh, wrinkle, or gray hair. Increasingly, both women and men

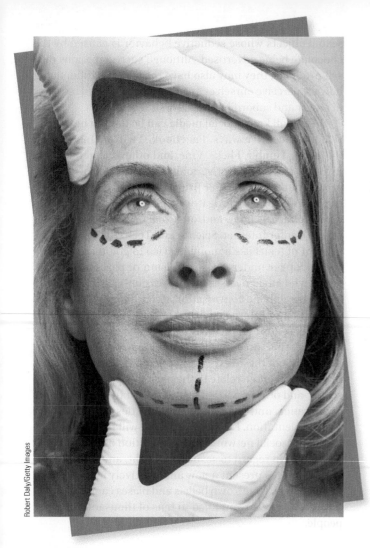

Robert Daly/Getty Images

FIGURE 11.9 Does the double standard of aging for women and men contribute to some women's desire to have surgical procedures that claim to restore their "youth" as they increase in chronological age?

have turned to "miracle" products, and sometimes to cosmetic surgery, to reduce the visible signs of aging. However, the vast majority (90.6 percent) of all cosmetic surgery is performed on female patients. In 2013 more than 10.3 million cosmetic procedures were performed on women in the United States, a 471 percent increase from 1997 (American Society for Aesthetic Plastic Surgery, 2014).

Knowledge of how we develop a gender-related self-concept and learn to feel, think, and act in feminine or masculine ways is important for an understanding of ourselves. Examining gender socialization makes us aware of the effect of our parents, siblings, teachers, friends, and the media on our perspectives about gender. However, the gender socialization perspective has been criticized on several accounts. Childhood gender-role socialization may not affect people as much as some analysts have suggested. For example, the types of jobs that people take as adults may have less to do with how they were socialized in childhood than with how they are treated in the workplace. From this perspective, women

and men will act in ways that bring them the most rewards and produce the fewest punishments. Also, gender socialization theories can be used to blame women for their own subordination by not taking into account structural barriers that perpetuate gender inequality. We will now examine a few of those structural forces.

LO6 **Discuss** ways in which the contemporary workplace reflects gender stratification.

Contemporary Gender Inequality

According to feminist scholars, women experience gender inequality as a result of past and present economic, political, and educational discrimination. Women's position in the U.S. workforce reflects the years of subordination that women have experienced in society.

Gendered Division of Paid Work in the United States

Where people are located in the occupational structure of the labor market has a major impact on their earnings. The workplace may be a gendered institution if jobs are often segregated by gender and by race/ethnicity (• Figure 11.10). In a comprehensive study, sociologists Kevin Stainback and Donald Tomaskovic-Devey (2012) describe how data from five million private-sector workplaces that they examined confirm that white men still dominate the management ranks and that workplace segregation, based on both gender and race, is increasing in many employment sectors. Consider, for example, that white men are 68 percent more likely to be in management positions than to be regular staffers, white women are 28 percent less likely to be in management, African American (black) men are 53 percent less likely to be in leadership positions, and African American (black) women are 73 percent less likely to be in management positions (Stainback and Tomaskovic-Devey, 2012).

Gender-segregated work refers to the concentration of women and men in different occupations, jobs, and places of work. Today, 93 percent of all secretaries in the United States are women while 91 percent of all mechanical engineers are men (U.S. Bureau of Labor Statistics, 2014). To eliminate gender-segregated jobs in the United States, more than half of all men or all women workers would have to change occupations. Moreover, women are severely underrepresented at the top of U.S. corporations. Out of the top S&P 500 companies (U.S. stock market index companies), only 23 have female CEOs (Catalyst, 2015). In Fortune 500 companies (the top 500 public corporations ranked by gross revenue), women of color are absent on most boards, making up only 2.8 percent of board directors. The overall share of board seats held by women of color is 3.1 percent, but this number is larger only because some of the same women hold more

Redsnapper/Alamy

Kevin Foy/Alamy

FIGURE 11.10 What stereotypes are associated with men in female-oriented positions? With women in male-oriented occupations? Do you think such stereotypes will change in the near future?

than one board seat. This figure shows that board selection committees tend to rely on the same women of color to fill board seats rather than seeking a larger pool of eligible women of color to appoint to the positions (Catalyst, 2015). When there are few, or no, women in top leadership roles in business, young women lack role models and mentors to encourage them to enter the business world.

Although the degree of gender segregation in the professional labor market (including physicians, dentists, lawyers, accountants, and managers) has declined since the 1970s, racial–ethnic segregation has remained deeply embedded in the social structure. Although some change has occurred in recent years, women of color are more likely than their white counterparts to be concentrated in public-sector employment (as public schoolteachers, welfare workers, librarians, public defenders, and faculty members at public colleges, for example) rather than in the private sector (for example, in large corporations, major law firms, and private educational institutions). And it appears that resegregation is occurring in the private sector. According to the study of fifty-eight industries by Stainback and Tomaskovic-Devey (2012), seven had a rise in gender segregation between 2001

and 2005. These include airlines, railroads, and mining. Similarly, the research found an increase in racial segregation in eighteen industries, including transportation and the lumber and leather industries. Across all categories of occupations, white women and all people of color are not evenly represented, as shown in • Table 11.3.

Labor market segmentation—the division of jobs into categories with distinct working conditions—results in women having separate and unequal jobs. Why does gender-segregated work matter? Although we look more closely at the issue of the pay gap in the following section, it is important to note here that the pay gap between men and women is the best-documented consequence of gender-segregated work. Most women work in lower-paying, less prestigious jobs, with less opportunity for advancement than their male counterparts.

Gender-segregated work affects both men and women. Men are often kept out of certain types of jobs. Those who enter female-dominated occupations often have to justify themselves and prove that they are "real men." Even if these concerns do not push men out of female-dominated occupations, they affect how the men manage their gender

TABLE 11.3

Percentage of the Workforce Represented by Women, African Americans, Hispanics, and Asian Americans in Selected Occupations

The U.S. Census Bureau accumulates data that show what percentage of the total workforce is made up of women, African Americans, Asian Americans, and Hispanics. As used in this table, *women* refers to females in all racial–ethnic categories, whereas *African Americans, Hispanics,* and *Asian Americans* refer to both women and men.

	Women	African Americans	Hispanics	Asian Americans
All occupations	46.9	11.4	16.1	5.7
Managerial, professional, and related occupations	51.6	8.8	8.7	7.5
Management occupations	38.6	6.7	9.1	5.4
Professional and related occupations	57.2	9.7	8.6	8.6
Architecture and engineering	15.4	5.2	8.2	11.7
Lawyers	32.9	5.7	5.6	4.4
Physicians and surgeons	36.7	5.5	6.3	21.0
Service occupations (all)	56.7	16.2	23.4	5.4
Food preparation and serving	55.1	12.6	24.9	6.0
Building and grounds cleaning	40.2	14.6	36.7	3.4
Health care support occupations	87.6	25.7	16.2	5.2
Grounds maintenance workers	6.3	6.3	43.6	1.7

Source: U.S. Bureau of Labor Statistics, 2015.

identity at work. For example, men in occupations such as nursing tend to emphasize their masculinity, attempt to distance themselves from female colleagues, and try to move quickly into management and supervisory positions.

Occupational gender segregation contributes to stratification in society. Job segregation is structural; it does not occur simply because individual workers have different abilities, motivations, and material needs. As a result of gender and racial segregation, employers are able to pay many men of color and all women less money, promote them less often, and provide fewer benefits.

Pay Equity (Comparable Worth)

Occupational segregation contributes to a ***pay gap***—the disparity between women's and men's earnings. The pay gap is calculated by dividing women's earnings by men's earnings to yield a percentage, also known as the earnings ratio. When the 1963 Equal Pay Act was passed, women who were classified as "full-time wage and salary workers" earned about 59 cents for every dollar her male counterpart earned. In 2014 women classified the same way earned about 78 percent (or 78 cents for every dollar) of the amount earned by men in the same category. Although some progress has been made, the gender pay gap has been persistent and has basically stalled over the past decade.

As • Figure 11.11 shows, women in all age categories also receive less pay than men, with the disparity growing wider in the older age brackets.

Earnings differences between women and men in various racial–ethnic categories are the widest for white Americans and Asian Americans. White (non-Hispanic) women's earnings were about 78 percent of their white male counterparts in 2013, while Asian American women earned about 79 percent as much as their male counterparts (see • Figure 11.12). By comparison, Hispanic women (Latinas) earned about 90 percent as much as their Hispanic male counterparts, American Indian and Alaska Native women earned 85 percent as much as their male counterparts, and African American women earned about 91 percent as much as African American men (AAUW, 2015). • Figure 11.13 shows women's median earnings as compared to men's median earnings in each of the fifty states, the District of Columbia, and Puerto Rico.

The gender gap increases as a person goes higher up the income ladder: Women near the top of the ladder earn 79 percent of wages for men at the same level. Among men and women with advanced degrees beyond a college diploma, women are paid about 74 percent (roughly $33 per hour) of what men make ($44 per hour). At the bottom of the income ladder, minimum-wage laws influence what people are paid, so there is less disparity in income.

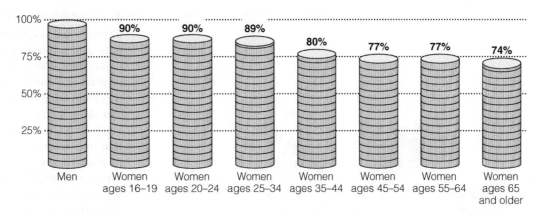

FIGURE 11.11 The Wage Gap, 2013
Source: AAUW, 2015

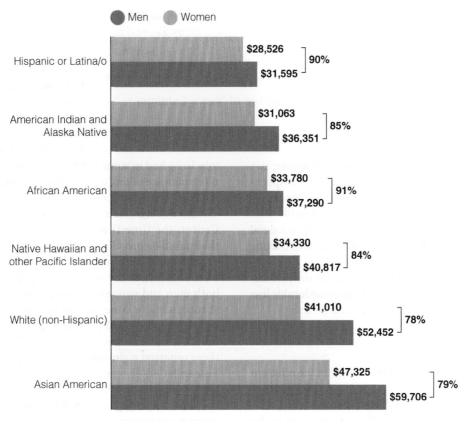

FIGURE 11.12 Women's Wages as a Percentage of Men's in Each Racial–Ethnic Category
Source: AAUW, 2015

However, even in low-wage jobs, males typically earn more than their female counterparts.

Pay equity or **comparable worth** is the belief that wages ought to reflect the worth of a job, not the gender or race of the worker. How can the comparable worth of different kinds of jobs be determined? One way is to compare the actual work of women's and men's jobs and see if there is a disparity in the salaries paid for each. To do this, analysts break a job into components—such as the education, training, and skills required, the extent of responsibility for

others' work, and the working conditions—and then allocate points for each (Lorber, 2005). For pay equity to exist, men and women in occupations that receive the same

pay gap
the disparity between women's and men's earnings.

comparable worth
(or *pay equity*) the belief that wages ought to reflect the worth of a job, not the gender or race of the worker.

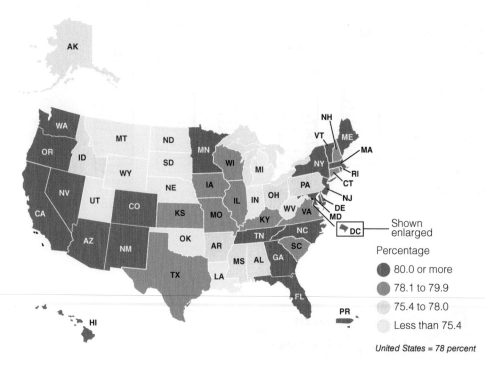

FIGURE 11.13 **Women's Earnings as a Percentage of Men's Earnings by State and Puerto Rico, 2013**

Source: U.S. Bureau of Labor Statistics, 2012e.

Percentage

● 80.0 or more
● 78.1 to 79.9
● 75.4 to 78.0
○ Less than 75.4

United States = 78 percent

number of points should be paid the same. However, pay equity exists for very few jobs. What are the prospects for the future? The Paycheck Fairness Act—proposed by the Obama Administration in 2010, 2012, and 2014—that would have extended pay-equity rules that apply to federal contractors to the entire U.S. workforce, while also making updates to the Equal Pay Act, was blocked from consideration by members of the U.S. Congress. So for the foreseeable future, women will continue to learn considerably less than men, even in similar occupational categories.

Paid Work and Family Work

As previously discussed, the first big change in the relationship between family and work occurred with the Industrial Revolution and the rise of capitalism. The cult of domesticity kept many middle- and upper-class women out of the workforce during this period. Primarily, working-class and poor women were the ones who had to deal with the work/family conflict. Today, however, the issue spans the entire economic spectrum. The typical married woman in the United States combines paid work in the labor force and family work as a homemaker. Although this change has occurred at the societal level, individual women bear the brunt of the problem.

Even with dramatic changes in women's workforce participation, the sexual division of labor in the family has remained essentially unchanged for many years. Most married women share responsibility for the breadwinner role, yet some men do not accept their full share of domestic responsibilities. Consequently, women may have a "double

day" or "second shift" because of their dual responsibilities for paid and unpaid work (Hochschild, 1989, 2003). Although the original work on the second shift was completed twenty-five years ago, the primary researcher, Arlie Hochschild, announced in 2014 that time-use research continues to show that women still do, on average, about twice the housework and child care as men even when the women are employed full time. According to Hochschild, women now make up half of the workforce, and they are earning more overall than in the past; however, they have found that the workplace does not provide them with the necessary flex time and parental leave to help them deal most effectively with both their work and family life. Some men find it more difficult to find work today because jobs are less certain and there are fewer jobs to be had in some fields, which may be one reason why more men are taking a larger role in maintaining the household and children (Schulte, 2014).

According to studies conducted by the Pew Research Center, the time that mothers and fathers spend with their families has changed significantly, with fathers now performing more housework and child-care activities and women being more involved in paid employment. For both men and women, juggling work and family life constitutes a major concern that may produce stress. Among working mothers, 56 percent reported that they found it difficult to balance work and family life; 50 percent of fathers reported a similar problem (Parker, 2013). As a Pew Research Center report states, "Fathers have by no means caught up with mothers in terms of time spent caring for children and doing household chores, but there has been some gender convergence in the way they divide their time between work and home" (Parker, 2013). Among dual-income couples, fathers spend about 42 hours each week on paid work, as compared to 31 hours of paid work for mothers. Housework takes up an average of 16 hours per week of mothers' time, as compared to 9 hours of fathers' time. Child care accounts for an average of 12 hours per week of mothers' time, as compared to 7 hours per week of fathers' time (Parker, 2013).

Problems from the past remain in many households: Working women have less time to spend on housework; if husbands do not participate in routine domestic chores, some chores simply do not get done or get done less often. Although the income that many women earn is essential to the economic survival of their families, they still must spend part of their earnings on family maintenance, such as day-care centers, fast-food restaurants, and laundries, in an attempt to keep up with their obligations.

Especially in families with young children, domestic responsibilities consume a great deal of time and energy. Although some kinds of housework can be put off, the needs of children often cannot be ignored or delayed. When children are ill or school events cannot be scheduled around work, parents (especially mothers) may experience stressful role conflicts ("Shall I be a good employee or a good mother?"). Many working women care not only for themselves, their husbands, and their children but also for elderly parents or in-laws. Some analysts refer to these women as the "sandwich generation"—caught between the needs of their young children and their elderly relatives. Many women try to solve their time crunch by forgoing leisure time and sleep. When Arlie Hochschild interviewed working mothers, she found that they talked about sleep "the way a hungry person talks about food" (1989: 9). Perhaps this is one reason that in later research, Hochschild (1997) learned that some married women with children found more fulfillment at work and that they worked longer hours because they liked work better than facing the pressures of home.

Perspectives on Gender Stratification

Sociological perspectives on gender stratification vary in their approach to examining gender roles and power relationships in society. Some focus on the roles of women and men in the domestic sphere; others note the inequalities arising from a gendered division of labor in the workplace. Still others attempt to integrate both the public and private spheres into their analyses.

 LO7 **Compare** functionalist and conflict perspectives on gender inequality.

Functionalist and Neoclassical Economic Perspectives

As seen earlier, functionalist theory views men and women as having distinct roles that are important for the survival of the family and society. The most basic division of labor is biological: Men are physically stronger, and women are the only ones able to bear and nurse children. Gendered belief systems foster assumptions about appropriate behavior for men and women and may have an impact on the types of work that women and men perform.

The Importance of Traditional Gender Roles According to functional analysts such as Talcott Parsons (1955), women's roles as nurturers and caregivers are even more pronounced in contemporary industrialized societies. While the husband performs the *instrumental* tasks of providing economic support and making decisions, the wife assumes the *expressive* tasks of providing affection and emotional support for the family. This division of family labor ensures that important societal tasks will be fulfilled; it also provides stability for family members.

This view has been adopted by a number of politically conservative analysts who assert that relationships between men and women are damaged when changes in gender roles occur, and family life suffers as a consequence. From this perspective, the traditional division of labor between men and women is the natural order of the universe.

The Human Capital Model Functionalist explanations of occupational gender segregation are similar to neoclassical economic perspectives, such as the human capital model. According to this model, individuals vary widely in the amount of human capital they bring to the labor market. *Human capital* is acquired by education and job training; it is the source of a person's productivity and can be measured in terms of the return on the investment (wages) and the cost (schooling or training).

From this perspective, what individuals earn is the result of their own choices (the kinds of training, education, and experience they accumulate, for example) and of the labor-market need (demand) for and availability (supply) of certain kinds of workers at specific points in time. For example, human capital analysts might argue that women diminish their human capital when they leave the labor force to engage in childbearing and child-care activities (• Figure 11.14). While women are out of the labor force, their human capital deteriorates from nonuse. When they return to work, women earn lower wages than men because they have fewer years of work experience and have "atrophied human capital" because their education and training may have become obsolete. One study found that over a fifteen-year period, women compared to men worked fewer years and fewer hours when the women were married and had dependent children. As a result, the women were more likely to work fewer hours in the labor market and be low earners (Rose and Hartman, 2008).

Evaluation of Functionalist and Neoclassical Economic Perspectives Although Parsons and other functionalists did not specifically endorse the gendered division of labor, their analysis suggests that it is natural and perhaps inevitable. However, critics argue that problems inherent in traditional gender roles, including the personal role strains of men and women and the social costs to society, are minimized by the functionalist approach. For example, men are assumed to be "money machines" for their families when they might prefer to spend more time in child-rearing activities. Also, the woman's place is assumed to be in the home, an assumption that ignores the fact that many women hold jobs because of economic necessity.

In addition, the functionalist approach does not take a critical look at the structure of society (especially the economic inequalities) that makes educational and occupational opportunities more available to some than to others.

Peter Dazeley/The Image Bank/Getty Images

FIGURE 11.14 According to the human capital model, women may earn less in the labor market because of their child-rearing responsibilities. What other sociological explanations are offered for the lower wages that women receive?

Furthermore, it fails to examine the underlying power relations between men and women or to consider the fact that the tasks assigned to women and to men are unequally valued by society. Similarly, the human capital model is rooted in the premise that individuals are evaluated based on their human capital in an open, competitive market where education, training, and other job-enhancing characteristics are taken into account. From this perspective, those who make less money (often men of color and all women) have no one to blame but themselves.

Critics note that instead of blaming people for their choices, we must acknowledge other realities. Wage discrimination occurs in two ways: (1) the wages are higher in male-dominated jobs, occupations, and segments of the labor market, regardless of whether women take time for family duties, and (2) in any job, women and people of color will be paid less.

Conflict Perspectives

According to many conflict analysts, the gendered division of labor within families and in the workplace results from male control of and dominance over women and resources. Differentials between men and women may exist in terms of economic, political, physical, and/or interpersonal power (• Figure 11.15). The importance of a male monopoly in any of these arenas depends on the significance of that type of power in a society. In hunting and gathering and horticultural societies, male dominance over women is limited because all members of the society must work in order to survive. In agrarian

societies, however, male sexual dominance is at its peak. Male heads of household gain a monopoly not only on physical power but also on economic power, and women become sexual property.

Although men's ability to use physical power to control women diminishes in industrial societies, men still remain the head of household and control the property. In addition, men gain more power through their predominance in the most highly paid and prestigious occupations and the highest elected offices. By contrast, women have the ability in the marriage market to trade their sexual resources, companionship, and emotional support for men's financial support and social status. As a result, women as a group remain subordinate to men.

All men are not equally privileged; some analysts argue that women and men in the upper classes are more privileged, because of their economic power, than men in lower-class positions and all people of color. In industrialized societies, persons who occupy elite positions in corporations, universities, the mass media, and government or who have great wealth have the most power. Most of these are men, however.

Conflict theorists in the Marxist tradition assert that gender stratification results from private ownership of the means of production; some men not only gain control over property and the distribution of goods but also gain power over women. According to Friedrich Engels and Karl Marx, marriage serves to enforce male dominance. Men of the capitalist class instituted monogamous marriage (a gendered institution) so that they could be certain of the paternity of their offspring, especially sons, whom they wanted to inherit their wealth. Feminist analysts have examined this theory, among others, as they have sought to explain male domination and gender stratification.

 LO8 **Describe** four feminist perspectives on gender inequality.

Feminist Perspectives

Feminism—the belief that women and men are equal and should be valued equally and have equal rights—is embraced by many men as well as women. It holds in common with men's studies the view that gender is a socially constructed concept that has important consequences for the

FIGURE 11.15 Although the demographic makeup of the U.S. Senate has been gradually changing in recent decades, men still dominate it, a fact that the conflict perspective attributes to a very old pattern in human societies.

lives of all people. According to sociologists, both women and men can be feminists and propose feminist theories because they have much in common as they seek to gain a better understanding of the causes and consequences of gender inequality (see "You Can Make a Difference").

Feminist theory seeks to identify ways in which norms, roles, institutions, and internalized expectations limit women's behavior. It also seeks to demonstrate how women's personal control operates even within the constraints of relative lack of power. In the twenty-first century, feminist theory focuses more on global issues such as how "fat stigma" among women has become globalized (see "Sociology in Global Perspective").

Liberal Feminism In liberal feminism, gender equality is equated with equality of opportunity. The roots of women's oppression lie in women's lack of equal civil rights and educational opportunities. Only when these constraints on women's participation are removed will women have the same chance for success as men. This approach notes the importance of gender-role socialization and suggests that changes need to be made in what children learn from

their families, teachers, and the media about appropriate masculine and feminine attitudes and behavior. Liberal feminists fight for better child-care options, a woman's right to choose an abortion, and the elimination of sex discrimination in the workplace (● Figure 11.16).

Radical Feminism According to radical feminists, male domination causes all forms of human oppression, including racism and classism. Radical feminists often trace the roots of patriarchy to women's childbearing and child-rearing responsibilities, which make them dependent on men. In the radical feminist view, men's oppression of women is deliberate, and ideological justification for this subordination is provided by other institutions such as the media and religion. For women's condition to improve, radical feminists claim, patriarchy must be abolished. If institutions are currently gendered, alternative institutions—such as

feminism
the belief that men and women are equal and should be valued equally and have equal rights.

YOU CAN MAKE A DIFFERENCE

"Love Your Body": Women's Activism on Campus and in the Community

Do You Love What You See When You Look in the Mirror?

Every day the beauty industry and media tell women and girls that being admired, envied, and desired based on their looks is a primary function of true womanhood. They provide them with a beauty template that is narrow, unrealistic, and most importantly ingrained into their brains leaving any woman who does not fit this template feeling inadequate. The Love Your Body campaign challenges the message that a woman's value is best measured through her willingness and ability to embody current beauty standards.

—promotion for "Love Your Body Day," sponsored by the NOW Foundation (2015)

Although this message appears to be for girls and women only, many boys and men are also concerned about their physical appearance, as well as how girls and women are represented in the media. Both men and women can make a difference by becoming involved in a campus or community organization that helps people gain a better understanding of body-image issues:

- Participate in the national Love Your Body Day, which is a day of action to speak out against ads and images of women that are offensive, dangerous, and disrespectful.

- Discourage sexist ads and media reporting about women (for example, a focus on weight or other physical attributes rather than on their accomplishments) by sending letters to the publications or encouraging journalists to rethink how they frame stories about girls and women.

- Think of on-campus traditions or events that promote negative body-image stereotypes, such as parties where students are encouraged to wear scant clothing. Actively encourage the organizers of such events to rethink "theme party" clothing or other kinds of dress that contribute to body-image problems.

- Promote positive body image on campus by encouraging your club or organization to host a "Friends Don't Let Friends Fat Talk" day. Have students write down on an index card their negative body-image thoughts such as "I hate my thighs." Then ask students to wad up the cards and throw those thoughts into trash cans.

Other opportunities for involvement exist through local, state, and national organizations. Here are two places to start:

- The National Organization for Women (NOW). NOW works to end gender bias and seeks greater representation of women in all areas of public life. On the Internet, NOW's website provides links to other feminist resources.

- The National Organization for Men Against Sexism (NOMAS). NOMAS has a *profeminist stance* that seeks to end sexism and an *affirmative stance* on the rights of gay men and lesbians.

Lorna Roberts/Alamy

"Love Your Body Day" and more-frequent use of plus-sized models (shown here) in fashion campaigns are two examples of how people send a positive message to girls and women about loving what they see in the mirror rather than feeling judgmental about their appearance. Are you aware of campus or community organizations that help individuals gain a better understanding of body-image issues?

---

women's organizations seeking better health care, day care, and shelters for victims of domestic violence and rape—should be developed to meet women's needs.

Socialist Feminism   Socialist feminists argue that the oppression of women results from their dual roles as paid *and* unpaid workers in a capitalist economy. In the workplace, women are exploited by capitalism; at home, they are exploited by patriarchy. Women are easily exploited in both sectors; they are paid low wages and have few economic resources. According to some feminist scholars, gender-segregated work is a central way in which men remain dominant over women in capitalist economies, primarily because most women have lower wages and fewer opportunities than men. As a result, women must do domestic labor either to gain a better-paid man's economic support or to stretch their own wages. According to socialist feminists, the only way to achieve gender equality is to eliminate capitalism and develop a socialist economy that would bring equal pay and rights to women.

Women's Body Size and the Globalization of "Fat Stigma"

Of all the things we could be exporting to help people around the world, really negative body image and low self-esteem are not what we hope is going out with public health messaging.

> —Alexandra Brewis, lead researcher for a *Current Anthropology* article on the globalization of fat stigma, describes how perceptions from the United States and the United Kingdom have contributed to negative beliefs about body size in dozens of developing countries (qtd. in Parker-Pope, 2011).

In past sociological and cultural ethnographic studies, people in nations and territories such as Fiji, Puerto Rico, and American Samoa were found to appreciate the "fuller figure" as the norm for women's body size. According to Professor Brewis and colleagues, "Plump bodies represented success, generosity, fertility, wealth, and beauty" (qtd. in Bates, 2011). In some cultures, weight has traditionally been associated with class position in society. For example, being overweight or obese in India can be considered to be a sign that the person is middle class or

WARNING

IT'S HARD TO BE A LITTLE GIRL IF YOU'RE NOT.

Stop childhood obesity.　strong4life.com

Many organizations in the United States and other nations use public health campaigns like the one shown here to encourage individuals to be concerned about health problems that are associated with being overweight or obese. However, some social analyst believe that certain health messages may contain negative moral messages about the worth of people as well. Do you think this is a valid concern? Why or why not?

wealthy. In Tahiti it was a custom to encourage young women to gain weight and to have rounded faces and bodies that made them more attractive for marriage. However, Professor Brewis's research team was surprised to discover in their eleven-country study that people in Mexico, Paraguay, American Samoa, and some other areas where people typically have been more favorable toward the fuller-figured norm, respondents had high scores for "fat stigma" based on twenty-three survey questions asked through in-person interviews or Internet surveys. Items included in the study represented socially credited or socially discrediting attributions related to body fat and obesity such as "People are overweight because they are lazy," "Being fat is prestigious," "People should be proud of their big bodies," and "Obese people should be ashamed of their bodies" (Brewis et al., 2011).

Although it is important for people to learn the detrimental effects of obesity on the individual's health and for public officials to view wide-scale obesity as a public health concern, fat stigma has become a troubling side effect of extensive global media and public health campaigns to make everyone more aware of the problems associated with being overweight or obese. Stigmatization of obesity generally often becomes a stigma against fat individuals specifically (Parker-Pope, 2011). Negative body image and self-deprecation follow when individuals are labeled as "lazy," "unattractive," and "undesirable." It is possible that negative health messages also contain negative moral messages about the worth of people as well. So the delicate balance in messaging for the future becomes how to have effective public health campaigns that help curb diabetes and high blood pressure worldwide but do not negatively stigmatize those individuals who are overweight or obese.

REFLECT & ANALYZE

What signs of fat stigma do you see in the United States or another country with which you are most familiar? How might the media and global health organizations more effectively send the message of the problematic health risks associated with being overweight or obese while, at the same time, encouraging people to be nonjudgmental about the body size of other individuals?

Multicultural Feminism Recently, academics and activists have been rethinking the experiences of women of color from a feminist perspective. The experiences of African American women and Latinas/Chicanas have been of particular interest to some social analysts. Building on the civil rights and feminist movements of the late 1960s and early 1970s, some contemporary black feminists have focused on the cultural experiences of African American

women. A central assumption of this analysis is that race, class, and gender are forces that simultaneously oppress African American women. The effects of these three statuses cannot be adequately explained as "double" or "triple" jeopardy (race + class + gender = a poor African American woman) because these ascribed characteristics are not simply added to one another. Instead, they are multiplicative in nature

in a position of privilege when compared to people of color (race) and men from lower socioeconomic positions (class), yet be in a subordinate position as compared with a white man (gender) from the capitalist class (Andersen and Collins, 2010). In order to analyze the complex relationship among these characteristics, the lived experiences of African American women and other previously "silenced people" must be heard and examined within the context of particular historical and social conditions.

A classic example of multicultural feminist studies is the work of the psychologist Aida Hurtado (1996), who explored the cultural identification of Latina/Chicana women. According to Hurtado, distinct differences exist between the worldviews of the white (non-Latina) women who participate in the women's movement and many Chicanas, who have a strong sense of identity with their own communities. From this perspective, women of color do not possess the "relational privilege" that white women have because of their proximity to white patriarchy through husbands, fathers, sons, and

FIGURE 11.16 In recent decades, more women have become doctors and lawyers than in the past. How has this affected the way that people "do gender" in settings that reflect their profession? Do professional women look and act more like their male colleagues, or have men changed their appearance and activities at work as a result of having female colleagues?

(race × class × gender); different characteristics may be more significant in one situation than another. For example, a well-to-do white woman (class) may be others. To change this situation, there must be a "politics of inclusion," which might create social structures that lead to positive behavior and bring more people into

CONCEPT QUICK REVIEW

Sociological Perspectives on Gender Stratification

Perspective	Focus	Theory/Hypothesis
Functionalist	Macrolevel analysis of women's and men's roles	Traditional gender roles ensure that expressive and instrumental tasks will be performed. Human capital model
Conflict	Power and economic differentials between men and women	Unequal political and economic power heightens gender-based social inequalities.
Feminist Approaches	Feminism should be embraced to reduce sexism and gender inequality.	1. Liberal feminism 2. Radical feminism 3. Socialist feminism 4. Multicultural feminism

a dialogue about how to improve social life and reduce inequalities.

Evaluation of Conflict and Feminist Perspectives Conflict and feminist perspectives provide insights into the structural aspects of gender inequality in society. These approaches emphasize factors external to individuals that contribute to the oppression of white women and people of color; however, they have been criticized for emphasizing the differences between men and women without taking into account the commonalities that they share. Feminist approaches have also been criticized for their emphasis on male dominance without a corresponding analysis of the ways in which some men may also be oppressed by patriarchy and capitalism. The Concept Quick Review outlines the key aspects of each sociological perspective on gender socialization.

Ken Howard/Alamy

FIGURE 11.17 Latinas have become increasingly involved in social activism for causes that they believe are important. This woman states her belief that we must "Fight Ignorance, Not Immigrants" in establishing policies and laws to protect the U.S. border.

twenty-first century. For example, recent national surveys have shown that the movement toward attitudes of greater gender equality in the United States has slowed and that more people are embracing a new cultural framework of "egalitarian essentialism," which is a blend of feminism equality and traditional motherhood roles (Hermsen, Cotter, and Vanneman, 2011).

In the labor force, gender segregation and the wage gap are still problems. As the United States attempts to climb out of the worst economic recession in decades, job loss has affected both women and men. However, data from the U.S. Bureau of Labor Statistics (2014) show that the wages of the typical woman who has a job have risen slightly faster than those of the typical man. Rather than this being considered a gain for women, some analysts suggest that it is a situation where everyone is losing but that men are simply losing more because of job insecurity or loss, declining real wages, and the loss of benefits such as health care and pension funds.

In the United States and other nations of the world, gender equity, political opportunities, education, and health care remain pressing problems for women. Gender issues and imbalances can contribute not only to individual problems but also to societal problems, such as the destabilization of nations in the global economy. Gender inequality is also an international problem because it is related to violence against women, sex trafficking, and other crimes against girls and women (see ● Figure 11.17). To bring about social change for women, it is important for them to be equal players in the economy and the political process (International Foundation for Electoral Systems, 2011). According to Hillary Rodham Clinton (2011), former U.S. secretary of state, and presidential candidate in 2016,

Governments and business leaders worldwide should view investing in women as a strategy for job creation and economic growth. And while many are doing so,

Looking Ahead: Gender Issues in the Future

Over the past century, women made significant progress in the labor force. Laws were passed to prohibit sexual discrimination in the workplace and school. Affirmative action programs helped make women more visible in education, government, and the professional world. More women entered the political arena as candidates and elected officials instead of as volunteers in the campaign offices of male candidates. And a woman ran for president of the United States in the 2016 election.

Many men joined movements to raise their consciousness, realizing that what is harmful to women may also be harmful to men. For example, women's lower wages in the labor force suppress men's wages as well; in a two-paycheck family, women who are paid less contribute less to the family's finances, thus placing a greater burden on men to earn more money. In the midst of these changes, however, many gender issues remain unresolved in the second decade of the

the pool of talented women remains underutilized, underpaid, and underrepresented overall in business and society. Worldwide, women do two-thirds of the work, yet they earn just one-third of the income and own less than 2 percent of the land.... If we invest in women's education and give them the opportunity to access credit or start a small business, we add fuel to a powerful engine for progress for women, their families, their communities and their countries.

As Clinton suggests, an investment in girls and women, whether in the United States or in other nations of the world, will strengthen other efforts to deal with social problems such as violence against women, inequality, and poverty. However, we must ask this: How will economic problems around the world affect gender inequality in the twenty-first century? What do you think might be done to provide more equal opportunities for girls and women in difficult political, economic, and social times?

CHAPTER REVIEW Q & A

LO1 How do sex and gender differ?

Sex refers to the biological categories and manifestations of femaleness and maleness; gender refers to the socially constructed differences between females and males. In short, sex is what we (generally) are born with; gender is what we acquire through socialization.

LO2 What kinds of prejudice and discrimination occur on the basis of sexual orientation?

Homophobia refers to extreme prejudice and sometimes discriminatory actions directed at gays, lesbians, bisexuals, and others who are perceived as not being heterosexual. Discrimination occurs in many forms, including marital and parenting rights, housing, health care, bank lending policies, and other rights and privileges taken for granted by heterosexual persons.

LO3 What are gender role, gender identity, body consciousness, and sexism?

Gender role encompasses the attitudes, behaviors, and activities that are socially assigned to each sex and that are learned through socialization. Gender identity is an individual's perception of self as either female or male. Body consciousness is how a person perceives and feels about his or her body. Sexism is the subordination of one sex, usually female, based on the assumed superiority of the other sex.

LO4 How does the division of labor between women and men differ in various kinds of societies?

In most hunting and gathering societies, fairly equitable relationships exist between women and men because neither sex has the ability to provide all of the food necessary for survival. In horticultural societies, a fair degree of gender equality exists because neither sex controls the food supply. In agrarian societies, male dominance is overt; agrarian tasks require more labor and physical strength, and females are often excluded from these tasks because they are viewed as too weak or too tied to child-rearing activities. In industrialized societies, a gap exists between

nonpaid work performed by women at home and paid work performed by men and women. A wage gap also exists between women and men in the marketplace.

LO5 What are the primary agents of gender socialization, and what is their role in socializing people throughout life?

Parents, peers, teachers and schools, sports, and the media are agents of socialization that tend to reinforce stereotypes of appropriate gender behavior. From birth, parents act differently toward children on the basis of the child's sex. Peers help children learn gender-role stereotypes, as well as gender-appropriate and gender-inappropriate behavior. Schools operate as gendered institutions, and teachers provide messages about gender through the formal content of assignments and informal interactions. In terms of sports, boys are socialized to participate in highly competitive, rule-oriented games, whereas girls have traditionally been socialized to participate in activities that involve less competitiveness. Recently, however, more girls have started to participate in sports formerly regarded as "male" activities.

LO6 In what ways does the contemporary workplace reflect gender stratification?

Many women work in lower-paying, less prestigious jobs than men. This occupational segregation leads to a disparity, or pay gap, between women's and men's earnings. Even when women are employed in the same job as men, on average they do not receive the same, or comparable, pay.

LO7 How do functionalists and conflict theorists differ in their perspectives on gender inequality?

According to functionalist analysts, women's roles as caregivers in contemporary industrialized societies are crucial in ensuring that key societal tasks are fulfilled. While the husband performs the instrumental tasks of economic support and decision making, the wife assumes the expressive tasks of providing affection and emotional support for the family. According to conflict analysts, the gendered division of labor within families and the workplace—particularly in agrarian and industrial societies—results from male control and dominance over women and resources.

LO8 What are the feminist perspectives on gender inequality?

Feminist perspectives provide insights into the structural aspects of gender inequality in society. In liberal feminism, gender equality is equated with equality of opportunity. Radical feminists often trace the roots of patriarchy to women's childbearing and child-rearing responsibilities, which make them dependent on men. Socialist feminists argue that the oppression of women results from their dual roles as paid and unpaid workers in a capitalist economy. Academics and activists have been rethinking the experiences of women of color from a feminist perspective. The experiences of African American women and Latinas/Chicanas have been of particular interest to some social analysts.

KEY TERMS

body consciousness 310
comparable worth 323
crossdresser 306
feminism 326
gender 309
gender bias 317
gender identity 309

gender role 309
homophobia 308
intersex person 304
matriarchy 312
patriarchy 312
pay gap 322
primary sex characteristics 304

secondary sex characteristics 304
sex 304
sexism 311
sexual orientation 306
sexualization 304
transgender person 305

QUESTIONS for CRITICAL THINKING

1 Do the media reflect societal attitudes on gender, or do the media determine and teach gender behavior? (As a related activity, watch television for several hours, and list the roles for women and men depicted in programs and those represented in advertising.)

2 Examine the various academic departments at your college. What is the gender breakdown of the faculty in selected departments? What is the gender breakdown of undergraduates and graduate students in those departments? Are there major differences among various academic areas of teaching and study? What hypothesis can you come up with to explain your observations?

ANSWERS to the SOCIOLOGY QUIZ

ON GENDER, SEXUAL ORIENTATION, AND WEIGHT BIAS

1 **False** Although both women and men are vulnerable to weight bias and discrimination, research has found that women typically experience higher levels of weight stigmatization than men, even when they have lower levels of excess weight than men.

2 **True** Some studies show that African American women are less likely to be stigmatized by other women of color on the basis of their body size. Other studies show that African American men experience lower levels of stigmatization from both African American men and white (Caucasian) men when compared to the stigmatization experienced by white (Caucasian) men. Additional research is needed before conclusions can be drawn about weight discrimination among people in various racial and ethnic categories.

3	True	More than half of all adult women in the United States are currently dieting, and over three-fourths of normal-weight women think they are "too fat." Recently, very young girls have developed similar concerns.
4	True	Women have been socialized to believe that being physically attractive is very important. Studies have found that weight and body shape are the central determinants of women's perception of their physical attractiveness.
5	False	Gay and bisexual men have been found to be more concerned about their weight and body image than heterosexual men as a category. Consequently, gay or bisexual men are three times more likely to develop eating disorders than their heterosexual counterparts.
6	False	The "ideal" body image for women has changed a number of times. A positive view of body fat has prevailed for most of human history; however, in the twentieth century in the United States, this view gave way to "fat aversion."
7	True	Discussions of eating disorders are more "taboo" among men than women because this illness is perceived as a weakness and a "female problem." However, gay men are up to three times more likely than heterosexual men to have an eating disorder, and many struggle with the problem throughout their lives.
8	True	Women in the United States are bombarded by advertising, television programs, and films containing images of women that typically represent an ideal that most real women cannot attain.

Sources: Based on American Psychological Association, 2010; Grogan, 2007; Obesity Action Coalition, 2015; prideagenda.org, 2014; and Zurbriggen and Roberts, 2012.

AGING AND INEQUALITY BASED ON AGE

Joe McBride/The Image Bank/Getty Images

LEARNING OBJECTIVES

1 **Distinguish** between functional age and chronological age, and explain why this distinction is important in understanding the social significance of age.

2 **Describe** how age has been a factor in determining individuals' roles and statuses in different types of societies.

3 **Explain** what is meant by age stratification and identify how age stratification affects people throughout the life course.

4 **Discuss** ageism and describe how some older people resist it.

5 **Explain** how age is intertwined with gender and race/ethnicity in patterns of social and economic inequality.

6 **Identify** the major forms of elder abuse.

7 **Compare** and contrast functionalist, conflict, and symbolic interactionist perspectives on aging.

8 **Describe** four widely used frameworks for explaining how people cope with the process of dying or with the loss of a loved one.

SOCIOLOGY & EVERYDAY LIFE

Facing Obstacles to Living a Long, Full Life

The women in my family, at least on my mother's side, seem to live long and well. My grandmother Pearl's "third act" was one of worldwide travel and voracious learning. . . . I am a good 20-something years from retirement, yet I find myself thinking often lately of my own third act. In part that is because I am watching my mother . . . tinker with her personal life's script. In keeping with her family legacy, Mom is not one to stand still, and over the years has accumulated a couple of master's degrees, a Ph.D. and a law degree, along with a kaleidoscope of work experience. After my father died two years ago, it seemed only logical that my mother would mourn, then take a few exotic trips and find a more challenging job.

But just because Mom was ready for her third act didn't mean the working world was ready for Mom. Unlike so many career shifts she had made over the years, this one did not go smoothly. Her calls to prospective employers often went unanswered. Her résumé did not always open doors. She looks as if she's in her 50's, but her résumé makes it clear that she's in her 60's, and suddenly the years of experience that have been her greatest strength somehow disqualified her.

Mom did not take this quietly. "I know more than I did 20 years ago, and my brain works as well as it did 30 years ago," she said. The problem, she points out, is not with her generation, but with mine. "Employers—who are your age—have dismissed people who are my age," she said.

How might employers' views on aging affect the lives of older people seeking to be a part of the workforce?

I f we apply our sociological imagination to the issue of aging and work as described by Lisa Belkin, we see that views on age and the problems that people experience in growing older are not just personal problems but also public issues of concern to everyone. In the second decade of the twenty-first century, about 7.9 million persons ages 65 and older are in the labor force, and many more, like Lisa Belkin's mother, are looking for work. A recent study by AARP (2015) found that half of older workers who experienced unemployment in the last five years (between 2009 and 2014) were still not working and that half of those who were able to find jobs were earning less than they did in their former positions. In addition, many older job seekers who were looking for full-time employment settled for part-time work (AARP, 2015).

Eventually, all of us will be affected by aging and possible problems such as finding work when we are considered to be "too old." **Aging** is the physical, psychological, and social processes associated with growing older. The United States has been described as having an aging population, which many younger people see as a looming economic and social burden. As a result, older people may be the targets of prejudice and discrimination based on myths about aging and the emphasis on the aging of the population. When used as a population term, *aging of the population* means that the proportion of people in the older age categories increases. Today, people ages 65 and older make up about 14.5 percent of the total population in the United States. It is estimated that people in this age category will constitute more than 20 percent of the total U.S. population by 2030 and that by 2060, people in the over-65 category will make up slightly more than one in five U.S. residents (Ortman, Velkoff, and Hogan, 2014).

In this chapter we examine the sociological aspects of aging and inequality based on age. We will also examine how older people seek dignity, autonomy, and empowerment in societies such as the United States that may devalue those who do not fit the ideal norms of youth, beauty, physical fitness, and self-sufficiency. Before reading on, test your knowledge about aging and age-based discrimination in the United States by taking the "Sociology and Everyday Life" quiz.

—Journalist Lisa Belkin (2006) describes the problem that many older workers, including her mother, encounter in finding new jobs. However, for Belkin's mother the story had a positive outcome: She eventually found the fulfilling job that she was seeking.

How Much Do You Know About Aging and Age-Based Discrimination?

TRUE	FALSE		
T	F	1	U.S. Supreme Court rulings have made it easier for individuals to show that they have been discriminated against based on their age.
T	F	2	Women in the United States have a longer life expectancy than do men.
T	F	3	Scientific studies have documented the fact that women age faster than men do.
T	F	4	Most older persons are economically secure today as a result of Social Security, Medicare, and retirement plans.
T	F	5	Studies show that advertising no longer stereotypes older persons.
T	F	6	People over age 85 make up one of the fastest-growing segments of the U.S. population.
T	F	7	Organizations representing older individuals have demanded the same rights and privileges as those accorded to younger persons.
T	F	8	The rate of elder abuse in the United States has been greatly exaggerated by the media.

Answers can be found at the end of the chapter.

 L01 **Distinguish** between functional age and chronological age, and explain why this distinction is important in understanding the social significance of age.

The Social Significance of Age

"How old are you?" This is a frequently asked question. Beyond indicating how old or young a person is, age is socially significant because it defines what is appropriate for or expected of people at various stages of life. For example, child development specialists have identified stages of cognitive development based on children's ages:

[W]e do not expect our preschool children, much less our infants, to have adult-like memories or to be completely logical. We are seldom surprised when a 4-year-old is misled by appearances; we express little dismay when our 2½-year-old calls a duck a chicken.... But we would be surprised if our 7-year-olds continued to think segmented routes were shorter than other identical routes or if they continued to insist on calling all reasonably shaggy-looking pigs "doggy." We expect some intellectual (or cognitive) differences between preschoolers and older children. (Lefrançois, 1996: 196)

At the other end of the age continuum, a 75-year-old grandmother wearing short shorts who goes skateboarding through her neighborhood will probably raise eyebrows for her actions because she is defying norms regarding age-appropriate behavior, such as spending time with her grandchild (see ● Figure 12.1).

When people say "Act your age," they are referring to *chronological age*—a person's age based on date of birth.

aging
the physical, psychological, and social processes associated with growing older.

chronological age
a person's age based on date of birth.

FIGURE 12.1 What can people across generations learn by spending time with each other? Can the learning process flow in both directions?

However, most of us actually estimate a person's age on the basis of *functional age*—observable individual attributes such as physical appearance, mobility, strength, coordination, and mental capacity that are used to assign people to age categories. Because we typically do not have access to other people's birth certificates to learn their chronological age, visible characteristics—such as youthful appearance or gray hair and wrinkled skin—may become our criteria for determining whether someone is "young" or "old." Appearance has been identified, more than any other factor, as how the objectification of aging occurs. Most people define someone as old because they "look old." Youthful-appearing celebrities, particularly women, are praised in popular magazines, the Internet, and social media around the world for being people who "defy their age." But is aging the same thing for women and men? Feminist scholars argue that functional age is subjective and that it is evaluated differently for women and men. From this perspective, as men age, people see them as becoming more distinguished or powerful, but as women grow older, they are thought to be "in need of cosmetic surgery," "over the hill," or "grandmotherly."

Trends in Aging in the United States

A major demographic shift is occurring in the United States as the population ages. Because of rising longevity and a drop in fertility following the Baby Boom era between 1946 and 1964, when more than 78 million people were born, the population will age substantially over the next 40 years. The median age (the age at which 50 percent of the population is younger than the median age and

50 percent are older than the median age) increased by more than 7 years—from 30 in 1980 to 37.6 in 2014. For males the estimated median age was 36.3 years, as compared with 39 years for females (*CIA World Factbook*, 2015). This change in median age was partly a result of the Baby Boomers moving into middle age and partly a result of more people living longer. The average length of life increased to 79.5 years in 2014, as compared to 47 years in 1900, because mortality rates have fallen and health at older ages has improved over the last 50 years as disability rates have fallen. Life expectancy for males is 77 years, as contrasted with nearly 82 years for females (*CIA World Factbook*, 2015).

As shown in • Figure 12.2a, the largest percentage change in population by age between 2000 and 2010 (the date of the most recently conducted U.S. national census) was for persons in the 60–64 category, which grew by 55.6 percent. By 2012, more than 81.9 million people were in the 45–64 age category, which equals more than one-quarter (26.5 percent) of the total U.S. population. The number of persons in the age 65+ category also continues to increase, with 13.4 percent of the U.S. population in this age category in 2012 (U.S. Census Bureau, Current Population Survey, Annual Social and Economic Supplement, 2012).

The increase in the number of older people living in the United States and other high-income nations resulted from an increase in life expectancy (greater longevity) combined with a stabilizing of birth rates. *Life expectancy* is the average number of years that a group of people born in the same year can expect to live. Based on the death rates in the year of birth, life expectancy shows the average length of life of a *cohort*—a group of people born within a specified period of time. Cohorts may be established on the basis of one-, five-, or ten-year intervals; they may also be defined by events taking place at the time of their birth, such as Depression-era babies, Baby Boomers, or Millennials. For the cohort born in 2014, as an example, life expectancy at birth was 79.6 years for the overall U.S. population—77.1 for males and 81.9 for females. However, as Figure 12.2b shows, there are significant racial–ethnic and sex differences in life expectancy. Compare, for example, life expectancy at birth for Native Americans and white Americans, as shown in the figure. Although the life expectancy of people of color has improved over the past 60 years, higher rates of illness and disability—attributed to poverty,

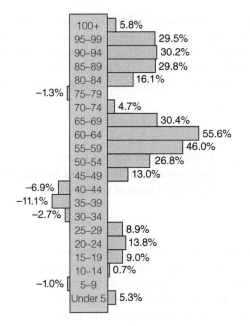

a. U.S. Population Growth, 2000–2010
The percentage of persons 65 years of age and above increased dramatically between 2000 and 2010.

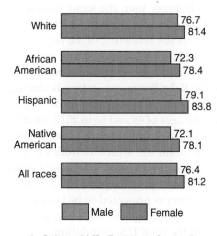

Male   Female

b. Selected Life Expectancies by Race, Ethnicity, and Sex, 2013
There are significant racial–ethnic and sex differences in life expectancy.

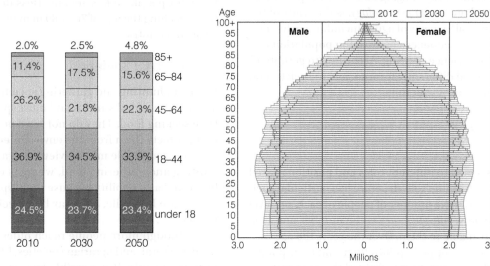

c. Percentage Distribution of U.S. Population by Age, 2000–2050 (projected)
Projections indicate that an increasing percentage of the U.S. population will be over age 65; one of the fastest-growing categories is persons age 85 and over.

d. Population by Age and Sex, 2012, 2030 and 2050
Declining birth rates and increasing aging of the population are apparent in this age pyramid.

FIGURE 12.2 **Trends in Aging and Life Expectancy**
Sources: Howden and Meyer, 2011; Ortman, Velkoff, and Hogan, 2014.

inadequate health care, and greater exposure to environmental risk factors—still persist.

Today, a much larger percentage of the U.S. population is over age 65 than in the past. In fact, one of the fastest-growing segments of the population is made up of people ages 85 and above. By 2050, as Figure 12.2c shows, the Census Bureau predicts that nearly 5 percent of the U.S. population will be made up of people ages 85 and older. Even more astonishing is the fact that the number of

functional age
observable individual attributes such as physical appearance, mobility, strength, coordination, and mental capacity that are used to assign people to age categories.

life expectancy
the average number of years that a group of people born in the same year can expect to live.

cohort
a group of people born within a specified period of time.

centenarians (persons 100 years of age and above) in this country will increase more than 12 times, from 66,000 in 1999 to about 834,000 in 2050.

The distribution of the U.S. population by age and sex in 2012, 2030, and 2050 is depicted in the "age–sex pyramid" in Figure 12.2d. If, every year, the same number of people are born as in the previous year and a certain number die in each age group, the rendering of the population distribution should be pyramid shaped. As you will note, however, Figure 12.2d is not a perfect pyramid but instead demonstrates how the Baby Boom generation is shaping the population now and in the future. The Baby Boom generation was between the ages of 48 and 66 in 2012. By 2030, all Baby Boomers will have moved into the older population, resulting in a shift in the age structure from 13.7 percent of the population in the 65 years of age and older category in 2012 to 20.3 percent of the population in 2030. By 2050, every age category is projected to be larger than it was in 2012, including those ages 85 and over, who will make up 4.5 percent of the total U.S. population, as compared to 1.9 percent in 2012.

As a result of changing population trends, research on aging has grown dramatically in the past sixty years. *Gerontology* is the study of aging and older people. A subfield of gerontology, *social gerontology*, is the study of the social (nonphysical) aspects of aging, including such topics as the societal consequences of an aging population and the personal experience of aging. According to gerontologists, age is viewed differently from society to society, and its perception changes over time.

Age in Historical and Contemporary Perspectives

People are assigned to different roles and positions based on the age structure and role structure in a particular society. *Age structure* is the number of people of each age level within the society; *role structure* is the number and type of positions available to them. Over the years the age continuum has been chopped up into finer and finer points. Two hundred years ago, people divided the age spectrum into "babyhood," a *very* short childhood, and then adulthood. What we would consider "childhood" today was quite different two hundred years ago, when agricultural societies needed a large number of strong arms and backs to work on the land to ensure survival. When 95 percent of the population had to be involved in food production, categories such as toddlers, preschoolers, preteens, teenagers, young adults, the middle-aged, and older persons did not exist.

As we will see in the next section of the chapter, if the physical labor of young people is necessary for society's survival, then young people are considered "little adults" and are expected to act like adults and do adult work. Older people are also expected to continue to be productive for the benefit of the society as long as they are physically able. In preindustrial societies, people of all ages help with the work, and little training is necessary for the roles that they fill. During the seventeenth and eighteenth centuries in the United States, for example, older individuals helped with the work and were respected because they were needed—and because few people lived that long.

Describe how age has been a factor in determining individuals' roles and statuses in different types of societies.

Age in Global Perspective

Physical and sociocultural environments have different effects on how people experience aging and old age. In fact, concepts such as *young* and *old* may vary considerably from culture to culture. Unlike the sophisticated data-gathering techniques used to determine the number of older people in high-income and middle-income nations, we know less about the life expectancies and the aging populations in hunting and gathering, horticultural, pastoral, and agrarian societies. However, reaching the age of 30 or 40 is less likely for people in low-income (less developed) nations than reaching the age of 70 or 80 in many high-income (developed) countries.

Preindustrial Societies

People in hunting and gathering societies are not able to accumulate a food surplus and must spend much of their time seeking food. They do not have permanent housing that protects them from the environment. In such societies, younger people may be viewed as a valuable asset in hunting and gathering food, whereas older people may be viewed as a liability because they typically move more slowly, are less agile, and may be perceived as being less productive.

Although more people reach older ages in horticultural, pastoral, and agrarian societies, life is still very hard for most people. It is possible to accumulate a surplus, so older individuals, particularly men, are often the most privileged in a society because they have the most wealth, power, and prestige. In agrarian societies, farming makes it possible for more people to live to adulthood and to more-advanced years. In these societies the proportion of older people living with other family members is extremely high, with few elderly living alone.

In recent years a growing number of people are reaching ages 60 and above in some middle- and lower-income nations that are in the preindustrial phase or are moving into early industrialization. Consider that India, for example, has about a billion people in its population. If only 5 percent of the population reaches age 60 or above, there will still be a significant increase in the number of older people in that country. Because so much of the world's population resides in India and other low-income nations, the proportion of older people in these countries will increase dramatically during the twenty-first century.

Industrial and Postindustrial Societies

In industrial societies, living standards improve, and advances in medicine contribute to greater longevity for more people. Although it is often believed that less industrialized countries accord greater honor, prestige, and respect to older people, some studies have found that the stereotypical belief that people in such nations will be taken care of by their relatives, particularly daughters and sons, is not necessarily true today.

In postindustrial societies, information technologies are extremely important, and a large proportion of the working population is employed in service-sector occupations in the fields of education and health care, both of which may benefit older people. Some more-affluent older people may move away from family and friends upon retirement in pursuit of recreational facilities or a better climate. Others may relocate to be closer to children or other relatives. The shift from a society that was primarily young to a society that is older will bring about major changes in societal patterns and in the needs of the population. Issues that must be addressed include the health care system, the Social Security system, transportation, housing, and recreation.

A Case Study: Aging in Japanese Society

The older population in Japan has increased significantly over the past 35 years, even though the rapid growth in the age 65+ population is almost done. Between 1980 and 2005, the elderly population more than doubled, but this rate of increase began a sharp decline in about 2010. People ages 65 and over now make up 25.1 percent of

Ryouchin/Taxi Japan/Getty Images

FIGURE 12.3 Recent studies suggest that sociocultural changes and population shifts are diminishing the social importance of older persons in Japan. Why might this be a significant problem for older individuals in a nation where elders previously have been revered?

the total population, and this percentage will continue to go up because the total population in Japan continues to shrink. Japan has a low birth rate, long life expectancy, and one of the highest proportions of people over age 65 in the world. By 2060, it is estimated that 40 percent of the total Japanese population will be age 65 or over. At the same time, the primary working-age population (between 15 and 64) is decreasing as persons in the Baby Boom generation have reached age 65, and many have retired. It is feared that this reduced workforce and the number of older people dependent on younger workers for their retirement and pension benefits will continue to create major problems for the Japanese economy in the future (Jiji, 2014).

What caused this growth of the elderly population? One reason is that people are living longer: Life expectancy in Japan is one of the highest worldwide, with men having an average life expectancy of 79.6 years and women having an average life expectancy of 86.4 years. Another factor is the declining birth rate: Fewer people are marrying, those who marry are waiting longer to have children, and married couples are having fewer children (International Longevity Center, 2015).

In the past it was widely assumed that older people in Japan were respected and revered; however, recent studies suggest that sociocultural changes and population shifts are producing a gradual change in the social importance of the elderly in that nation (● Figure 12.3). This change might be attributed to contemporary demographic patterns, but it may also be related to younger and middle-aged couples working many hours per day and feeling that they have little time to socialize with, or take care of, their parents.

gerontology
the study of aging and older people.

LO3 **Explain** what is meant by age stratification and identify how age stratification affects people throughout the life course.

Age and the Life Course in Contemporary Society

During the twentieth century, life expectancy steadily increased as industrialized nations developed better water and sewage systems, improved nutrition, and made tremendous advances in medical science. However, in the twenty-first century, children are often viewed as an economic liability. Younger children cannot contribute to the family's financial well-being and must be supported. In industrialized and postindustrial societies, the skills necessary for many roles are more complex and the number of unskilled positions is more limited. Consequently, children are expected to attend school and learn the necessary skills for future employment rather than perform unskilled labor. Further, older people are typically expected to retire so that younger people can assume their economic, political, and social roles. However, when economic crises occur and many jobs are lost, age-based inequality tends to increase.

In the United States, age differentiation is typically based on categories such as infancy, childhood, adolescence, young adulthood, middle adulthood, and later adulthood. Chapter 4 examines the socialization process that occurs when people are in various stages of the life course. However, these narrowly defined age categories have had a profound effect on our perceptions of people's capabilities, responsibilities, and entitlement. In this chapter we will look at what is considered appropriate for or expected of people at various ages. These expectations are somewhat arbitrarily determined and produce *age stratification*—inequalities, differences, segregation, or conflict between age groups. We will now examine some of those age groups and the unique problems associated with each one.

Infancy and Childhood

Infancy (birth to age 2) and childhood (ages 3 to 12) are typically thought of as carefree years; however, children are among the most powerless and vulnerable people in society. Historically, children were seen as the property of their parents, who could do with them as they pleased. In fact, whether an infant survives the first year of life depends on a wide variety of parental factors, as a community health scholar explains:

> All infants are not created equal. Those born to teenage mothers or to mothers who smoke

cigarettes, drink alcohol, or take drugs are at higher risk for death in their first year. Those born in very rural areas or in inner cities are more likely to die as infants. . . . But surviving the first year is only one piece of the equation. Quality of life is another. Infants who survive the first year can have lives so compromised that their future is seriously limited. . . . We cannot always predict which infants will survive, and we certainly cannot predict who will be happy. (Schneider, 1995: 26)

Moreover, early socialization plays a significant part in children's experiences and their quality of life (• Figure 12.4). Many children are confronted with an array of problems in their families because of marital instability, an increase in the number of single-parent households, and the percentage of families in which both parents are employed full time. These factors have heightened the need for high-quality, affordable child care for infants and young children.

FIGURE 12.4 People in all cultures understand their responsibility to socialize the next generation. However, the physical arrangements associated with rearing children, including how they are moved from place to place, vary from culture to culture.

However, many parents have few options regarding who will take care of their children while they work. These statistics from the Children's Defense Fund (2014) point out potential problems of infancy and childhood:

> Each day in the United States, 4 children are killed by abuse or neglect; 5 children or teens commit suicide; 7 children or teens are killed by firearms; 1,837 children are confirmed as abused or neglected; 2,723 babies are born into poverty; and 4,028 children are arrested.

As these statistics show, childhood has many perils. In fact, two-thirds of all childhood deaths are caused by injuries. (Cancers, birth defects, heart disease, pneumonia, and HIV/AIDS cause the other third.) Although many previous childhood killers such as polio, measles, and diphtheria are better controlled through immunizations and antibiotics, motor vehicle accidents have become a major source of injury and death for infants and children. Despite laws and protective measures implemented to protect infants and children, far too many lose their lives at an early age because of the abuse, neglect, or negligence of adults.

Adolescence

In contemporary industrialized countries, adolescence roughly spans the teenage years between 13 and 19. Before the twentieth century, adolescence did not exist as an age category. Today, it is a period in which young people are expected to continue their education and perhaps hold a part-time job if they are able to find one. Most states have compulsory school-attendance laws requiring young people between certain ages, usually 6 to 16 or 18 years old, to attend school regularly; however, students who believe that they will receive little or no benefit from attending school or who believe that the money they make working is more important may find themselves labeled as "dropouts" or "juvenile offenders" for missing school. Juvenile laws define behaviors such as truancy and running away from home as forms of delinquency—which would not be offenses if an adult committed them. Despite labor laws implemented to control working conditions for younger employees, many adolescents are employed in settings with hazardous working conditions, low wages, no benefits, and long work hours. Overall, the most significant concerns of teenagers from low-income and/or minority families are the lack of opportunity for education and future employment. Teens from low-income and minority families constitute the majority of young people in jails, prisons, and detention facilities—locations where some lawmakers and law enforcers believe that "out-of-control" adolescents will be less able to harm others.

Why do adolescents experience inequalities based on age in the United States? Adolescents are not granted full status as adults, but they are held more accountable than younger children. Early teens are considered too young to do "adult" things, such as stay out late at night, vote, drive a motor vehicle, use tobacco, or consume alcoholic beverages. In national surveys, teens consistently list the following as potential problems they face: drug and alcohol abuse; sexually transmitted diseases; teen pregnancy; eating disorders such as anorexia, bulimia, and binging; obesity; the presence of gangs in their school and neighborhood; feelings of tiredness and depression; fear of bullying; and excessive peer pressure. Of course, the United States is a diverse nation in terms of class, race/ethnicity, and other social attributes, causing it to be difficult to make generalizations about the problems of all adolescents.

What are teenagers known for today? Teens are often known for their use of smartphones, social media, and other digital technology. Rather than having face-to-face conversations with others, many teens prefer texting or talking on their phone (Lenhart, 2015). According to a recent study, nearly 75 percent of teens own or have access to a smartphone. The most widely used social media sites for teens are Facebook, Instagram, and Snapchat. Overall, teenage girls use social media sites more than their male counterparts. Males are more likely to play video games, and a higher percentage of boys than girls own or have access to a game console. Both boys (84 percent in the survey) and girls (59 percent in the survey) play video games online or on their phones.

In 2015 about 91 percent of teens with cell phones used texting as a routine way to communicate with their family and friends. Messaging apps such as Kik and Whats App have become very popular among teens, particularly Hispanic and African American young people. On average, teens with cell phones receive and/or send about 30 text messages each day, with girls sending more—typically about 40 per day—as compared with boys (Lenhart, 2015). Texting and extensive use of Facebook and Twitter may become areas of conflict between teens and parents. Texting while driving is of great concern, and more states are passing laws prohibiting this behavior. Extensive social media use has produced concerns about teen safety and the growing problem of bullying by friends and acquaintances. However, many parents want their children to have a cell phone because they believe that it makes them safer and because parents can more easily stay in touch with them. Overall, teens who are actively involved with their families and friends and who give their time to volunteer work indicate that they are more satisfied with their lives and have higher self-esteem than those who are uninvolved. The teen years are supposed to lead adolescents toward a happier and more productive young adulthood (• Figure 12.5).

age stratification
inequalities, differences, segregation, or conflict between age groups.

Syda Productions/Shutterstock.com

FIGURE 12.5 The college years serve as an important bridge between adolescence and young adulthood for many young people.

Young Adulthood

Young adulthood, which follows adolescence and lasts to about age 39, is socially significant. As shown in • Figure 12.6, the U.S. median age was 37.2 in 2010; however, distribution of the U.S. population based on age varies from state to state. This is the age span in which many people are healthier and more energetic, have goals in life to which they aspire, and express enthusiasm for the future. During the earlier years of young adulthood, identity formation is important. Many young adults seek satisfactory sexual relations and hope to find a life partner. Self-growth is an important concern for many young adults today.

In the past the standard expectation for these years was that individuals would get married, have children, and hold down a meaningful job. People who did not fulfill these activities during young adulthood were viewed negatively. However, this is not necessarily true today. Problems in the larger society and global economy have made it more difficult for young adults to find permanent, well-paid employment and to have the necessary resources to get married, purchase a residence independent of one's parents, and have children. Although the middle to late thirties have been considered a time for "settling down," this is not universally true today, as a growing number of media reports tell of young adults moving back home with their parents or making arrangements to "double up" with other relatives or friends because of high rates of job loss, mortgage foreclosures, and other economic crises reflecting national and international patterns.

High levels of debt and financial stress are common among young adults. Those who do not attend or graduate from college are faced with limited employment opportunities. Some who complete college find that they owe large student loans but have limited financial resources to pay them off. Lack of affordable housing and expensive food and transportation costs also contribute to financial stress among young adults. For some, finding a job is more difficult than for others. Race/ethnicity, class, and gender influence people's career opportunities. People who are unable to earn income have both present and future problems: They have problems living in the here and now, but they also cannot save money or pay into retirement plans and Social Security, which leaves them further disadvantaged as they enter middle and late adulthood.

Middle Adulthood

Prior to the twentieth century, life expectancy in the United States was only about 47 years, so the concept of middle adulthood—people between the ages of 40 and 65—did not exist until fairly recently. Middle adulthood for some people represents the time during which (1) they have the highest levels of income and prestige, (2) they leave the problems of child rearing behind them and are content with their spouse of many years, and (3) they may have grandchildren, who give them another tie to the future. Job satisfaction and commitment to work are relatively high, and individuals enjoy what they are doing. Some people have adjusted their aspirations to goals that are attainable; others may have a feeling of sadness and frustration over unaccomplished goals. For many in middle adulthood, social stability is based on enduring relationships with spouses, friends, and workplace colleagues.

A major characteristic of middle adulthood is the physical and biological changes that occur in the body. Normal changes in appearance occur during these years, and although these changes have little relationship to a person's health or physical functioning, they are socially significant to many people. As people progress through middle adulthood, they experience *senescence* (primary aging) in the form of molecular and cellular changes in the body. Wrinkles and gray hair are visible signs of senescence.

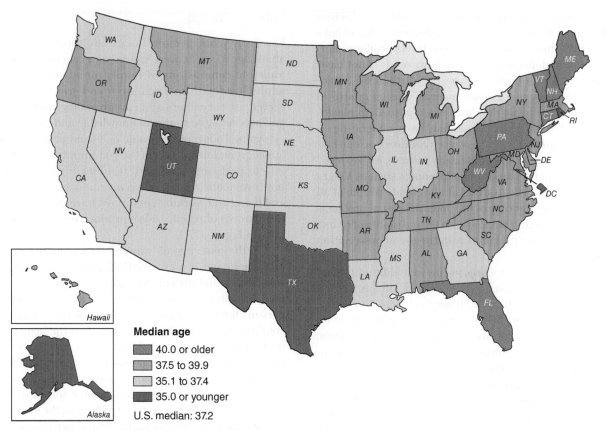

FIGURE 12.6 Median Age by State, 2013
Source: U.S. Census Bureau Newsroom, 2014.

Median age
- 40.0 or older
- 37.5 to 39.9
- 35.1 to 37.4
- 35.0 or younger

U.S. median: 37.2

Less visible signs include arthritis and a gradual dulling of the senses of taste, smell, touch, and vision. Typically, reflexes begin to slow down, but the actual nature and extent of change vary widely from person to person. And stereotypes and self-stereotypes about aging may often be important in determining changes in a person's work ability.

Consider this question: Is work ability strictly related to a person's age? The Norwegian sociologist Per Erik Solem (2008) explored the influence of the work environment on how age-related subjective changes occurred in people's work ability. He found that although age and physical health are obviously associated with decline in the work ability of older individuals, stereotypes and self-stereotypes (negative assumptions about themselves that are embraced by individuals who are being stereotyped) about aging are also important factors in producing a decline in a person's work ability. According to Solem (2008: 44), objective capacities such as physical strength and endurance may be important in performing certain tasks, such as lifting heavy equipment or nursing bedridden patients; however, "what workers believe they are able to do . . . influences to what extent they use their potential of objective abilities." In other words, if older individuals *subjectively* define themselves as less able than they actually are, given their *objective* potential, then they may be less likely to perform a task that might be quite within their capabilities to do. Although physical changes associated with aging

will always be a factor in some types of job performance, particularly those requiring quick reaction times or heavy physical labor, many occupations could benefit from older workers' experience and expertise.

People also experience a change of life in middle adulthood. Women undergo *menopause*—the cessation of the menstrual cycle caused by a gradual decline in the body's production of the "female" hormones estrogen and progesterone. Menopause typically occurs between the midforties and the early fifties and signals the end of a woman's childbearing capabilities. Some women may experience irregular menstrual cycles for several years, followed by hot flashes, retention of body fluids, swollen breasts, and other aches and pains. Other women may have few or no noticeable physical symptoms. The psychological aspects of menopause are often as important as any physical effects. In one study, Anne Fausto-Sterling (1985) concluded that many women respond negatively to menopause because of negative stereotypes associated with menopausal and postmenopausal women. These stereotypes make the natural process of aging in women appear abnormal when compared with the aging process of men. Actually, many women experience a new interest in sexual activity because they no longer have to worry about the possibility of becoming pregnant. On the other hand, more women have recently chosen to produce children using new medical technologies even after they have gone through menopause.

Men undergo a *climacteric,* in which the production of the "male" hormone testosterone decreases. Some have argued that this change in hormone levels produces nervousness and depression in men; however, it is not clear whether these emotional changes are attributable to biological changes or to a more general "midlife crisis," in which men assess what they have accomplished. Ironically, even as such biological changes may have a liberating effect on some people, they also reinforce societal stereotypes of older people, especially women, as "sexless." Recently, intensive marketing of products such as Viagra and Cialis for erectile dysfunction has made people of all ages more aware not only of the potential sexual problems associated with aging but also with the possibility of reducing or solving these problems with the use of prescription drugs.

Along with primary aging, people in middle adulthood also experience *secondary aging,* which occurs as a result of environmental factors and lifestyle choices. For example, smoking, drinking heavily, and engaging in little or no physical activity are factors that affect the aging process. People who live in regions with high levels of environmental degradation and other forms of pollution are also at greater risk of aging more rapidly and having chronic illnesses and diseases associated with these external factors.

In middle adulthood, many people face family-related issues such as the empty nest syndrome, divorce, death of a spouse, remarriage, or learning that they are part of the "sandwich generation," which means that they not only have responsibility for their children but also for taking care of their aging parents. The term "club sandwich generation" was coined to describe people who are sandwiched between their aging parents, their children, and their grandchildren (Abaya, 2011). The burden of these responsibilities disproportionately falls on women: "It's not like your husband is going to wash his own mother," a female forty-two-year-old former lawyer noted. "There's just no amount of feminism that is going to change that" (Vincent, 2004).

Middle adulthood has both high and low points. Given society's current structure, many people know that their status may begin to change significantly when they reach the end of this period of their lives. Those who had few opportunities available earlier in life tend to become increasingly disadvantaged as they grow older.

Late Adulthood

For many years, persons in late adulthood, beginning at about age 65 or 70, were viewed as a crucial link between multiple generations of family members. Many older adults lived with their children and grandchildren and served as storytellers and caregivers for young children and as a valuable source of knowledge and experience for everyone. Today, more people in late adulthood are living longer, remaining active in the community or employed in full- or part-time work, and residing in their own home. Improvements in health care, nutrition, exercise, and general living and working conditions have greatly increased the chances of people to live productively into late adulthood (● Figure 12.7).

In the past, age 65 was referred to as the "normal" retirement age; however, with changes in Social Security regulations that provide for full retirement benefits to be paid only after a person reaches 67 years of age, many older persons have either chosen, or been forced by economic conditions, to retire later. *Retirement* is the institutionalized separation of an individual from an occupational position, with continuation of income through a retirement pension based on prior years of service. Retirement means the end of a status that has long been a source of income and a means of personal identity. Perhaps the loss of a valued status explains why many retired persons introduce themselves by saying "I'm retired now, but I was a (banker, lawyer, plumber, supervisor, and so on) for 40 years."

Some gerontologists subdivide late adulthood into three categories: (1) the "young-old" (ages 65 to 74), (2) the "old-old" (ages 75 to 85), and (3) the "oldest-old"

FIGURE 12.7 Exercise is one way to offset the natural process of aging throughout life. What other activities might keep older individuals physically and mentally fit?

(over age 85). Although these are somewhat arbitrary divisions, the "young-old" are less likely to suffer from disabling illnesses, whereas some of the "old-old" are more likely to suffer such illnesses. The 2010 U.S. Census (the latest available) counted more than 53,000 people age 100 and older in the United States, so it is possible that the age categories previously identified in late adulthood will be changed. More than 80 percent of all centenarians are women: For every 100 centenarian women, there are only 20.7 centenarian men (Meyer, 2012). Many persons ages 100 and over live in nursing homes or other residential-care facilities, primarily located in urban areas in states such as California, New York, Florida, and Texas. A newer category, *supercentenarians*—persons who have lived to age 110 or more—has been identified by researchers, but only one in every thousand centenarians reaches this advanced-age category.

So how important do you think chronological age is? The rate of biological and psychological changes in older persons may be as important as their chronological age in determining how they and others perceive them. As adults grow older, they actually become shorter, partly because bones that have become more porous with age develop curvature. A loss of three inches in height is not uncommon. As bones become more porous, they also become more brittle; simply falling may result in broken bones that take longer to heal. With age, arthritis increases, and connective tissue stiffens joints. Wrinkled skin, "age spots," gray (or white) hair, and midriff bulge appear; however, people sometimes use "miracle" creams and cosmetic surgery in the hope of avoiding looking older.

Older persons also have increased chances of heart attacks, strokes, and cancer, and some diseases affect virtually only persons in late adulthood. Alzheimer's disease (a progressive and irreversible deterioration of brain tissue) is an example. One in three older Americans dies with Alzheimer's or another dementia. It currently is the sixth-leading cause of death in the United States, and almost two-thirds of people diagnosed with Alzheimer's disease are women (Alzheimer's Association, 2015). Persons with this disease have an impaired ability to function in everyday social roles; eventually, they cease to be able to recognize people they have always known and lose all sense of their own identity. Finally, they may revert to a speechless, infantile state such that others must feed them, dress them, sit them on the toilet, and lead them around. The disease can last up to 20 years; currently, there is no cure. An estimated 5.3 million people of all ages have Alzheimer's disease, and 5.1 million of these are people ages 65 and older. About one in nine people ages 65 and older (11 percent) has Alzheimer's disease, and the percentage increases to 32 percent of people ages 85 and over (Alzheimer's Association, 2015). Fortunately, many older individuals do not suffer from Alzheimer's and are not incapacitated by their physical condition.

Although older people experience some decline in strength, flexibility, stamina, and other physical capabilities, much of that decline does not result simply from the aging process and is avoidable; with proper exercise, some of it is even reversible. With the physical changes come changes in the roles that older adults are expected (or even allowed) to perform. For example, people may lose some of the abilities necessary for driving a car safely, such as vision or reflexes. The issue of older drivers has been widely debated in the media and political arenas after serious accidents were caused by a few older drivers (see "Sociology and Social Policy").

The physical and psychological changes that come with increasing age can cause stress. According to Erik Erikson (1963), older people must resolve a tension of "integrity versus despair." They must accept that the life cycle is inevitable, that their lives eventually will end, and that achieving inner harmony requires accepting both one's past accomplishments and past disappointments. And this is what the later years mean for many older adults: a chance to enjoy life and participate in leisure activities.

Will the life stages as we currently understand them accurately reflect aging in the future? Research continues to show that there are limited commonalities between those who are age 65 and those who are centenarians; however, many people tend to place everyone from 65 upward in categories such as "old," "elderly," or "senior citizen." In the future we will probably see such categorizations revised as growing numbers of older people reject such labels as forms of "ageism." Some analysts believe that the existing life-course and life-stages models will be modified to reflect a sense of "old age" beginning at age 75 or 80 while new stages will be added for those who reach 100 and over.

Inequalities Related to Aging

In previous chapters we have seen how prejudice and discrimination may be directed toward individuals based on ascribed characteristics—such as race/ethnicity or gender—over which they have no control. The same holds true for age.

 LO4 **Discuss** ageism and describe how some older people resist it.

Ageism

Stereotypes regarding older persons reinforce *ageism*, defined in Chapter 4 as prejudice and discrimination against people on the basis of age, particularly against older persons. Ageism against older persons is rooted in the assumption that as people grow older, they become unattractive, unintelligent, asexual, unemployable, and mentally incompetent.

Ageism is reinforced by stereotypes, whereby people have narrow, fixed images of certain groups. One-sided and exaggerated images of older people are used repeatedly in everyday life. Older persons are often stereotyped

Driving While Elderly: An Update on Policies Pertaining to Mature Drivers

You are 84 years old and yesterday you killed my son. I don't know if you're even sound enough to understand what you've done but I have no sympathy for you or any of your family. And I pray EVERY day that you remain on this planet you see my son when you close your eyes. . . . We very much support a mandatory limit on the driving age for seniors.

—Amanda Wesling, the mother of an eight-year-old boy killed when an elderly driver drove her car into the boy's elementary school, expressing her anger and frustration that the woman was still allowed to drive (qtd. in Suhr, 2007).

Ironically, the driver involved in this deadly accident was on her way to a driving class at a senior citizens' center because it was time to renew her driver's license. When an accident such as this occurs, journalists and social media bloggers often focus on the age of the driver and make generalizations about how older drivers are unsafe on the roads. To update this feature for the eleventh edition, I Googled "elderly driver AND accident," and a number of articles appeared. I was greeted by headlines such as "After Fatal Crash, Questions Arise About Elderly Driver," "Elderly Driver Kills Three After Church in Car Accident," and "Accident Puts Spotlight on Senior Drivers." Each article suggested that the older driver was at fault, or at least was negligent, in the fatal crash.

It is true that drivers over age 70 are keeping their licenses longer and driving more miles than earlier generations. This trend has led to dire predictions that aging Baby Boomers will have higher rates of motor vehicle crashes. However, research from the Insurance Institute for Highway Safety shows that fatal car accidents involving older drivers have actually declined in the past decade.

Most of us can recall the time when we anxiously awaited our next birthday so that we would be old enough to get our first driver's license. All states in the United States have a minimum age for getting a learner's permit or a first driver's license, but far fewer states have policies regarding drivers over the age of 65. However, this is changing with the aging of the U.S. population. By 2020, nearly 50 million Americans over

What issues are involved in social policies pertaining to older drivers?

Pavel L Photo and Video/Shutterstock.com

as thinking and moving slowly; as being bound to themselves and their past, unable to change and grow; as being unable to move forward and often moving backward. They are viewed as cranky, sickly, and lacking in social value; as egocentric and demanding; as shallow, enfeebled, aimless, and absentminded.

As previously discussed, popular media and the marketing of products created for "senior citizens" contribute to negative images of older persons, many of whom are portrayed as being wrinkled, suffering from erectile dysfunction, needing better bladder control, and having chronic depression. Fortunately, some members of the media are making an effort to draw attention to the positive contributions, talents, and physical stamina of many older persons rather than focusing primarily on negative portrayals. For example, some media sources highlight artistic, literary, and athletic accomplishments of older people, such as individuals over age 80 who participate in marathons and other sporting events.

Despite some changes in media coverage of older people, many younger individuals still hold negative stereotypes of "the elderly." In a now-classic study, William C. Levin (1988) showed photographs of the same man (disguised to appear as ages 25, 52, and 73 in various photos) to a group of college students and asked them to evaluate these (apparently different) men for employment purposes. Based purely on the photographs, the "73-year-old" was viewed by many of the students as being less competent, less intelligent, and less reliable than the "25-year-old" and the "52-year-old." Although this study was conducted more than two decades ago, the findings remain consistent with contemporary studies of ageism and age-based stereotypes. Social psychologist and gerontologist Becca Levy's (2009) research also shows that ageist stereotypes persist and that a causal link may exist between such stereotypes and various outcomes for older people. A self-fulfilling prophecy may occur in regard to memory loss, cardiac reactivity to stress, and decreased longevity because older people have been bombarded by negative images of aging.

Many older people resist ageism by continuing to view themselves as being in middle adulthood long after their

age 65 will be eligible for driver's licenses, and half of them will be age 75 or older. Mature drivers sometimes face impairments in three functions that may affect their driving abilities:

1. *Vision*. Adequate visual acuity and field of vision are necessary for safe driving but tend to decline with age. Glare, impaired contrast sensitivity, and increased time needed to adjust to changes in light levels are problems commonly experienced by mature drivers.

2. *Cognition*. Driving requires a variety of high-level cognitive skills, including memory, visual processing, and attention. Medical conditions (such as Alzheimer's) and medications often prescribed for the older population may affect cognitive levels.

3. *Motor function*. Motor abilities such as muscle strength, endurance, and flexibility are necessary for operating vehicle controls and turning. Changes related to aging such as arthritis or medical occurrences such as strokes can decrease an individual's ability to drive safely. (Governors Highway Safety Association, 2011)

Although licensing and renewal policies vary throughout the United States, more states are implementing special renewal procedures for older drivers. Some states require one or more of the following for license renewal if a driver is over a specific age (usually 65 or 70):

- Shorter renewal intervals will be required for driver's licenses.

- Renewals must be done in person rather than electronically or by mail.

- Vision, reaction, and/or road tests (not routinely required of younger drivers) may be required for renewal.

Clearly, many issues are involved in social policies pertaining to older drivers. On the one hand, older drivers point out that most cities lack adequate public transportation and that not being able to drive makes them a burden on other people. On the other hand, pedestrians, other drivers, and the general public have a vested interest in not having individuals of *any* age drive who are unsafe and who might, for whatever reason, constitute a threat to others.

As we think about social policy pertaining to older people, particularly in regard to the right to drive, it is important to have empathy for people who will be affected by rules requiring them to prove their competence, but it is also important to remember that if we are fortunate to live long enough, each of us will also face such questions as we reach later stages in our life.

REFLECT & ANALYZE

What safety issues are similar for young drivers and older drivers? How can a nation balance the rights of individuals to engage in activities such as driving a motor vehicle with the need to protect the safety of others? What part do social rules play in dealing with issues such as these?

actual chronological age would suggest otherwise (• Figure 12.8). In one study of people ages 60 and over, 75 percent of the respondents stated that they thought of themselves as middle-aged and only 10 percent viewed themselves as being old. When the same people were interviewed again 10 years later, one-third still considered themselves to be middle-aged. Even at age 80, one out of four men and one out of five women said that the word *old* did not apply to them; this lack of willingness to acknowledge having reached an older age is a consequence of ageism in society.

Tina Stallard/Getty Images

FIGURE 12.8 Many older people turn to cosmetic surgery and Botox injections to reduce the effects of aging on their appearance. The woman shown here had a face-lift two months before this photo was taken and is now admiring her new image in the mirror. Do negative stereotypes about growing older contribute to the desire to defy the aging process? What do you think?

can afford plastic surgery, exercise classes, and social activities such as ballroom dancing or golf, and that they have available time and facilities to engage in pursuits that will "keep them young." However, many older people with meager incomes, little savings, and poor health, as well as those who are isolated in rural areas or high-crime sections of central cities, do not have the same opportunities to follow popular recommendations about successful aging. In fact, for many older people of color, aging is not so much a matter of seeking to defy one's age but rather of attempting to survive in a society that devalues both old age and minority status.

If we compare wealth (all economic resources of value, whether they produce cash or not) with income (available money or its equivalent in purchasing power), we find that older people tend to have more wealth but less income than younger people. The real median income of households with householders ages 65 and older was $35,611 in 2013. Older people are more likely to own a home that has increased substantially in market value; however, some may not have the available cash to pay property taxes, to buy insurance, and to maintain the property. Among older persons, a wider range of assets and income is seen than in other age categories. For example, many people on the *Forbes* list of the richest people in this country (see Chapter 8) are over 65 years of age. On the other hand, about 9.5 percent of all people 65 and over live in poverty. • Figure 12.10 shows the percentage of persons ages 65 and over who live below the poverty level, which has ranged from about $11,770 for a one-person household to somewhere in the $24,250 area for a family of four in 2015. Poverty rates are also higher for women ages 65 and older (11.6 percent) than for men (6.8 percent). This makes gender differences in poverty rates more pronounced for those ages 65 and older than in other age categories (DeNavas-Walt and Proctor, 2014).

FIGURE 12.9 How have tablets, smartphones, and other communications technologies helped older people keep in touch with family, friends, and the social world around them? Can social networking help older people resist ageism?

Older people may also resist ageism by maintaining positive relationships with others and keeping up with newer technologies that help them communicate with children, grandchildren, and friends (• Figure 12.9). Studies have found that social networking use among people ages 50 and older has nearly doubled in recent years. Although social media use has grown dramatically across all age groups, older users are particularly enthusiastic about social media networking. Social media use among Internet users ages 65 and older grew 150 percent between 2009 and 2012 as older users found that this form of communication provided them with a way to connect with children, grandchildren, other relatives, and longtime friends. When older people share photos, videos, and status updates with others, they feel more in touch with others even if they live alone. But other seniors remain relatively unattached from online and mobile life, according to a recent Pew Research Center survey which found that 41 percent of persons ages 65 and older do not use the Internet at all, 53 percent do not have broadband access at home, and 23 percent do not use cell phones (Smith, 2014). Among younger seniors who have higher incomes and are more highly educated, there is much greater use of Internet, broadband, and social media, showing once again a linkage between socioeconomic status and other aspects of how people of all ages live their lives.

 Explain how age is intertwined with gender and race/ethnicity in patterns of social and economic inequality.

Age, Gender, and Inequality

Age, gender, and poverty are intertwined. Men and women over the age of 50 have experienced problems because of the recent economic crisis in the United States and other nations linked to the global economy. Many older employees have lost jobs as businesses have closed or the number

Wealth, Poverty, and Aging

Many of the positive images of aging and suggestions on how to avoid the most negative aspects of ageism are based on an assumption of class privilege, meaning that people

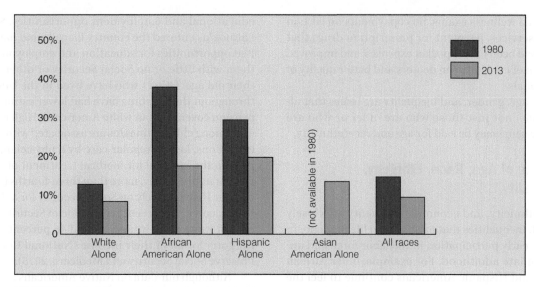

FIGURE 12.10 Percentage of Persons Age 65+ Below the Poverty Level

Source: U.S. Census Bureau, Historical Poverty Tables, 2015.

of employees has been reduced. Despite many years of experience, older workers find it more difficult to get new positions. It is often assumed that they are overqualified, that they are too expensive to hire, or that insurance and other benefits will be more costly for older workers.

While some women have made gains in recent decades, older women in the workplace tend to earn less than men their age, and they are more likely to work in gender-segregated jobs that provide less income and fewer opportunities for advancement (see Chapter 11). As a result, women do not garner economic security for their retirement years at the same rate that many men do. These factors have contributed to the economic marginality of the current cohort of older women. Many women who are over age 70 spent their early adult lives as financial dependents of husbands or as working, nonmarried women trying to support themselves in a culture that did not see women as the heads of households or as sole providers of family income. Women who stayed home to raise a family or to care for an aging parent or sick child typically accumulated far less in Social Security credits toward retirement than did their male counterparts.

Women also have a greater risk of poverty in their later years; statistically, women tend to marry men who are older than themselves, and women live longer than men. As a result, women over age 65 are about three times as likely to be widowed as men, 40 percent as compared to 13 percent. Many widowed persons are living on a fixed income from sources such as Social Security or a spouse's retirement benefits. As shown in Figure 12.10, the percentage of persons ages 65 and older living below the poverty line decreased significantly between 1980 and 2013, but major discrepancies still exist across racial–ethnic categories. This decrease over three decades largely resulted from increasing benefit levels of *entitlements*—certain benefit payments made by the government, including Social

Security, Supplemental Social Income (SSI), Medicare, Medicaid, and civil service pensions, which are the primary source of income for many persons over age 65. Ninety percent of all people in the United States over age 65 receive Social Security benefits, and Social Security is the major source of income for most of the older population. Among elderly Social Security beneficiaries, nearly one-fourth of married couples and slightly under half of all unmarried persons rely on Social Security for 90 percent or more of their income. However, Social Security benefits are not as high as many people believe: In 2015 the average Social Security retirement benefit for a retired worker was $1,331 a month, or $15,972 a year (Social Security Administration, 2015).

Medicare, the other of the two largest entitlement programs, is a nationwide health care program for persons ages 65 and older who are covered by Social Security or who are eligible to "buy into" the program by paying a monthly premium. Different parts of Medicare help cover specific health services. Part A is Hospital Insurance, which helps cover inpatient care in hospitals. It also helps cover skilled-nursing facilities, hospice, and home health care. Part B is Medical Insurance, which helps cover doctors' services, hospital outpatient care, and home health care. Part D is Prescription Drug Coverage, which is an option run by Medicare-approved private insurance companies to help cover the cost of prescription drugs. Some sections of Medicare have changed as a result of the 2010 Affordable Care Law. The federal government is taking strong action to reduce payment errors, waste, fraud, and abuse in Medicare. In addition to reforming abuses in Medicare, the Affordable Care Law provides benefits to seniors such

entitlements
certain benefit payments made by the government.

as free annual wellness exams, no copayments on certain preventive services, payment for prescription drugs that previously had been out-of-pocket expenses, and improved coordination of care between doctors and better quality of care at hospitals.

In sum, age, gender, and inequality are issues that affect everyone, not just those who are older or who are women. The same may be said for age and race/ethnicity.

Intertwining of Age, Race/Ethnicity, and Inequality

Age, race/ethnicity, and economic inequality are closely intertwined. Inequalities that exist later in life originate in individuals' early participation in the labor force and are amplified in late adulthood. For example, older African Americans and Hispanic Americans continue to feel the impact of unequal educational opportunities, income inequality, and historical patterns of job discrimination. African Americans account for 17.6 percent of older Americans in poverty, while older Hispanics make up nearly one-fifth (19.8 percent) of older Americans in poverty. By contrast, white Americans make up 8.4 percent of persons ages 65 and over in poverty.

As previously discussed, gender also intersects with age and race/ethnicity in regard to levels of inequality. Older white American (non-Hispanic) women and African American women have higher poverty rates than their male counterparts. Race and ethnicity are particularly related to poverty among men ages 65 and over. For example, older white (non-Hispanic) men are less likely to live in poverty than older men who are African American, Hispanic, or Asian American. For example, 5 percent of white American men ages 65 and over live in poverty, as sharply contrasted with 14 percent each for older African American, Hispanic, and Asian American men. The data for women are more disparate: Eight percent of white (non-Hispanic) women ages 65 and older live in poverty, as compared with more than 21 percent each for older African American and Hispanic (Latina) women, and 27.1 percent for older Native American women (National Women's Law Center, 2013).

Why do some persons in some older racial–ethnic minority categories have higher rates of poverty than others? As previously noted, the primary reason for the lower income status of many older African Americans can be traced to a pattern of limited employment opportunities and periods of unemployment throughout their lives, combined with their concentration in secondary-sector jobs, which pay lower wages, are sporadic, have few benefits, and were not covered by Social Security prior to the 1950s. Moreover, health problems may force some African Americans out of the labor force earlier than other workers because of a higher rate of chronic diseases such as hypertension, diabetes, and kidney failure.

Similarly, older Hispanics (Latinas/os) have higher rates of poverty than whites (Anglos) because of a lack of educational and employment opportunities. Some older Latinos/as entered the country illegally and have had limited opportunities for education and employment, leaving them with little or no Social Security or other benefits in their old age. Others who have lived in the United States throughout their lifetime have had lower earnings and less pension coverage than white Americans. High rates of poverty among older Latinas/os are associated with poor health conditions, lack of regular care by a physician, and fewer trips to the hospital for medical treatment of illness, disease, or injury. Today, more than three-fourths (77 percent) of older Hispanics rely on Social Security for at least half of their income, 55 percent rely on Social Security for 90 percent or more of their income, and 46 percent rely on this program for all of their income (National Committee to Preserve Social Security and Medicare, 2015).

Although our data on Native Americans are more limited, research has shown that older Native Americans are among the most disadvantaged of all categories. Nationwide, the highest poverty rates are for American Indians (Native Americans), and older Native Americans are more likely to live in high-poverty, rural areas than are other minority older populations. In addition to experiencing educational and employment discrimination similar to that of African Americans and Latinos/as, older Native Americans were also the objects of historical oppression and federal policies toward the native nations that exacerbated patterns of economic impoverishment, as discussed in Chapter 10.

Shifting our focus to older Asian Americans, many of whom migrated to the United States prior to and during World War II (between 1939 and 1945) or during the Vietnam War era (1955 to 1975), we find that many in this category have fared less well in their old age than Asian Americans who were either born in the United States or have arrived more recently. As discussed in Chapter 10, many older Asian Americans from Japan and China received less education and experienced more economic deprivation than did later cohorts of Japanese Americans and Chinese Americans. Today, many of the oldest Asian Americans remain in ethnic enclaves such as Chinatown, Japantown, and Koreatown, where others speak their language and provide goods and services that help them maintain their culture, and where mutual-aid and benevolent societies and recreational clubs provide them with social contacts and delivery of services. For Asian Americans who arrived as refugees from Southeast Asia, either during or after the Vietnam conflict, and for Pacific Islanders, who often reside in Hawaii, low-wage jobs and social marginality originally contributed to higher rates of poverty and greater reliance on Social Security income in retirement. As is evident from this discussion, the economic needs and concerns of persons ages 65 and older have both commonalities and differences in regard to income, poverty, and benefit programs such as Social Security. Much more research is needed on the unique needs of older people from diverse racial and ethnic categories.

Alison Wright/Encyclopedia/Corbis

Clive Brunskill/Getty Images

FIGURE 12.11 Age, race, and inequality are intertwined in both rural and urban settings when we examine the lifestyles and opportunities of people through a sociological lens. What are some of the unique problems experienced by older individuals who reside in rural areas?

Older People in Rural Areas

The lives of many older people differ based on whether they reside in urban or rural areas (● Figure 12.11). Rural areas are those counties that are outside the boundaries of metropolitan areas, which include central counties with one or more urban areas together with outlying counties economically tied to the central counties as measured by work commuting. The terms *rural* and *nonmetro* are often used interchangeably today (Farrigan and Parker, 2012). Some older individuals who reside in rural or nonmetro areas are middle- and upper-income people who have chosen to retire away from the noise and fast-paced lifestyle of urban areas. Others have lived in rural areas throughout their lives and have limited resources on which to survive. Despite the stereotypical image of the rural elderly living in a pleasant home located in an idyllic country setting, many rural elders, as compared with older urban residents, typically have lower incomes, are more likely to be poor, and have fewer years of schooling.

Nearly one-third (32.4 percent) of the rural elderly live in high-poverty counties and tend to be in poorer health. Many of these people are less likely to receive adequate medical care because rural areas typically lack adequate health and long-term care facilities. Specific health issues

among the rural elderly include inadequate nutrition, obesity, and illnesses such as type 2 diabetes and Alzheimer's. Although health-related problems such as these are experienced by many older individuals in both metro and nonmetro areas, the concerns of elderly rural residents are often intensified because of inaccessibility to high-tech medical facilities and lack of transportation for routine doctors' appointments or other health checkups (usda.gov, 2012).

Whether or not they are poor, the rural elderly receive a higher proportion of their income from Social Security payments than do the urban elderly. Housing also differs between rural and urban elderly, with older rural residents being more likely to own their own homes than older residents in urban settings. However, the homes of the elderly in rural areas are more likely to have a lower value in the real estate market and to be in greater need of repair than are those owned by urban elderly residents.

The economic recessions and the slow pace of recovery have further contributed to the problems of older people in rural areas, particularly among persons in minority groups who are exposed to high poverty. Nonmetro elderly African Americans, Hispanic Americans, and Native Americans have been especially hard hit by the migration of their children and grandchildren to metro areas in hopes of finding jobs; environmental challenges, including lack of proper sanitary conditions and facilities for disposal of sewage and garbage; and extreme weather conditions that have become more routine in some areas of the country.

 LO6 **Identify** the major forms of elder abuse.

Elder Abuse

Elder abuse refers to physical abuse, psychological abuse, financial exploitation, and medical abuse or neglect of people age 65 or older. How prevalent is the problem of elder abuse? We do not know for certain because relatively few cases are identified. According to the National Center for Victims of Crime (2011), as many as 1.6 million older people in the United States are the victims of physical or mental abuse each year. Just as with violence against children or women, it is difficult to determine exactly how much abuse against older people occurs. Many victims are understandably reluctant to talk about it. Studies have shown that elder abuse may occur from age 50 on, but victims of elder abuse tend to be concentrated among those over age 75.

elder abuse
physical abuse, psychological abuse, financial exploitation, and medical abuse or neglect of people age 65 or older.

Most of the victims are white, middle- to lower-middle-class Protestant women, ages 75 to 85, who suffer some form of impairment. Sons, followed by daughters, are the most frequent abusers of older persons.

Neglect is the most frequently reported type of elder abuse, with 58 percent of all reported cases being in this category; however, there are various kinds of elderly abuse. Physical abuse involves the use of force to threaten or injure an older person. It may be identified by injuries such as bruises, welts, sprains, and dislocations, or problems such as malnutrition. In contrast, emotional or psychological abuse is made up of verbal attacks, threats, rejection, isolation, or belittling acts that cause or could cause mental anguish, pain, or distress to a senior. Financial exploitation typically involves theft, fraud, or misuse of the older person's money or property by another person. Medical abuse occurs when a person withholds, or improperly administers, medications or health aids such as dentures, glasses, or hearing aids. Neglect is a caregiver's failure or refusal to provide care sufficient for the older person to maintain physical and mental health (Hooyman and Kiyak, 2011). Other types of elder abuse include sexual abuse, including sexual contact that is forced, tricked, threatened, or otherwise coerced, and abandonment, which occurs when a person with a duty of care deserts a frail or vulnerable older person (National Center on Elder Abuse, 2013).

Today, almost every state has enacted mandatory reporting laws regarding elder abuse or has provided some type of governmental protection for older persons. Cases of abuse and neglect of older people are highly dramatized in the media because they offend very central values in the United States—respect for and consideration of older persons.

Living Arrangements for Older Adults

Where older people live is linked to their income, health status, and the availability of caregivers. As shown in ● Figure 12.12, older men are more likely to live with their spouse than older women are. Among older men in 2012, 71 percent lived with their spouse, as compared with only 45 percent of older women. Older women were almost twice as likely (35 percent) as older men (19 percent) to live alone. Differences have also been identified in living arrangements when comparisons are made across racial and ethnic categories: White (non-Hispanic) women and African American women were more likely than women of other racial–ethnic categories to live alone. More than one-third of older white American women (41 percent) and African American women (39 percent) lived alone in 2012, as compared with about 18 percent of older Asian American women and 26 percent of Hispanic (Latina) women (U.S. Census Bureau, Families and Living Arrangements, 2014).

Support Services, Homemaker Services, and Day Care

Support services help older individuals cope with the problems in their day-to-day lives. These services are often expensive even when they are provided through state or federally funded programs, hospitals, or community organizations. For older persons, homemaker services perform basic chores (such as light housecleaning and laundry); other services deliver meals to homes (such as Meals on Wheels) or provide transportation for medical appointments. Some programs provide balanced meals at set locations, such as churches, synagogues, or senior centers.

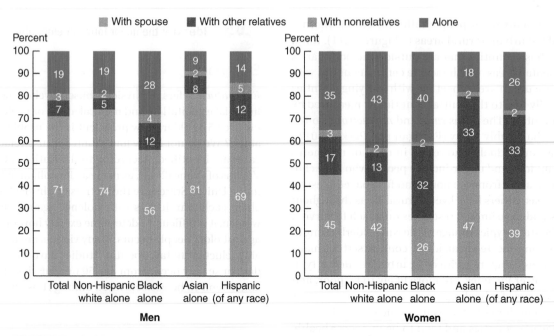

FIGURE 12.12 Living Arrangements of the Population Ages 65 and Over, by Sex and Race and Hispanic Origin, 2013

Source: U.S. Census Bureau, Families and Living Arrangements, 2014.

Day-care centers have also been developed to provide meals, services, and companionship to older people and in the process to help them maintain as much dignity and autonomy as possible. Numbering more than 4,600 nationwide, these centers typically provide transportation, activities, some medical personnel (such as a licensed practical nurse) on staff, and nutritious meals. The cost averages about $64 per day, adding up to $1,000 to $1,500 per month for a five-day week (Helpguide.org, 2015). Medicare does not cover the cost of adult day care; however, in some states, Medicaid covers the fees of older persons living below the poverty line.

Housing Alternatives

Some older adults remain in the residence where they have lived for many years—a process that gerontologists refer to as *aging in place*. Remaining in a person's customary residence is a symbol that he or she is able to maintain independence and preserve ties to his or her neighborhood and surrounding community. However, either by choice or necessity, some older adults move to smaller housing units or apartments.

This type of relocation frequently occurs when older people are living in a residence that does not meet their current needs. Moving to another location may be desirable or necessary when their family has grown smaller, the cost and time necessary to maintain the existing residence have become a strain on them, or individuals experience illness or disabilities that make it difficult for them to remain in the same location.

More housing alternatives are available to middle-income and upper-income older people than to low-income individuals. These include *retirement communities* such as Sun City, where residents must be age 55 or above. (Sun City is the name of several privately owned retirement communities located in such states as Arizona and Texas.) Residents of retirement communities purchase their housing units and in some instances pay additional fees for the upkeep of shared areas and amenities such as a golf course, swimming pool, or other recreational facilities. Typically, residents of planned retirement communities are similar to one another with respect to race and ethnic background and social class.

People needing assistance with daily activities or desiring the regular companionship of other people may move to an *assisted-living* facility. Some of these facilities offer fully independent apartments and provide residents with support services such as bathing, help with dressing, food preparation, and taking medication. Some centers provide residents with transportation to medical appointments, beauty salons and barbershops, and social events in the community. However, cost is a compelling factor. Assisted-living centers with more-luxurious amenities are primarily available to retired professionals and other upper-middle- and upper-income people. A number of assisted-living facilities provide residents with the opportunity to age in place by also having long-term care facilities available if and when the need arises (• Figure 12.13). Some of these facilities are well managed and carefully maintained; however, others are a major source of concern to residents, their families, and state regulators who are responsible for inspecting these facilities and responding to complaints filed against them. Some lower-income older people live in planned housing projects that are funded by federal, state, or local government agencies or by religious groups. Other lower-income individuals receive care and assistance from relatives, neighbors, or members of religious congregations.

Nursing Homes

It sometimes becomes necessary for an older person to move into a long-term care facility or nursing home as a result of major physical and/or cognitive problems that prevent the person from living in any other setting. There are two types of care provided in nursing homes: *skilled care* and *long-term care*. Skilled care provides services that

Dennis MacDonald/AGE Fotostock

FIGURE 12.13 Many people in assisted-living facilities enjoy planned social activities and gain a feeling of family ties with other residents. Why is companionship so important for our well-being regardless of our age?

are rendered by a doctor, licensed nurse, physical therapist, occupational therapist, social worker, respiratory therapist, or other specialist. This type of care typically follows a hospital visit and involves short-term nursing care or therapy until the patient can return home. By contrast, long-term care is provided for individuals who can no longer take care of themselves and are likely to remain in the nursing home facility for the remainder of their lifetime.

About 15,700 nursing homes operate in the United States, with a total of more than 1.7 million beds and 1.4 million residents. At any given time, about 80 percent of the beds are occupied. Nearly 70 percent of all nursing homes are owned by for-profit organizations. There has been some downward fluctuation in the total number of people living in nursing homes in recent years, largely because of the availability of home health care options and assisted-living facilities. Women make up about three-fourths of all residents and have relatively high rates of chronic illness or disability as they reach ages 80 and over, making it necessary for them to receive assistance with various activities such as bathing, dressing, walking, and eating (U.S. Centers for Disease Control and Prevention, 2015).

Some nursing home residents have mental disturbances or have been diagnosed with dementia—a group of conditions that all gradually destroy brain cells and lead to a progressive decline in mental function. As previously discussed, Alzheimer's disease is the most common form of dementia, and a growing number of nursing home residents have this progressive brain disorder, which gradually destroys a person's memory and ability to learn new information, make judgments, and communicate with others. Some nursing home facilities are designed exclusively for residents with Alzheimer's disease because of the physical and behavioral changes that typically accompany this condition.

Nursing home care is paid for in a variety of ways, including Medicare, Medicaid, and long-term-care insurance. Medicare pays for skilled nursing or rehabilitative services but typically does not pay for long-term care. Medicaid, a joint state and federal program, pays most nursing home costs for people with limited income and assets if they are in a nursing home facility certified by the government. Some people have long-term-care insurance or managed-care plans through their health insurance policies; however, about half of all nursing home residents pay nursing home costs out of their own savings. This expense often has devastating consequences on their finances, and after these savings are depleted, they must turn to Medicaid for their long-term care.

How satisfactory are nursing homes in meeting the needs of older individuals? Although there have been frequent criticisms in the media and lawsuits brought against some nursing homes for problems such as neglect or mistreatment of patients, there has been a greater emphasis in both the public and private sector on quality care in nursing facilities. Regulatory changes have mandated that nursing homes must meet specific training guidelines and minimum staffing requirements. Many nursing homes have adopted codes that specify the rights of residents, including the right to be treated with dignity and respect; full disclosure about the services and costs associated with living at the nursing home; the right to manage one's own money or choose the person who will do this; the right to privacy and to keep personal belongings and property unless they constitute a health or safety hazard; the right to be informed about one's medical condition and medications and to see one's own doctor; and the right to refuse medications and treatments. Clearly, there is a difference in *providing care* and in *caring*, for the latter involves offering all residents the emotional support they need in a homelike setting where they can live out the remainder of their lives with a sense of well-being and dignity.

 LO7 **Compare** and contrast functionalist, conflict, and symbolic interactionist perspectives on aging.

Sociological Perspectives on Aging

Sociologists and social gerontologists have developed a number of explanations regarding the social effects of aging. Some of the early theories were based on a microlevel analysis of how individuals adapt to changing social roles. More-recent theories have used a macrolevel approach to examine the inequalities produced by age stratification at the societal level.

Functionalist Perspectives on Aging

Functionalist explanations of aging focus on how older persons adjust to their changing roles in society. According to the mid-twentieth-century sociologist Talcott Parsons (1960), the roles of older persons need to be redefined by society. He suggested that devaluing the contributions of older persons is dysfunctional for society; older persons often have knowledge and wisdom to share with younger people.

How does society cope with the disruptions resulting from its members growing older and dying? According to *disengagement theory*, older persons make a normal and healthy adjustment to aging when they detach themselves from their social roles and prepare for their eventual death (Cumming and Henry, 1961). Gerontologists Elaine C. Cumming and William E. Henry (1961) noted that disengagement can be functional for both the individual and society. For example, the withdrawal of older persons from the workforce provides employment opportunities for younger people. Disengagement also aids a gradual and orderly transfer of statuses and roles from one generation to the next; an abrupt change would result in chaos. Retirement, then, can be thought of as recognition for years of service and the acknowledgment that the person no longer fits into the world of paid work. The younger workers who move into the vacated positions have received more up-to-date training—for example, the computer skills that are taught to most young people today.

Critics of this perspective object to the assumption that all older persons want to disengage while they are still productive and still gain satisfaction from their work. Disengagement may be functional for organizations but not for individuals. A corporation that encourages retirement by paying "retirement bonuses" or uses other means to encourage older workers to retire may be able to replace higher-paid, older workers with lower-paid, younger workers, but retirement is not always beneficial for older workers, particularly if they were not ready to retire. Contrary to disengagement theory, a number of studies have found that activity in society is *more* important with increasing age.

FIGURE 12.14 According to the symbolic interactionist perspective, older people who invest their time and energy in volunteer work or in other enjoyable activities tend to be healthier, be happier, and live longer than those who disengage from society.

Symbolic Interactionist Perspectives on Aging

Symbolic interactionist perspectives examine the connection between personal satisfaction in a person's later years and a high level of activity. *Activity theory* states that people tend to shift gears in late middle age and find substitutes for previous statuses, roles, and activities (Havighurst, Neugarten, and Tobin, 1968). From this perspective, older people have the same social and psychological needs as middle-aged people and thus do not want to withdraw unless restricted by poor health or disability.

Whether retired persons invest their energies in grandchildren, traveling, hobbies, new work roles, or volunteering in the community, their social activity is directly related to longevity, happiness, health, and overall social well-being (● Figure 12.14). Healthy people who remain active have a higher level of life satisfaction than do those who are inactive or in ill health. Among those whose mental capacities decline later in life, deterioration is most rapid in people who withdraw from social relationships and activities (see "You Can Make a Difference").

A variation on activity theory is the concept of *continuity*—that people are constantly attempting to maintain their self-esteem and lifelong principles and practices, and that they simply adjust to the feedback from and needs of others as they grow older. From this perspective, aging is a continuation of earlier life stages rather than a separate and unique period. Thus, values and behaviors that have previously been important to an individual will continue to be so as the person ages. Although physical changes associated with aging will always be a factor in some types of job performance, particularly those requiring quick reaction times or heavy physical labor, many occupations could benefit from older workers' experience and expertise. For this reason, Solem (2008) suggests that older workers should be provided with new learning opportunities and a chance to maintain their subjective work ability throughout their careers.

Another interactionist approach, the *social constructionist perspective on aging*, seeks to understand how individual processes of aging are influenced by social definitions and social structures. The social constructionist approach recognizes that individuals are active participants in their daily lives as they create and maintain social meanings for themselves and their acquaintances. In other words, through a person's relationships with others, the individual creates a "reality" that then structures his or her life. For example, one study examined how "frailty" is not necessarily a specific physical condition as much as it is a social construct produced through the interactions of

disengagement theory
the proposition that older persons make a normal and healthy adjustment to aging when they detach themselves from their social roles and prepare for their eventual death.

activity theory
the proposition that people tend to shift gears in late middle age and find substitutes for previous statuses, roles, and activities.

Getting Behind the Wheel to Help Older People: Meals on Wheels

[Meals on Wheels volunteers] make you feel like you are part of the outside world. For me, it's just hard to explain how much that means to me to have them come by here every day and have a few words of "How are you?" and back and forth. That they bring you something that is good for you—I just can't say enough about how much I appreciate that work on their part.

—a woman explaining how much Meals on Wheels of San Antonio, Texas, means to her (qtd. in MOWAA, 2011)

Like this grateful woman, thousands of people across the United States who are elderly, homebound, disabled, or frail have nutritious meals delivered to their residence by volunteers from Meals on Wheels, an organization that asks recipients to contribute what they can toward the cost of their meals but relies primarily on donations of time and money to cover the cost of the service that it provides. Although the majority of Meals on Wheels recipients are over age 60, some younger disabled or ill persons also benefit from this service.

What do "Wheels" volunteers do? On their designated service days, volunteers pick up prepared meals packed in insulated carriers at a neighborhood meal site and deliver the food to the people on their "route," which usually serves about ten people and takes no more than about an hour, even allowing for time to visit briefly with each food recipient. Volunteers serve as a safety check to ensure that the older person is doing all right. In orientation sessions, volunteers learn what to do if they get to a residence and find that a person is in need of assistance (MOWAA, 2011).

What does it take to become a Meals on Wheels volunteer? "Wheels" volunteers must have a valid driver's license and proof of insurance. They must be eighteen years of age or older, or if they are younger, they must be accompanied by a person over age eighteen who meets the licensing and insurance requirements. Meals on Wheels organizations in various communities have different guidelines on how frequently they want volunteers to work, but a minimum of two days per month is a general requirement. Obviously, many local "Wheels" organizations would like for volunteers to work more often than that and attempt to assign volunteers to an area near their own home or workplace in order to make it easier to meet that goal (MOWAA, 2011).

What can you do to help provide food for elderly, homebound residents? Providing a little of our time and/or some financial support (even a small donation) helps programs such as Meals on Wheels. Although the Meals on Wheels Association of America is a national organization, it is largely a grassroots operation that relies on local volunteers to make a difference in the lives of the most vulnerable individuals in their community. If you are interested in helping, contact your local Meals on Wheels organization or visit the website of the Meals on Wheels Association of America for more information.

Meals on Wheels provides a valuable service for older persons throughout the United States. What other support services may be of special importance in meeting the daily needs of these individuals?

---

older people, their caretakers, and their health professionals. Through these interactions the subjective experience of frailty becomes interpreted and defined in a "medical/social idiom" that is framed in terms of surveillance ("keeping an eye on them") and independence (how much freedom they should have to move around on their own without falling, for example). Then rules (made by caregivers and/or medical professionals) for what the person should or should not do become facts and a social reality apart from the individual's actual physical condition. Theories based on a symbolic interactionist tradition provide new understandings about the problems of aging and how interactions with others create social realities that become important in social–structural contexts.

Conflict Perspectives on Aging

Conflict theorists view aging as especially problematic in contemporary capitalistic societies. As people grow older, their power tends to diminish unless they are able to maintain wealth. Consequently, those who have been disadvantaged in their younger years become even more so in late adulthood. Women ages 75 and over are among the most disadvantaged because they often must rely solely on Social Security, having outlived their spouses and sometimes their children.

Underlying the capitalist system is an ideology that assumes that all people have equal access to the means of gaining wealth and that poverty results from individual weakness. When older people are in need, they may be viewed as not having worked hard enough or planned adequately for their retirement. The family and the private sector are seen as the "proper" agents to respond to their needs. To minimize the demand for governmental assistance, these services are made punitive and stigmatizing to those who need them. Class-based theories of inequality assert that government programs for older persons stratify society on the basis of class. Feminist approaches claim that these programs perpetuate inequalities on the basis of gender and race in addition to class. A political economy approach seeks to explain how economic and political forces come together to determine how resources are allocated in a society and why there is so often in high-income nations a variation in the treatment and status of older people. To better understand where these inequalities originated and how they are perpetuated, researchers using a political economy perspective examine public policies, economic trends, and social–structural factors.

Conflict analysis draws attention to the diversity in the older population. Differences in social class, gender, and race/ethnicity divide older people just as they do everyone else. Wealth cannot forestall aging indefinitely, but it can soften the economic hardships faced in later years. The conflict perspective adds to our understanding of aging by focusing on how capitalism devalues older people, especially women. However, critics assert that this approach ignores the fact that industrialization and capitalism have greatly enhanced the longevity and quality of life for many older persons.

The major sociological perspectives regarding aging are summarized in the Concept Quick Review.

LO8 **Describe** four widely used frameworks for explaining how people cope with the process of dying or with the loss of a loved one.

Death and Dying

Historically, death was a common occurrence at all stages of the life course. Until the twentieth century, the chances that a newborn child would live to adulthood were very small. Poor nutrition, infectious diseases, accidents, and natural disasters took their toll on men and women of all ages. But in contemporary, industrial societies, death is looked on as unnatural because it has been largely removed from everyday life. Most deaths now occur among older persons, and the association of death with the aging process has contributed to ageism in our society; if people can deny aging, they believe that they can deny death.

How do people cope with dying? (See • Figure 12.15.) There are four widely known frameworks for explaining how people cope with the process of dying or with the loss of a loved one: the *stage-based approach*, the *trajectories of grief approach*, the *dying trajectory*, and the *task-based approach*.

The *stage-based approach* was popularized by psychiatrist Elisabeth Kübler-Ross (1969), who proposed five stages in the dying process: (1) denial and isolation ("Not me!"), (2) anger and resentment ("Why me?"), (3) bargaining and an attempt to postpone ("Yes me, but . . ."—negotiating for divine intervention), (4) depression and sense of loss, and (5) acceptance. She pointed out that these stages are not the same for all people; some of the stages may exist at the same time. Kübler-Ross (1969: 138) also stated that "the one thing that usually persists through all these stages is hope." Kübler-Ross's stages were attractive to the general public and the media because they provided common responses to a difficult situation. On the other hand, her stage-based model also generated a great deal of criticism. Some have pointed out that these stages have never been conclusively demonstrated or comprehensively explained.

The *trajectories of grief approach* was introduced by the clinical psychologist George Bonanno (2009, 2010), who claims that bereavement studies have

FIGURE 12.15 Sharing experiences with other people is very important in coping with death and dying. Support groups, both in person and online, provide individuals with many opportunities to discuss their own health problems and mortality or to deal with the illness or death of a loved one.

CONCEPT QUICK REVIEW

Theoretical Perspectives on Aging

	Focus	Theory/Hypothesis
Functionalist Perspectives	How older people adjust to changing roles in society	Disengagement theory suggests that detachment and preparation for death are normal and healthy adjustments for older individuals.
Symbolic Interactionist Perspectives	Why microlevel contacts between individuals are particularly important for older people	Activity theory is based on the assumption that people are more satisfied in old age if they remain active and find new statuses, roles, and activities. Social constructionist approaches examine how individual processes of aging are influenced by social definitions and social structures.
Conflict Perspectives	How aging is difficult in a capitalist economy and why race, class, and gender are factors that make a difference in the well-being of older people	Inequality follows people across the life course, and poor and middle-income individuals often live on fixed incomes and must rely on Social Security and Medicare benefits.

disproved the stage-based approach and have helped researchers identify four common trajectories of grief:

- *Resilience*—the ability of people to maintain a relatively stable, healthy level of psychological and physical functioning while they are dealing with a highly disruptive event such as the death of a relative or a life-threatening situation.

- *Recovery*—a gradual return to previous levels of normal functioning after experiencing a period of psychological stress such as depression or posttraumatic stress disorder.

- *Chronic dysfunction*—lengthy suffering (sometimes several years or more) and the inability to function after experiencing grief.

- *Delayed grief or trauma*—experiencing what appears to be a normal adjustment after the loss of a loved one but then having an increase in distress and other symptoms months later. (based on Bonanno, 2010)

To explain his views on grief-related behavior, Bonanno coined the term "coping ugly" to describe how some seemingly inappropriate or counterintuitive behavior (such as telling jokes or laughing) may seem odd at the time but help a person move on after a loss. Although Bonanno's ideas about persistent resilience have been embraced by some gerontologists and bereavement specialists, others disagree and see this condition as a form of denial that might necessitate counseling or other treatment.

A third approach is referred to as the *dying trajectory*, which focuses on the perceived course of dying and the expected time of death. For example, a dying trajectory may be sudden, as in the case of a heart attack, or it may be slow, as in the case of lung cancer. According to the dying-trajectory approach, the process of dying involves three phases: the acute phase, characterized by the expression of maximum anxiety or fear; the chronic phase, characterized by a decline in anxiety as the person confronts reality; and the terminal phase, characterized by the dying person's withdrawal from others (Glaser and Strauss, 1968).

Finally, the *task-based approach* is based on the assumption that the dying person can and should go about daily activities and fulfill tasks that make the process of dying easier on family members and friends, as well as on the dying person. Physical tasks can be performed to satisfy bodily needs, whereas psychological tasks can be done to maximize psychological security, autonomy, and richness of experience. Social tasks sustain and enhance interpersonal attachments and address the social implications of dying. Spiritual tasks help people to identify, develop, or reaffirm sources of spiritual energy and to foster hope (Corr, Nabe, and Corr, 2003).

In the final analysis, however, how a person dies or experiences the loss of a loved one is shaped by many social and cultural factors. These endeavors are influenced by an individual's personality and philosophy of life, as well as the social context in which these events occur.

In recent years the process of dying has become an increasingly acceptable topic for public discussion. Such discussions helped further the hospice movement in the 1970s. A **hospice** is an organization that provides a homelike facility or home-based care (or both) for people who are terminally ill. The hospice philosophy asserts that people should participate in their own care and have control over as many decisions pertaining to their life as possible. Pain and suffering should be minimized, but artificial measures should not be used to sustain life. This approach is family based and provides support for family members and friends, as well as for the person who is dying (see Corr, Nabe, and Corr, 2003). Over time, hospice care has moved toward hospital standards because of the need for hospices to work with the federal government and the American Hospital Association to gain accreditation. Medicare funding and the resultant federal regulations have further changed hospices.

Looking Ahead: Aging in the Future

The size of the older population in the United States will continue to increase dramatically in the early decades of the twenty-first century. By the year 2050, there will be an estimated 88.5 million people ages 65 and older (20 percent of the total population). Thus, combined with decreasing birth rates, most of the population growth will occur in the older age cohorts during the next 50 years. More people will survive to age 85, and more will reach 100 and older to become centenarians (Meyer, 2012).

As the previous statement indicates, the demographics of the older population will undergo a dramatic transformation in the future. By 2050, the number of African Americans over age 65 will triple, moving them from 8 to 10 percent of all Americans ages 65 and above. Older Latinos/as (Hispanics) will increase from fewer than 4 percent of all people over age 65 to nearly 16 percent of older adults, for an 11 percent gain. This demographic shift means that many older Americans will have increased needs for health and social services: African Americans and Latinos/as have special concerns regarding the onset of chronic illnesses at an earlier age than do white (non-Hispanic) Americans, higher incidences of obesity and type 2 diabetes, and social problems that contribute to mental health concerns such as higher rates of poverty, racially segregated communities, poor schools, high unemployment rates, and limited access to health care. In regard to access to health care for all older individuals, the effects of the 2010 Affordable Care Act will become more apparent as no-cost preventive medical services will become available and the closing of the income-based gap in Medicare prescription drug coverage, known as the "doughnut hole," is completed by 2020. It will be important to develop better and more comprehensive ways of assisting people from all racial–ethnic and nationality groupings so that they may be able to live full and productive lives.

In the future, who will assist older people with needs they cannot meet themselves? Family members in the future may be less willing or able to serve as caregivers. Women, the primary caregivers in the past, are faced with *triple* workdays if they attempt to combine working full time with caring for their children and assisting older relatives. More caregivers in the future will probably be paid employees who work in the individual's home or who are employed by independent or assisted-living facilities or nursing homes.

As biomedical research on aging continues, new discoveries in genetics may eliminate life-threatening diseases and make early identification of other diseases possible. Technological advances in the diagnosis, prevention, and treatment of Alzheimer's disease may revolutionize people's feelings about growing older. Advances in medical technology may lead to a more positive outlook on aging.

If these advances occur, will they help everyone or just some segments of the population? This is a very important question for the future. As we have seen, many of the benefits and opportunities of living in a highly technological, affluent society are not available to all people. Classism, racism, sexism, and ageism all serve to restrict individuals' access to education, medical care, housing, employment, and other valued goods and services in society.

For older persons, the issues discussed in this chapter are not merely sociological abstractions; they are an integral part of their everyday lives.

hospice
an organization that provides a homelike facility or home-based care (or both) for people who are terminally ill.

CHAPTER REVIEW

LO1 What is the difference between functional age and chronological age, and why is this distinction important in understanding the social significance of age?

Aging refers to the physical, psychological, and social processes associated with growing older. Chronological age refers to a person's age based on date of birth. However, most of us actually estimate a person's age on the basis of functional age—observable individual attributes such as physical appearance, mobility, strength, coordination, and mental capacity that are used to assign people to age categories.

LO2 How has age been a factor in determining individuals' roles and statuses in different types of societies?

In preindustrial societies, people of all ages are expected to share the work, and the contributions of older people are valued. In industrialized societies, however, older people are often expected to retire so that younger people may take their place. In postindustrial societies, information technologies are extremely important, and a large proportion of the working population is employed in service-sector occupations in the fields of education and health care, both of which may benefit older people.

LO3 What is meant by age stratification, and how does age stratification affects people throughout the life course?

In the United States, age differentiation is typically based on categories such as infancy, childhood, adolescence, young adulthood, middle adulthood, and later adulthood. These age categories have had a profound effect on our perceptions of people's capabilities, responsibilities, and entitlement. What is considered appropriate for or expected of people at various ages is somewhat arbitrarily determined and produces age stratification—inequalities, differences, segregation, or conflict between age groups.

LO4 What is ageism, and how do some older people resist it?

Ageism is prejudice and discrimination against people on the basis of age, particularly against older persons. Ageism is reinforced by stereotypes of older people. Many older people resist ageism by continuing to view themselves as being in middle adulthood long after their actual chronological age would suggest otherwise. Older people may also resist ageism by maintaining positive relationships with others and keeping up with newer technologies that help them communicate with children, grandchildren, and friends.

LO5 How is age intertwined with gender and race/ethnicity in patterns of social and economic inequality?

Age, gender, and poverty are intertwined. Men and women over the age of 50 have experienced problems in the recent economic crisis in the United States and other nations linked to the global economy. Older women in the workplace tend to earn less than men their age, and they are more likely to work in gender-segregated jobs that provide less income and fewer opportunities for advancement. Age, race/ethnicity, and economic inequality are also closely intertwined. Inequalities that exist later in life originate in individuals' early participation in the labor force and are amplified in late adulthood.

LO6 What are the major forms of elder abuse?

Elder abuse includes physical abuse, psychological abuse, financial exploitation, and medical abuse or neglect of people age 65 or older. Passive neglect is the most common form of abuse. Today, almost every state has enacted mandatory reporting laws regarding elder abuse or has provided some type of governmental protection for older persons.

LO7 How do functionalist, conflict, and symbolic interactionist perspectives on aging differ?

Functionalist explanations of aging focus on how older persons adjust to their changing roles in society; the gradual transfer of statuses and roles from one generation to the next is necessary for the functioning of society. Conflict theorists link the loss of status and power experienced by many older persons to their lack of ability to produce and maintain wealth in a capitalist economy. Activity theory, a part of the symbolic interactionist perspective, states that people change in late middle age and find substitutes for previous statuses, roles, and activities. This theory asserts that people do not want to withdraw unless restricted by poor health or disability.

LO8 What are four widely used frameworks for explaining how people cope with the process of dying or with the loss of a loved one?

Four widely known frameworks for explaining how people cope with the process of dying or with the loss of a loved one are the stage-based approach, the trajectories of grief approach, the dying trajectory, and the task-based approach.

KEY TERMS

activity theory 359	cohort 340	functional age 340
age stratification 344	disengagement theory 358	gerontology 342
aging 338	elder abuse 355	hospice 362
chronological age 339	entitlements 353	life expectancy 340

QUESTIONS for CRITICAL THINKING

1 Why does activity theory contain more positive assumptions about older persons than disengagement theory does? Analyze your grandparents (or other older persons whom you know well or even yourself if you are older) in terms of disengagement theory and activity theory. Which theory seems to provide the most insights? Why?

2 How are race, class, gender, and aging related?

3 Is it necessary to have a mandatory retirement age? Why or why not?

4 How will the size of the older population in the United States affect society and programs such as Social Security in the future?

ANSWERS to the SOCIOLOGY QUIZ

ON AGING AND AGE-BASED DISCRIMINATION

1	**False**	In 2009 the U.S. Supreme Court ruled that persons claiming age discrimination must prove that *age* was the determining factor in adverse job decisions, and few employers state "I won't hire you because of your age" or "I'm letting you go because of your age."
2	**True**	For the cohort born in 2014, as an example, life expectancy at birth was 79.6 years for the overall U.S. population—77.1 for males and 81.9 for females. These figures vary by race and ethnicity.
3	**False**	No studies have documented that women actually age faster than men. However, some scholars have noted a "double standard" of aging that places older women at a disadvantage with respect to older men because women's worth in the United States is often defined in terms of physical appearance.
4	**False**	Although some older persons are economically secure, persons who rely solely on Social Security, Medicare, and/or pensions tend to live on low, fixed incomes that do not adequately meet their needs and often place them below the official poverty line.
5	**False**	Studies have shown that advertisements frequently depict older persons negatively—for example, as chronically ill or absentminded.
6	**True**	The U.S. population is growing older, and persons ages 85 and over constitute one of the fastest-growing segments of the population.
7	**True**	Organizations such as the Gray Panthers and AARP have been instrumental in the enactment of legislation beneficial to older persons.
8	**False**	Although cases of abuse and neglect of older persons are highly dramatized in the media, most coverage pertains to problems in hospitals, nursing homes, or other long-term care facilities. We know very little about the nature and extent of abuse that occurs in private homes.

THE ECONOMY AND WORK
IN GLOBAL PERSPECTIVE

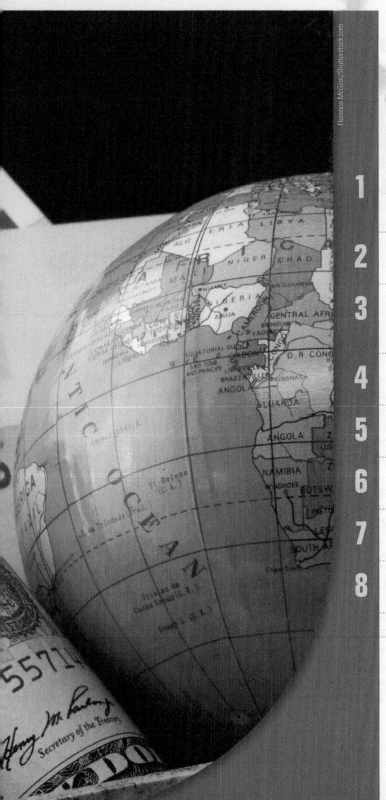

Florence McGinn/Shutterstock.com

CHAPTER

13

LEARNING OBJECTIVES

1 **Describe** the primary function of the economy and distinguish among preindustrial, industrial, and postindustrial economies.

2 **Discuss** the key characteristics of capitalism, socialism, and mixed economies.

3 **State** the key features of functionalist, conflict, and symbolic interactionist perspectives on the economy and work in the United States.

4 **Describe** professions and discuss how they differ from other occupations.

5 **Compare** primary and secondary labor markets, and identify jobs that fit into each category.

6 **Explain** problems that are associated with contingent work and the underground (informal) economy.

7 **Describe** different kinds of unemployment and discuss current trends in U.S. unemployment.

8 **Identify** ways in which workers attempt to gain control over their work situation.

Defining the Twenty-First-Century Workplace: Office, Home, or Neither?

Over the past few months, we have introduced a number of great benefits and tools to make us more productive, efficient and fun. . . . To become the absolute best place to work, communication and collaboration will be important, so we need to be working side-by-side. That is why it is critical that we are all present in our offices. . . . Speed and quality are often sacrificed when we work from home. We need to be one Yahoo! And that starts with physically being together.

—JACKIE RESES, director of Yahoo's Human Relations Department, informing employees that CEO Marissa Mayer had canceled work-from-home arrangements and wanted all employees to work in the office (qtd. in Swisher, 2013).

When the newly hired CEO Marissa Mayer kicked off her turnaround at Yahoo, she banned employees from working from home. When Andy Mattes [CEO of an ATM manufacturing plant in Canton, Ohio] kicked off his

turnaround at Diebold, the first thing the new CEO did was look for employees who wanted to work remotely. . . . "We wanted the brightest people on the planet. We were fishing in a small fishing pond" (qtd. in Peck, 2015).

When Yahoo called remote workers back to the office in 2013 in order to improve worker collaboration and communication, there was a lot of hand wringing about the future of

Justin Sullivan/Getty Images

Yahoo CEO Marissa Mayer surprised many when she announced that all Yahoo employees would no longer be allowed to work from home. What might employees gain from working in the office rather than at home or other locations such as the local Starbucks?

When the original Yahoo memo came out, advocates for telecommuting argued that going into the office every day is a bad idea in high-traffic areas such as the Silicon Valley and that working from home is much more efficient. As one sociologist noted, "Yahoo will not get more out of their employees by watching them like hawks and monitoring their every move" (Glass, 2013). People who supported the decision to ban telecommuting at Yahoo argued that employees were more innovative and productive if they had frequent face-to-face interactions with one another. By contrast, when Andy Mattes was trying to lure workers to Diebold, the conversation had shifted the other way, and many companies, including Yahoo, were encouraging some employees to work from home (known as WFH). Although only a limited number of employees WFH all the time, at least 67 percent of all full-time workers engage in some remote work, which is more common among employees at the top and at the lower ends of the pay scale (Peck, 2015).

Regardless of which side of the WFH argument we might find most compelling, the idea of working from home as opposed to performing our workplace duties at the office raises a number of important sociological questions, such as "What

is the connection between the contemporary workplace and telecommuting?" and "How is the U.S. economy related to the global economy?" In this chapter we discuss the economy as a social institution and explain why some of our social and economic problems are linked to the larger world of work. By the world of work, we mean what people are paid for their labor, who gains and who loses in difficult economic times, how people feel about their work, and what impact that changes in the economy may have on your own future. Before reading on, test your knowledge about work in the United States in the 2010s by taking the "Sociology and Everyday Life" quiz.

Comparing the Sociology of Economic Life with Economics

Perhaps you are wondering how a sociological perspective on the economy differs from the study of economics. Although aspects of the two disciplines overlap, each provides a unique perspective on economic institutions. Economists focus on the complex workings of economic systems (such as monetary policy, inflation, and the national debt),

the workplace. Was it the office? At home? Now, two years later, it is clear: Telecommuting has won.

—journalist EMILY PECK (2015) discussing the seemingly opposing strategies of two CEOs regarding telecommuting for workers

How Much Do You Know About Work in the United States in the 2010s?

TRUE	FALSE		
T	F	1	Telecommuting has decreased in recent years because employers want to have more control over their employees.
T	F	2	Studies have shown that working from home leads to a decrease in workers' performance and overall productivity.
T	F	3	Workers with bachelor's degrees or higher have much lower rates of unemployment than workers with less than a high school education.
T	F	4	Corporations use the term *employee engagement* to refer to employee happiness or workers' satisfaction with their jobs.
T	F	5	Health-related occupations and service positions are among the fastest-growing occupational fields in the second decade of the twenty-first century.
T	F	6	Most U.S. adults with disabilities are in the labor force today.
T	F	7	Physicians and surgeons and dentists in specialties such as oral and maxillofacial surgery and orthodontia are in the highest-paying occupations based on median pay.
T	F	8	A direct linkage exists between parents' education level and income and their children's scores on college admissions tests, which often serve as gatekeepers for higher education and, subsequently, occupations.

Answers can be found at the end of the chapter.

whereas sociologists focus on interconnections among the economy, other social institutions, and the social organization of work. At the macrolevel, sociologists may study the impact of transnational corporations on industrialized and low-income nations. At the microlevel, sociologists might study people's satisfaction with their jobs. To better understand these issues, we will examine how economic systems came into existence and how they have changed over time.

 Describe the primary function of the economy and distinguish among preindustrial, industrial, and postindustrial economies.

Economic Systems in Global Perspective

The *economy* is the social institution that ensures the maintenance of society through the production, distribution, and consumption of goods and services. *Goods* are tangible objects that are necessary (such as food, clothing, and shelter) or desired (such as robots for the home,

smartphones, and tablets). *Services* are intangible activities for which people are willing to pay (such as dry cleaning or medical care). In high-income nations today, many of the goods and services that we consume are information goods. Examples include databases and surveys ("intermediate products") and the mass media, computer software, and the Internet ("information goods").

Some goods and services are produced by human labor (the plumber who unstops your sink, for example); others are primarily produced by capital (such as Internet and high-definition-TV access available through a cable or dish-service provider). *Labor* refers to the group of people who contribute their physical and intellectual services to the production process in return for wages that they are paid by firms. *Capital* is wealth (money or property) owned or used in business by a person or corporation. Obviously, money, or financial capital, is needed to invest in

economy
the social institution that ensures the maintenance of society through the production, distribution, and consumption of goods and services.

the physical capital (such as machinery, equipment, buildings, warehouses, and factories) used in production. For example, a person who owns a thousand shares of stock in a high-tech company owns financial capital, but these shares also represent an ownership interest in that corporation's physical capital.

To better understand the economy in the United States today, let's briefly look at three broad categories of economies: preindustrial, industrial, and postindustrial economies (two of which are shown in ● Figure 13.1).

Preindustrial Economies

Preindustrial economies include hunting and gathering, horticultural and pastoral, and agrarian societies. Most workers engage in *primary sector production*—the

FIGURE 13.1 Although agriculture and manufacturing still exist in contemporary high-income countries such as the United States, the work process continues to change in industrial and postindustrial economies as new technologies are introduced in agribusiness or this uniquely designed manufacturing plant.

extraction of raw materials and natural resources from the environment. These materials and resources are typically consumed or used without much processing. The production units in hunting and gathering societies are small; family members produce most goods. The division of labor is by age and gender. The potential for producing surplus goods increases as people learn to domesticate animals and grow their own food.

In horticultural and pastoral societies, the economy becomes distinct from family life. The distribution process becomes more complex, with the accumulation of a *surplus* such that some people can engage in activities other than food production. In agrarian societies, production is primarily related to producing food. However, workers have a greater variety of specialized tasks, such as warlord or priest; for example, warriors are necessary to protect the surplus goods from plunder by outsiders. Once a surplus is accumulated, more people can also engage in trade. Initially, the surplus goods are distributed through a system of *barter*—the direct exchange of goods or services considered of equal value by the traders. However, bartering is limited as a method of distribution; equivalencies are difficult to determine (how many fish equal one rabbit?) because there is no way to assign a set value to the items being traded. As a result, *money,* a medium of exchange with a relatively fixed value, came into use in order to aid the distribution of goods and services in society.

What was the U.S. economy like in the preindustrial era? In the economy of the colonial period (from the 1600s to the early 1700s), white men earned a livelihood through agricultural work or as small-business owners who ran establishments such as inns, taverns, and shops. During this period, white women worked primarily in their homes, doing such tasks as cooking, cleaning, and child care. Some also developed *cottage industries*—producing goods in their homes that could be sold to nonfamily members. However, a number of white women also worked outside their households as midwives, physicians, nurses, teachers, innkeepers, and shopkeepers.

However, the experiences of people of color were quite different in preindustrial America. According to the sociologists Sharlene Hesse-Biber and Gregg Lee Carter (2000), slavery, which came about largely as a result of the demand for cheap agricultural labor, was a major force in the exploitation of people of color, particularly women of color who suffered a double burden of both sexism and racism. By contrast, Native American women in some agricultural communities held greater power because they were able to maintain control over land, tools, and surplus food (Hesse-Biber and Carter, 2000).

Do preindustrial forms of work still exist in contemporary high-income nations? In short, yes. For example, portions of contemporary sub-Saharan Africa have a relatively high rate of exports of primary commodities, and foreign direct investment is concentrated in mineral extraction. Even in high-income nations such as the United States, entire families work in the agricultural sector of the economy, performing tasks such as picking fruits and vegetables. Some parts of the agricultural sector coexist beside industrial and postindustrial sectors. For example, workers who help grow alfalfa, almonds, and pistachios on the West Coast are employed in the same region as high-tech information workers who are employed by Microsoft and other software developers or computer manufacturers.

Industrial Economies

Industrial economies result from sweeping changes to the system of production and distribution of goods and services. Drawing on new forms of energy (such as steam, gasoline, and electricity) and machine technology, factories proliferate as the primary means of producing goods. Most workers engage in **secondary sector production**— the processing of raw materials (from the primary sector) into finished goods. For example, steelworkers process metal ore; autoworkers then convert the ore into automobiles, trucks, and buses. In industrial economies, work becomes specialized and repetitive, activities become bureaucratically organized, and workers primarily work with machines instead of with one another. With the emergence of mass production, larger surpluses are generated, typically benefiting some people and organizations but not others.

In sum, the typical characteristics of industrial economies include the following:

1. New forms of energy, mechanization, and the growth of the factory system.

2. Increased division of labor and specialization among workers.

3. Universal application of scientific methods to problem solving and profit making.

4. Introduction of wage labor, time discipline, and workers' deferred gratification, which means that employees should be diligent at work and pursue personal activities on their own time only.

5. Strengthening of bureaucratic organizational structure and the enforcement of rules, policies, and procedures to make the workplace more efficient and profitable.

All these characteristics contribute to the development of industrial economies, greater productivity in the workplace, and a dramatic increase in consumption because many more goods are available at affordable prices.

Although industrialization brought about an increased standard of living for many people, the sociologist Thorstein Veblen (1857–1929) criticized industrialism in his famous book *The Theory of the Leisure Class* (1967/1899). According to Veblen, the idle rich, who made vast sums of money through factory ownership and the hard work of their employees, represented a conspicuously consuming, parasitic leisure class. *Conspicuous consumption* is the ostentatious display of symbols of wealth, such as owning numerous mansions and expensive works of art, wearing extravagant jewelry and clothing, or otherwise flaunting the trappings of great wealth. By contrast, *conspicuous leisure* involves wasteful and highly visible leisure activities such as casino gambling or expensive sporting events that require costly gear or excessive travel expenses (such as going on a safari in Africa). If Veblen were alive today, do you think he would feel the same way about conspicuous consumption and perhaps incorporate the spending habits of Wall Street bankers, hedge fund managers, or billionaire entrepreneurs?

Postindustrial Economies

A postindustrial economy is based on **tertiary sector production**—the provision of services rather than goods— as a primary source of livelihood for workers and profit for owners and corporate shareholders. Tertiary sector production includes a wide range of activities, such as fast-food service, transportation, communication, education, real estate, advertising, sports, and entertainment. A majority of U.S. jobs are in tertiary sector employment, as contrasted with primary or secondary sector employment, and some of the employment sectors are segregated by gender and/or race/ethnicity.

Five characteristics are central to the postindustrial economy:

1. Service industries dominate over manufacturing.

2. Information and technological innovation displace property as the central preoccupations in the economy.

3. Professional and technical classes grow more predominant, and workplace culture shifts from factories to diversified work settings.

4. Traditional boundaries between work and home (public and private life) no longer exist because digital technologies such as cell phones and computers make global communication possible twenty-four hours per day.

5. High levels of urbanization occur, along with a decline in population in many rural areas.

primary sector production
the sector of the economy that extracts raw materials and natural resources from the environment.

secondary sector production
the sector of the economy that processes raw materials (from the primary sector) into finished goods.

tertiary sector production
the sector of the economy that is involved in the provision of services rather than goods.

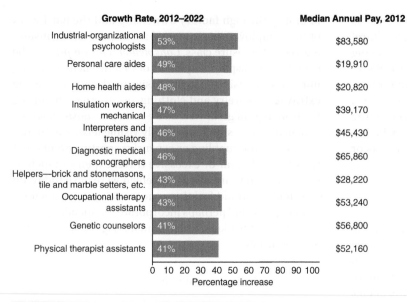

Growth Rate, 2012–2022	Median Annual Pay, 2012
Industrial-organizational psychologists — 53%	$83,580
Personal care aides — 49%	$19,910
Home health aides — 48%	$20,820
Insulation workers, mechanical — 47%	$39,170
Interpreters and translators — 46%	$45,430
Diagnostic medical sonographers — 46%	$65,860
Helpers—brick and stonemasons, tile and marble setters, etc. — 43%	$28,220
Occupational therapy assistants — 43%	$53,240
Genetic counselors — 41%	$56,800
Physical therapist assistants — 41%	$52,160

0 10 20 30 40 50 60 70 80 90 100
Percentage increase

FIGURE 13.2   Top Ten Fastest-Growing Occupations, 2012–2022

Source: U.S. Bureau of Labor Statistics, 2014.

How far has the United States moved into postindustrialization? Although this question is difficult to answer, projections by the Census Bureau indicate that some of the fastest-growing occupations of this century are in health care and the service sector, such as personal care aides and home health aides, as shown in ● Figure 13.2.

Discuss the key characteristics of capitalism, socialism, and mixed economies.

Contemporary World Economic Systems

Capitalism and socialism are the principal economic models in industrialized countries. Sociologists use two criteria—property ownership and market control—to distinguish between types of economies.

Capitalism

Capitalism is an economic system characterized by private ownership of the means of production, from which personal profits can be derived through market competition and without government intervention. Most of us think of ourselves as "owners" of private property because we own a car, a television, or other possessions. However, most of us are not capitalists; we *spend money* on the things we own rather than *make money* from them. Only a relatively few people own income-producing property from which a profit can be realized by producing and distributing goods and services. Everyone else is a consumer. "Ideal" capitalism has four distinctive features: (1) private ownership of the means of production, (2) pursuit of personal profit, (3) competition, and (4) lack of government intervention.

Private Ownership of the Means of Production   Capitalist economies are based on the right of individuals to own income-producing property, such as land, water, mines, and factories, and the right to "buy" people's labor.

In the early stages of industrial capitalism (1850–1890), virtually all the capital for investment in the United States was individually owned—prior to the Civil War, an estimated 200 families controlled all major trade and financial organizations. By the 1890s, individual capitalists, including Andrew Carnegie, Cornelius Vanderbilt, and John D. Rockefeller, controlled most of the capital in commerce, agriculture, and industry (Feagin, Baker, and Feagin, 2006).

As workers grew tired of toiling for the benefit of capitalists instead of for themselves, some of them banded together to form the first national labor union, the Knights of Labor, in 1869. A *labor union* is a group of employees who join together to bargain with an employer or a group of employers over wages, benefits, and working conditions. The Knights of Labor included both skilled and unskilled laborers, but the American Federation of Labor (AFL), founded in 1886, targeted groups of skilled workers such as plumbers and carpenters; each of these craft unions maintained autonomy under the "umbrella" of the AFL.

Under early monopoly capitalism (1890–1940), most ownership rapidly shifted from individuals to huge *corporations*—large-scale organizations that have legal powers, such as the ability to enter into contracts and buy and sell property, separate from their individual owners. During this period, major industries, including oil, sugar, and grain, came under the control of a few corporations owned by shareholders. (Shareholders hold or own shares of stock and cannot be personally blamed for the actions of the corporation.) As automobile and steel plants shifted to mass production, the Congress of Industrial Organizations (CIO) was established in 1935 to represent both skilled and unskilled workers in industries such as automobile manufacturing. In 1937 General Motors workers held their first *sit-down strike* by refusing to work and paralyzing production, a move that came to dominate U.S. labor activism.

In advanced monopoly capitalism (1940–present), ownership and control of major industrial and business sectors have become increasingly concentrated, and many corporations have become more global in scope. *Transnational corporations* are large corporations that are headquartered in one or a few countries but sell and produce goods and services in many countries. Also referred to as *multinational corporations,* these entities play a major role in the economies and governments of many nations. Revenues for the top twenty public and private corporations worldwide are shown in ● Table 13.1, where their concentration in only a few industries, such as oil and gas, is evident.

Transnational corporations are not dependent on the labor, capital, or technology of any one country and may

TABLE 13.1

Revenues of the World's 20 Largest Public and Private Corporations (2014)

Company and Location	Revenues/GDP (in billions)
Wal-Mart Stores, Inc.—U.S. (Retail)	$485
Sinopec Group—China (Oil and gas)	$471
China National Petroleum—China (Oil and gas)	$438
Royal Dutch Shell—UK (Oil and gas)	$421
Exxon Mobil—U.S. (Oil and gas)	$394
Saudi Aramco—Saudi Arabia (Oil and gas)	$400 (2011)
BP—UK (Oil and gas)	$358
State Grid Corporation of China—China (Electric utility)	$338
Vitol—Netherlands (Commodities)	$307
Volkswagen Group—Germany (Automotive)	$263
Total—France (Oil and gas)	$253
Toyota—Japan (Automotive)	$249
Glencore Xstrata—Switzerland (Commodities)	$239
Chevron—U.S. (Oil and gas)	$220
Samsung Electronics—South Korea (Electronics)	$216
Apple Inc.—U.S. (Electronics)	$182
Berkshire Hathaway—U.S. (Conglomerate)	$182
China Railway Corp.—China (Transport)	$172
Phillips 66—U.S. (Oil and gas)	$171
E.ON—Germany (Electric utility)	$163

Sources: Compiled by the author and by wikipedia.org, 2015.

Francesco Dazzi/Shutterstock.com

FIGURE 13.3 Apple Inc. is one of the most successful transnational corporations on the planet, making billions from its innovative computers and mobile communications technologies.

move their operations to countries where wages and taxes are lower and the potential profits are higher (● Figure 13.3). Corporate considerations of this kind help explain why many jobs formerly located in the United States have shifted to lower-income nations where few employment opportunities

capitalism
an economic system characterized by private ownership of the means of production, from which personal profits can be derived through market competition and without government intervention.

labor union
a group of employees who join together to bargain with an employer or a group of employers over wages, benefits, and working conditions.

corporations
large-scale organizations that have legal powers, such as the ability to enter into contracts and buy and sell property, separate from their individual owners.

transnational corporations
large corporations that are headquartered in one or a few countries but sell and produce goods and services in many countries.

FIGURE 13.4 The high-tech sector of the global economy has changed the way in which ownership and control are exercised in twenty-first century capitalism. The entrepreneurs shown here are speaking at TechCrunch Disrupt, an event where technology startups launch their products and services so that they can compete for instant prize money, media coverage, and longer-term investments from venture capitalists.

generations, descendants of some of the early industrial capitalists have benefited from the economic deeds (and misdeeds) of their ancestors. In early monopoly capitalism, some stockholders derived massive profits from companies that held near-monopolies on specific goods and services.

In advanced (late) monopoly capitalism, profits have become more concentrated and technology plays a much more significant part in economic outcomes (● Figure 13.4). Although some people own *some* stock, they do not own *control*; in other words, they are unable to participate in establishing the policies that determine the size of the profit or the rate of return on investments, which affects the profits they derive.

Competition In theory, competition acts as a balance to excessive profits. When producers vie with one another for customers, they must be able to offer innovative goods and services at competitive prices. However, from the time of early industrial capitalism, the trend has been toward less, rather than more, competition among companies. In early monopoly capitalism, competition was diminished by increasing concentration *within* a particular industry, a classic case being the virtual monopoly on oil held by John D. Rockefeller's Standard Oil Company (Tarbell, 1925/1904; Lundberg, 1988). Today, Microsoft Corporation so dominates certain areas of the computer software industry that it has virtually no competitors in those areas.

In other situations, several companies may dominate certain industries. An **oligopoly** exists when several companies overwhelmingly control an entire industry. An example is the music industry, in which a few giant companies are behind many of the labels and artists (see ● Table 13.2). In fact, what was

exist and workers can be paid significantly less than their U.S. counterparts. This appears to be a fact of life whether workers are reviewing legal documents, producing automobiles and computers, or cooking hamburgers in fast-food restaurants owned by transnational corporations. It has become more widely accepted that economic crises and a growing global economy under capitalism have produced a rapidly shrinking middle class in the United States.

Pursuit of Personal Profit A tenet of capitalism is the belief that people are free to maximize their individual gain through personal profit; in the process, the entire society will benefit from their activities (Smith, 1976/1776). Economic development is assumed to benefit both capitalists and workers, and the general public also benefits from public expenditures (such as for roads, schools, and parks) made possible through an increase in business tax revenues.

During the period of industrial capitalism, however, specific individuals and families (not the general public) were the primary recipients of profits. For many

TABLE 13.2

The Music Industry's Big Three		
Company	**Country**	**Leading Artists**
Universal Music Group (EMI Recorded Music)	France	Taylor Swift
		Lady Gaga
		Shawn Mendes
		Maroon 5
		Nick Jonas
Sony Music Entertainment (EMI Music Publishing)	United States	Garth Brooks
		CAM
		5th Harmony
		Kelly Clarkson
Warner Music Group (Atlantic, Elektra)	United States	Red Hot Chili Peppers
		Josh Groban
		Wiz Khalifa
		Ed Sheeran
		Jason Derulo
		Muse

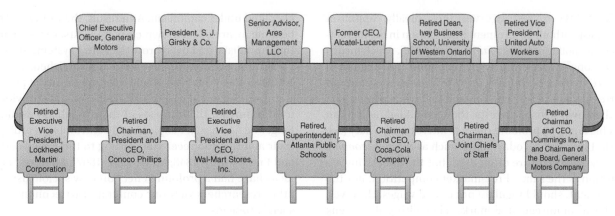

FIGURE 13.5 **The General Motors Board of Directors**

The 2015 General Motors Board of Directors shows the nature of interlocking directorates and the synergy that exists when people from various corporate and educational backgrounds come together to work for another corporation. On the chair representing each of the directors is the name of another entity each director is connected with, and his or her position with that entity.

Source: General Motors, 2015.

formerly known as the Big Four—Universal, Sony, Warner, and EMI—further consolidated into the Big Three in 2011, leaving only Universal, Sony, and Warner. This would be considered a *shared monopoly*—when four or fewer companies supply 50 percent or more of a particular market. Other industries that are dominated by just a few companies include those that manufacture automobiles, breakfast cereals, cigarettes, oil, and personal computers and tablets. However, as we have seen in recent years, even some corporations, such as automobile manufacturers, that have been considered "too big to fail" have indeed had problems serious enough that government bailouts have been required to keep them in business and some of their workers employed.

In advanced monopoly capitalism, mergers also occur *across* industries: Corporations gain near-monopoly control over all aspects of the production and distribution of a product by acquiring both the companies that supply the raw materials and the companies that are the outlets for their products. For example, an oil company may hold leases on the land where the oil is pumped out of the ground, own the plants that convert the oil into gasoline, and own the individual gasoline stations that sell the product to the public.

Corporations with control both within and across industries are often formed by a series of mergers and acquisitions across industries. These corporations are referred to as *conglomerates*—combinations of businesses in different commercial areas, all of which are owned by one holding company. Media ownership is a case in point; companies such as Time Warner have extensive holdings in radio and television stations, cable television companies, book publishing firms, and film production and distribution companies, to name only a few.

Competition is reduced over the long run by *interlocking corporate directorates*—members of the board of directors of one corporation who also sit on the board(s) of other corporations. Although the Clayton Antitrust Act of 1914 made it illegal for a person to sit simultaneously on the boards of directors of two corporations that are in *direct* competition with each other, a person may serve simultaneously on the board of a financial institution (a bank, for example) and the board of a commercial corporation (a computer manufacturing company or a furniture store chain, for example) that borrows money from the bank. Directors of competing corporations may also serve together on the board of a third corporation that is not in direct competition with the other two. An example of interlocking directorates is depicted in • Figure 13.5. Compensation for members of the boards of top corporations can be a million dollars or more per person per year when stock, stock options, and pensions are taken into account. To deflect public scrutiny, in recent years some corporate board members have "retired" from previous executive positions in high-powered corporations, banks, or law firms and thus have no visible conflict of interest.

Interlocking directorates diminish competition by producing interdependence. Individuals who serve on multiple boards are often able to forge cooperative arrangements that benefit their corporations but not necessarily the general public. When the same financial interests control several corporations, they are more likely to cooperate with one another than to compete.

oligopoly
a condition that exists when several companies overwhelmingly control an entire industry.

shared monopoly
a condition that exists when four or fewer companies supply 50 percent or more of a particular market.

conglomerate
a combination of businesses in different commercial areas, all of which are owned by one holding company.

interlocking corporate directorates
members of the board of directors of one corporation who also sit on the board(s) of other corporations.

Lack of Government Intervention Ideally, capitalism works best without government intervention in the marketplace. The policy of *laissez-faire* (les-ay-FARE, which means "leave alone") was advocated by economist Adam Smith in his 1776 treatise *An Inquiry into the Nature and Causes of the Wealth of Nations*. Smith argued that when people pursue their own selfish interests, they are guided "as if by an invisible hand" to promote the best interests of society (see Smith, 1976/1776). Today, terms such as *market economy* and *free enterprise* are often used, but the underlying assumption is the same: Free-market competition, not the government, should regulate prices and wages. However, the "ideal" of unregulated markets benefiting all citizens has seldom been realized. Individuals and companies in pursuit of higher profits have run roughshod over weaker competitors, and small businesses have grown into large, monopolistic corporations. Accordingly, government regulations were implemented in an effort to curb the excesses of the marketplace brought about by laissez-faire policies.

However, much of what is referred to as government intervention has been in the form of aid to business. Between 1850 and 1900, corporations received government assistance in the form of public subsidies and protection from competition by tariffs, patents, and trademarks. The federal government also gave large tracts of land to the privately owned railroads to encourage their expansion across the nation. Antitrust laws originally intended to break up monopolies were used instead against labor unions that supported workers' interests.

Government intervention in the twenty-first century has included the Emergency Economic Stabilization Act, which created the $700 billion Troubled Assets Relief Program (TARP). This program allowed the government to purchase failed bank assets that resulted primarily from the subprime mortgage crisis (in which people were encouraged to purchase homes that many of them could not afford). General Motors, Bank of America, and American International Group (AIG) were among the companies that received taxpayers' money from the TARP bailout. Overall, most corporations have gained much more than they have lost as a result of government involvement in the economy.

Socialism

Socialism is an economic system characterized by public ownership of the means of production, the pursuit of collective goals, and centralized decision making. Like "pure" capitalism, "pure" socialism does not exist. Karl Marx described socialism as a temporary stage en route to an ideal communist society. Although the terms *socialism* and *communism* are associated with Marx and are often used interchangeably, they are not identical. Marx defined communism as an economic system characterized by common ownership of all economic resources. In the *Communist Manifesto* and *Das Kapital*, he predicted that the working class would become increasingly impoverished and alienated under capitalism. As a result, the workers would become aware of their own class interests, revolt against the capitalists, and overthrow the entire system. After the revolution, private property would be abolished, and collectives of workers who would own the means of production would control capital. The government (previously used to further the interests of the capitalists) would no longer be necessary. People would contribute according to their abilities and receive according to their needs (Marx and Engels, 1967/1848; Marx, 1967/1867). Many of Marx's ideas have had a profound effect on how sociologists and other researchers view our contemporary economic and social problems.

"Ideal" socialism has three distinctive features: (1) public ownership of the means of production, (2) pursuit of collective goals, and (3) centralized decision making.

Public Ownership of the Means of Production In a truly socialist economy, the means of production are owned and controlled by a collectivity or the state, not by private individuals or corporations. For example, prior to the early 1990s the state owned all the natural resources and almost all the capital in the Soviet Union. At least in theory, goods were produced to meet the needs of the people. Access to housing and medical care was considered to be a right.

The leaders of what was then called the Soviet Union and some Eastern European nations decided to abandon government ownership and control of the means of production because the system was unresponsive to the needs of the marketplace and offered no incentive for increased efficiency. Since the 1990s, Russia and other states in the former Soviet Union have attempted to privatize ownership of production. Economic reforms in the 1990s privatized most industries, with the exceptions of the energy and defense-related sectors. Today, the state-owned Russian oil company Rosneft makes billions of dollars annually from the sale of oil.

China—previously the world's other major communist economy—has privatized many state industries. In *privatization*, resources are converted from state ownership to private ownership; the government takes an active role in developing, recognizing, and protecting private property rights. In the second decade of the twenty-first century, China has a hybrid political economy made up of both capitalism and an autocratic form of Communist Party governance. Economic growth brought about an increase in annual urban income, life expectancy increased by more than six years, and the rate of illiteracy dropped significantly. With these improvements, it is likely that the combination of communism and a modified form of capitalism will remain for the foreseeable future. By 2015, however, China was experiencing a slower economy with a lopsided job market that reduced many workers' opportunities to find the jobs they had hoped for, given the gradual transformation of their nation's economy (see "Sociology in Global Perspective").

Lopsided Job Market in China: A Mismatch Between Workers and Jobs

News Item: Wang Junping, a high school graduate who was previously employed as a farmer and then a coal miner, was being instructed at a Beijing employment agency on how to use a broom and mop to clean the local subway system. Ultimately, Junping decided not take the $320 a month position, believing the salary was too low and that it should be easy to find a good job in China's capital city. However, he was mistaken on this last point. (Gough, 2015)

With recent discussions about vast economic growth in China and a rapid shift from agriculture and rural living to manufacturing and urban living, it is no surprise that Wang Junping and many others like him believed that good jobs would be plentiful. Although agriculture jobs have been declining for years and rural residents have been moving in large numbers to urban areas, the manufacturing and other sectors are not thriving as much as analysts had predicted. One problem is that many companies have rising debt and *overcapacity*—a situation where the capacity to produce exceeds the demand for a specific product. In other words, businesses have produced more than customers are buying. So good jobs are not as readily available as Mr. Junping initially thought. Positions in the service sector are more readily available, but these involve tasks like subway cleaning (sanitation), retail, and fast-food service that are low-wage jobs. Nearly 40 percent of the workforce in China is employed in the service sector: This is approximately 300 million people! For a period of time, migrant workers performed many of these unskilled and semiskilled jobs, but as the

Chinese economy slowed down, migration also slowed, leaving the jobs for people like Mr. Junping.

Here is the other mismatch in the lopsided job market: China produces nearly 7 million new university graduates each year, and the supply greatly outpaces the demand for workers with a college degree. Each year, for every 88 jobs that require a college degree, there are about 100 university-graduate job seekers (Gough, 2015). According to a spokesperson for the Asian Development Bank, it is a fact that "a taxi driver in Beijing, unskilled, can make more money than a new university graduate" (qtd. in Gough, 2015: B8). Of course, young people are shocked to find this out after passing ultracompetitive exams to get into the finest institutions of higher education and then working diligently to earn top grades.

As we look at the workplace worldwide, we may see comparisons to our own nation in regard to the employment outlook for people with various levels of education. These examples show the intertwining of a nation's economy and the job opportunities for its residents as they attempt to fit into the existing social and economic structures of a society.

REFLECT & ANALYZE

Does China's slowing economy and changes in workers' opportunities have any effect on our lives? Do you believe the United States has a lopsided job market? Is there a mismatch between workers and the jobs that are available? Why or why not?

Pursuit of Collective Goals Socialism is based on the pursuit of collective goals rather than on personal profits. Equality in decision making replaces hierarchical relationships (such as between owners and workers or political leaders and citizens). Everyone shares in the goods and services of society, especially necessities such as food, clothing, shelter, and medical care, based on need, not on ability to pay. In reality, in nations such as China, members of the Communist Party are able to obtain low-interest loans from state-owned and state-operated banks as long as they play by party rules. In sum, even though pursuit of collective goals is one of the ideals of socialism, few societies can or do pursue purely collective goals.

Centralized Decision Making Another tenet of socialism is centralized decision making. In theory, economic decisions are based on the needs of society; the government is responsible for aiding the production and distribution of goods and services. Central planners set wages and prices to

ensure that the production process works. When problems such as shortages and unemployment arise, they can be dealt with quickly and effectively by the central government.

Mixed Economies

As we have seen, no economy is truly capitalist or socialist; most economies are mixtures of both. A **mixed economy** combines elements of a market economy (capitalism) with elements of a command economy (socialism). Sweden, Great Britain, France, and a number of other countries have mixed economies, sometimes referred to as

socialism
an economic system characterized by public ownership of the means of production, the pursuit of collective goals, and centralized decision making.

mixed economy
an economic system that combines elements of a market economy (capitalism) with elements of a command economy (socialism).

democratic socialism—an economic and political system that combines private ownership of some of the means of production, governmental distribution of some essential goods and services, and free elections (● Figure 13.6). For example, government ownership in Sweden is limited primarily to railroads, mineral resources, a public bank, and liquor and tobacco operations. Compared with capitalist economies, however, the government in a mixed economy plays a larger role in setting rules, policies, and objectives.

The government is also heavily involved in providing services such as medical care, child care, and transportation. In Sweden, for example, all residents have health insurance, housing subsidies, child allowances, paid parental leave, and day-care subsidies. Recently, some analysts have suggested that the United States has assumed many of the characteristics of a *welfare state*, a state in which there is extensive government action to provide support and services to the citizens, as it has attempted to meet the basic needs of older persons, young children, unemployed people, and persons with a disability.

FIGURE 13.6 For many decades, Russia had a state-controlled economy. However, beginning in the 1990s the government privatized many sectors of the economy, and new companies made vast profits in sectors such as oil exploration and similar endeavors. Here, workers in Siberia are setting pipe for the Yukos Oil Company.

LO3 **State** the key features of functionalist, conflict, and symbolic interactionist perspectives on the economy and work in the United States.

Perspectives on Economy and Work in the United States

Functionalists, conflict theorists, and symbolic interactionists view the economy and the nature of work from a variety of perspectives. We first examine functionalist and conflict views of the economy; then we focus on the symbolic interactionist perspective on job satisfaction and alienation.

Functionalist Perspective

Functionalists view the economy as a vital social institution because it is the means by which needed goods and services are produced and distributed. When the economy runs smoothly, other parts of society function more effectively. However, if the system becomes unbalanced, such as when demand does not keep up with production, maladjustment occurs (in this case, a surplus). Some problems can be easily remedied in the marketplace (through "free enterprise") or through government intervention (such as paying farmers *not* to plant when there is an oversupply of a crop).

However, other problems, such as periodic *peaks* (high points) and *troughs* (low points) in the business cycle, are more difficult to resolve. The *business cycle* is the rise and fall of economic activity relative to long-term growth in the economy. From this perspective, peaks occur when "business" has confidence in the country's economic future. During a peak, or *expansion period,* the economy thrives, and upward social mobility for workers and their families becomes possible.

The American Dream of upward mobility is linked to peaks in the business cycle. Once the peak is reached, however, the economy turns down because too large a surplus of goods has been produced. In part, this is because of *inflation*—a sustained and continuous increase in prices. Inflation erodes the value of people's money, and they are no longer able to purchase as high a percentage of the goods that have been produced. Because of this lack of demand, fewer goods are produced, workers are laid off, credit becomes difficult to obtain, and people cut back on their purchases even more, fearing unemployment. Eventually, this produces a distrust of the economy, resulting in a *recession*—a decline in an economy's total production that lasts six months or longer. To combat a recession, the government lowers interest rates (to make borrowing easier and to get more money back into circulation) in an attempt to spur the beginning of the next expansion period.

Conflict Perspective

Conflict theorists have a different view of business cycles and the economic system. From a conflict perspective, business cycles are the result of capitalist greed. In order to maximize profits, capitalists suppress the wages of workers

FIGURE 13.7 Fast-food employees in New York City and elsewhere led the charge in demanding an increase in the minimum wage paid by employers in the United States. Here, the signs of McDonald's workers focus on how an increase in pay will improve their lives and help lift up their families.

(• Figure 13.7). As the prices of products increase, workers are not able to purchase them in the quantities that have been produced. The resulting surpluses cause capitalists to reduce production, close factories, and lay off workers, thus contributing to the growth of the reserve army of the unemployed, whose presence helps reduce the wages of the remaining workers. In some situations, workers are replaced with machines or nonunionized workers.

Karl Marx referred to the propensity of capitalists to maximize profits by reducing wages as the *falling rate of profit*, which he believed to be one of the inherent contradictions of capitalism that would produce its eventual downfall. According to the political sociologist Michael Parenti, business *is* the economic system. Parenti believes that political leaders treat the health of the capitalist economy as a necessary condition for the health of the nation and that the goals of big business (rapid growth, high profits, and secure markets at home and abroad) become the goals of government (Parenti, 2007). To some conflict theorists, capitalism is the problem; to some functionalist theorists, however, capitalism is the solution to society's problems.

Symbolic Interactionist Perspective

Sociologists who focus on microlevel analyses are interested in how the economic system and the social organization of work affect people's attitudes and behavior. In particular, symbolic interactionists have examined the factors that contribute to job satisfaction. According to these analysts, work is an important source of self-identity for many people; it can help people feel positive about themselves, or it can cause them to feel alienated.

Alienation occurs when workers' needs for self-identity and meaning are not met and when work is done strictly for material gain, not a sense of personal satisfaction. According to Marx, workers are resistant to having very little power and no opportunities to make workplace decisions. This lack of control contributes to an ongoing struggle between workers and employers. Job segmentation, isolation of workers, and the discouragement of any type of pro-worker organizations (such as unions) further contribute to feelings of helplessness and frustration. Some occupations may be more closely associated with high levels of alienation than others.

Job Satisfaction The term *job satisfaction* is used to refer to people's attitudes toward their work, based on (1) their job responsibilities, (2) the organizational structure in which they work, and (3) their individual needs and values. Earlier studies have found that worker satisfaction is highest when employees have some degree of control over their work, when they are part of the decision-making process, when they are not too closely supervised, and when they believe that they play an important part in the outcome (Kohn et al., 1990). A study by the Heldrich Center for Workplace Development at Rutgers University (Godofsky, Zukin, and Van Horn, 2011) found that overall job satisfaction and job security declined significantly in the first decade of the twenty-first century. The Society for Human Resource Management (2012) found that on certain dimensions of job satisfaction, less than 30 percent of workers in their study indicated they were "very satisfied" with compensation/pay, communication between employees and senior management, benefits, and career advancement opportunities, among other things.

Job satisfaction is often related to both intrinsic and extrinsic factors. Intrinsic factors pertain to the nature of the work itself, whereas extrinsic factors include such things as vacation and holiday policies, parking privileges, on-site day-care centers, and other amenities that

democratic socialism
an economic and political system that combines private ownership of some of the means of production, governmental distribution of some essential goods and services, and free elections.

welfare state
a state in which there is extensive government action to provide support and services to the citizens.

Key Concept	
Functionalist Perspective	The economy is a vital social institution because it is the means by which needed goods and services are produced and distributed.
Conflict Perspective	The capitalist economy is based on greed. In order to maximize profits, capitalists suppress the wages of workers, who, in turn, cannot purchase products, making it necessary for capitalists to reduce production, close factories, lay off workers, and adopt other remedies that are detrimental to workers and society.
Symbolic Interactionist Perspective	Many workers experience job satisfaction when they like their job responsibilities and the organizational structure in which they work and when their individual needs and values are met. Alienation occurs when workers do not gain a sense of self-identity from their jobs and when their work is done completely for material gain and not for personal satisfaction.

FIGURE 13.8 What workplace amenities, such as nursery facilities or diaper-changing stations, may be of value to working mothers?

contribute to workers' overall perception that their employer cares about them (● Figure 13.8). Some studies of job satisfaction link this issue to employee engagement.

Employee Engagement The term *employee engagement* refers to the emotional commitment that the employee has to the organization and its goals (Kruse, 2012). Based on this definition, employee engagement does not mean the same thing as employee happiness or satisfaction: Engagement means that the workers have a genuine commitment to their work and the company or organization that employs them. According to some business professionals, engaged employees help create higher profits, better workplace morale, and overall better business outcomes (Kruse, 2012). From a sociological

perspective, the question remains about whether having engaged employees who provide better service, higher customer satisfaction, increased sales, higher levels of profit, and higher shareholder returns also brings about workers who are highly satisfied with their work environment and their work life.

The Concept Quick Review summarizes the three major sociological perspectives on the economy and work in the United States.

Work in the Contemporary United States

As we have seen, the kind of work that people perform has changed dramatically, and the distribution of persons in the labor force has also changed dramatically. The term *labor force* refers to the number of people ages 16 and over who are either employed or actively looking for work. It does not include active-duty military personnel or persons who are institutionalized, such as prison inmates. Let's look specifically at recent trends in labor force participation.

Trends in Labor Force Participation

Beginning with the Great Recession in 2007, there was a decrease in the share of the U.S. population (ages 16 and over) that participated in the U.S. labor force. Even between March 2014 and March 2015, the civilian labor force participation rate continued to slide slightly, from 63.2 percent to 62.7 percent (U.S. Bureau of Labor

Statistics, 2015b). Analysts believe that these factors contribute to shifts in overall labor force participation:

- **Aging of the U.S. population** as more of the Baby Boomer population reaches retirement age and becomes eligible for Social Security benefits.

- **Later retirement** as a larger share of older workers are remaining in the labor force, partly because of healthier aging and partly become of lost retirement resources during the Great Recession.

- **Schooling among the young** as more 16- to 24-year olds decide to attain more education.

- **Disability** as a higher percent of the population in all age categories is identified as too sick to work or as disabled.

- **"Don't want a job"** as a reason given by more individuals between the ages of 35–54 (prime work-age category) for not having or wanting a job. This sometimes relates to disability; other times it refers to school attendance, retirement, taking care of the house or family, and various other reasons.

As this list indicates, the U.S. labor force participation rate may vary based on many factors, not just the availability of employment opportunities in a specific location. Participation rate by age is an important consideration when analyzing work in the contemporary United States. For example, if older workers delay retirement, this may put upward pressure on labor force participation rates because younger people are less likely to find certain kinds of employment opportunities. Gender is another factor that varies in labor force participation.

Trends in Gender and Race/Ethnicity in Employment

In the 1950s, more than 80 percent of employable men were in the U.S. labor force, as contrasted with less than 40 percent of employable adult women. However, the number of women, both single and married, in the labor force has increased significantly since 1970, when about 53 percent of single women and 41 percent of married women held paid positions. After women's participation in the labor market reached an all-time high of 60 percent in 1999, the percentage was slightly lower at 57 percent in 2014 (U.S. Bureau of Labor Statistics, 2015c).

Throughout the years, distinct differences have existed in labor force participation between single and married women. About 53 percent of single women and 41 percent of married women held paid positions in the 1970s. Today, the labor force participation rate is 74.8 percent for mothers who have never married or who are widowed, divorced, separated, or married but living apart from their spouse. The participation rate for married mothers with a spouse present still remains lower (67.8 percent) than the rate of mothers in the other categories (U.S. Bureau of Labor Statistics, 2015b).

Does the presence of children make a difference in whether women are employed outside their household?

The answer is both yes and no. Yes, many women with children are employed, but mothers with younger children are less likely to be in the labor force than those with older children. The participation rate of mothers with infants under a year old was 57.1 percent, as compared to 74.7 percent for mothers whose youngest child was in the 6 to 17 age category (U.S. Bureau of Labor Statistics, 2015c).

Women's wages are important for their families, but women are nearly twice as likely as men to work part time and to be employed in near-minimum-wage jobs. Women account for nearly 30 percent of all part-time workers, while men make up slightly more than 15 percent of all part-time workers. Part-time employment typically does not have the same benefits as full-time work, including sick leave, health insurance, retirement contributions by employers, or paid vacation time. Some people work part time by choice, but others, particularly women, find it necessary to do so in order to take care of children and other household responsibilities. Most part-time jobs pay an hourly wage often at or below the federal minimum wage, so many women hold more than one job. This is particularly true for women who are the single heads of households with one or more children present.

Looking at the intersection of gender and race/ethnicity as it relates to labor force participation, African American women ages 16 and older have the highest workforce participation rate at 59.1 percent, as compared to white (non-Hispanic) women (56.7 percent), Hispanic women (56 percent), and Asian American women (55.8 percent). Data are not available for Native American women, but among both men and women in specific nations, the Chippewa and the Pueblo have the highest labor force participation rates (roughly 59 percent), while the Navajo and the Cherokee have rates in the 52–53 percent range (Institute for Women's Policy Research, 2015). In a later section, we will look at the problem of unemployment as it intersects with race/ethnicity and gender in the United States.

 LO4 **Describe** professions and discuss how they differ from other occupations.

The Kinds of Work We Do

The economy in the United States and other contemporary societies is partially based on the work (purposeful activity, labor, or toil) that people perform. However, work in high-income nations is highly differentiated and often fragmented because people have many kinds of occupations. Some occupations are referred to as *professions*.

Professions

Although sociologists do not always agree on exactly which occupations are professions, most of them agree that the term *professionals* includes most doctors, natural scientists, engineers, computer scientists, certified public accountants, economists, social scientists, psychotherapists, lawyers, policy experts of various sorts, professors, at least some journalists and editors, some clergy, and some artists and writers. (See • Figure 13.9 to find the 2014 average annual pay for some occupations and professions.)

Characteristics of Professions   *Professions* are high-status, knowledge-based occupations that have five major characteristics (Freidson, 1970, 1986; Larson, 1977):

1. *Abstract, specialized knowledge.* Professionals have abstract, specialized knowledge of their field based on formal education and interaction with colleagues.

2. *Autonomy.* Professionals are autonomous in that they can rely on their own judgment in selecting the relevant knowledge or the appropriate technique for dealing with a problem.

3. *Self-regulation.* In exchange for autonomy, professionals are theoretically self-regulating. All professions have licensing, accreditation, and regulatory associa-

tions that set professional standards and that require members to adhere to a code of ethics as a form of public accountability.

4. *Authority.* Because of their authority, professionals expect compliance with their directions and advice. Their authority is based on mastery of the body of specialized knowledge and on their profession's autonomy.

5. *Altruism.* Ideally, professionals have concern for others that goes beyond their self-interest or personal comfort so that they can help a patient or client.

Social Reproduction of Professionals   Although higher education is one of the primary qualifications for a profession, the emphasis on education gives children whose parents are professionals a disproportionate advantage early in life. There is a direct linkage between parental education/income and children's scores on college admissions tests such as the SAT, as shown in • Figure 13.10. In turn, test scores are directly related to students' ability to

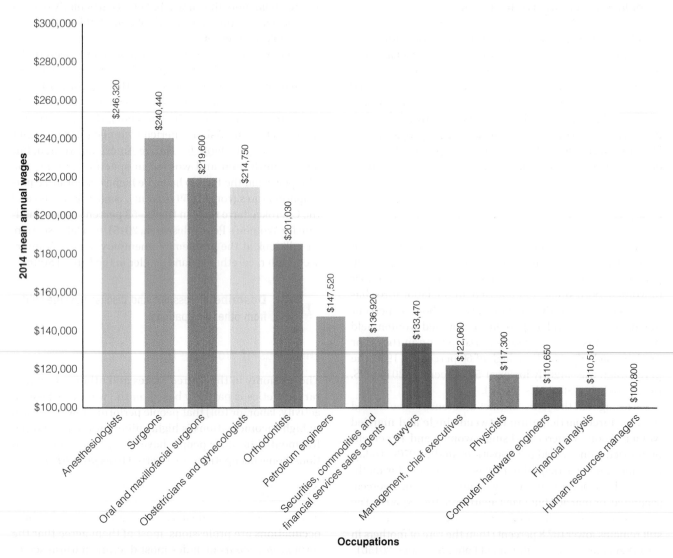

FIGURE 13.9   Selected Highest-Paying Occupations, 2014

Source: U.S. Bureau of Labor Statistics, 2015d.

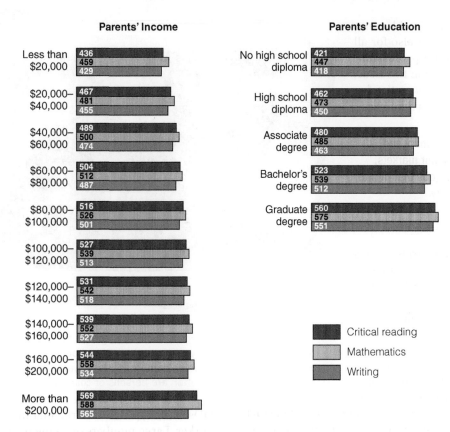

Parents' Income

	Critical reading	Mathematics	Writing
Less than $20,000	436	459	429
$20,000–$40,000	467	481	455
$40,000–$60,000	489	500	474
$60,000–$80,000	504	512	487
$80,000–$100,000	516	526	501
$100,000–$120,000	527	539	513
$120,000–$140,000	531	542	518
$140,000–$160,000	539	552	527
$160,000–$200,000	544	558	534
More than $200,000	569	588	565

Parents' Education

	Critical reading	Mathematics	Writing
No high school diploma	421	447	418
High school diploma	462	473	450
Associate degree	480	485	463
Bachelor's degree	523	539	512
Graduate degree	560	575	551

- ■ Critical reading
- ■ Mathematics
- ■ Writing

FIGURE 13.10 SAT Scores by Parents' Income and Education, 2014
Source: SAT, 2014.

gain admission to colleges and universities, which serve as springboards to most professions. Race and gender are also factors in access to the professions.

Deprofessionalization Certain professions are undergoing a process of *deprofessionalization*, in which some of the characteristics of a profession are eliminated. Occupations such as pharmacist have already been *deskilled,* as Nino Guidici explains:

> In the old days [people] took druggists as doctors. . . . All we do [now] is count pills. Count out twelve on the counter, put 'em in here, count out twelve more. . . . Doctors used to write out their own formulas and we made most of these things. Most of the work is now done in the laboratory. The real druggist is found in the manufacturing firms. They're the factory workers and they're the pharmacists. We just get the name of the drugs and the number and the directions. It's a lot easier. (qtd. in Terkel, 1990/1972)

However, colleges of pharmacy in many universities have fought against deprofessionalization by upgrading the degrees awarded to pharmacy graduates from the traditional B.S. in pharmacy to a Pharm.D. This upgrading of degrees has also occurred over the past two decades in law schools, where the Bachelor of Laws (LL.B.) has been changed to the Juris Doctor (J.D.) degree.

Today, deprofessionalization has expanded into higher education. In her article "Remaking the Public U's Professorate," Jennifer Ruth (2015) of Portland State University voices a common concern heard on many college and university campuses nationwide about how far fewer teaching positions are tenured or tenure-track appointments than in the past. Increasing numbers of faculty members are being hired as full-time nontenure-track faculty (NTT), leaving them without any assurance that they will have a continuing academic appointment at the institution or the security of "academic freedom" and "shared governance" allegedly accorded to tenure-track and tenured faculty members. Although NTT faculty members may have similar academic qualifications as tenure-track faculty who were hired on the research/teaching tenure track, NTT positions have been deprofessionalized because persons in these academic appointments are employed on a year-to-year basis without any long-range plan for their career success, and they are not guaranteed the similar income and status within the university as the tenure-track and tenured faculty. Some believe that a way to reprofessionalize these positions is to create a full-time teaching-intensive tenure track composed of higher course loads than research/teaching faculty but with comparable salaries and equal benefits (Ruth, 2015). Some tenure-track and tenured faculty would argue that deprofessionalization

professions
high-status, knowledge-based occupations.

in higher education is a problem not only for NTT faculty but also for all research and teaching professionals. If you are interested in this issue, *The Chronicle of Higher Education* frequently has discussions on this topic.

LO5 **Compare** primary and secondary labor markets, and identify jobs that fit into each category.

Other Occupations

Occupations are categories of jobs that involve similar activities at different work sites. More than 600 different occupational categories and 35,000 occupation titles, ranging from motion-picture cartoonist to drop-hammer operator, are currently listed by the U.S. Census Bureau. Historically, occupations were classified as blue collar and white collar. Blue-collar workers were primarily factory and craft workers who did manual labor; white-collar workers were office workers and professionals. However, contemporary workers in the service sector do not easily fit into either of these categories; neither do the so-called pink-collar workers, primarily women, who are employed in occupations such as preschool teacher, dental assistant, secretary, and clerk. (The term refers to an era when some female restaurant employees were required to wear uniforms with a pink collar.)

Sociologists establish broad occupational categories by distinguishing between employment in the primary labor market and in the secondary labor market. The *primary labor market* consists of high-paying jobs with good benefits that have some degree of security and the possibility of future advancement. By contrast, the *secondary labor market* consists of low-paying jobs with few benefits and very little job security or possibility for future advancement.

Upper-Tier Jobs: Managers and Supervisors
Managers are essential in contemporary bureaucracies, where work is highly specialized and authority structures are hierarchical. Workers at each level of the hierarchy take orders from their immediate superiors and perhaps give orders to a few subordinates. Upper-level managers are typically responsible for coordination of activities and control of workers.

Lower-Tier and Marginal Jobs
Positions in the lower tier of the service sector are part of the secondary labor market, characterized by low wages, little job security, few chances for advancement, higher unemployment rates, and very limited (if any) unemployment benefits. Typical lower-tier positions include janitor, waitress, messenger,

Hongqi Zhang/Alamy

FIGURE 13.11 Occupational segregation by race and gender is visible in the construction industry where architects and engineers are more likely to be white (non-Hispanic) men as compared with lower-level supervisors and construction workers who most often are persons of color or recent immigrants.

sales clerk, typist, file clerk, migrant laborer, and textile worker. Large numbers of young people, people of color, recent immigrants, and white women are employed in this sector (• Figure 13.11).

Marginal jobs differ from the employment norms of the society in which they are located; examples in the U.S. labor market include jobs in personal service industries such as eating and drinking places, hotels, and laundries, as well as private household workers. Marginal jobs are frequently not covered by government work regulations—such as minimum standards of pay, working conditions, and safety standards—or do not offer sufficient hours of work each week to provide a living. Marginal jobs are often associated with the immigrant labor force because some employers believe they can hire willing workers for less pay and little governmental oversight regarding wages and safety standards.

LO6 **Explain** problems that are associated with contingent work and the underground (informal) economy.

Contingent Work

Contingent work is part-time work, temporary work, or subcontracted work that offers advantages to employers but that can be detrimental to the welfare of workers. Contingent work is found in every segment of the labor force. The federal government is part of this trend, as is private enterprise. In the health care field, physicians, nurses, and other workers are increasingly employed through temporary agencies. Employers benefit by hiring workers on a part-time or temporary basis; they are able to cut costs, maximize profits, and have workers available only when they need them. Temporary workers are the fastest-growing segment of the contingent workforce, and agencies that "place" them have increased dramatically in number in the last decade.

Subcontracted work is another form of contingent work that often cuts employers' costs at the expense of workers. Instead of employing a large workforce, many companies have significantly reduced the size of their payrolls and benefit plans by **subcontracting**—an agreement in which a corporation contracts with other (usually smaller) firms to provide specialized components, products, or services to the larger corporation. Hiring and paying workers become the responsibility of the subcontractor, not of the larger corporation.

New York City/Alamy

FIGURE 13.12 How have media representations glorifying luxury items such as designer handbags contributed to an underground economy? Shown here are temporary display racks that are quickly set up by unscrupulous vendors who remove them and run when law enforcement officials arrive on the scene.

The Underground (Informal) Economy

Some social analysts make a distinction between the legitimate and the underground (informal) economies in the United States. For the most part the occupations previously described in this chapter operate within the *legitimate economy:* Taxes on income are paid by employers and employees, and individuals who hold jobs requiring a specialized license (such as craftspeople or taxi drivers) possess the appropriate credentials for their work. By contrast, the *underground economy* is made up of a wide variety of activities through which people make money that they do not report to the government or through endeavors that may involve criminal behavior (Venkatesh, 2006). Sometimes referred to as the "informal" or "shadow economy," one segment of the underground economy is made up of workers who are paid "off the books," which means that they are paid in cash, their earnings are not reported, and no taxes are paid. Lawful jobs, such as nannies, construction workers, and landscape/yard workers, are often part of the shadow economy because workers and bosses make under-the-table deals so that both can gain through the transaction: Employers pay less for workers' services, and workers have more money to take home than if they paid taxes on their earnings.

The underground economy also involves the selling of lawful goods that are purchased "off the books" so that no taxes are paid, as well as the sale of "designer alternative fashion" products that may be counterfeit ("knockoff") merchandise (● Figure 13.12). Demand for such products is strong at all times but frequently increases in difficult economic times

because many people retain a desire for luxury goods that are widely publicized by the media even when people have fewer resources to allocate to such purchases.

According to one way of thinking, operating a business in the underground economy reveals capitalism at its best because it shows how the "free market" might work if there were no government intervention. However, from another perspective, selling goods or services in the underground economy borders on—or moves into—criminal behavior. For some individuals the underground economy offers the only way to purchase certain goods or to overcome unemployment,

occupations
categories of jobs that involve similar activities at different work sites.

primary labor market
the sector of the labor market that consists of high-paying jobs with good benefits that have some degree of security and the possibility of future advancement.

secondary labor market
the sector of the labor market that consists of low-paying jobs with few benefits and very little job security or possibility for future advancement.

marginal jobs
jobs that differ from the employment norms of the society in which they are located.

contingent work
part-time work, temporary work, or subcontracted work that offers advantages to employers but that can be detrimental to the welfare of workers.

subcontracting
an agreement in which a corporation contracts with other (usually smaller) firms to provide specialized components, products, or services to the larger corporation.

particularly in low-income and poverty areas where people may feel alienated from the wider world and believe that they must use shady means to survive (see Venkatesh, 2006).

Describe different kinds of unemployment and discuss current trends in U.S. unemployment.

Unemployment

Who is considered to be unemployed? Based on a definition established by the Bureau of Labor Statistics, part of the U.S. Department of Labor, people are classified as "unemployed" if they currently do not have a job, have been actively looking for a job within the four weeks prior to the Labor Department's survey, and are currently available for work. Actively looking for work means that the person must be involved in these activities: contacting employers directly or having a job interview, submitting résumés or filling out applications, answering job ads, checking union and professional registers that show open positions, or engaging in some other active job-search methods.

There are three major types of unemployment: cyclical, seasonal, and structural. *Cyclical unemployment* occurs as a result of lower rates of production during recessions in the business cycle; a recession is a decline in an economy's total production that lasts at least six months. Although massive

layoffs initially occur, some of the workers will eventually be rehired, largely depending on the length and severity of the recession. *Seasonal unemployment* results from shifts in the demand for workers based on conditions such as the weather (in agriculture, the construction industry, and tourism) or the season (holidays and summer vacations). Both of these types of unemployment tend to be relatively temporary.

By contrast, structural unemployment may be permanent. *Structural unemployment* arises because the skills demanded by employers do not match the skills of the unemployed or because the unemployed do not live where the jobs are located. This type of unemployment often occurs when a number of plants in the same industry are closed or when new technology makes certain jobs obsolete. Structural unemployment often results from capital flight—the investment of capital in foreign facilities, as previously discussed. Today, many workers fear losing their jobs, exhausting their unemployment benefits (if any), and still not being able to find another job.

The ***unemployment rate*** is the percentage of unemployed persons in the labor force actively seeking jobs. The second decade of the twenty-first century has seen significant change in the unemployment rate. The U.S. unemployment rate in 2000 was 4.0 percent. By 2011, the overall rate hovered around 9 percent before falling to 5.4 percent in April 2015. About 8.5 million individuals were classified as unemployed in April 2015, but the rate of unemployment varied by state. (● Figure 13.13 shows

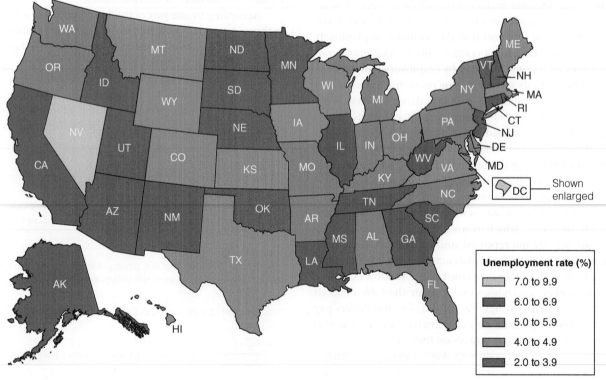

p = preliminary.

NOTE: Rates shown are a percentage of the labor force. Data refer to place of residence. Estimates for the current month are subject to revision the following month.

FIGURE 13.13 U.S. Unemployment Rates by State, 2015

Source: U.S. Bureau of Labor Statistics Local Area Unemployment Statistics, April 2015.

how unemployment rates vary by state.) In 2015 the overall U.S. unemployment rate for adult men (20 years and over) was 5.0 percent, as compared to 4.9 percent for adult women. However, the breakdown across race/ethnic and age categories tells a more complete story: African Americans of all ages had a 9.6 percent unemployment rate, Hispanics (Latinos/as) had a rate of 6.9 percent, white Americans (non-Hispanics) had a rate of 4.7 percent, and Asian Americans had a rate of 4.4 percent.

Among teenagers (ages 16 to 19 years), however, the story is more disparate. Teenagers of all racial–ethnic categories (and both sexes) had a 17.5 percent unemployment rate, but African Americans in the same category had an unemployment rate of 25 percent. Although lower than the rate for African American youths, the unemployment rate among Hispanics (Latinos and Latinas) was 21.1 percent. Among white Americans (non-Hispanics) of both sexes, the unemployment rate was 15.7 percent. Data were not available for Asian Americans in this age category but previously had been as high as 20 percent (U.S. Bureau of Labor Statistics, 2015c).

Unemployment is also related to educational attainment. The unemployment rate for persons with less than a high school diploma was 8.6 percent in 2015. For high school graduates with no college, the unemployment rate was 5.4 percent, as compared with 4.7 for persons with some college or an associate's degree, and a low of 2.7 percent for individuals with a bachelor's degree and higher (U.S. Bureau of Labor Statistics, 2015c).

How reliable are unemployment statistics? This is a widely debated topic among politicians and media analysts. We must consider that, like other types of "official" statistics, unemployment rates may not provide us with the whole story. Individuals who become discouraged in their attempt to find work and no longer actively seek employment are not counted as unemployed. Some analysts believe that unemployment rates may drop for a period of time because people either do not seek work or they accept part-time or temporary jobs when they cannot find full-time employment. According to the U.S. Bureau of Labor Statistics (2015c), 2.1 million people were marginally attached to the labor force in 2015, but they were seeking employment and had looked for a job at some time during the prior 12 months. Among the marginally attached were 756,000 "discouraged workers" who were not currently looking for work because they believed that no jobs were available for them or had other problems in

AFP/Getty Images

FIGURE 13.14   In many countries, workers protest and go on strike in an effort to improve their wages and working conditions. Employees who attempt to gain control over their work situation have varying degrees of success in achieving their goal. How prevalent is worker resistance and activism in the United States today?

seeking employment at that time (U.S. Bureau of Labor Statistics, 2015c).

 LO8   **Identify** ways in which workers attempt to gain control over their work situation.

Worker Resistance and Activism

In their individual and collective struggles to improve their work environment and gain some measure of control over their work-related activities, workers around the world have used a number of methods to resist workplace alienation (• Figure 13.14). Many have joined labor unions to gain strength through collective action.

Labor Unions

U.S. labor unions came into being in the mid-nineteenth century. Unions have been credited with gaining an eight-hour workday, a five-day workweek, health and retirement benefits, sick leave and unemployment insurance, and workplace health and safety standards for many employees. As one bumper sticker reads, "Unions: The folks who brought you the weekend."

Most of these gains have occurred through *collective bargaining*—negotiations between employers and labor union leaders on behalf of workers. However, some states

unemployment rate
the percentage of unemployed persons in the labor force actively seeking jobs.

have passed laws making it harder for workers to organize or to engage in collective bargaining because of state and local budget shortfalls. For example, Wisconsin passed a law that bans collective bargaining by unionized government workers for benefits and pensions but allows them to bargain as a union for pay as long as their raises do not exceed the rate of inflation.

A 2012 study by the Pew Research Center found that nearly two-thirds (64 percent) of Americans agree that labor unions are necessary to protect the working person. However, 33 percent of respondents disagreed with this assessment, meaning that there is not a clear consensus about labor unions. The disagreement is particularly strong based on political party lines: Fifty-four percent of Republicans in the Pew study indicated that they did not believe labor unions were necessary to protect workers, as compared with 43 percent who agreed. By contrast, Democrats continue to be strong supporters of labor unions, with 80 percent responding that they believe labor unions are necessary (Pew Research Center, 2012b).

Although views about labor unions appear to have stabilized in recent years, some analysts attribute diminished support for unions to the fact that more people may have had a better understanding of unions in the past because more individuals belonged to unions or had family members who did, while unions have now largely disappeared from the private sector.

In the past, more union leaders called for strikes to force employers to accept the union's position on wages and benefits. The number of workers involved in the actions declined from a peak of more than 2.5 million in 1971 to 34,000 in 2014, when there were 11 major strikes and lockouts that involved 1,000 or more workers (see • Figure 13.15). In 2014 there were 200,000 days that were idle from major work stoppages, which was much lower than 2012, when 1.13 million days were idle (U.S. Bureau of Labor Statistics, 2015a). Work stoppages in 2014 were primarily in the health care industry, educational services, construction, and manufacturing.

Union membership has also been shrinking over the past three decades. In 2014 only 11.1 percent of wage and salary workers were union members, compared with 20.1 percent in 1983, the first year for which the federal government compiled such data (U.S. Bureau of Labor Statistics, 2015a). The total number of wage and salary workers belonging to unions was 14.6 million in 2014, down from 17.7 million in 1983. Union membership is higher for public-sector workers (35.7 percent) than for private-sector employees (6.6 percent). Workers in education, training, library, and protective-service occupations had the highest unionization rate, at 35.3 percent. More men (11.7 percent) are union members than women (10.5 percent). In 2014, among major racial and ethnic groups, African American workers were more likely to be union members (13.2 percent) than workers who were white American (10.8 percent), Asian American (10.4 percent), or Hispanic (9.2 percent).

The highest rates of union membership were among workers between the ages of 45 to 64. In the 45–54 age category, the rate was 13.8 percent; in the 55–64 age category, the rate was 14.1 percent. If unions continue to "age" in their membership ranks and decline in overall membership and involvement in the workplace, should we anticipate that this will have an even more detrimental effect on the hard-earned gains of workers in the

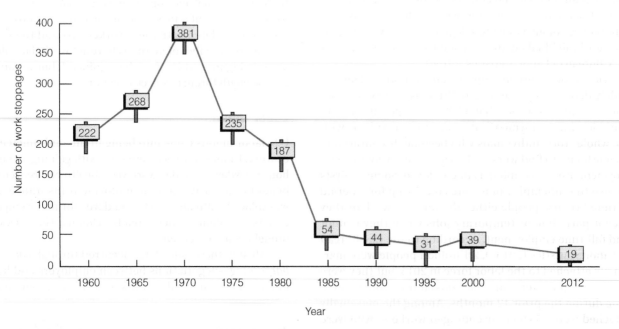

FIGURE 13.15 **Major Work Stoppages in the United States, 1960–2012**

Source: U.S. Bureau of Labor Statistics, 2013c.

United States? Or is the era of labor union influence on the U.S. workplace and economy effectively reaching its end? These questions remain to be more fully answered in the future.

Employment Opportunities for Persons with a Disability

In the past and to some extent today, people with disabilities have been steered away from certain occupations by teachers, parents, prospective employers, and others who tended to focus more on what they could not do rather than what they were capable of doing. In 1990 the United States became the first nation to formally address the issue of equality for persons with a disability when Congress passed the Americans with Disabilities Act (ADA) to prohibit discrimination on the basis of disability. Combined with previous disability rights laws (such as those that provide for the elimination of architectural barriers from new, federally funded buildings and for the maximum integration of schoolchildren with disabilities), the ADA is a legal mandate for the full equality of people with disabilities. The federal law defines a person with a disability as an individual with at least one of the following conditions: He or she is deaf or has serious difficulty hearing; is blind or has serious difficulty seeing even when wearing glasses; has serious difficulty concentrating, remembering, or making decisions because of a physical, mental, or emotional condition; has serious difficulty walking or climbing stairs; has difficulty dressing or bathing; or has difficulty doing errands alone such as visiting a doctor's office or shopping because of a physical, mental, or emotional condition.

Despite the ADA and other laws regarding disability rights, many persons with a disability remain unemployed or have been tracked into disability-related service roles (such as helping other persons with a disability). The economic recession and the slow recovery thereafter also hit workers with disabilities hard, particularly those individuals who have mobility impairments or difficulty performing routine daily activities. In 2015 only 19.3 percent of persons ages 16 years and over with one or more disabilities were employed (as compared with 68.4 percent of people without disabilities). The 2015 unemployment rate for people with disabilities was 10.0 percent, as compared with 4.9 percent for people without disabilities. However, many other persons with disabilities were not in the labor force at all, nor were they actively seeking employment (U.S. Bureau of Labor Statistics, 2015c).

Many persons with a disability believe that they could work if they were offered the opportunity. However, even when persons with a disability are able to find jobs, they earn less than persons without a disability. On average, workers with a disability receive lower pay, have less job security, and have less access to health insurance, pension plans, and training than their nondisabled counterparts

(Society for Human Resource Management, 2010). In the second decade of the twenty-first century, there is some cause for hope because increasing numbers of people are becoming involved in the campaign for disability employment (see "You Can Make a Difference").

Looking Ahead: The Global Economy and Work in the Future

How will the U.S. economy and work look in the future? What about the global economy? Although sociologists do not have a crystal ball with which to predict the future, some general trends can be suggested.

The U.S. Economy

Many of the trends we examined in this chapter will continue to produce dramatic changes in the organization of the economy and work in the twenty-first century. Some U.S. workers may find themselves fighting for a larger piece of the economic pie. Millennials between the ages of 25 and 34 are entering the work world in rapidly increasing numbers, combined with job growth in 2015. New, nonfarm job gains were higher in 2015 than they have been in the previous 17 years. It remains to be seen whether this growth will be sustained, but it appears that the U.S. economy might be stronger than it has been since the Great Recession of 2007–2009.

Of course, one of the greatest concerns for the future is employment for U.S. workers. Well-paid jobs have been the backbone of the middle class in this country, as well as a major source of upward mobility for people throughout history. In the mid-to-late twentieth century, permanent, stable jobs were the norm. These jobs provided opportunities for advancement and employee benefits such as health insurance and a retirement plan. In the second decade of the twenty-first century, the norm has become a job market divided between well-paid, stable employment and lower-paying jobs or temporary and contingent work that provides little job security and a feeling of stagnation. Thus, workers are increasingly fragmented into two major labor market divisions: (1) those who work in the more innovative, primary sector and (2) those whose jobs are located in the growing secondary, marginal sector. In the innovative sector, increased productivity will be the watchword as corporations respond to heightened international competition. In the marginal sector, alienation continues to grow as temporary workers, sometimes including professionals, look for a chance to increase their earnings and find job security.

Where will the greatest problems be in the workforce? Workers with less formal education are losing ground on wages faster than other workers because the kinds of jobs that are available have changed. Men with fewer years of formal education hold fewer jobs in well-paid manufacturing positions than they did in the past. More men with

YOU CAN MAKE A DIFFERENCE

Focusing on What People with Disabilities Can Do, Not What They Can't Do!

Sometimes employers hesitate to hire an adult with a disability because of a misguided sense that the person may not be able to handle the job when the going gets tough. I've used a wheelchair for most of my life, and I would argue that people with disabilities are in fact better equipped for acute problem-solving than their peers without disabilities. That's mainly because we're experts in finding creative ways to perform tasks that others may take for granted.

—Ralph Braun (2009), CEO of BraunAbility, explaining why he believes that employers' fears about hiring workers with a disability are unfounded

If you are a person with a disability or a person who plans to hire and supervise employees—some of whom may have a disability—you can make a difference in the workplace by learning more about disability employment and by promoting positive employment outcomes for people with disabilities. Here are some important job-search tips from the Massachusetts One-Stop Career Centers (Commonwealth of Massachusetts, 2011) for job seekers who have a disability:

- *Focus on your abilities and not disabilities.* While an employer cannot use the interview of the application

When they are provided with employment opportunities, people with disabilities often become outstanding employees in the labor force. What can be done to dispel commonly held myths about workers with disabilities?

fewer years of education are employed in low-wage jobs, such as food service, groundskeeping, and maintenance. Even manufacturing employees are earning less (when adjusted for inflation) than workers who held similar positions a generation ago (Irwin, 2015). According to a Brookings Institution report, the median earnings of working men between the ages of 30 and 45 without a high school diploma fell 20 percent from 1990 to 2013 (Irwin, 2015). So as we look toward the future, the economy looks most bleak for those with limited years of formal education who work in low-wage positions with no opportunity for advancement or who are long-term unemployed.

Some analysts attribute work-related problems, particularly in manufacturing jobs, to a long, downward slide in union organization and membership. Employees have lost bargaining power, globalization and technological advances have eliminated many jobs, and less educated workers are forced to compete for a limited number of low-wage jobs in food service and maintenance. These factors contribute to a further depression of wages.

What do you think will happen to work in the future? Some analysts believe that additional worker activism will take place as employees in the service sector demand higher wages and the right to unionize. For example,

process to inquire about a potential disability, it is helpful to bring up your disability if it relates to your ability to perform the job.

- *Practice interviewing before the interview.* Use mock interview sessions with trusted friends and relatives, be sure that clothing is clean and pressed, and make eye contact with the interviewer.

- *Be realistic about what types of positions will fit best for your disability.*

- *Read up on how to conduct a successful job search and interview.*

- *Learn as much as you can about the company or organization where you are interviewing.*

- *Become an in-house expert.* Add new job skills and become an expert in a particular area so that you will be more valuable to an employer.

Although these are just a few suggestions for having a positive employment outcome when seeking a job, each of us can make a difference, regardless of our ability/disability status, if we recognize the value and talent that persons with disabilities bring to the workplace. According to the Campaign for Disability Employment (2011), "Whether good economic times or bad, it's the organizations that know how to identify and recognize talent that are most likely to succeed." If you are interested in becoming involved, search the Internet for sites such as the Campaign for Disability Employment, which explains how to promote positive employment outcomes for people with disabilities or how to find your own opportunities, if you are a person with a disability.

Finally, each of us can make a difference by actively working to dispel these commonly held myths about workers with disabilities (U.S. Department of Labor, 2013):

Myth 1: Employees with disabilities have a higher absentee rate than employees without disabilities.

Fact: Studies show that employees with disabilities are not absent any more than employees without disabilities.

Myth 2: Persons with disabilities are unable to meet performance standards, thus making them a bad employment risk.

Fact: Studies have shown that the vast majority of employees with disabilities rated average or better in job performance.

Myth 3: Persons with disabilities have problems getting to work.

Fact: Persons with disabilities are as successful at supplying their own transportation as are persons without a disability.

Myth 4: Employees with disabilities are more likely to have accidents on the job than are employees without disabilities.

Fact: Safety records are virtually identical for workers with and without disabilities.

Lack of familiarity with disability issues and lack of involvement with individuals with a disability contribute to negative attitudes about employing persons with disabilities. You can make a difference by helping to dissipate myths and groundless fears that exclude many people from becoming fully productive members of the U.S. workforce.

Organizations such as Think Beyond the Label work to help employers find candidates with a disability. This organization has a national TV and print advertising campaign to make employers and others aware that what gets in the way of employment is the *label,* not the *disability* (Entwisle, 2015). If you are interested, you can visit the organization's website for additional information. As the website states, "Think Beyond the Label is a public–private partnership that delivers information, outreach and resources to businesses, job seekers and the public workforce system to ensure greater recruiting and hiring opportunities for job candidates with disabilities."

McDonald's has experienced frequent protests by employees. In November 2012, 200 New York workers walked off their jobs at McDonald's restaurants in protests organized by the Service Employees International Union. The protests expanded to nearly 200 cities by 2015. The workers were part of the "Fight for $15" campaign to increase the minimum wage not only at McDonald's but also at state and federal levels. Subsequently, McDonald's Corporation announced an increase in the minimum wage paid to employees who worked in fast-food outlets owned by the corporation. Ironically, the vast majority of McDonald's employees nationwide work in independently owned franchises, so this wage increase would not be applicable to them. Other occupations are also affected by the state and federal minimum-wage issue, including rental car companies, hotels, auto maintenance, tax preparation services, and many others (Scheiber, 2015).

How will the decline in labor unions affect workers in the future? According to a 2015 Pew Research Center study, the decline of organized labor has already affected *all* occupation groups, but the problem will be most intense in private-sector occupations that formerly were unionized. Although the Pew study found that labor union membership had declined overall, there was a slight increase in labor organizing in "management occupations, even by some workers who formerly opposed union activism. Persons in management occupations, such as construction foreman, food-service manager, school administrator, and other mid-level supervisory personnel, have experienced an inflation-adjusted decline in annual wages in recent years, thus making them more responsive to labor union participation (DeSilver, 2015).

Despite problems in the workplace and the U.S. economy, some economists and political leaders suggest that we can be cautiously optimistic about the future of the economy. Some researchers and politicians stress that we need reform in workforce policies such as these:

- Providing additional tax credits to businesses to hire new workers.

- Providing long-term education and training programs to help people make career changes that will help them find work.

- Requiring people to enroll in training programs to help them find jobs if they are going to receive unemployment insurance.

- Creating more government-funded jobs for unemployed people.

- Establishing longer and higher unemployment insurance benefits or changing unemployment insurance to "reemployment insurance" that is coordinated with job search and career counseling (based on Van Horn, 2013).

Another factor in providing opportunities in the workplace is better coordination between the social institutions of education and the economy so that there will be more understanding and alignment among schools, students, and employers. Of course, problems in regard to the U.S. economy and work are not confined to our own national boundaries: Our lives are also intertwined with the global economy.

Global Economic Interdependence and Competition

Global economic interdependence refers to the mutual dependence that various countries have on one another with regard to importing and exporting goods and services. Commodities such as oil are produced in some countries and purchased by people in other countries that do not have a sufficient supply to meet existing demand. Over time, global interdependence emerges as countries come to rely on one another for specific products. Conflict may arise when competition over scarce goods and services occurs or when multiple suppliers flood the market with a product, such as a glut of Middle Eastern oil on the U.S. market.

What effect does global economic interdependence have on the United States? For one thing, the global economy is affecting the value of the U.S. dollar, which has been slipping in recent years. This affects the price of gas in this country, as well as the cost of most consumer goods, because it means that a dollar today is not worth as much as it was previously in purchasing power. Overall, the U.S. dollar is shrinking as a percentage of the world's currency supply. This means that in the future the dollar may not be the denomination that is most often used in international transactions and may not be one of the world's dominant reserve currencies—monies that are held in significant quantities by many governments and institutions as part of their foreign exchange reserves. Historically, the U.S. dollar has been the international pricing currency for products traded on a global market and for commodities such as oil and gold (● Figure 13.16). In 2013 the dollar reached a fifteen-year low, as compared to global currencies. However, by 2015 the U.S. dollar had reached a twelve-year high against the euro and an eight-year high versus the Japanese yen.

This example helps us to see how global economic interdependence and competition can affect countries such as the United States and bring about international shifts in economic conditions. Changes in the global economy require people in all nations—including the United States—to make changes in the way that things are done on the individual, regional, and national levels (see "Sociology and Social Policy"). Will we be able to make sufficient changes to strengthen our economy and workforce now and in the future?

FIGURE 13.16 The global economy is deeply embedded in the oil industry worldwide. How might fluctuations in oil demand affect workers across the globe?

SOCIOLOGY & SOCIAL POLICY

How Globalization Changes the Nature of Social Policy

In 1492 Christopher Columbus set sail for India, going west. He had the Niña, the Pinta and the Santa Maria. He never did find India, but he called the people he met "Indians" and came home and reported to his king and queen: "The world is round." I set off for India 512 years later. I knew just which direction I was going. I went east. I had Lufthansa business class, and I came home and reported only to my wife and only in a whisper: "The world is flat."

—author and newspaper columnist Thomas L. Friedman (2005) describing how the global economy is changing all areas of economic, political, and social life, including how we formulate social policy

What does Thomas Friedman mean when he states that "the world is flat"? As discussed in Friedman's (2007) best-selling book, *The World Is Flat 3.0: A Brief History of the Twenty-First Century,* flat means "level" or "connected" because (in his opinion) there is a more level global playing field in business (and almost any other endeavor) in the twenty-first century. According to Friedman, global telecommunications and the lowering of many trade and political barriers have brought about a new global era driven by *individuals,* not just by major corporations or giant trade organizations such as the World Bank. These individuals include entrepreneurs who create startup ventures around the world and computer freelancers whose work knows no boundaries (based on the idea of the older nation-state borders) when it comes to the transfer of information. Other factors that Friedman also believes have contributed to the "flattening" of the world include the streaming of the supply chain (Walmart, for example) and the organization of information on the Internet by Google and Yahoo!, among others.

Worldwide, many freelancers and business entrepreneurs are not in the United States: They reside in nations such as India and China, where it is now possible to do more than merely compete in low-wage manufacturing and routine information labor (such as workers in call centers), but also be in the top levels of research and design work and in professions such as law.

If Friedman's assertions are correct that the world is becoming more flat and the United States is losing some of its dominance in the global political and economic arena, where does this leave us in regard to social policy? Friedman suggests that if the United States is to remain competitive in the global economy, we must not continue to do things as they have previously been done. He believes that we should have a thoughtful national discussion about what globalization means in all of our lives. Because there has been a shift from large-scale corporate players in the global economy to individual entrepreneurs and freelancers, we must look to each *individual* in our country to see how we can best play the economic game in the twenty-first century. In addition to nationwide policies, Friedman (2007) believes that our social policy regarding globalization must begin at home; children and young adults must be encouraged to rise to the economic challenge that faces them.

REFLECT & ANALYZE

When social policy becomes personal (as Friedman believes), are we willing to engage in the changes that it requires? Are Friedman's assumptions about the changing world order accurate? What do you think? What other arguments might be presented?

CHAPTER REVIEW Q & A

LO1 What is the primary function of the economy, and how do preindustrial, industrial, and postindustrial economies compare?

The economy is the social institution that ensures the maintenance of society through the production, distribution, and consumption of goods and services. Preindustrial economies include hunting and gathering, horticultural and pastoral, and agrarian societies. Most workers engage in primary sector production—the extraction of raw materials and natural resources from the environment. Industrial societies engage in secondary sector production, which is based on the processing of raw materials (from the primary sector) into finished goods. Postindustrial societies engage in tertiary sector production by providing services rather than goods.

LO2 What are the key characteristics of capitalism, socialism, and mixed economies?

Capitalism is characterized by private ownership of the means of production, the pursuit of personal profit, competition, and limited government intervention. By contrast, socialism is characterized by public ownership of the means of production, the pursuit of collective goals, and centralized decision making. In mixed economies, elements of a capitalist, market economy are combined with elements of a command, socialist economy. These mixed economies are often referred to as democratic socialism.

LO3 What are the functionalist, conflict, and symbolic interactionist perspectives on the economy and work in the United States?

According to functionalists, the economy is a vital social institution because it is the means by which needed goods and services are produced and distributed. Conflict theorists suggest that the capitalist economy is based on greed. In order to maximize profits, capitalists suppress the wages of workers, who, in turn, cannot purchase products, making it necessary for capitalists to reduce production, close factories, lay off workers, and adopt other remedies that are detrimental to workers and society. Symbolic interactionists focus on the microlevel of the economic system, particularly on the social organization of work and its effects on workers' attitudes and behavior. Many workers experience job satisfaction when they like their job responsibilities and the organizational structure in which they work and when their individual needs and values are met. Alienation occurs when workers do not gain a sense of self-identity from their jobs and when their work is done completely for material gain and not for personal satisfaction.

LO4 What are the characteristics of professions, and how do professions differ from other occupations?

Professions are high-status, knowledge-based occupations characterized by abstract, specialized knowledge; autonomy; self-regulation; authority over clients and subordinate occupational groups; and a degree of altruism. Occupations are categories of jobs that involve similar activities at different work sites.

LO5 What is the difference in primary and secondary labor markets, and what jobs fit into each category?

The primary labor market consists of high-paying jobs with good benefits and some degree of job security and a chance for future advancement (examples: managers and supervisors). By contrast, the secondary labor market is made up of low-wage jobs with few benefits and very little job security or possibility for future advancement (examples: janitors and waitresses). Marginal jobs are part of the secondary labor market because they differ in some manner from mainstream employment norms that jobs should be legal, be covered by government regulations, be relatively permanent, and provide adequate hours and pay in order to make a living.

LO6 What problems are associated with contingent work and the underground (informal) economy?

Contingent work is part-time work, temporary work, or subcontracted work that offers advantages to employers but may be detrimental to workers. Through the use of contingent workers, employers are able to cut costs and maximize profits, but workers have little or no job security. The underground economy is made up of a wide variety of activities through which people make money that they do not report to the government or through endeavors that may involve criminal behavior.

LO7 What are the different kinds of unemployment, and what are the current trends in U.S. unemployment?

Unemployment may be cyclical, meaning that it is not constant but occurs as a result of lower rates of production and/or problems such as an economic recession. It may also be seasonal, which results from shifts in the demands for workers based on conditions such as weather or the season. Finally, structural unemployment is the most difficult to overcome because it may be permanent: The skills of workers do not match the skills demanded by employers, or no jobs exist in a specific city or region. The second decade of the twenty-first century has seen a significant increase in unemployment.

LO8 How do workers attempt to gain control over their work situation?

Some workers have engaged in resistance or activism in an effort to overcome workplace alienation and gain control over their environment. For some this has meant joining labor unions and participating in strikes; however, such activism has been in a period of decline in the second decade of the twenty-first century.

KEY TERMS

capitalism 372
conglomerate 375
contingent work 384
corporations 372
democratic socialism 378

economy 369
interlocking corporate
   directorates 375
labor union 372
marginal jobs 384

mixed economy 377
occupations 384
oligopoly 374
primary labor market 384
primary sector production 370

QUESTIONS for CRITICAL THINKING

1 If you were the manager of a computer software division, how might you encourage innovation among your technical employees? How might you encourage efficiency? If you were the manager of a fast-food restaurant, how might you increase job satisfaction and decrease job alienation among your employees?

2 Using Chapter 2 as a guide, design a study to determine the degree of altruism in certain professions. What might be your hypothesis? What variables would you study? What research methods would provide the best data for analysis?

3 What types of occupations will have the highest prestige and income in 2030? The lowest prestige and income? What, if anything, does your answer reflect about the future of the U.S. economy?

ANSWERS to the SOCIOLOGY QUIZ

ON WORK IN THE UNITED STATES IN THE 2010s

1 **False** According to U.S. Census Bureau data, telecommuting has increased 79 percent during the 2010s. Now, about 3 percent of the workforce (approximately 3.2 million employees) work at least half the time at home.

2 **False** Although studies are limited, most research tends to show that working at home leads to performance increases, as well as to improved work satisfaction and less turnover among employees.

3 **True** This widely held assumption has been proven true by statistics on unemployment by education level. In early 2015 the unemployment rate for bachelor's degree or higher was 2.5 percent, while the rate for less than high school was 8.4 percent.

4 **False** Human resources departments in corporations and other workplaces distinguish between happiness or satisfaction and engagement at work. *Employee engagement* refers to the emotional commitment that the employee has to the organization and its goals.

5 **True** Among the fastest-growing occupations between 2012 and 2022 in the United States are industrial–organizational psychologists, personal care aides, and home health aides. Annual incomes in these top three categories range from above $80,000 to less than $20,000.

6 **False** Only 19.3 percent of persons with disabilities were in the U.S. labor force in April 2015, as compared to 68.4 percent of people without disabilities.

7 **True** According to the latest statistics available, physicians, surgeons, and specialists in oral and maxillofacial surgery and orthodontia earned mean wages between $166,810 and $246,320 in 2014. These figures place them among the highest-paying occupations listed by the U.S. Bureau of Labor Statistics.

8 **True** Based on the tests currently given, the College Board, which administers the SAT, documents the fact that students' scores on critical reading, mathematics, and writing are highest among students whose parents have the highest levels of income and education. There are plans for changing the exams.

Sources: Based on Bloom et al., 2013; College Board, 2012; Glass, 2013; Kruse, 2012; U.S. Bureau of Labor Statistics, 2015d; and U.S. Department of Labor, 2015.

POLITICS AND GOVERNMENT
IN GLOBAL PERSPECTIVE

iStockphoto.com/Best-photo

LEARNING OBJECTIVES

1 **Define** the terms *power* and *authority*, and distinguish among traditional, charismatic, and rational–legal authority.

2 **Compare** the major political systems found around the world.

3 **Describe** the pluralist model of the U.S. power structure and note the part that special interest groups and political action committees play in this approach.

4 **Discuss** the elite model of power and identify the approaches of C. Wright Mills and G. William Domhoff.

5 **Explain** how political parties shape the government and identify recent trends in political participation and voter apathy.

6 **Identify** key factors that contribute to the power of government bureaucracies and explain how bureaucratic power tends to take on a life of its own.

7 **Explain** what is meant by the iron triangle and the military–industrial complex.

8 **Discuss** the effects of militarism, terrorism, and war on citizens and the U.S. political system.

SOCIOLOGY & EVERYDAY LIFE

Experiencing Politics and History at a Presidential Inauguration

Tens of thousands of college students were present in Washington, D.C., when Barack Obama was publicly sworn in for his second term as president of the United States on January 21, 2013. Millions of others watched the inauguration live on TV, computers, or smartphones and tablets that were using the latest apps. George Washington University senior Jamie Blynn blogged about how much more exciting it was to attend the event in person, as compared to watching the 2009 inauguration on an outdated TV in her twelfth-grade English classroom:

> Luckily, we pushed our way through the hordes of people . . . [to] the entrances to the National Mall, and somehow stumbled upon the perfect location just minutes before Sen. Chuck Schumer kicked off the ceremony. . . . In the crowd, we stood among hundreds of thousands of people from [all] around the country, all of us waving the same American flags proudly in the air. [The patriotism] was like nothing I have ever experienced before; we all came from different races, ethnicities, and all-around different backgrounds, but we could all find common ground behind the flag. . . . This was a great end to . . . my college career.

—JAMIE BLYNN (2013)

Student media representatives were also present in Washington: Multimedia reporter Kevin Kucharshi, Yale class of 2015, provided students with eyewitness accounts of the inauguration and students' perceptions about the event and the future of the United States in his YouTube documentary *A Part of History: College Students and the 2013 Inauguration*.

Thousands of students attended the 2013 presidential inauguration. What do you think college students like you might learn about the political process by observing such a historic event?

A re you interested in politics and the political process taking place around you? College students Jamie Blynn and Kevin Kucharshi both believe that politics is interesting and that it was important to be at a presidential inauguration—to be a part of history and to see the results of the democratic process of electing public officials. Student involvement in politics and government stands in sharp contrast to what we often hear about young people's voter apathy and lack of engagement. However, many college students care about their nation and what their own future holds in the United States and the larger global community.

The political landscape has been vastly modified in a very short period of time as a result of instantaneous social networking and websites. Only a few years ago, voters learned about political candidates from stump speeches, newspapers, and the evening news on television. In the second decade of the twenty-first century, millions of people of all ages sign on to Facebook, Twitter, and other social media sites to follow their favorite candidates and learn more about their messages. Do people actually vote for specific candidates or engage in political activism because they have been influenced by the media? This is an even more

significant question in light of the nature of both traditional and social media. And the line between traditional and more-contemporary media has blurred rapidly. Most traditional media outlets, including television, radio, newspaper, and other print media, have adopted social media into their reporting so that the two flow together and are often intertwined. Television anchors give play-by-play accounts of a political event, while viewers comment on Twitter and/or Facebook.

Our major social institution of politics and government is both reflective of and influenced by the social institution of media, which constitutes another powerful social force in its own right. We live in an age of political and economic uncertainty and constant discord regarding most decisions made by our nation's political and business leaders. Sociologists are concerned about how the social institution of politics and government operates and the extent to which people are (or are not) influenced by media sources. In this chapter we discuss the intertwining nature of contemporary politics, government, and media. Before reading on, test your knowledge of politics and the media by taking the "Sociology and Everyday Life" quiz.

How Much Do You Know About Politics and the Media?

TRUE	FALSE		
T	F	1	More than 80 percent of American adults use the Internet, and many of them also participate in social networking sites where political information is available.
T	F	2	Studies have found that most users of social networking sites find these sites to be "very important" or "somewhat important" in keeping up with political news or discussing political issues with others.
T	F	3	People who describe their political beliefs as conservative are more likely to use social networking sites than those who describe themselves as "moderate" or "liberal."
T	F	4	People using social networking sites are often surprised about what their friends have posted about politics because they assumed their friends' political beliefs were different from what their statements indicate.
T	F	5	Young adults and African Americans are more likely than white Americans (non-Hispanic) and Hispanics to see social networking sites as important for keeping up with political news, recruiting people to get involved, finding others with similar views, and debating or discussing political issues with others.
T	F	6	Many social networking site users indicate that they have changed their political views based on reading posts about specific issues.
T	F	7	Because more media users have turned to social media for information about politics, traditional print and television journalists have taken a more active role in conducting investigations into the backgrounds and political platforms of candidates.
T	F	8	The number of viewers of local TV news programming has decreased each year since 2007, even in presidential election years.

Answers can be found at the end of the chapter.

Define the terms *power* and *authority,* and distinguish among traditional, charismatic, and rational–legal authority.

Politics, Power, and Authority

Politics is the social institution through which power is acquired and exercised by some people and groups. In contemporary societies the government is the primary political system. *Government* is the formal organization that has the legal and political authority to regulate the relationships among members of a society and between the society and those outside its borders. Some social analysts refer to the government as the *state*—the political entity that possesses a legitimate monopoly over the use of force within its territory to achieve its goals.

Whereas political science focuses primarily on power and its distribution in different types of political systems, *political sociology* is the area of sociology that examines the nature and consequences of power within or between societies, as well as the social and political conflicts that

lead to changes in the allocation of power. Political sociology primarily focuses on the *social circumstances* of politics and explores how the political arena and its actors are intertwined with social institutions such as the economy, religion, education, and the media.

What is the relationship between politics and media? Recent research suggests that this relationship is

politics
the social institution through which power is acquired and exercised by some people and groups.

government
the formal organization that has the legal and political authority to regulate the relationships among members of a society and between the society and those outside its borders.

state
the political entity that possesses a legitimate monopoly over the use of force within its territory to achieve its goals.

political sociology
the area of sociology that examines the nature and consequences of power within or between societies, as well as the social and political conflicts that lead to changes in the allocation of power.

dramatically changing in an era where various forms of media are omnipresent twenty-four hours a day, seven days a week, or "24/7." As early as 1998, sociologist Michael Parenti argued that the media distort—either intentionally or unintentionally—the information they provide to citizens. According to Parenti, the media use their power to influence public opinion in favor of management over labor, corporations over those who criticize them, affluent whites over racial and ethnic minorities, political officials over protestors, and free-market capitalists over those who are in favor of public-sector development. Nearly two decades later, if we look at the overall relationship among politics, government, and both the "traditional" media and the rapidly expanding social media, we can see that some of Parenti's assertions remain relevant regarding the use, distribution, and possible abuse of power. According to a Pew Research Center Project study (2013c), the role of social media such as Twitter and Facebook grew exponentially in influencing voters in the 2012 U.S. presidential election (see ●Figure 14.1), while the part journalists played was much smaller than in previous elections.

FIGURE 14.1   Facebook has changed the face of politics, including how candidates run for office and how people respond to candidates and the political process. What social media sites do you prefer to use to learn more about current political and social events?

Power and Authority

Power is the ability of persons or groups to achieve their goals despite opposition from others (Weber, 1968/1922). Through the use of persuasion, authority, or force, some people are able to get others to acquiesce to their demands. Consequently, power is a *social relationship* that involves both leaders and followers. Power is also a dimension in the structure of social stratification. Persons in positions of power control valuable resources of society—including wealth, status, comfort, and safety—and are able to direct the actions of others while protecting and enhancing the privileged social position of their class (Domhoff, 2014). And according to the sociologist G. William Domhoff (2002), the media tend to reflect "the biases of those with access to them—corporate leaders, government officials, and policy experts." For example, research by the Pew Research Center (2013c) concluded that in the 2012 presidential election, campaign reporters acted primarily as "megaphones," rather than "investigators," in reporting the assertions that were put forward by candidates and other political participants. As a result, there was less effort on the part of journalists to interpret, contextualize, and sometimes verify statements that were made by the opposing parties. Reports such as this raise an important question: What effects might power shifts in the media have on our perceptions of what is taking place in U.S. politics and government?

Shifting to the bigger picture: What about power on a global basis? Although the most basic form of power is physical violence or force, most political leaders do not want to base their power on force alone. Instead, they seek to legitimize their power by turning it into ***authority***—power that people accept as legitimate rather than coercive.

Ideal Types of Authority

Who is most likely to accept authority as legitimate and adhere to it? People have a greater tendency to accept authority as legitimate if they are economically or politically dependent on those who hold power. They may also accept authority more readily if it reflects their own beliefs and values. Weber's outline of three *ideal types* of authority—traditional, charismatic, and rational–legal—shows how different bases of legitimacy are tied to a society's economy.

Traditional Authority   According to Weber, ***traditional authority*** is power that is legitimized on the basis of long-standing custom (●Figure 14.2). In preindustrial societies the authority of traditional leaders, such as kings, queens, pharaohs, emperors, and religious dignitaries, is usually grounded in religious beliefs and custom. For example, British kings and queens historically traced their authority from God. Members of subordinate classes obey a traditional leader's edicts out of economic and political dependency and sometimes personal loyalty. However, as societies industrialize, traditional authority is challenged by a more complex division of labor and by the wider

diversity of people who now inhabit the area as a result of high immigration rates.

Gender, race, and class relations are closely intertwined with traditional authority. Political scientist Zillah R. Eisenstein (1994) suggests that *racialized patriarchy*—the continual interplay of race and gender—reinforces traditional structures of power in contemporary societies. According to Eisenstein (1994: 2), "Patriarchy differentiates women from men while privileging men. Racism simultaneously differentiates people of color from whites and privileges whiteness. These processes are distinct but intertwined."

Charismatic Authority   *Charismatic authority* is power legitimized on the basis of a leader's exceptional personal qualities or the demonstration of extraordinary insight and accomplishment that inspire loyalty and obedience from followers (Figure 14.2). Charismatic leaders may be politicians, soldiers, religious leaders, and entertainers, among others.

Charismatic authority tends to be temporary and relatively unstable; it derives primarily from individual leaders (who may change their minds, leave, or die) and from an administrative structure usually limited to a small number of faithful followers. For this reason, charismatic authority often becomes routinized. The ***routinization of charisma*** occurs when charismatic authority is succeeded by a bureaucracy controlled by a rationally established authority or by a combination of traditional and bureaucratic authority (Turner, Beeghley, and Powers, 2007). According to Weber (1968/1922: 1148), "It is the fate of charisma to recede . . . after it has entered the permanent structures of social action."

Rational–Legal Authority   According to Weber, ***rational–legal authority*** is power legitimized by law or written rules and regulations (Figure 14.2). Rational–legal authority—also known as *bureaucratic authority*—is based

FIGURE 14.2   Max Weber's three types of global authority are shown here in global perspective. Pope Francis is an example of traditional authority sanctioned by custom. Mother Teresa exemplifies charismatic authority, for her leadership was based on personal qualities. The U.S. Supreme Court represents rational–legal authority, which depends upon established rules and procedures.

power
according to Max Weber, the ability of persons or groups to achieve their goals despite opposition from others.

authority
power that people accept as legitimate rather than coercive.

traditional authority
power that is legitimized on the basis of long-standing custom.

charismatic authority
power legitimized on the basis of a leader's exceptional personal qualities or the demonstration of extraordinary insight and accomplishment that inspire loyalty and obedience from followers.

routinization of charisma
the process by which charismatic authority is succeeded by a bureaucracy controlled by a rationally established authority or by a combination of traditional and bureaucratic authority.

rational–legal authority
power legitimized by law or written rules and procedures. Also referred to as *bureaucratic authority*.

Weber's Three Types of Authority

Type	Description	Examples
Traditional	Legitimized by long-standing custom	Patrimony (authority resides in traditional leader supported by larger social structures, as in old British monarchy)
	Subject to erosion as traditions weaken	Patriarchy (rule by men occupying traditional positions of authority, as in the family)
Charismatic	Based on leader's personal qualities	Napoleon Adolf Hitler
	Temporary and unstable	Martin Luther King Jr. César Chávez Mother Teresa
Rational–Legal	Legitimized by rationally established rules and procedures	Modern British Parliament
	Authority resides in the office, not the person	U.S. presidency, Congress, federal bureaucracy

on an organizational structure that includes a clearly defined division of labor, hierarchy of authority, formal rules, and impersonality. Power is legitimized by procedures; if leaders obtain their positions in a procedurally correct manner (such as by election or appointment), they have the right to act.

Rational–legal authority is held by elected or appointed government officials and by officers in a formal organization. However, authority is invested in the *office,* not in the *person* who holds the office. For example, although the U.S. Constitution grants rational–legal authority to the office of the presidency, a president who fails to uphold the public trust may be removed from office. In contemporary society the media may play an important role in bringing to light allegations about presidents or other elected officials, including President Richard M. Nixon in the 1970s Watergate investigation, the late-1990s sex scandal involving President Bill Clinton, and the more-recent ethics charges of money laundering involving members of Congress during the Obama administration.

In a rational–legal system, the governmental bureaucracy is the apparatus responsible for creating and enforcing rules in the public interest. Weber believed that rational–legal authority was the only means to attain efficient, flexible, and competent regulation under a rule of law. Weber's three types of authority are summarized in the Concept Quick Review.

 LO2 **Compare** the major political systems found around the world.

Political Systems in Global Perspective

Political systems as we know them today have evolved slowly. In the earliest societies, politics was not an entity separate from other aspects of life. Political institutions first emerged in agrarian societies as they acquired surpluses and developed greater social inequality. Elites took control of politics and used custom or traditional authority to justify their position. When cities developed circa 3500–3000 B.C.E., the *city-state*—a city whose power extended to adjacent areas—became the center of political power.

Nation-states as we know them began to develop in Europe between the twelfth and fifteenth centuries (see Tilly, 1975). A *nation-state* is a unit of political organization that has recognizable national boundaries and whose citizens possess specific legal rights and obligations. Nation-states emerge as countries develop specific geographic territories and acquire greater ability to defend their borders. Improvements in communication and transportation make it possible for people in a larger geographic area to share a common language and culture. As charismatic and traditional forms of authority are superseded by rational–legal authority, legal standards come to prevail in all areas of life, and the nation-state claims a monopoly over the legitimate use of force.

The U.S. State Department recognizes 195 independent nation-states throughout the world (U.S. Department of State, 2015). Today, everyone is born, lives, and dies under the auspices of a nation-state. Four main types of political systems are found in nation-states: monarchy, authoritarianism, totalitarianism, and democracy.

Monarchy

Monarchy is a political system in which power resides in one person or family and is passed from generation to generation through lines of inheritance (●Figure 14.3). Monarchies are most common in agrarian societies and are associated with traditional authority patterns. However, the relative power of monarchs has varied across nations, depending on religious, political, and economic conditions.

FIGURE 14.3 Through its many ups and downs, the British royal family has remained a symbol of Great Britain's monarchy—today headed by Queen Elizabeth. Monarchies typically pass power from generation to generation, and Queen Elizabeth's grandson, Prince William and Catherine (Kate), Duchess of Cambridge represent the future of the royal family's rule.

Absolute monarchs claim a hereditary right to rule (based on membership in a noble family) or a divine right to rule (a God-given right to rule that legitimizes the exercise of power). In limited monarchies, rulers depend on powerful members of the nobility to retain their thrones. Unlike absolute monarchs, *limited monarchs* are not considered to be above the law. In *constitutional monarchies,* the royalty serve as symbolic rulers or heads of state, while actual authority is held by elected officials in national parliaments. In present-day monarchies such as the United Kingdom, Sweden, Spain, and the Netherlands, members of royal families primarily perform ceremonial functions. In the United Kingdom, for example, the media often focus large amounts of time and attention on the royal family but concentrate on the personal lives of its members.

Authoritarianism

Authoritarianism is a political system controlled by rulers who deny popular participation in government. A few authoritarian regimes have been absolute monarchies whose rulers claim a hereditary right to their position and where the final decision on issues rests with the monarch. Today, Saudi Arabia, Qatar, and Swaziland are examples of absolute monarchies. In *dictatorships,* power is gained and held by a single individual. Pure dictatorships are rare; all rulers need the support of the military and the backing of business elites to maintain their position. *Military juntas* result when military officers seize power from the government, as has happened over time in Argentina, Chile, and Haiti. In recent decades, authoritarian regimes have existed in Cuba and the People's Republic of China. Political analysts question whether authoritarian regimes will remain into the twenty-first century if these nations' leaders continue to engage in discussions about normalization of relations with the United States. Reestablishment of diplomatic relations, closely watched by international media sources, may deter the typical efforts of an authoritarian regime to control the media and suppress coverage of topics that do not reflect positively on the regime.

monarchy
a political system in which power resides in one person or family and is passed from generation to generation through lines of inheritance.

authoritarianism
a political system controlled by rulers who deny popular participation in government.

Totalitarianism

Totalitarianism is a political system in which the state seeks to regulate all aspects of people's public and private lives. Totalitarianism relies on modern technology to monitor and control people; mass propaganda and electronic surveillance are widely used to influence people's thinking and to control their actions. One example of a totalitarian regime was the Nazi Party in Germany during World War II; military leaders there sought to control all aspects of national life, not just government operations. Other examples include the former Soviet Union, with vestiges of this approach remaining in contemporary Russian leadership.

To keep people from rebelling, totalitarian governments enforce conformity: People are denied the right to assemble for political purposes, access to information is strictly controlled, and secret police enforce compliance, creating an environment of constant fear and suspicion.

Many nations do not recognize totalitarian regimes as being the legitimate government of a particular country. Afghanistan in the year 2001 was an example. As the war on terrorism began in the aftermath of the September 11 terrorist attacks on the United States, many people developed a heightened awareness of the Taliban regime, which ruled most of Afghanistan and was engaged in fierce fighting to capture the rest of the country. The Taliban regime maintained absolute control over the Afghan people in most of that country, including requiring that all Muslims take part in prayer five times each day and that women wear the *hijab* (veil). Since U.S. military action commenced in Afghanistan, most of what U.S. residents know about the Taliban, the war, and its aftermath has been based on social media accounts and "expert opinions" expressed on television and the Internet.

Democracy

Democracy is a political system in which the people hold the ruling power either directly or through elected representatives. The literal meaning of *democracy* is "rule by the people" (from the Greek words *demos*, meaning "the people," and *kratein*, meaning "to rule"). In an ideal-type democracy, people would actively and directly rule themselves. *Direct participatory democracy* requires that citizens be able to meet together regularly to debate and decide the issues of the day. Because there are approximately 320.9 million people in the United States today, it would be impossible for everyone to come together in one place for a meeting.

In countries such as the United States, Canada, Australia, and the United Kingdom, people have a voice in the government through ***representative democracy***, whereby citizens elect representatives to serve as bridges between themselves and the government. The U.S. Constitution requires that each state have two senators and a minimum of one member in the House of Representatives. The number of voting representatives in the House (435 seats) has not changed since the apportionment

following the 1910 census; however, those 435 seats are reapportioned based on an increase or decrease in a state's population as shown in the census data gathered every ten years.

In a representative democracy, elected representatives are supposed to convey the concerns and interests of those they represent, and the government is expected to be responsive to the wishes of the people. Elected officials are held accountable to the people through elections. However, representative democracy is not always equally accessible to all people in a nation. Throughout U.S. history, members of subordinate racial–ethnic groups have been denied full participation in the democratic process. Gender and social class have also limited some people's democratic participation.

Even representative democracies are not all alike. As compared to the winner-takes-all elections in the United States, which are usually decided by the candidate who wins the most votes, the majority of European elections are based on a system of proportional representation, meaning that each party is represented in the national legislature according to the proportion of votes that party received. For example, a party that won 40 percent of the vote would receive 40 seats in a 100-seat legislative body, and a party receiving 20 percent of the votes would receive 20 seats.

Perspectives on Power and Political Systems

Is political power in the United States concentrated in the hands of the few or distributed among the many? Sociologists and political scientists have suggested many different answers to this question; however, two prevalent models of power have emerged: pluralist and elite.

 LO3 **Describe** the pluralist model of the U.S. power structure and note the part that special interest groups and political action committees play in this approach.

Functionalist Perspectives: The Pluralist Model

The pluralist model is rooted in a functionalist perspective, which assumes that people share a consensus on central concerns, such as freedom and protection from harm, and that the government serves important functions that no other institution can fulfill. According to Emile Durkheim (1933/1893), the purpose of government is to socialize people to be good citizens, to regulate the economy so that it operates effectively, and to provide necessary services for citizens. Contemporary functionalists state the four main functions as follows: (1) maintaining law and order, (2) planning and directing society,

FIGURE 14.4 **Government from a Functionalist Perspective**
From the functionalist perspective, government serves important functions that no other institution can fulfill. Contemporary functionalists identify four main functions: (a) maintaining law and order, (b) planning and directing society, (c) meeting social needs, and (d) handling international relations, including warfare.

(3) meeting social needs, and (4) handling international relations, including warfare (see ● Figure 14.4).

But what happens when people do not agree on specific issues or concerns? Functionalists suggest that divergent viewpoints lead to a system of political pluralism in which the government functions as an arbiter between competing interests and viewpoints. According to the *pluralist model*, power in political systems is widely dispersed throughout many competing interest groups.

In the pluralist model the diverse needs of women and men, people of all religions and racial–ethnic backgrounds, and the wealthy, middle class, and poor are met by political leaders who engage in a process of bargaining, accommodation, and compromise. Competition among leadership groups in government, business, labor, education, law, medicine, and consumer organizations, among others, helps prevent abuse of power by any one group. Everyday people can influence public policy by voting in elections, participating in existing special interest groups, or forming new ones to gain access to the political system. In sum, power is widely dispersed, and leadership groups that wield influence on some decisions are not the same groups that may be influential in other decisions.

Special Interest Groups *Special interest groups* are political coalitions made up of individuals or groups which share a specific interest that they wish to protect or advance with the help of the political system. Examples of special interest groups include the AFL-CIO (representing the majority of labor unions) and public interest or citizens' groups such as the American

totalitarianism
a political system in which the state seeks to regulate all aspects of people's public and private lives.

democracy
a political system in which the people hold the ruling power either directly or through elected representatives.

representative democracy
a form of democracy whereby citizens elect representatives to serve as bridges between themselves and the government.

pluralist model
an analysis of political systems that views power as widely dispersed throughout many competing interest groups.

special interest groups
political coalitions made up of individuals or groups which share a specific interest that they wish to protect or advance with the help of the political system.

Conservative Union and Population Connection (formerly Zero Population Growth).

What purpose do special interest groups serve in the political process? According to some analysts, special interest groups help people advocate their own interests and further their causes. Broad categories of special interest groups include banking, business, education, energy, the environment, health, labor, persons with a disability, religious groups, retired persons, women, and those espousing a specific ideological viewpoint; obviously, many groups overlap in interests and membership. Special interest groups are also referred to as *pressure groups* (because they put pressure on political leaders) or *lobbies*. Lobbies are often referred to in terms of the organization they represent or the single issue on which they focus—for example, the "gun lobby" and the "dairy lobby." The people who are paid to influence legislation on behalf of specific clients are referred to as *lobbyists*.

Over the past fifty years, special interest groups have become more involved in "single-issue politics," in which political candidates are often supported or rejected solely on the basis of their views on a specific issue—such as abortion, gun control, LGBTQ rights, or the environment. Single-issue groups derive their strength from the intensity of their beliefs; leaders have little room to compromise on issues.

Political Action Committees For many years the funding of lobbying efforts has been a hotly debated issue. Numerous attempts have been made to limit campaign contributions and expenditures to ensure that wealthy and influential individuals and organizations are not able to silence the voices of people who do not have equal resources. Reforms in campaign finance laws in the 1970s set limits on direct contributions to political candidates and led to the creation of *political action committees* (PACs)—organizations of special interest groups that solicit contributions from donors and fund campaigns to help elect (or defeat) candidates based on their stances on specific issues. As the cost of running for political office has skyrocketed, candidates have relied more on PACs for financial assistance. Advertising, staff, direct-mail operations, telephone banks, computers, consultants, travel expenses, office rentals, and other expenses incurred in political campaigns make PAC money vital to candidates.

Some PACs represent the "public interest" and ideological interest groups such as LGBTQ rights or the National Rifle Association. Other PACs represent the capitalistic interests of large corporations. Realistically, PACs do not represent members of the least-privileged sectors of society, such as food stamp PACs.

Various efforts have been made to curb excessive spending in political elections. As an outgrowth of record-setting campaign spending in the 1996 national election, Congress passed the 2002 McCain–Feingold campaign finance law prohibiting soft money contributions (which are made outside the limits imposed by federal election law). This law pertains to federal elections only and does not include state or local elections.

But a shift occurred in 2010, when the U.S. Supreme Court ruled in *Citizens United v. Federal Election Commission* that corporations and other organizations could bypass existing spending limits by giving unlimited amounts to "independent groups" that support candidates (but not to the candidates themselves). In this decision the Court struck down a provision of the McCain–Feingold Act that prohibited both for-profit and not-for-profit corporations and unions from broadcasting "electioneering communications," defined as a broadcast, cable, or satellite communication that mentioned a candidate within sixty days of a general election or thirty days of a primary. The Court's decision was made in terms of First Amendment rights, based on the assumption that a decision to spend money in support of a political cause or candidate was similar to giving a speech or carrying a campaign sign and thus protected by the First Amendment (Toobin, 2011). This controversial decision was criticized for granting lobbyists for special interests even more power than they previously held in Washington, while average Americans were further downgraded in their efforts to support a political candidate. Clearly, the Court's decision contributed to the rise of *super PACs,* which were major contributors in the 2012 elections.

Rise of the Super PACs Super PACs came into existence following two important legal cases. In *SpeechNow .org v. Federal Election Commission,* a federal court ruled that it was a violation of the First Amendment to establish limitations on individual contributions to independent organizations that seek to influence elections. In the second case, as previously discussed, the U.S. Supreme Court decided in *Citizens United v. Federal Election Commission* that it was a violation of the U.S. Constitution to establish spending limits for corporations and unions. This decision opened the door for corporations, unions, and other organizations to bypass existing spending limits and give unlimited amounts of money to various "independent groups" that support candidates, but not to candidates or specific political parties.

What is the difference between a PAC and a super PAC? Simply stated, PACs have restrictions on who may contribute and how much they can give, whereas super PACs are unrestricted in who can give and how much they may donate. As a result, large sums of money flow into candidates' campaign coffers from entities that have specific agendas. Super PACs can contribute money to candidates who will uphold the best interests of specific interest groups, corporations, or unions, or they can provide funds to defeat candidates whose political platforms will not benefit the major contributors of the super PAC. The presidential election in 2012 was the first election in which super PACs were involved in supporting campaigns, as shown in • Figure 14.5. It remains to be seen what the longer-term effects of these massive contributions from corporations, unions, and other major donors will be on U.S. elections and the political process.

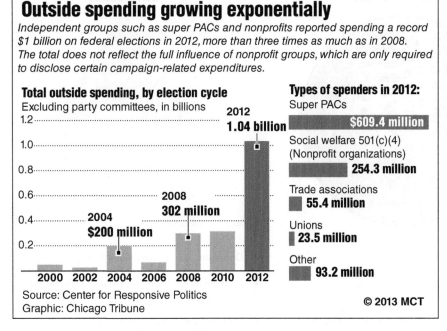

Outside spending growing exponentially

Independent groups such as super PACs and nonprofits reported spending a record $1 billion on federal elections in 2012, more than three times as much as in 2008. The total does not reflect the full influence of nonprofit groups, which are only required to disclose certain campaign-related expenditures.

Total outside spending, by election cycle
Excluding party committees, in billions

2012
1.04 billion

Types of spenders in 2012:
Super PACs
$609.4 million

Social welfare 501(c)(4)
(Nonprofit organizations)
254.3 million

Trade associations
55.4 million

Unions
23.5 million

Other
93.2 million

2000 2002 2004 2006 2008 2010 2012

2004
$200 million

2008
302 million

Source: Center for Responsive Politics
Graphic: Chicago Tribune

© 2013 MCT

FIGURE 14.5 Influence of Super Pacs in the 2012 Presidential Election
Source: Center for Responsive Politics, 2013.

Based on a functionalist approach, some special interest groups and political action committees continue to represent the broader needs and interests of many individuals, groups, and organizations in society. However, what happens when they become narrowly focused on the issues that benefit only a very small percentage of elites at the top of the pyramid, as has recently happened with some super PACs? At this juncture we must turn to conflict perspectives and elite models for a better explanation of how politics and government operate when power is more concentrated.

LO4 **Discuss** the elite model of power and identify the approaches of C. Wright Mills and G. William Domhoff.

Conflict Perspectives: Elite Models

Although conflict theorists acknowledge that the government serves a number of useful purposes in society, they assert that government exists for the benefit of wealthy or politically powerful elites who use it to impose their will on the masses. According to the *elite model*, power in political systems is concentrated in the hands of a small group of elites, and the masses are relatively powerless. The pluralist model and the elite model are compared in • Figure 14.6.

Contemporary elite models are based on the assumption that decisions are made by the elites, who agree on the basic values and goals of society. However, the needs and concerns of the masses are not often given consideration by those in the elite. According to this approach, power is highly concentrated at the top of a pyramid-shaped social hierarchy, and public policy reflects the values and preferences of the elite, not the preferences of the people.

C. Wright Mills and the Power Elite Who makes up the U.S. power elite? According to the sociologist C. Wright Mills (1959a), the *power elite* is made up of leaders at the top of business, the executive branch of the federal government, and the military. Of these three, Mills speculated that the "corporate rich" (the highest-paid officers of the biggest corporations) were the most powerful because of their unique ability to parlay the vast economic resources at their disposal into political power. At the middle level of the pyramid, Mills placed the legislative branch of government, special interest groups, and local opinion leaders. The bottom (and widest layer) of the pyramid is occupied by the unorganized masses, who are relatively powerless and are vulnerable to economic and political exploitation.

G. William Domhoff and the Ruling Class Sociologist G. William Domhoff (2002) asserts that this nation in fact has a *ruling class*—the wealthiest persons in the upper class and the corporate rich, who make up less than 1 percent of the U.S. population. Domhoff uses the term *ruling class* to signify a relatively fixed group of privileged people who

political action committees (PACs)
organizations of special interest groups that solicit contributions from donors and fund campaigns to help elect (or defeat) candidates based on their stances on specific issues.

elite model
a view of society that sees power in political systems as being concentrated in the hands of a small group of elites whereas the masses are relatively powerless.

power elite
C. Wright Mills's term for the group made up of leaders at the top of business, the executive branch of the federal government, and the military.

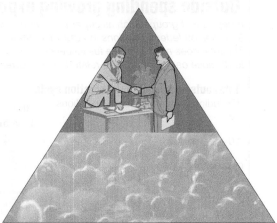

- Decisions are made on behalf of the people by leaders who engage in bargaining, accommodation, and compromise.

- Competition among leadership groups makes abuse of power by any one group difficult.

- Power is widely dispersed, and people can influence public policy by voting.

- Public policy reflects a balance among competing interest groups.

- Decisions are made by a small group of elite people.

- Consensus exists among the elite on the basic values and goals of society.

- Power is highly concentrated at the top of a pyramid-shaped social hierachy.

- Public policy reflects the values and preferences of the elite.

FIGURE 14.6   Pluralist and Elite Models

wield power sufficient to constrain political processes and serve underlying capitalist interests. Although the power elite controls the everyday operation of the political system, who *governs* is less important than who *rules*.

According to Domhoff (2005), the upper class and the corporate rich influence politics in the following ways:

- "The rich" coalesce into a social upper class that has developed institutions by which the children of its members are socialized into an upper-class worldview and newly wealthy people are assimilated.

- Members of this upper class control corporations, which have been the primary mechanisms for generating and holding wealth in the United States for more than 150 years.

- In a network of nonprofit organizations, members of the upper class and hired corporate leaders not yet in the upper class shape policy debates.

- Members of the upper class, with the help of their high-level employees in for-profit and nonprofit institutions, are able to dominate the federal government.

- The rich, and corporate leaders, nonetheless claim to be relatively powerless.

- Working people have less power than in many other democratic countries, so they have little chance to influence the political process.

However, Domhoff emphasizes that people in the United States do have options because of their rights to free speech and to vote. Although the U.S. government may not always be responsive to the will of the majority of voters, citizens have been able to place restraints on some actions of wealthy elites to keep them from having an even greater influence on policy than they already possess.

Power elite models call our attention to a central concern in contemporary U.S. society: the ability of democracy and its ideals to survive in the context of the increasingly concentrated power held by capitalist oligarchies such as the media giants previously discussed in Chapter 13.

LO5 **Explain** how political parties shape the government and identify recent trends in political participation and voter apathy.

The U.S. Political System

The U.S. political system is made up of formal elements, such as the legislative process and the duties of the president, and informal elements, such as the role of political parties in the election process. We now turn to an examination of these informal elements, including political parties, political socialization, and voter participation.

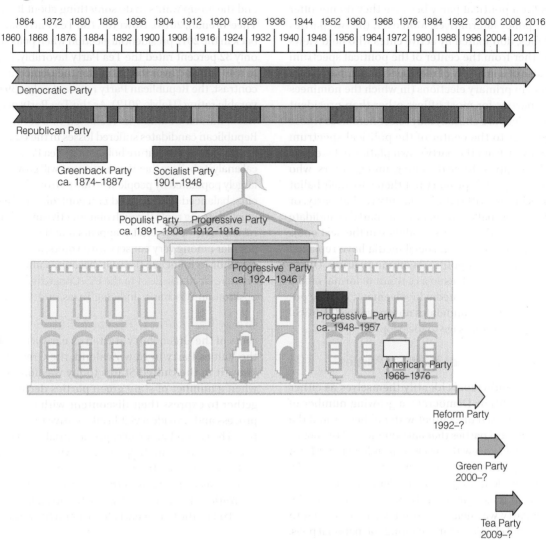

FIGURE 14.7   Major U.S. Political Parties

Despite recurring attempts by other groups to organize third parties, the Democratic and Republican parties have dominated national politics in the United States. Control of the presidency has alternated between these two parties since the Civil War.

Note: Three different "third parties" have used the name Progressive Party.

Political Parties and Elections

A **political party** is an organization whose purpose is to gain and hold legitimate control of government; it is usually composed of people with similar attitudes, interests, and socioeconomic status. A political party (1) develops and articulates policy positions, (2) educates voters about issues and simplifies the choices for them, and (3) recruits candidates who agree with those policies, helps those candidates win office, and holds the candidates responsible for implementing the party's policy positions. In carrying out these functions, a party may try to modify the demands of special interests, build a consensus that could win majority support, and provide simple and identifiable choices for the voters on election day. Political parties create a

platform, a formal statement of the party's political positions on various social and economic issues.

Since the Civil War, the Democratic and Republican parties have dominated the U.S. political system (see • Figure 14.7). Although one party may control the presidency for several terms, at some point the voters elect the other party's nominee, and control shifts.

How well do the parties measure up to the ideal-type characteristics of a political party? Although both parties have been successful in getting their candidates elected at

political party
an organization whose purpose is to gain and hold legitimate control of government.

various times, they generally do not meet the ideal characteristics for a political party because they do not offer voters clear policy alternatives. Moreover, the two parties are oligarchies, dominated by active elites who hold views that are further from the center of the political spectrum than are those of a majority of members of their party. As a result, voters in primary elections (in which the nominees of political parties for most offices other than president and vice president are chosen) may select nominees whose views are closer to the center of the political spectrum and further away from the party's own platform. Likewise, party loyalties appear to be declining among voters, who may vote in one party's primary but then cast their ballot in general elections without total loyalty to that party, or cast a "split-ticket" ballot (voting for one party's candidate in one race and another party's candidate in the next one).

Finally, mainstream and social media have replaced the party as a means of political communication. Often, the candidate who wins does so as a result of formal media presentations and social media involvement, not the political party's platform. Candidates no longer need political parties to carry their message to the people.

Discontent with Current Political Parties

Although many individuals identify themselves as either Republicans (GOP) or Democrats, a growing number of people have expressed discontent with politicians and the existing U.S. political parties that have dominated the political system, and may view themselves as independent from the traditional parties. In one national survey, for example, persons who self-identified as "independents" accounted for 39 percent of all those surveyed (Holyk, 2012). In 2012 the Democratic Party was near a record low, as compared to prior presidential election years, in ratings on national polls, and the GOP was even further behind in ratings. Although the Tea Party movement initially came into the political arena with gusto, when its leaders tried to turn it into a full-fledged third political party, the Tea Party continued to slip in favorability in national polls (Holyk, 2012). Let's look briefly at the background of the Tea Party and the Green Party as case studies in discontent with current political parties.

Tea Party: Past, Present, Future   In 2009 the Tea Party movement emerged to support more constitutionally limited government and to oppose various stimulus and bailout programs that use federal monies. The protesters refer to themselves as the "Tea Party" based on the Boston Tea Party, a 1773 protest by American colonists against "taxation without representation" by the British government because the colonists were not represented in the British Parliament but were required to pay taxes to that government. A demographic analysis reveals that the typical Tea Party supporter is white, male, married, and over forty-five years of age. Most are registered Republicans, but they disagree with party leadership about various issues. Clearly, the Tea Party movement is pessimistic about the direction that its members think the

United States has taken under the Obama administration, and the group wants to do something about it (Zernike and Thee-Brenan, 2010). However, the Tea Party suffered a setback in the 2012 presidential election when polls showed that only 32 percent rated the Tea Party favorably, as compared to a 49 percent favorable rating for the Democratic Party. By contrast, the Republican Party received only a 39 percent favorable rating (Holyk, 2012). As the Tea Party drifted downward, Barack Obama was reelected to a second term, and Republican candidates suffered losses in the U.S. Senate.

What does the future hold for the Tea Party? Some political analysts believe that the movement will grow in the future, largely populated by people who have strong beliefs about tax cuts, balanced budgets, and gun control. The Tea Party has created strong roots through online activism and strong fundraising activities. Overall, it appears that the Tea Party is most popular among very conservative voters, and this group will continue to influence the GOP because some of its key players were elected in 2014 to the U.S. Congress and to statewide positions in Wisconsin, Indiana, Texas, and Nevada.

Green Party: Past, Present, Future   Although the roots of the Green Party started as early as the 1970s with the Values Party in New Zealand, the Green Party as we know it in the United States today was founded in 1996, when a number of state green parties decided to join together to express their discontent with the U.S. political process and provide a viable alternative to the existing parties. The Green Party's first presidential candidate was the consumer advocate Ralph Nader, who ran unsuccessfully in 1996 and 2000. In 2001 the Green Party of the United States was formed out of the older association of state organizations to help establish parties in all fifty states.

Today, the Green Party focuses on these ten key values:

1. Grassroots democracy
2. Social justice and equal opportunity
3. Ecological wisdom
4. Nonviolence
5. Decentralization of wealth and power in social, political, and economic institutions
6. Community-based economics and economic justice
7. Feminism and gender equity
8. Respect for diversity
9. Personal and global responsibility
10. Future focus and sustainability to protect valuable natural resources and develop sustainable economies that are not dependent on continual expansion for survival. (GreenParty.org, 2015)

The stated purposes of the organization are to renew democracy without the support of corporate donors, to take a stand against super PACs and the big-money system of government, and to find viable solutions for real-world problems such as health care, corporate globalization, alternative energy, election reform, and wages for workers (GreenParty.org, 2015).

What does the future hold for the Green Party? In national elections the party has not fared well as of 2015, receiving very few votes for its candidate, Jill Stein (• Figure 14.8), in the 2012 presidential election and currently holding no seats in the U.S. Congress. In 2015 Greens held about 130 positions in local offices, ranging from mayor or city council member to school board members. In the future the Green Party, like other third parties, will probably continue to struggle in overcoming ballot access laws, such as the requirement of many states that candidates get petitions signed by a certain percentage of registered voters before they can run for office.

Political Participation and Voter Apathy

Why do some people vote and others not? How do people come to think of themselves as being conservative, moderate, or liberal? Key factors include individuals' political socialization and attitudes.

Political socialization is the process by which people learn political attitudes, values, and behavior. For young children the family is the primary agent of political socialization, and children tend to learn and hold many of the same opinions as their parents. By the time children reach school age, they typically identify with the political party (if any) of their parents. As we grow older, other agents of socialization begin to influence our political beliefs, including our peers, teachers, and the media. If we grow up around family members and friends who vote and discuss politics, we are more likely to be interested in the political process and to vote.

In addition to the socialization process, people's socioeconomic status affects their political attitudes, values, and beliefs. For example, individuals who are very poor or are unable to find employment tend to believe that society has failed them and therefore tend to be indifferent toward the political system. Believing that casting a ballot would make no difference to their own circumstances, they do not vote.

Democracy in the United States has been defined as a government "of the people, by the people, and for the people." Accordingly, it would stand to reason that "the people"

FIGURE 14.8 Jill Stein was the Green Party candidate for president in the most recent U.S. presidential election. How does the Green Party differ from other political parties?

would actively participate in their government at any or all of four levels: (1) voting, (2) attending and taking part in political meetings, (3) actively participating in political campaigns, and (4) running for and/or holding political office. At most, about 10 percent of the voting-age population in this country participates at a level higher than simply voting, and over the past fifty years less than half of the voting-age population has voted in nonpresidential elections. In the 2014 midterm elections, for example, voter turnout was the lowest it had been since World War II. Of the voting-eligible population, slightly over one-third (36.4 percent) cast ballots (DelReal, 2014). Even in presidential elections, voter turnout is often relatively low, but it is higher than in the midterm elections.

Voter Turnout and Political Preferences In the 2012 presidential election, 58.7 percent of the 219 million eligible voters cast ballots. This percentage was down from the 2008 presidential election, when 62.3 percent of the 208.3 million eligible voters actually voted. The number of ballots cast in 2008 was the highest in history, at 131 million votes, as compared to 126 million in 2012. However, more than half (57.4 percent) of the drop-off in voter turnout was in three states: New York and New Jersey had experienced the devastation of Hurricane Sandy just prior to the election, and California continued a downward decline in voter turnout, from ranking thirty-third in turnout in the 2008 election to forty-first in turnout in 2012. Overall, Minnesota had the highest voter turnout (76.1 percent), while Hawaii had the lowest (44.5 percent).

Among voters between the ages of eighteen and twenty-nine, 23 million people cast ballots. The presidential election in 2012 was the third election in a row in which about half of the eligible younger voting population actually voted, and their vote was a determining factor in the swing states of Florida, Ohio, Pennsylvania, and Virginia. These voters were the most racially and ethnically

political socialization
the process by which people learn political attitudes, values, and behavior.

Scott Olson/Getty Images News/Getty Images

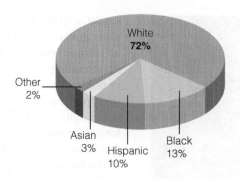

White
72%

Other
2%

Asian
3%

Hispanic
10%

Black
13%

FIGURE 14.9 Voter Participation in the 2012 Presidential Election by Race and Ethnicity

Source: Based on Taylor, 2012.

diverse: Latinos/as, African Americans, and other so-called nonwhite voters made up 42 percent of the voting-eligible population, as compared to only 24 percent among voters ages thirty and older (Pillsbury and Johannesen, 2012).

For the first time in 2012, Latinos/as (across age categories) constituted one in ten voters (10 percent of the electorate). This increase is attributed to changing demographics; it is estimated that the percentage of Latino/a voters will double in size by 2030, at which time they will account for 40 percent of the growth in the eligible electorate in the United States. By contrast, increases in the numbers of African American voters in the 2012 election are attributed by political analysts to higher voter turnout among African Americans rather than demographic shifts in the population. In 2012 African Americans voted at a higher rate than in the past and accounted for an estimated 13 percent of all votes cast (Taylor, 2012). Barack Obama's candidacy in 2008 and 2012 is believed to be a key reason why African Americans constituted a higher percentage of all voters than in the past, but other factors may also be significant, such as increasing diversity of the electorate and a declining white (non-Hispanic) turnout rate (Taylor, 2012). • Figure 14.9 shows the percentage of voter participation by race and ethnicity in the 2012 presidential election.

Voter Turnout in Swing States   In presidential elections, we frequently hear about "swing states." What are swing states? They are those states in which no single candidate or political party appears to have overwhelming support in securing the state's electoral college votes. This is important because electoral college votes determine who will be the next president. Consequently, swing states are very important. For example, in 2012 presidential campaign officials spent most of their time and money in swing states such as Utah, Colorado, Wisconsin, Massachusetts, and Iowa, along with the District of Columbia, trying to convince voters to decide for their candidate and tip their state in that party's favor.

Red and Blue States   State-by-state differences in voting preferences are also highly visible in what political analysts refer to as the "red states" and the "blue states." These terms refer to those states whose residents predominantly vote for Republican Party (red) or Democratic Party (blue) presidential candidates. Today, the terms are often used to indicate that voters residing there are either conservative or liberal but not necessarily to indicate their political party affiliation (see • Figure 14.10 for a breakdown of the 2012 presidential election).

Voter Apathy or Something Else?   Why is it that so many eligible voters in this country stay away from the polls? During any election, millions of voting-age persons do not go to the polls because of illness, disability, lack of transportation, lack of registration, or absenteeism. However, these explanations do not account for why many other people do not vote. According to some analysts, people may not vote because they are satisfied with the status quo or because they are apathetic and uninformed—they lack an understanding of both public issues and the basic processes of government.

By contrast, others argue that people stay away from the polls because they feel alienated from politics at all levels of government—federal, state, and local—because of political corruption and influence peddling by special interests, large corporations, and now the power of super PACs. Participation in politics is influenced by gender, age, race/ethnicity, and, especially, socioeconomic status (SES). One explanation for the higher rates of political participation at higher SES levels is that advanced levels of education may give people a better understanding of government processes, a belief that they have more at stake in the political process, and greater economic resources to contribute to the process. Some studies suggest that during their college years, many people develop assumptions about political participation that continue throughout their lives.

Much remains to be learned about voter participation and voter apathy. With increasingly sophisticated techniques for studying who goes (or does not go) to the polls and who remains apathetic about the political process, we will be able to know much more in the future about why the United States has some of the lowest voter turnout rates of any democracy worldwide (• Figure 14.11).

Identify key factors that contribute to the power of government bureaucracies and explain how bureaucratic power tends to take on a life of its own.

Governmental Bureaucracy

When most people think of political power, they overlook one of its major sources—the governmental bureaucracy. As previously discussed, Weber's rational–legal authority finds its contemporary embodiment in bureaucratic organizations. Negative feelings about bureaucracy are perhaps strongest when people are describing the "faceless bureaucrats" and "red tape" with which they must deal in government. But who are these "faceless bureaucrats," and what do they do?

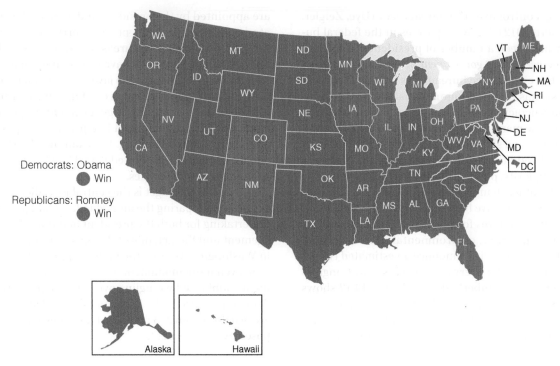

Democrats: Obama
● Win

Republicans: Romney
● Win

Alaska   Hawaii

FIGURE 14.10   2012 Presidential Election: State by State

Source: Staff/MCT/Newscom

Bureaucratic power tends to take on a life of its own. During the nineteenth century, the government had a relatively limited role in everyday life. In the 1930s, however, the scope of government was extended greatly during the Great Depression to deal with labor–management relations, public welfare, and the regulation of the securities markets. With dramatic increases in technology and increasing demands that the federal government "do something" about a wide variety of problems facing the United States, such as fighting terrorism and providing homeland security, the government has continued to grow in many areas in recent decades. Today, the federal bureaucracy employs about 2.03 million people in civilian jobs, despite a series of hiring freezes and a *sequestration*—the practice of using mandatory spending cuts in the federal budget if the cost of running the government exceeds either an arbitrary amount or the gross revenue that the government brings in during the fiscal year. For example, a 2013 sequester temporarily reduced the number of government positions available, cut federal spending in both defense and nondefense categories,

FIGURE 14.11   Voting is one of the many rights that people gain when they become naturalized citizens. Some studies find that naturalized citizens are more likely to vote and engage in civic participation than persons born in the United States. Why do you think that this might be true?

furloughed federal employees, and made other cuts because of lack of funds.

Because much of the actual functioning of the government is carried on by its bureaucracy, even the president, the White House staff, and cabinet officials have difficulty establishing control over the bureaucracy (Dye, Zeigler, and Schubert, 2012). Many employees in the federal bureaucracy have seen a number of presidents "come and go," but the *permanent government* in Washington, made up of top-tier civil service bureaucrats who have a major power base, has remained more constant. The governmental bureaucracy has been able to perpetuate itself and expand because many of its employees have highly specialized knowledge and skills and cannot be replaced easily by "outsiders." In addition, as the United States has grown in size and complexity, public policy is increasingly made by bureaucrats rather than by elected officials. For example, offices and agencies have been established to create rules, policies, and procedures for dealing with complex issues such as nuclear power, environmental protection, and drug safety; bureaucracies announce an estimated twenty rules or regulations for every one law passed by Congress (Dye, Zeigler, and Schubert, 2012). • Figure 14.12 shows characteristics of the "typical" federal civilian employee.

The executive branch is also highly bureaucratized, as shown in • Figure 14.13. Cabinet-level secretaries, who are appointed by the president and approved by the Senate, head fifteen departments that carry out governmental functions. Like all other areas of governmental bureaucracy, these departments have grown in number and size over the years. Adding to the layers of bureaucracy are the bureaus and agencies that are subdivisions within cabinet departments, as well as government corporations—agencies that are organized like private companies and operate in a market setting. For example, the U.S. Postal Service competes with United Parcel Service (UPS), FedEx, and other delivery services.

The federal budget is the central ingredient in the bureaucracy. Preparing the annual federal budget is a major undertaking for both the president and the Office of Management and Budget, one of the most important agencies in Washington. Getting the budget approved by Congress is an even more monumental task; however, even with the highly publicized wrangling over the budget by the president and Congress, the final congressional appropriations are usually within 2–3 percent of the budget originally proposed by the president.

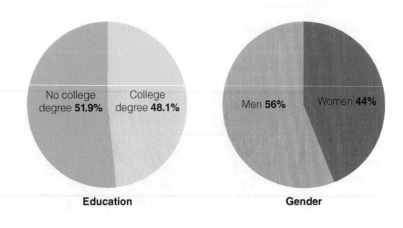

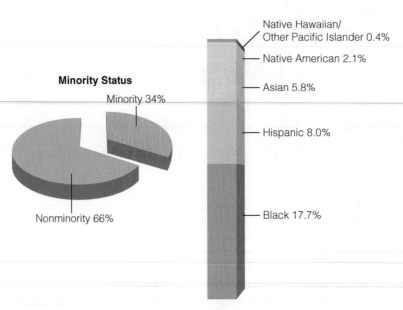

FIGURE 14.12   The "Typical" Federal Civilian Employee
Source: U.S. Office of Personnel Management, 2013.

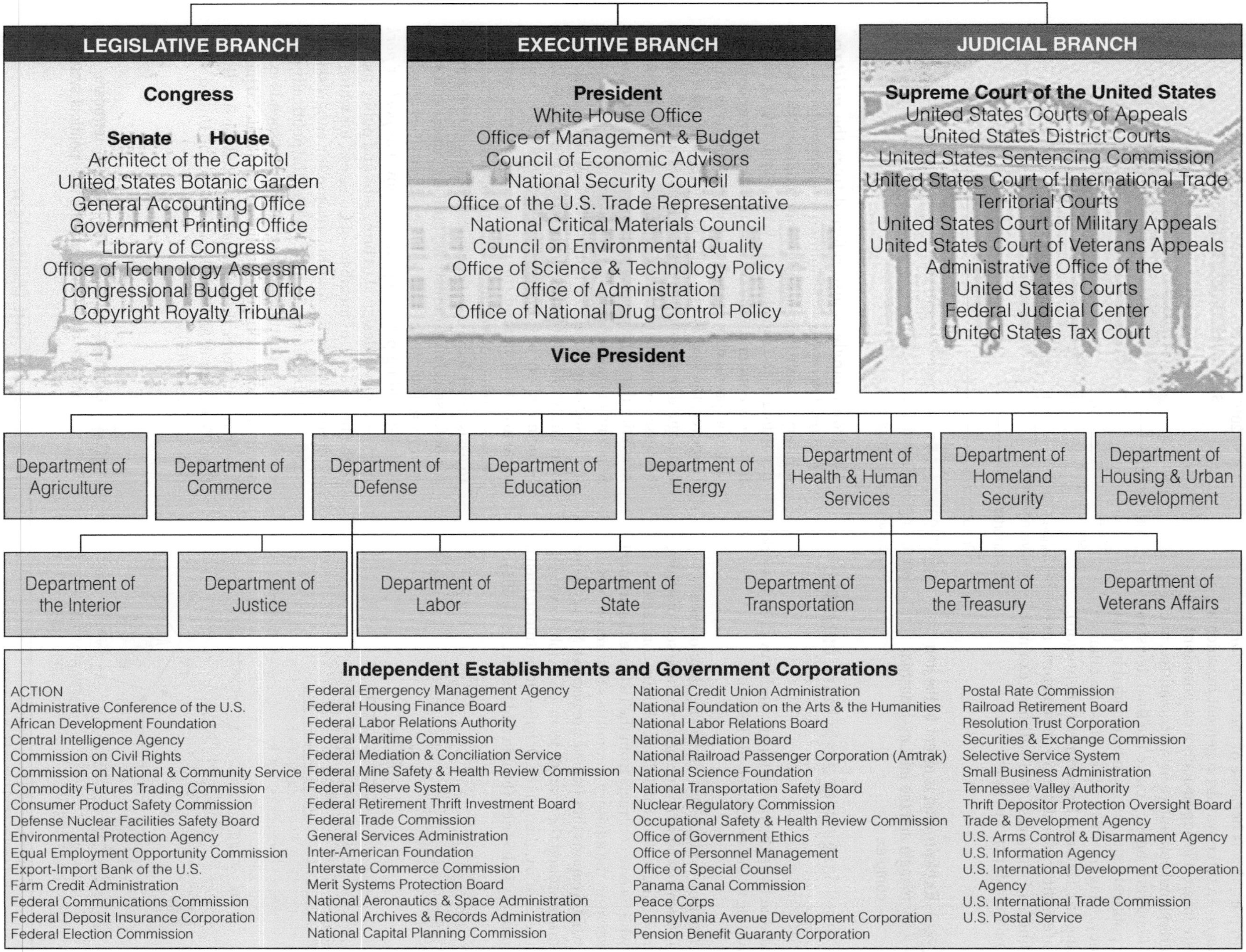

LEGISLATIVE BRANCH

Congress

Senate **House**

Architect of the Capitol
United States Botanic Garden
General Accounting Office
Government Printing Office
Library of Congress
Office of Technology Assessment
Congressional Budget Office
Copyright Royalty Tribunal

EXECUTIVE BRANCH

President

White House Office
Office of Management & Budget
Council of Economic Advisors
National Security Council
Office of the U.S. Trade Representative
National Critical Materials Council
Council on Environmental Quality
Office of Science & Technology Policy
Office of Administration
Office of National Drug Control Policy

Vice President

JUDICIAL BRANCH

Supreme Court of the United States

United States Courts of Appeals
United States District Courts
United States Sentencing Commission
United States Court of International Trade
Territorial Courts
United States Court of Military Appeals
United States Court of Veterans Appeals
Administrative Office of the
United States Courts
Federal Judicial Center
United States Tax Court

Department of Agriculture
Department of Commerce
Department of Defense
Department of Education
Department of Energy
Department of Health & Human Services
Department of Housing & Urban Development

Department of the Interior
Department of Justice
Department of Labor
Department of State
Department of Transportation
Department of Homeland Security
Department of the Treasury

Department of Veterans Affairs

Independent Establishments and Government Corporations

ACTION
Administrative Conference of the U.S.
African Development Foundation
Central Intelligence Agency
Commission on Civil Rights
Commission on National & Community Service
Commodity Futures Trading Commission
Consumer Product Safety Commission
Defense Nuclear Facilities Safety Board
Environmental Protection Agency
Equal Employment Opportunity Commission
Export-Import Bank of the U.S.
Farm Credit Administration
Federal Communications Commission
Federal Deposit Insurance Corporation
Federal Election Commission

Federal Emergency Management Agency
Federal Housing Finance Board
Federal Labor Relations Authority
Federal Maritime Commission
Federal Mediation & Conciliation Service
Federal Mine Safety & Health Review Commission
Federal Reserve System
Federal Retirement Thrift Investment Board
Federal Trade Commission
General Services Administration
Inter-American Foundation
Interstate Commerce Commission
Merit Systems Protection Board
National Aeronautics & Space Administration
National Archives & Records Administration
National Capital Planning Commission

National Credit Union Administration
National Foundation on the Arts & the Humanities
National Labor Relations Board
National Mediation Board
National Railroad Passenger Corporation (Amtrak)
National Science Foundation
National Transportation Safety Board
Nuclear Regulatory Commission
Occupational Safety & Health Review Commission
Office of Government Ethics
Office of Personnel Management
Office of Special Counsel
Panama Canal Commission
Peace Corps
Pennsylvania Avenue Development Corporation
Pension Benefit Guaranty Corporation

Postal Rate Commission
Railroad Retirement Board
Resolution Trust Corporation
Securities & Exchange Commission
Selective Service System
Small Business Administration
Tennessee Valley Authority
Thrift Depositor Protection Oversight Board
Trade & Development Agency
U.S. Arms Control & Disarmament Agency
U.S. Information Agency
U.S. International Development Cooperation Agency
U.S. International Trade Commission
U.S. Postal Service

FIGURE 14.13 Organization of the U.S. Government

In part, this is because of the *iron triangle of power*—a three-way arrangement in which a private interest group (usually a corporation), a congressional committee or subcommittee, and a bureaucratic agency make the final decision on a political issue that is to be decided by that agency. • Figure 14.14 illustrates the alliance among the Defense Department (Pentagon), private military (or defense) contractors, and members of Congress. We will now examine this relationship more closely.

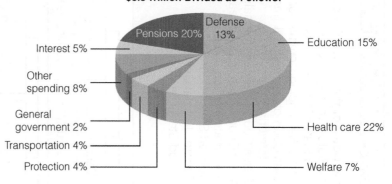

2015 Overall U.S. Budget Approximately $3.8 Trillion Divided as Follows:

Pensions 20%
Defense 13%
Interest 5%
Other spending 8%
General government 2%
Transportation 4%
Protection 4%
Education 15%
Health care 22%
Welfare 7%

FIGURE 14.15   Categories and Percentages of U.S. Federal Spending in Fiscal Year 2015
Source: usgovernmentspending.com, 2015.

L07 **Explain** what is meant by the iron triangle and the military–industrial complex.

The Iron Triangle and the Military–Industrial Complex

What exactly is the iron triangle, and how does it work? According to the sociologist Joe Feagin,

> The Iron Triangle has a revolving door of money, influence, and jobs among these three sets of actors, involving trillions of dollars. Military contractors who receive contracts from the Defense Department serve on the advisory committees that recommend what weapons they believe are needed. Many people move around the triangle from job to job, serving in the military, then in the Defense Department, then in military industries. (Feagin and Feagin, 1994: 405)

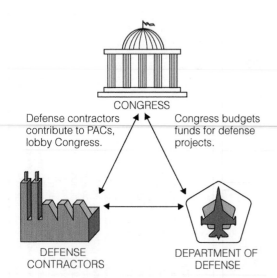

CONGRESS

Defense contractors contribute to PACs, lobby Congress.

Congress budgets funds for defense projects.

DEFENSE CONTRACTORS

DEPARTMENT OF DEFENSE

Defense Department awards contracts to contractors; personnel move back and forth between employment by Defense Department and by defense contractors.

FIGURE 14.14   Example of the Iron Triangle of Power

The iron triangle is also referred to as the ***military–industrial complex***—the mutual interdependence of the military establishment and private military contractors. According to the sociologist C. Wright Mills (1976), this alliance of economic, military, and political power could result in a "permanent war economy" or "military economy." However, the economist John Kenneth Galbraith (1985) has argued that the threat of war is good for the economy because government money spent on military preparedness stimulates the private sector of the economy, creates jobs, and encourages consumer spending.

Issues regarding the military–industrial complex are intricately linked to other problems in the U.S. economy. Some members of Congress have been willing to support measures that expand the military–industrial complex because doing so provides a unique opportunity for them to assist their local constituencies by authorizing funding for defense-related industries, military bases, and space centers in their home districts. These activities constitute a long-standing political practice known as *pork* or *pork barrel*—projects designed to bring jobs and public monies to the home state of members of Congress, for which they can then take credit. Another reason for congressional support of the defense industry is that the politically powerful military industry employs millions and spends extravagantly on contributions to candidates and political parties.

In recent years, defense has accounted for slightly less than 25 percent of all federal spending in the United States (see • Figure 14.15).

L08 **Discuss** the effects of militarism, terrorism, and war on citizens and the U.S. political system.

The Military and Militarism

Militarism is a societal focus on military ideals and an aggressive preparedness for war. Core U.S. values such as patriotism, courage, reverence, loyalty, obedience, and faith in authority help to support militarism, which has become central to our national identity, in the words of the historian and Vietnam war veteran Andrew J. Bacevich (2013: Introduction):

> Today as never before in their history Americans are enthralled with military power. The global military supremacy that the United States presently enjoys—and is bent on perpetuating—has become central to our national identity. More than America's matchless material abundance or even the effusions of its pop culture, the nation's arsenal of high-tech weaponry and the soldiers who employ that arsenal have come to signify who we are and what we stand for.

As Bacevich suggests, many people have become engrossed in the combination of unprecedented military strength and faith in American values, which inevitably encourages endless involvement in wars and the continually increasing militarization of U.S. policy and federal government expenditures.

The term *militarism* is not synonymous with *war*; rather, it involves an ideology that perpetuates societal activities directed toward military preparedness and enormous spending on weapons production and "defense," as compared with other national priorities.

FIGURE 14.16   For many centuries, "warrior culture" has been the province of men. However, women now make up a growing percentage of soldiers in the U.S. armed services, and they have faced numerous obstacles in the process.

Explanations for Militarism

Sociologists have proposed several reasons for militarism. One is the economy. According to Cynthia H. Enloe (1987: 527), people who see capitalism as the moving force behind the military's influence "believe that government officials enhance the status, resources, and authority of the military in order to protect the interests of private enterprises at home and overseas." From this perspective, the origin of militarism is found in the boardroom, not the war room. Because government and military funding is also a prime source of support for research in the physical, biological, and social sciences, administrators and faculty in some institutions of higher learning feel the need to support war-related spending to obtain grants for research.

Gender and the Military

In recent years, some significant changes have occurred in the policies and composition of the U.S. armed forces. Especially with the introduction of the all-volunteer force and the end of the draft in the 1970s, the focus of the military shifted from training "good citizen-soldiers" to recruiting individuals who would enlist in the military in the same way that another person might take a job in the private sector. Now, in return for joining the military a person

is rewarded with a reasonable salary, medical and dental care, educational opportunities, and other benefits.

Since the introduction of the volunteer forces, considerable pressure has been placed on the military to recruit women (● Figure 14.16). In 2015 about 201,400 women served in active-duty military positions. In addition, 2,600 women were in military academies or served as midshipmen. Of the 71,000 military personnel deployed in 2015, about 9,200 were women.

The ban on women serving in direct ground combat was lifted in 2013, potentially making an additional 91,000 positions available for qualified women. In the meantime, however, U.S. military spending continued to fall as ongoing budget-deficit-reduction measures were implemented and troop reductions occurred. In spite of some changes in the gender composition of the military and revised rules involving their service, many women in the military continue to face sexual harassment and physical and verbal assault in a traditionally "male-dominated world."

Terrorism and War

Terrorism is the use of calculated, unlawful physical force or threats of violence against a government, organization,

military–industrial complex
the mutual interdependence of the military establishment and private military contractors.

militarism
a societal focus on military ideals and an aggressive preparedness for war.

terrorism
the use of calculated, unlawful physical force or threats of violence against a government, organization, or individual to gain some political, religious, economic, or social objective.

or individual to gain some political, religious, economic, or social objective. Terrorist tactics include bombing, kidnapping, hostage taking, hijacking, assassination, and extortion. Although terrorists sometimes attack government officials and members of the military, they more often target civilians as a way of pressuring the government.

In the twenty-first century, people in the United States have become acutely aware of the problem of terrorism on their own soil, a concern that people in nations around the world have faced for many years. Today, people remember the terrorist attacks of 2001 when they face heightened security at airports or when the nation commemorates each year the anniversary of that horrific event on September 11. The nation was shaken once again in 2013, when two youthful bombers allegedly set their sights on the Boston Marathon, where they placed homemade "pressure-cooker bombs" at the race's finish line that caused deadly blasts. This act killed three spectators and seriously injured 286 other people. In the subsequent chase to apprehend the two bombing suspects, a university police officer was killed and a transit worker was injured.

Around the world, acts of terrorism extract a massive toll by producing rampant fear, widespread loss of human life, and extensive destruction of property. One form of terrorism—political terrorism—is actually considered a form of unconventional warfare. *Political terrorism* uses intimidation, coercion, threats of harm, and other forms of violence that attempt to bring about a significant change in or overthrow an existing government.

In 2013, the latest year for which global terrorism data were available at the time of this writing, terrorist activity had substantially increased, with the total number of deaths rising from 11,133 in 2012 to 17,958 in 2013 (Global Terrorism Index, 2014). The number of countries that experienced more than 50 deaths also increased from 15 to 24. Just five countries accounted for 82 percent of all lives lost to terrorist activities: Iraq, Afghanistan, Pakistan, Nigeria, and Syria (Global Terrorism Index, 2014). Only four terrorist organizations—ISIS, Boko Haram, the Taliban, and al-Qaida—took credit for most of these terrorist attacks. According to experts who study terrorism, various motivations exist for terrorism, with religious ideology being only one of them. Others include political or nationalistic and separatist movements. However, since 2000, religious ideology driving terrorism has increased dramatically (Global Terrorism Index, 2014).

History of Terrorism in the United States

In the United States, the years 1995 and 2001 stand out as the periods when this nation suffered from the two worst terrorist attacks that have ever occurred in the continental United States. Prior to the September 11, 2001, attacks, the April 1995 bombing of the Federal Building in Oklahoma City took the lives of 168 adults and children, and injured others. Many other people were left with psychological scars. The Oklahoma City bombing was described as an act of domestic terrorism because it was attributed to two men who had no apparent connection to external enemies of the United States.

By contrast, external enemies (who had planted perpetrators in the United States) were believed to be responsible for the attacks on New York's World Trade Center, first in 1993 in the form of a bombing that killed six people and injured more than a thousand, and then in the 2001 destruction of the Trade Center's twin towers (and damage to the Pentagon, in Washington, D.C., that same morning) by suicide hijackers using passenger airplanes as lethal weapons, resulting in the deaths of almost 3,000 people.

Prior to the events of 9/11, terrorism in the form of biological and chemical attacks was considered by most analysts to be a remote possibility in the United States. However, with the deaths of five people and twenty-two others infected from the delivery of anthrax, suspicions deepened that for the first time, terrorists—whether homegrown or international—might use germ warfare to achieve their subversive political goals. In the second decade of twenty-first century, people in the United States still fear terrorism and problems associated with war.

War

War is organized, armed conflict between nations or distinct political and/or religious factions. Social scientists define war to include both *declared wars* between nations or parties and *undeclared wars:* civil and guerrilla wars, covert operations, and some forms of terrorism. War is an institution that involves *violence*—behavior intended to bring pain, physical injury, and/or psychological stress to people or to harm or destroy property. As such, war is a form of *collective violence* by people who are seeking to promote their cause or resist social policies or practices that they consider oppressive.

The direct effects of war are loss of human life and serious physical and psychological effects on some survivors. Although it is impossible to determine how many human lives have been lost in wars throughout human history, World War I took the lives of approximately 8 million combatants and 1 million civilians. In World War II, more than 50 million people (17 million combatants and 35 million civilians) lost their lives. During World War II, U.S. casualties alone totaled almost 300,000, and more than 600,000 were wounded. Later military actions in Korea, Vietnam, Iraq, Afghanistan, Libya, and other regions have brought about the deaths of millions of other military personnel and civilians, but estimates of those numbers range so widely that it is difficult, if not impossible, to come up with an accurate number. As of 2015, it was estimated that the United States had 4,478 military fatalities in Iraq and 2,215 in

Afghanistan; however, these figures do not include military personnel from other nations or civilian casualties in the countries in which the wars occurred.

Even more frightening is the possibility of the use of chemical weapons in warfare. In 2013 the U.S. government received reports that the Syrian government had been using chemical weapons in its civil war, particularly in attacks in Aleppo and near Damascus. This posed a grave political concern for the Obama administration because the president had previously issued statements indicating that the United States would intervene in Syria if chemical weapons were used in that nation's warfare, even as he was trying to get the United States out of Afghanistan. Lack of stability in various regions of the world is a cause of concern for political leaders in the more-stable nations because they realize that the destabilization of some other region not only is harmful for the people who live in that region but also for the citizens of their own nation as well.

In recent years, problems associated with war have increased as terrorism and the potential for even more terrorism have increased in various regions. For example, the Islamic State, also known as ISIS, has dramatically increased its hold on Iraq and Syria in recent years. Through a combination of persuasion and violence, this radical group has become a powerful social and political movement that alleges it protects the interest of Sunni Muslims (• Figure 14.17). Sunni Muslims believe they are under attack by Shiite-backed governments in Iraq and Syria. Ultimately, the constant conflict between the Sunni and Shiite branches of Islam fuels political battles and perpetuates war in the Middle East.

Egypt became more authoritarian and repressive. Civil war seemed increasingly possible in Libya and Yemen.

Boko Haram insurgents increased their terrorist attacks in northern Nigeria, and Russia annexed Crimea (Guehenno, 2015). These aggressive activities create greater potential for international conflict and war because relationships between various organizations and countries have become much more antagonistic. Relations among the United States, Europe, and Russia are more hostile than they have been in the recent past. Rivalry between nations and regional powers creates the potential for additional instability and wars, which is of concern to all people who hope to see greater stability in the vulnerable areas of the world.

Looking Ahead: Politics and Government in the Future

Thinking about U.S. politics and government in the future is very much like the old story about optimists and pessimists: An eight-ounce cup containing exactly four ounces of water is placed on a table. The optimist comes in, sees the cup, and says, "The cup is half full." The pessimist comes in, sees the cup, and says, "The cup is half empty." Clearly, both the pessimist and the optimist are looking at the same cup containing the same amount of water, but their perspective on what they see is quite different. For some analysts, looking at the future of the U.S. government and its political structure is very much like this.

Views of the future of politics and government relate to specific concerns about the United States:

- What will be the future of political parties? What have recent elections told us about the nature of these institutions?

  - Are global corporate interests and the concerns of the wealthy in this nation and elsewhere overshadowing the needs and interests of everyday people?

  - Is it possible to prevent future terrorist attacks in the United States (and other nations as well) through tightening organizational intelligence and reorganizing some governmental bureaucracies?

  - Will we be able to balance the need for national security with the individual's right to privacy and freedom of movement within this country?

  - How will elected politicians and appointed government officials handle the challenges

FIGURE 14.17 Global conflict and terrorism have further destabilized various regions of the world. In the Middle East, for example, Sunni Muslim groups like Al Qaeda and ISIS have become powerful social and political movements that frequently come into conflict with Shiite branches of Islam such as Hezbollah. International power struggles and extreme violence are likely to continue far into the future.

war
organized, armed conflict between nations or distinct political and/or religious factions.

YOU CAN MAKE A DIFFERENCE

Keeping an Eye on the Media

Do we get all of the news that we should about how our government operates and about the pressing social problems of our nation? Consider this list of the top five news stories that media analysts such as Project Censored (2015) believe were *not* adequately covered by the U.S. media in 2013–2014:

1. **Ocean Acidification Increasing at Unprecedented Rate.** Carbon dioxide is rapidly changing the ocean's chemistry, with potentially devastating consequences for ocean life and humans who depend on fisheries for protein and their livelihood. Carbon dioxide is released into the air and absorbed by oceans when we burn fossil fuels in the form of coal, oil, and natural gas.

2. **Top Ten U.S. Aid Recipients All Practice Torture.** These top ten nations that were supposed to receive U.S. assistance in 2014 all practice torture and engage in human rights violations: Israel, Afghanistan, Egypt, Pakistan, Nigeria, Jordan, Iraq, Kenya, Tanzania, and Uganda.

3. **WikiLeaks Revelations on Trans-Pacific Partnership Ignored by Corporate Media.** A trade agreement designed to improve trade among twelve countries, including the United States, Vietnam, Australia, Canada, Japan, and Mexico, has been largely hidden from public view as it has been written by corporate advisors who represent the pharmaceutical industry, entertainment companies, and the oil industry. Corporations may be the big winners in this agreement because major sections of the proposal involve corporate law and intellectual property rights, not free trade. Potentially, 800 million people and one-third of all world trade may be affected by this largely unpublished trans-Pacific partnership.

4. **Corporate Internet Providers Threaten Net Neutrality.** The Federal Communications Commission (FCC) has published new rules for Internet traffic and created much corporate media attention about net neutrality. However, underlying stories about how companies such as Verizon challenge the FCC regarding its authority to regulate Internet service providers have received much less scrutiny. Will service providers be able to charge varying prices or give priority to users who access certain websites? May some Internet service providers be able to pay more for faster speeds compared to their competitors? These issues have largely been downplayed in the media.

5. **Bankers Back on Wall Street Despite Major Crimes.** Some bankers have been charged with engaging in illegal business practices, such as rigging auctions

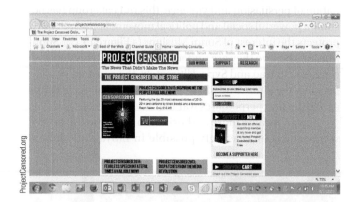

ProjectCensored.org

Project Censored is an organization of students and professors who produce the annual "Top 25 Censored Stories" list at Sonoma State University's Sociology Department. What can we learn from their research on news stories that are undervalued by the media?

regarding ongoing budget crises and the changing demographics of the United States?

- Will immigration and employment policies be based on the best interests of the largest number of people, or will these policies be based on the best interests of elites and major transnational corporations that are major contributors to political campaigns?

- Do the media accurately report what is going on at all levels of government? To what extent can individuals and grassroots organizations influence the media and the political process? (See "You Can Make a Difference.")

In the second decade of the twenty-first century, these are a few of the many questions regarding politics and the government that face us and people in other nations. How we (and our elected officials) answer these (and related) questions will in large measure determine the future of politics and government in the United States. Our answers will also have a profound influence on people and governments in other countries, whether they are in high-, middle-, or low-income nations, around the world.

In regard to global politics, one of the most compelling problems is how to reduce terrorism and lessen the risk of large-scale wars. According to some foreign policy

of municipal bonds by paying kickbacks to brokers and manipulating bids. These municipal bonds were intended to build schools, hospitals, libraries, and nursing homes throughout the United States. However, three former General Electric bankers involved in these nefarious activities had their convictions reversed on a technicality because the statute of limitations had run out. Although the media have investigated such crimes, the issue of why white-collar perpetrators of this magnitude usually serve no jail time is seldom discussed.

According to Project Censored, an organization of students and professors who produce the annual "Top Censored Stories" list at Sonoma State University's Sociology Department, many important stories are either missing from the news altogether or do not receive the attention they deserve. (To view the entire list, visit Project Censored's website.)

What should be the role of the media in keeping us informed? The media are referred to as the "Fourth Estate" or the "Fourth Branch of the Government" because they are supposed to provide people with relevant information on important topics regarding how the government operates in a democratic society. This information can then be used by citizens to decide how they will vote on candidates and issues presented for their approval or disapproval on the election ballot.

According to studies by the Pew Research Center and Project Censored, the media have not been fulfilling their duties to the public. Concern has increased that the corporate news media are blurring the facts on numerous issues and supporting elite ideology on issues such as global warming, how technology will ultimately save us, and how militarization is the right course of action for acquiring everything from oil to ensuring global economic stability. In sum, many important news stories are ignored, misreported, or simply censored by "mainstream" media, which are controlled by elite corporate interests (Huff and Roth, with Project Censored, 2012).

Can we become more aware of how the media influence our thinking on pressing issues? Yes, and the first step in keeping an eye on the news is to become more analytical about the "news" that we do receive. How can we evaluate the information that we receive from the media? In *How to Watch TV News*, media analysts Neil Postman and Steve Powers (2008) suggest the following:

1. We should keep in mind that television news shows are called "shows" for a reason. They are not a public service or a public utility.

2. We should never underestimate the power of commercials, which tell us much about our society.

3. We should learn about the economic and political interests of those who run television stations or own a controlling interest in a media conglomerate.

4. We should pay attention to the *language* of newscasts, not just the visual imagery. For example, a *question* may reveal as much about the *questioner* as the person answering the question.

Becoming aware of the media's role in influencing people's opinions about how our government is run is the first step toward becoming an informed participant in the democratic political process. The second step in keeping an eye on the news is becoming aware of national and international events that should receive more coverage than they do or that might not be reported in a fair and unbiased manner. With traditional media, these steps are somewhat easier to follow; however, with newer social media, we must look even closer to distinguish what may be nothing more than individual opinions from concrete information and facts. How do you think we might go about this endeavor?

analysts, it is important for the United States and other leading nations to have a coherent policy for dealing with global problems. An example of an international problem requiring a coherent political strategy is how to deal with the Islamic State. From this perspective, military action alone will not solve, or effectively reduce, many problems associated with global conflict. Instead, military action exacerbates power inequalities, creates greater problems with global underdevelopment, and produces new and dangerous terrorist factions. Can dialogue reduce the world's greatest political problems? Is there room for international policy makers to be more flexible and to seek other avenues of accord? Answers to these questions remain problematic when groups with exclusionary agendas and/or criminal motives are unwilling or unable to come to the negotiating table to talk about their agendas and concerns. Exclusion contributes to international conflict and wars, while political inclusion will be necessary to bring about more stability in fragile countries (Guehenno, 2015). Overall, preventing crises in the global political arena is more desirable than having to contain these crises later (Guehenno, 2015). Improved approaches to dealing with international political crises are long overdue. What do you think the future will hold in this regard?

CHAPTER REVIEW `Q & A`

LO1 What are power and authority, and what are the differences among traditional, charismatic, and rational–legal authority?

Power is the ability of persons or groups to carry out their will even when opposed by others. Authority is power that people accept as legitimate rather than coercive. Max Weber identified these types of authority: traditional, charismatic, and rational–legal. Traditional authority is based on long-standing custom. Charismatic authority is power based on a leader's personal qualities. Rational–legal authority is based on law or written rules and regulations, as found in contemporary bureaucracies.

LO2 What are the main types of political systems around the world?

The main types of political systems are monarchies, authoritarian systems, totalitarian systems, and democratic systems. In a monarchy, one person is the hereditary ruler of the nation. In authoritarian systems, rulers tolerate little or no public opposition and generally cannot be removed from office by legal means. In totalitarian systems, the state seeks to regulate all aspects of society and to monopolize all societal resources in order to exert complete control over both public and private life. In democratic systems, the powers of government are derived from the consent of all the people.

LO3 How is the pluralist model of power in the United States, and what part do special interest groups and political action committees play in this approach?

According to the pluralist (functionalist) model, power is widely dispersed throughout many competing interest groups. People influence policy by voting, joining special interest groups and political action campaigns, and forming new groups. Special interest groups are political coalitions made up of individuals or groups that share a specific interest they wish to protect or advance with the help of the political system. Political action committees (PACs) are organizations of special interest groups that solicit contributions from donors and fund campaigns to help elect (or defeat) candidates based on their stances on specific issues.

LO4 How does the power elite model view power in the United States?

According to the elite (conflict) model, power is concentrated in a small group of elites, whereas the masses are relatively powerless. According to C. Wright Mills, the power elite is composed of influential business leaders, key government leaders, and the military. The elites possess greater resources than the masses, and public policy reflects their preferences. According to G. William Domhoff, the United States has a ruling class—a relatively fixed group of privileged people who wield power sufficient to constrain political processes and serve underlying capitalist interests.

LO5 How do political parties shape the government, and what are some recent trends in political participation and voter apathy?

A political party (1) develops and articulates policy positions, (2) educates voters about issues and simplifies the choices for them, and (3) recruits candidates who agree with those policies, helps those candidates win office, and holds the candidates responsible for implementing the party's policy positions. Some people respond when they are displeased with current political parties and form their own organization; however, others simply do not participate and contribute to high rates of voter apathy in many elections. In the 2012 presidential election, 58.7 percent of the 219 million eligible voters cast ballots. This percentage was down from the 2008 presidential election, when 62.3 percent of the 208.3 million eligible voters actually voted.

LO6 What are key factors that contribute to the power of government bureaucracies, and how does bureaucratic power tend to take on a life of its own?

With dramatic increases in technology and increasing demands that the federal government "do something" about a wide variety of problems facing the United States, the government has continued to grow in many areas in recent decades. Because much of the actual functioning of the government is carried on by its bureaucracy, even the president, the White House staff, and cabinet officials have difficulty establishing control over the bureaucracy. The governmental bureaucracy has been able to perpetuate itself and expand because many of its employees have highly specialized knowledge and skills and cannot be replaced easily by "outsiders."

LO7 What is the iron triangle and the military–industrial complex?

The iron triangle is also referred to as the military–industrial complex—the mutual interdependence of the government military establishment and private military contractors.

LO8 What are the effects of militarism, terrorism, and war on citizens and the U.S. political system?

Militarism is a societal focus on military ideals and an aggressive preparedness for war. The term *militarism* is not synonymous with *war*; rather, it involves an ideology that perpetuates societal activities directed toward military preparedness and enormous spending on weapons production and "defense," as compared with other national priorities. Terrorism is the use of calculated, unlawful physical force or threats of violence against a government, organization, or individual to gain some political, religious, economic, or social objective. Around the world, acts of terrorism extract a massive toll by producing rampant fear, widespread loss of human life, and extensive destruction of property.

KEY TERMS

authoritarianism 403
authority 400
charismatic authority 401
democracy 404
elite model 407
government 399
militarism 416
military–industrial complex 416
monarchy 402

pluralist model 405
political action committee (PAC) 406
political party 409
political socialization 411
political sociology 399
politics 399
power 400
power elite 407
rational–legal authority 401

representative democracy 404
routinization of charisma 401
special interest group 405
state 399
terrorism 417
totalitarianism 404
traditional authority 400
war 418

QUESTIONS for CRITICAL THINKING

1 Who is ultimately responsible for decisions and policies that are made in a democracy such as the United States—the people or their elected representatives?

2 How would you design a research project to study the relationship between campaign contributions to elected representatives and their subsequent voting records? What would be your hypothesis? What kinds of data would you need to gather? How would you gather accurate data?

3 How does your school (or workplace) reflect a pluralist or elite model of power and decision making?

4 Can democracy survive in a context of rising concentrated power in the capitalist oligarchies? Why or why not?

ANSWERS to the SOCIOLOGY QUIZ

ON POLITICS AND THE MEDIA

1 **True** Despite the "digital divide" (which means unequal access to the Internet based on the interactivity of race and income), more than 80 percent of adults in the United States indicate in studies that they use the Internet, and many are also social network users.

2 **False** Although social networking sites do contain political news, only about one-third of social network users indicated that such sites were "very" or "somewhat" important in helping them keep up with information about the 2012 presidential election; only about one-quarter of users surveyed indicated that they used social networks to debate or discuss political issues with others.

3	False	Persons who describe their political beliefs as conservative are less likely to use social networking than those who describe themselves as "moderate" or "liberal." Only 49 percent of conservatives use such sites, as compared to 60 percent of liberals and 61 percent of moderates. Although the figures change slightly, the differences hold true for the overall U.S. population generally and Internet users specifically.
4	True	About 38 percent of social network users indicate that they were surprised about their friends' political beliefs, and 18 percent have blocked, unfriended, or hidden someone on a site because of what the other person posted about politics.
5	True	Young adults between the ages of 18 and 29 and African Americans are more likely to indicate that social networking sites are important for politics than are white Americans (non-Hispanics) and Hispanics.
6	False	The vast majority of social network users do not indicate that they have changed their political views or activity based on information from the sites. However, some who listed themselves as Democrats or liberals stated that they became more active on a political issue as a result of social network involvement.
7	False	Partly as a result of substantial cuts in newsroom personnel, media campaign reporters are now less involved in investigative journalism and instead primarily broadcast statements and assertions set forth by political candidates and their managers without interpreting or contextualizing them.
8	True	Audiences for local TV news programs have continued to decline over the past six years, even in years (such as 2012) in which a presidential election was held.

Source: Based on PRCPEJ, 2015.

FAMILIES AND INTIMATE RELATIONSHIPS

© Patrick Foto/Shutterstock.com

LEARNING OBJECTIVES

1   **Explain** these key concepts: families, kinship, family of orientation, family of procreation, extended family, and nuclear family.

2   **Describe** the differences among the following marriage patterns—monogamy, polygamy, polygyny, and polyandry—and the differences among these patterns of descent—patrilineal, matrilineal, and bilateral.

3   **Identify** the authority figure(s) in each of the following kinds of families: patriarchal, matriarchal, and egalitarian.

4   **Compare** functionalist, conflict/feminist, symbolic interactionist, and postmodernist perspectives on the family as a social institution.

5   **Discuss** issues that many contemporary couples face when thinking of developing intimate relationships and establishing families.

6   **Describe** child-related family issues that are of concern to many people in the twenty-first century.

7   **Identify** some of the key stressors that contribute to family violence and to the need for foster care for children.

8   **Discuss** divorce and how it affects remarriage patterns and blended families in the United States.

SOCIOLOGY & EVERYDAY LIFE

Diverse Family Landscapes in the Twenty-First Century

It definitely felt a little weird to be back [living at home]. I enjoy spending time with my mom, but I also felt like by now I should be established and not depending on her to help me.

—AMANDA SEALS, age thirty, describing how it feels to be part of the "boomerang generation"—adult children living at home with their parents while looking for a job or otherwise getting back on their feet (qtd. in Trounson, 2012).

Kelly is 33, and if dreams were winds, you might say that hers have shifted. She believes that every household needs one primary caretaker, that women are, broadly speaking, better at that job than men, and that no amount of professional success could possibly console her if she felt her two children were not being looked after the right way. The maternal instinct is a real thing, Kelly argues.

—In a *New York* magazine article, "The Retro Wife," Kelly Makino's life at a stay-at-home mom is described by journalist Lisa Miller (2013: 22).

Contemporary families are more diverse than in the past, including an increasing number of households made up of young people who, at least temporarily, have returned to live in their parents' homes. What larger societal factors contributed to this living arrangement?

I hit the breaking point as a parent a few years ago. It was the week of my extended family's annual gathering in August, and we were struggling with assorted crises. . . . Sure enough, one night all the tensions boiled over. At dinner, I noticed my nephew texting under the table. I knew I shouldn't say anything, but I couldn't help myself and asked him to stop. Ka-boom! My sister snapped at me to not discipline her child. My dad pointed out that my girls were the ones balancing spoons on their noses.

Iakov Filimonov/Shutterstock.com

Although the personal narratives about the families of Amanda, Kelly, and Bruce may initially seem unrelated, a common thread weaves through them: Contemporary families are diverse, ranging from "boomerang" families in which adult children return home until they can find jobs, to families with a stay-at-home parent who takes care of the children, and to extended families in which individuals may not always see things in the same way. The United States is a highly diverse nation with many family patterns. Parents socialize and interact with their children in various ways, often linked to their cultural background, socioeconomic status, race/ethnicity, religion, national origin, and/or other personal characteristics, attributes, and preferences.

In this chapter we examine the increasing complexity of family life in the United States and other nations. Pressing family and societal issues such as interpersonal communications, financial hardships, teenage childbearing, divorce, and child-care issues will be used as examples of how families and intimate relationships continue to change over time and place. Before reading on, test your knowledge about trends in U.S. family life by taking the "Sociology and Everyday Life" quiz.

LO1 **Explain** these key concepts: families, kinship, family of orientation, family of procreation, extended family, and nuclear family.

Families in Global Perspective

As the nature of family life has changed in high-, middle-, and low-income nations, the issue of what constitutes a "family" continues to be widely debated. In the "Universal Declaration of Human Rights," Article 16, adopted by the United Nations (1948), the family is defined as follows:

- Men and women of full age, without any limitation due to race, nationality, or religion, have the right to marry and to found a family. They are entitled to equal rights as to marriage, during marriage and at its dissolution.

- Marriage shall be entered into only with the free and full consent of the intending spouses.

- The family is the natural and fundamental group unit of society and is entitled to protection by society and the States.

My mom said none of the grandchildren had manners. Within minutes, everyone had fled to separate corners.

—Journalist Bruce Feiler (2013) describes how easily tensions can mount at extended-family gatherings. He further argues that families need a shared narrative about their history across generations so each person will develop an "intergenerational self" to provide personal identity and help in difficult times.

How Much Do You Know About Contemporary Trends in U.S. Family Life?

TRUE	FALSE		
T	F	1	Most U.S. adults view having a baby outside of marriage as "morally wrong."
T	F	2	More than four in ten U.S. adults have at least one step-relative in their family.
T	F	3	The percentage of children living with two parents, regardless of their marital status, differs by race and Hispanic origin.
T	F	4	Couples who cohabit before they get married are more likely to stay married.
T	F	5	The birth rate among U.S. teenagers is much lower than the birth rate among teens in other Western industrialized countries.
T	F	6	Stay-at-home parents have become less common since the 1990s, when more two-paycheck families were established to help pay the bills.
T	F	7	The percentage of U.S. households that contain only one person has continued to increase since the 1970s.
T	F	8	Among U.S. married couples, it is very rare for the wife to earn more money than the husband.

Answers can be found at the end of the chapter.

According to this declaration, the social institution of family must be protected in all societies because family is the "natural" and "fundamental" group unit of society. Although families differ widely around the world, they also share certain common concerns in their everyday lives. Food, clothing, shelter, and child care are necessities important to all people.

In the United States the Census Bureau defines a family as consisting of two or more people who are related by birth, marriage, or adoption, and residing in the same housing unit. (The Census Bureau specifies that one person in the household unit will be identified as the "householder.") For many years the standard sociological definition of *family* has been a group of people who are related to one another by bonds of blood, marriage, or adoption and who live together, form an economic unit, and bear and raise children. Some people believe that this definition should not be expanded—that social approval should not be extended to other relationships simply because the persons in those relationships wish to consider themselves to be a family. However, other people challenge this definition because it simply does not match the reality of family life in contemporary society, particularly at a time when only about half of adults ages eighteen and older are married in the legal usage of the term.

Today's families include many types of living arrangements and relationships, including single-parent households, unmarried couples, LGBTQ couples with or without children, and multiple generations (such as grandparent, parent, and child) living in the same household (Figure 15.1). To accurately reflect these changes in family life, some sociologists believe that we need a more encompassing definition of what constitutes a family. Accordingly, *families* are relationships in which people live together with commitment, form an economic unit and care for any young, and consider their identity to be significantly attached to the group. Sexual expression and parent–child relationships are a part of most, but not all, family relationships.

families
relationships in which people live together with commitment, form an economic unit and care for any young, and consider their identity to be significantly attached to the group.

FIGURE 15.1 Contemporary families are more diverse than in the past, including an increasing number of households made up of young people who, at least temporarily, have returned to live in their parents' homes. What larger societal factors contribute to this living arrangement?

How do sociologists approach the study of families? In our study of families we will use our sociological imagination to see how our personal experiences are related to the larger happenings in society. At the microlevel, each of us has a "biography," based on our experience within our family; at the macrolevel, our families are embedded in a specific culture and social context that has a major effect on them. We will examine the institution of the family at both of these levels, starting with family structure and characteristics.

Family Structure and Characteristics

In preindustrial societies the primary form of social organization is through kinship ties. **Kinship** refers to a social network of people based on common ancestry, marriage, or adoption. Through kinship networks, people cooperate so that they can acquire the basic necessities of life, including food and shelter. Kinship systems can also serve as a means by which property is transferred, goods are produced and distributed, and power is allocated.

In industrialized societies, other social institutions fulfill some of the functions previously taken care of by the

kinship network. For example, political systems provide structures of social control and authority, and economic systems are responsible for the production and distribution of goods and services. Consequently, families in industrialized societies serve fewer and more-specialized purposes than do families in preindustrial societies. Contemporary families are primarily responsible for regulating sexual activity, socializing children, and providing affection and companionship for family members.

Families of Orientation and Procreation During our lifetime, many of us will be members of two different types of families—a family of orientation and a family of procreation. The *family of orientation* is the family into which a person is born and in which early socialization usually takes place. Although most people are related to members of their family of orientation by blood ties, those who are adopted have a legal tie that is patterned after a blood relationship (• Figure 15.2). The *family of procreation* is the family that a person forms by having, adopting, or otherwise creating children. Both legal and blood ties are found in most families of procreation.

FIGURE 15.2 Whereas the relationship between spouses is based on legal ties, relationships between parents and children may be established by either blood or legal ties.

The relationship between a husband and wife is based on legal ties; however, the relationship between a parent and child may be based on either blood ties or legal ties, depending on whether the child has been adopted.

Some sociologists have emphasized that "family of orientation" and "family of procreation" do not encompass all types of contemporary families. Instead, many gay, lesbian, transsexual, bisexual, and transgender persons have *families we choose*—social arrangements that include intimate relationships between couples and close familial relationships among other couples and other adults and children. According to the sociologist Judy Root Aulette (1994), "families we choose" include blood ties and legal ties, but they also include *fictive kin*—persons who are not actually related by blood but who are accepted as family members.

Extended and Nuclear Families Sociologists distinguish between extended families and nuclear families based on the number of generations that live within a household. An **extended family** is a family unit composed of relatives in addition to parents and children who live in the same household. These families often include grandparents, uncles, aunts, or other relatives who live close to the parents and children, making it possible for family members to share resources. In horticultural and agricultural societies, extended families are extremely important; having a large number of family members participate in food production may be essential for survival. Today, extended-family patterns are found in Latin America, Africa, Asia, and some parts of Eastern and Southern Europe. With the advent of industrialization and urbanization, maintaining the extended-family pattern becomes more difficult. Increasingly, young people move from rural to urban areas in search of employment in the industrializing sector of the economy. At that time, some extended families remain, but the nuclear family typically becomes the predominant family form in the society.

A **nuclear family** is a family composed of one or two parents and their dependent children, all of whom live apart from other relatives. A traditional definition specifies that a nuclear family is made up of a "couple" and their dependent children; however, this definition became outdated when a significant shift occurred in the family structure. A comparison of Census Bureau data from 1970 and 2015 shows that there has been a significant decline in the percentage of U.S. households comprising a married couple with their own children under eighteen years of age, so we will look at what some social analysts refer to as the contemporary, diverse family.

The Contemporary Family—Family Diversity in the Twenty-First Century In the second decade of the twenty-first century, researchers have found that there is no such thing as a typical family. In the past the typical family comprised two married, heterosexual parents in their first marriage and their children under 18 years of age. In the 1960s, this was the norm for 73 percent of children living in the United States. However, by 1980, only 61 percent of children lived in such families, and the percentage reached

a new low at less than one-half (46 percent) in 2014 (Livingston, 2014). In the words of a *Time* magazine article, "Pretty much everyone agrees that the era of the nuclear family, with a dad who went to work, and the mom who stayed at home, has declined to the point of no return" (Luscombe, 2014). Of course, the question remains: "What is taking the place of the nuclear family?" And more family researchers are finding that the answer is *diversity*—a wider variety of family living arrangements has become the norm. According to the sociologist Philip Cohen (2014), three major factors have contributed to this dramatic change in family structure in the United States: (1) a decline in marriage rates; (2) a rise in the number of women who are employed in the paid workforce, and (3) a shift from the majority living in a nuclear family to a wider variety of living arrangements, such as blended families, cohabitation, and more-extensive patterns of remarriage (discussed later in this chapter).

Describe the differences among the following marriage patterns—monogamy, polygamy, polygyny, and polyandry—and the differences among these patterns of descent—patrilineal, matrilineal, and bilateral.

Marriage Patterns

Across cultures, different forms of marriage characterize families. **Marriage** is a legally recognized and/or socially approved arrangement between two or more individuals that carries certain rights and obligations and usually involves sexual activity. In most societies, marriage involves a mutual commitment by each partner, and linkages between two individuals and families are publicly demonstrated.

In the United States the only legally sanctioned form of marriage is **monogamy**—the practice or state of being married to one person at a time. For some people, marriage is

kinship
a social network of people based on common ancestry, marriage, or adoption.

family of orientation
the family into which a person is born and in which early socialization usually takes place.

family of procreation
the family that a person forms by having, adopting, or otherwise creating children.

extended family
a family unit composed of relatives in addition to parents and children who live in the same household.

nuclear family
a family composed of one or two parents and their dependent children, all of whom live apart from other relatives.

marriage
a legally recognized and/or socially approved arrangement between two or more individuals that carries certain rights and obligations and usually involves sexual activity.

monogamy
the practice or state of being married to one person at a time.

FIGURE 15.3 Polygamy is the concurrent marriage of a person of one sex with two or more persons of the opposite sex. Although most people do not practice this pattern of marriage, some men are married to more than one wife. Shown here is a polygamist family made up of Kody Brown and his four wives, who have been featured on the TLC reality television series, *Sister Wives*.

a lifelong commitment that ends only with the death of a partner. For others, marriage is a commitment of indefinite duration. Through a pattern of marriage, divorce, and remarriage, some people practice *serial monogamy*—a succession of marriages in which a person has several spouses over a lifetime but is legally married to only one person at a time.

Polygamy is the concurrent marriage of a person of one sex with two or more members of the opposite sex. The most prevalent form of polygamy is **polygyny**—the concurrent marriage of one man with two or more women. Polygyny has been practiced in a number of societies, including parts of Europe until the Middle Ages. More recently, some marriages in Islamic societies in Africa and Asia have been polygynous; however, the cost of providing for multiple wives and numerous children makes the practice impossible for all but the wealthiest men. In addition, because roughly equal numbers of women and men live in these areas, this nearly balanced sex ratio tends to limit polygyny. Contemporary cable TV shows have portrayed several U.S. families whose members live the polygamous lifestyle (● Figure 15.3).

The second type of polygamy is **polyandry**—the concurrent marriage of one woman with two or more men. Polyandry is very rare; when it does occur, it is typically found in societies where men greatly outnumber women because of high rates of female infanticide.

Patterns of Descent and Inheritance

Even though a variety of marital patterns exist across cultures, virtually all forms of marriage establish a system of descent so that kinship can be determined and inheritance rights established. In preindustrial societies, kinship is usually traced through one parent (unilineally). The most common pattern of unilineal descent is **patrilineal descent**—a system of tracing descent through the father's side of the family. Patrilineal systems are set up in such a manner that a legitimate son inherits his father's property and sometimes his position upon the father's death. In nations such as India, where boys are seen as permanent patrilineal family members but girls are seen as only temporary family members, girls tend to be considered more expendable than boys.

Even with the less common pattern of **matrilineal descent**—a system of tracing descent through the mother's side of the family—women may not control property. However, inheritance of property and position is usually traced from the maternal uncle (mother's brother) to his nephew (mother's son). In some cases, mothers may pass on their property to daughters.

By contrast, kinship in industrial societies is usually traced through both parents (bilineally). The most common form is **bilateral descent**—a system of tracing descent through both the mother's and father's sides of the family. This pattern is used in the United States for the purpose of determining kinship and inheritance rights; however, children typically take the father's last name.

 LO3 **Identify** the authority figure(s) in each of the following kinds of families: patriarchal, matriarchal, and egalitarian.

Power and Authority in Families

Descent and inheritance rights are intricately linked with patterns of power and authority in families. The most prevalent forms of familial power and authority are patriarchy, matriarchy, and egalitarianism. A **patriarchal family** is a family structure in which authority is held by the eldest male (usually the father). The male authority figure acts as head of the household and holds power and authority over the women and children, as well as over other males. A **matriarchal family** is a family structure in which authority is held by the eldest female (usually the mother). In this case the female authority figure acts as head of the household. Although there has been a great deal of discussion about matriarchal families, scholars have found no historical evidence to indicate that true matriarchies ever existed.

The most prevalent pattern of power and authority in families is patriarchy. Across cultures, men are the primary (and often sole) decision makers regarding domestic, economic, and social concerns facing the family. The existence of patriarchy may give men a sense of power over their own lives, but it can also create an atmosphere in which some men feel greater freedom to abuse women and children.

An *egalitarian family* is a family structure in which both partners share power and authority equally. Recently, a trend toward more-egalitarian relationships has been evident in a number of countries as women have sought changes in their legal status and increased educational and employment opportunities. Some degree of economic independence makes it possible for women to delay marriage or to terminate a problematic marriage. Recent cross-national studies have found that larger increases in the proportion of women who have higher levels of education, who hold jobs with higher wages, who have more commitment to careers outside the family, and who have greater interest in gender equality all contribute to the support of egalitarian gender values in the larger society as these ideas eventually spread to others.

Residential Patterns

Residential patterns are interrelated with the authority structure and the method of tracing descent in families. *Patrilocal residence* refers to the custom of a married couple living in the same household (or community) as the husband's parents. Across cultures, patrilocal residency is the most common pattern. Patrilocal residency can be found in countries where it is to the distinct advantage of young men to remain close to their parents' household.

Few societies have residential patterns known as *matrilocal*—the custom of a married couple living in the same household (or community) as the wife's parents. In industrialized nations such as the United States, most couples hope to live in a *neolocal residence*—the custom of a married couple living in their own residence apart from both the husband's and the wife's parents.

Up to this point, we have examined a variety of marriage and family patterns found around the world. Even with the diversity of these patterns, most people's behavior is shaped by cultural rules pertaining to endogamy and exogamy. *Endogamy* is the practice of marrying within one's own group. In the United States, for example, most people practice endogamy: They marry people who come from the same social class, racial–ethnic group, religious affiliation, and other categories considered important within their own social group. *Exogamy* is the practice of marrying outside one's own group. Depending on the circumstances, exogamy may not be noticed at all, or it may result in a person being ridiculed or ostracized by other members of the "in" group. The three most important sources of positive or negative sanctions for intermarriage are the family, the church, and the state. Participants in these social institutions may look unfavorably on the marriage of an in-group member to an "outsider" because of the belief that it diminishes social cohesion in the group. However, educational attainment is also a strong indicator of marital choice. Higher education emphasizes individual achievement, and college-educated people may be less likely than others to identify themselves with their social or cultural roots and thus more willing to marry outside their own social group or category if their potential partner shares a similar level of educational attainment.

 LO4 **Compare** functionalist, conflict/feminist, symbolic interactionist, and postmodernist perspectives on the family as a social institution.

Theoretical Perspectives on Family

The *sociology of family* is the subdiscipline of sociology that attempts to describe and explain patterns of family life and variations in family structure. Functionalist

---

polygamy
the concurrent marriage of a person of one sex with two or more members of the opposite sex.

polygyny
the concurrent marriage of one man with two or more women.

polyandry
the concurrent marriage of one woman with two or more men.

patrilineal descent
a system of tracing descent through the father's side of the family.

matrilineal descent
a system of tracing descent through the mother's side of the family.

bilateral descent
a system of tracing descent through both the mother's and father's sides of the family.

patriarchal family
a family structure in which authority is held by the eldest male (usually the father).

matriarchal family
a family structure in which authority is held by the eldest female (usually the mother).

egalitarian family
a family structure in which both partners share power and authority equally.

patrilocal residence
the custom of a married couple living in the same household (or community) as the husband's parents.

matrilocal residence
the custom of a married couple living in the same household (or community) as the wife's parents.

neolocal residence
the custom of a married couple living in their own residence apart from both the husband's and the wife's parents.

endogamy
the practice of marrying within one's own group.

exogamy
the practice of marrying outside one's own group.

sociology of family
the subdiscipline of sociology that attempts to describe and explain patterns of family life and variations in family structure.

perspectives emphasize the functions that families perform at the macrolevel of society, whereas conflict and feminist perspectives focus on families as a primary source of social inequality. Symbolic interactionists examine microlevel interactions that are integral to the roles of different family members. Postmodern analysts view families as being permeable, capable of being diffused or invaded so that their original purpose is modified.

Functionalist Perspectives

Functionalists emphasize the importance of the family in maintaining the stability of society and the well-being of individuals. According to Emile Durkheim, marriage is a microcosmic replica of the larger society; both marriage and society involve a mental and moral fusion of physically distinct individuals. Durkheim also believed that a division of labor contributes to greater efficiency in all areas of life—including marriages and families—even though he acknowledged that this division imposes significant limitations on some people.

In the United States, Talcott Parsons was a key figure in developing a functionalist model of the family. According to Parsons (1955), the husband/father fulfills the *instrumental role* (meeting the family's economic needs, making important decisions, and providing leadership), whereas the wife/mother fulfills the *expressive role* (running the household, caring for children, and meeting the emotional needs of family members).

Contemporary functionalist perspectives on families derive their foundation from Durkheim. Division of labor makes it possible for families to fulfill a number of functions that no other institution can perform as effectively. In advanced industrial societies, families serve four key functions:

1. *Sexual regulation.* Families are expected to regulate the sexual activity of their members and thus control reproduction so that it occurs within specific boundaries. At the macrolevel, incest taboos prohibit sexual contact or marriage between certain relatives. For example, virtually all societies prohibit sexual relations between parents and their children and between brothers and sisters.
2. *Socialization.* Parents and other relatives are responsible for teaching children the necessary knowledge and skills to survive. The smallness and intimacy of families make them best suited for providing children with the initial learning experiences they need.

3. *Economic and psychological support.* Families are responsible for providing economic and psychological support for members. In preindustrial societies, families are economic production units; in industrial societies, the economic security of families is tied to the workplace and to macrolevel economic systems. In recent years, psychological support and emotional security have been increasingly important functions of the family.
4. *Provision of social status.* Families confer social status and reputation on their members. These statuses include the ascribed statuses with which individuals are born, such as race/ethnicity, nationality, social class, and sometimes religious affiliation.

One of the most significant and compelling forms of social placement is the family's class position and the opportunities (or lack thereof) resulting from that position. Examples of class-related opportunities are access to quality health care, higher education, and a safe place to live.

Conflict and Feminist Perspectives

Conflict and feminist analysts view functionalist perspectives on the role of the family in society as idealized and inadequate. Rather than operating harmoniously and for the benefit of all members, families are sources of social inequality and conflict over values, goals, and access to resources and power (• Figure 15.4).

According to some classical conflict theorists, families in capitalist economies are similar to the work environment

FIGURE 15.4   Functionalist theorists believe that families serve a variety of functions that no other social institution can adequately fulfill. In contrast, conflict and feminist theorists believe that families may be a source of conflict over values, goals, and access to resources and power. Children in upper-class families have many advantages and opportunities that are not available to other children.

of a factory: Men in the home dominate women in the same manner that capitalists and managers in factories dominate their workers (Engels, 1970/1884). Although childbearing and care for family members in the home contribute to capitalism, these activities also reinforce the subordination of women through unpaid (and often devalued) labor. Other conflict analysts are concerned with the effect that class conflict has on the family. The exploitation of the lower classes by the upper classes contributes to family problems such as high rates of divorce and overall family instability.

Some feminist perspectives on inequality in families focus on patriarchy rather than class. From this viewpoint, men's domination over women existed long before capitalism and private ownership of property. Women's subordination is rooted in patriarchy and men's control over women's labor power. According to one scholar, "Male power in our society is expressed in economic terms even if it does not originate in property relations; women's activities in the home have been undervalued at the same time as their labor has been controlled by men" (Mann, 1994: 42). In addition, men have benefited from the privileges they derive from their status as family breadwinners.

Symbolic Interactionist Perspectives

Early symbolic interactionists such as Charles Horton Cooley and George Herbert Mead provided key insights on the roles that we play as family members and how we modify or adapt our roles to the expectations of others—especially significant others such as parents, grandparents, siblings, and other relatives. How does the family influence the individual's self-concept and identity? In order to answer questions such as this one, contemporary symbolic interactionists examine the roles of husbands, wives, and children as they act out their own parts and react to the actions of others. From such a perspective, what people think, as well as what they say and do, is very important in understanding family dynamics.

Some symbolic interactionist theorists focus on how interaction between marital partners contributes to a shared reality (Berger and Kellner, 1964). Although newlyweds bring separate identities to a marriage, over time they construct a shared reality as a couple. In the process, the partners redefine their past identities to be consistent with new realities. Development of a shared reality is a continuous process, taking place not only in the family but in any group in which the couple participates together. Divorce is the reverse of this process; couples may start with a shared reality and, in the process of uncoupling, gradually develop separate realities (• Figure 15.5).

Symbolic interactionists explain family relationships in terms of the subjective meanings and everyday interpretations that people give to their lives. As the sociologist Jessie Bernard (1982/1973) pointed out, women and men experience marriage differently. Although the husband may see *his* marriage very positively, the wife may feel less

FIGURE 15.5 Marriage is a complicated process involving rituals and shared moments of happiness. When marriage is followed by divorce, couples must abandon a shared reality and then reestablish individual ones.

positive about *her* marriage, or vice versa. Researchers have found that husbands and wives may give very different accounts of the same event and that their "two realities" frequently do not coincide.

Postmodernist Perspectives

According to postmodern theories, we have experienced a significant decline in the influence of the family and other social institutions. As people have pursued individual freedom, they have been less inclined to accept the structural constraints imposed on them by institutions. Given this assumption, how might a postmodern perspective view contemporary family life? For example, how might this approach answer the question "How is family life different in the digital age where many of us are surrounded by our technological gadgets?"

The postmodern family has been described as *permeable*—a more fluid and pliable form of the nuclear family that is characterized by larger variations in family structures. These variations are generated by divorce, remarriage, cohabitation, single-parent family structures, and families in which one or more grandchildren live with

Theoretical Perspectives on Families

	Focus	Key Points	Perspective on Family Problems
Functionalist	Role of families in maintaining stability of society and individuals' well-being	In modern societies, families serve the functions of sexual regulation, socialization, economic and psychological support, and provision of social status.	Family problems are related to changes in social institutions such as the economy, religion, education, and law/government.
Conflict/ Feminist	Families as sources of conflict and social inequality	Families both mirror and help perpetuate social inequalities based on class and gender.	Family problems reflect social patterns of dominance and subordination.
Symbolic Interactionist	Family dynamics, including communication patterns and the subjective meanings that people assign to events	Interactions within families create a shared reality.	How family problems are perceived and defined depends on patterns of communication, the meanings that people give to roles and events, and individuals' interpretations of family interactions.
Postmodernist	Permeability of families	In postmodern societies, families are diverse and fragmented. Boundaries between workplace and home are blurred.	Family problems are related to cyberspace, consumerism, and the hyperreal in an age increasingly characterized by high-tech "haves" and "have-nots."

their grandparents. In the postmodern family, traditional gender roles are much more flexible. Younger people are much less constrained by the hierarchy and power relations of more-traditional families, sometimes to the displeasure of parents and other adult caregivers. In the postmodern era, the nuclear family is now only one of many family forms. Similarly, the idea of romantic love has given way to the idea of consensual love: Some individuals agree to have sexual relations with others whom they have no intention of marrying or, if they marry, do not necessarily see the marriage as having permanence. Maternal love has also been transformed into shared parenting, which includes not only mothers and fathers but also caregivers who may either be relatives or nonrelatives.

Urbanity is another characteristic of the postmodern family. The boundaries between the public sphere (the workplace) and the private sphere (the home) are becoming much more open and flexible. In fact, family life may be negatively affected by the decreasing distinction between what is work time and what is family time. As more people are becoming connected "24/7" (twenty-four hours per day, seven days per week), the boss who in the past would not call at 11:30 P.M. may send a text or e-mail asking for an immediate response to some question that has arisen while the person is away from the workplace

The Concept Quick Review summarizes sociological perspectives on the family. Taken together, these perspectives on the social institution of families reflect various ways in which familial relationships may be viewed in contemporary societies. Now we shift our focus to love, marriage, intimate relationships, and family issues in the United States.

Discuss issues that many contemporary couples face when thinking of developing intimate relationships and establishing families.

Developing Intimate Relationships and Establishing Families

The United States has been described as a "nation of lovers"; it has been said that we are "in love with love." Why is this so? Perhaps the answer lies in the fact that our ideal culture emphasizes *romantic love*, which refers to a deep emotion, the satisfaction of significant needs, a caring for and acceptance of the person we love, and involvement in an intimate relationship (Lamanna, Riedmann, and Stewart, 2015) (• Figure 15.6). In the United States the notion of romantic love is deeply intertwined with our beliefs about how and why people develop intimate relationships and establish families. Not all societies share this concern with romantic love. However, in this country the number of opportunities for romance is sometimes increased by online matching services, such as FarmersOnly.com, which caters to farmers and people who love nature.

Love and Intimacy

In the late nineteenth century, during the Industrial Revolution, people came to view work and home as separate spheres in which different feelings and emotions were appropriate. The public sphere of work—men's sphere—emphasized self-reliance and independence. By contrast, the private sphere of the home—women's

FIGURE 15.6 FarmersOnly.com is one of many online matchmaking services that cater to individuals of various ages, occupations, interests, and geographic locations. What might be gained by using an online dating service? What limitations may be caused by this approach?

sphere—emphasized the giving of services, the exchange of gifts, and love. Accordingly, love and emotions became the domain of women, and work and rationality became the domain of men (Lamanna, Riedmann, and Stewart, 2015). Although the roles of women and men have changed dramatically over the past one hundred years, women and men may still not share the same perceptions about romantic love today, and women may express their feelings verbally, whereas men may express their love through non-verbal actions; however, in other cases, the situation may be just the opposite.

Love and intimacy are closely intertwined. Intimacy may be psychic ("the sharing of minds"), sexual, or both. Although sexuality is an integral part of many intimate relationships, perceptions about sexual activities vary from one culture to the next and from one time period to another. For example, kissing has traditionally been found primarily in Western cultures; many African and Asian cultures have viewed kissing negatively, although some change has occurred among younger people in recent years.

For decades, the work of the biologist Alfred C. Kinsey was considered to be the definitive research on human sexuality, even though some of his methodology had serious limitations. However, in the 1990s the work of Kinsey and his associates was superseded by the National Health and Social Life Survey conducted by the National Opinion Research Center at the University of Chicago (see Laumann et al., 1994; Michael et al., 1994). Based on interviews with more than 3,400 men and women ages 18 to 59, this random survey tended to reaffirm the significance of the dominant sexual ideologies. Most respondents reported that they engaged in heterosexual relationships, although 9 percent of the men said they had had at least one homosexual encounter resulting in orgasm. Although 6.2 percent of men and 4.4 percent of women said that they were at least somewhat attracted to others of the same gender, only 2.8 percent of men and 1.4 percent of women identified themselves as gay or lesbian. According to the study, persons who engaged in extramarital sex found their activities to be more thrilling than those with a marital partner, but they also felt more guilt. Persons in sustained relationships such as marriage or cohabitation found sexual activity to be the most satisfying emotionally and physically.

Today, research on human sexuality continues at numerous universities and other research facilities. Some studies are funded by corporations, such as condom manufacturers or pharmaceutical companies, which might benefit from the research findings. If you wish to look at findings from one such study, the National Survey of Sexual Health and Behavior (NSSHB), conducted by researchers from the Center for Sexual Health Promotion at Indiana University's School of Health, Physical Education and Recreation, is available online. An area of investigation in this study is the effects of condom use because it is believed that more information will not only be beneficial for individuals but also that it will assist medical and public health professionals who address issues such as HIV, sexually transmitted infections, and unintended pregnancy. A few key findings from this research are as follows (National Survey of Sexual Health and Behavior, 2015):

- Enormous variability exists in the sexual behavior of U.S. adults, with more than 40 combinations of sexual activity reported by respondents in the study.

- Many older adults have active, pleasurable sex lives and engage in a range of behaviors and partner types.

- Although 85 percent of men in the study reported that their partner had an orgasm at their most recent sexual event, only 64 percent of women reported having an orgasm at their most recent sexual event.

- About 7 percent of adult women and 8 percent of men identified as gay, lesbian, or bisexual; however, the proportion of individuals in the United States who reported that they had had same-gender sexual inter-actions at some point in their lives was much higher.

- Despite popular media representations, most adolescents in the study did not report that they were engaged in partnered sexual behavior.

Cohabitation and Domestic Partnerships

Attitudes about cohabitation have changed dramatically over the past five decades. ***Cohabitation*** refers to two people who live together, and think of themselves as a couple, without being legally married. The U.S. Census Bureau uses the terms *unmarried partner, cohabiting partner,* and *cohabiter* interchangeably when referring to individuals who cohabit (Vespa, Lewis, and Kreider, 2013). The number of unmarried opposite-sex partners living together has increased from about 2.9 million in 1996 to 7.9 million in 2014 (see • Figure 15.7). About 3.1 million cohabiting couples (unmarried) had one or more children younger than age 18 residing in their household in 2014, an increase from approximately 1.2 million cohabitating partners with children under 18 in the late 1990s (Child Trends Data Bank, 2015).

As compared with children who reside with married parents, cohabiting couples with children are typically younger, have fewer years of formal education, hold lower-income positions, and have less secure sources of employment. Often this contributes to economic disadvantages for the couples and their children. Those who are statistically most likely to cohabit are younger individuals (typically under age 45), people who have been married before, or older persons who do not want to lose retirement benefits that are contingent upon not remarrying.

How common is cohabitation among women? A study by the National Center for Health Statistics found that nearly one-half (48 percent) of all U.S. women between the ages of 15 and 44 had cohabited before marriage. Between 1995 and 2010, the percentage of women who cohabited as a first union increased to 57 percent of Hispanic (Latina) women, 43 percent of white (non-Hispanic) women, and 39 percent of African American (black) women (Copen, Daniels, and Mosher, 2013). According to researchers in this study, 40 percent of first premarital cohabitations among women transitioned to marriage within 3 years, about 33 percent of these cohabitating relationships continued, and 27 percent dissolved (Copen, Daniels, and Mosher, 2013). This study looked only at *first premarital cohabitation,* so it did not include those women who had cohabited with other partners.

To what extent do cohabiting couples have children together? Over the past decade, childbearing has become more frequent among cohabiting couples. Among all U.S. women ages 15–44, about 23 percent of births occurred within cohabiting relationships, up from 14 percent in 2002. According to research participants, about 50 percent of these births were unintended (Copen, Daniels, and Mosher, 2013).

Does cohabitation serve as a step toward marriage, or is it an alternative to marriage? Research shows that cohabitation is more likely to serve as a transition into marriage among women with higher levels of education and income than for cohabiting women with lower levels (Copen, Daniels, and Mosher, 2013).

Among heterosexual couples, many reasons continue to exist for cohabitation; for LGBTQ couples, however, no alternatives to cohabitation existed for many years until the 2015 U.S. Supreme Court ruling in *Obergefell v. Hodges* legalized same-sex marriage (see Chapter 11). For that reason, many lesbians, gays, bisexuals, and transgender persons often sought recognition of their civil unions or created ***domestic partnerships***—household partnerships in which an unmarried couple lives together in a committed, sexually intimate relationship and is granted some of the same rights and benefits as those accorded to married heterosexual couples. Civil unions—which have been available in some states to both same-sex and opposite-sex couples—provide legal recognition of the couple's relationship and afford legal rights to the partners similar to those accorded to spouses in marriage. It should be noted that domestic partnerships vary from state to state and that they remain in a period of fluctuation following the Supreme Court's decision that legal same-sex marriages provide the same rights and benefits to partners as legal marriages entered into by heterosexual couples. Rights typically granted to those domestic partners who meet state requirements include the following: hospital visitation rights, child custody and visitation rights, health insurance coverage if it is provided for heterosexual spouses by the partner's employer, and similar family-related rights that are granted to opposite-sex families. State and federal laws are in a period of transition regarding civil unions and domestic partnerships. Interesting questions have also arisen for same-sex partners currently in domestic partnerships or civil unions: Do we now choose to enter into a legally binding same-sex marriage? Some of the many issues involved include laws related to inheritance rights, tax-related benefits and liabilities, and provisions of divorce laws in various states.

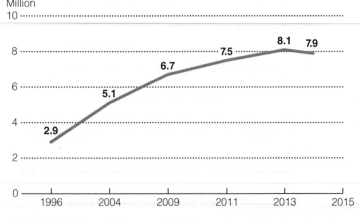

Million

FIGURE 15.7 Estimated Number of Opposite-Sex Couples Cohabiting in the United States in Selected Years, 1996–2014
Source: National Vital Statistics Report, 2015.

Marriage

Why do people get married? Couples get married for a variety of reasons. Some do so because they are "in love," desire companionship and sex, want to have children, feel social pressure, are attempting to escape from a bad situation in their parents' home, or believe that they will have more money or other resources if they get married. These factors notwithstanding, the selection of a marital partner is actually fairly predictable. As previously discussed, most people in the United States tend to choose marriage partners who are similar to themselves. *Homogamy* refers to the pattern of individuals marrying those who have similar characteristics, such as race/ethnicity, religious background, age, education, or social class. However, homogamy provides only the general framework within which people select their partners; people are also influenced by other factors. For example, some researchers claim that people want partners whose personalities match their own in significant ways. Thus, people who are outgoing and friendly may be attracted to other people with those same traits. However, other researchers claim that people look for partners whose personality traits differ from but complement their own.

The number of married households in the United States has been in a downward trend since the 1970s, when married couples made up 71 percent of all households, to 2014, when less than half of all U.S. households were composed of married couples. The median age at first marriage continued to increase over the past four decades. The median age at first marriage in 2014 was 29.3 for men and 27 for women, as compared to 23.2 for men and 20.8 for women in 1970 (U.S. Census Bureau Marital Status Data, 2015).

Same-Sex Marriages

As discussed in Chapter 11, controversy continues over the legal status of gay and lesbian couples, particularly those who seek to make their relationship a legally binding commitment through marriage (● Figure 15.8). Let's take a brief look at the history of how same-sex marriage law evolved in the U.S. Supreme Court prior to the 2015 decision in *Obergefell v. Hodges.* One case regarding same-sex marriage, *United States v. Windsor,* challenged Section 3 of the 1996 Defense of Marriage Act (DOMA), which explicitly defined marriage for all purposes under federal law as the legal union of one man and one woman as husband and wife. The issue before the Court was whether DOMA deprives same-sex couples, who are lawfully married in states that permit it, of the equal protection of the laws guaranteed by the Fifth Amendment to the Constitution. In 2011 President Obama declared this act to be unconstitutional and ordered the U.S. Justice Department to stop defending the law in court. His highly controversial decision was applauded by gay and lesbian rights advocates but was sharply denounced by conservative political leaders. The case moved through various courts until it reached the highest court in the nation.

FIGURE 15.8 The issue of same-sex marriage has been in the headlines because of two major legal challenges before the U.S. Supreme Court and wider public awareness of the importance of various civil rights issues involved.

cohabitation
a situation in which two people live together, and think of themselves as a couple, without being legally married.

domestic partnerships
household partnerships in which an unmarried couple lives together in a committed, sexually intimate relationship and is granted some of the same rights and benefits as those accorded to married heterosexual couples.

homogamy
the pattern of individuals marrying those who have similar characteristics, such as race/ethnicity, religious background, age, education, or social class.

The U.S. Supreme Court ruled that DOMA is unconstitutional because it amounts to a deprivation of the equal liberty of persons that is protected by the Fifth Amendment. This ruling struck down the central provisions of DOMA that denied federal benefits to same-sex couples who were married in jurisdictions that permit same-sex unions. However, this decision did not address the larger issue of whether there was a nationwide right of all same-sex couples to marry regardless of where they live.

In the second Supreme Court case, *Hollingsworth v. Perry*, the Court examined the issue of whether the Equal Protection Clause of the Fourteenth Amendment prohibited the state of California from enforcing Proposition 8, a voter-approved measure defining marriage as the union of a man and woman and banning same-sex marriage. Prior to passage of this proposition, in a five-month period in 2008 same-sex marriages were legally performed in California. Ultimately, the Supreme Court ruled that the proponents who intervened to defend Proposition 8 did not have legal standing based on the U.S. Constitution and that their appeal should be dismissed. The Court's decision cleared the way for same-sex marriages to resume in California. It also meant that same-sex couples married in jurisdictions that permit such unions could not be denied federal benefits.

Although legal, political, and social support for same-sex marriage continued to increase over the past decade, many questions remained unanswered, including (1) how state-by-state laws would be applied in various situations, (2) what happens when a same-sex couple legally married in one state moved across state lines to another state where same-sex marriage was not recognized, (3) how the patchwork of state laws affected divorce, and many other issues that the Supreme Court did not address when it reached decisions in these two cases. The 2015 Supreme Court decision, issued in a 5–4 ruling by the Justices, served to settle at least some of the disputes that existed regarding same-sex marriage and to acknowledge that marriage and equal dignity in the eyes of the law are a Constitutional right in the United States. Although some of the Supreme Court Justices strongly disagreed with the Court's decision and wrote scathing dissents, the nation's marriage laws were changed as the United States became the twenty-first country to legalize same-sex marriage nationwide. Married same-sex couples now have the same legal rights and obligations as married heterosexual couples and to be recognized on birth and death certificates.

This decision is in keeping with surveys by the Pew Research Center that indicate 57 percent of persons responding to a 2015 survey supported same-sex marriage while only 39 percent opposed it. (Pew Research Center, 2015). Some researchers attribute this change in attitude to the Millennial generation (born after 1980), which is made up of young adults who are more open to gay rights than previous generations. The most frequently cited reason for changing attitudes on same-sex marriage is having friends, family members, or acquaintances who are lesbian or gay (Pew Research Center, 2015).

Some people assume that there would be uniform acceptance by all LGBTQ Americans of same-sex marriage, but Pew Research Center studies show that additional factors—such as political preference, age, and religion—are involved in how individuals view this issue. According to Pew Research Center findings, support for same-sex marriage among LGBTQ adults differs slightly along political party lines, with 45 percent of those who identified as Republicans stating they "strongly favor" same-sex marriage as compared with 81 percent among Democrats. Younger LGBTQ individuals also were more likely to "strongly favor" same-sex marriage (82 percent) than persons age 30 or older (71 percent). Religious affiliation is another factor: Sixty-seven percent of the religiously affiliated strongly favored same-sex marriage, as compared with 82 percent of the religiously unaffiliated. Overall, even with differences in political party, age, and/or religion, at least nine in ten LGBTQ adults "strongly favor" or "favor" same-sex marriage in this Pew Research Center (2015) survey.

Housework and Child-Care Responsibilities

Taking care of housework, child-care responsibilities, and paid employment is a continuing challenge for families (● Figure 15.9). In 2014, among the 34.4 million families with children in the United States, 88.7 percent of these families had at least one employed parent. In families with children maintained by a woman with no spouse present, the mother was employed in 69.4 percent of the households. But in families with children maintained by a man with no spouse present, 81.9 percent of the fathers were employed. And in married-couple heterosexual families with children, 96.6 percent of the households had at least one employed parent. Both parents work in 60.2 percent of all heterosexual married-couple families where children under age 18 reside (U.S. Bureau of Labor Statistics, 2015). What this means is that many marriages in the United States are *dual-earner marriages*—marriages in which both spouses are in the labor force. Today, even when their children are infants, 57.1 percent of mothers are employed in the labor force. For children under age 6, 64.2 percent of mothers are employed. For youths between the ages of 6 and 17, 74.7 percent of mothers are employed in the paid labor force (U.S. Bureau of Labor Statistics, 2015).

So, as we can see, many mothers are working all or part of the day in paid employment and then returning to their household, where additional work awaits them. Sociological research shows that many women, after leaving their paid employment at the end of the day, go home to perform hours of housework and child care. Sociologist Arlie Hochschild (2012) refers to this as the *second shift*—the domestic work that employed women perform at home after they complete their workday on the job. Thus, many married women contribute to the economic well-being of their families and also meet many of the domestic needs of family members by cooking, cleaning, shopping, taking care of children, and managing household routines.

According to Hochschild, the unpaid housework that women do on the second shift amounts to an extra month of work each year. For fathers with children and no spouse present, a similar second shift may also exist.

But haven't things changed? Aren't men doing much more around the household these days? In recent years, married and cohabiting couples with more-egalitarian ideas about women's and men's roles have tended to share more equally in food preparation, housework, and child-care responsibilities (Lachance-Grzela and Bouchard, 2010). Although some studies show that when husbands share household responsibilities, they spend less time in these activities than do their wives, other studies shown that the higher a woman's educational resources and earnings potential, the more help from her partner that she actually gets with child care and housework (Sullivan, 2011). However, certain kinds of activities, such as night duty with young children, appear to be unevenly distributed. More mothers report sustained sleep deprivation: The American Time Use Survey shows that women in dual-earner couples are still three times more likely than men to report interrupted sleep patterns if they are the parent of a child under the age of one. Even more telling, stay-at-home mothers are six times more likely to get night duty where they are up with their children as are stay-at-home fathers (Senior, 2014).

In marriages with same-sex partners, negotiations about child-rearing tasks are also prevalent. If one partner is employed full time outside the household while the other is a stay-at-home parent or is employed part time, child-rearing duties and school-related meetings and tasks are often the responsibility of the one who fulfills the "daytime" parenting duties. However, when both partners are employed full-time, particularly in professional occupations, each parent may have variable family duties depending on external scheduling demands and priorities.

In the United States, millions of parents rely on outside child care so that they can work. Relatives (father,

Ariel Skelley/The Image Bank/Getty Images

FIGURE 15.9 Juggling housework, child care, and a job in the paid workforce is all part of the average day for many women. Why does sociologist Arlie Hochschild believe that many women work a "second shift"?

grandparent, sibling, or other relative) take care of almost half of all children between birth and four years of age who have full-time employed mothers. Another 25 percent of children with full-time employed mothers spend most of their time away from home in a center-based arrangement, such as day care, nursery school, preschool, or Head Start. An additional 13 percent are primarily cared for by a nonrelative, such as a babysitter, nanny, au pair, or family day-care provider, in a home-based environment (Federal Interagency Forum on Child and Family Statistics, 2013). Research is very limited on how same-sex partners deal with outside child care or the extent to which they employ in-house day or evening care for their children. The cost of child-care programs often makes it difficult for families to find a high-quality environment for their children. Church-sponsored programs, such as "Mothers' Day Out," have become increasingly popular with young working mothers who use these facilities for one or more days per week as a form of relatively inexpensive day care for their children while they are at work. In the future, some of these programs may be referred to as "Parents' Day Out" instead.

Although organized after-school programs have become more numerous, the percentage of children staying home alone has remained steady in recent years. About 25 percent of school-age children stay alone after the school day ends until a parent returns home from work. Child-care specialists are concerned about this because children need productive and safe activities to engage in while their parents are working, but many home-alone children spend time eating junk food, watching television, talking on their

dual-earner marriages
marriages in which both spouses are in the labor force.

second shift
Arlie Hochschild's term for the domestic work that employed women perform at home after they complete their workday on the job.

Wombs-for-Rent: Commercial Surrogacy in India

I wanted to be a surrogate mother because I wanted to deposit money into an account for my children for their future. I also wanted to help parents who cannot have children. I am proud to have given birth to a beautiful baby. . . . I feel like part of the family.

> —Thapa, a 31-year-old Indian woman who became a surrogate for an Australian couple, explains why, in addition to her own children, she is willing to help other people become parents (qtd. in AFP, 2013). Thapa works with a New Delhi, India, surrogacy center, where she and other surrogate mothers earn about $6,000 for carrying a child for a foreign couple.

Why do some infertile couples in the United States, Britain, and elsewhere want to "hire" a woman in India to have

their child? Most couples who engage in this practice have made numerous attempts to have a child through in vitro fertilization and other assisted reproductive technologies.

A surrogate mother (left) has delivered a baby for Karen Kim (center), with the help of infertility specialist Dr. Nayna Patel (right). This practice, sometimes called "rent-a-womb," remains controversial.

cell phone, or playing computer and video games. Many of these children are under the supervision of an older brother or sister who may not be particularly interested in taking care of them. Older children are more likely than younger ones to care for themselves: Thirty-three percent of children ages 12–14 are regularly in self-care situations, as compared to 10 percent of children ages 9–11 and 2 percent of children ages 5–8 (Federal Interagency Forum on Child and Family Statistics, 2013).

LO6 **Describe** child-related family issues that are of concern to many people in the twenty-first century.

Child-Related Family Issues and Parenting

Not all couples become parents. Those who decide not to have children often consider themselves to be "child-free," whereas those who do not produce children through no choice of their own may consider themselves "childless."

Deciding to Have Children

Cultural attitudes about having children and about the ideal family size began to change in the late 1950s and have continued to decline in the decades following in the United States.

In 2013, the latest year for which comprehensive data are available, the birth rate continued to decline slightly for white (non-Hispanic) and Hispanic women. However, no appreciable change was recorded for African American (black/non-Hispanic) women (National Vital Statistics Reports, 2015).

Age is a factor in deciding to have children. Birth rates for teenagers between the ages of 15 and 19 decreased by 10 percent between 2012 and 2013, with 26.5 births per 1,000 teenagers in that age category. This was a historical low for the United States and included teenagers in nearly all racial and Hispanic-origin groups. Birth rates declined to record lows for women in their twenties. Birth rates rose for women in their thirties and late forties but remained relatively unchanged for women in their early forties. So this means that birth rates declined for all women under age thirty and rose for women ages 30–39 and 45–49. The mean (average) age of "mother at first birth" rose to 26.0 years in 2013, up from 25.8 years in 2012 (National Vital Statistics Reports, 2015).

Today, the percentage of births to unmarried women is 40.6 percent of all births, and this number declined by 1 percent from 2012 to 2013, the latest year for which data are available (National Vital Statistics Reports, 2015).

Advances in birth control techniques over the past four decades—including the birth control pill and contraceptive patches and shots—now make it possible for people to decide whether they want to have children, how many they wish to have, and to determine (at least somewhat) the spacing of the children's births. However,

If they have been unsuccessful in their efforts, the couple may first attempt to find a surrogate in the United States, but they quickly learn that a U.S. gestational surrogate costs more than $50,000—far more than they would pay for a surrogate in India (Kohl, 2007; United Press International, 2007). According to infertility specialist Dr. Nayna Patel, earning money through surrogacy helps uplift Indian women: It provides money for their household and makes them more independent. For example, the typical woman might earn more for one surrogate pregnancy than she would earn in 15 years from other kinds of employment (*CBS News*, 2007).

Are there any problems with global "rent-a-womb"? If there is an agreement between a surrogate mother and a couple who badly wants a child, some analysts believe that "offer and acceptance" is nothing more than capitalism at work—where there is a demand (for infants by infertile couples), there will be a supply (from low-income surrogate mothers). However, some ethicists raise troubling questions about the practice of commercial surrogacy: A mother should give birth to her child because it is hers and she loves it, not because she is being paid to give birth to someone else's baby. Other

social critics are concerned about the potential mistreatment of low-income women who may be exploited or may suffer long-term emotional damage from functioning as a surrogate (Dunbar, 2007). For the time being, in clinics such as the one in India, hopeful parents just provide the egg, the sperm, and the money, and all the rest is done for them by the clinic and the surrogates, who live in a spacious house where they are taken care of by maids, cooks, and doctors.

REFLECT & ANALYZE

What are your thoughts on surrogacy? Is there any difference between surrogacy when it occurs in high-income nations such as the United States and Britain as compared to situations in which the parents live in a high-income nation and the surrogate mother lives in a lower-income nation? How might we relate the specific issue of outsourced surrogacy to some larger concerns about families and intimate relationships that we have discussed in this chapter?

sociologists suggest that fertility is linked not only to reproductive technologies but also to women's beliefs that they do or do not have other opportunities in society that are viable alternatives to childbearing (Lamanna, Riedmann, and Stewart, 2015).

Today, the concept of reproductive freedom includes both the desire *to have* or *not to have* one or more children. According to some sociologists, many U.S. women spend up to one-half of their life attempting to control reproduction while other women and men choose to be child-free. Among U.S. women ages 15 to 44, 6 percent are voluntarily childless. The percentage of women who are voluntarily childless (and have never given birth) is higher among women who hold professional degrees (a master's or equivalent and higher). One in four women with a master's degree and nearly that many women with doctorates have no biological children by the time the women reach the ages of 40 to 44. Of course, some of these women may have children through adoption, by marriage to a partner who already had one or more children, or other factors (Wade, 2012). When some people decide not to have children, their wishes come into conflict with our society's *pronatalist bias,* which assumes that having children is the norm and can be taken for granted, whereas those who choose not to have children believe they must justify their decision to others (Lamanna, Riedmann, and Stewart, 2015).

Some couples experience the condition of *involuntary infertility*, whereby they want to have a child but

find that they are physically unable to do so. *Infertility* is defined as an inability to conceive after a year of unprotected sexual relations. Women who are able to get pregnant but who are not able to stay pregnant may also be defined as infertile. Research suggests that fertility problems originate in females in approximately one-third of the cases, with males in about one-third of the cases, and with both the male and female in about one-third of the cases (Mayo Clinic, 2011). Leading causes of male infertility are abnormal sperm production or function, sexual problems, general health and lifestyle issues, overexposure to certain environmental factors (such as pesticides and other chemicals or heat), and age. The most common causes of female infertility include fallopian tube damage or blockage, endometriosis, ovulation disorders, early menopause, and other health-related disorders (Mayo Clinic, 2011). It is estimated that about half of infertile couples who seek treatments such as medication, behavioral approaches, fertility drugs, artificial insemination, and surgery can be helped; however, some are unable to overcome infertility despite expensive treatments such as *assisted reproductive technology (ART),* which includes medical procedures such as *in vitro fertilization,* whereby medical professionals help infertile couples achieve pregnancy (Mayo Clinic, 2011). Some people who are involuntarily childless may choose surrogacy or adoption as an alternate way to become a parent (see "Sociology in Global Perspective").

Adoption

Adoption is a legal process through which the rights and duties of parenting are transferred from a child's biological and/or legal parents to a new legal parent or parents. This procedure gives the adopted child all the rights of a biological child. In most adoptions a new birth certificate is issued, and the child has no future contact with the biological parents; however, some states have "right-to-know" laws under which adoptive parents must grant the biological parents visitation rights. In 2014 approximately 67 percent of all adoptive households were made up of married couples, followed by single women at 27 percent, single men at 3 percent, and unmarried couples at 3 percent (U.S. Department of Health and Human Services, 2015).

Matching children who are available for adoption with prospective adoptive parents can be difficult. The available children have specific needs, and the prospective parents often set specifications on the type of child they want to adopt. Some adoptions are by relatives of the child; others are by infertile couples (although many fertile couples also adopt). Increasing numbers of LGBTQ persons and individuals who are single are adopting children. Some prospective parents seek out children in nations such as China, Ethiopia, Ukraine, Haiti, Uganda, and Russia (U.S. Department of Homeland Security, 2014).

Teenage Childbearing

Teenage childbearing is a popular topic in the media and political discourse, and U.S. teen pregnancy rates are among the highest of all Western industrialized nations. However, as shown in • Figure 15.10, the U.S. teen birth rate (between the ages of 15 and 19) has generally been in a decline from 1960 to 2013, briefly increased in periods such as 2005–2007, and then resumed its long-term downward trend. In 2013 the birth rate for teenagers between the ages of 15 and 19 dropped 10 percent (to 26.5 births per 1,000 teenagers ages 15–19), as compared with 2012. This was the lowest rate ever reported in the United States. Rates were also down for ages 15 through 19 and for nearly all races and Hispanic-origin groups (National Vital Statistics Report, 2015). Although teen pregnancy rates have continued to decline, concern remains about the number of younger teenagers (ages 15–17) who are producing children. But the good news is that the steepest decline in nonmarital birth rates in 2013 was in this age category, falling 13 percent from 13.7 per 1,000 to 11.9 per 1,000 (National Vital Statistics Report, 2015).

What are the primary reasons for teenage pregnancy? At the microlevel, several issues are most important: (1) many sexually active teenagers do not use contraceptives; (2) teenagers—especially those from low-income families and/or subordinate racial and ethnic groups—may receive little accurate information about the use of, and problems associated with, contraception; (3) some teenage males (because of a double standard based on the myth that sexual promiscuity is acceptable among males but not females) believe that females should be responsible for contraception; and (4) some teenagers view pregnancy as a sign of male prowess or as a way to gain adult status (• Figure 15.11).

At the macrolevel, structural factors also contribute to teenage pregnancy rates. Lack of education and employment opportunities may discourage young people's thoughts of upward mobility. Likewise, religious and political opposition has resulted in issues relating to reproductive responsibility not being dealt with as openly in the United States as in some other nations. Finally, advertising, films, television programming, magazines, music, and other forms of media often flaunt the idea of being sexually active without showing the possible consequences of such behavior.

Teen pregnancies are of concern to analysts who argue that teenage mothers and their children experience strong socioeconomic disadvantages (Mollborn and Dennis, 2011). Studies show that teen mothers may be less skilled at parenting, are less likely to complete high school than their counterparts without children, and possess few economic and social supports other than their relatives. In addition, these births may have negative long-term consequences for these mothers and their children, who may also have limited educational and employment opportunities and a high likelihood of living in poverty.

Teenage fathers have largely been left out of the picture in most studies of teen pregnancy and parenting. How does having a teen father affect a child's health and development? One study found that the father–child relationship does not

FIGURE 15.10   Birth Rates for Teenagers Ages 15–19 Years, by Age, United States, 1960–2013

FIGURE 15.11 Although the rates of teen pregnancy have been declining in the United States, the number of pregnant teens in this country remains high. Among high-income nations of the world, the United States has one of the highest rates of teen pregnancy and teen parenthood. What are the effects of teen pregnancy and parenthood on the lives of young mothers and fathers?

differ significantly between having a teen or adult father when factors such as marital status and economic disadvantage are held constant. For example, in families where the father, regardless of his age, is nonresidential (lives elsewhere) or cohabits occasionally, children are placed at a social disadvantage when compared to children whose father is married to the teen mother and resides in a more permanent family arrangement (Mollborn and Lovegrove, 2010). What about the effects of teen parenting on the father? According to a study on the effects of teenage fatherhood, teen pregnancy led to a decrease in the number of years of schooling among teen fathers who sought early full-time employment, enrolled in the military, or acquired high school equivalency diplomas once they became fathers (ScienceDaily, 2011).

Single-Parent Households

Since 1970, there has been a significant increase in single-parent or one-parent households with children under age 18 because of divorce and births outside of marriage. In 2014, 24 percent of children lived in mother-only families, and 4 percent lived in father-only families, for a total of 28 percent of children in single-parent households. Seven percent of all children lived in the home of their grandparents, and in two-thirds of those families, one or both parents were also present (Child Trends Data Bank, 2015). As shown in • Figure 15.12, the percentage of mother-only family groups continues to rise while the percentage of father-only family groups with children under age 18 remains consistently much lower.

Does living in a single-parent family put children at risk? Single-parent households tend to have much lower incomes than two-parent families. Cohabiting families fall

in between those two. Income is only one factor, however. Health, educational attainment, behavior problems, and psychological well-being are also factors that some researchers associate with living outside of a married, two-parent family.

According to some researchers, the increase in the proportion of single-parent households tends to place children in situations where they experience a lower standard of living, receive less-effective parenting, experience less cooperative co-parenting, are less emotionally close to both parents, and are subjected to more stressful events and circumstances than children who grow up in stable, two-parent families. Why does this occur? Because of factors such as economic hardships that force single-parent families to do without books, without computers, and without homes in better neighborhoods and school districts. When this problem is coupled with the lack of time for parenting while the single parent struggles to make ends meet, and the fact that many children lose contact with their fathers after separation or divorce, the quality of parenting is often less than that found in supportive, co-parenting relationships. Even for a person with a stable income and a network of friends and family to help with child care, raising a child alone can be an emotional and financial burden.

Because of the nature of marriage laws in some states, LGBTQ partners are counted in some studies as single

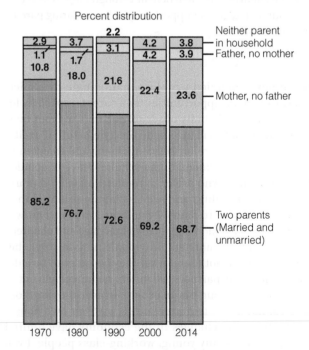

FIGURE 15.12 Living Arrangements of Children Under 18 Years Old for Selected Years, 1970–2014
Source: U.S. Census Bureau, 2015.

parents even when they share parenting responsibilities with their partner. More research has been conducted in recent years on parenting by lesbian, gay, bisexual, transgender, and queer couples, and the results are generally favorable; living in these families appears in some respects to be better for children than living in heterosexual families. One of the areas that appears to be better is the division of parenting and household labor, which has a distinct pattern of equality and sharing among LGBTQ couples as compared with heterosexual parents. It also appears that lesbian and gay parents tend to be more responsive to their children and more child-oriented in their outlook.

Two-Parent Households

Between 1970 and 2014, the share of all U.S. households comprising married couples with children under age 18 halved from 40 percent to slightly less than 20 percent. During that time period, the percentage of children living in two-parent households dropped from 85.2 percent to 64.4 percent, while the percentage living with a single parent increased. In computing these statistics for "parents," the U.S. Census Bureau includes not only biological parents but also stepparents who adopt their children. However, foster parents are considered nonrelatives.

For families in which a couple truly shares parenting, children have two primary caregivers. Some parents share parenting responsibilities by choice; others share out of necessity because both hold full-time jobs. Some studies have found that men's taking an active part in raising the children is beneficial not only for mothers (who then have a little more time for other activities) but also for the men and the children. The men benefit through increased access to children and greater opportunity to be nurturing parents (Coltrane, 2010).

Remaining Single

Some never-married people remain single by choice. Reasons include opportunities for a career (especially for women), the availability of sexual partners without marriage, the belief that the single lifestyle is full of excitement, and the desire for self-sufficiency and freedom to change and experiment. Some scholars have concluded that individuals who prefer to remain single hold more-individualistic values and are less family-oriented than those who choose to marry. Friends and personal growth tend to be valued more highly than marriage and children.

Other never-married individuals remain single because they have not found what they consider to be a desirable marriage partner; still others remain single out of necessity. Being single is an economic necessity for those who cannot afford to marry and set up their own household. Structural changes in the economy have limited the options of many young, working-class people. Even some college graduates have found that they cannot earn enough money to set up a household separate from that of their parents.

The proportion of singles varies significantly by racial and ethnic group, as shown in • Figure 15.13. Among persons ages 18 and over in 2014, nearly 44 percent of African Americans had never married, compared with 35.9 percent of Latinos/as, nearly 27 percent of Asian and Pacific Islander Americans, and slightly more than 25 percent of non-Hispanic whites (ProQuest Statistical Abstract of the U.S. 2015 Online Edition, 2015).

 Identify some of the key stressors that contribute to family violence and to the need for foster care for children.

Transitions and Problems in Families

Families go through many transitions and experience a wide variety of problems, ranging from high rates of divorce and teen pregnancy to domestic abuse and family violence. These all-too-common experiences highlight two important facts about families: (1) for good or ill, families are central to our existence, and (2) the reality of family life is far more complicated than the idealized image of families found in the media and in many political discussions. Moreover, as people grow older, transitions inevitably occur in family life.

Family Violence

Family violence refers to various forms of abuse that take place among family members, including child abuse, spousal abuse, and elder abuse. We will primarily focus on domestic violence—also referred to as spousal abuse or intimate-partner violence—and elder abuse. *Domestic violence* refers to any intentional act or series of acts—whether physical, emotional, or sexual—by one or both partners in an intimate relationship that causes injury to either person. An intimate relationship might include marriage or cohabitation, as well as people who are separated or living apart from a former partner or spouse. Domestic violence is a way in which some individuals seek to establish power and control over others through the use of fear and intimidation. This type of intimate-partner violence often includes the threat or use of violence, relationship abuse, and various kinds of bullying and battering.

There are numerous causes of domestic violence, and many factors are interrelated. Factors contributing to unequal power relations in families include economic inequality, legal and political sanctions that deny girls and women equal rights, and cultural sanctions that dictate appropriate sex roles and reinforce the belief that males are inherently superior to females. Cultural factors that perpetuate domestic violence include gender-specific socialization that establishes dominant–subordinate sex roles. Economic factors include poverty or limited financial resources within families that contribute to tension and sometimes to violence. Economic factors are intertwined

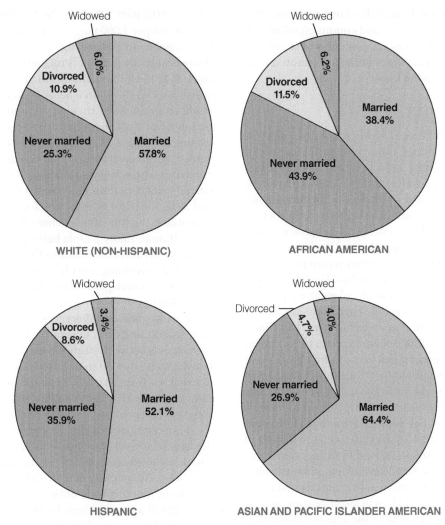

FIGURE 15.13 Marital Status of U.S. Population Ages 18 and Over by Race/Ethnicity

Source: Proquest 2015.

with women's limited access to education, employment, and sufficient income so that they can take care of themselves and their children. Regardless of the factors that contribute to domestic violence, control is central to all forms of abuse: Gaining and maintaining control over the victim is the key factor in abuse. As a result, family violence often involves a cycle of abuse that goes on for extended periods of time.

How much do we really know about family violence? Women, as compared with men, are more likely to be the victims of violence perpetrated by intimate partners. Recent statistics indicate that one in four women, or more than 25 million women in the United States, will experience domestic violence at some time during their lives. It is estimated that men are the victims of nearly 3 million physical assaults in the household. However, we cannot know the true extent of family violence because much of it is not reported to police. More than 60 percent of all domestic violence incidents occur at home. Consequently, it is estimated that only about one-half of the intimate-partner

violence against women is reported. The reasons that many victims of family violence do not report the violence to police include (1) belief that such violence is a private or personal matter, (2) fear of retaliation, (3) view of the violence as a "minor" crime, (4) desire to protect the offender, and (5) belief that the police will not help or will be ineffective.

Although everyone in a household where family violence occurs is harmed psychologically, whether or not they are the victims of violence, children are especially affected by household violence. It is estimated that between 3 million and 10 million children witness some form of domestic violence in their homes each year, and there is evidence to suggest that domestic violence and child maltreatment often take place in the same household.

In some situations, family violence can be reduced or eliminated through counseling, the removal of one parent from the household, or other steps that are taken either by the family or by social service or law enforcement officials. However, children who witness violence in the home may display certain emotional and behavioral problems that adversely affect their school life and communication with other people. In some families the problems of family violence are great enough that the children are removed from the household and placed in foster care.

Children in Foster Care

A special problem in families is when children must be placed in foster care either voluntarily or involuntarily. Many foster children have been in dysfunctional homes where parents or other relatives lack the ability to meet the children's daily needs. *Foster care* refers to institutional settings or residences where adults other than a child's own parents or biological relatives serve as caregivers. States provide financial aid to foster parents, and the intent of such programs is that the children will either return to their own families or be adopted by other families. However, this is often not the case for "difficult to place" children, particularly those who are over ten years of age, have illnesses or disabilities, or are perceived as suffering from "behavioral problems." More than 402,000 children are in foster care at any given time. This number had been declining each year since 2005 but took an upswing in 2013. More than one-half of all children in foster care are in just nine states: California, Florida, Illinois, Indiana, Michigan, New York, Ohio, Pennsylvania, and Texas. Forty-five percent or more of all children in foster care in one-fourth of the states and the District of Columbia are African American, and in twenty-one states the percentage of African American children in foster care is more than twice their proportion in the general child population. About one-third of children in foster care are age five or younger, one-third are between the ages of six and thirteen, and one-third are age fourteen or older. Those who enter foster care at a younger age tend to remain in the system for a longer period of time (Children's Defense Fund, 2013).

Problems in the family contribute to the large numbers of children who are in foster care. Such factors include parents' illness, unemployment, or death; violence or abuse in the family; and high rates of divorce.

 LO8 **Discuss** divorce and how it affects remarriage patterns and blended families in the United States.

Divorce

Divorce is the legal process of dissolving a marriage that allows former spouses to remarry if they so choose. Most divorces today are granted on the grounds of *irreconcilable differences*, meaning that there has been a breakdown of the marital relationship for which neither partner is specifically blamed. Prior to the passage of more-lenient divorce laws, many states required that the partner seeking the divorce prove misconduct on the part of the other spouse. Under *no-fault divorce laws*, however, proof of "blameworthiness" is generally no longer necessary.

Over the past 100 years, the U.S. divorce rate (number of divorces per 1,000 population) has varied from a low of 0.7 in 1900 to an all-time high of 5.3 in 1981; by 2012, it had decreased to 3.4 percent (ProQuest Statistical Abstract of the U.S. 2015 Online Edition, 2015). Divorce statistics may vary based on the source because some organizations include annulments in their count, and a number of states no longer provide divorce statistics to national reporting agencies. ● Figure 15.14 shows the latest available U.S. divorce rates for each state so that you can see how your state compares with others in the nation. Overall, a decrease occurred in divorce rates in a number of states between 2011 and 2012, the latest year for which comprehensive data are available at the time this is being written.

Although many people believe that marriage should last for a lifetime, others believe that marriage is a commitment that may change over time. Between 40 and 50 percent of new U.S. marriages will end in divorce (Cherlin, 2010). According to various studies on divorce in the United States, there are significant differences in the rates of divorce for first, second, and third marriages: The divorce rate for first marriages is between 40 and 50 percent, the rate for second marriages is from 60 to 67 percent, and the rate for third marriages is from 73 to 75 percent. Research has also found that couples with children have a slightly lower rate of divorce as compared to couples without children.

Financial stressors are a contributing factor to some divorces (National Marriage Project, 2010). During times of a national recession such as the one that the United States experienced beginning in 2007–2008, some people decide to remain married only until conditions change so that they can sell a house or have better financial stability when they part. However, others gain a deeper commitment to their marriage as they struggle through adversity (National Marriage Project, 2010).

As previously stated, we often hear that 50 percent of all marriages end in divorce. But this figure is misleading in some ways, even though it is true for the U.S. population as a whole. Although there is a 40–50 percent chance that a first marriage will end in either divorce or separation before one partner dies, divorce rates vary widely based on certain factors. There is a decrease in risk of divorce for people in the following categories:

- Making over $50,000 annually (as compared with under $25,000)
- Having graduated from college (as opposed to not completing high school)
- Having a baby 7 months or more after marriage (as opposed to before marriage)
- Marrying when partners are over 25 years of age (as opposed to younger age categories)
- Coming from an intact family of origin (as opposed to having divorced parents)
- Having a religious affiliation (as opposed to having none) (National Marriage Project, 2010)

	Rates per 1,000 population[a, b]						Rates per 1,000 population[a, b]			
	1990	2000	2005	2012			1990	2000	2005	2012
United States[c]	**4.7**	**4.0**	**3.6**	**3.4**		Missouri	5.1	4.5	3.6	3.7
Alabama	6.1	5.5	4.9	3.6		Montana	5.1	4.2	4.5	3.9
Alaska	5.5	3.9	4.3	4.5		Nebraska	4.0	3.7	3.3	3.4
Arizona	6.9	4.6	4.2	4.3		Nevada	11.4	9.9	7.4	5.5
Arkansas	6.9	6.4	6.0	5.3		New Hampshire	4.7	4.8	3.9	3.6
California	4.3	---	---	---		New Jersey	3.0	3.0	2.9	2.8
Colorado	5.5	4.7	4.4	4.3		New Mexico	4.9	5.1	4.6	3.0
Connecticut	3.2	3.3	3.0	2.7		New York	3.2	3.0	2.9	2.9
Delaware	4.4	3.9	3.8	3.5		North Carolina	5.1	4.5	4.1	3.7
District of Columbia	4.5	3.2	2.0	2.9		North Dakota	3.6	3.4	2.9	3.1
Florida	6.3	5.1	4.6	4.2		Ohio	4.7	4.2	3.5	3.4
Georgia	5.5	3.3	---	---		Oklahoma	7.7	---	5.6	4.8
Hawaii	4.6	3.9	---	---		Oregon	5.5	4.8	4.2	3.8
Idaho	6.5	5.5	5.0	4.7		Pennsylvania	3.3	3.1	2.3	2.8
Illinois	3.8	3.2	2.6	2.4		Rhode Island	3.7	2.9	3.0	3.2
Indiana	---	---	---	---		South Carolina	4.5	3.8	2.9	3.2
Iowa	3.9	3.3	2.7	2.2		South Dakota	3.7	3.5	2.8	3.0
Kansas	5.0	3.6	3.1	3.4		Tennessee	6.5	5.9	4.6	4.2
Kentucky	5.8	5.1	4.6	4.1		Texas	5.5	4.0	3.3	3.0
Louisiana	---	---	---	---		Utah	5.1	4.3	4.1	3.3
Maine	4.3	5.0	4.1	3.9		Vermont	4.5	4.1	3.6	3.5
Maryland	3.4	3.3	3.1	2.8		Virginia	4.4	4.3	4.0	3.7
Massachusetts	2.8	2.5	2.2	2.7		Washington	5.9	4.6	4.3	3.9
Michigan	4.3	3.9	3.4	3.3		West Virginia	5.3	5.1	5.1	4.7
Minnesota	3.5	3.2	---	---		Wisconsin	3.6	3.2	2.9	2.9
Mississippi	5.5	5.0	4.4	4.0		Wyoming	6.6	5.8	5.2	4.4

[a]Based on total population residing in area; population enumerated as of April 1 for 1990; estimated as of July 1 for all other years.
[b]Includes annulments.
[c]U.S. totals for the numbers of divorces are an estimate that includes states not reporting, such as California, Colorado, Indiana, and Louisiana.

FIGURE 15.14   U.S. Divorce Rate by State, 1990–2012
Source: ProQuest Statistical Abstract of the United States, 2015.

Consequences of Divorce   In some families, divorce may have a dramatic economic and emotional impact on family members. In others, the effect may be more marginal. Overall, in families where one or more children are present, the children remain with their mothers and live in a single-parent household for a period of time. In recent years, there has been a debate over whether children who live with their same-sex parent after divorce are better off than their peers who live with an opposite-sex parent. However, virtually no evidence has been found to support the belief that children are better off living with a same-sex parent.

Although divorce decrees provide for parental joint custody of many children, this arrangement may create unique problems. Furthermore, some children experience more than one divorce during their childhood because one or both of their parents may remarry and subsequently divorce again.

But divorce does not have to be always negative. For some people, divorce may be an opportunity to terminate destructive relationships. For others, it may represent a means to achieve personal growth by managing their lives and social relationships and establishing their own social identity. Still others choose to remarry one or more times.

Remarriage

Many people who divorce or are widowed get remarried one or more times. Although 50 percent of all men and 54 percent of women ages 15 and over in 2012 had been married only once, the proportion of adults that had married only once decreased for both men (from 54 percent to 50 percent) and women (from 60 percent to 54 percent) between 1996 and 2012. In 2012 an increasing share of women ages 50 and older, and men ages 60 and older, reported that they had been married two, three, or more times (Lewis and Kreider, 2015).

Age is an important factor in remarriage. People who are older have had more time to see a marriage conclude and to remarry. The proportion of women and men who have married twice is about 20 percent higher for persons between the ages of 50 and 69. Overall, 13 percent of men and 14 percent of women reported that they had been married twice, and 4 percent reported that they had been married three or more times in the latest U.S. Census Bureau study, which incorporated the years 2008–2012.

Education levels are also a factor in both divorce and remarriage patterns. Sixty-four percent of married persons with a bachelor's degree had been married only once, as compared to 52 percent of adults in the general population. Because there is a lower risk of divorce for those who have earned a bachelor's degree or more, there is less possibility of them having second or third marriages. As previously discussed, lower rates of divorce among those with more years of formal education may be related to a tendency to delay marriage and later ages at first marriage, which are associated with lower rates of marital instability.

Employment status and income are important factors in remarriage. Persons who are employed are more likely to be married once and to stay married to the same partner than those who are not in the labor force. Individuals who are unemployed have a slightly higher risk of being married three or more times than those in other employment categories. Of course, unemployment often brings financial hardship, which is related to strain in marriage, higher rates of divorce, and fewer prospects of remarriage if a marriage ends. Similarly, income is related to marriage and remarriage. Adults with incomes of $100,000 and above are more likely to have married only once. At the other end of the income spectrum, persons living below the poverty line, as well as those receiving public assistance, are more likely to have never married and thus are less likely to have remarried (Lewis and Kreider, 2015).

As a result of divergent marital histories, many marriages include one or both spouses who have previously been married and/or have children and other commitments from previous marital unions. As a result of these divergent marriage patterns, complex family relationships are often created. Some people become part of stepfamilies or **blended families**, which consist of a husband and wife, children from previous marriages, and children (if any) from the new marriage (● Figure 15.15). At least initially, levels of family stress may be fairly high because of rivalry among the children and hostilities directed toward stepparents or babies born into the family. In spite of these problems, however, many blended families succeed. The family that results from divorce and remarriage is typically a complex, binuclear family in which children may have a biological parent and a stepparent, biological siblings and stepsiblings, and an array of other relatives, including aunts, uncles, and cousins.

The norms governing divorce and remarriage are ambiguous. Because there are no clear-cut guidelines, people must make decisions about family life (such as whom to invite for a birthday celebration or wedding) based on their beliefs and feelings about the people involved. This adds to individuals' insecurity and confusion about how to interact with others on social occasions and has contributed to the use of informal social networking technology to communicate rather than having face-to-face discussions about plans that must be made involving children, finances, or other shared concerns.

Kayte Deioma/PhotoEdit

FIGURE 15.15   Remarriage and blended families create new opportunities and challenges for parents and children alike.

Looking Ahead: Family Issues in the Future

As we have seen, families and intimate relationships changed dramatically as we moved into the twenty-first century. Some people believe that the family as we know it is doomed. Others believe that a return to traditional family values will save this important social institution and create greater stability in society.

Current economic conditions constitute a threat for many families in the United States and other nations. Throughout good and bad economic times, families are important to people because the family is a vital social institution in society and often serves as the major source of support for individuals. However, in periods like the 1930s depression or the recent recession, families are particularly affected by problems in the economy. Worldwide, the economies in various nations have been undergoing transformation, and in the United States alone, since the onset of the 2007 recession, there has been a net loss of 7.5 million jobs (Zuckerman, 2011). From 2011 to 2013, the rate of unemployment averaged about 9 percent, after having peaked at over 10 percent in parts of 2010, and this counts only individuals who are actively seeking employment, not those who have become discouraged and have given up looking for work.

Many families experience long-term repercussions from a national or global economic crisis and the financial and emotional havoc that such crises wreak on workers and their families. For example, studies have found that individuals are feeling more tension and having more arguments with other family members because of personal fears that people have regarding unemployment, loss of housing values, mortgage foreclosures, and other personal and financial problems (Luo and Thee-Brenan, 2009). It is not surprising to sociologists that families and personal relationships are affected by financial insecurity: Individuals are forced to deal with problems of low self-esteem associated with job loss and seeking new employment, and they must also focus on how to take care of their family in tough economic times.

People in the lower tiers of the U.S. class structure are confronted with problems such as these all of the time; however, this level of anxiety is somewhat new to many individuals in the middle and upper-middle classes in this country. Family problems related to economic crises are not limited to the United States: People in many countries are affected by changes in the global economy. The present economy casts a shadow over family issues in the future. How national and global political and business leaders deal with economic issues will no doubt have an important effect on the future of families worldwide.

As previously discussed, how we view the concept of "family" is continuing to evolve in the twenty-first century. Some of these changes are already becoming evident. For example, many men are taking an active role in raising their children and helping with household chores. More individuals rely on others who are not necessarily immediate family for friendship, emotional support, and help in time of emergencies. As well, more people are cohabiting, living in domestic partnerships, or are same-sex partners who now have the benefit of legally formalized marriages. The number of gay couples with children has doubled in the past decade, and more than 100,000 same-sex couples are actively parenting children.

Regardless of problems facing families today, many people still demonstrate their faith in the future by establishing families, in both the traditional and less traditional sense of the term. It will be interesting to see what people in the future decide about family relationships. What will family life be like in 2030 or 2040? What will your own family be like?

blended family
a family consisting of a husband and wife, children from previous marriages, and children (if any) from the new marriage.

CHAPTER REVIEW Q & A

LO1 How are the following key concepts defined: families, kinship, family of orientation, family of procreation, extended family, and nuclear family?

Today, families may be defined as relationships in which people live together with commitment, form an economic unit and care for any young, and consider their identity to be significantly attached to the group. Through kinship networks, people cooperate so that they can acquire the basic necessities of life, including food and shelter. Kinship systems can also serve as a means by which property is transferred, goods are produced and distributed, and power is allocated.

Although most people are related to members of their family of orientation by blood ties, those who are adopted have a legal tie that is patterned after a blood relationship. The family of orientation is the family into which a person is born; the family of procreation is the family that a person forms by having or adopting children. An extended family is a family unit composed of relatives in addition to parents and children who live in the same household. A traditional definition specifies that a nuclear family is made up of a "couple" and their dependent children; however, this definition became outdated when a significant shift occurred in the family structure.

LO2 What are the differences among the following marriage patterns—monogamy, polygamy, polygyny, and polyandry—and the differences among these patterns of descent—patrilineal, matrilineal, and bilateral?

Marriage is a legally recognized and/or socially approved arrangement between two or more individuals that carries certain rights and obligations and usually involves sexual activity. In the United States, the only legally sanctioned form of marriage is monogamy—the practice or state of being married to one person at a time. Polygamy is the concurrent marriage of a person of one sex with two or more members of the opposite sex. The most prevalent form of polygamy is polygyny—the concurrent marriage of one man with two or more women. The second type of polygamy is polyandry—the concurrent marriage of one woman with two or more men. Virtually all forms of marriage establish a system of descent so that kinship can be determined and inheritance rights established. In preindustrial societies, kinship is usually traced through one parent (unilineally). The most common pattern of unilineal descent is patrilineal descent. Matrilineal descent is the less common pattern. Kinship in industrial societies is usually traced through both parents (bilineally). The most common form of descent is bilateral.

LO3 What are the kinds of authority figures in patriarchal, matriarchal, and egalitarian families?

Forms of familial power and authority that have been identified are patriarchy, matriarchy, and egalitarianism. A patriarchal family is a family structure in which authority is held by the eldest male (usually the father). A matriarchal family is a family structure in which authority is held by the eldest female (usually the mother). An egalitarian family is a family structure in which both partners share power and authority equally.

LO4 What are the primary sociological perspectives on the family as a social institution?

Functionalists emphasize the importance of the family in maintaining the stability of society and the well-being of individuals. Conflict and feminist perspectives view the family as a source of social inequality and an arena for conflict. Symbolic interactionists explain family relationships in terms of the subjective meanings and everyday interpretations that people give to their lives. Postmodern analysts view families as being permeable, capable of being diffused or invaded so that their original purpose is modified.

LO5 What issues do many contemporary couples face when thinking of developing intimate relationships and establishing families?

Families are changing dramatically in the United States. Cohabitation has increased significantly in the past three decades. Among heterosexual couples, many reasons exist for cohabitation; for gay and lesbian couples, however, no alternatives to cohabitation existed in many U.S. states before the 2015 U.S. Supreme Court decision legalizing same-sex marriage. For that reason, many lesbian and gay couples sought recognition of their domestic partnerships—household partnerships in which an unmarried couple lives together in a committed, sexually intimate relationship and is granted some of the same rights and benefits as those accorded to married heterosexual couples. With the increase in dual-earner marriages, women have become larger contributors to the financial well-being of their families, but some have become increasingly burdened by the second shift—the domestic work that employed women perform at home after they complete their workday on the job.

LO6 What child-related family issues are of concern to many people in the twenty-first century?

Cultural attitudes about having children and about the ideal family size have changed dramatically in the United States. Today, the concept of reproductive freedom includes both the desire *to have* or *not to have* one or more children. Issues of concern include teenage childbearing. Many single-parent families also exist today.

LO7 What are some of the key stressors that contribute to family violence and the need for foster care for children?

Factors contributing to unequal power relations in families include economic factors, legal and political sanctions that deny girls and women equal rights, and cultural factors that perpetuate domestic violence. Regardless of the factors that contribute to domestic violence, control is central to all forms of abuse. Women and children are most strongly affected by family violence, although domestic violence is also perpetrated against men. However, everyone in a household where family violence occurs is harmed psychologically, and children are especially harmed. Foster care is often used as a safe place for children who have been in dysfunctional families, some of which are the sites of family violence, others of which are not.

LO8 What is divorce, and how does it affect remarriage patterns and blended families in the United States?

Divorce is the legal process of dissolving a marriage. At the macrolevel, changes in social institutions may contribute to an increase in divorce rates; at the microlevel, factors contributing to divorce include age at marriage, length of acquaintanceship, economic resources, education level, and parental marital happiness.

KEY TERMS

bilateral descent 432
blended family 450
cohabitation 438
domestic partnerships 438
dual-earner marriages 440
egalitarian family 433
endogamy 433
exogamy 433
extended family 431
families 429

family of orientation 430
family of procreation 430
homogamy 439
kinship 430
marriage 431
matriarchal family 432
matrilineal descent 432
matrilocal residence 433
monogamy 431
neolocal residence 433

nuclear family 431
patriarchal family 432
patrilineal descent 432
patrilocal residence 433
polyandry 432
polygamy 432
polygyny 432
second shift 440
sociology of family 433

QUESTIONS for CRITICAL THINKING

1 In your opinion, what constitutes an ideal family? How might functionalist, conflict, feminist, and symbolic interactionist perspectives describe this family?

2 Suppose that you wanted to find out about people's perceptions about love and marriage. What specific issues might you examine? What would be the best way to conduct your research?

3 You have been appointed to a presidential commission on child-care problems in the United States. How to provide high-quality child care at affordable prices is a key issue for the first meeting. What kinds of suggestions would you take to the meeting? How do you think your suggestions should be funded? How does the future look for children in high-, middle-, and low-income families in the United States?

ANSWERS to the SOCIOLOGY QUIZ

ON CONTEMPORARY TRENDS IN U.S. FAMILY LIFE

1	**False**	A Gallup Values and Beliefs poll found that more than half (54 percent) of U.S. adults say that having a baby outside of marriage is "morally acceptable." Slightly less than half (41 percent) view out-of-wedlock childbearing as "morally wrong" (Saad, 2011).
2	**True**	A Pew Research report revealed that more than four in ten U.S. adults have at least one step-relative in their family, such as a stepparent, a step- or half-sibling, or a stepchild.
3	**True**	Among persons reporting a single race on the U.S. Census, 85 percent of Asian American children lived with two parents, compared with 77 percent of white (non-Hispanic) children, 66 percent of Hispanic (Latino/a) children, and 38 percent of black (African American) children.

4	**False**	Recent research indicates that couples who cohabit before they get married are less likely to stay married; however, their chances of remaining married improve if they were already engaged when they began living together.
5	**False**	The U.S. teen birth rate is up to nine times *higher* than it is in other Western industrialized countries.
6	**False**	There has been an *increase* in the number of stay-at-home parents, particularly mothers. In 2012, among married-couple families with children younger than 15, the percentage of stay-at-home mothers rose to 24 percent, up from 20 percent in the 1990s.
7	**True**	The percentage of U.S. households that consist of one person living alone has risen from 17 percent in 1970 to 27 percent in 2012.
8	**False**	According to a study by the Pew Research Center, more than one in five (22 percent) of U.S. husbands are in marriages in which the wife earns more than the husband—up from only 4 percent in 1970.

Sources: Based on Newport, 2011; Parker, 2011; Roberts, 2010; Saad, 2011; U.S. Census Bureau, 2012a; and U.S. Centers for Disease Control and Prevention, 2011d.

EDUCATION

16

LEARNING OBJECTIVES

1   **Define** *education* and trace how the social institution of education has changed throughout history.

2   **Identify** the key assumptions of functionalist, conflict, symbolic interactionist, and postmodernist perspectives on education.

3   **Describe** major problems in elementary and secondary schools, such as unequal funding of schools, high dropout rates, and problems of inequality based on race or disability.

4   **Explain** how options such as school vouchers, charter schools, and homeschooling differ from traditional educational approaches, and identify strengths and weaknesses of each approach.

5   **Discuss** the educational opportunities and challenges found in community colleges in the twenty-first century.

6   **Describe** the economic problems facing many four-year colleges and universities, and discuss how these fit into larger patterns of state funding.

7   **Identify** problems in racial and ethnic diversity in higher education, including student enrollment issues, lack of faculty diversity, and continuing controversy over affirmative action policies and laws.

8   **Discuss** future trends in education and explain why education will remain an important social institution in the future.

457

SOCIOLOGY & EVERYDAY LIFE

Learning the Value of Education

My parents came to the United States from Egypt 30 years ago. Today, they still can barely speak English and haven't been educated beyond high school. My parents, my three older brothers and I live in a two-bedroom apartment, which forces me to share a room with my parents. My household has always consisted of constant yelling, fighting, putting others down, and negative forces. . . . Slowly, I started to care less about my school work and failed my classes. . . . My only alternative was to spend time at my aunt's house. She and her husband are both college educated . . . [and] lived happily and comfortably in comparison to those who weren't educated (my parents), who lived a life of hard physical labor and financial troubles. It came down to one simple question: What kind of lifestyle do I want to live? . . .

As the end of my senior year [of high school] approached, I began to get response letters from the Cal State universities. . . . I had been rejected from every Cal State to which I applied. . . . I faced a crossroads. I could follow in my brothers' footsteps and begin working at a minimum-wage

Achieving a college degree has long been a part of the American Dream. An increasingly popular way to start this process is to begin at the community college level.

job instead of going to college, or I could prove my mother wrong by starting at community college to begin the path of building the lifestyle I wanted in the future.

College it was! . . . [After completing two years of college,] I can't explain in words exactly how I feel inside because I don't think words can describe the level of happiness I now feel. . . . I have completely changed as a person, daughter, friend, leader, employee, and most importantly, as a student since I decided to attend Glendale Community College. If I

Sally Morgan is one of millions of people who have attended a community college to gain knowledge and skills that will benefit them and enhance their opportunities in the future. In fact, a substantial proportion of all college students start their higher education at a community college.

From prekindergarten through postgraduate studies, education is one of the most significant social institutions in the United States and other high-income nations. Although most social scientists agree that schools are supposed to be places where people acquire knowledge and skills, not all of them agree on how a large number of factors—including class, race, gender, national origin, age, religion, and family background—affect individuals' access to educational opportunities or to the differential rewards that accrue at various levels of academic achievement.

In this chapter we discuss education as a key social institution and analyze some of the problems that affect contemporary elementary, secondary, and higher education. Before reading on, test your knowledge about U.S. education by taking the "Sociology and Everyday Life" quiz.

LO1 Define *education* and trace how the social institution of education has changed throughout history.

An Overview of Education

Education is the social institution responsible for the systematic transmission of knowledge, skills, and cultural values within a formally organized structure. As a social institution, education imparts values, beliefs, and knowledge considered essential to the social reproduction of individual personalities and entire cultures. Education grapples with issues of societal stability and social change, reflecting society even as it attempts to shape it. Education serves an important purpose in all societies. At the microlevel, people must acquire the basic knowledge and skills they need to survive in society. At the macrolevel, the social institution of education is an essential component in maintaining and perpetuating the culture of a society across generations. *Cultural transmission*—the process by which children and recent immigrants become acquainted with the dominant cultural beliefs, values, norms,

had let those Cal State rejections stop me from this path, I wouldn't be as educated, well-rounded, happy and the Sally Morgan I am today. I graduate with my associate's degree in a few weeks, and I was accepted to several Cal States. . . . Re-

member that you have the power to live the lifestyle you want.

—SALLY MORGAN (2010) describing how completing a community college education changed her life

How Much Do You Know About U.S. Education?

TRUE	FALSE		
T	F	1	Public education in the United States dates back more than 160 years.
T	F	2	Equality of opportunity is a vital belief in the U.S. educational system.
T	F	3	Each year in the United States, men earn more doctoral degrees than women do.
T	F	4	Although there is much discussion about gangs in schools, very few students report that gangs are present at their school.
T	F	5	Young people who bully are more likely to smoke, drink alcohol, and get into fights.
T	F	6	School dropout rates increase each year, and there is little hope for change in the foreseeable future.
T	F	7	In public schools in the United States, core classes such as history and mathematics were never taught in any language other than English before the 1960s civil rights movement.
T	F	8	The federal government has only limited control over how funds are spent by individual school districts because most of the money comes from the state and local levels.

Answers can be found at the end of the chapter.

and accumulated knowledge of a society—occurs through informal and formal education. However, the process of cultural transmission differs in preliterate, preindustrial, and industrial nations.

The earliest education in *preliterate societies,* which existed before the invention of reading and writing, was informal in nature. People acquired knowledge and skills through ***informal education***—learning that occurs in a spontaneous, unplanned way—from parents and other group members who provided information on survival skills such as how to gather food, find shelter, make weapons and tools, and get along with others. Formal education for elites first came into being in *preindustrial societies,* where few people knew how to read and write. ***Formal education*** is learning that takes place within an academic setting such as a school, which has a planned instructional process and teachers who convey specific knowledge, skills, and thinking processes to students.

Perhaps the earliest formal education occurred in ancient Greece and Rome, where philosophers such as Socrates, Plato, and Aristotle taught elite males the skills required to become thinkers and orators who could engage in the art of persuasion (Ballantine and Hammack,

2012). During the Middle Ages the first colleges and universities were developed under the auspices of the Catholic church. In the Renaissance the focus of education shifted to the importance of developing well-rounded and liberally educated people. With the rapid growth of industrial capitalism and factories during the Industrial Revolution, it became necessary for workers to have basic skills in reading, writing, and arithmetic, and pressure to provide formal education for the masses increased significantly.

education
the social institution responsible for the systematic transmission of knowledge, skills, and cultural values within a formally organized structure.

cultural transmission
the process by which children and recent immigrants become acquainted with the dominant cultural beliefs, values, norms, and accumulated knowledge of a society.

informal education
learning that occurs in a spontaneous, unplanned way.

formal education
learning that takes place within an academic setting such as a school, which has a planned instructional process and teachers who convey specific knowledge, skills, and thinking processes to students.

FIGURE 16.1   Some early forms of mass education took place in one-room schoolhouses such as the one shown here, where children in various grades were all taught by the same teacher. How do changes in the larger society bring about changes in education?

In the United States, Horace Mann started the free public school movement in 1848 when he declared that education should be the "great equalizer." By the mid-1850s, the process of mass education had begun in the United States as all states established free, tax-supported elementary schools that were readily available to children throughout the country (● Figure 16.1). **Mass education** refers to providing free, public schooling for wide segments of a nation's population. As industrialization and bureaucratization intensified, managers and business owners demanded that schools educate students beyond the third or fourth grade so that well-qualified workers would be available for rapidly emerging "white-collar" jobs in management and clerical work.

Today, schools attempt to meet the needs of industrial and postindustrial society by teaching a wide diversity of students a myriad of topics, including history and science, computer skills, how to balance a checkbook, and how to avoid contracting AIDS. According to sociologists, many functions performed by other social institutions in the past are now under the auspices of the public schools.

LO2

Identify the key assumptions of functionalist, conflict, symbolic interactionist, and postmodernist perspectives on education.

Sociological Perspectives on Education

Sociologists have divergent perspectives on education in contemporary society. Here, we examine functionalist, conflict, symbolic interactionist, and postmodernist approaches to analyzing schooling.

Functionalist Perspectives

Functionalists view education as one of the most important components of society. According to Emile Durkheim, education is crucial for promoting social solidarity and stability in society: Education is the "influence exercised by adult generations on those that are not yet ready for social life" (Durkheim, 1956: 28). Durkheim asserted that moral values are the foundation of a cohesive social order and that schools are responsible for teaching a commitment to the common morality. In analyzing the values and functions of education, sociologists using a functionalist framework distinguish between manifest functions and latent functions, which are compared in ● Figure 16.2.

Manifest Functions of Education   Some functions of education are *manifest functions*—previously defined as open, stated, and intended goals or consequences of activities within an organization or institution. Education serves six major manifest functions in society:

1. *Socialization.* From kindergarten through college, schools teach students the student role, specific academic subjects, and political socialization.

2. *Transmission of culture.* Schools transmit cultural norms and values to each new generation and play an active part in the process of assimilation of recent immigrants.

3. *Multicultural education.* Schools promote awareness of and appreciation for cultural differences so that students can work and compete successfully in a diverse society and a global economy.

4. *Social control.* Schools teach values such as discipline, respect, obedience, punctuality, and perseverance. Schools teach conformity by encouraging young people to be good students, conscientious future workers, and law-abiding citizens.

5. *Social placement.* Schools identify the most-qualified people to fill the positions available in society. As a result, students are channeled into programs based on individual ability and academic achievement. Graduates receive the appropriate credentials to enter the paid labor force.

6. *Change and innovation.* Schools are a source of change and innovation to meet societal needs. Faculty members are responsible for engaging in research and passing on their findings to students, colleagues, and the general public.

Manifest functions—open, stated, and intended goals or consequences of activities within an organization or institution. In education, these are:

- socialization
- transmission of culture
- multicultural education
- social control
- social placement
- change and innovation

Latent functions—hidden, unstated, and sometimes unintended consequences of activities within an organization. In education, these include:

- matchmaking and production of social networks
- restricting some activities
- creating a generation gap

FIGURE 16.2   Manifest and Latent Functions of Education

Latent Functions of Education   Education serves at least three *latent functions,* which we have previously defined as hidden, unstated, and sometimes unintended consequences of activities within an organization or institution:

1. *Restricting some activities.* States have *mandatory education laws* that require children to attend school until they reach a specified age (usually age sixteen) or complete a minimum level of formal education (generally the eighth grade). Out of these laws grew one latent function of education: keeping students off the streets and out of the full-time job market until they are older.

2. *Matchmaking and production of social networks.* Because schools bring together people of similar ages, social class, and race/ethnicity, young people often meet future marriage partners and develop lasting social networks.

3. *Creation of a generation gap.* Students learn information and develop technological skills that may create a generation gap between them and their parents, particularly as the students come to embrace a newly acquired perspective.

Dysfunctions of Education   Functionalists acknowledge that education has certain dysfunctions. Some analysts argue that U.S. education is not promoting the high-level skills in reading, writing, science, and

mathematics that are needed in the workplace and the global economy. For example, mathematics and science education in the United States does not compare favorably with that found in many other industrialized countries. Are U.S. schools dysfunctional as a result of lower test scores? Analysts do not agree on what the exam score differentials mean. For many functionalist thinkers, lagging test scores are a sign that dysfunctions exist in the nation's educational system. According to this approach, improvements will occur only when more-stringent academic requirements are implemented for students and when teachers receive sufficient training. Overall, functionalists typically advocate the importance of establishing a more rigorous academic environment in which students are required to learn the basics that will make them competitive in school and job markets.

Conflict Perspectives

Conflict theorists emphasize that schools solidify the privileged position of some groups at the expense of others by perpetuating class, racial–ethnic, and gender inequalities (Ballantine and Hammack, 2012). Contemporary conflict theorists also focus on how politics and corporate interests dominate schools, particularly higher education.

mass education
the practice of providing free, public schooling for wide segments of a nation's population.

Cultural Capital and Class Reproduction   Although many factors—including intelligence, motivation, and previous accomplishments—are important in determining how much education a person will attain, conflict theorists argue that access to quality education is closely related to social class. From this approach, education is a vehicle for reproducing existing class relationships. According to the French sociologist Pierre Bourdieu, the school legitimates and reinforces the social elites by engaging in specific practices that uphold the patterns of behavior and the attitudes of the dominant class. Bourdieu asserts that students from diverse class backgrounds come to school with different amounts of *cultural capital*—social assets that include values, beliefs, attitudes, and competencies in language and culture (Bourdieu and Passeron, 1990). Cultural capital involves "proper" attitudes toward education, socially approved dress and manners, and knowledge about books, art, music, and other forms of high and popular culture (• Figure 16.3). Middle-and upper-income parents endow their children with more cultural capital than do working-class and poverty-level parents. Because cultural capital is essential for acquiring an education, children with less cultural capital have fewer opportunities to succeed in school. For example, standardized tests that are used to group students by ability and to assign them to classes often measure students' cultural capital rather than their "natural" intelligence or aptitude. Thus, a circular effect occurs: Students with dominant cultural values are more highly rewarded by the educational system. In turn, the educational system teaches and reinforces values that sustain the elite's position in society.

Tracking and Detracking   Closely linked to the issue of cultural capital is how tracking in schools is related to social inequality. *Tracking* refers to the practice of assigning students to specific curriculum groups and courses on the basis of their test scores, previous grades, or other criteria. Conflict theorists believe that tracking seriously affects many students' educational performance and their overall academic accomplishments. Tracking first came into practice in the early twentieth century, when a large influx of immigrant children entered U.S. schools for the first time and were sorted by ability and past performance. In elementary schools, tracking is often referred to *ability grouping* and is based on the assumption that it is easier to teach a group of students who have similar abilities. However, class-based factors also affect which children are most likely to be placed in "high," "middle," or "low" groups, often referred to by such innocuous terms as "Blue Birds," "Red Birds," and "Yellow Birds." This practice is described by the well-known journalist Ruben Navarrette Jr. (1997: 274–275), who tells us about his own childhood experience with tracking:

> One fateful day, in the second grade, my teacher decided to teach her class more efficiently by dividing it into six groups of five students each. Each group was assigned a geometric symbol to differentiate it from the others. There were the Circles. There were the Squares. There were the Triangles and Rectangles.
>   I remember being a Hexagon.... The Hexagons were the smartest kids in the class. These distinctions are not lost on a child of seven.... And on the day on which we were assigned our respective shapes, we knew that our teacher knew, too. As Hexagons, we would wait for her to call on us, then answer by hurrying to her with books and pencils in hand. We sat around a table in our "reading group," chattering excitedly to one another and basking in the intoxication of positive learning. We did not notice, did not care to notice, over our shoulders, the frustrated looks on the faces of Circles and Squares and Triangles who sat quietly at their desks, doodling on scratch paper or mumbling to one another. We knew also that, along with our geometric shapes, our books were different and that each group had different amounts of work to do.... The Circles had the easiest books and were assigned to read only a few pages at a time.... Not surprisingly, the Hexagons had the most difficult books

Cleve Bryant/PhotoEdit

FIGURE 16.3   Children who are able to visit museums, libraries, and musical events may gain cultural capital that other children do not possess. What is cultural capital? Why is it important in the process of class reproduction?

FIGURE 16.4 As Ruben Navarrette Jr. so powerfully describes, school is extremely tedious for underachieving students, who may find themselves "tracked" in such a way as to deny them upward mobility in the future.

of all, those with the biggest words and the fewest pictures, and we were expected to read the most pages.

The result of all of this education by separation was exactly what the teacher had imagined that it would be: Students could, and did, learn at their own pace without being encumbered by one another. Some learned faster than others. Some, I realized only [later], did not learn at all.

As Navarrette suggests, tracking does make it possible for students to work together based on their perceived abilities and at their own pace; however, it also takes a toll on students who are labeled as "underachievers" or "slow learners." Today, Navarrette is a nationally recognized journalist and blogger who writes about many important issues facing our nation and the world. However, as he points out in his discussion of tracking, race, class, language, gender, and many other social categories may determine the placement of children in elementary tracking systems as much as or more than their actual academic abilities and interests.

The practice of tracking continues in middle school/ junior high and high school. Although schools in some communities bring together students from diverse economic and racial–ethnic backgrounds, the students do not necessarily take the same courses or move on the same academic career paths, or have the same opportunities even when they attend the same school (Gilbert, 2014). Overall, many scholars have documented in their research that tracking does not improve student achievement but does intensify educational inequality, particularly along racial–ethnic and class-based lines.

The *detracking movement*—which emphasizes that students should be deliberately placed in classes of mixed ability to improve their academic performance and test scores—has influenced a growing number of educators. Detracking is based on the assumption that intensifying the secondary school curricula may help close the achievement gap among students, particularly those dimensions that are based on class or race/ethnicity. For example, a growing proportion of U.S. students are being enrolled in higher-level academic math courses than in the past so that they will be exposed to more-complex course content and be influenced by students who perform at a higher level in such courses.

Detracking is a major concern for parents of high-achieving students: They often believe their children are losing out because lower-achieving students are in their courses. According to their perspective, high-achieving students should have classes that maximize their potential rather than hold them back with less able or less talented students. According to the sociologist Maureen Hallinan (2005), who has extensively studied detracking, tracking is not the answer: Schools should provide more-engaging lessons for all students, alter teachers' assumptions about students, and raise students' performance requirements (● Figure 16.4).

cultural capital
Pierre Bourdieu's term for people's social assets, including values, beliefs, attitudes, and competencies in language and culture.

tracking
the practice of assigning students to specific curriculum groups and courses on the basis of their test scores, previous grades, or other criteria.

The Hidden Curriculum   According to conflict theorists, the **hidden curriculum** is the transmission of cultural values and attitudes, such as conformity and obedience to authority, through implied demands found in the rules, routines, and regulations of schools. In other words, through the experience of being in school, students pick up on subtle messages about attitudes, beliefs, values, and behavior that are either "appropriate" or "inappropriate" from teachers and other school personnel. These messages are not part of the official curriculum or the school's mission to educate students for the future.

Although all students are subjected to the hidden curriculum, students who are from low-income families and/or are African American or Hispanic (Latino/a) may be affected the most adversely by educational settings that have been established on the basis of upper- and middle-class white (non-Hispanic) values, attitudes, and behavior (AAUW, 2008). When teachers from middle- and upper-middle-class backgrounds instruct students from lower-income families, the teachers often establish a more structured classroom and a more controlling environment for students (•Figure 16.5). These teachers may also have lower expectations for students' academic achievements.

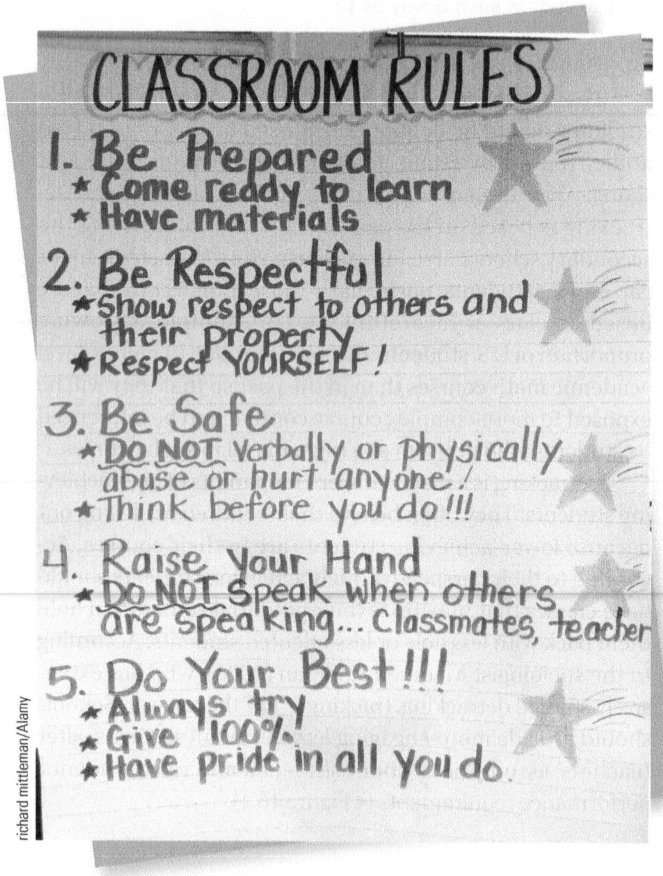

richard mittleman/Alamy

FIGURE 16.5   Signs such as this found in elementary classrooms list the rules, and sometimes the rewards and consequences, of different types of student behavior. According to conflict theorists, schools impose rules on working-class and poverty-level students so that they will learn to follow orders and to be good employees in the workplace.

Schools with many students from low-income families often emphasize procedures and rote memorization without focusing on decision making and choice, or on providing explanations of why something is done a particular way. Schools for middle-class students stress the processes (such as figuring and decision making) involved in getting the right answer. Schools for affluent students focus on creative activities in which students express their own ideas and apply them to the subject under consideration, as well as building students' analytical and critical-thinking skills.

Over time, low-income students become frustrated with the educational system and drop out or become very marginal students, making it even more difficult for them to attend college and gain the appropriate credentials for gaining better-paying jobs. Educational credentials are extremely important in a nation such as ours that emphasizes **credentialism**—a process of social selection in which class advantage and social status are linked to the possession of academic qualifications. Credentialism is closely related to *meritocracy*, a social system in which status is assumed to be acquired through individual ability and effort. Persons who acquire the appropriate credentials for a job are assumed to have gained the position through what they know, not who they are or whom they know. According to conflict theorists, the hidden curriculum determines in advance that the most valued credentials will primarily stay in the hands of the middle and upper classes, so the United States is not actually as meritocratic as some might claim.

The hidden curriculum is also related to gender bias. For many years the focus in education was on how gender bias harmed girls and women: Reading materials, classroom activities, and treatment by teachers and peers contributed to a feeling among many girls and young women that they were less important than male students. The accepted wisdom was that, over time, differential treatment undermines females' self-esteem and discourages them from taking certain courses, such as math and science, that have been dominated by male teachers and students. In the 1990s the American Association of University Women issued *The AAUW Report: How Schools Shortchange Girls*, which highlighted inequalities in women's education and started a national debate on gender equity (AAUW, 1995). Over the past twenty years, improvements have occurred in girls' educational achievement, as females have attended and graduated from high school and college at a higher rate than their male peers. More females have enrolled in advanced-placement or honors courses and in academic areas, such as math and science, where they had previously lagged. However, some traditional gender differences persist at some grade levels, with boys generally outscoring girls on standardized math tests by a small margin and girls outscoring boys on standardized reading tests by a small margin.

Ironically, after many years of discussion about how the hidden curriculum and other problems in schools served to disadvantage female students, the emphasis has now shifted to the question of whether girls' increasing

FIGURE 16.6 According to some conflict theorists, a persistent problem in education is the large racial–ethnic discrepancy in test scores from the early grades through high school and college. How might issues of racism, unequal funding of schools, and similar concerns contribute to this problem?

accomplishments from elementary school to college and beyond have come at the expense of boys and young men. But this is not true, according to research by the AAUW (2008: 2): "Educational achievement is not a zero-sum game, in which a gain for one group results in a corresponding loss for the other. If girls' success comes at the expense of boys, one would expect to see boys' scores decline as girls' scores rise, but this has not been the case."

Regardless of gender, large differences remain in scores on academic tests among students by race/ethnicity (●Figure 16.6). Studies have shown that white children are more likely to graduate from high school and college than are their African American and Hispanic peers. Likewise, children from higher-income families are more likely to graduate from high school than are children from lower-income families, who are also less likely to attend college (AAUW, 2008).

The conflict theorists' focus on the hidden curriculum calls our attention to the fact that students learn far more—both positively and negatively—than just the subject matter being taught in the classroom. Students are exposed to a wide range of beliefs, values, attitudes, and behavioral expectations that are not directly related to specific subject matter.

Symbolic Interactionist Perspectives

Unlike functionalist analysts, who focus on the functions and dysfunctions of education, and conflict theorists, who focus on the relationship between education and inequality, symbolic interactionists focus on classroom communication patterns and educational practices, such as labeling, which affect students' self-concept and aspirations.

Labeling and the Self-Fulfilling Prophecy According to symbolic interactionists, the process of labeling is directly related to the power and status of those persons who do the labeling and those who are being labeled. Chapter 7 explains that *labeling* is the process whereby others identify a person as possessing a specific characteristic or exhibiting a certain pattern of behavior (such as being deviant). In schools, teachers and administrators are empowered to label children in various ways, including grades, written comments on classroom behavior, and placement in classes. For some students, labeling amounts to a *self-fulfilling prophecy*—an unsubstantiated belief or prediction resulting in behavior that makes the originally false belief come true (first defined by Merton, 1968).

A classic form of labeling and the self-fulfilling prophecy has occurred for many years through the use of various IQ (intelligence quotient) tests, which claim to measure a person's inherent intelligence, apart from any family or school influences on the individual. Schools have used IQ tests as one criterion in determining student placement in classes and ability groups (see ● Figure 16.7). The way in which IQ test scores may become a self-fulfilling prophecy was revealed in the 1960s when two social scientists conducted an experiment in an elementary school during which they intentionally misinformed teachers that some of the students had extremely high IQ test scores whereas others had average to below-average scores. As the researchers observed, the teachers began to teach "exceptional" students in a different manner from other students. In turn, the "exceptional" students began to outperform their "average" peers and to excel in their classwork. This study called attention to the labeling effect of IQ scores.

hidden curriculum
the transmission of cultural values and attitudes, such as conformity and obedience to authority, through implied demands found in the rules, routines, and regulations of schools.

credentialism
a process of social selection in which class advantage and social status are linked to the possession of academic qualifications.

Question 2: Consider the following two statements: all farmers who are also ranchers cannot come near town; and most of the ranchers who are also farmers cannot surf. Which of the following statements MUST be true?

- ○ Most of the farmers who cannot come near town can surf
- ○ Only some farmers who ranch can surf near town
- ○ A surfer who ranches and farms cannot surf near town
- ○ Some ranchers who farm can come to town to learn to surf
- ○ Any farmer who cannot surf also ranches

FIGURE 16.7 IQ Test Sample Question
IQ tests containing items such as this are often used to place students in ability groups. Such placement can set the course of a person's entire education.

Today, so-called IQ fundamentalists continue to label students and others on the basis of IQ tests, claiming that these tests measure some identifiable trait that predicts the quality of people's thinking and their ability to perform. Critics of IQ tests continue to argue that these exams measure a number of factors—including motivation, home environment, type of socialization at home, and the quality of schooling—not intelligence alone (Yong, 2011).

Postmodernist Perspectives

How might a postmodern approach describe higher education? One of the major postmodern theorists is Jean-Francois Lyotard (1984), who described how knowledge has become a commodity that is exchanged between producers and consumers. "Knowledge" is now an automated database, and teaching and learning are primarily about data presentation, stripped of their former humanistic and spiritual associations.

In the postmodern era an emphasis in higher education is on how to make colleges and universities more efficient and how to bring these institutions into the service of business and industry. A major objective is looking for the best way to transform these schools into corporate entities such as the "McUniversity," which refers to a means of educational consumption that allows students to consume educational services, to eventually obtain "goods" such as degrees and credentials, and to think of themselves and their parents as consumers. The rapidly increasing cost of higher education has contributed to the perception of "McUniversity" and to the idea of students as consumers.

Savvy college and university administrators are aware of the permeability of higher education and the "students-as-consumers" model. To attract new students and enhance current students' opportunities for consumption, most campuses have amenities such as spacious food courts with many franchise choices, ATMs, video games on gigantic HDTV screens, Olympic-sized swimming pools, and massive rock-climbing walls (• Figure 16.8). Wi-Fi–enabled campuses are also a major attraction for student consumers, and virtual classrooms make it possible for some students to earn college credit without having to look for a parking place at the traditional brick-and-mortar campus.

Based on a postmodern approach, what do you believe will be the dominant means by which future students will consume educational services and goods at your college or university? For many in the second decade of the twenty-first century, the answer becomes that the digital age will continue to rapidly transform what we think of as knowledge and the social institution of education in which people consume new information. College students and many other tech-savvy persons find information on the Web by searching in a purposeful but somewhat random and sporadic manner because of the way in which hyperlinked sources send users nomadically searching from site to site. This postmodern approach to learning has been referred to as the "rhizomatic model of learning," which refers to a rhizomatic plant: a plant that has no center or defined boundaries but instead has a number of semi-independent nodes that are capable of growing and spreading individually within the boundaries of a specific habitat (Cormier, 2008). Based on this analogy, knowledge is increasingly nonhierarchical, is open ended, and involves the "wisdom of the crowds," in which large communities of Web users find meaning and identify what is important to learn from what initially might appear to be random searching in cyberspace.

If this approach to knowledge and the process of education becomes the norm, a specific curriculum determined by experts becomes irrelevant, and people begin to identify and legitimatize for themselves what is important to know

AP Images/Newspaper Member/Laramie Daily Boomerang/Jeremy Martin

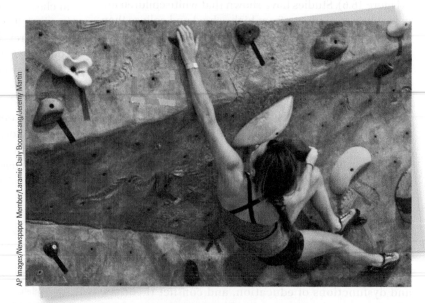

FIGURE 16.8 To attract new students, some college campuses have amenities such as rock-climbing walls.

Sociological Perspectives on Education

	Key Points
Functionalist Perspectives	Education is one of the most important components of society: Schools teach students not only content but also to put group needs ahead of the individual's.
Conflict Perspectives	Schools perpetuate class, racial–ethnic, and gender inequalities through what they teach to whom.
Symbolic Interactionist Perspectives	Labeling and the self-fulfilling prophecy are an example of how students and teachers affect each other as they interpret their interactions.
Postmodernist Perspectives	Knowledge has become a commodity, and students and their parents are consumers of education in the twenty-first century.

and the significance of the work they are doing. As one analyst stated, "The community, then, has the power to create knowledge within a given context and leave that knowledge as a new node connected to the rest of the network" (Cormier, 2008). Do you think that this postmodern approach is more useful in analyzing how education in the future may be in traditional fields of study with long-accepted knowledge and accepted "experts" or in looking at new and developing fields that are now emerging in the digital age?

The Concept Quick Review summarizes the major theoretical perspectives on education.

LO3 **Describe** major problems in elementary and secondary schools, such as unequal funding of schools, high dropout rates, and problems of inequality based on race or disability.

Problems in Elementary and Secondary Schools

Education in kindergarten through high school is a microcosm of many of the concerns facing the United States. The problems we examine in this section include unequal funding of public schools, dropout rates, racial segregation and resegregation, equalizing opportunities for students with disabilities, and competition for public schools in the form of school choice, charter schools, and homeschooling.

Unequal Funding of Public Schools

Why does unequal funding exist in public schools? Most educational funds come from state legislative appropriations and local property taxes (see • Figure 16.9). Some legislatures provide far fewer funds for schools in their state because they have fewer resources or because they have other political priorities. The same is true for local

communities. Some cities have properties that are more expensive than others. Other cities have large amounts of land that is under the control of various levels of the government and thus not subject to school taxation. All of this adds up to unequal funding for public schools. As shown in Figure 16.9, in the 2014–2015 school year, state sources contributed 46.3 percent of public elementary–secondary school system revenue, 44.1 percent came from local sources, and 9.6 percent came from federal sources (National Education Association, 2015b). Much of the money from federal sources is earmarked for special programs for students who are disadvantaged (e.g., the Head Start program) or who have a disability. Expenditures per pupil for public and secondary public schools vary from state to state, with the U.S. average being $11,355. Per-pupil expenditures were predicted to range from a high of $21,263 in Vermont to a low of $8,632 in North Carolina (National Education Association, 2015b).

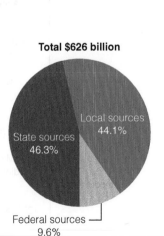

Total $626 billion

Local sources 44.1%
State sources 46.3%
Federal sources 9.6%

FIGURE 16.9   Percentage Distribution of Total Public Elementary–Secondary School System Revenue, 2014–2015

Source: National Education Association, 2015b.

School Dropouts

High dropout rates are a major problem facing contemporary schools. Dropout rates are computed in various ways, but one of the most telling is the *status dropout rate*—the percentage of people in a specific age range who are not currently enrolled in high school and who do not have a high school degree or its equivalent. In recent years, slightly more than three million 16- to 24-year-olds were not enrolled in high school and had not earned a high school diploma or its equivalent.

Status dropout rates vary by gender, race/ethnicity, and region of the country (see • Figure 16.10). Males (7.0 percent) have a higher status dropout rate than females (6.0 percent). Status dropout rates also vary by race/ethnicity: Hispanics/Latinos/as (12.0 percent), American Indian/Alaska natives (13.0 percent), and blacks/African Americans (7.0 percent have higher status dropout rates than whites (5.0 percent), Asian/Pacific Islanders (3.0 percent), and persons reporting two or more races (6.0 percent) (National Center for Educational Statistics, 2015a). Finally, region is also an issue in status dropout rates: The Northeastern United States has the lowest status dropout rates, while the South and West have the highest.

Using a second approach to determine dropout rates, the *event dropout rate*—which estimates the percentage of both public and private high school students who left high school between the beginning of one school year and the beginning of the next without earning a high school diploma or an alternative credential such as a GED—we find that every school day, at least 7,000 U.S. students (on average) leave high school and never return. What this means is that, on average, 3.5 percent of students who were enrolled in public or private high schools in October 2008 left school before October 2009 without completing a high school program. (These data are the latest available at the time of this writing, but the trends have remained relatively constant in the 2000s.) Perhaps the most telling statistic when using the event dropout rate is that students living in low-income families are about four-and-one-half times more likely to drop out in any given year than students living in high-income families (National Center for Educational Statistics, 2012a).

Dropping out of school produces serious economic and social consequences for individuals and nations. There is substantial evidence that education levels are related to earnings. Young people without a high school

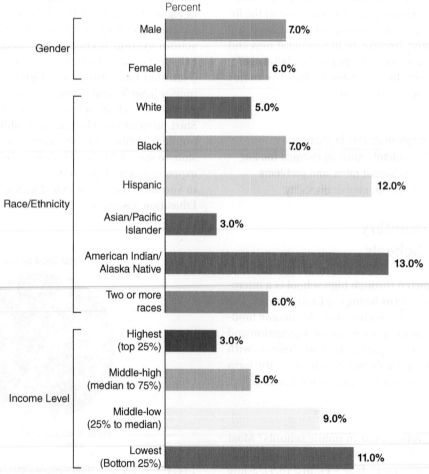

FIGURE 16.10   Status Dropout Rates for 16- to 24-Year-Olds, by Race/Ethnicity, Gender, and Region

Source: National Center for Educational Statistics, 2015a.

education not only have a difficult time finding work that will pay a living wage, but when they are employed, they make far less than individuals with at least a high school diploma. The median weekly earnings of full-time workers ages 25 and over who had not completed high school (in 2012) was $478, as compared to earnings of $647 for persons with a high school diploma, GED, or equivalent (no college) and $1,168 for those holding at least a bachelor's degree (U.S. Bureau of Labor Statistics, 2013e).

A higher percentage of dropouts (ages 25 and over) are also unemployed or holding temporary or part-time jobs while seeking full-time work. Cities and states suffer because tax revenues are lower when many people are unemployed, and the societal costs for public assistance, crime control, and health care are higher.

Why do students drop out of school? Some students believe that their classes are boring; others are skeptical about the value of schooling and think that completing high school will not increase their job opportunities. Upon leaving school, many dropouts have high hopes of making money and enjoying newfound freedom; however, many find that few jobs are available and that they do not have the minimum education required for any "good" jobs that exist.

Racial Segregation and Resegregation

Although some people believe that the issue of racial segregation has long been solved in America's schools, in many areas of the United States schools remain racially segregated or have become resegregated. In 1954 the U.S. Supreme Court ruled (in *Brown v. Board of Education of Topeka, Kansas*) that "separate but equal" segregated schools are unconstitutional because they are inherently unequal. Today, racial segregation remains a fact of life in education: Efforts to bring about *desegregation*—the abolition of legally sanctioned racial–ethnic segregation—or *integration*—the implementation of specific action to change the racial–ethnic and/or class composition of the student body—have failed in many districts. Some school systems have bused students across town to achieve racial integration. Others have changed school attendance boundaries or introduced magnet schools with specialized programs such as science or the fine arts to change the racial–ethnic composition of schools. But school segregation does not exist in isolation: Racially segregated housing patterns contribute to high rates of school segregation.

Resegregation is also an issue because some school districts have abandoned programs that had produced greater racial integration in local schools. Raleigh, North Carolina, is a case in point: A local school board decided to end consideration of race and socioeconomic status in determining school assignments and stopped the district's busing-for-diversity program. Those who opposed this return to the "neighborhood school" concept argued that resegregation would quickly occur throughout the district (Mooney, 2011).

How segregated are U.S. schools? Here are a few facts: More than half of all African American public school students in Illinois, Michigan, and New York state attend predominantly black schools. For example, half of all African American students in the Chicago metro area and one-third of all African American students in New York attend "apartheid schools," where white Americans make up 0 to 1 percent of the total enrollment (Orfield, Kucsera, and Siegel-Hawley, 2012). In Maryland, Alabama, Mississippi, Tennessee, Georgia, and Texas, approximately 30 percent of African American public school students attend schools that have at least a 95 percent black/African American population (●Figure 16.11). Children of color now constitute more than half of public school students in a number of states because white (non-Hispanic) children are more often enrolled in charter schools, suburban school districts, or private schools with a high percentage of white students (Mack, 2010).

According to one recent study by the Civil Rights Project at the University of California, approximately 40 percent of African American and Latino/a students are enrolled in U.S. public schools where less than 10 percent of their classmates are white (non-Hispanic) Americans. As these statistics show, despite declining residential

FIGURE 16.11   Although many people believe that the United States is a racially integrated nation, a look at schools throughout the country reveals that many of them remain segregated or have become largely resegregated in recent decades.

segregation for African Americans and extensive movement to the suburbs, school segregation remains very high for African American students. The problem involves double segregation by both race and poverty: The typical African American or Latina/o student attends a school with almost twice as many low-income students as a school attended by the typical white American or Asian American student (Orfield, Kucsera, and Siegel-Hawley, 2012).

Equalizing Opportunities for Students with Disabilities

Another concern in education has been how to provide better educational opportunities for students with a disability—any physical and/or mental condition that limits students' access to, or full involvement in, school life (•Figure 16.12). The U.S. Department of Education's Office of Special Education Programs (OSEP) collects information on students with disabilities as part of the implementation of the Individuals with Disabilities Education Act (IDEA) and uses categories of disabilities that include autism, deafness/blindness, developmental delay, emotional disturbance, hearing impairment, intellectual disability, multiple disabilities, orthopedic impairment, other health impairment, specific learning disabilities, speech or language impairments, traumatic brain injury, visual impairments, and preschool disability.

Along with other provisions, the Americans with Disabilities Act of 1990 requires schools to make their facilities, services, activities, and programs accessible to people with disabilities. Many schools have attempted to *mainstream* children with disabilities by providing *inclusion programs,* under which the special education curriculum is integrated with the regular education program and each child receives an *individualized education plan* that provides annual educational goals. *Inclusion* means that children with disabilities work with a wide variety of people; over the course of a day, a child may interact with his or her regular education teacher, the special education teacher, a speech therapist, an occupational therapist, a physical therapist, and a resource teacher, depending on the child's individual needs. Although much remains to be done, measures to enhance education for children with disabilities have increased the inclusion of many young people who were formerly excluded or marginalized in the educational system. Among these is IDEA (the Individuals with Disabilities Education Act), a 1975 law enacted by Congress to ensure that children with disabilities have the opportunity to receive a free appropriate public education like other children. The law has been revised numerous times, and in September 2011 it was widened to include babies and toddlers until the third birthday.

Traditional public schools have the highest rate (34 percent) of enrollment for students with disabilities, as compared to 23 percent in charter schools. Some analysts believe that this difference remains persistent because public schools may have more resources to assist students with learning disabilities; however, some critics of charter schools argue that some of these schools have discouraged enrollment of special-needs students or have indicated their inability to continue serving such students after the school year started (Resmovits, 2012). Among the students served by public schools, specific learning disabilities are the most common concern (about 38 percent), followed by speech or language impairments (nearly 22 percent), other health impairments (10.6 percent), and intellectual disability (7.1 percent) (National Center for Education Statistics, 2012a). Only about 15 percent of students with disabilities (who are enrolled in regular school settings) spend more than 60 percent of their school time outside the general ("mainstream") classrooms: Nearly 60 percent of students with disabilities (who are enrolled in regular school settings) spend at least 80 percent of their school time in regular classrooms. About 3 percent of all public school students with disabilities attend a separate school for students with disabilities. Top categories served in these

Angela Hampton/Bubbles Photolibrary/Alamy

FIGURE 16.12 What steps can schools take to provide better educational opportunities for students with a disability?

schools are students with multiple disabilities, deafness/blindness, and emotional disturbances (National Center for Education Statistics, 2012a).

Explain how options such as school vouchers, charter schools, and homeschooling differ from traditional educational approaches, and identify strengths and weaknesses of each approach.

Competition for Public Schools

Public schools do not have a monopoly on K–12 education. Today, parents have more choices of where to send their children than in the past.

School Choice and School Vouchers   School choice is a persistent issue in education. Much of the discussion about school choice focuses on school *voucher programs* in which public funds (tax dollars) are provided to parents so that they can pay their child's tuition at a private school of their choice. Many parents praise the voucher system because it provides them with options for schooling their children. Some political leaders applaud vouchers and other school-choice policies for improving public school performance. However, voucher programs are controversial: Some critics believe that giving taxpayer money to parents so that they can spend it at private (often religious) schools violates constitutional requirements for the separation of church and state. However, the U.S. Supreme Court ruled in *Zelman v. Simmons-Harris,* a case involving a Cleveland, Ohio, school district, that voucher policies are constitutional because parents have a choice and are not required to send their children to church-affiliated schools. Other critics claim that voucher programs are less effective in educating children than public schools. According to studies in the District of Columbia, Milwaukee, and Cleveland, public school students outperformed voucher students in both reading and math on state proficiency tests; however, neither group reached state proficiency requirements (Ott, 2011). In sum, advocates like the choice factor in voucher programs, while critics believe that vouchers undermine public education, lack accountability, and may contribute to the collapse of the public school system.

Charter Schools   Charter schools (or "schools of choice") are primary or secondary schools that receive public money but are free from some of the day-to-day bureaucracy of a larger school district that may limit classroom performance. These schools operate under a charter contract negotiated by the school's organizers (often parents or teachers) and a sponsor (usually a local school board, a state board of education, or a university) that oversees the provisions of the contract. Some school districts "contract out" by hiring for-profit companies on a contract basis to manage charter schools, but the schools themselves are nonprofit. Among the largest educational management organizations are Imagine Schools, National Heritage Academies, the Leona Group, Edison Learning, White Hat Management, and Mosaica Education.

In the 2012–2013 school year, more than 500 new public charter schools opened; however, in the same time period, about 150 public charter schools closed their doors because of low enrollment, financial concerns, and low academic performance. Advocates for charter schools suggest that these figures show that there is both demand for charter schools and evidence that schools that do not meet the needs of their students will be closed (National Alliance for Public Charter Schools, 2013).

What are the unique claims for charter schools? Charter schools are supposed to provide more autonomy for individual students and teachers, and to provide a large number of minority students with a higher-quality education than they would receive in the public schools in their area. Public charter schools enroll a greater percentage of low-income students (46 percent as compared to 41 percent in traditional public schools), African American students (27 percent versus 15 percent), Hispanic/Latino/a students (26 percent versus 22 percent), and students with lower standardized assessments before their transfer to charters schools (National Alliance for Public Charter Schools, 2013). Charter schools attempt to maintain an organizational culture that motivates students and encourages achievement rather than having a negative school environment where minority students are ridiculed for "acting white" or making good grades. Some charter schools offer college-preparatory curriculums and help students of color achieve their goal of enrolling in the college or university of their choice.

However, charter schools have numerous challenges. Some schools have high turnover rates, perhaps partly because of family instability, students' socioeconomic status, or other factors not under the direct control of the schools. A number of charter-school officials have been accused of misappropriating school funds or other financial irregularities, but many analysts believe that the positives seem to outweigh the negatives when it comes to charter schools addressing the academic gap among minority students.

Homeschooling   Another alternative, homeschooling, has been chosen by some parents who hope to avoid the problems of public schools while providing a quality education for their children (● Figure 16.13). It is estimated that about 1.5 million children are homeschooled in grades K through 12 (Kerkman, 2011). This is a significant increase from the estimated 1.1 million students who were homeschooled in 2003. The primary reasons that parents indicated for preferring to homeschool their children are (1) concern about the school environment, (2) the desire to provide religious or moral instruction, and (3) dissatisfaction with the academic instruction available at traditional schools. Typically, the parents of homeschoolers are better educated, on average, than other parents; however, their

FIGURE 16.13 Homeschooling has grown in popularity in recent decades as parents have sought to have more control over their children's education. Although some homeschool settings may resemble a regular classroom, other children learn in more-informal settings, such as the family kitchen.

income is about the same. Researchers have found that boys and girls are equally likely to be homeschooled.

Parents who educate their children at home believe that their children are receiving a better education at home because instruction can be individualized to the needs and interests of their children. Some parents also indicate religious reasons for their decision to homeschool their children. An association of homeschoolers now provides communication links for parents and children, and technological advances in computers and the Internet have made it possible for homeschoolers to gain information and communicate with one another. In some states, parents organize athletic leagues, proms, and other social events so that their children will have an active social life without being part of a highly structured school setting. According to advocates, homeschooled students typically have high academic achievements and high rates of employment.

Critics of homeschooling question how much parents know about school curricula and how competent they are to educate their own children at home, particularly in rapidly changing subjects such as science and computer technology. Some states have passed accountability laws that must be met by parents who teach their children at home.

School Safety and Violence at All Levels

Today, officials in schools from the elementary level to two-year colleges and four-year universities are focusing on how to reduce or eliminate violence. In many schools, teachers and counselors are instructed in anger management and peer mediation, and they are encouraged to develop classroom instruction that teaches values such as respect and responsibility. Some schools create partnerships with local law enforcement agencies and social service organizations to link issues of school safety to larger concerns about safety in the community and the nation.

In fact, 85 percent of all public schools, grades K through 12, recorded one or more incidents of violence, theft, or other crimes, amounting to nearly 2 million crimes each year. The most frequent incidents reported are physical attack or fight with a weapon, threat of physical attack without a weapon, vandalism, theft/larceny, possession of a knife or other sharp object, and distribution, possession, or use of illegal drugs (National Center for Education Statistics, 2014). However, statistics related to school safety continue to show that U.S. schools are among the safest places in the world for young people. According to "Indicators of School Crime and Safety," jointly released by the National Center for Education Statistics and the U.S. Department of Justice's Bureau of Justice Statistics, young people are more likely to be victims of violent crime at or near their home, on the streets, at commercial establishments, or at parks than they are at school (National Center for Educational Statistics, 2014). However, these statistics do not keep many people from believing that schools are becoming more dangerous with each passing year and that all schools should have high-tech surveillance equipment and police officers in every school to help maintain a safe environment.

Even with safety measures in place, violence and fear of violence continue to be pressing problems in schools throughout the United States. This concern extends from kindergarten through grade 12 because violent acts have resulted in deaths in a number of communities throughout the United States. The Newtown, Connecticut, school massacre occurred on December 14, 2012, when Adam Lanza fatally shot twenty children and six adult staff members at an elementary school before committing suicide (• Figure 16.14). Considered the second deadliest mass murder at an elementary school in U.S. history (following the 1927 Bath School bombings in Michigan), this violent attack prompted extensive debate about gun control and

Gina Jacobs/Shutterstock.com

FIGURE 16.14 In the aftermath of the 2012 mass school shooting in Newtown, Connecticut, many people sought more-stringent gun laws to prevent future occurrences of similar horrendous events. Do you believe the impetus remains to bring about stronger gun-control legislation in the United States?

college and university campuses. At the time of this writing, the state of Texas was in the process of creating a law to take effect in August 2016 that legalizes carrying concealed handguns into classrooms, dorms, and buildings of public colleges and universities. The only exception this law allows is for limited "gun-free campus zones" where school administrators can justify reasons why guns should not be allowed, such as facilities where biohazardous materials are stored. The law exempts private universities from "concealed carry" because legislators have assumed that the state does not have a right to tell private property owners how to handle their property. Most college and university professors and administrators are strongly opposed to this legislation.

whether school officials should be armed to prevent future occurrences of this kind or whether police officers should be stationed in every school.

So far, the evidence is unclear on the effect of placing police officers in every school. Do they deter crime or provide safety if an armed person seeks to harm students and school personnel? Advocates say that the answer is an unequivocal "yes." However, critics note that the recent increase in police officers and armed guards in schools has also brought about an increase in the number of students who are arrested for minor behavior problems, thus pushing children, whose detrimental behavior might best be handled by the school's discipline system, into the criminal courts for relatively minor offenses (Eckholm, 2013).

Like public elementary and secondary schools, college and university campuses are not immune to violence, as deranged individuals have engaged in acts of personal terrorism at the expense of students, professors, and other victims. In the aftermath of tragedies such as the one that occurred at Virginia Tech in 2007, in which thirty-two people were killed by a student, there was a massive outpouring of public sympathy and calls for greater campus security. As usual, gun-control advocates called for greater control over the licensing and ownership of firearms and for heightened police security on college campuses, whereas pro-gun advocates argued that people should be allowed to carry firearms on campus for their own protection. Lawmakers in a number of states introduced measures seeking to relax concealed-weapons restrictions on

LO5 **Discuss** the educational opportunities and challenges found in community colleges in the twenty-first century.

Opportunities and Challenges in Higher Education

Who attends college? What sort of college or university do they attend? We will explore these and other questions in this section.

Community Colleges

One of the fastest-growing areas of U.S. higher education today is the community college; however, the history of two-year colleges goes back more than a century, with the establishment of Joliet Junior College in Illinois (• Figure 16.15). Later, following World War II, the G.I. Bill of Rights provided the opportunity for more people to attend college, and in 1948 a presidential commission report called for the establishment of a network of public community colleges that would charge little or no tuition, serve as cultural centers, be comprehensive in their program offerings, and serve the area in which they were located.

Hundreds of community colleges were opened across the nation during the 1960s, and the number of such institutions has steadily increased since that time

four-year degrees, sometimes in conjunction with four-year colleges located in the same state.

Community colleges educate about half of the nation's undergraduates. According to the American Association of Community Colleges (2013), the 1,132 community colleges (including public, private, and tribal colleges) in the United States enroll about 13 million students in credit and noncredit courses. Community college enrollment accounts for 45 percent of all U.S. undergraduate students. Women make up more than half (57 percent) of community college students, and for working women and mothers of young children, these schools provide a unique opportunity to attend classes on a part-time basis as their schedule permits. Men also benefit from flexible scheduling because they can work part time or full time while enrolled in school. About 59 percent of all community college students are enrolled part time, while 41 percent are full-time students (taking 12 or more credit hours each semester). Community colleges are also important for underrepresented minority student enrollment: Fifty-five percent of all Native American college students attend a community college, as do 56 percent of all Hispanic students, 49 percent of African American students, and 44 percent of Asian American/Pacific Islanders (American Association of Community Colleges, 2013).

One of the greatest challenges facing community colleges today is money. Across the nation, state and local governments struggling to balance their budgets have slashed funding for community colleges. In a number of regions, these cuts have been so severe that schools have been seriously limited in their ability to meet the needs of their students. In some cases, colleges have terminated programs, slashed course offerings, reduced the number of faculty, and eliminated essential student services. Many people were hopeful that President Obama's "American Graduation Initiative" of 2009 would strengthen community colleges, offer greater financial support for students, and increase a high college graduation rate for the nation by 2020. However, the $12 billion that President Obama called for to launch this initiative was not realized because of a tradeoff to get the Patient Protection and Affordable Care Act of 2010 passed (see Chapter 18). Instead, $2 billion was pledged for a job training and workforce development program in community colleges, administered by the U.S. Department of Labor, to help economically dislocated workers who are changing careers.

FIGURE 16.15   Joliet Junior College (Illinois) is the oldest two-year college in the United States, having opened its doors in 1901. The bottom photo shows a scene from graduation day at the nation's largest two-year school, Miami Dade College (Florida), which recently expanded into offering four-year degrees, part of a national trend among two-year colleges. Today, Joliet, Miami Dade, and other schools like them fulfill many needs in the competitive world of higher education.

as community colleges have responded to the needs of their students and local communities. Community colleges offer a variety of courses, some of which are referred to as "transfer courses" in which students earn credits that are fully transferable to a four-year college or university. Other courses are in technical/occupational programs, which provide formal instruction in fields such as nursing, emergency medical technology, plumbing, carpentry, and computer information technology. Many community colleges are now also offering

By September 2012, $1 billion of this money had been distributed in grants to community colleges and universities around the country for the development and expansion of innovative training programs. The grants are part of the Trade Adjustment Assistance Community College and Career Training initiative, which promotes skills development and employment opportunities in fields such as advanced manufacturing, transportation, and health care, as well as science, technology, engineering, and math careers, through partnerships between training providers and local employers (U.S. Department of Labor, 2012). Although the community college grants are not as large as the Obama administration had originally intended, they may help many students who otherwise might not have had an opportunity to develop life and workplace skills, acquire additional years of schooling, and gain employment opportunities that will give them a stronger place in the community and a better future.

 Describe the economic problems facing many four-year colleges and universities, and discuss how these fit into larger patterns of state funding.

Four-Year Colleges and Universities

About 19.9 million undergraduate and graduate students attend public or private degree-granting colleges or universities in the United States (*Chronicle of Higher Education,* 2014). Four-year schools typically offer a general education curriculum that gives students exposure to multiple disciplines and ways of knowing, along with more in-depth study (known as a "major") in at least one area of concentration. However, many challenges are faced by four-year institutions, ranging from the cost of higher education to racial and ethnic differences in enrollment and lack of faculty diversity.

The High Cost of a College Education   What does a college education cost? According to the College Board (2014), the average published tuition and fee prices for full-time students living on campus and paying in-state tuition was $18,943 at public four-year schools and $42,419 at private nonprofit institutions in the 2013–2014 academic year. The published rate for out-of-state students at four-year public institutions was $32,762. At for-profit postsecondary schools the published rate averaged about $15,230. The lowest total costs were for students at public two-year institutions ($3,347) (College Board, 2014). Although public institutions such as community colleges and state colleges and universities typically have lower tuition and overall costs—because they are funded primarily by tax dollars—than private colleges have, the cost of attending public institutions has increased significantly over the past two decades (●Figure 16.16).

According to some social analysts, a college education is a bargain and a means of upward mobility. However,

FIGURE 16.16   Soaring costs of both public and private institutions of higher education are a pressing problem for today's college students and their parents. What factors have contributed to the higher overall costs of obtaining a college degree?

other analysts believe that the high cost of a college education reproduces the existing class system: Students who lack money may be denied access to higher education, and those who are able to attend college receive different types of education based on their ability to pay. For example, a community college student who receives an associate's degree or completes a certificate program may be prepared for a position in the middle of the occupational status range, such as a dental assistant, computer programmer, or welder. In contrast, university graduates with four-year degrees are more likely to find initial employment with firms where they stand a chance of being promoted to high-level management and executive positions.

Given the necessity of getting a college education in the twenty-first century, is any financial assistance available? The Obama administration passed a 2010 student-loan bill to aid colleges and students. The legislation was designed to "cut out the middle person" by ending the bank-based system of distributing federally subsidized student loans and instead have the Department of Education give loan money directly to colleges and their students. With the savings from this approach, more money was to be put into the Pell Grant program. Unlike a loan, a federal Pell Grant does not have to be repaid. The maximum award for the 2015–2016 academic year is $5,775, but the average amount awarded is $3,541.

Many questions remain about student loans and the possible long-term effects of high student debt on individuals after they complete their college education. More than $1.2 trillion is owed on student loans in the United States today, up from $509 billion in 2006. The problem with the ratio of student borrowing to overall debt really began in the 1980s, when four-year college tuition began to rise faster than incomes. The booming of for-profit schools in the 1990s further contributed to the problem of excessive student loan debt because many students took out larger loans than they could repay with the types of jobs available to them based on the education they received. Between 2001 and 2015, state and local financing per student for higher education declined by more than 25 percent nationwide, while tuition and fees at many public state colleges and universities increased by 75 percent. Among those most harmed by spiraling college costs, rising student debt, and less availability of grant money that does not have to be repaid are lower- and middle-income students of all racial–ethnic categories.

Slashed Budgets at State Colleges and Universities

The problem of increasing costs of higher education for students is compounded by slashed funding for public higher education as states have encountered budget shortfalls and contentious debates over competing priorities for public funds. Declining state and federal support has become a major concern for colleges and universities that must find new sources of revenue, sharply reduce current operating expenses, and rework their budgetary priorities. Recent studies have found that spending per student (adjusted for inflation) decreased by 20 percent (approximately $1,805) in 47 out of 50 states between the economic downturn beginning in 2007–2008 and fiscal year 2015. During that same period, students' tuition rose about 29 percent, or an average of $2,068 (Chokshi, 2015).

Why do states cut funding for higher education when it appears that more years of formal education would be worthwhile not only for students but also for the economy of the states? One answer involves the kinds of problems that states faced when the recession of late 2007 and early 2008 occurred, because states' tax revenues declined rapidly, leaving many states with steep budget shortfalls. To reduce the problem, many political leaders slashed budgets and cut funding wherever possible: Funding for higher education was an obvious target. Since the Great Recession, politicians in many states have been unwilling, or unable, to implement the necessary tax hikes and spending cuts that would be required to provide sufficient funding for higher education (Chokshi, 2015). Members of the Boards of Regents and school administrators at many public college and universities have been left to scramble for funds, including looking increasingly toward private contributors and other revenue sources to fund their institutions.

LO7 **Identify** problems in racial and ethnic diversity in higher education, including student enrollment issues, lack of faculty diversity, and continuing controversy over affirmative action policies and laws.

Racial and Ethnic Differences in Enrollment

How does college enrollment differ by race and ethnicity? People of color (who are more likely than the average white student to be from lower-income families) are underrepresented in higher education. White Americans make up nearly 55 percent of all college students at both two-year and four-year public and private institutions, as compared to African American enrollment at 13.1 percent, Hispanic/Latina/o enrollment at 13.6 percent, Asian American and Pacific Islander enrollment at 5.8 percent, and American Indian (Native American)/Alaska Native at 0.8 percent. Students who reported two or more races accounted for 2.3 percent of students, persons whose race was unknown made up 6.1 percent, and nonresident students constituted 3.8 percent (*Chronicle of Higher Education*, 2014).

Although gaps in college enrollment rates by race and ethnicity have been reduced in recent years and greater access has occurred for some African American and Hispanic American students, distinct differences exist in the kinds of schools that students attend, producing increasing stratification by race and ethnicity (Lipka, 2014). More-selective institutions, such as public or private research universities, are more likely to enroll white Americans (60 percent). African American students are more highly represented (28.2 percent) at two-year private nonprofit institutions than at other types of institutions. Hispanic American students are primarily enrolled (22.3 percent) at two-year, for-profit institutions (*Chronicle of Higher Education*, 2014).

Native American/Alaska Native enrollment rates have remained stagnant at less than 1.0 percent for more than a decade. However, tribal colleges have experienced growth in student enrollment. Founded to overcome racism experienced by Native American students in traditional colleges and to shrink the high dropout rate among Native American college students, thirty-seven tribal colleges are now chartered and run by the Native American nations. Tribal colleges receive no funding from state and local governments and, as a result, are often short of funds to fulfill their academic mission. Various organizations seek to raise funds for academic endeavors and scholarships for students.

The proportionately low number of people of color enrolled in colleges and universities in the past is reflected in the educational achievement of people ages 25 and over in 2014, as shown in • Figure 16.17. If we focus on persons who receive advanced degrees (such as the master's, doctoral, and professional degrees), the underrepresentation of persons of color is even more striking. According to the *Chronicle of Higher Education* (2014), of the 51,008 doctoral degrees conferred in 2012, African Americans earned 6.3 percent, Hispanics earned 6.5 percent, and American Indian or Alaska Natives earned 0.3 percent. By contrast, whites (non-Hispanic) earned 73.5 percent of the total number of degrees awarded,

476 • **PART 4** Social Institutions

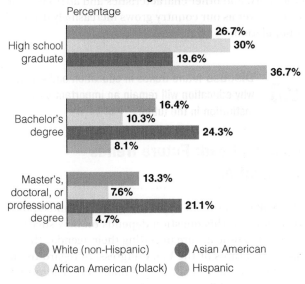

Educational Achievement of Persons Ages 25 and Over

Percentage

High school graduate
- 26.7%
- 30%
- 19.6%
- 36.7%

Bachelor's degree
- 16.4%
- 10.3%
- 24.3%
- 8.1%

Master's, doctoral, or professional degree
- 13.3%
- 7.6%
- 21.1%
- 4.7%

● White (non-Hispanic) ● Asian American
● African American (black) ● Hispanic

FIGURE 16.17 Educational Achievement of Persons Ages 25 and Over

Source: Author's compilation of U.S. Census Bureau Educational Attainment Table 3, 2014.

and Asian Americans earned 9.1 percent. Temporary visa holders in the United States earned 29 percent of all doctoral degrees, with more than half of them in engineering.

Underrepresentation is not the only problem faced by students of color: Problems of prejudice and discrimination continue on some college campuses, as discussed in previous chapters. Some problems are overt and highly visible; others are more covert and hidden from public view. Examples of overt racism include mocking Black History Month or a Latino celebration on campuses, referring to individuals by derogatory names, tying nooses on doorknobs of dorm rooms or faculty offices, and having "parties" where guests dress in outfits that ridicule people from different cultures or nations. A study by the sociologists Leslie Houts Picca and Joe R. Feagin (2007) found that many blatantly racist events, ranging from private jokes and conversations to violent incidents, occurred in the presence of 600 white students at 28 colleges and universities across the country who were asked to keep diaries and record any racist events that they observed. In addition to overt patterns of discrimination, other signs of racism included numerous conversations that took place "backstage" (in whites-only spaces where no person of color was present) and involved derogatory comments, skits, or jokes about persons of color. According to Picca and Feagin, most of the racial events were directed at African Americans, but Latinos/as and Asian Americans were also objects of some negative comments.

E-mail and social networking sites offer additional avenues for college students and others to engage in backstage racism. However, racist statements made by white persons in "private" digital communications are sometimes made public and shared in the front stage by individuals who do not share their views. As a result, cyber racism

brings embarrassment not only to the perpetrators but also to persons of color who experience emotional distress as a result of the behavior of others (Daniels, 2009, 2010).

The Lack of Faculty Diversity and Equity

Despite the widely held assumption that there has been a significant increase in the number of minority professors, the latest comprehensive data available indicate that white Americans make up nearly 74 percent of all full-time faculty members, as compared with African Americans at slightly less than 7 percent, Asian Americans at 6.2 percent, Hispanics (Latinos/as) at 4.3 percent, and American Indians/Alaska Natives at 0.5 percent (*Chronicle of Higher Education,* 2014).

Gender is also a factor in faculty diversity. Across all racial and ethnic categories, women are underrepresented at the level of full professor and overrepresented at the lower, assistant professor and instructor levels. At the full-professor level at doctoral universities, for example, there is one woman for every three men, and the women, on average, earn 90 cents per dollar paid to their male counterparts (*Chronicle of Higher Education,* 2014). By contrast, women make up more than half of all full-time faculty members at two-year-colleges. In regard to underrepresentation of women in certain academic fields, the problem is greatest among STEM (science, technology, engineering, and mathematics) faculty. For example, the percentage of tenured and tenure-track faculty who are women in engineering remains at less than 15 percent. The percentage of women who are full professors in engineering reached an all-time high of 9.4 percent in 2013 (Yoder, 2014). Law schools and business schools have similar patterns of underrepresentation of women and people of color, particularly in tenure-track and tenured faculty positions.

Faculty diversity along lines of race/ethnicity and gender is an important issue, and it is linked to another problem in higher education: Colleges and universities are experiencing a long-term trend toward more contingent faculty appointments. Data from 2014 indicate that part-time faculty and graduate-student employees make up more than 75 percent of the total instructional staff in higher education and that nearly half of undergraduate courses are taught by nontenure-track instructors or graduate students. The growth in full-time nontenure-track and part-time faculty positions continues to outstrip the increase in tenure-line positions (*Chronicle of Higher Education,* 2014).

Affirmative Action

Affirmative action has been a controversial issue for many years. Why does affirmative action generate such a controversy among people? And what is affirmative action anyway? ***Affirmative action*** is a term that describes policies or

affirmative action
policies or procedures that are intended to promote equal opportunity for categories of people deemed to have been previously excluded from equality in education, employment, and other fields on the basis of characteristics such as race or ethnicity.

procedures that are intended to promote equal opportunity for categories of people deemed to have been previously excluded from equality in education, employment, and other fields on the basis of characteristics such as race or ethnicity.

Education was one of the earliest targets of social policy pertaining to civil rights in the United States. Increased educational opportunity has been a goal of many subordinate-group members because of the widely held belief that education is the key to economic and social advancement. Beginning in the 1970s, most public and private colleges developed guidelines for admissions, financial aid, scholarships, and faculty hiring that took race, ethnicity, and gender into account. These affirmative action policies were challenged in a number of lawsuits, especially when the policies involved public colleges. Critics of affirmative action often assert that these policies amount to *reverse discrimination,* a term that describes a situation in which a person who is better qualified is denied enrollment in an educational program or employment in a specific position as a result of another person receiving preferential treatment as a result of affirmative action.

In 2003 the U.S. Supreme Court ruled in *Grutter v. Bollinger* (involving the admissions policies of the University of Michigan's law school) and *Gratz v. Bollinger* (involving the undergraduate admissions policies of the same university) that race can be a factor for universities in shaping their admissions programs, but only within carefully defined limits. Following the *Grutter* decision, many colleges and universities have used race and ethnicity as one of a number of factors in determining admissions. This case led to significantly more African American and Hispanic (Latina/o) students being admitted to selective colleges and universities than would have occurred if decisions had been made strictly on the basis of test scores and high school grade point average (GPA).

A more recent case regarding affirmative action is *Fisher v. University of Texas,* brought by Abigail Fisher, a white student who claimed that the University of Texas denied her admission because of her race. Fisher's attorneys argued that the university's use of race in admissions decisions violated her right to equal protection under the Fourteenth Amendment. The attorneys further stated that diversity should be eliminated as a factor that can be used to justify any use of race in decisions regarding college admission. The Supreme Court heard the case but sent it back to the Fifth Circuit for review regarding the specific techniques that the school used to accomplish its goal of diversity. The university needed to prove, for example, that it could not accomplish its goal of diversity without having race-based preferences. The Fifth Circuit ruled in favor of UT and in 2014 declined to rehear the case. Fisher's attorneys then asked the Supreme Court to review the case again, and that request was granted in June 2015. The Supreme Court will hear the case in the fall of 2015, and you can learn how the Court ruled by searching for *Fisher v. University of Texas* online.

No matter the final outcome of this case, one thing remains clear: Discussions regarding affirmative action and access to higher education—particularly regarding the way

that access is influenced by income, race/ethnicity, gender, nationality, and other characteristics and attributes—are far from over as our country grows increasingly diverse in its population.

 LO8 **Discuss** future trends in education and explain why education will remain an important social institution in the future.

Looking Ahead: Future Trends in Education

What will the future of education be in the United States? The answer to this question depends on how successful that elected officials are in getting their agendas through state legislatures or the U.S. Congress. In regard to public elementary and secondary schools, provisions of the Elementary and Secondary Education Act—or No Child Left Behind Act—implemented in 2001 by the Bush administration remain in effect as of this writing. The Obama administration subsequently outlined in 2011 how states might get relief from some of its provisions by implementing state-led efforts to close academic achievement gaps. Overall, however, the effects of No Child Left Behind (NCLB) are still being felt in the nation's schools.

A primary purpose of NCLB was to close the achievement gap between rich and poor students by holding schools accountable for students' learning. The law required states to test every student's progress toward meeting established standards. School districts were required to report students' results to demonstrate they were making progress toward meeting these standards. Schools that closed the education gap received additional federal dollars, but schools that did not show adequate progress lost funding and pupils: In some school districts, parents were able to move their children from lower- to higher-performing schools.

During the NCLB era, improvements occurred in fourth- and eighth-grade reading and math scores nationwide, particularly among African American and Hispanic students. However, critics believed that NCLB did not accurately define the main problem facing U.S. education: What the schools really needed was more money and incentives for teaching and learning, not more testing. As testing became the focal point in schools and pressure increased on teachers and schools to improve test scores, investigations in a number of states uncovered scandals in which educators had tampered with children's standardized tests in an effort to improve both scores and their own performance reviews. Although scandals such as these have been rare overall, they pointed to the kind of pressure that the federal No Child Left Behind law put on students, teachers, and school administrators as performance requirements moved higher annually and students and schools were penalized if they were unable to reach the level of educational attainment that was expected.

Efforts continue to implement educational reform for the future. One federal government initiative is known as "Race to the Top," which offers incentives to states that implement systemic reform to improve teaching and learning in schools. Among the goals of this reform are to raise standards and align policies and structures to the goals of college and career readiness. As of this writing, more than $4 billion has been given to nineteen states to create plans that address four key areas of K–12 educational reform: (1) development of rigorous standards and better assessments; (2) adoption of better data systems to provide schools, teachers, and parents with information about student progress; (3) support for teachers and school leaders to become more effective; and (4) increased emphasis and resources for the rigorous interventions needed to turn around the lowest-performing schools (Whitehouse.gov, 2015). Other issues addressed by this measure include attracting outstanding teachers, creating conditions in schools that support effective teaching and learning, and modernizing outdated schools. However, some critics point out that such initiatives reward a few schools but leave many without the necessary resources to improve. Other critics argue that many plans have been drawn up for education but that without the necessary financial backing at local, state, and federal levels, it is unlikely that many of the idealistic goals of plans such as Race to the Top will be fully implemented.

Other future areas for reform include improving STEM (science, technology, engineering, and math) education to help students become more competitive in the global marketplace. Encouraging innovation and ensuring opportunity for all are also priorities at the federal level. Having a top-tier educational system and providing high-quality job-training opportunities will provide people with the ability to be innovative and will offer greater opportunities that will help narrow the achievement gap. Improving the quality of underperforming schools and strengthening the teaching profession are other goals for the future.

Looking ahead to higher education, the United States has been outpaced internationally, and federal officials believe that it is important for our nation to regain its position as the first in the world in four-year-degree attainment among 25- to 34-year-olds. Today, the United States ranks twelfth, after being number one in 1990 (Whitehouse.gov, 2015). A goal for the future is increasing participation of students from all levels of family income in higher education. Although most students from wealthy families attend college, slightly over half of high school graduates from families in the bottom 25th percentile of families in income attend college, and the completion rate for those who do attend is about 25 percent (Whitehouse.gov, 2015). So one major concern is how to help middle- and lower-income families afford college, and the federal administration has made suggestions such as free college tuition for community college students. However, it remains to be seen if funding will follow goals such as this.

In the meanwhile, colleges and universities continue to expand their focus while, at the same time, undergoing strenuous budget cuts coupled with increasing demands to meet the needs of diverse student populations. The tightening of financial resources available to colleges and universities will lead to even more schools seeking alternative ways to fund their operations. Some will further raise tuition; others will seek different sources of funding. Some will move beyond the United States to find ways to expand their base of operation. For example, some U.S. universities are expanding their educational operations to emerging nations where demand is high for certain kinds of curricula, such as advanced business and petroleum engineering courses in Qatar and other Middle Eastern countries. Experts suggest that "university globalization" is here to stay, with both the export of students from countries such as India and China to other countries to study and the development of top-tier research universities in countries, including China, Singapore, and Saudi Arabia, where students may study without living abroad (● Figure 16.18).

FIGURE 16.18 The term "university globalization" refers to the export of students from countries such as India and China to other nations where they study and immerse themselves in another culture. How does this process affect higher education in the United States and other countries?

SOCIOLOGY & SOCIAL POLICY

Cultural Lag and Social Policy: Should We "Control" MOOCs?

If you are a student, can you imagine what it would be like to be in a class of 10,000 or more students? If you are a professor, can you imagine what it would be like to teach 10,000 or more students at the same time?

These are questions facing faculty and students at some universities where MOOCs, or *massive open online courses,* are either currently being offered or are in the planning stage for implementation in the near future. Indeed, MOOCs are trending now, particularly after professors at prestigious universities such as Stanford, Harvard, and MIT have drawn hundreds of thousands of students to their online courses in computer science and similar fields.

Originally, practically all learning in established colleges and universities took place in brick-and-mortar buildings with live professors and students engaged in the teaching–learning process. With the advent of television and cable TV channels, instructional television became a means by which students could take courses without physically attending them. The introduction of the Internet and the dawning of the digital age opened up higher education to many more people who had access to a computer and an Internet connection. In the future, some analysts believe that the process of moving teaching out of the classroom will go one step further with wider use of MOOCs, allowing for large-scale participation of students and open access via the Web. The interactive user forums available with MOOCs make it possible for students to engage not only with the material but also with other students and professors in a manner that does not require them to have face-to-face meetings in real time. Although MOOCs are similar to an older teaching method known as distance learning or distance education, today's MOOCs are unique in that many are taught by well-known professors in elite universities that are currently allowing open access to the courses.

From a social policy perspective, many questions remain about the role of MOOCs in the future of higher education. What should be the role of virtual teaching in higher education? Who should fund MOOCs? Should anyone regulate MOOCs? What state or federal agency will control how credit is granted to students who successfully complete MOOCs? How will students receive academic credit from universities where they are not officially enrolled, and can they use those credits toward graduation at another institution? From a practical standpoint, there are also questions about how well students learn in MOOCs. Can students concentrate on video lectures for extended periods of time? Are these courses set up for the interactive aspects of engaging with the materials and other students, or are they primarily digitized versions of the older distance-learning courses?

At least for now, it appears that MOOCs are moving forward. It has been reported that there are more than 3,800 institutions offering MOOCs worldwide and that this number grew 201 percent in 2014 alone. MOOCs have gained momentum because they take what has always been the noncommercial realm of teaching at public and nonprofit private institutions and make this realm accessible to twenty-first-century entrepreneurs. Venture capitalists and start-ups view MOOCs as cash cows that will help them build company value, and if, in the process, they help educate students, this too is a good thing.

Some analysts believe that MOOCs, or massive open online courses, will forever change the nature of higher education. Other analysts believe that the role of MOOCs in higher education will be much more limited. What effect do you believe MOOCs will have on more traditional brick-and-mortar colleges in the future?

Christian Science Monitor/Getty Images

REFLECT & ANALYZE

Should state higher education coordinating boards and other entities that regulate colleges and universities be designing and implementing policies pertaining to MOOCs, or should professors and/or institutions that wish to offer such course listings be able to do whatever they wish as long as the market supports their endeavors? What do you think?

Increasing numbers of U.S. students may go to school in these countries if their institutions offer opportunities, possibly at a lower cost, than do schools in the United States.

As discussed throughout this chapter, one of the major issues—present and future—is the part that education will play in reducing or maintaining and perpetuating social inequality. Education is an important social institution that must be sustained and enhanced because we know that simply having new initiatives and arguing over spending larger sums of money on education does not guarantee that the problems facing our schools will be resolved.

CHAPTER REVIEW Q & A

LO1 What is education, and how has the social institution of education changed throughout history?

Education is the social institution responsible for the systematic transmission of knowledge, skills, and cultural values within a formally organized structure. Perhaps the earliest formal education occurred in ancient Greece and Rome, where philosophers taught elite males the skills required to become thinkers and orators. By the mid-1850s, the process of mass education had begun in the United States as all states established free, tax-supported elementary schools that were readily available to children throughout the country. Today, schools attempt to meet the needs of society by teaching a wide diversity of students a myriad of topics.

LO2 What are the key assumptions of functionalist, conflict, symbolic interactionist, and postmodernist perspectives on education?

According to functionalists, education has both manifest functions (socialization, transmission of culture, multicultural education, social control, social placement, and change and innovation) and latent functions (keeping young people off the streets and out of the job market, matchmaking and producing social networks, and creating a generation gap). From a conflict perspective, education is used to perpetuate class, racial–ethnic, and gender inequalities through tracking, ability grouping, and a hidden curriculum that teaches subordinate groups conformity and obedience. Symbolic interactionists examine classroom dynamics and study ways in which practices such as labeling may become a self-fulfilling prophecy for some students. Some postmodernists suggest that in a consumer culture, education has become a commodity that is bought by students and their parents.

LO3 What are some major problems in elementary and secondary schools?

Most educational funds come from state legislative appropriations and local property taxes. In difficult economic times, this means that schools must do without the necessary funds to provide students with teachers, supplies, and the best educational environment for learning. High dropout rates, racial segregation and resegregation, and how to equalize educational opportunities for students are also among the many pressing issues facing U.S. public education today.

LO4 How do options such as school vouchers, charter schools, and homeschooling differ from traditional educational approaches?

In the voucher program, tax dollars are provided to parents so that they can pay their child's tuition at a private school of their choice. Charter schools are schools that operate under a charter contract negotiated by the school's organizers and a sponsor that oversees the provisions of the contract. Homeschooling has been chosen by some parents who hope to avoid the problems of public schools while providing a quality education for their children.

LO5 What are the educational opportunities and challenges found in community colleges in the twenty-first century?

One of the fastest-growing areas of U.S. higher education today is the community college. Community colleges educate about half of the nation's undergraduates and offer a variety of courses. One of the greatest challenges facing community colleges today is money. Across the nation, state and local governments struggling to balance their budgets have slashed funding for community colleges.

LO6 What are the economic problems facing many four-year colleges and universities?

The problem of increasing costs of higher education for students is compounded by state budget shortfalls, which have caused funding for public higher education to be slashed. Declining state and federal support has become a major concern for colleges and universities because as enrollments drop, along with financial support, these institutions will have to find new sources of revenue, sharply reduce expenses, and rework administrative costs.

LO7 What problems exist in higher education related to racial and ethnic diversity?

Among the most pressing problems are the underrepresentation of minorities as students, a lack of faculty diversity, and continuing controversy over affirmative action policies and laws. People of color (who are more likely than the average white student to be from lower-income families) are underrepresented in higher education. Underrepresentation is not the only problem faced by students of color: Problems of prejudice and discrimination continue on some college campuses. Despite the widely held assumption that there has been a significant increase in the number of minority professors, the latest figures indicate that this is not the case. Gender is also a factor in faculty diversity. In all ranks and racial and ethnic categories, men make up nearly 60 percent of the full-time faculty, while women account for roughly 40 percent.

LO8 What may the future hold with regards to education?

The future of education will depend in part on how successful that elected officials are in getting their agendas through state legislatures or the U.S. Congress. The tightening of financial resources available to colleges and universities will lead to even more schools seeking alternative ways to fund their operations. Another issue facing higher education in the future is how academic instruction will take place.

KEY TERMS

affirmative action 477
credentialism 464
cultural capital 462
cultural transmission 458

education 458
formal education 459
hidden curriculum 464
informal education 459

mass education 460
tracking 462

QUESTIONS for CRITICAL THINKING

1 What are the major functions of education for individuals and for societies?

2 Why do some theorists believe that education is a vehicle for decreasing social inequality whereas others believe that education reproduces existing class relationships?

3 Why is there so much controversy over what should be taught in grades K–12 and in colleges and universities?

4 How are the values and attitudes that you learned from your family reflected in your beliefs about education?

ANSWERS to the SOCIOLOGY QUIZ
ON U.S. EDUCATION

1	True	As far back as 1848, free public education was believed to be important in the United States because of the high rates of immigration and the demand for literacy so that the country would have an informed citizenry that could function in a democracy.
2	True	Despite the fact that equality of educational opportunity has not been achieved, it remains a goal of many people in this country. A large number subscribe to the belief that this country's educational system provides equal educational opportunities and that it is up to each individual to make the most of them.
3	True	In 2012, the latest year for which comprehensive data are available, men earned 53.7 percent of all doctoral degrees conferred. In the physical sciences and engineering, men earned 71.4 percent of all doctoral degrees. Women earned more than 50 percent of all doctoral degrees in education, social sciences, life sciences, and humanities (*Chronicle of Higher Education,* 2014).
4	False	Approximately 20 percent of students report the presence of gangs at their school.
5	True	Government data show that young people who bully are more likely to engage in other problematic behavior, such as smoking, drinking alcohol, and getting into fights.

6	**False**	An encouraging educational statistic is that the status dropout rate has decreased between 1990 and 2013 (from 12 to 7 percent), with the sharpest decline being for Hispanic students. The Hispanic status dropout rate declined from 32 to 13 percent during that time period.
7	**False**	Late in the nineteenth century and early in the twentieth century, some classes were conducted in Italian, Polish, German, and other languages of recent immigrants.
8	**True**	Most funding for public education comes from state and local property taxes, and similar sources of revenue.

Sources: *Chronicle of Higher Education,* 2014; and National Center for Educational Statistics, 2014.

RELIGION

Bill Pugliano/Getty Images

LEARNING OBJECTIVES

1 **Define** *religion* and identify its key components.

2 **Identify** the four main categories of religion based on their dominant belief.

3 **Discuss** the central beliefs of major Eastern religions such as Hinduism, Buddhism, and Confucianism.

4 **Describe** the key beliefs associated with major Western religions such as Judaism, Islam, and Christianity.

5 **Compare** and contrast four sociological perspectives on religion.

6 **Describe** the types of religious organizations.

7 **Identify** major trends in religion in the United States in the twenty-first century.

8 **Discuss** the future of religion as a social institution.

SOCIOLOGY & EVERYDAY LIFE

An Ongoing Debate

As someone who has spent her career in science, I have no delusion that the debate over teaching evolution in public schools will go away . . . [, but] I hope we can put at least one issue to rest. Intelligent design is not science. . . . Scientific theories must change in light of new data. . . . The lack of this characteristic in intelligent design is well-illustrated by a conversation I had with the tour guide at a creationist museum. . . . I posited to the fellow: "Look, you and I are both trying to learn the truth. But in science, theories must be modified to account for new data, and you are never going to change your theory. So creationism—by any name—is not science!" He calmly replied, "But I already know the truth and the truth doesn't change." . . . That statement explains precisely why creationism, intelligent

design, or whatever other label comes along 10 years from now, is not science. The theory refuses to change.

—ELIZABETH COOPER (2011), a college lecturer, stating why she believes that intelligent design, or creationism, should not be taught in public-school science courses

We are teaching our children a theory [evolution] that most of us don't believe in. I don't think God creates everything on a day-to-day basis, like the color of the sky. But I do believe he created Adam and Eve—instantly.

For many years people—including the parents shown here at a local PTA meeting—have argued about what should (or should not) be taught in U.S. public schools. One issue of key concern has been the teaching of creationism or intelligent design, as contrasted with evolution. What do you think these controversies are really about? Why do some debates about school curriculum eventually become lawsuits?

Enigma/Alamy

What is all the controversy about? Why do lawsuits about teaching intelligent design in schools arise so often? The argument over intelligent design—an assertion that the universe is so complex that an intelligent, supernatural power must have created it—as an alternative to the theory of evolution is the latest debate in a lengthy battle over the teaching of creationism versus evolutionism in public schools, and it is only one of many arguments that will continue to take place regarding the appropriate relationship between public education and religion in the United States. Nearly a century ago, evolutionism versus creationism was hotly debated in the famous "Scopes monkey trial," so named because of Charles Darwin's assertion that human beings had evolved from lower primates. In this case, John Thomas Scopes, a substitute high school biology teacher in Tennessee, was found guilty of teaching evolution, which denied the "divine creation of man as taught in the Bible." Although an appeals court later overturned Scopes's conviction and fine (on the grounds that the fine was excessive), teaching evolution in Tennessee's public schools remained illegal until 1967. By contrast, more-recent U.S. Supreme Court rulings have looked unfavorably on the teaching of creationism in public schools, based on a provision in the Constitution that

requires a "wall of separation" between church (religion) and state (government). Initially, this wall of separation was erected to protect religion from the state, not the state from religion. In the latest national case at the time of this writing, the U.S. Supreme Court in 2014 rejected the appeal of an Ohio public-school science teacher who was fired for promoting the theory of creationism and refusing to remove religious materials from his classroom. The justice allowed an Ohio Supreme Court ruling to stand that found the school district had grounds to fire the teacher and that his free speech rights had not been violated. No doubt other cases will arise in the future regarding teaching of creationism, intelligent design, and evolution in the public schools of the United States.

As these examples show, religion can be a highly controversial topic. One group's deeply held beliefs or cherished religious practices may be a source of irritation to another. Today, religion is a source of both stability and conflict not only in the United States but also throughout the world. In this chapter we examine how religion influences life in the United States and in other areas of the world. Before reading on, test your knowledge about how religion affects public education in this country by taking the "Sociology and Everyday Life" quiz.

—S<small>TEVE</small> F<small>ARRELL</small>, a Dover, Pennsylvania, resident explaining why he approved of the Dover school board's decision to require biology teachers to teach intelligent design, before

a U.S. District Court ruled that such teaching violates the Establishment Clause of the First Amendment to the U.S. Constitution (qtd. in Powell, 2004: A1)

How Much Do You Know About the Effect of Religion on U.S. Education?

TRUE	FALSE		
T	F	1	The U.S. Constitution originally specified that religion should be taught in the public schools.
T	F	2	Virtually all sociologists have advocated the separation of moral teaching from academic subject matter.
T	F	3	Private, religiously based schools have decreased in enrollment as interest in religion has waned in the United States.
T	F	4	The number of children from religious backgrounds other than Christianity has grown steadily in public schools over the past four decades.
T	F	5	Debates over evolution versus creationism are primarily in elementary education because of the vulnerability of young children.
T	F	6	School prayer is no longer a political issue in the United States.
T	F	7	Buddhism is one of the fastest-growing Eastern religions in the United States.
T	F	8	Early social theorists agreed that religion restricted social change in societies.

Answers can be found at the end of the chapter.

 Define *religion* **and identify its key components.**

The Sociological Study of Religion

What is religion? ***Religion*** is a social institution composed of a unified system of beliefs, symbols, and rituals—based on some sacred or supernatural realm—that guides human behavior, gives meaning to life, and unites believers into a community. Based on this definition, religion is a stable institution that exists independently from individuals who attend religious services or officials (such as priests, pastors, or other clergy) in the administrative hierarchy. Religion is sometimes thought of as a platform for the expression of ***spirituality***—the relationship between the individual and something larger than oneself, such as a broader sense of connection with the surrounding world. As such, spirituality involves the individual's *inner,* subjective feelings and experiences rather than the act of giving devotion to *external* beliefs, rituals, and deities that are set forth in established creeds or religious communities.

In the final analysis, both religion and spirituality require that persons engage in a leap of ***faith***—a confident belief that cannot be proven or disproven but is accepted

as true. Religious beliefs require faith because religion provides answers for seemingly unanswerable questions that underlie human existence. According to the sociologist Peter Berger (1967), these questions are *Who am I? Why am I here? How should I live? What happens when I die?* Berger suggests that religion provides a system of meaning that connects people to society and provides them with a sense of purpose that transcends the ordinary realm of life (• Figure 17.1). Consequently, religious beliefs bind people together and establish rites of passage through various stages of life, such as birth, marriage, and death. People with similar religious beliefs and practices gather together

religion
a social institution composed of a unified system of beliefs, symbols, and rituals—based on some sacred or supernatural realm—that guides human behavior, gives meaning to life, and unites believers into a community.

spirituality
the relationship between the individual and something larger than oneself, such as a broader sense of connection with the surrounding world.

faith
a confident belief that cannot be proven or disproven but is accepted as true.

FIGURE 17.1 Hanukkah, a major holiday in Judaism, provides worshippers with the opportunity to come together and worship their Creator and celebrate their community.

in a moral community (such as a church, mosque, temple, or synagogue), where they engage in religious beliefs and practices with similarly minded people.

Given the diversity and complexity of contemporary religion, how is it possible for sociologists to study this social institution? Most sociologists studying religion are committed to the pursuit of "disinterested scholarship," meaning that they do not seek to make value judgments about religious beliefs or to determine whether particular religious bodies are "right" or "wrong." However, many acknowledge that it is impossible to completely rid themselves of those values and beliefs into which they were socialized.

Religion and the Meaning of Life

Because religion seeks to answer important questions such as why we exist and why people suffer and die, Peter Berger (1967) referred to religion as a *sacred canopy*—a sheltering fabric hanging over individuals that provides them with security and answers for the most difficult questions of life. Whereas science and medicine typically rely on existing scientific evidence to respond to questions of suffering, death, and injustice, religion seeks to answer these questions by referring to the sacred. According to Emile Durkheim (1995/1912), *sacred* refers to those aspects of life that exist beyond the everyday, natural world that we cannot experience with our senses—in other words, those things that are set apart as "holy" (see ● Figure 17.2). People feel a sense of awe, reverence, deep respect, or fear for that

which is considered sacred. Across cultures and in different eras, many things have been considered sacred, including invisible gods, spirits, specific animals or trees, altars, crosses, holy books, and special words or songs that only the initiated could speak or sing.

Those things that people do not set apart as sacred are referred to as *profane*—the everyday, secular, or "worldly" aspects of life that we know through our senses. Thus, whereas sacred beliefs are rooted in the holy or supernatural, secular beliefs have their foundation in scientific knowledge or everyday explanations. In the debate between creationists and evolutionists, for example, advocates of creationism view their beliefs as founded in sacred (Biblical) teachings, whereas advocates of evolutionism assert that their beliefs are based on provable scientific facts.

In addition to beliefs, religion also comprises symbols and rituals. According to the now-classical works of anthropologist Clifford Geertz (1966), religion is a set of cultural symbols that establishes powerful and pervasive moods and motivations to help people interpret the meaning of life and establish a direction for their behavior. People often act out their religious beliefs in the form of *rituals*—regularly repeated and carefully prescribed forms of behaviors that symbolize a cherished value or belief. Rituals range from songs and prayers to offerings and sacrifices that worship or praise a supernatural being, an ideal, or a set of supernatural principles. For example, Muslims bow toward Mecca, the holy city of Islam, five times a day at fixed times to pray to God, whereas Christians participate in the celebration of communion (or the "Lord's Supper") to commemorate the life, death, and resurrection of Jesus. Rituals differ from everyday actions in that they involve very strictly determined behavior. The rituals involved in praying or in observing communion are carefully orchestrated and must be followed with precision. According to the sociologist Randall Collins (1982: 34), "In rituals, it is the forms that count. Saying prayers, singing a hymn, performing a primitive sacrifice or a dance, marching in a procession, kneeling before an idol or making the sign of the cross—in these, the action must be done the right way."

Not all sociologists believe that the "sacred canopy" metaphor suggested by Berger accurately describes contemporary religion. Some analysts believe that a more accurate metaphor for religion in the global village is that of the *religious marketplace*, in which religious institutions and traditions compete for adherents, and worshippers

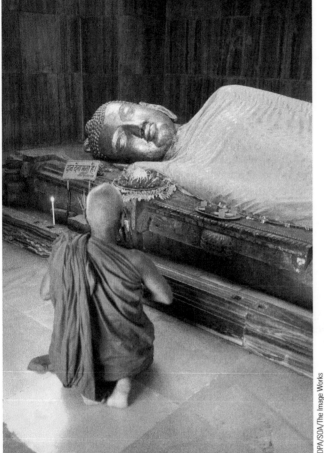

iStockphoto.com/Joel Carillet

Renaud Visage/Digital Vision/Getty Images

Richard A. Cooke/Documentary/Corbis

DPA/SOA/The Image Works

Ammar Awad /Reuters/Landov

FIGURE 17.2 One of the rituals in most religions is the identification of a holy place. For Jews (clockwise from top left), one such place is the Western Wall in Jerusalem. For Muslims, an example of a holy place is the Blue Mosque in Turkey. Historical Native Americans believed in the ritual of creating burial mounds. For Christians, Bethlehem is a historic and spiritual pilgrimage. And for Buddhists, the Mahaparinirvana Stupa, a shrine in India with a reclining, sleeping golden Buddha, is a holy place.

shop for a religion in much the same way that consumers decide what goods and services they will purchase in the marketplace. However, other analysts do not believe that moral and ethical beliefs are bought and sold like groceries, shoes, or other commodities, and they note that many of the world's religions have persisted from earlier eras to advanced technological societies. But this poses another question: When did the earliest religions begin?

sacred
those aspects of life that exist beyond the everyday, natural world that we cannot experience with our senses.

profane
the everyday, secular, or "worldly" aspects of life that we know through our senses.

rituals
regularly repeated and carefully prescribed forms of behaviors that symbolize a cherished value or belief.

Categories of Religion

Although it is difficult to establish exactly when religious rituals first began, anthropologists have concluded that all known groups over the past hundred thousand years have had some form of religion (Haviland et al., 2014). The original locations of the world's major religions are shown in • Figure 17.3.

Religions have been classified into four main categories based on their dominant belief: simple supernaturalism, animism, theism, and transcendent idealism. In very simple preindustrial societies, religion often takes the form of *simple supernaturalism*—the belief that supernatural forces affect people's lives either positively or negatively. This type of religion does not acknowledge specific gods or supernatural spirits but focuses instead on impersonal forces that may exist in people or natural objects. For example, simple supernaturalism has been used to explain mystifying events of nature, such as sunrises and thunderstorms, and ways that some objects may bring a person good or bad luck. By contrast, *animism* is the belief that plants, animals, or other elements of the natural world are endowed with spirits or life forces that have an impact on events in society. Although these spirits or life forces play a part in everyday life, they are not worshipped by humans.

Animism is associated with early hunting and gathering societies and with many Native American societies, in which everyday life is not separated from the elements of the natural world (Albanese, 2013).

The third category of religion is *theism*—a belief in a god or gods who shape human affairs. Horticultural societies were among the first to practice *monotheism*—a belief in a single, supreme being or god who is responsible for significant events such as the creation of the world. Three of the major world religions—Christianity, Judaism, and Islam—are monotheistic. By contrast, Shinto and a number of the indigenous religions of Africa are forms of *polytheism*—a belief in more than one god. The fourth category of religion, *transcendent idealism*, is *nontheistic* because it does not focus on worship of a god or gods. *Transcendent idealism* is a belief in sacred principles of thought and conduct. Principles such as truth, justice, affirmation of life, and tolerance for others are central tenets of transcendent idealists, who seek an elevated state of consciousness in which they can fulfill their true potential.

World Religions

Although there are many localized religions throughout the world, those religions classified as *world religions* cover vast expanses of the Earth and have millions of followers. These six religions are compared in • Table 17.1.

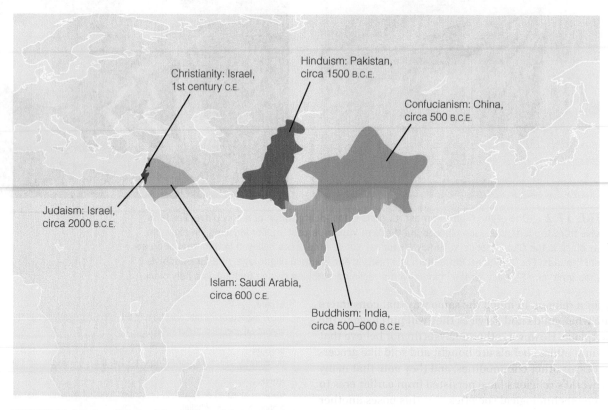

Christianity: Israel, 1st century C.E.

Hinduism: Pakistan, circa 1500 B.C.E.

Confucianism: China, circa 500 B.C.E.

Judaism: Israel, circa 2000 B.C.E.

Islam: Saudi Arabia, circa 600 C.E.

Buddhism: India, circa 500–600 B.C.E.

FIGURE 17.3   Original Locations of the World's Major Religions

TABLE 17.1

Major World Religions

		Current Followers	Founder/Date	Beliefs
✝	**Christianity**	2.2 billion	Jesus/first century C.E.	Jesus is the Son of God. Through good moral and religious behavior (and/or God's grace), people achieve eternal life with God.
☪	**Islam**	1.6 billion	Muhammad/ca. 600 C.E.	Muhammad received the Qur'an (scriptures) from God. On Judgment Day, believers who have submitted to God's will, as revealed in the Qur'an, will go to an eternal Garden of Eden.
ॐ	**Hinduism**	1 billion	No specific founder/ca. 1500 B.C.E.	Brahma (creator), Vishnu (preserver), and Shiva (destroyer) are divine. Union with ultimate reality and escape from eternal reincarnation are achieved through yoga, adherence to scripture, and devotion.
☸	**Buddhism**	500 million	Siddhartha Gautama/500 to 600 B.C.E.	Through meditation and adherence to the Eightfold Path (correct thought and behavior), people can free themselves from desire and suffering, escape the cycle of eternal rebirth, and achieve nirvana (enlightenment).
✡	**Judaism**	14 million	Abraham, Isaac, and Jacob/ca. 2000 B.C.E.	God's nature and will are revealed in the Torah (Hebrew scripture) and in His intervention in history. God has established a covenant with the people of Israel, who are called to a life of holiness, justice, mercy, and fidelity to God's law.
☯	**Confucianism**	6.3 million	K'ung Fu-Tzu (Confucius)/ca. 500 B.C.E.	The sayings of Confucius (collected in the *Analects*) stress the role of virtue and order in the relationships among individuals, their families, and society.

Discuss the central beliefs of major Eastern religions such as Hinduism, Buddhism, and Confucianism.

Hinduism

We begin with Hinduism because it is believed to be one of the world's oldest current religions, having originated along the banks of the Indus River in Pakistan between 3,500 and 4,500 years ago. Hinduism began before written records were kept, so modern scholars have only limited information about its earliest leaders and their teachings. Hindu beliefs and practices have been preserved through an oral tradition and expressed in texts and hymns known as the *Vedas* (meaning "knowledge" or "wisdom"); however, this religion does not have a "sacred" book, such as the Judeo-Christian Bible or the Islamic Qur'an. Consequently, Hindu beliefs and practices emerged over the centuries across the subcontinent of India in a variety of forms, reflecting the influence of the various regional cultures. Because Hinduism has no scriptures that are thought to be inspired by a god or gods and is not based on the teachings of any one person, religion scholars refer to it as an *ethical religion*—a system of beliefs that calls upon adherents to follow an ideal way of life. For most Hindus, this is partly achieved by adhering to the expectations of the caste system (see Chapter 8).

Central to Hindu teachings is the belief that individual spirits or living beings (*jivas*) enter the world and roam the universe until they break free into the limitless atmosphere of illumination (*moska*) by discovering their own *dharma*—duties or responsibilities. According to Hinduism, individual *jivas* pass through a sequence of bodies over time as they undergo a process known as reincarnation (*samsara*)—an endless passage through cycles of life, death, and rebirth until the soul earns liberation. The soul's

simple supernaturalism
the belief that supernatural forces affect people's lives either positively or negatively.

animism
the belief that plants, animals, or other elements of the natural world are endowed with spirits or life forces that have an impact on events in society.

theism
a belief in a god or gods who shape human affairs.

monotheism
a belief in a single, supreme being or god who is responsible for significant events such as the creation of the world.

polytheism
a belief in more than one god.

transcendent idealism
a belief in sacred principles of thought and conduct.

acquisition of each new body is tied to the law of *karma* (deed or act), which is a doctrine of the moral law of cause and effect. The present condition of the soul—how happy or unhappy it is, for example—is directly related to what it has done in the past, and its present thoughts and decisions are the ultimate determinants of what its future will be. The final goal of Hindu existence is entering the state of *nirvana*—becoming liberated from the world by uniting the individual soul with the universal soul (*Brahma*).

Hinduism has been devoid of some of the social conflict experienced by other religions. Because Hinduism is based on the assumption that there are many paths to the "truth" and that the world's religions are alternate paths to that goal, Hindus typically have not engaged in religious debates or "holy wars" with those holding differing beliefs. One of the best-known Hindu leaders of modern times was Mohandas ("Mahatma") Gandhi, the champion of India's independence movement, who was devoted to the Hindu ideals of nonviolence, honesty, and courage (● Figure 17.4). However, some social analysts note that Hinduism has been closely associated with the perpetuation of the caste system in India. Although people in the lower castes are taught to live out their lives with dignity even in the face of poverty and despair, they may also come to believe that their lowly position is the acceptable and appropriate place for them to be—which allows the upper castes to exploit them.

The Hindu religion is almost as diverse as the wide array of people who adhere to its teachings. About one billion people are identified as Hindus worldwide, representing 15 percent of the global population. Most Hindus live in countries where they constitute a majority, including Nepal, India, and Mauritius (Pew Forum on Religion and Public Life, 2012a).

Since the 1960s, the number of Hindus in North America has increased significantly, particularly as Asian Indian immigration grew rapidly in the late twentieth and early twenty-first centuries. Today, approximately 2.3 million Hindus reside in North America. In the United States, nearly half of all Asian Indian immigrants reside in California, New Jersey, New York, and Texas—also the states with the largest number of Hindu temples. Most Asian Indian immigrants to the United States have been well-educated professionals who have joined the ranks of the U.S. middle and upper-middle classes and have been active in supporting Hindu temples in the nation. For most people of Asian Indian descent in the United States, these temples are sites of worship and gathering places where they can maintain a sense of community. They are also ritual centers where language, arts, and practices from their ethnic past can be preserved. In the future the influence of Hindus in North America and particularly the United States may become more profound as their number continues to increase and their influence in STEM fields (science, technology, engineering, and mathematics) reaches an all-time high.

Buddhism

In the second decade of the twenty-first century, research suggests that there are nearly 500 million Buddhists worldwide who reside primarily in Asia and the Pacific, especially China, Thailand, and Japan. However, when Buddhism first emerged in India some twenty-five hundred years ago, it was thought of as a "new religious movement," arising as it did around the sixth century B.C.E., after many earlier religions had become virtually defunct.

Buddhism's founder, Siddhartha Gautama of the Sakyas (also known as Gautama Buddha), was born about 563 B.C.E. into the privileged caste. His father was King Suddhodana (who was more like a feudal lord than a king because many kingdoms existed on the Indian subcontinent during that era). According to historians, the king attempted to keep his young son in the palace at all times so that he would neither see how poor people lived nor experience the suffering present in the outside world. As a result, Siddhartha was oblivious to social inequality until he began to make forays beyond the palace walls and into the "real" world, where he observed how other people lived. On one excursion, he saw a monk with a shaven head and became aware that some people withdraw from the secular world and live a life of strict asceticism. Later, Siddhartha engaged in intense meditation underneath a bodhi tree in what is now Nepal, eventually declaring that he had obtained Enlightenment—an awakening to the true nature of reality. From that day forward, Siddhartha was referred to as *Buddha*, meaning "the Enlightened One" or the "Awakened One," and spent his life teaching others how to reach nirvana.

FIGURE 17.4 One of the best-known Hindu leaders of modern times was Mohandas ("Mahatma") Gandhi, the champion of India's independence movement.

SZ Photo/Scherl/The Image Works

Because of the efforts of a series of invaders, Buddhism had ceased to exist in India by the thirteenth century but had already expanded into other nations in various forms. *Theravadin Buddhism,* which focuses on the life of the Buddha and seeks to follow his teachings, gained its strongest foothold in Southeast Asia. *Mahayana Buddhism* is centered in Japan, China, and Korea, and primarily focuses on meditation and the Four Noble Truths:

1. Life is *dukkha*—physical and mental suffering, pain, or anguish that pervades all human existence.

2. The cause of life's suffering is rooted in *tanha*—grasping, craving, and coveting.

3. One can overcome *tanha* and be released into Ultimate Freedom in Perfect Existence (nirvana).

4. Overcoming desire can be accomplished through the Eightfold Path to Nirvana. This path is a way of living that avoids extremes of indulgence and suggests that a person can live in the world but not be worldly. The path's eight steps are *right view* (proper belief), *right intent* (renouncing attachment to the world), *right speech* (not lying, slandering, or using abusive talk), *right action* (avoiding sexual indulgence), *right livelihood* (avoiding occupations that do not enhance spiritual advancement), *right effort* (preventing potential evil from arising), *right mindfulness,* and *right concentration* (overcoming sensuous appetites and evil desires).

The third major branch of Buddhism—*Vajrayana*—incorporates the first two branches along with some aspects of Hinduism; it emerged in Tibet in the seventh century (Albanese, 2013). Like Hinduism, the teachings of this type of Buddhism—and specifically those of the Dalai Lama, the Tibetan Buddhist leader—emphasize the doctrine of *ahimsa,* or nonharmfulness, and discourage violence and warfare. In the aftermath of the 2011 U.S. raid that killed al-Qaida leader Osama bin Laden, who was responsible for the September 11, 2001, terrorist attack on the United States, the Dalai Lama reaffirmed his lifelong commitment to nonviolence and emphasized that dialogue is the only way to resolve differences because violence can get out of control easily and cause resentments that breed additional discord (Karnowski, 2011).

When did Buddhism first arrive in the United States? According to most scholars, some branches appeared as early as the 1840s, when Chinese immigrants arrived on the West Coast. Shortly thereafter, temples were erected in San Francisco's Chinatown; however, ethnic Buddhism in the United States expanded after the Civil War when Japanese immigrants arrived first in Hawaii (then a U.S. possession) and a decade later in California. The branch of Pure Land Buddhism brought by the Japanese probably had the greatest chance of succeeding in the United States because it most closely resembled Christianity (Albanese, 2013). Buddhism went through the process of Americanization, which is reflected in the contemporary use of terms such as *church, bishop,* and *Sunday school*—terms previously unknown in this religion.

FIGURE 17.5 Confucianism is based on the ethical teachings formulated by Confucius, shown here in a portrait created by a Manchu prince in 1735.

Confucianism

Some scholars view Confucianism as a philosophy; others view it as a religion. Confucianism—which means the "family of scholars"—started as a school of thought or a tradition of learning before its eventual leader, Confucius, was born (Wei-ming, 1995). Confucius (the Latinized form of K'ung Fu-tzu) lived in China between 551 and 479 B.C.E. and emerged as a teacher at about the same time that the Buddha became a significant figure in India (● Figure 17.5).

Confucius—whose sayings are collected in the *Analects*—taught that people must learn the importance of *order* in human relationships and must follow a strict code of moral conduct, including respect for others, benevolence, and reciprocity (Kurtz, 1995). A central teaching of Confucius was that humans are by nature good and that they learn best by having an example or a role model. As a result, he created a *junzi* ("chun-tzu"), or model person, who has such attributes as being upright regardless of outward circumstances, being magnanimous by expressing forgiveness toward others, being directed by internal principles rather than external laws, being sincere in speech and action, and being earnest and benevolent. Confucius wanted to demonstrate these traits, and he believed that he should be a role model for his students. The *junzi's* behavior is to be based on the Confucian principle of *Li,* meaning righteousness or propriety, which refers both to ritual and to correct conduct in public. One of the central attributes exhibited by the *junzi* is *ren* (*jen*), which means having deep empathy or compassion for other humans.

Confucius established the foundation for social hierarchy—and potential conflict—when he set forth his Five Constant Relationships: *ruler–subject, husband–wife, elder brother–younger brother, elder friend–junior friend,* and *father–son*. In each of these pairs, one person is unequal to the other, but each is expected to carry out specific responsibilities to the other. Confucius taught that due authority is not automatic; it must be earned. The subject does not owe loyalty to the ruler or authority figure if that individual does not fulfill his or her end of the bargain.

Confucianism is based on the belief that Heaven and Earth are not separate places but rather a continuum in which both realms are constantly in touch with each other. According to this approach, those who inhabit Heaven are ruled over by a supreme ancestor and are the ancestors of those persons who are on Earth. Those who are currently on Earth eventually join these forefathers; therefore, death is nothing more than the promotion to a more honorable estate.

How prominent is Confucianism? Some analysts place the number at about 6.3 million worldwide. However, it is difficult to provide an accurate answer to this question because, as previously mentioned, some people view Confucianism as a set of ethical teachings rather than as a religion. However, many immigrants from China and Southeast Asia adhere to the teachings of Confucius, although perhaps mixed with those of other great Eastern religious philosophers and teachers. Neo-Confucianism, which emerged on the West Coast, is heavily influenced by Buddhism and Taoism.

We now move from the Eastern religions (Hinduism, Buddhism, and Confucianism), which tend to be based on ethics or values more than on a deity or Supreme Being, to the Western religions (Judaism, Christianity, and Islam), which are founded on the Abrahamic tradition and place an emphasis on God and a relationship between human beings and a Supreme Being. • Figure 17.6 analyzes world religions in terms of estimated percentage of adherents.

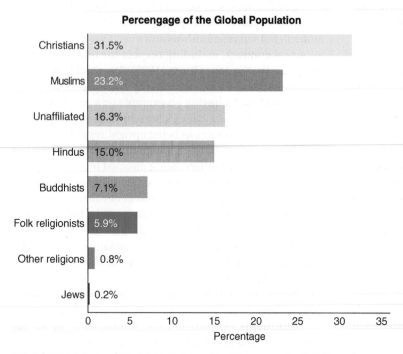

Percengage of the Global Population

Religion	Percentage
Christians	31.5%
Muslims	23.2%
Unaffiliated	16.3%
Hindus	15.0%
Buddhists	7.1%
Folk religionists	5.9%
Other religions	0.8%
Jews	0.2%

FIGURE 17.6   World Religions by Percentage of Adherents
Source: Pew Research Center on Religion and the Public Life, 2014.

Describe the key beliefs associated with major Western religions such as Judaism, Islam, and Christianity.

Judaism

Although Judaism has fewer adherents worldwide than some other major religions, its influence is deeply felt in Western culture. A worldwide estimate of the Jewish population stands at about 14 million people, or 0.2 percent of the entire population. The majority of Jewish adherents live in Israel, the United States, France, and Russia. The latest available estimates indicate that about 6 million adherents to Judaism live in North America (Pew Forum on Religion and Public Life, 2012a).

Central to contemporary Jewish belief is monotheism, the idea of a single god, called Yahweh, the God of Abraham, Isaac, and Jacob. The Hebrew tradition emerged out of the relationship of Abraham and Sarah, a husband and wife, with Yahweh. According to Jewish tradition, the God of the Jews made a covenant with Abraham and Sarah—His chosen people—that He would protect and provide for them if they swore Him love and obedience. When God appeared to Abraham in about the eighteenth century B.C.E., He encouraged Abraham to emigrate to the area near the Sea of Galilee and the Dead Sea (what is now Israel), leaving behind the ancient fertile crescent of the Middle East (present-day Iraq).

The descendants of Abraham and Sarah migrated to Egypt, where they became slaves of the Egyptians. In a vision, God's chosen leader, Moses, was instructed to liberate His chosen people from the bondage and slavery imposed upon them by the pharaoh. After experiencing a series of ten plagues, the pharaoh decided to free the slaves. The tenth and final plague had involved killing all of the firstborn in the land of Egypt—human beings and lower animals, as well—except the firstborn children of the Hebrews, who had put the blood of a lamb on the doorposts of their houses so that they would be passed over. This practice inspired the Jewish holiday Passover, which commemorates God's deliverance of the Hebrews from slavery in Egypt during the time of Moses. It is believed that the first Passover took place in Egypt in about 1300–1200 B.C.E.

Wandering in the desert after their release, the Hebrews established a covenant with God, Who promised that if they would serve Him exclusively, He would give the Hebrews a promised land and make them a great nation. Known as the Ten Commandments (or *Decalogue*), the covenant between God and human beings was given to Moses on top of Mount Sinai. The Ten Commandments and discussions of moral, ceremonial, and cultural laws are contained in the Torah, the Hebrew word for "law." Torah has several meanings in contemporary life. One meaning is the first five books of the Bible: Genesis, Exodus, Leviticus, Numbers, and Deuteronomy (also known as the Pentateuch). Another meaning is the entire Jewish Bible or the whole body of Jewish law and teachings.

The Jewish people believe that they have a unique relationship with God, affirmed on the one hand by His covenant and on the other by His law. Judaism has three key components: God (the deity), Torah (God's teachings), and Israel (the community or holy nation). Although God guides human destiny, people are responsible for making their own ethical choices in keeping with His law; when they fail to act according to the law, they have committed a sin. Also fundamental to Judaism is the belief that one day the Messiah will come to Earth, ushering in an age of peace and justice for all.

Today, Jews worship in synagogues in congregations led by a *rabbi*—a teacher or ordained interpreter and leader of Judaism. The Sabbath is observed from sunset Friday to sunset Saturday, based on the story of Creation in Genesis, especially the belief that God rested on the seventh day after He had created the world. Worship services consist of readings from scripture, prayer, and singing. Jews celebrate a set of holidays distinct from U.S. dominant cultural religious celebrations. The most important holidays in the Jewish calendar are Rosh Hashanah (New Year), Yom Kippur (Day of Atonement), Hanukkah (Festival of Lights), and Pesach (Passover).

Throughout their history, Jews have been the object of prejudice and discrimination. The Holocaust, which took place in Nazi Germany (and several other nations that the Germans occupied) between 1933 and 1945, remains one of the saddest eras in history. After the rise of Hitler in Germany in 1933 and the Nazi invasion of Poland, Jews were singled out with special registrations, passports, and clothing. Many of their families were separated by force, and some family members were sent to slave labor camps while others were sent to "resettlement." Eventually, many Jews were imprisoned in death camps, where six million were killed. Anti-Semitism has been a continuing problem in the United States. Like other forms of prejudice and discrimination, anti-Semitism has extracted a heavy toll on multiple generations of Jewish Americans.

Today, Judaism has three main branches—Orthodox, Reform, and Conservative. Orthodox Judaism follows traditional practices and teachings, including eating only kosher foods prepared in a designated way, observing the traditional Sabbath, segregating women and men in religious services, and wearing traditional clothing. Reform Judaism, which began in Germany in the nineteenth century, is based on the belief that the Torah is binding only in its moral teachings and that adherents should no longer be required to follow all of the Talmud, the compilation of Jewish law setting forth the strict rabbinic teachings on practices such as food preparation, rituals, and dress. In some Reform congregations, gender-segregated seating is no longer required. In the United States, services are conducted almost entirely in English, a Sunday Sabbath is observed, and less emphasis is placed on traditional Jewish holidays (Albanese, 2013).

Conservative Judaism emerged between 1880 and 1914 with the arrival of many Jewish immigrants in the United States from countries such as Russia, Poland, Romania, and Austria. Seeking freedom and an escape from persecution, these new arrivals settled in major cities such as New York and Chicago, where they primarily became factory workers, artisans, and small shopkeepers. Conservative Judaism, which became a middle ground between Orthodox and Reform Judaism, teaches that the Torah and Talmud must be followed and that *Zionism*—the movement to establish and maintain a Jewish homeland in Israel—is crucial to the future of Judaism. In Conservative synagogues, worship services are typically performed in Hebrew. Men are expected to wear head coverings, and women have roles of leadership in the congregation; some may become ordained rabbis. Despite centuries of religious hatred and discrimination, Judaism persists as one of the world's most influential religions.

Islam (Muslims)

Like Judaism, Islam is a religion in the Abrahamic tradition; both religions arose through sons of Abraham—Judaism through Isaac and Islam through Ishmael. Islam, whose followers are known as Muslims, is based on the teachings of its founder, Muhammad, who was born in Mecca (now in Saudi Arabia) in about 570 C.E. According to Muhammad, followers must adhere to the five Pillars of Islam: (1) believing that there is no god but Allah, (2) participating in five periods of prayer each day, (3) paying taxes to help support the needy, (4) fasting during the daylight hours in the month of Ramadan, and (5) making at least one pilgrimage to the Sacred House of Allah in Mecca.

The Islamic faith is based on the Qur'an—the holy book of the Muslims—as revealed to the Prophet Muhammad through the Angel Gabriel at the command of God. According to the Qur'an, it is up to God, not humans, to determine which individuals are deserving of punishment and what kinds of violence are justified under various conditions.

The Islamic notion of *jihad*—meaning "struggle"—is a core belief. The Greater Jihad is believed to be the internal struggle against sin within a person's heart, whereas the Lesser Jihad is the external struggle that takes place in the world, including violence and war. The term *jihad* is typically associated with religious fundamentalism. Despite the fact that fundamentalism is found in most of the world's religions, some social analysts believe that Islamic fundamentalism is uniquely linked to the armed struggles of groups such as Hamas, an alleged terrorist organization, and the militant Islamic Jihad, which is believed to engage in continual conflict. In the second decade of the twenty-first century, when violence anywhere in the world is allegedly perpetrated by individuals or groups thought to be identified with Muslim or Islamic fundamentalist beliefs and values, an immediate outcry goes up to bring the perpetrators to justice. An effort is made to identify specifically how their conduct is related to Islamic fundamentalism and possibly other "sleeper cells" that might bring even further harm to people and physical structures. Although some terrorist attacks are perpetrated by angry religious fundamentalists, many others may have roots in terrorism that is not related to Islamic or other forms of religious fundamentalism.

Today, about 23.2 percent of the world's population considers itself to be Muslim. The majority (63 percent) of the 1.6 billion adherents worldwide of this religion reside in the Asia–Pacific region. However, the Middle East–North African region has an overwhelmingly Muslim population (93 percent). Other large populations of Muslims are located in Indonesia, India, Pakistan, and Bangladesh.

Islam is one of the fastest-growing religions in North America. Driven by recent waves of migration and a relatively high rate of conversion, there has been a significant increase in the number of Muslims, making the estimated population in North America about 3.5 million (Pew Forum on Religion and Public Life, 2012a). In the United States the states with the largest Muslim populations include California, New York, Illinois, New Jersey, Indiana, Michigan, Virginia, Texas, Ohio, and Maryland. Muslims in the United States have experienced prejudice and discrimination because of fears that they are terrorists. The 1993 bombing of New York's World Trade Center, the 2001 terrorist attacks on the United States, and other successful or attempted terrorist attacks around the world have intensified distrust regarding people who appear to be Muslim or from countries associated with Islam (● Figure 17.7). Overall, however, many Muslims have fared well in the United States because of their level of education, professional status, and higher-than-average household income levels.

Christianity

Today, about 2.2 billion people (or about one in three persons worldwide) are identified as Christians (Pew Forum on Religion and Public Life, 2012a). Along with Judaism and Islam, Christianity follows the Abrahamic tradition, tracing its roots to Abraham and Sarah. Although Jews and Christians share common scriptures in the portions of the Bible known to Christians as the "Old Testament," they interpret them differently. The Christian teachings in the "New Testament" present a worldview in which the old covenant between God and humans, as found in the Old Testament, is obsolete in light of God's offer of a new covenant to the followers of Jesus, whom Christians believe to be God's only son.

As described in the New Testament, Jesus was born to the Virgin Mary and her husband, Joseph. After a period of youth in which He prepared himself for the ministry, Jesus appeared in public and went about teaching and preaching, including performing a series of miracles—events believed to be brought about by divine intervention—such as raising people from the dead. The central themes in the teachings of Jesus are the kingdom of God and standards of personal conduct for adherents of Christianity. Jesus emphasized the importance of righteousness before God and of praying to the Supreme Being for guidance in the daily affairs of life.

One of the central teachings of Christianity is linked to the unique circumstances surrounding the death of Jesus. Just prior to His death, Jesus and His disciples held a special supper, now referred to as "the last supper," which is commemorated in contemporary Christianity in the sacrament of Holy Communion. Afterward, Jesus was arrested by a group sent by the priests and scribes for claiming to be king of the Jews. After being condemned to death by political leaders, Jesus was executed by crucifixion, which made the cross a central symbol of the Christian religion. According to the New Testament, Jesus died, was placed in a tomb, and on the third day was resurrected—restored to life—establishing that He is the son of God. Jesus then remained on Earth for forty days, after which He ascended into heaven on a cloud. Two thousand years later, many Christian churches teach that one day Jesus will "come again in glory" and that His second coming will mark the end of the world as we know it.

FIGURE 17.7 Many Muslims show their patriotism and work to let other people know that Islam is against terrorism. Why do individuals like those shown here feel the need to make this public affirmation?

Whereas Judaism is basically an inherited religion and most adherents are born into the community, Christianity has universalistic criteria for membership—meaning that it does not have ethnic or tribal qualifications—based on acceptance of a set of beliefs. Becoming a Christian requires personal belief that Jesus is the son of God; that He died, was buried, and on the third day rose from the dead; and that the Supreme Being is a sacred *trinity*—the Christian belief of "God the Three in One," comprising God the Father, Jesus the Son, and the Holy Spirit—a presence that lives within those who have accepted Jesus as savior. To become members of the religious community, believers must affirm their faith and go through a rite of passage of baptism. According to the teachings of Christianity, those who believe in Jesus as their savior will be resurrected from death and live eternally in the presence of God, whereas those who are wicked will endure an eternity in hell.

LondonPhotos-Homer Sykes/Alamy

FIGURE 17.8 Throughout recorded history, churches and other religious bodies have provided people with a sense of belonging and of being part of something larger than themselves. Members of this congregation show their unity by clapping and singing together.

Christianity was first introduced to North America when Roman Catholic missionaries established missions and converted Native Americans to Christianity before the early Protestant settlers arrived on this continent. From the earliest days of the British colonies in this country, a variety of religions were represented, including Anglicans (the forerunners of the Protestant Episcopal church), Baptists, Quakers, Presbyterians, Methodists, and Lutherans. Freedom of religion provided people with the opportunity to establish other denominations and sects and generally to worship as they pleased.

The African American church was the center of community life first for slaves and freed slaves, and then for generations of blacks who experienced ongoing prejudice and discrimination based on their race (Figure 17.8). The theory of nonviolent protest and civil disobedience used in the civil rights movement in the 1960s, largely orchestrated by the Rev. Dr. Martin Luther King Jr., emerged from the African American church in the South. Over the years, members of other minority groups, including Latinos/as and Asian Americans, have benefited from religious and social ties to various Christian denominations, including the Roman Catholic church.

Worldwide, about half of all Christians are Catholic (50 percent), followed by 37 percent who identify with the Protestant tradition. The largest proportion of the Christian population is found in the Americas (37 percent), followed by Europe (26 percent), sub-Saharan Africa (24 percent), and Asia and the Pacific (13 percent). Although the majority of the world's Christians reside in the Americas and Europe,

extensive growth has occurred in the Christian population in sub-Saharan Africa and the Asia–Pacific region over the past century (Pew Forum on Religion and Public Life, 2011). In the United States the percentage of American adults identifying themselves as Christians dropped from a high of 86 percent in 1990 to 77 percent in 2012 (Newport, 2012). By 2014, a Pew Research Center (2015) survey found a further drop in the Christian share of the U.S. population to 70.6 percent, or approximately 173 million people.

For centuries, Christians, Jews, and Muslims have lived together in peace and harmony in some areas of the world; in others, however, they have engaged in strife and fighting. The wounds and animosities between Muslims and Christians have remained since the early Christian era; nevertheless, some religion scholars believe that there is hope for a genuine Christian–Muslim dialogue in the future. In fact, it has been said that the contemporary challenge to Christianity is not from other religions but rather from a rejection of all forms of organized religion.

LO5 **Compare** and contrast four sociological perspectives on religion.

Sociological Perspectives on Religion

Religion as a social institution is a powerful, deeply felt, and influential force in human society. Sociologists study the social institution of religion because of the importance that religion holds for many people; they also want to know more about the influence of religion on society, and vice versa.

SOCIOLOGY & SOCIAL POLICY

Prayer in Public Schools? The Issue of Separation of Church and State

ALL THINGS ARE POSSIBLE!

"Only Believe"

The faculty and staff are working hard to prepare our students for the MSA.

Let's come together, as one, in prayer (teachers, students and parents) and ask God to bless our school to pass the MSA (Jeremiah 33:3). He will do it again! Everyone is invited.

Time: Saturday, 10:00–10:30 a.m.

Where: Tench Tilghman

> —This poster announcing a Baltimore, Maryland, principal's invitation to students, teachers, and parents to join together in prayer at a public school for success on the Maryland School Assessments test for third through eighth graders created a stir and brought an attorney for the American Civil Liberties Union forward to argue that such activity was a violation of the U.S. Constitution (Green, 2011).

Why is the issue of prayer or other religious observances in public schools an ongoing concern for many people? As we think about this question, it is necessary to understand how social policy has been historically used to establish a division between church and state. When the U.S. Constitution was ratified, in 1789, the colonists who made up the majority of the population of the original states were of many different faiths. Because of this diversity, there was no mention of religion in the original Constitution. However, in 1791 the First Amendment added the following

Rob Crandall/Stock Connection Blue/Alamy

Should prayer be permitted in public-school classrooms? On the school grounds? At school athletic events? Given the diversity of beliefs that U.S. people hold about separation of church and state, arguments and court cases over activities such as prayer around the school flagpole will no doubt continue in the future.

For example, some people believe that the introduction of prayer or religious instruction in public schools would have a positive effect on the teaching of values such as honesty, compassion, courage, and tolerance because these values could be given a moral foundation. However, society has strongly influenced the practice of religion in the United States as a result of court rulings and laws that have limited religious activities in public settings, including schools (see "Sociology and Social Policy").

The major sociological perspectives have different outlooks on the relationship between religion and society. Functionalists typically emphasize the ways in which religious beliefs and rituals can bind people together. Conflict explanations suggest that religion can be a source of false consciousness in society. Symbolic interactionists focus on the meanings that people give to religion in their everyday lives. Rational choice theorists view religion as a competitive marketplace in which religious organizations (suppliers) offer a variety of religions and religious products to potential followers (consumers), who shop around for the religious theologies, practices, and communities that best suit them.

Functionalist Perspectives on Religion

The functionalist perspective on religion finds its roots in the works of early sociologist Emile Durkheim, who emphasized that religion is essential to the maintenance of society. He suggested that religion is a cultural universal found in all societies because it meets basic human needs and serves important societal functions.

For Durkheim, the central feature of all religions is the presence of sacred beliefs and rituals that bind people together in a collectivity. In his studies of the religion of the Australian aborigines, Durkheim found that each clan had established its own sacred totem, which included kangaroos, trees, rivers, rock formations, and other animals or natural creations. To clan members, their totem was sacred; it symbolized some unique quality of their clan. People developed a feeling of unity by performing ritual dances around their totem, causing them to abandon individual self-interest. Durkheim suggested that the correct performance of the ritual gives rise to religious conviction. Religious beliefs and rituals are *collective representations*—group-held meanings

provision (to this day, the only reference to religion in that document): "Congress shall make no law respecting an establishment of religion, or prohibiting the free exercise thereof." The ban was binding only on the federal government; however, in 1947 the U.S. Supreme Court held that the Fourteenth Amendment ("No State shall make or enforce any law which shall abridge the privileges or immunities of citizens") had the effect of making the First Amendment's separation of church and state applicable to state governments as well.

Historically, the Supreme Court has been called on to define the boundary between permissible and impermissible governmental action with regard to religion. In 1947 the Court held that the ban on establishing a religion made it unconstitutional for a state to use tax revenues to support an institution that taught religion. In 1962 and 1963 the Court expanded this ruling to include many types of religious activities at schools or in connection with school activities, including group prayer, invocations at sporting events, and distribution of religious materials at school. However, in 1990 the Court ruled that religious groups could meet on school property if certain conditions were met, including that attendance must be voluntary, the meetings must be organized and run by students, and the activities must occur outside of regular class hours.

Compromises regarding prayer in schools have been attempted in some states as legislators have passed laws that either permit or mandate a daily moment of silence in public schools; however, the effect of such compromises has often been that neither side in the debate is appeased. Advocates of allowing more religious activities in public education believe that the constitutional dictate prohibiting "establishment of religion" was not intended to keep religion out of the public schools and that students need greater access to religious and moral training. Opponents believe that any entry of religious training and religious observances into public education and taxpayer-supported school facilities or events violates the Constitution and might be used by some people to promote their religion over others, or over a person's right to have no religion at all. No doubt, this debate will continue in the future as some political leaders in state legislatures look for creative ways to authorize school prayer or other religious functions in public schools that will pass constitutional muster while, at the same time, staunch opponents of such measures will continue to argue that any form of religious expression should not be allowed in this nation's public schools.

REFLECT & ANALYZE

Where should the line be drawn on religious observances and public education? What do you think?

Sources: Based on Albanese, 2013; and Green, 2011.

that express something important about the group itself. Because of the intertwining of group consciousness and society, functionalists suggest that religion has three important functions in any society:

1. *Meaning and purpose.* Religion offers meaning for the human experience. Some events create a profound sense of loss on both an individual basis (such as injustice, suffering, or the death of a loved one) and a group basis (such as famine, earthquake, economic depression, or subjugation by an enemy). Inequality may cause people to wonder why their own situation is no better than it is. Most religions offer explanations for these concerns. Explanations may differ from one religion to another, yet each tells the individual or group that life is part of a larger system of order in the universe. Some (but not all) religions even offer hope of an afterlife for persons who follow the religion's tenets of morality in this life. Such beliefs help make injustices easier to endure.

2. *Social cohesion and a sense of belonging.* By emphasizing shared symbolism, religious teachings and practices help promote social cohesion. An example is the Christian ritual of communion, which not only commemorates a historical event but also allows followers to participate in the unity ("communion") of themselves with other believers. All religions have some form of shared experience that rekindles the group's consciousness of its own unity.

3. *Social control and support for the government.* All societies attempt to maintain social control through systems of rewards and punishments. Sacred symbols and beliefs establish powerful, pervasive, long-lasting motivations based on the concept of a general order of existence. In other words, if individuals consider themselves to be part of a larger order that holds the ultimate meaning in life, they will feel bound to one another (and to past and future generations) in a way that might not be possible otherwise. Religion also helps maintain social control in society by conferring supernatural legitimacy on the norms and laws of a society. In some societies, social control occurs as a result of direct collusion between the dominant classes and the dominant religious organizations.

In the United States the separation of church and state reduces religious legitimation of political power. Nevertheless, political leaders often use religion to justify their decisions, stating that they have prayed for guidance in deciding what to do. This informal relationship between religion and the state has been referred to as *civil religion*—the set of beliefs, rituals, and symbols that makes sacred the values of the society and places the nation in the context of the ultimate system of meaning. Civil religion is not tied to any one denomination or religious group; it has an identity all its own. For example, many civil ceremonies in the United States have a marked religious quality. National values are celebrated on "high holy days" such as Memorial Day and the Fourth of July. Political inaugurations and courtroom trials often require people to place their hand on a Bible while swearing to do their duty or tell the truth, as the case may be. The U.S. flag is the primary sacred object of our civil religion, and the Pledge of Allegiance includes the phrase "one nation under God." U.S. currency bears the inscription "In God We Trust."

Some critics have attempted to eliminate all vestiges of civil religion from public life. However, sociologist Robert Bellah (1967), who has studied civil religion extensively, argues that civil religion is not the same thing as Christianity; rather, it is limited to affirmations of loyalty and patriotism that adherents of any religion can accept. However, Bellah's assertion does not resolve the problem for those who do not believe in the existence of God or for those who believe that *true* religion is trivialized by civil religion.

Conflict Perspectives on Religion

Many functionalists view religion, including civil religion, as serving positive functions in society, but some conflict theorists view religion negatively. From a conflict perspective, religion tends to promote conflict between groups and societies. According to conflict theorists, conflict may be *between* religious groups (for example, anti-Semitism), *within* a religious group (for example, when a splinter group leaves an existing denomination), or between a religious group and the *larger society* (for example, the conflict over religion in the classroom). Conflict theorists assert that in attempting to provide meaning and purpose in life while at the same time promoting the status quo, religion is used by the dominant classes to impose their own control over society and its resources.

Karl Marx on Religion   For Marx, *ideologies*—systematic views of the way the world ought to be—are embodied in religious doctrines and political values. These ideologies serve to justify the status quo and restrict social change. The capitalist class uses religious ideology as a tool of domination to mislead the workers about their true interests. For this reason, Marx wrote his now-famous statement that religion is the "opiate of the masses." People become complacent because they have been taught to believe in an afterlife in which they will be rewarded for their suffering

and misery in this life. Although these religious teachings soothe the masses' distress, any relief is illusory. According to Marx, religion unites people in a "false consciousness" that they share common interests with members of the dominant class.

Max Weber's Response to Marx   Whereas Marx believed that religion restricts social change, Weber argued just the opposite. For Weber, religion could be a catalyst to produce social change. In *The Protestant Ethic and the Spirit of Capitalism* (1976/1904–1905), Weber asserted that the religious teachings of John Calvin are directly related to the rise of capitalism. Calvin emphasized the doctrine of *predestination*—the belief that even before they are born, all people are divided into two groups, the saved and the damned, and only God knows who will go to heaven (the elect) and who will go to hell. Because people cannot know whether they will be saved, they tend to look for earthly signs that they are among the elect. According to the Protestant ethic, those who have faith, perform good works, and achieve economic success are more likely to be among the chosen of God. As a result, people work hard, save their money, and do not spend it on worldly frivolity; instead, they reinvest it in their land, equipment, and labor.

The spirit of capitalism grew in the fertile soil of the Protestant ethic. Even as people worked ever harder to prove their religious piety, structural conditions in Europe led to the Industrial Revolution, free markets, and the commercialization of the economy—developments that worked hand in hand with Calvinist religious teachings. From this viewpoint, wealth is an unintended consequence of religious piety and hard work. With the contemporary secularizing influence of wealth, people often think of wealth and material possessions as the major (or only) reason to work. Although the "Protestant ethic" is rarely invoked today, many people still refer to the "work ethic" in somewhat the same manner that Weber did.

Like Marx, Weber was acutely aware that religion could reinforce existing social arrangements, especially the stratification system (• Figure 17.9). The wealthy can use religion to justify their power and privilege: It is a sign of God's approval of their hard work and morality. As for the poor, if they work hard and live a moral life, they will be richly rewarded in another life.

Feminist Perspectives on Religion   Like other approaches in the conflict tradition, feminist perspectives focus on the relationship between religion and women's inequality. Some feminist perspectives highlight the patriarchal nature of religion and seek to reform religious language, symbols, and rituals to eliminate elements of patriarchy. As you will recall, *patriarchy* refers to a hierarchical system of social organization that is controlled by men. In virtually all religions, male members predominate in positions of power in the religious hierarchy, and women play subordinate roles in the hierarchy and in everyday life. For example, an Orthodox Jewish man may focus on *public*

FIGURE 17.9 According to Marx and Weber, religion serves to reinforce social stratification in a society. For example, according to Hindu belief, a person's social position in his or her current life is a result of behavior in a former life.

ritual roles and discussions of sacred texts, while an Orthodox Jewish woman may have few, if any, ritual duties and focus on her *private* responsibilities in the home. Orthodox Judaism does not permit women to become rabbis; however, more-liberal Jewish movements have granted more women this opportunity in recent years.

According to feminist theorists, religious symbolism and language consistently privilege men over women. Religious symbolism depicts the higher deities as male and the lower deities as female. Women are also more likely than men to be depicted as negative or evil spiritual forces in religion. In the Jewish and Christian traditions, for example, Eve in the Book of Genesis is a temptress who contributes to the Fall of Man. In the Hindu tradition, the goddess Kali represents men's eternal battle against the evils of materialism. Until recently, the language used in various religious texts made women virtually nonexistent. Phrases such as *for all men* in Catholic and Episcopal services have gradually been changed to *for all people*; however, some churches have retained the traditional liturgy, where all of the language is male centered. As women have become increasingly aware of their subordination, more of them have fought to change existing rules and have a voice in their religious community. The fact that women make up between 56 percent and 58 percent of all adherents among Pentecostals, Baptists, and mainline Christians (such as Methodists, Lutherans, Presbyterians, and Episcopalians/

Anglicans) may contribute to additional changes in the roles and statuses of women in organized religion.

Symbolic Interactionist Perspectives on Religion

Thus far, we have been looking at religion primarily from a macrolevel perspective. Symbolic interactionists focus their attention on a microlevel analysis that examines the meanings people give to religion in their everyday lives.

Religion as a Reference Group For many people, religion serves as a reference group to help them define themselves. For example, religious symbols have meaning for large bodies of people. The Star of David holds special significance for Jews, just as the crescent moon and star do for Muslims and the cross does for Christians. For individuals as well, a symbol may have a certain meaning beyond that shared by the group. For instance, a symbolic gift given to a child may have special meaning when he or she grows up and faces war or other crises. It may not only remind the adult of a religious belief but also create a feeling of closeness with a relative who is now deceased.

civil religion
the set of beliefs, rituals, and symbols that makes sacred the values of the society and places the nation in the context of the ultimate system of meaning.

Sociological Perspectives on Religion

	Key Points
Functionalist Perspectives	Sacred beliefs and rituals bind people together and help maintain social control.
Conflict Perspectives	Religion may be used to justify the status quo (Marx) or to promote social change (Weber). Feminist approaches focus on the relationship between religion and women's inequality.
Symbolic Interactionist Perspectives	Religion may serve as a reference group for many people, but because of race, class, and gender, people may experience it differently.
Rational Choice Perspectives	Religious persons and organizations, interacting within a competitive market framework, offer a variety of religions and religious products to consumers, who shop around for religious theologies, practices, and communities that best suit them.

Religion and Social Meaning   Religion provides social meaning for individuals as they learn about beliefs, rituals, and religious ideas from others. This learning process contributes to personal identity, which, in turn, helps people adjust to their surroundings. For example, children may learn appropriate conduct during the Christian sacrament of Communion or the Lord's Supper by attending instruction classes for prospective church members or by observing how their parents and other adults quietly and reverently participate in this ceremony.

Social meaning in religion emerges through the process of socialization and from interaction with others in a religious setting. This social meaning then develops into personal meaning that provides a religious identity for the individual. According to a symbolic interactionist approach, members of a religious group expect each member to perform certain normative religious behaviors and to adhere to certain normative religious beliefs. In other words, a specific kind of *role performance* is expected of members of a religious group, and this role performance contributes to religion becoming a *master status* for some people, despite the fact that each person performs numerous roles and has many identities in everyday life.

Rational Choice Perspectives on Religion

In terms of religion, *rational choice theory* is based on the assumption that religion is essentially a rational response to human needs; however, the theory does not claim that any particular religious belief is necessarily true or more rational than another. The rational choice perspective views religion as a competitive marketplace in which religious organizations (suppliers) offer a variety of religions and religious products to potential followers (consumers), who shop around for the religious theologies, practices, and communities that best suit them.

According to this approach, people need to know that life has a beneficial supernatural element, such as that there is meaning in life or that there is life after death, and they seek to find these *rewards* in various religious organizations. The rewards include explanations of the meaning of life and reassurances about overcoming death. However, because

religious organizations cannot offer religious certainties, they instead offer *compensators*—a body of language and practices that compensates for some physical lack or frustrated goal. According to some sociologists, all religions offer compensators, such as a belief in heaven, personal fulfillment, and control over evil influences in the world, to offset the fact that they cannot offer certainty of an afterlife or other valued resources that potential followers and adherents might desire (Stark, 2007). Rational choice theory focuses on the process by which actors—individuals, groups, and communities—settle on one optimal outcome out of a range of possible choices (a cost–benefit analysis). These compensators provide a range of possible choices for people in the face of a limited (or nonexistent) supply of the choice (certainty, for example) that they truly desire.

Sociologists of religion have applied rational choice theory to an examination of the very competitive U.S. religious marketplace and found that people are actively shopping around for beliefs, practices, and religious communities that best suit them. For example, some religious followers have been attracted to churches that preach the so-called *prosperity gospel*, which is based on the assumption that if you give your money to God, He will bless you with more money and other material possessions (such as a larger house and a luxury vehicle) that you desire. Several megachurches, including Joel Osteen's Lakewood in Houston, T. D. Jakes's Potter's House in south Dallas, and Creflo Dollar's World Changers near Atlanta, are partly based on teaching that suggests God wants people to be prosperous if they are "right" with Him.

Based on the diverse teaching and practices of various religious bodies, adherents and prospective followers move among various religious organizations, with every major religious group simultaneously gaining and losing adherents. Although some find the religious home they seek, others decide to consider themselves unaffiliated with any specific faith. Based on the movement of possible adherents, religious groups challenge one another for followers, emphasize specific moral values, and create a civil society that offers followers a religious faith that does not unduly burden them.

The Concept Quick Review provides an overview of the four sociological perspectives on religion that we have discussed.

FIGURE 17.10   One of the largest places of worship in the United States is Joel Osteen's Lakewood Church, in Houston, Texas. This church broadcasts its message worldwide.

LO6   **Describe** the types of religious organizations.

Types of Religious Organization

Religious groups vary widely in their organizational structure. Although some groups are small and have a relatively informal authority structure, others are large and somewhat bureaucratically organized (● Figure 17.10). Some require the total commitment of their members; others expect members to have only a partial commitment. Sociologists have developed typologies or ideal types of religious organization to enable them to study a wide variety of religious groups. The most common categorization sets forth five types: *ecclesia*, *church*, *sect*, *denomination*, and *cult* (also referred to as a "new religious movement").

Ecclesia

An *ecclesia* is a religious organization that is so integrated into the dominant culture that it claims as its membership all members of a society. Membership in the ecclesia occurs as a result of being born into the society, rather than by any conscious decision on the part of individual members. The linkages between the social institutions of religion and government are often very strong in such societies. Although no true ecclesia exists in the contemporary world, the Anglican church (the official church of England), the Lutheran church in Sweden and Denmark, the Roman Catholic church in Italy and Spain, and the Islamic mosques in Iran and Pakistan come fairly close.

Churches, Sects, and Denominations

Unlike an ecclesia, a church is not considered to be a state religion; however, it may still have a powerful influence on political and economic arrangements in society. A *church* is a large, bureaucratically organized religious organization that tends to seek accommodation with the larger society in order to maintain some degree of control over it. Church membership is largely based on birth; typically, children of

ecclesia
a religious organization that is so integrated into the dominant culture that it claims as its membership all members of a society.

church
a large, bureaucratically organized religious organization that tends to seek accommodation with the larger society in order to maintain some degree of control over it.

church members are baptized as infants and become life-long members of the church. Older children and adults may choose to join the church, but they are required to go through an extensive training program that culminates in a ceremony similar to the one that infants go through. Churches have a bureaucratic structure, and leadership is hierarchically arranged. Usually, the clergy have many years of formal education. Churches have very restrained services that appeal to the intellect rather than the emotions. Religious services are highly ritualized; they are led by clergy who wear robes, enter and exit in a formal processional, administer sacraments, and read services from a prayer book or other standardized liturgical format. The Lutheran church and the Episcopal church are two examples.

A *sect* is a relatively small religious group that has broken away from another religious organization to renew what it views as the original version of the faith. Unlike churches, sects offer members a more personal religion and an intimate relationship with a supreme being, depicted as taking an active interest in the individual's everyday life. Sects have informal prayers composed at the time they are given, whereas churches use formalized prayers, often from a prayer book.

According to the church–sect typology, as members of a sect become more successful economically and socially, they tend to focus more on this world and less on the next. However, sect members who do not achieve financial success often believe that they are being left behind as the other members, and sometimes the minister, shift their priorities to things of this world. Eventually, this process weakens some religious groups, and the dissatisfied or downwardly mobile split off to create new, less worldly versions of the original group that will be more committed to "keeping the faith." Those who defect to form a new religious organization may start another sect or form a cult. (See ● Table 17.2 for a summarized description of churches and sects.)

Midway between the church and the sect is the *denomination*—a large organized religion characterized by accommodation to society but frequently lacking in ability or intention to dominate society. Denominations have a trained ministry, and although involvement by lay members is encouraged more than in the church, their participation is usually limited to particular activities, such as readings or prayers. Denominations tend to be more tolerant and are less likely than churches to expel or excommunicate members. This form of organization is most likely to thrive in societies characterized by *religious pluralism*—a situation in which many religious groups exist because they have a special appeal to specific segments of the population. Perhaps because of its diversity, the United States has more denominations than any other nation.

Cults (New Religious Movements)

Previously, sociologists defined a *cult* as a loosely organized religious group with practices and teachings outside the dominant cultural and religious traditions of a society. Because the term *cult* has assumed a negative and sometimes offensive meaning because of the beliefs and actions of a few highly publicized cults, some researchers now use the term *new religious movement* (NRM) and point out that a number of major world religions (including Judaism, Islam, and Christianity) and some denominations (such as the Mormons) started as cults. Also, most cults or NRMs do not exhibit the bizarre behavior or have the unfortunate ending that a few notorious groups have had in the past.

NRMs usually have a leader who exhibits charismatic characteristics (personal magnetism or mystical leadership) and possesses a unique ability to communicate and form attachments with others. An example was the late Rev. Sun Myung Moon, a former Korean electrical engineer who founded the Unification church, or "Moonies," claiming God revealed to him that the Judgment Day was rapidly approaching. This former cult identified itself as "comprised of families striving to embody the ideal of true love and to establish a world of peace and unity among all peoples, races, and religions as envisioned by Rev. Sun Myung Moon. Members of the Unification Church accept and follow Rev. Moon's particular religious teaching, the Divine Principle" (*Unification Church News*, 2011). Rev. Moon died at the age of 92 in 2012. Initially, his movement flourished because it recruited new members through their personal attachments to present members (● Figure 17.11). In the twenty-first century, it became more institutionalized.

Other cult leaders did not fare so well for so long, including Jim Jones, whose ill-fated cult ended up committing mass suicide in Guyana in 1978, and Marshall Herff Applewhite ("Do"), who led his thirty-eight Heaven's Gate

TABLE 17.2

Characteristics of Churches and Sects

Characteristic	Church	Sect
Organization	Large, bureaucratic organization, led by a professional clergy	Small, faithful group, with high degree of lay participation
Membership	Open to all; members usually from upper and middle classes	Closely guarded membership, usually from lower classes
Type of Worship	Formal, orderly	Informal, spontaneous
Salvation	Granted by God, as administered by the church	Achieved by moral purity
Attitude Toward Other Institutions and Religions	Tolerant	Intolerant

FIGURE 17.11 Mass wedding ceremonies of thousands of brides and grooms brought widespread media attention to the late Rev. Sun Myung Moon and the Unification Church, which many people viewed as a religious cult.

followers to commit mass suicide at their Rancho Santa Fe, California, mansion after convincing them that the comet Hale-Bopp, which swung by Earth in late March 1997, would be their celestial chariot taking them to a higher level.

What eventually happens to cults or NRMs? Over time, some disappear; however, others gradually transform into other types of religious organizations, such as sects or denominations. An example is Mary Baker Eddy's Christian Science church, which started as a cult but became an established denomination with mainstream methods of outreach, such as Christian Science Reading Rooms placed in office buildings or shopping malls, where individuals can learn of the group's beliefs while going about their routine activities. Other cults or new religious movements lose their "newness" as they are embraced, as Scientology has been, by celebrities such as Tom Cruise, John Travolta, Kirstie Alley, and Brandy.

LO7 **Identify** major trends in religion in the United States in the twenty-first century.

Trends in Religion in the United States

Religion in the United States is very diverse. Pluralism and religious freedom are among the cultural values most widely espoused, and no state church or single denomination predominates. As shown in ● Figure 17.12, Protestants constitute the largest religious body in the United States, followed by Roman Catholics, Mormons, Jews, Eastern churches, and others.

In this section we examine trends that have most influenced religion in the United States over the past two centuries. One of the most important has been the debate over secularization.

The Secularization Debate

Secularization is the process by which religious beliefs, practices, and institutions lose their significance in society and nonreligious values, principles, and institutions take their place. Secularization has two components: (1) a decline in religious values and institutions in everyday life and (2) a corresponding increase in nonreligious values or principles and greater significance given to secular institutions. Although secularization has been widely studied and hotly debated, many scholars argue that levels of religiosity are not declining in the United States and other high-income nations. In fact, sociologist Peter Berger, who proposed the secularization theory, reconsidered his theory when he observed that modernity, rather than secularizing society, had contributed to a counter-secularization movement. Berger also found that higher levels of religiosity accompanied modernity—the way that people are influenced by religious beliefs and shape their social reality accordingly. *Religiosity* may also be referred to as religious commitment or "religiousness." Social scientists use various measures to determine religiosity, including the extent to which a person does one or more of the following: (1) believes in and "feels" or experiences certain aspects of religion, (2) becomes involved in religious activities such as attending church or reading sacred texts, (3) believes in the teachings of the church, and (4) lives in accordance with those teachings and beliefs.

So is secularization occurring in the twenty-first century? Although the U.S. religious landscape has fluid and

sect
a relatively small religious group that has broken away from another religious organization to renew what it views as the original version of the faith.

denomination
a large organized religion characterized by accommodation to society but frequently lacking in ability or intention to dominate society.

cult
(also known as *new religious movement* or *NRM*) a loosely organized religious group with practices and teachings outside the dominant cultural and religious traditions of a society.

secularization
the process by which religious beliefs, practices, and institutions lose their significance in society and nonreligious values, principles, and institutions take their place.

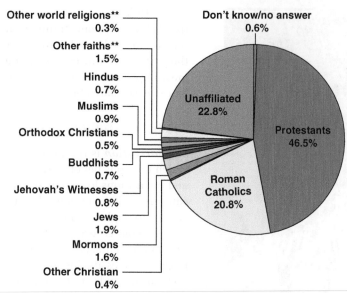

Other world religions** 0.3%

Other faiths** 1.5%

Hindus 0.7%

Muslims 0.9%

Orthodox Christians 0.5%

Buddhists 0.7%

Jehovah's Witnesses 0.8%

Jews 1.9%

Mormons 1.6%

Other Christian 0.4%

Don't know/no answer 0.6%

Unaffiliated 22.8%

Protestants 46.5%

Roman Catholics 20.8%

**The "other world religions" category includes Sikhs, Baha'is, Taoists, Jains, and a variety of other world religions. The "other faiths" category includes Unitarians, New Age religions, Native American religions, and a number of other non-Christian faiths.

FIGURE 17.12   U.S. Religious Traditions' Membership
Sources: Pew Research Center, 2014; Religious Landscape Study, 2015.

diverse patterns, some trends have been identified that might support the secularization thesis. First, the number of people in the United States who do not identify with any religion has grown at a rapid rate. In 2012 one-fifth of the U.S. public, and one-third of adults under age thirty, indicated that they were religiously unaffiliated (Pew Forum on Religion and Public Life, 2012b). When the Pew Research Center looked at trends from 2007 to 2012, it found that over the five-year period, the number of people who considered themselves to be unaffiliated increased from slightly over 15 percent to just under 20 percent of all U.S. adults. These percentages included nearly 33 million individuals who stated that they had no particular religious affiliation and 13 million who self-described as atheists and agnostics (Pew Forum for Religion and Public Life, 2012b).

However, these figures do not necessarily mean that Americans are losing their religion. According to recent surveys, many of the 46 million unaffiliated adults say that they are religious or spiritual in some way (● Figure 17.13). For example, 68 percent state that they believe in God, and more than 50 percent indicate that they feel a deep connection with nature and the Earth. In addition, one in five of the so-called "unaffiliated" say that they pray every day (Pew Forum on Religion and Public Life, 2012b). One of the reasons that many unaffiliated persons indicate that they are not looking for a church or other religious institution to join is that they believe these organizations are too concerned with money and power, as well as overly emphasizing rules and becoming too involved

in politics (Pew Forum on Religion and Public Life, 2012b). Issues of secularization and the growing number of unaffiliated people in major national religion surveys will continue to be of concern to persons who identify themselves as "fundamentalists."

The Rise of Religious Fundamentalism

Fundamentalism is a traditional religious doctrine that is conservative, is typically opposed to modernity, and rejects "worldly pleasures" in favor of otherworldly spirituality. In the past, traditional fundamentalism primarily appealed to people from lower-income, rural, southern backgrounds; however, newer fundamentalist movements have had a much wider appeal to people from all socioeconomic levels, geographical areas, and occupations in the United States. One reason for the rise of fundamentalism has been a reaction against modernization and secularization. Around the world, those who adhere to fundamentalism—whether they are Muslims, Christians, or followers of one of the other world religions—believe that sacred traditions must be revitalized. For example, public education in the United States has been the focus of some who follow the tenets of Christian fundamentalism. Various religious and political leaders vow to bring the Christian religion "back" into the public life of this country. They have been especially critical of educators who teach what they perceive to be *secular humanism*—the belief that human beings can become better through their own efforts rather than through belief in God and a religious conversion. But how might students and teachers who come from diverse religious heritages feel about Christian religious instruction or organized prayer in public schools? Some social analysts believe that such practices would cause conflict and perhaps discrimination on the basis of religion.

FIGURE 17.13 Although a growing number of U.S. adults are not affiliated with a particular religion, many of them still feel that they are spiritual or religious in some way.

Trends in Race, Class, Gender, and Religion

One of the most recent comprehensive surveys on contemporary religion identified a number of trends regarding religion and race, class, and gender (Pew Forum on Religion and Public Life, 2012b). In regard to race and religious affiliation, white Americans (non-Hispanic) made up 68 percent of the religiously affiliated, as compared to 66 percent of the U.S. population in 2012. African Americans are the most likely of all the so-called "minority" racial and ethnic groups in the United States to report that they have a formal religious affiliation. To determine this, we compare their percentage in religious organizations as compared to their percentage in the U.S. population. In 2012 African Americans accounted for 11 percent of the U.S. population, but they constituted 12 percent of the religiously affiliated. In the same year, Hispanic Americans accounted for 15 percent of the U.S. population but only 14 percent of the religiously affiliated. Likewise, Asian Americans made up 5 percent of the U.S. population but only 2 percent of the religiously affiliated (Pew Forum on Religion and Public Life, 2012b).

The linkage between race and religion has been very important throughout this nation's history. Most racial and cultural minorities who have felt overpowered by a lack of resources have been drawn to churches or other places of worship that help them establish a sense of dignity and self-worth that is otherwise missing in daily life.

In regard to religion and gender, the religiously affiliated population is more female than the general public. There are more women than men in the U.S. population, partly because women live longer than men. Specifically, the adult population is made up of about 48 men to every 52 women, and this pattern is reflected in religious adherents. Men make up 47 percent of the religiously affiliated, and women make up 53 percent. An interesting trend is seen among the religiously unaffiliated in regard to which gender is most likely to identify with the atheist/agnostic category: Men make up 64 percent of those who self-identify as atheists or agnostics, as compared to only 36 percent of women (Pew Forum on Religion and Public Life, 2012b). The most common trend in research on gender and religion is the finding that women consistently have higher rates of religiosity and are more involved in the life of the religious community than are men.

Finally, social science research continues to identify significant patterns involving social class and religious affiliation. In the Pew study, Hindus and Jews in the United States reported higher incomes than people in other categories. Researchers attribute this higher income level to the fact that Hindus and Jews also have the highest levels of education. Historically, other religious groups with the most members making more than $100,000 per year have been the mainline liberal churches, such as the Episcopalians, the Presbyterians, and the Congregationalists. However, Mormons, Buddhists, and Orthodox Christians also tend to have higher income levels. But looking at the other end of the income spectrum, members of evangelical churches, historically black churches, Jehovah's Witnesses, and Muslims typically earn less than $50,000 per year. For example, only 3 percent of members of historically black churches and 20 percent of members of evangelical churches are in the top income bracket in the Pew study. The intertwining relationship among race, class, gender, and religion is complicated and remains an important topic for additional research in the sociology of religion (● Figure 17.14).

FIGURE 17.14   Churches in converted buildings such as this seek to win new religious followers and to offer solace and hope to people in low-income areas.

fundamentalism
a traditional religious doctrine that is conservative, is typically opposed to modernity, and rejects "worldly pleasures" in favor of otherworldly spirituality.

Looking Ahead: Religion in the Future

What significance will religion have in the future? In the twenty-first century, organized religion has seen a significant rise in power and importance in many regions of the world. It is interesting to note that religion has been empowered by three trends that secularization theory believed would be the cause of its death: (1) modernization, (2) democratization, and (3) globalization. Rather than modernization spelling the end of religion, more than 80 percent of people around the world claim that religion is important to them. Even for advanced industrialized nations, the figure is more than 60 percent (Toft, Philpott, and Shah, 2011).

With the growth of democracies in numerous countries, including those Arab nations currently seeking change, religious actors have played a leading part in the political arena that was prohibited by former authoritarian regimes. Finally, globalization has produced a surge in the number of people and ideas that now travel around the world at much faster speeds than in the past, and this has made it possible for transnational actors such as religious groups to make themselves known to others and to use newer technologies such as the Internet, text messaging, and Twitter to bring people together quickly for collective action. How will these three trends play out in the future? According to Toft, Philpott, and Shah (2011),

> Religion is far from being the only or even the most decisive factor in global politics. But it has played—and will continue to play—a key role. The 21st century has brought us a world radically different from the one secularization theory promised. We have no choice but to build new theories and devise fresh policy strategies for the religious age we live in, God's Century, not the secular age that never came.

As discussed in this chapter, religion can be a unifying force in societies and around the world, or it can be a source of conflict and violence. It remains to be seen what the relationship between religion and peace and/or conflict will be in this century. Throughout history, people with religious beliefs have contributed to both peace and terrorism, democracy and authoritarianism, and reconciliation and civil war (Toft, Philpott, and Shah, 2011).

Debates over religion around the world, particularly issues such as secularization, fundamentalism, and violence perpetrated by religious extremists, will no doubt continue. Some studies show that the more-conservative or fundamental expressions of religious traditions are the most likely to present major challenges to peaceful accords among individuals and nations. Why? It is because conservative or fundamentalist factions primarily focus on those issues that bring about discord, conflict, and even the use of violence to resolve perceived differences. It is important to consider ways in which the leaders and teachings of the world's religions can best help cope with the opportunities and challenges that we face in the twenty-first century. However, the challenge is greatest when we acknowledge that some peoples of the world may believe that their religion will be furthered only if they are able to remain aloof from, and perhaps even banish, Western civilization as we know it.

Regarding the future of religion in America, studies have shown that we have become more polarized religiously over the past fifty years but that this may not remain true in the future. According to the political scientists Robert Putnam and David Campbell (2010), the United States is not only religiously devout and diverse, but it is also more religiously tolerant than we initially might think. To support this point, they point to intermarriage across religious lines, as well as to the extent that people today are far more likely to change religions or list themselves as "unaffiliated" or "none" when asked about religious preference than they were fifty years ago. In the future, more people may turn away from organized religions, creating a drop in American religiosity, but that does not necessarily mean that these individuals have no religion at all: As mentioned earlier, many people who indicate that they have no religious affiliation do state that they believe in God or provide other indicators of their religious beliefs. The extent to which religion is linked to conservative politics and/or discussions of sexual morality and rights regarding abortion, same-sex marriage, and other "hot topics" will play a central part in whether younger people become more (or less) involved in organized religion in the future.

In the twenty-first century we are likely to see not only the creation of new religious forms but also a dramatic revitalization of traditional forms of religious life. Religious fundamentalism has found a new audience among young people between the ages of eighteen and twenty-four. Some are Christian fundamentalists who will continue to emphasize biblical literalism, which means that every word of the Bible is literally true (inerrancy). Other Christians are creating their own modified version of what it means to be a Christian, a fundamentalist, or an evangelical (a person actively involved in sharing the "Good News" of his or her religion with others). Ultimately, one thing we probably can expect is considerable religious tension between Americans with differing beliefs and people around the globe who do not share our worldview. All of these factors are going to make it imperative that we become more tolerant of others and hope that they will do likewise toward us.

YOU CAN MAKE A DIFFERENCE

Understanding and Tolerating Religious and Cultural Differences

[The small strip of kente cloth that hangs on the marble altar and the flags of a dozen West Indies nations, including Trinidad, Jamaica, and Barbados,] represent all of the people in this parish. It's an opportunity for us to celebrate everyone's culture. You won't find your typical Episcopal church looking like that.

—Rev. Cannon Peter P. Q. Golden, then serving as rector of St. Paul's Episcopal Church in Brooklyn, New York (qtd. in Pierre-Pierre, 1997: A11)

One part of the history of St. Paul's Episcopal Church in Flatbush, Brooklyn, is etched into its huge stained-glass windows, which tell the story of the prominent descendants of English and Dutch settlers who founded the church. However, the church's more recent history is told in the faces of its parishioners—most of whom are Caribbean immigrants from the West Indies who labor as New York City's taxi drivers, factory workers, accountants, and medical professionals.

Traveling only a short distance, we find a growing enclave of Muslims in Brooklyn. It is their wish that people will recognize that Islam shares a great deal with Christianity and Judaism, including the fact that all three believe in one God, are rooted in the same part of the world, and share some holy sites. Some Muslim adherents also hope that more people in this country will learn greater tolerance toward those who have a different religion and celebrate different holidays. For example, in some years the Islamic holy season, Ramadan, falls at about the same time as Christmas and Hanukkah, but many Muslim children and adults are disparaged by their neighbors because they do not celebrate the same holidays as others. As one person observed, "As Muslims, we have to respect all religions" (Sengupta, 1997: A12). Left unsaid was the belief that other people should do likewise.

How can each of us—regardless of our race, color, creed, or national origin—help to bring about greater tolerance of religious diversity in this country? Here is one response to this question, from Huston Smith (1991: 389–390), a historian of religion:

Whether religion is, for us, a good word or bad; whether (if on balance it is a good word) we side with a single religious tradition or to some degree open our arms to all: How do we comport ourselves in a pluralistic world that is riven by ideologies, some sacred, some profane? . . . We listen. . . . If one of the [world's religions] claims us, we begin by listening to it. Not uncritically, for new occasions teach new duties and everything finite is flawed in some respects. Still, we listen to it expectantly, knowing that it houses more truth than can be encompassed in a single lifetime.

But we also listen to the faith of others, including the secularists. We listen first because . . . our times require it. The community today can be no single tradition; it is the planet. Daily, the world grows smaller, leaving understanding the only place where peace can find a home. . . . Those who listen work for peace, a peace built not on ecclesiastical or political hegemonies but on understanding and mutual concern.

Perhaps this is how each of us can make a difference—by *learning* more about our own beliefs and about the diverse denominations and world religions represented in the United States and around the globe, and *listening* to what other people have to say about their own beliefs and religious experiences.

Will religious tolerance increase in the United States? In the world? What steps can you take to help make a difference?

As we look to the future, we can expect even more fluidity in religion and the expansion of religious pluralism. In the words of Putnam and Campbell (2010: 550),

How has America solved the puzzle of religious pluralism—the coexistence of religious diversity and devotion? And how has it done so in the wake of growing religious polarization? By creating a web of interlocking personal relationships among people of many different faiths. This is America's grace.

This is an interesting and optimistic view of the future of religion. As we have seen in this chapter, the debate continues over what religion is, what it should do, and what its relationship to other social institutions such as education should be. It will be up to your generation to understand other religions and to work for greater understanding among the diverse people who make up our nation and the world (see "You Can Make a Difference").

CHAPTER REVIEW

LO1 What is religion, and what are its key components?

Religion is a social institution composed of a unified system of beliefs, symbols, and rituals, based on some sacred or supernatural realm, that guides human behavior, gives meaning to life, and unites believers into a community.

LO2 What are the four main categories of religion based on their dominant belief?

Religions have been classified into four main categories based on their dominant belief: simple supernaturalism, animism, theism, and transcendent idealism.

LO3 What are the central beliefs of major Eastern religions such as Hinduism, Buddhism, and Confucianism?

Hindus believe that Brahma (creator), Vishnu (preserver), and Shiva (destroyer) are divine. Union with ultimate reality and escape from eternal reincarnation are achieved through yoga, adherence to scripture, and devotion. Buddhists believe that through meditation and adherence to the Eightfold Path (correct thought and behavior), people can free themselves from desire and suffering, escape the cycle of eternal rebirth, and achieve nirvana (enlightenment). The sayings of Confucius (collected in the *Analects*) stress the role of virtue and order in the relationships among individuals, their families, and society.

LO4 What are the central beliefs associated with major Western religions such as Judaism, Islam, and Christianity?

Judaism holds that God's nature and will are revealed in the Torah (Hebrew scripture) and in His intervention in history. Jewish adherents also believe that God has established a covenant with the people of Israel, who are called to a life of holiness, justice, mercy, and fidelity to God's law. According to Muslims, Muhammad received the Qur'an (scriptures) from God. On Judgment Day, believers who have submitted to God's will, as revealed in the Qur'an, will go to an eternal Garden of Eden. Christians believe that Jesus is the Son of God and that through good moral and religious behavior (and/or God's grace), people achieve eternal life with God.

LO5 What are sociological perspectives on religion?

According to functionalists, religion has three important functions in any society: (1) providing meaning and purpose to life, (2) promoting social cohesion and a sense of belonging, and (3) providing social control and support for the government. From a conflict perspective, religion can have negative consequences in that the capitalist class uses religion as a tool of domination to mislead workers about their true interests. However, Max Weber believed that religion could be a catalyst for social change. Symbolic interactionists focus on a microlevel analysis of religion, examining the meanings that people give to religion and that they attach to religious symbols in their everyday life. According to the rational choice perspective, religious persons and organizations offer a variety of religions and religious products to consumers who shop around for religious theologies, practices, and communities that best suit them.

LO6 What are the types of religious organization?

Religious organizations can be categorized as ecclesia, churches, denominations, sects, and cults (now frequently referred to as new religious movements or NRMs).

LO7 What are some major trends in religion in the United States in the twenty-first century?

One of the most important has been the debate over secularization—the process by which religious beliefs, practices, and institutions lose their significance in society and nonreligious values, principles, and institutions take their place. Fundamentalism has emerged in many religions because people do not like to see social changes taking place that affect their most treasured beliefs and values. In some situations, people have viewed modernity as a threat to their traditional beliefs and practices. The latest-available comprehensive survey on contemporary religion identified a number of trends regarding religion and race, class, and gender, which are explained in detail in the chapter.

LO8 What is the future of religion as a social institution?

Because religion meets many needs that no other institution can provide, it will continue as a major social institution. Organized religion has grown and changed in many regions of the world because of modernization, democratization, and globalization. In the United States, religious organizations have become much more fluid and more diverse. Congregations in the future will hold an even wider variety of beliefs, and adherents will be drawn from many diverse cultures.

KEY TERMS

animism 490
church 503
civil religion 500
cult 504
denomination 504
ecclesia 503
faith 487

fundamentalism 506
monotheism 490
polytheism 490
profane 488
religion 487
rituals 488
sacred 488

sect 504
secularization 505
simple supernaturalism 490
spirituality 487
theism 490
transcendent idealism 490

QUESTIONS for CRITICAL THINKING

1 Why do some people who believe they have "no religion" subscribe to civil religion? How would you design a research project to study the effects of civil religion on everyday life?

2 How is religion a force for social stability? How is it a force for social change?

3 What is the relationship among race, class, gender, and religious beliefs?

4 If Durkheim, Marx, and Weber were engaged in a discussion about religion, on what topics might they agree? On what topics would they disagree?

5 How does the rational choice perspective differ from other theoretical explanations of religion in contemporary societies?

ANSWERS to the SOCIOLOGY QUIZ

ON RELIGION AND U.S. EDUCATION

1   **False**   Because of the diversity of religious backgrounds of the early settlers, no mention of religion was made in the original Constitution. Even the sole provision that currently exists (the establishment clause of the First Amendment) does not speak directly about the issue of religious learning in public education.

2   **False**   Obviously, contemporary sociologists hold strong beliefs and opinions on many subjects; however, most of them do not think that it is their role to advocate specific stances on a topic. Early sociologists were less inclined to believe that they had to be "value free." For example, Durkheim strongly advocated that education should have a moral component and that schools had a responsibility to perpetuate society by teaching a commitment to the common morality.

3   **False**   Just the opposite has happened. As parents have felt that their children were not receiving the type of education they desired in public schools, private, religiously based schools have flourished.

4	True	Although most Americans describe their religion as some Christian denomination, there has still been a significant increase in those who adhere either to no religion or who are Jewish, Muslim/Islamic, Unitarian–Universalist, Buddhist, or Hindu.
5	False	Debates over evolution versus creationism take place not only in elementary schools but in high schools and colleges as well because of the larger societal disagreement over the role of religion and science in education.
6	False	School prayer periodically reemerges as a political issue in this country, and for some it remains a hotly contested subject.
7	True	Adherents to Buddhism have been increasing rapidly, partly because of intense interest on the part of younger, middle-class people living primarily in New York and in western states such as California.
8	False	Although social theorists such as Karl Marx believed that religion kept social change from occurring, other theorists such as Max Weber argued that religion was a catalyst for social change.

HEALTH, HEALTH CARE,
AND DISABILITY

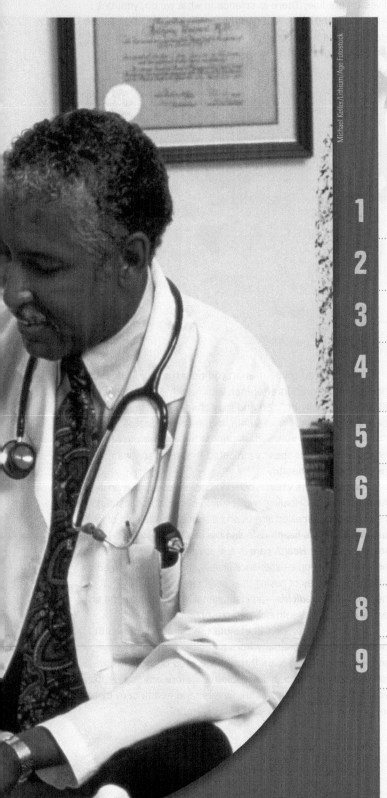

CHAPTER

18

LEARNING OBJECTIVES

1 **Discuss** the relationship between the social environment and health and illness.

2 **Define** *social epidemiology* and identify key demographic factors that are studied by social epidemiologists.

3 **Explain** how the profession of medicine emerged in the United States.

4 **Discuss** how U.S. health care was paid for in the past and how this is in the process of change since passage of the Affordable Care Act (Obamacare).

5 **Compare** how the United States pays for health care with how other nations provide health services for their citizens.

6 **Describe** how advanced medical technology has changed the practice of medicine and the cost of health services.

7 **Explain** how functionalist, conflict, symbolic interactionist, and postmodern approaches differ in their analysis of health and health care.

8 **Discuss** what is meant by the term *mental illness* and explain why it is a difficult topic for sociological research.

9 **Discuss** the concept of disability and identify key sociological perspectives on disability.

SOCIOLOGY & EVERYDAY LIFE

Medicine as a Social Institution

Medicine is, I have found, a strange and in many ways disturbing business. The stakes are high, the liberties taken tremendous. We drug people, put needles and tubes into them, manipulate their chemistry, biology, and physics, lay them unconscious and open their bodies up to the world. We do so out of an abiding confidence in our know-how as a profession. What you find when you get in close, however—close enough to see the furrowed brows, the doubts and missteps, the failures as well as the successes—is how messy, uncertain, and also surprising medicine turns out to be.

The thing that still startles me is how fundamentally human an endeavor it is. Usually, when we think about medicine and its remarkable abilities, what comes to mind is the science and all it has given us to fight sickness and misery: the tests, the machines, the drugs, the procedures. And without question, these are at the center of virtually everything medicine achieves. But we rarely see how it all actually works. You have a cough that won't go away—and then? It's not science you call upon but a doctor. A doctor with good

days and bad days. A doctor with a weird laugh and a bad haircut. A doctor with three other patients to see and, inevitably, gaps in what he knows and skills he's still trying to learn. . . . We look for medicine to be an orderly field of knowledge and procedure. But it is not. It is an imperfect science, an enterprise of constantly changing knowledge, uncertain information, fallible individuals, and at the same time lives on the line. There is science in what we do, yes, but also habit, intuition, and sometimes plain old guessing. The gap between what we know and what we aim for persists. And this gap complicates everything we do.

Dr. Atul Gawande (center) has written movingly about the differences between people's expectations of physicians and the medical establishment and the realities that they find in health care today. Sociologists study these contradictions to better understand a very complex and important part of U.S. social life.

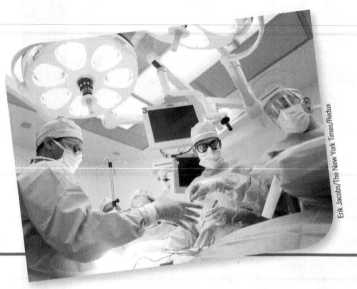

Erik Jacobs/The New York Times/Redux

The everyday life of a doctor like Atul Gawande is filled with its high points and low points: Some patients benefit from medical treatments they receive from physicians, whereas others have sustained injuries or developed illnesses that are too severe or are beyond the scope of current knowledge and practice in the health care system to be successfully resolved. Physicians are human beings just like the patients they treat; however, much more is expected of them because of the availability of health care in the United States and other high-income nations and because the dominant role of doctors in modern high-tech medicine has led many individuals to believe that virtually anything should be possible when it comes to one's health and longevity. However, this assumption is often not an accurate reflection of how health, illness, and health care actually work.

What does the concept of health mean to you? At one time, health was considered to be simply the absence of disease. However, the World Health Organization defines **health** as a state of complete physical, mental, and social well-being. According to this definition, health involves not only the absence of disease but also a positive sense of wellness. In other words, health is a multidimensional phenomenon: It includes physical, social, and psychological factors.

What do you think of when you hear the word "illness?" *Illness* refers to an interference with health; like health, illness is socially defined and may change over time and between cultures. For example, in the United States and Canada, obesity is viewed as unhealthy, whereas in other times and places, obesity indicated that a person was prosperous and healthy.

What happens when a person is perceived to have an illness or disease? Healing involves both personal and institutional responses to perceived illness and disease. One aspect of institutional healing is health care and the health care delivery system in a society. **Health care** is any activity intended to improve health. When people experience illness, they often seek medical attention in hopes of having their health restored. A vital part of health care is **medicine**—an institutionalized system for the scientific diagnosis, treatment, and prevention of illness.

In this chapter we will explore the dynamics of health, health care, and disability from a sociological perspective, as well as look at issues through the eyes of those who have experienced medical problems. Before reading on, test your knowledge about health, illness, and health care by taking the "Sociology and Everyday Life" quiz.

—Atul Gawande, M.D. (2002: 4, 5, 7), now a professor at Harvard University School of Public Health and a surgeon at Brigham and Women's Hospital in Boston, was a surgical resident when he wrote these words describing how he feels about the power and the limits of medicine.

How Much Do You Know About Health, Illness, and Health Care?

TRUE	FALSE		
T	F	1	The idea that everyone should have guaranteed health insurance coverage is accepted by nearly all Americans.
T	F	2	Worldwide, most HIV infections are spread by homosexual contact.
T	F	3	The primary reason that African Americans have shorter life expectancies than whites is the high rate of violence in central cities and the rural South.
T	F	4	The most common cause of death among Americans ages 15 to 24 is unintentional injuries.
T	F	5	Health care in most high-income, developed nations is organized on a fee-for-service basis as it is in the United States.
T	F	6	The medical–industrial complex has operated in the United States with virtually no regulation, and allegations of health care fraud have largely been overlooked by federal and state governments.
T	F	7	Media coverage of chronic depression and other mental conditions focuses primarily on these problems as "women's illnesses."
T	F	8	About two-thirds of U.S. adults are either overweight or obese.

Answers can be found at the end of the chapter.

 Discuss the relationship between the social environment and health and illness.

Health in Global Perspective

Why is it important to know about health and health care worldwide? Studying health and health care issues around the world offers each of us important insights on illness and how political and economic forces shape health care in nations. Disparities in health are glaringly apparent between high-income and low-income nations when we examine factors such as the prevalence of life-threatening diseases, rates of life expectancy and infant mortality, and access to health services. Statistics on global health typically run a year or two behind when you are reading this, but it is important to see trends in illness and health care. One important trend in HIV/AIDS data, for example, is that some progress is being made in reducing this disease but much more remains to be done. For example, approximately 2.1 million people worldwide became newly infected with HIV in 2013, and about 1.5 million died from

AIDS-related causes (UNAIDS, 2014). Between 2001 and 2013, however, new HIV infections decreased by 38 percent. Also, AIDS-related deaths worldwide decreased by 35 percent, to 1.5 million people, after a peak number of 2.4 million deaths in 2005. Tragically, since the start of the HIV/AIDS epidemic, more than 78 million people have become infected with HIV, and 39 million have died of AIDS-related illnesses (UNAIDS, 2014). However, AIDS is not the only disease reducing life expectancy in many nations.

Life expectancy refers to a statistical estimate of the average number of years that a person born in a specific

health
a state of complete physical, mental, and social well-being.

health care
any activity intended to improve health.

medicine
an institutionalized system for the scientific diagnosis, treatment, and prevention of illness.

life expectancy
a statistical estimate of the average number of years that a person born in a specific year will live.

year will live. Sometimes life expectancy is calculated from birth; other times it is calculated based on how long an individual at a given age can expect to live given present mortality rates. Worldwide, life expectancy at birth ranges from a low of 49.4 years in Chad to 89.6 years in Monaco (*CIA World Factbook,* 2015). Life expectancy in low-and middle-income nations is often reduced by problems such as infectious and parasitic diseases that are now rare in high-income, industrialized nations. Among these diseases are tuberculosis, polio, measles, diphtheria, meningitis, hepatitis, malaria, and leprosy.

The ***infant mortality rate*** is the number of deaths of infants under 1 year of age per 1,000 live births in a given year. In 2014 the infant mortality rate ranged from a low of 1.8 in high-income nations such as Monaco to a high of 117 in low-income nations such as Afghanistan (• Figure 18.1). Consider the vast difference in a nation where 117 infants out of every 1,000 born in a specific year die before their first birthday, as compared with another nation in which slightly less than 2 infants (statistically speaking) out of every 1,000 infants born in the same year die before their first birthday. A large proportion of these deaths occur during the infants' *first month* of life.

There are many reasons for these differences in life expectancy and infant mortality. Many people in low-income countries have insufficient or contaminated food; lack access to pure, safe water; and do not have adequate sewage and refuse disposal. Added to these hazards is a lack of information about how to maintain good health. Many of these nations also lack qualified physicians and health care facilities with up-to-date equipment and medical procedures.

Nevertheless, tremendous progress has been made in saving the lives of children and adults over the past 25 years. The average life expectancy at birth is more than 70 years in the following regions: the Arab States, East Asia and the Pacific, Europe and Central Asia, and Latin America and the Caribbean. By contrast, life expectancy at birth is 56.8 years in Sub-Saharan Africa (United Nations Development Programme, 2015). One of the key indicators of the level of human development in a nation or region is the rate at which children survive. Referred to as "child survival," this is often measured by the child's fifth birthday. However, even for children who survive past this birthday, poor health is still a major concern because it can permanently damage their cognitive development and ability to fully function as an adult (United Nations Development Programme, 2015).

World health statistics for 2015, the latest year for which comprehensive data are available, are informative about both our progress and our most pressing challenges in regard to health and illness:

- Worldwide, life expectancy at birth has increased six years for both men and women since 1990.
- Two-thirds of deaths worldwide are caused by non-communicable diseases.
- In low- and middle-income countries, only two-thirds of pregnant women with HIV receive antiretrovirals to prevent transmission to their baby.
- Over one-third of adult men smoke tobacco.
- Only one in three African children with suspected pneumonia receives antibiotics.
- Fifteen percent of women worldwide are obese.
- One-quarter of men have elevated blood pressure.
- In some countries, less than 5 percent of total government expenditure is on health.

Despite progress that has been made in life expectancy and medical care, many challenges remain. Although elevated blood pressure is a known risk factor for deaths from stroke and coronary heart disease, many people worldwide have undiagnosed or untreated hypertension. In Africa it is estimated that more than one-third of the population has high blood pressure, and this problem is growing worse. Being overweight or obese is a problem that increases the risk of coronary heart disease, ischemic stroke, type 2 diabetes, and some cancers, and the worldwide prevalence of these conditions has doubled over the past two decades. Worldwide, the amount of money spent on health care varies widely, and the extent to which persons derive improved medical attention or quality of life through higher expenditures (such as in the United States) is not easily measured. However, lack of adequate funds to pay for appropriate health care is also a pressing concern.

Natee K.Jindakum/Shutterstock.com

FIGURE 18.1   Japan has one of the lowest infant mortality rates in the world. Why? Finding the answer to this question would benefit people in every nation.

A key issue in health care is the cost and availability of advanced diagnostic and surgical technologies and lifesaving drugs around the world (• Figure 18.2). Advanced technologies such as the da Vinci surgery system (physician-directed robotic surgery) and new MRI machines cost in the range of $1.5 million to $3 million, not including expenses involved in creating specially designed suites to house the equipment.

Similarly, drugs to reduce pain and suffering or to save lives are very costly for individuals and medical institutions. The major pharmaceutical companies that hold patents on the most widely used drugs believe that their name-brand products need to be protected by law, whereas people in international human relief agencies believe that the most important concern is providing needed medication to the one-third of the world's population that does not have access to essential medicines. Transnational pharmaceutical companies fear that if they provide their name-brand drugs (if these are the only ones available to treat a specific condition) at lower prices in low-income

countries, this will undercut their major sales base in high-income countries. For this reason the companies are reluctant to provide inexpensive or free drugs to people in low-income countries while, at the same time, they have been aggressively marketing name-brand drugs to patients in high-income nations through television, the Internet, and social media. If we are to see a significant improvement in life expectancy and health among people of all nations, improvements are needed in accessibility to advanced medical technologies and effective drugs.

How about improvements in health and health care within one nation? Is there a positive relationship between the amount of money that a society spends on health care and the overall physical, mental, and social well-being of its people? Not necessarily. If there were such a relationship, people in the United States would be among the healthiest and most physically fit people in the world. Some estimates suggest that we spend as much as $2.9 trillion annually on health care. Moreover, health care spending continues to account for more than 17 percent of the gross domestic product (GDP) in the United States. Although these figures have not increased as rapidly as earlier estimates predicted, particularly with initial implementation of the Affordable Care Act and wider health coverage for more people, other factors may also be in play.

Some initial analyses have found that out-of-pocket medical expenses have increased for many patients as newer insurance plans, including those available under the Affordable Care Act, now have higher co-payments, deductibles, and other incidental costs that are passed on to patients along the way. In addition, when some patients find out how large their out-of-pocket expenses and deductibles will be, they do not elect to get the medical treatment or drugs that might improve their condition. Although the Affordable Care Act specifies that certain screening services must be provided at no cost to patients, the law does not always mandate that payments for any treatments that ensue from the test results will be paid for by patients' insurance plans. By contrast, in nations where residents are provided with universal health coverage, the goal is to provide all people with the preventive, curative, rehabilitative, and palliative health services they need while ensuring that the services do not create financial hardship for patients. *Preventive care* seeks to reduce the rate of illness by proactive measures before a person becomes ill. An example is a wellness program that offers diet and exercise plans to help promote fitness. *Curative services* involve treatment to alleviate illness or promote healing after an injury. *Rehabilitative services* seek to restore some or all of the patient's physical, sensory, and mental capabilities that were reduced or lost because of illness, disease, or injury. *Palliative services* are medical care

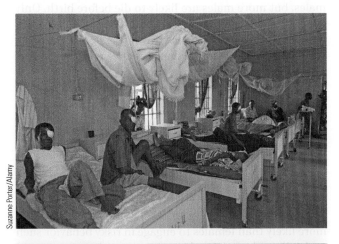

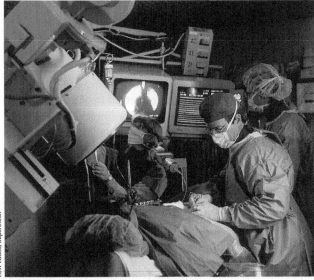

FIGURE 18.2 Access to quality health care is much greater for some people than for others. The factors that are involved vary not only for people within one nation but also across the nations of the world.

infant mortality rate
the number of deaths of infants under 1 year of age per 1,000 live births in a given year.

that helps reduce pain, symptoms, and stress associated with very serious health conditions. This type of treatment is directed toward improving a patient's quality of life but not necessarily extending the person's life.

Health in the United States

Looking at the United States specifically, why do you think that some of us are healthier than others? Is it biology—our genes—that accounts for this difference? Does the environment within which we live have an effect? How about our own individual lifestyle? Social epidemiology attempts to answer questions such as these.

 LO2 **Define** *social epidemiology* and identify key demographic factors that are studied by social epidemiologists.

Social Epidemiology

Social epidemiology is the study of the causes and distribution of health, disease, and impairment throughout a population (Weiss and Lonnquist, 2014). Typically, the target of social epidemiologists' investigations is disease agents, the environment, and the human host. *Disease agents* include biological agents such as insects, bacteria, and viruses that carry or cause disease; nutrient agents such as fats and carbohydrates; chemical agents such as gases and pollutants in the air; and physical agents such as temperature, humidity, and radiation. The *environment* includes the physical (geography and climate), biological (presence or absence of known disease agents), and social (socioeconomic status, occupation, and location of home) environments. The *human host* takes into account demographic factors (age, sex, and race/ethnicity), physical condition, habits and customs, and lifestyle (Weiss and Lonnquist, 2014). Let's look briefly at some of these factors.

Age Rates of illness and death are highest among the old and the young. Mortality rates drop shortly after birth and begin to rise significantly during middle age. As many people reach their late sixties or seventies, rates of chronic diseases and mortality increase. *Chronic diseases* are illnesses that are long term or lifelong and that develop gradually or are present from birth; in contrast, *acute diseases* are illnesses that strike suddenly and cause dramatic incapacitation and sometimes death (Weitz, 2013). Two of the most common sources of chronic disease and premature death are tobacco use, which increases mortality among both smokers and the people who breathe the tobacco smoke of others, and alcohol abuse, both of which are discussed later in this chapter.

The fact that rates of chronic diseases increase as people grow older has obvious implications for individuals, their families, and the entire nation. As the older population continues to grow in the United States, health care costs

will rise as the use of medical care services expands. Population aging has national implications, as we have already seen in patterns of health care spending, and institutional long-term care is particularly costly for older age groups. As the U.S. population continues to age in the twenty-first century, additional upward pressure will be placed on health care facilities and overall medical spending.

Sex Prior to the twentieth century, women had lower life expectancies than men because women had high mortality rates because of pregnancy and childbirth. Preventive measures have greatly reduced this cause of female mortality, and women now live longer than men. For babies born in the United States in 2015, for example, the estimated life expectancy at birth was 79.6 years. Males had a life expectancy of 77.1 years as compared with 81.9 years for females (*CIA World Factbook*, 2015). Even from the beginning of life, women have a slight biological advantage over males: Females have lower mortality rates than males in the prenatal stage and the first month of life. About 124 males are conceived for every 100 females, but more males are likely to die before birth. Only 105 boys are born for every 100 girls, and boys are more likely to be born prematurely and to have neonatal conditions that affect them. Boys are also about 18 percent more likely to die before their first birthday than girls.

Do you think gender roles and gender socialization might also contribute to differences in life expectancy? Yes, the kinds of work that men perform and societal pressures about what it means to be a "man" or a "woman" may contribute to differences in life expectancy by gender. Traditionally, more men have been employed in dangerous occupations such as commercial fishing, mining, construction, and public safety/firefighting (• Figure 18.3). As a result of gender roles and socialization, males may be more likely than females to engage in risky behavior such as drinking alcohol, smoking cigarettes (there is more social pressure on women not to smoke), using drugs, driving dangerously, and engaging in fights. Finally, women are more likely to use the health care system, with the result that health problems are identified and treated earlier (while there is a better chance of a successful outcome), whereas many men are more reluctant to consult doctors.

Because women on average live longer—have longer life expectancy at every age—than men, it is easy to jump to the conclusion that women are healthier than men. However, although men at all ages have higher rates of fatal diseases, women have higher rates of chronic illness and are more likely to extensively use health care services throughout their life.

Race/Ethnicity and Social Class Racial–ethnic differences are also visible in statistics pertaining to life expectancy. Not taking into account differences in education levels or other important variables, life expectancy for African American males is estimated at 71.6 years as compared to 78.9 for Latino (Hispanic) males and 76.4 for white (non-Hispanic) males. For African American females, life

FIGURE 18.3 Occupation and life expectancy may be related. Men are overrepresented in high-risk jobs, such as construction, that may affect their life expectancies.

FIGURE 18.4 Can your neighborhood be bad for your health? According to recent research, it can indeed, especially if it predominantly contains fast-food restaurants, liquor stores, and similarly unhealthy lifestyle options.

show that income and the neighborhood in which a person lives may be equally or more significant than race or ethnicity with respect to these issues (● Figure 18.4). How is it possible that the neighborhood you live in may significantly affect your risk of dying during the next year? Numerous studies have found that people have a higher survival rate if they live in better-educated or wealthier neighborhoods than if the neighborhood is low income and has low levels of education. Among the reasons that researchers believe that neighborhoods make a difference are the availability (or lack thereof) of safe areas to exercise, grocery stores with nutritious foods, and access to transportation, education, and good jobs. Many low-income neighborhoods are characterized by fast-food restaurants, liquor stores, and other facilities that do not afford residents healthy options.

As discussed in prior chapters, people of color are more likely to have incomes below the poverty line, and the poorest people typically receive less preventive care and less-optimal management of chronic diseases than do other people. People living in central cities, where there are high levels of poverty and crime, or in remote rural areas generally have greater difficulty in getting health care because most doctors prefer to locate their practice in a "safe" area, particularly one with a patient base that will produce a high income. Although rural Americans make up 20 percent of the U.S. population, only about 10 percent of the nation's physicians practice in rural areas, and fewer specialists such as cardiologists are available in these areas.

social epidemiology
the study of the causes and distribution of health, disease, and impairment throughout a population.

chronic diseases
illnesses that are long term or lifelong and that develop gradually or are present from birth.

acute diseases
illnesses that strike suddenly and cause dramatic incapacitation and sometimes death.

expectancy is 77.8 years as compared to 83.7 for Latina (Hispanic) females and 81.1 for white females (Minino, 2013). For both sexes, the widest life expectancy gap in 2013 was 6.6 years, and this was found between Hispanics at 81.4 years and African Americans (non-Hispanic) at 74.8 years.

Although race/ethnicity and social class are related to issues of health and mortality, research continues to

Medical Crises in the Aftermath of Disasters: From Oklahoma to Nepal

- Moore, Oklahoma (USA): Oklahoma City–Area Hospitals Report Hundreds of Patients After Tornado (KJRH.com, 2013)

- Kathmandu, Nepal: More Aid Arriving to Help Nepal Survivors Face Monsoon Season and Aftershocks (Americares.org, 2015)

Although Moore, Oklahoma, is thousands of miles away from Kathmandu, Nepal, both of these cities have had similar problems: how to take care of sick and injured patients quickly and effectively in the aftermath of deadly natural disasters. In Oklahoma and other U.S. states, devastating tornadoes have continued to rip through cities and the countryside in recent years, leaving hundreds dead and many others injured and sometimes homeless. Hospitals have been bombarded with injured patients whom rescue workers dug from the rubble or found in the streets. Although some of the injured had minor scrapes, bruises, or cuts, others had

multiple fractures, contusions, and potentially life-threatening internal injuries that required surgery.

In Nepal, a powerful earthquake rocked the country, set off Mount Everest avalanches, and created chaos throughout the region (Americares.org, 2015). This natural disaster left at least 8,600 people dead and many thousands more injured or missing. Shortages of water, food, clothing, shelter, and medicine further contributed to the hardship that many people experienced.

Shortly after the Nepal disaster, medical teams from India and other nations treated many patients in temporary

Major disasters such as the tornadoes that continue to rip through U.S. cities and the recent major earthquake in Nepal produce medical crises for health care workers and hospitals. Many injured patients, whom rescue workers often have dug from the rubble or found in the streets, must be transported to medical facilities and treated quickly.

Another factor is occupation. People with lower incomes are more likely to be employed in jobs that expose them to danger and illness—working in the construction industry or around heavy equipment in a factory, for example, or holding a job as a convenience store clerk or other position that exposes a person to the risk of armed robbery. Finally, people of color and poor people are more likely to live in areas that contain environmental hazards.

However, although Latinas/os are more likely than non-Latino/a whites to live below the poverty line, they have lower death rates from heart disease, cancer, accidents, and suicide, and they have a higher life expectancy. One explanation may be dietary factors and the strong family life and support networks found in many Latina/o families (Weiss and Lonnquist, 2014). Obviously, more research is needed on this point, for the answer might be beneficial to all people.

Health Effects of Disasters

When we hear about disease or impairment, most of us think about health problems associated with acute or

chronic conditions, ranging from colds and flu to diabetes and coronary disease. However, disasters also have a detrimental effect on people's health and well-being, and they also contribute to higher rates of disability and mortality. The World Health Organization defines a *disaster* as a "sudden ecological phenomenon of sufficient magnitude to require external assistance" (Goolsby, 2011). This statement means that a disruption of such magnitude has occurred that the community, state, or nation is unable to return to a normal condition following the event without outside assistance.

Disasters are commonly classified as natural or technological (human-made) disasters. Natural disasters include tornadoes, earthquakes, hurricanes, floods, volcano eruptions, tsunamis, and other potentially lethal conditions that originate in nature. By contrast, technological disasters include toxic spills, fires, and nuclear crises. Such a technological disaster occurred in 2013 in West, Texas, when a fire in a fertilizer plant produced a massive explosion, resulting in the loss of 15 lives, injuries to more than 200 other people, and many millions

medical camps and in the ruins of their communities. AmeriCare team members and other relief workers distributed medications and supplies, and workers coordinated relief efforts across agencies and government officials.

Given the horrific nature of major disasters such as these, what kind of national and global planning should be done to provide the best possible medical care for people? This question deserves careful consideration because the global impact of natural disasters has taken a turn for the worse in recent years. In 2013 alone, natural disasters worldwide killed a total of 29,163 people, and 78 percent (22,875) of the victims were in Asia (International Federation of Red Cross and Red Crescent Societies, 2014).

The World Health Organization is one of the organizations leading the drive for more-effective disaster risk management for health-related concerns. According to this organization (2011a), disaster-related injuries, diseases, deaths, disabilities, and psychosocial problems can be avoided or reduced by effective disaster risk management. Unlike traditional approaches of the health sector that respond to emergencies only after they happen, disaster risk management offers a proactive approach that emphasizes prevention, or at least reduction, of the problem. Although many factors go into creating a health-related risk-management plan, here are a few priorities:

1. Provide direction and support for disaster risk management at all levels, particularly local communities.

2. Assess potential risks to health and health systems, particularly from biological, natural, and technological sources, to enable early detection and warning to prompt action by the public and health workers.

3. Identify individuals, populations, infrastructure, and other community elements that are most vulnerable to harm in disasters and their aftermath (examples include young children, people over age 65, low-income persons, individuals with chronic illness or disability, and those who are socially isolated).

4. Evaluate the system's capacity to manage health risks when responding to, or recovering from, a major disaster.

Although these priorities are abstract, identifying them is the first step in disaster preparedness. It is important to manage health risks so that fewer lives will be lost and fewer injuries sustained in natural and technological disasters that are both hazardous to and frightening for all of us.

REFLECT & ANALYZE

How do inequality and poverty contribute to some people's vulnerability in a major disaster? Can physicians and other health care providers become better prepared for disasters? How might they accomplish this goal?

---

of dollars in damages to the surrounding area. In some ways an event such as this has commonalities with terrorist attacks and war, which also have a devastating effect on human life because of the potential for mass casualties and the physical and psychological trauma that follows such events.

Both physical and mental health effects from disasters are important concerns. The risk of physical injury during and after natural and technological disasters is high. Individuals who sustain wound injuries are at a high risk for tetanus, a serious, often fatal toxic condition. However, this condition is virtually 100 percent preventable with the appropriate vaccination. Similarly, infection is a potential problem with any wound or rash sustained in a disaster. For this reason, organizations such as the Centers for Disease Control and Prevention issue guidelines for health care professionals so that they will be aware of special precautions that are needed for emergency wound management. Often, the effects of disasters on mental health are not known for some period of time. The stress and trauma of survivors and those

who are seriously injured may linger for extended periods. In fact, some medical specialists compare the psychological effects of being a survivor of a deadly disaster such as an earthquake, tornado, hurricane, or terrorist attack to the anxiety or posttraumatic stress disorders exhibited by some wartime combat survivors.

How does treatment differ in disasters as compared to medical emergencies? Emergency medical services typically provide *maximal resources* to a *small number* of people, while disaster medical services are designed to direct *limited resources* to the *greater number* of individuals. Although most of us do not like to think about these issues in regard to health and illness, in the United States and around the world a disaster occurs somewhere almost daily, and some of these are of sufficient magnitude that individuals and nations are devastated for extended periods of time. For this reason, social epidemiologists have intensified research on this phenomenon, and the World Health Organization has created plans for disaster risk management for health (see "Sociology in Global Perspective").

Lifestyle Factors

As noted previously, social epidemiologists also examine lifestyle choices as a factor in health, disease, and impairment. We will examine three lifestyle factors as they relate to health: drugs, sexually transmitted diseases, and diet and exercise.

Drug Use and Abuse   What is a drug? There are many different definitions, but for our purposes a **drug** is any substance—other than food and water—that, when taken into the body, alters its functioning in some way. Drugs are used for either therapeutic or recreational purposes. *Therapeutic* use occurs when a person takes a drug for a specific purpose such as reducing a fever or controlling a cough. In contrast, *recreational* drug use occurs when a person takes a drug for no purpose other than achieving a pleasurable feeling or psychological state. Alcohol and tobacco are examples of drugs that are primarily used for recreational purposes; their use by people over a fixed age (which varies from time to time and place to place) is lawful. Other drugs—such as some antianxiety drugs or tranquilizers—may be used legally only if prescribed by a physician for therapeutic use but are frequently used illegally for recreational purposes.

Alcohol   The use of alcohol is considered an accepted part of the culture in the United States. According to a Gallup survey, more than six in ten adults say they consume alcoholic beverages. About 32 percent of Gallup survey respondents indicated that they had consumed between one and seven drinks in the past week while another nine percent stated that they had consumed eight or more drinks in that period. But not everyone consumes alcohol to that extent: Twenty-two percent of respondents indicated that they drink only occasionally, and slightly more than one-third stated that they do not drink at all (Saad, 2015).

Although some respondents indicated that they sometimes drink too much, particularly on weekends, most people who drink in the United States consume about four drinks per week (Saad, 2015).

Age, gender, and race are factors in regard to drinking behavior: Men tend to drink more than women and are more likely to drink beer while women are more likely to drink wine. Younger men between the ages of 18 to 49 are the heaviest drinkers. White Americans are more likely to drink than African Americans and Latinos/as (Hispanic Americans). When white Americans drink, they typically consume more alcohol than do nonwhite Americans (Saad, 2015).

Is alcohol consumption a problem? Even short-term alcohol use may have negative effects, including motor vehicle crashes, violence against others, sexual aggression, spread of HIV and sexually transmitted diseases (STDs), unplanned pregnancy, and binge drinking and the risky behavior associated with such activity. *Binge drinking* is defined as men drinking five or more alcoholic drinks within a short period of time or women drinking four or more drinks within a short period of time (U.S. CDC, 2012b). More than 38 million U.S. adults engage in binge drinking, but the extent of binge drinking varies from state to state (see • Figure 18.5). The largest number of drinks consumed within a short period of time among binge drinkers ranges from six drinks in the District of Columbia to nine drinks in Wisconsin (see • Figure 18.6).

Binge drinking is especially problematic because it can lead to death from alcohol poisoning. According to a study by the U.S. Centers for Disease Control and Prevention (2015), an average of six individuals die of alcohol poisoning each day in the United States. More than three-fourths of alcohol-poisoning deaths occur among adults between the ages of 35 and 64, and about 76 percent of those who die from alcohol poisoning are men. Although the largest number of alcohol-poisoning deaths are among white

FIGURE 18.5   Percentage of Adults Who Binge Drink
Source: U.S. CDC, 2012.

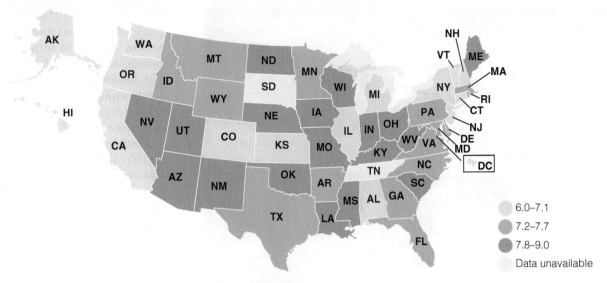

6.0–7.1	
7.2–7.7	
7.8–9.0	
Data unavailable	

FIGURE 18.6 The Average Largest Number of Drinks Consumed by Binge Drinkers on a Single Occasion
Source: U.S. CDC, 2012a.

Americans (non-Hispanic), Native Americans and Alaska Natives have the highest rate of alcohol-poisoning deaths per million people of any race because of the fact that they are a much smaller proportion of the total U.S. population.

Many long-term problems are associated with alcohol consumption as well. Chronic heavy drinking or alcoholism can cause permanent damage to the brain or other parts of the body. For alcoholics, the long-term negative health effects include *nutritional deficiencies* resulting from poor eating habits (chronic heavy drinking contributes to high caloric consumption but low nutritional intake); *cardiovascular problems* such as inflammation and enlargement of the heart muscle, high blood pressure, and stroke; and eventually *alcoholic cirrhosis*—a progressive development of scar tissue that chokes off blood vessels in the liver and destroys liver cells by interfering with their use of oxygen. The social consequences of heavy drinking are not always limited to the person doing the drinking. For example, abuse of alcohol and other drugs by a pregnant woman can damage her unborn fetus.

Can alcoholism be overcome? Many people get and stay sober. Some rely on organizations such as Alcoholics Anonymous or church support groups to help them overcome their drinking problem. Others undergo medical treatment and therapy sessions to learn more about the root causes of their problem.

Nicotine (Tobacco)   The nicotine in tobacco is a toxic, dependency-producing psychoactive drug that is more addictive than heroin. It is classified as a stimulant because it stimulates central nervous system receptors and activates them to release adrenaline, which raises blood pressure, speeds up the heartbeat, and gives the user a temporary sense of alertness. Although the overall proportion of smokers in the general population has declined somewhat since

the 1964 Surgeon General warning that smoking is linked to cancer and other serious diseases, tobacco is still the single most preventable cause of disease, disability, and death in the United States. Tobacco is responsible for about one in every five deaths in this country, and it is estimated that more than 450,000 people die prematurely from smoking or exposure to secondhand smoke each year. Even people who never light up a cigarette can be harmed by *environmental tobacco smoke*—the smoke in the air inhaled by nonsmokers as a result of other people's tobacco smoking and the residue of smoke on garments and furniture, for example. Researchers have found that environmental smoke is especially hazardous for nonsmokers who carpool or work with heavy smokers. Another 8.5 to 9 million people have a serious illness that is related to smoking (U.S. CDC, 2013c).

Illegal Drugs   In the United States, marijuana is the most extensively used drug that is illegal in most states. As of 2015, four states—Colorado, Washington, Alaska, and Oregon—had legalized the sale of recreational marijuana (as compared to marijuana for medicinal purposes). (See • Figure 18.7.) The District of Columbia also had legalized possession of small amounts of marijuana. By contrast, other states have passed medical marijuana laws that permit limited use of cannabis for certain medical conditions. The laws vary from state to state.

The use of marijuana has grown exponentially among young people in the nation's middle and high schools. Researchers in a 2012 study found that a high rate of marijuana use by eighth-, tenth-, and twelfth-grade students, combined with a drop in perceptions about the potential harm of such

drug
any substance—other than food and water—that, when taken into the body, alters its functioning in some way.

FIGURE 18.7   Although this is an unfamiliar sight throughout most of the United States, customers at The Clinic in Denver, Colorado, are able to legally purchase marijuana for recreational purposes as a result of the passage of Colorado Amendment 64, which allowed for the commercial sale of cannabis to the general public at licensed establishments in that state.

drug use, had contributed to regular or daily use of marijuana by more young people. According to the survey, as teens get older, their perception about the risk involved in smoking marijuana diminishes despite the fact that studies funded by the National Institutes of Health (NIH) have found that persons who used marijuana heavily in their teens and continued through their adulthood showed a significant drop in IQ between the ages of 13 and 38 (NIH, 2012). Moreover, even young people who used marijuana heavily before age 18 (when the brain is still developing) and quit taking it showed impaired mental abilities even after they no longer took the drug.

High doses of marijuana smoked during pregnancy can disrupt the development of a fetus and result in congenital abnormalities and neurological disturbances. Furthermore, some studies have found an increased risk of cancer and other lung problems associated with marijuana because its smokers are believed to inhale more deeply than tobacco users.

Another widely used illegal drug is cocaine. Cocaine may be either inhaled, injected intravenously, or smoked ("crack cocaine"). According to the National Institute on Drug Abuse, nearly five million Americans ages twelve and older have used cocaine, and one million have used crack at least once. People who use cocaine over extended periods of time have higher rates of infection, heart problems, internal bleeding, hypertension, stroke, and other neurological and cardiovascular disorders than do nonusers. Intravenous cocaine users who share contaminated needles are also at risk for contracting HIV.

Each of the drugs discussed in these few paragraphs represents a lifestyle choice that affects health. Whereas age, race/ethnicity, sex, and—at least to some degree—social class are ascribed characteristics, taking drugs is a voluntary action on a person's part.

Sexually Transmitted Diseases

The circumstances under which a person engages in sexual activity constitute another lifestyle choice with health implications. Although most people find consensual sexual activity enjoyable, it can result in transmission of certain *sexually transmitted diseases (STDs)*, including HIV/AIDS, chlamydia, gonorrhea, and syphilis. The latest estimates from the U.S. Centers for Disease Control and Prevention (CDC) show that there are about 20 million new STD infections in the United States each year. These new infections alone cost the health care system nearly $16 billion in direct medical costs.

Prior to 1960, the incidence of STDs in this country had been reduced sharply by barrier-type contraceptives (e.g., condoms) and the use of penicillin as a cure. However, in the 1960s and 1970s the number of cases of STDs increased rapidly with the introduction of the birth control pill, which led to women having more sexual partners and couples being less likely to use barrier contraceptives. Since that time, many people have become more aware of STDs, and organizations such as the CDC have mounted aggressive public service campaigns to make individuals aware of choices that contribute to sexual health.

Chlamydia   Chlamydia is the most commonly reported STD in terms of the number of cases (more than 1.4 million) reported annually to the U.S. Centers for Disease Control and Prevention (2014a). However, the CDC estimates that more than half of new cases remain undiagnosed and unreported each year. This is an issue of great concern because chlamydial infections in women, which are usually asymptomatic, can result in pelvic inflammatory disease (PID)—a major cause of infertility, ectopic pregnancy, and chronic pelvic pain. Like other STDs, chlamydia can also contribute to the transmission of HIV. Women are at higher risk of chlamydia, as shown in ● Figure 18.8. However, in recent years more men have also been tested for chlamydial infection, resulting in an increase in the number of male cases reported. Rates of this disease vary widely among racial and ethnic populations: The rate for African Americans in 2012 was seven times the rate for white Americans (U.S. CDC, 2014a).

Gonorrhea   Gonorrhea, the second most commonly reported STD in the United States, is caused by a bacterium that can grow and multiply easily in the warm, moist areas of the reproductive tract. It can also grow in the mouth, throat, eyes, and anus. Untreated gonorrhea may lead to serious outcomes such as tubal infertility, ectopic pregnancy, and chronic pelvic pain. Gonorrhea can help the

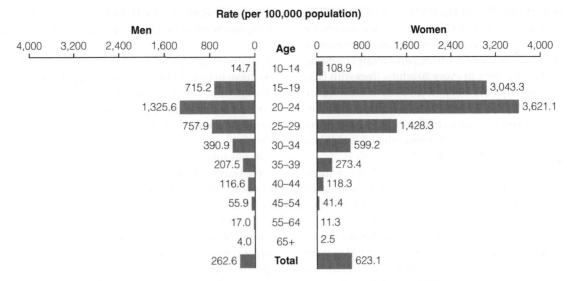

Rate (per 100,000 population)

Men	Age	Women
14.7	10–14	108.9
715.2	15–19	3,043.3
1,325.6	20–24	3,621.1
757.9	25–29	1,428.3
390.9	30–34	599.2
207.5	35–39	273.4
116.6	40–44	118.3
55.9	45–54	41.4
17.0	55–64	11.3
4.0	65+	2.5
262.6	**Total**	623.1

FIGURE 18.8 Chlamydia—Rates by Age and Sex, United States, 2012
Source: U.S. CDC, 2014a.

transmission of HIV infection, and in advanced cases it can spread to the blood or joints and be life threatening.

Specific populations are at higher risk for gonorrhea. Among African Americans, rates are nearly fifteen times higher than rates among white Americans. The South has the highest gonorrhea rate among the four regions of the country (U.S. CDC, 2014b). Although antibiotics can cure gonorrhea, some strains of the infection have developed resistance to each of the antibiotics used for treatment.

Syphilis Syphilis is a genital ulcerative disease that can be acquired not only by sexual intercourse but also by kissing or coming into intimate bodily contact with an infected person. Untreated early syphilis in pregnant women results in perinatal death in up to 40 percent of cases and may lead to infection of the fetus in up to 80 percent of cases if syphilis was acquired during the four years prior to the pregnancy (U.S. CDC, 2014c). Untreated syphilis can, over time, cause cardiovascular problems, brain damage, and even death. Penicillin can cure most cases of syphilis as long as the disease has not spread.

Like gonorrhea, syphilis is a major health problem in the South and in urban areas in other regions. High-risk populations include young African American men, all men who have sex with other men, and men who have sex with both men and women (U.S. CDC, 2012c).

AIDS AIDS (acquired immunodeficiency syndrome), which is caused by HIV (human immuno-deficiency virus), is among the most significant health problems that this nation—and the world—faces today. Although AIDS almost inevitably ends in death, no one actually dies *of* AIDS. Rather, AIDS reduces the body's ability to fight diseases, making a person vulnerable to many diseases—such as pneumonia—that result in death.

AIDS was first identified in 1981, and the total number of AIDS-related deaths in the United States through 1985 was only 12,493; however, the numbers rose rapidly and precipitously after that. In recent years the CDC has developed new estimates of HIV prevalence, meaning the total number of people living with HIV. Current estimates suggest that 1.14 million people in the United States ages 13 years and older are in this category. Nearly 16 percent (180,900 people) are not aware that they are infected with HIV. The majority of new HIV infections are attributed to male-to-male sexual behavior, followed by heterosexual transmission and intravenous drug use. Gay, bisexual, and other men who have sex with men (MSM) of all races and ethnicities are the population most profoundly affected by HIV (U.S. CDC, 2013b). Although African Americans make up about 12 percent of the total U.S. population, they account for nearly half (44 percent) of new HIV infections each year. Hispanics/Latinos/as represent 16 percent of the U.S. population but account for 21 percent of new HIV infections. Heterosexuals make up about 25 percent of estimated new HIV infections each year. Most new HIV infections among women are attributed to heterosexual contact or injection drug use (U.S. CDC, 2013b).

According to the U.S. Centers for Disease Control and Prevention (2012a), racial disparities in HIV/AIDS are driven by factors such as the higher HIV prevalence (proportion of people living with HIV) in many African American and Latino communities, which means that individuals who reside in these communities have a higher risk of infection with every sexual encounter. Other factors include the stigma associated with HIV/AIDS and homophobia, which may prevent people from seeking prevention or medical attention.

Worldwide, the number of people with HIV/AIDS continues to increase, but progress has been made in some countries in addressing this epidemic. An estimated 35.3 million people worldwide are living with HIV/AIDS, including 2.1 million between the ages of 10 and 19 (AIDS. gov, 2013). HIV is the world's leading infectious killer.

According to the World Health Organization (WHO), about 2.3 million new cases of HIV were reported worldwide in 2012. Almost all (95 percent) of new infections are reported in low- and middle-income countries, particularly sub-Saharan Africa, where nearly 1 in every 20 adults lives with HIV. WHO also reports that 3.34 million children worldwide are infected. Most of the children were infected by their HIV-positive mothers during pregnancy, childbirth, or breast feeding. Even more alarming is the fact that more than 700 children per day are newly infected with HIV (AIDS.gov, 2013).

HIV is transmitted through unprotected (or inadequately protected) sexual intercourse with an infected partner (either male or female), by sharing a contaminated hypodermic needle with someone who is infected, by exposure to blood or blood products (usually from a transfusion), or, as previously stated, by an HIV-inflected woman who passes the virus on to her child. It is not transmitted by casual contact such as shaking hands.

Over the past decade, global efforts have been increased to reduce the HIV/AIDS epidemic. More people are receiving standard antiretroviral therapy (ART), which is a combination of at least three antiretroviral (ARV) drugs to maximally suppress the HIV virus and stop the progression of the disease. Massive reductions in rates of death and suffering have occurred when the potent ART regimen is used, particularly in early stages of the disease. ART is also used for the prevention of HIV infection, particularly among pregnant women, children, and other high-risk populations.

Staying Healthy: Diet and Exercise   Up to this point we have primarily been looking at problematic aspects of lifestyle that have a negative effect on health; however, lifestyle choices also include positive actions such as a healthy diet and good exercise. Over the past several decades, a dramatic improvement in our understanding of food and diet has taken place, and many people in the United States have begun to improve their dietary habits. A significant portion of the population now eats larger amounts of vegetables, fruits, and cereals, substituting unsaturated fats and oils for saturated fats. However, many people in the United States still eat diets that are deficient in fruits and vegetables, and their meals contain more fats and added sugars than are recommended by current dietary guidelines. Eating behavior such as this contributes to obesity, which may shorten the average U.S. life expectancy by 2 to 5 years, reversing the steady increase in life expectancy that has occurred over the past 200 years. ● Figure 18.9 shows adult obesity rates across the United States.

Obesity increases the risk of health problems such as hypertension, type 2 diabetes, and certain types of cancer. It is estimated that more than one-third of all U.S. adults are obese and that many others are overweight. One of the important

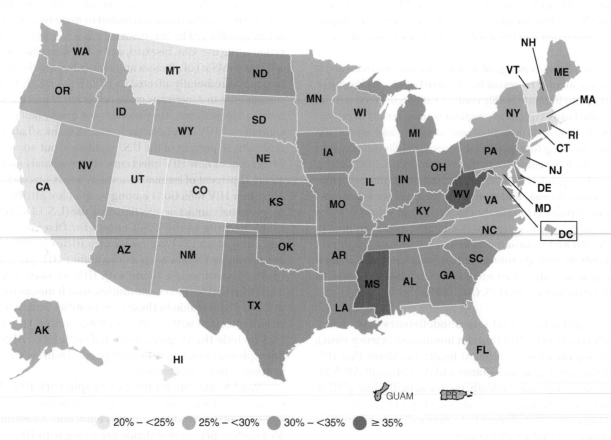

20% – <25%   25% – <30%   30% – <35%   ≥ 35%

FIGURE 18.9   Adult Obesity in the United States, 2013
Source: U.S. CDC, 2013.

ways to reduce obesity is through exercise. Regular exercise (at least three times a week) keeps the heart, lungs, muscles, and bones in good health and slows the aging process.

Health Care in the United States

Understanding health care as it exists in the United States today requires a brief examination of its history. During the nineteenth century, people became doctors in this country either through apprenticeships, purchasing a mail-order diploma, completing high school and attending a series of lectures, or obtaining bachelor's and M.D. degrees and studying abroad for a number of years. At that time, medical schools were largely proprietary institutions, and their officials were often more interested in acquiring students than in enforcing standards. The state licensing boards established to improve medical training and stop the proliferation of "irregular" practitioners failed to slow the growth of medical schools, and their number increased from 90 in 1880 to 160 in 1906. Medical school graduates were largely poor and frustrated because of the overabundance of doctors and quasi-medical practitioners, so doctors became highly competitive and anxious to limit the number of new practitioners. The obvious way to accomplish this was to reduce the number of medical schools and set up licensing laws to eliminate unqualified or irregular practitioners (Kendall, 1980).

LO3 **Explain** how the profession of medicine emerged in the United States.

The Rise of Scientific Medicine and Professionalism

Although medicine had been previously viewed more as an art than as a science, several significant discoveries during the nineteenth century in areas such as bacteriology and anesthesiology began to give medicine increasing credibility as a science. At the same time that these discoveries were occurring, the ideology of science was being advocated in all areas of life, and people came to believe that almost any task could be done better if the appropriate scientific methods were used. To make medicine in the United States more scientific (and more profitable), the Carnegie Foundation (at the request of the American Medical Association and the forerunner of the Association of American Medical Colleges) commissioned an official study of medical education. The "Flexner report" that resulted from this study has been described as the catalyst of modern medical education but has also been criticized for its lack of objectivity.

The Flexner Report To conduct his study, Abraham Flexner met with the leading faculty at the Johns Hopkins University School of Medicine to develop a model of how

medical education should take place; he next visited each of the 155 medical schools then in existence, comparing them with the model. Included in the model was the belief that a medical school should be a full-time, research-oriented, laboratory facility that devoted all of its energies to teaching and research, not to the practice of medicine (Kendall, 1980). It should employ "laboratory men" to train students in the "science" of medicine, and the students should then apply the principles they had learned in the sciences to the illnesses of patients (Brown, 1979). Only a few of the schools Flexner visited were deemed to be equipped to teach scientific medicine; nonetheless, his model became the standard for the profession.

As a result of the Flexner report (1910), all but two of the African American medical schools then in existence were closed, and only one of the medical schools for women survived. As a result, white women and people of color were largely excluded from medical education for the first half of the twentieth century. Until the civil rights movement and the women's movement of the 1960s and 1970s, virtually all physicians were white, male, and upper or upper-middle class.

The Professionalization of Medicine Despite its adverse effect on people of color and women who might desire a career in medicine, the Flexner report did help *professionalize* medicine (• Figure 18.10). When we compare post-Flexner medicine with the characteristics of professions, we find that it meets those characteristics:

1. *Abstract, specialized knowledge.* Physicians undergo a rigorous education that results in a theoretical understanding of health, illness, and medicine. This education provides them with the credentials, skills, and training associated with being a professional.

2. *Autonomy.* Physicians are autonomous and (except as discussed subsequently in this chapter) rely on their own judgment in selecting the appropriate technique for dealing with a problem. They expect patients to respect that autonomy.

3. *Self-regulation.* Theoretically, physicians are self-regulating. They have licensing, accreditation, and regulatory boards and associations that set professional standards and require members to adhere to a code of ethics as a form of public accountability.

4. *Authority.* Because of their authority, physicians expect compliance with their directions and advice. They do not expect clients to argue about the advice rendered (or the price to be charged).

5. *Altruism.* Physicians perform a valuable service for society rather than acting solely in their own self-interest. Many physicians go beyond their self-interest or personal comfort so that they can help a patient.

However, with professionalization, licensed medical doctors gained control over the entire medical

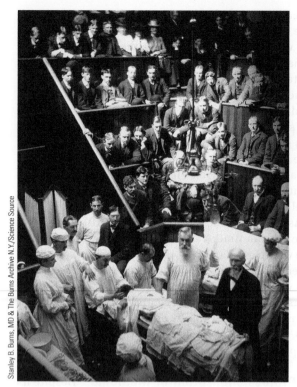

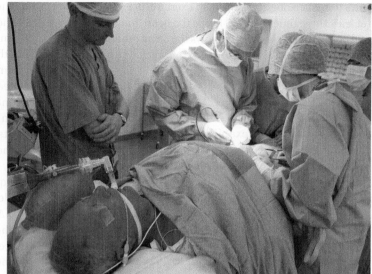

FIGURE 18.10 A sight similar to the one on the left might have greeted Abraham Flexner as he prepared his report on medical education in the United States: students observing while their professor demonstrates surgical techniques. And although today's health care facilities look very different, many of the same teaching techniques are still employed.

establishment, a situation that has continued until the present and—despite current efforts at cost control by insurance companies and others—may continue into the future.

Medicine Today

Throughout its history in the United States, medical care has been on a *fee-for-service* basis: Patients have been billed individually for each service they receive, including treatment by doctors, laboratory work, hospital visits, prescriptions, and other health-related expenses. Fee for service is an expensive way to deliver health care because there are few restrictions on the fees charged by doctors, hospitals, and other medical providers. For this reason the Affordable Care Act, passed in 2010, proposed to change how health care is paid for and the ways in which some services are provided.

There are both good and bad sides to the fee-for-service approach. The good side is that in the "true spirit" of capitalism, coupled with the hard work and scholarship of many people, this approach has resulted in remarkable advances in medicine. The bad side of fee-for-service medicine is its inequality of distribution. In effect, the United States has a two-tier system of medical care. For the most part, those who can afford it are able to get good medical treatment and comfortable surroundings.

However, top-tier medical care is not within the budget of most people. The annual cost of health care per person in the United States rose from $141 in 1960 to $8,233 in 2012 (pbs.org, 2012). (• Figure 18.11 reflects how expenditures for health care have increased from 1993 to 2014.)

Keeping in mind the issues that have been raised throughout this text regarding income disparity in the United States, the questions to be considered at this point are "Who pays for medical care, and how?" and "What about the people who simply cannot afford adequate medical care?"

Discuss how U.S. health care was paid for in the past and how this is in the process of change since passage of the Affordable Care Act (Obamacare).

Paying for Medical Care in the United States

Until recently, the United States was the only high-income nation without some form of universal health coverage for all citizens. As previously mentioned, the U.S. Congress passed a sweeping health care reform bill in 2010 that would gradually bring about some changes in how health care is funded. Let's look first at the health reform legislation and then compare its provisions with previous and current methods of funding of health care in the United States.

The Affordable Care Act of 2010 After a lengthy struggle in the U.S. Congress, a major health care reform bill was signed into law in 2010. One of the central tenets in the law was the creation of a new insurance marketplace that made it possible for individuals and families without coverage and small business owners to pool their resources to increase their buying power in order to make health insurance more

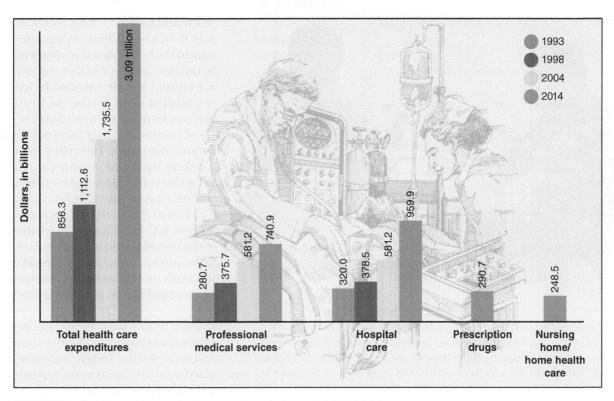

FIGURE 18.11   Increase in Cost of Health Care, 1993–2014
Source: Plunkett Research, 2014.

affordable (● Figure 18.12). Private insurance companies would compete for their business based on cost and quality. Advocates of the law believed that it might be a first step in curbing abuses in the insurance industry.

Health care reform was originally scheduled to occur in the following stages:

● **2010:** Adults who had been unable to get coverage because of a preexisting condition could join a high-risk insurance pool (as a stopgap measure until the competitive health insurance marketplace was scheduled to begin in 2014). Insurance companies would have to cover children with preexisting conditions. Policies could not be revoked when people got sick. Preventive services would be fully covered without co-pays or deductibles. Dependent children could remain on their parents' insurance plans until they reached the age of twenty-six.

● **2011:** Medicare recipients would have access to free annual wellness visits with no cost for preventive care, and those recipients who had to pay out of pocket for prescription drugs would receive substantial discounts.

● **2012:** The federal government would provide additional money for primary-care services, and new incentives would be offered to encourage doctors to join together in accountability-care organizations. Hospitals with high readmission rates would face stiff penalties.

● **2013:** Households with incomes above $250,000 would be subject to higher taxes to help pay for health care reform. Medicare would launch "payment bundling" so that hospitals, doctors, and other health care providers could be paid on the basis of patient outcome, not services provided.

● **2014:** Most people would be required to buy health insurance or pay a penalty for not having it. Insurance companies could not deny a policy to anyone based on health status, nor could they refuse to pay for treatment on the basis of preexisting health conditions. Annual limits on health care coverage would be abolished. Each state would have to open a health insurance exchange, or marketplace, so that individuals and small businesses without coverage could comparatively shop for health packages. Tax credits would make insurance and health care more affordable for those who earn too much to qualify for Medicaid.

● **2018:** Insurance companies and plan administrators would pay a 40 percent excise tax on all family plans costing more than $27,500 per year.

● **2019:** The health reform law should have reduced the number of uninsured people by 32 million, leaving about 23 million uninsured. About one-third of the uninsured would be immigrants residing in the country without legal documentation.

Joe Raedle/Getty Images

FIGURE 18.12 Following passage of the Affordable Care Act, many people enrolled either in person or online for health insurance coverage. It was estimated in 2015 that "Obamacare" provided health insurance to nearly 17 million formerly uninsured people. However, the Affordable Care Act has undergone many challenges and may face additional ones from political leaders.

These are a few of the highlights of the plan, which is a document of more than 2,000 pages in length. However, in 2012, after a series of legal challenges and an appeal of a decision of the U.S. Court of Appeals of the Eleventh Circuit, in Atlanta, the U.S. Supreme Court heard a challenge to the 2010 health care law on the issue of whether Congress had overstepped its constitutional authority by requiring most people in the United States to purchase health insurance or be assessed a penalty. According to the Obama administration, Congress was justified in this action by the constitutional power granted to this legislative body "to regulate commerce" and "to lay and collect taxes." After extensive deliberations, the U.S. Supreme Court ruled that Congress did not exceed its constitutional authority, and the Affordable Care Act was further implemented following the Court's decision.

The latest challenge to the Affordable Care Act came in 2015, when the U.S. Supreme Court heard *King vs. Burwell,* a case that involved a specific section of the law that says federal subsidies (tax credits) would be available for people who purchase insurance through exchanges set up by the state. Some states did set up their own exchanges, or marketplaces; however, other states opted to let the federal government set up the exchange rather than establishing their own. In *King vs. Burwell,* the plaintiffs argued that the language of the law means that *only* people enrolled in the state-run marketplaces—but not those in federally run marketplaces—are eligible to get subsidies from the federal government. Across all 50 states, about 85 percent of insurance marketplace customers were eligible for the subsidies based on their income. However, if the subsidies had been taken away for those who enrolled in federally established marketplaces, many people who had signed up for insurance under the Affordable Care

Act would have no longer been able to afford it. In a 6–3 ruling the Supreme Court upheld the health care law when a majority of the Justices decided that the estimated 6.4 million people enrolled in federally established plans run by the Healthcare .gov marketplace were entitled to keep the tax subsidies that help them pay their insurance premiums. This ruling was clearly a victory for the Affordable Care Act and encouraged President Obama to state that "the Affordable Care Act is here to stay."

To better understand how and why the Affordable Care Act has been so controversial in the United States, let's look at how heath insurance has been funded through private insurance, public insurance, health maintenance organizations, and managed care.

Private Health Insurance   Private health insurance is largely paid for by businesses and households. Today, employer-sponsored insurance plans cover approximately 149 million persons below the age of 65.

Looking back to the 1960s, medical insurance programs began to expand, and third-party providers (public and private insurers) started picking up large portions of doctor and hospital bills for insured patients. With third-party fee-for-service payment, patients pay premiums into a fund that in turn pays doctors and hospitals for each treatment the patient receives.

Private health insurance premiums have continued to increase for both individuals and families. The average annual premium for an individual in an employer-sponsored health insurance plan was $6,025 in 2014, up from $5,615 in 2012. The average annual premium for family coverage was $16,834 in 2014, up from $15,745 in 2012. In 2014 workers contributed (on average) $4,823, and employers paid (on average) $12,011 of the cost (Kaiser Family Foundation, 2015). These costs continue to rise annually, and most analysts believe they will be largely unaffected by provisions in the Affordable Care Act.

Over the years, some health care advocates have argued that a third-party fee-for-service approach is the best and most cost-efficient method of delivering medical care. Others have argued that fee for service is outrageously expensive and a very cost-ineffective way in which to provide for the medical needs of people in this country, particularly those who are without health insurance coverage. According to critics, third-party fee for service contributes greatly to medical inflation because it gives doctors and hospitals an incentive to increase medical services. In other words, the more services they provide, the more fees they charge and the more money they make. Patients have no incentive to limit their visits to doctors or hospitals because they have already paid their premiums and feel entitled to medical care, regardless of the cost. This is one of the spiraling costs that advocates of health care reform hope will be reduced.

Public Health Insurance   Since the 1960s, the United States has had two nationwide public health insurance programs, Medicare and Medicaid. In 1965 Congress enacted Medicare, a federal program for people ages 65 and over (who are covered by Social Security or railroad retirement insurance or who have been permanently and totally disabled for two years or more). This program was primarily funded through Social Security taxes paid by current workers. Medicare has several components, referred to as Part A, Part B, Part C, and Part D. Medicare Part A (hospital insurance) helps cover inpatient care in hospitals, skilled nursing facilities, hospice, and home health care. Part B (medical insurance) helps cover doctors' services and outpatient care. It covers some of the services of physical and occupational therapists, and some home health care. Most people pay a monthly premium for Part B. Medicare Part C is an insurance plan run by private companies approved by Medicare. Medicare Part D provides prescription drug coverage.

Medicaid is the federal government's health care program for low-income and disabled persons and certain groups of seniors in nursing homes. Medicaid is jointly funded by federal–state–local monies, and various factors are taken into account when determining whether a person is eligible for Medicaid. Among these are age, disability, blindness, and pregnancy. Income and resources are also taken into consideration, as well as whether the person is a U.S. citizen or a lawfully admitted immigrant. Each state has its own rules regarding who may be covered under Medicaid, and some provide time-limited coverage for specific categories of individuals, such as uninsured women with breast or cervical cancer or those individuals diagnosed with TB (tuberculosis).

As compared to Medicare, Medicaid has had a more tarnished image throughout its history. Unlike Medicare recipients, who are often seen as "worthy" of their health care benefits, Medicaid recipients have been stigmatized by politicians and media outlets for their participation in a "welfare program." For a number of years, many physicians refused to take Medicaid patients because the administrative paperwork is burdensome and reimbursements are low—typically less than one-half of what private insurance companies pay for the same services. The Affordable Care Act was designed to improve health care for low-income individuals, particularly if the state in which they resided participated in Medicaid expansion. By 2015, only thirty states and the District of Columbia had approved this expansion, with eighteen states declining to do so. However, even in some states with the expansion plans, all adults are not covered, or they are not covered at the same cost-sharing formula as other states. Consequently, many low-income persons living in states that did not expand Medicaid coverage remain without health insurance coverage today.

Preventive Health Care Services   Under the Affordable Care Act, some preventive services must be provided for individuals enrolled in employer-funded health plans or individual health insurance policies that were created after March 23, 2010. Among the preventive services covered at no cost are screening for hypertension (high blood pressure), diabetes, high cholesterol, and cancer. There is also counseling on quitting smoking, losing weight, eating healthfully, treating depression, and reducing alcohol use. Provision is made for certain types of routine vaccinations and flu and pneumonia shots, as well as well-baby and well-child visits to the physician. It is hoped that early detection and preventive measures will cut down the overall cost of health care.

Health Maintenance Organizations (HMOs)   Created in an effort to provide workers with health coverage by keeping costs down, *health maintenance organizations* **(HMOs)** provide, for a set monthly fee, total care with an emphasis on prevention to avoid costly treatment later. About 13 percent of workers in employer-sponsored insurance plans are enrolled in HMOs. In these plans, doctors do not work on a fee-for-service basis, and patients are encouraged to get regular checkups and to practice good health practices (e.g., exercise and eat right). However, research shows that preventive care is good for the individual's health but does not necessarily lower total costs. As long as patients use only the doctors and hospitals that are affiliated with their HMO, they pay no fees, or only small co-payments, beyond their insurance premiums.

Recent concerns about physicians being used as gatekeepers who might prevent some patients from obtaining referrals to specialists or from getting needed treatment have resulted in changes in the policies of some HMOs, which now allow patients to visit health care providers outside an HMO's network or to receive other previously unauthorized services by paying a higher co-payment. However, critics charge that those HMOs whose primary-care physicians are paid on a capitation basis—meaning that they receive only a fixed amount per patient whom they see, regardless of how long they spend with that patient—in effect encourage doctors to undertreat patients.

Managed Care   Another approach to controlling health care costs in the United States is known as *managed care*: any system of cost containment that closely monitors and controls health care providers' decisions about medical procedures, diagnostic tests, and other services that should be provided to patients. One type of managed care in the United States is a *preferred provider organization* (PPO), which is an organization of medical doctors, hospitals, and

---

health maintenance organizations (HMOs)
companies that provide, for a set monthly fee, total care with an emphasis on prevention to avoid costly treatment later.

managed care
any system of cost containment that closely monitors and controls health care providers' decisions about medical procedures, diagnostic tests, and other services that should be provided to patients.

other health care providers that enter into a contract with an insurer or a third-party administrator to provide health care at a reduced rate to patients who are covered under specific insurance plans. In most managed-care programs, patients choose a primary-care physician from a list of participating doctors. Unlike many of the HMOs, when a patient covered under a PPO plan needs medical services, he or she may contact any one of a number of primary-care physicians or specialists who are "in-network" providers. Like HMOs, most PPO plans do contain a pre-certification requirement in which scheduled (nonemergency) hospital admissions and certain kinds of procedures must be approved in advance. Through measures such as this, these insurance plans have sought unsuccessfully to curb the rapidly increasing costs of medical care and to reduce the extensive paperwork and bureaucracy involved in the typical medical visit. About 58 percent of workers in employer-sponsored health care programs participate in PPO plans.

For the foreseeable future, HMOs and PPOs are supposed to remain somewhat the same. After the passage of the health care reform measure, the Obama administration widely publicized a statement that people in plans such as these would not be affected by the new law. Rather, they would have assurance that they could get health care coverage even if they lost their job, changed jobs, moved out of state, got divorced, or were diagnosed with a serious illness.

The Uninsured   Despite existing public and private insurance programs, about 41.3 million people in the United States did not have health insurance coverage in 2013, the latest year for which comprehensive data are available. Although full-time workers are more likely to have health insurance than those who work less than full time or persons who are unemployed, having a full-time job is still no guarantee of health insurance coverage. Most of the uninsured were in working families but did not have access to or could not afford employer-sponsored health insurance coverage. In a worst-case scenario, some who are uninsured (despite being employed full time) make too little to afford health insurance but too much to qualify for Medicaid. Even if their employers offer group health insurance, some employees cannot afford their share of the premiums or the out-of-pocket expenditures such as co-payments.

Although children under age nineteen are the most likely to have health insurance coverage because many qualify for certain government health care programs, • Figure 18.13 shows that the uninsured rate for children in poverty (9.9 percent) is greater than the rate for children who do not live in poverty (7.5 percent). Children who are more likely to be uninsured are ones who live in poverty-level or in lower-income families. Less than 5 percent of children in households with income above $100,000 are uninsured. The highest rate of uninsured children by race is for Hispanics (any race) and by place of birth for those who are not U.S. citizens, particularly those whose families are more-recent immigrants to this country. The rate of uninsured children would be even higher if there were not

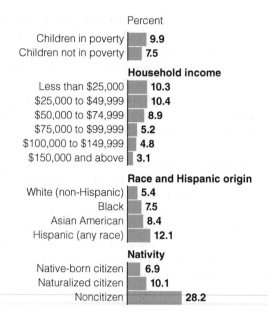

FIGURE 18.13   Uninsured Children by Poverty Status, Household Income, Age, Race, Hispanic Origin, and Nativity, 2013
Source: Smith and Medalia, 2014.

state-sponsored programs such as CHIP, a health insurance plan for children that is offered on a sliding scale with premiums (down to zero) based on family income.

In all age categories, race and ethnicity are important factors in health insurance coverage. Although every racial and ethnic group is affected, Hispanics (Latinos/as) and African Americans are more likely to be uninsured than are white (non-Hispanic) Americans. For Hispanic Americans the uninsured rate in 2013 was 24.3 percent, as compared to African Americans at 15.9 percent, Asian Americans at 14.5 percent, and white (non-Hispanic) Americans at 9.8 percent. These figures show a decline in percentage over earlier years, perhaps because of increased coverage through the Affordable Care Act.

As previously stated, under the 2010 Affordable Care Act the health insurance marketplace was implemented in 2014. This system was designed to make buying health coverage easier and more affordable by allowing individuals and small businesses to compare health plans, get answers to questions, find out if they are eligible for tax credits for insurance plans bought in the new marketplace or programs such as the Children's Health Insurance Program (CHIP), and enroll in the health plan that best meets their needs. Every health insurance plan in the new marketplace will offer comprehensive coverage, from doctors to medications to hospital visits. Persons wishing to purchase health insurance will be able to compare prices, benefits, quality, and other features that may be of importance to them so that they can make an informed choice about which plan they want to purchase.

At the time of this writing (June 2015), about 11.7 million people were enrolled in state and federal marketplaces. In addition, 5.7 million young people were able to stay on their parents' plan because of a provision in the Affordable Care Act. Similarly, over 10.8 million had enrolled in Medicaid or CHIP. Overall, if these estimates are accurate, the uninsured rate in the United States fell from 18 percent in 2013 to slightly less than 12 percent in 2015. This change translates to about 16.4 million fewer uninsured persons in the United States in a two-year time period (Obamacare Facts, 2015). Clearly, these enrollment figures are applauded by those who approve of the goals of the Affordable Care Act, but the program still has many political detractors. It will be interesting to see how further implementation takes place and what its effects are on the quality and cost of health care in the future.

LO5 **Compare** how the United States pays for health care with how other nations provide health services for their citizens.

Paying for Medical Care in Other Nations

Other nations have various ways in which they provide health care for their citizens. Let's examine how other nations pay for health care.

Canada Health care in Canada is delivered through a publicly funded health care system. Services are provided by private entities and are mostly free to patients. Information remains private between physician and patient, and the government is not involved in patient care. Each citizen receives a health card, and all patients receive the same level of care. As long as a person's premiums are paid up, health coverage is not affected by losing or changing jobs.

How did the contemporary Canadian health care system get started? In 1962 the government of the province of Saskatchewan implemented a health insurance plan despite opposition from doctors, who went on strike to protest the program. The strike was not successful, as the vast majority of citizens supported the government, which maintained health services by importing doctors from Great Britain. The Saskatchewan program proved itself viable in the years following the strike, and by 1972 all Canadian provinces and territories had coverage for medical and hospital services (Murray, Linden, and Kendall, 2012). As a result, Canada has *universal health care*—a health care system in which all citizens receive medical services paid for by tax revenues. In Canada these revenues are supplemented by insurance premiums paid by all taxpaying citizens.

One major benefit of the Canadian system is a significant reduction in administrative costs. Whereas more than 20 percent of the U.S. health care dollar represents administrative costs, in Canada the corresponding figure is 10 percent (Weiss and Lonnquist, 2014). However, the system is not without its critics, who claim that it is costly and often wasteful.

The Canadian health care system does not constitute what is referred to as *socialized medicine*—a health care system in which the government owns the medical care facilities and employs the physicians. Rather, Canada has maintained the private nature of the medical profession. Although the government pays most health care costs, the physicians are not government employees and have much greater autonomy than do physicians in the health care system in Great Britain.

Great Britain Great Britain has a centralized, single-payer health care system that is funded by general revenues. The National Health Service Act of 1946 provides for all health care services to be available at no charge to the entire population. Although physicians work out of offices or clinics—as in the United States or Canada—the government sets health care policies, raises funds and controls the medical care budget, owns health care facilities, and directly employs physicians and other health care personnel (Weiss and Lonnquist, 2014).

Unlike the Canadian model, the health care system in Great Britain *does* constitute socialized medicine. Physicians receive capitation payments from the government: a fixed annual fee for each patient in their practice regardless of how many times they see the patient or how many procedures they perform. They also receive supplemental payments for each low-income or elderly patient in their practice, to compensate for the extra time such patients may require; bonus payments if they meet targets for providing preventive services, such as immunizations against disease; and financial incentives if they practice in medically underserved areas. Physicians may accept private patients, but such patients rarely constitute more than a small fraction of a physician's practice; hospitals reserve a small number of beds for private patients (Weiss and Lonnquist, 2014). Why would anyone want to be a private patient who pays for her or his own care or hospital bed? The answer is primarily found in the desire to avoid the long waits ("queues") that the general population encounters and the fact that private patients can enter the hospital for surgery at times convenient to the consumer rather than wait upon the convenience of the system (• Figure 18.14).

People's Republic of China In recent years the People's Republic of China has seen many changes in the delivery and financing of medical care. When most U.S. people speak of "Chinese medicine," they are referring to treatments such as acupuncture, herbal remedies, or massage. However, health care in China is much more complex than the use of alternative remedies.

During the past two decades the health care system in China has become a complex mix of market-driven capitalism, communism, and massive government spending. The profit-driven system is based on fee-for-service practice by physicians and the sale of highly expensive pharmaceutical

universal health care
a health care system in which all citizens receive medical services paid for by tax revenues.

socialized medicine
a health care system in which the government owns the medical care facilities and employs the physicians.

FIGURE 18.14 Although the National Health Service (NHS) in England provides free health care services, funded by tax monies, for anyone who is a permanent UK resident, private health care is also available for those who are willing to pay. The private Lindo Wing suite of St. Mary's Hospital, where both of the children of Prince William and Princess Kate, Duchess of Cambridge, were born, is an example of the luxurious accommodations that are available to those who are willing to pay for hotel-like amenities.

products that are marketed by doctors and hospitals. Much of the profit is through drug sales, particularly intravenous (IV) treatments, which are used much more widely than in the United States. Until reforms occurred in about 2008–2009, pharmaceutical products were very costly; however, they have become somewhat more affordable since that time.

Ironically, as the Chinese government stepped up funding for health care, social disease epidemics, such as obesity and/or illnesses related to excessive use of tobacco, alcohol, and salt, took their toll on many in the population: Type 2 diabetes, hypertension, and environmentally induced heart and lung diseases have remained on the upswing for a number of years (Mills, 2010).

What is the history of health care delivery and funding in China? After a lengthy civil war, in 1949 the Communist Party won control of mainland China but found itself in charge of a vast nation with a population of one billion people, most of whom lived in poverty and misery. Malnutrition was prevalent, life expectancies were short, and infant and maternal mortality rates were high. In the cities, only the elite could afford medical care; in the rural areas, where most of the population resided, Western-style health care barely existed (Weitz, 2013). With a lack of financial

resources and not enough trained health care personnel, China adopted a policy to create a large number of physician extenders who could educate the public about health and the treatment of illness and disease. Referred to as *street doctors* in urban areas and *village doctors* (formerly "barefoot doctors") in the countryside, these individuals had little formal training and worked under the supervision of trained physicians (Weitz, 2013).

Over the past six decades, medical training has become more rigorous. All doctors receive training in both Western and traditional Chinese medicine. Most doctors who work in hospitals receive a salary; all other doctors now work on a fee-for-service basis. In urban areas, about 94 percent of the working population has health insurance coverage paid for by employers, but individuals often pay large out-of-pocket fees because of gaps in health care coverage. Today, 60 percent of the population of mainland China lives in rural areas, where large numbers of migrant farm workers must provide for their own health care. New government initiatives established in the 2010s call for the establishment of a clinic in each rural community; however, many more physicians will be needed to provide adequate care for the millions of people living in these areas.

In urban areas the latest reports show that it is difficult to get into a hospital even with insurance. Increasing demands for services have placed the already overburdened system under greater stress, and many people line up outside the more-prestigious hospitals early in the morning on the day before they want treatment to get their name on lengthy waiting lists. Although local doctors might be able to take care of their needs, many Chinese patients want to be treated in the more-prestigious hospitals, where they believe they will receive higher-quality care. To help meet the growing demands on the health care delivery system, the Chinese government is attempting to improve thousands of medical centers and to focus on preventive care, especially for infants, children, pregnant women, and those in need of mental health care. Despite these efforts, both services and funding remain quite different when comparing urban and rural areas: The health care disparities remain great, and more money and medical personnel are needed throughout the system, but particularly in rural areas, to meet the needs of a rapidly growing population.

Regardless of which approach to health care delivery and financing that a nation uses, health care providers, hospitals, governmental agencies, political leaders, and the general public all face many difficult issues about the best way to meet patients' needs and not bankrupt the medical system and the nation. One area that may increase costs and generate complex concerns is advanced medical technology.

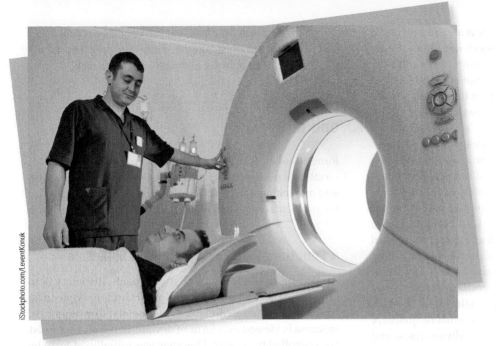

FIGURE 18.15 High-tech medical imaging devices such as the one shown here are very expensive. Some analysts question whether an expenditure of this size could be better used instead to help more people in different ways.

L06 **Describe** how advanced medical technology has changed the practice of medicine and the cost of health services.

Social Implications of Advanced Medical Technology

Advances in medical technology are occurring at a speed that is almost unbelievable; however, sociologists and other social scientists have identified specific social implications of some of the new technologies (see Weiss and Lonnquist, 2014):

1. *Advanced technologies create options for people and for society, but these options alter human relationships.* An example is the ability of medical personnel to sustain a life that in earlier times would have ended as the result of disease or an accident. Although this can be beneficial, technologically advanced equipment (that can sustain life after consciousness is lost and there is no likelihood that the person will recover) can create a difficult decision for the family of that person if he or she has not left a *living will*—a document stating the person's wishes regarding the medical circumstances under which his or her life should be terminated. Federal law requires all hospitals and other medical facilities to honor the terms of a living will.

2. *Advanced technologies increase the cost of medical care.* For example, the computerized axial tomography (CT or CAT) scanner—which combines a computer with X-rays that are passed through the body at different angles—produces clear images of the interior of the body that are invaluable in investigating disease. The cost of such a scanner is around $1 million.

Magnetic resonance imaging (MRI) equipment that allows pictures to be taken of internal organs ranges in cost from $1 million to $2.5 million (• Figure 18.15). Can the United States afford such equipment in every hospital for every patient? The money available for health care is not unlimited, and when it is spent on high-tech equipment and treatment, it is being reallocated from other health care programs that might be of greater assistance to more people.

3. Advanced technologies such as cloning and stem cell research raise provocative questions about the very nature of life. In 1997 Dr. Ian Williams and his associates in Scotland cloned a lamb (that they named Dolly) from the DNA of an adult sheep. Subsequently, scientists have cloned other animals in the same manner, raising a number of profound questions: If scientists can duplicate mammals from adult DNA, is it possible to clone a human being? If it is possible, would it be ethical? Like cloning, stem cell research has been an important and controversial issue in medicine. Stem cell research is important because stem cells—which are accessible in the skin and through extraction from umbilical cord blood and human embryos—can be used to generate virtually any type of specialized cell in the human body and to replace diseased or damaged human tissue. However, opponents of embryonic stem cell research believe that a human life is taken when a human embryo is destroyed in this research. Proponents of stem cell research respond that these studies do not always require the use of embryos.

At the same time that high-tech medicine is becoming a major part of overall health care, many people are turning to holistic medicine and alternative healing practices.

Holistic Medicine and Alternative Medicine

When examining the subject of medicine, it is easy to think only in terms of conventional (or mainstream) medical treatment. By contrast, *holistic medicine* is an approach to health

holistic medicine
an approach to health care that focuses on prevention of illness and disease and is aimed at treating the whole person—body and mind—rather than just the part or parts in which symptoms occur.

care that focuses on prevention of illness and disease and is aimed at treating the whole person—body and mind—rather than just the part or parts in which symptoms occur. Under this approach it is important that people not look solely to medicine and doctors for their health but that people also engage in health-promoting behavior. Likewise, medical professionals must not only treat illness and disease but also work with the patient to promote a healthy lifestyle and self-image.

Many practitioners of *alternative medicine*—healing practices inconsistent with dominant medical practice—take a holistic approach, and today many people are turning to alternative medicine, such as the use of herbal therapies, to supplement or replace traditional medicine (• Figure 18.16). However, some medical doctors are opposed to alternative medicine. In understanding the medical establishment's reaction to alternative medicine, it is important to keep in mind the philosophy of scientific medicine—that medicine is a science, not an art. Thus, to the extent to which alternative medicine is "nonscientific," it must be quackery and therefore something that is undoubtedly worthless and possibly harmful. Undoubtedly, self-interest is also involved in mainstream medicine's reaction to alternative medicine: If the public can be persuaded that scientific medicine is the only legitimate healing practice, fewer health care dollars will be spent on a form of medical treatment that is (at least to some extent) in competition with the medical establishment. But if all forms of alternative medicine (including chiropractic, massage, and spiritual) are taken into account, people spend more money on unconventional therapies than they do for all hospitalizations.

Xie Guang Hui/Redlink/Terra/Corbis

FIGURE 18.16   The use of herbal therapies is a form of alternative medicine that is increasing in popularity in the United States. How does this approach to health care differ from a more traditional medical approach?

LO7 **Explain** how functionalist, conflict, symbolic interactionist, and postmodern approaches differ in their analysis of health and health care.

Sociological Perspectives on Health and Medicine

Functionalist, conflict, symbolic interactionist, and postmodernist perspectives focus on different aspects of health and medicine; each provides us with significant insights on the problems associated with these pressing social concerns.

A Functionalist Perspective: The Sick Role

According to the functionalist approach, if society is to function as a stable system, it is important for people to be healthy and to contribute to their society. Consequently, sickness is viewed as a form of deviant behavior that must be controlled by society. This view was initially set forth by the sociologist Talcott Parsons (1951) in his concept of the *sick role*—the set of patterned expectations that defines the norms and values appropriate for individuals who are sick and for those who interact with them. According to Parsons, the sick role has four primary characteristics:

1. People who are sick are not responsible for their condition. It is assumed that being sick is not a deliberate and knowing choice of the sick person.

2. People who assume the sick role are temporarily exempt from their normal roles and obligations. For example, people with illnesses are typically not expected to go to school or work.

3. People who are sick must want to get well. The sick role is considered to be a temporary one that people must relinquish as soon as their condition improves sufficiently. Those who do not return to their regular activities in a timely fashion may be labeled as hypochondriacs or malingerers.

4. People who are sick must seek competent help from a medical professional to hasten their recovery.

As these characteristics show, Parsons believed that illness is dysfunctional for both individuals and the larger society. Those who assume the sick role are unable to fulfill their necessary social roles, such as being parents or employees. Similarly, people who are ill lose days from their productive roles in society, thus weakening the ability of groups and organizations to fulfill their functions.

According to Parsons, it is important for the society to maintain social control over people who enter the sick role. Physicians are empowered to determine who may enter this role and when patients are ready to exit it. Because physicians spend many years in training and have specialized knowledge about illness and its treatment, they are certified by the society to be "gatekeepers" of the sick role. When patients seek the advice of a physician, they enter into the doctor–patient relationship, which does not contain equal power for both parties. The patient is expected to follow the "doctor's orders" by adhering to a treatment regime, recovering from the malady, and returning to a normal routine as soon as possible.

What are the major strengths and weaknesses of Parsons's model and, more generally, of the functionalist view of health and illness? Parsons's analysis of the sick role was pathbreaking when it was introduced. Some social analysts believe that Parsons made a major contribution to our knowledge of how society explains illness-related behavior and how physicians have attained their gatekeeper status. In contrast, other analysts believe that the sick-role model does not take into account racial–ethnic, class, and gender variations in the ways that people view illness and interpret this role. For example, this model does not take into account the fact that many individuals in the working class may choose not to accept the sick role unless they are seriously ill—because they cannot afford to miss time from work and lose a portion of their earnings. Moreover, people without health insurance may not have the option of assuming the sick role.

A Conflict Perspective: Inequalities in Health and Health Care

Unlike the functionalist approach, conflict theory emphasizes the political, economic, and social forces that affect health and the health care delivery system. Among the issues of concern to conflict theorists are the ability of all people to obtain health care; how race, class, and gender inequalities affect health and health care; power relationships between doctors and other health care workers; the dominance of the medical model of health care; and the role of profit in the health care system.

FIGURE 18.17 According to the conflict perspective, problems in U.S. health care delivery are rooted in the capitalist economy, which views medicine as a commodity that is produced and sold by the medical–industrial complex.

Who is responsible for problems in the U.S. health care system? According to many conflict theorists, problems in U.S. health care delivery are rooted in the capitalist economy, which views medicine as a commodity that is produced and sold by the medical–industrial complex (• Figure 18.17). The *medical–industrial complex* encompasses local physicians, local hospitals, and global health-related industries such as insurance companies and pharmaceutical and medical supply companies.

The United States is one of the few industrialized nations that has relied almost exclusively on the medical–industrial complex for health care delivery and has not had any form of universal health coverage to provide some level of access to medical treatment for all people. Consequently, access to high-quality medical care has been linked to people's ability to pay and to their position within the class structure. Those who are affluent or have good medical insurance may receive high-quality, state-of-the-art care in the medical–industrial complex because of its elaborate technologies and treatments. However, people below the poverty level and those just above it have greater difficulty gaining access to medical care. Referred to as the *medically indigent*, these individuals do not earn enough to afford private medical care but earn just enough money to keep them from qualifying for Medicaid. In the

sick role
the set of patterned expectations that defines the norms and values appropriate for individuals who are sick and for those who interact with them.

medical–industrial complex
local physicians, local hospitals, and global health-related industries such as insurance companies and pharmaceutical and medical supply companies.

profit-oriented capitalist economy, these individuals are said to "fall between the cracks" in the health care system.

Who benefits from the existing structure of medicine? According to conflict theorists, physicians—who hold a legal monopoly over medicine—benefit from the existing structure because they can charge inflated fees. Similarly, clinics, pharmacies, laboratories, hospitals, supply manufacturers, insurance companies, and many other corporations derive excessive profits from the existing system of payment in medicine. In recent years, large drug companies and profit-making hospital corporations have come to occupy a larger and larger part of health care delivery. As a result, medical costs have risen rapidly, and the federal government and many insurance companies have placed pressure for cost containment on other players in the medical–industrial complex.

Conflict theorists increase our awareness of inequalities of race, class, and gender as these statuses influence people's access to health care. They also inform us about the problems associated with health care becoming "big business." However, some analysts believe that the conflict approach is unduly pessimistic about the gains that have been made in health status and longevity—gains that are at least partially the result of large investments in research and treatment by the medical–industrial complex.

A Symbolic Interactionist Perspective: The Social Construction of Illness

Symbolic interactionists attempt to understand the specific meanings and causes that we attribute to particular events. In studying health, symbolic interactionists focus on the meanings that social actors give their illness or disease and how these affect people's self-concept and relationships with others. According to symbolic interactionists, we socially construct "health" and "illness" and how both should be treated. For example, some people explain disease by blaming it on those who are ill. If we attribute cancer to the acts of a person, we can assume that we will be immune to that disease if we do not engage in the same behavior. Nonsmokers who learn that a lung cancer victim had a two-pack-a-day habit feel comforted that they are unlikely to suffer the same fate. Similarly, victims of AIDS are often blamed for promiscuous sexual conduct or intravenous drug use, regardless of how they contracted HIV. In this case the social definition of the illness leads to the stigmatization of individuals who suffer from the disease.

Although biological characteristics provide objective criteria for determining medical conditions such as heart disease, tuberculosis, or cancer, there is also a subjective component to how illness is defined. This subjective component is very important when we look at conditions such as childhood hyperactivity,

mental illness, alcoholism, drug abuse, cigarette smoking, and overeating, all of which have been medicalized. The term **medicalization** refers to the process whereby nonmedical problems become defined and treated as illnesses or disorders. Medicalization may occur on three levels: (1) the conceptual level (e.g., the use of medical terminology to define the problem), (2) the institutional level (e.g., physicians are supervisors of treatment and gatekeepers to applying for benefits), and (3) the interactional level (e.g., when physicians treat patients' conditions as medical problems). For example, the sociologists Deborah Findlay and Leslie Miller (1994: 277) explain how gambling was medicalized (● Figure 18.18):

> Habitual gambling . . . has been regarded by a minority as a sin, and by most as a leisure pursuit—perhaps wasteful but a pastime nevertheless. Lately, however, we have seen gambling described as a psychological illness—"compulsive gambling." It is in the process of being medicalized. The consequences of this shift in discourse (that is, in the way of thinking and talking) about gambling are considerable for doctors, who now have in gamblers a new market for their services or "treatment"; perhaps for gambling halls, which may find themselves subject to new regulations, insofar as they are deemed to contribute to the "disease"; and not least, for gamblers themselves, who are no longer treated as sinners or wastrels, but as patients, with claims on our sympathy, and to our medical insurance plans as well.

Sociologists often refer to this form of medicalization as the *medicalization of deviance* because it gives physicians and other medical professionals greater authority to determine what should be considered "normal" and "acceptable"

Stockbyte/Jupiter Images

FIGURE 18.18 Is gambling a moral issue or a medical one? According to sociologists, the recent trend toward viewing compulsive gambling as a health care issue is an example of the medicalization of deviance.

Sociological Perspectives on Health and Medicine

A Functionalist Perspective: The Sick Role	People who are sick are temporarily exempt from normal obligations but must want to get well and seek competent help.
A Conflict Perspective: Inequalities in Health and Health Care	Problems in health care are rooted in the capitalist system, exemplified by the medical–industrial complex.
A Symbolic Interactionist Perspective: The Social Construction of Illness	People socially construct both "health" and "illness," and how both should be treated.
A Postmodernist Perspective: The Clinical Gaze	Doctors gain power through observing patients to gather information, thus appearing to speak "wisely."

behavior and to establish the appropriate mechanisms for controlling "deviant behaviors."

According to symbolic interactionists, medicalization is a two-way process: Just as conditions can be medicalized, so can they be demedicalized. **Demedicalization** refers to the process whereby a problem ceases to be defined as an illness or a disorder. Examples include the removal of certain behaviors (such as homosexuality) from the list of mental disorders compiled by the American Psychiatric Association and the deinstitutionalization of mental health patients. The process of demedicalization also continues in women's health as advocates seek to redefine childbirth and menopause as natural processes rather than as illnesses.

In addition to how health and illness are defined, symbolic interactionists examine how doctors and patients interact in health care settings. Some physicians may hesitate to communicate certain kinds of medical information to patients, such as why they are prescribing certain medications or what side effects or drug interactions may occur.

Symbolic interactionist perspectives on health and health care provide us with new insights on the social construction of illness and how health and illness cannot be strictly determined by medical criteria. Symbolic interactionists also make us aware of the importance of communication between physicians and patients, including factors that may reduce effective medical treatment for some individuals. However, these approaches have been criticized for suggesting that few objective medical criteria exist for many illnesses and for overemphasizing microlevel issues without giving adequate recognition to macrolevel issues such as the effects on health care of managed care, health maintenance organizations, and for-profit hospital chains.

A Postmodernist Perspective: The Clinical Gaze

In *The Birth of the Clinic* (1994/1963), postmodern theorist Michel Foucault questioned existing assumptions about medical knowledge and the power that doctors have gained over other medical personnel and everyday people. Foucault asserted that truth in medicine—as in all other areas of life—is a social construction, in this instance one that doctors have created. Foucault believed that doctors gain power through the *clinical* (or "observing") *gaze,* which they use to gather information. Doctors develop the clinical gaze through their observation of patients; as the

doctors begin to diagnose and treat medical conditions, they also start to speak "wisely" about everything. As a result, other people start to believe that doctors can "penetrate illusion and see . . . the hidden truth" (Shawver, 1998).

According to Foucault, the prestige of the medical establishment was further enhanced when it became possible to categorize all illnesses within a definitive network of disease classification under which physicians can claim that they know why patients are sick. Moreover, the invention of new tests made it necessary for physicians to gaze upon the naked body, to listen to the human heart with an instrument, and to run tests on the patient's body fluids. Patients who objected were criticized by doctors for their "false modesty" and "excessive restraint" (Foucault, 1994/1963: 163). As the new rules allowed for the patient to be touched and prodded, the myth of the doctor's diagnostic wisdom was further enhanced, and "medical gestures, words, gazes took on a philosophical density that had formerly belonged only to mathematical thought" (Foucault, 1994/1963: 199). For Foucault, the formation of clinical medicine was merely one of the more-visible ways in which the fundamental structures of human experience have changed throughout history.

Foucault's work provides new insights on medical dominance, but it has been criticized for its lack of attention to alternative viewpoints. Among these is the possibility that medical breakthroughs and new technologies actually help physicians become wiser and more scientific in their endeavors. Another criticism is that Foucault's approach is based on the false assumption that people are passive individuals who simply comply with doctors' orders—he does not take into account that people (either consciously or unconsciously) may resist the myth of the "wise doctor" and not follow "doctors' orders" (Lupton, 1997). The Concept Quick Review summarizes the major sociological perspectives on health and medicine.

Foucault's analysis (1988/1961) was not limited to doctors who treat bodily illness; he also critiqued psychiatrists and the treatment of insanity.

medicalization
the process whereby nonmedical problems become defined and treated as illnesses or disorders.

demedicalization
the process whereby a problem ceases to be defined as an illness or a disorder.

Discuss what is meant by the term *mental illness* and explain why it is a difficult topic for sociological research.

Mental Illness

Mental illness affects many people; however, it is a difficult topic for sociological research. An earlier social analyst, Thomas Szasz (1984), argued that mental illness is a myth. According to Szasz, "mental illnesses" are actually individual traits or behaviors that society deems unacceptable, immoral, or deviant. As a result, labeling individuals as "mentally ill" harms them because they often come to accept the label and are then treated accordingly by others.

Is mental illness a myth? After decades of debate on this issue, social analysts are no closer to reaching a consensus than they were when Szasz originally introduced his ideas. However, many scholars believe that mental illness is a reality that has biological, psychological, environmental, social, and other factors involved (• Figure 18.19). Many medical professionals distinguish between a *mental disorder*—a condition that makes it difficult or impossible for a person to cope with everyday life—and *mental illness*—a condition in which a person has a severe mental disorder requiring extensive treatment with medication, psychotherapy, and sometimes hospitalization.

How many people are affected by mental illness? Data from the Substance Abuse and Mental Health Services Administration suggest that 43.8 million adults age 18 or older experienced a diagnosable mental illness in 2013 (SAMHSA, 2014). Of that number, about 10 million adults experienced a serious mental illness, which means that a person has serious functional impairment that substantially interferes with or limits life activities. An additional 15.7 million adults experienced a major depressive episode. Among young people between the ages of 12 to 17, major depressive episodes affected approximately one in ten, or 2.6 million youths. However, a recent study found that only 38.1 percent of youths who experienced a major depressive episode in the past year had received treatment for depression. In 2013 approximately 34.6 million adults (constituting 14.6 percent of all adults) received some form of mental health care, such as inpatient or outpatient services and/or prescription medication. The recent study found that among adults, 44.7 percent of those with any mental illness and 68.5 percent of adults with serious mental illness received mental health services (SAMHSA, 2014).

The most widely accepted classification of mental disorders is the American Psychiatric Association's (2013) *Diagnostic and Statistical Manual of Mental Disorders (DSM–5)*, which is illustrated in • Table 18.1. Mental disorders are very costly to the nation. Direct costs associated with mental disorders include the price of medication, clinic visits, and hospital stays. However, many indirect costs are incurred as well. These include the loss of earnings by individuals, the costs associated with homelessness and incarceration, and other indirect costs that exist but are difficult to document.

The Treatment of Mental Illness

In *Madness and Civilization*, Michel Foucault (1988/1961) examined the "archaeology of madness" from 1500 to 1800 to determine how ideas of mental illness have changed over time and to describe the "birth of the asylum." According to Foucault, early in this period insanity was considered part of everyday life, and people with mental

Tom Farmer/REX/Newscom

FIGURE 18.19 Many hospitals around the world have facilities designated for helping patients diagnosed with a mental disorder or mental illness. In the United States, the Affordable Care Act expanded mental health and substance use disorder services for millions of Americans who previously did not have this benefit.

TABLE 18.1

Criteria for Psychiatric Disorders Identified by the American Psychiatric Association

Neurodevelopmental disorders	Intellectual disability (intellectual developmental disorder), communication disorders, and autism spectrum disorder
Schizophrenia spectrum and other psychotic disorders	Disorders with symptoms such as delusions or hallucinations
Bipolar and related disorders	Mood swings that range from depression to mania
Depressive disorders	Includes major depressive disorder (clinical depression), disruptive mood dysregulation disorder in children up to age 18 years, and premenstrual dysphoric disorder
Anxiety disorders	Disorders characterized by anxiety that is manifested in phobias or panic attacks
Obsessive-compulsive and related disorders	Behavior that is obsessive and compulsive in nature, including excoriation (skin-picking) disorder, hoarding disorder, substance/medication-induced obsessive-compulsive and related disorder, and obsessive-compulsive and related disorder caused by another medical condition
Trauma- and stressor-related disorders	Including posttraumatic stress disorder (PTSD), reactive attachment disorder, and disinhibited social engagement disorder
Dissociative disorders	Depersonalization/derealization disorder and other problems involving disassociation with normal consciousness such as dissociative amnesia
Somatic symptom and related disorders	Including psychological problems that present themselves as symptoms of physical disease
Feeding and eating disorders	Including avoidant/restrictive food intake disorder, anorexia nervosa, bulimia nervosa, and binge eating disorder
Sleep–wake disorders	Including insomnia disorder, narcolepsy, some breathing-related sleep disorders, and other problems associated with sleep
Sexual dysfunctions	Gender-specific sexual dysfunctions of a duration of 6 months or more
Gender dysphoria	Intense anxiety about one's birth gender as compared to one's perceived gender
Disruptive, impulse-control, and conduct disorders	Including the inability to control disruptive behavior, lack of impulse control, and conduct disorders such as such as kleptomania and pyromania
Substance-related and addictive disorders	Disorders resulting from abuse of alcohol and/or drugs; gambling disorder and tobacco use disorder
Neurocognitive disorders (NCDs)	Including dementia and amnestic disorder and substantive/medication-induced NCDs
Paraphilic disorders	Sexual masochism disorder rooted in atypical sexual interests that cause persons to feel distressed about their interest and/or to cause another person to have psychological distress, injury, or death as a result of this disorder

Source: Adapted from American Psychiatric Association, 2013.

illnesses were free to walk the streets; however, beginning with the Renaissance and continuing into the seventeenth and eighteenth centuries, the mentally ill were viewed as a threat to others. During that time, asylums were built, and a clear distinction was drawn between the "insane" and the rest of humanity. According to Foucault (1988/1961: 252), people came to see "madness" as a minority status that does not have the right to autonomy:

> Madness is childhood. Everything at the [asylum] is organized so that the insane are transformed into minors. They are regarded as children who have an overabundance of strength and make dangerous

use of it. They must be given immediate punishments and rewards; whatever is remote has no effect on them.

As previously stated, many youths and adults with mental disorders do not receive professional treatment today. U.S. adults who receive care for mental health problems are more likely to receive such care from primary-care and general-care providers who are not specialists in mental health care. The lack of available mental health providers, as well as inadequate insurance coverage, means that many people with mental health problems fall through the cracks.

Many people seeking psychiatric assistance are treated with medications or psychotherapy—which is believed to help patients understand the underlying reasons for their problem—and sometimes treated in psychiatric wards of local hospitals or in private psychiatric hospitals. However, the introduction of new psychoactive drugs to treat mental disorders and the deinstitutionalization movement in the 1960s have created dramatic changes in how people with mental disorders are treated. ***Deinstitutionalization*** refers to the practice of rapidly discharging patients from mental hospitals into the community. Originally devised as a solution for the problem of "warehousing" mentally ill patients in large, prison-like mental hospitals in the first half of the twentieth century, deinstitutionalization is now viewed as a problem by many social scientists. The theory behind this process was that patients' rights were being violated because many patients experienced involuntary commitment (i.e., without their consent) to the hospitals, where they remained for extended periods of time. Instead, some professionals believed that the patients' mental disorders could be controlled with proper medications and treatment from community-based mental health services. Advocates of deinstitutionalization also believed that this practice would relieve the stigma attached to mental illness and hospitalization. However, critics of deinstitutionalization argue that this process exacerbated long-term problems associated with inadequate care for people with mental illness.

Admitting people to psychiatric hospitals on an involuntary basis ("involuntary commitment") has always been controversial; however, it remains the primary method by which police officers, judges, social workers, and other officials deal with people—particularly the homeless—whom they have reason to believe are mentally ill and imminently dangerous to others or themselves if not detained. State psychiatric hospitals continue to provide most of the chronic inpatient care for poor people with mental illnesses; these institutions tend to serve as a revolving door to poverty-level board-and-care homes, nursing homes, or homelessness, as contrasted with the situation of patients who pay their bills at private psychiatric facilities through private insurance coverage or Medicare.

According to the sociologist Erving Goffman's (1961a) classical work on this topic, mental hospitals are an excellent example of a *total institution,* previously defined as a place where people are isolated from the rest of society for a period of time and come under the complete control of the officials who run the institution.

LO9 **Discuss** the concept of disability and identify key sociological perspectives on disability.

Disability

What is a disability? ***Disability*** refers to a physical or mental impairment that substantially limits one or more major activities that a person would normally do at a given stage of life and that may result in stigmatization or discrimination against the person with a disability. Some disabilities involve physical conditions, while others involve mental abilities. However, according to disability rights advocates, social attitudes and the social and physical environments in which people live are contributing factors to the extent to which a person may be considered "disabled." An example of a disabling environment is a school or office building in which elevator buttons and faucets on public restroom sinks are located beyond the reach of a person using a wheelchair. In such a setting the person's disability derives from the fact that necessary objects in everyday life have been made inaccessible. According to advocates, disability must be thought of in terms of how society causes or contributes to the problem—not in terms of what is "wrong" with the person with a disability (Albrecht, Seelman, and Bury, 2001).

A second crucial factor in understanding disability is how a person with a disability is viewed and treated by other people. Many examples could be given of persons with a disability who have been made uncomfortable by "able-bodied" individuals who think they are "helping" the person with a disability when they make some comment about the disability or the perceived inadequacies it causes. However, Dr. Adrienne Asch (2001), director of the Center for Ethics at Yeshiva University, speaks volumes about what her life has been like as an outstanding academic and as a person with a visual impairment—blindness:

> [A] growing number of professionals with disabilities, including myself, can point to professional recognition and the joys of doing work we love as well as its relative financial security and social status. Yet like . . . others, we have all-too-frequent reminders that we are unanticipated participants in workshops or conferences or unexpected guests at social gatherings. Sitting beside a stranger waiting for a lecture to begin at an academic conference, the stranger whispers loudly not "Hello, my name is Carol," but "Let me know how I can help you." What help do I need while waiting for the speaker to begin? Why not introduce herself, rather than assume that the only sociability I could possibly want is her help? When I respond by saying that she can let me know if I can help her, she does not get the point and I am all too well aware that the point is subtle; instead she needs to be thanked for her offer and reassured that I will accept it—and then many pleasantries later perhaps we can discuss why we are at the lecture and whether we like it and what workshop we will attend that afternoon.

Asch emphasizes that the incident she described is not an isolated one in her life: Many people who have known her for years do not feel comfortable treating her the same way they would a person without a disability. Even more important, countless other persons with a disability have shared similar experiences in which they have been stigmatized or marginalized because of how another individual responded to them based on their disability.

Some estimates suggest that about 56.7 million Americans (about 18.7 percent of the civilian noninstitutionalized U.S. population) have some level of disability. Other estimates are lower, placing the overall rate of disability in the U.S. population at about 12.7 percent (Stoddard, 2014). These figures vary greatly by state and by age of persons. Of all persons in the United States with disabilities, children under age 5 account for only 0.4 percent, and children and youths between the ages of 5 and 17 account for 7.4 percent. More than half (51.9 percent) of persons with disabilities are between the ages of 18 and 64, and about 40.3 percent of those with disabilities are age 65 and older. More than one-third (36.6 percent) of all persons ages 65 and older had a disability in 2013 (Stoddard, 2014).

Although anyone can become disabled, some people are more likely to be or to become disabled than others. People who work in hazardous settings have higher rates of disability than workers in seemingly safer occupations. African Americans have higher rates of disability than whites, especially more-serious disabilities; persons with lower incomes also have higher rates of disability (Weitz, 2013). As age increases, the chances of disability also rise. But no one is immune to disability. Many people believe that most disabilities are caused by catastrophic events such as serious accidents; however, illness causes more than 90 percent of all disabilities.

Environment, lifestyle, and working conditions may all contribute to either temporary or chronic disability. For example, air pollution in automobile-clogged cities leads to a higher incidence of chronic respiratory disease and lung damage, which may result in severe disability for some people. Eating certain types of food and smoking cigarettes increase the risk for coronary and cardiovascular diseases. In contemporary industrial societies, workers in the second tier of the labor market (primarily recent immigrants, white women, and people of color) are at the greatest risk for certain health hazards and disabilities. Employees in data processing and service-oriented jobs may also be affected by work-related disabilities. The extensive use of computers has been shown to harm some workers' vision; to produce joint problems such as arthritis, low-back pain, and carpal tunnel syndrome; and to place employees under high levels of stress that may result in neuroses and other mental health problems.

As shown in • Table 18.2, more than one out of five people in the United States (21.3 percent) has a chronic health condition that, given the physical, attitudinal, and financial barriers built into the social system, makes it difficult to perform one or more activities generally considered appropriate for persons of their age. This percentage refers only to the noninstitutionalized population. If people living in nursing homes and other institutional settings are considered, the percentage is significantly higher.

TABLE 18.2

Percentage of Noninstitutionalized U.S. Population with Disabilities

Characteristic	Percentage
With a disability	21.3
Severe	14.8
Not severe	6.5
Has difficulty or is unable to:	
See	4.8
Hear	3.1
Have speech understood	1.2
Lifting	7.1
Using stairs	9.2
Walking	9.9
Has difficulty or needs assistance with:	
Getting around	1.9
Getting into bed	2.5
Taking a bath or shower	2.3
Dressing	1.8
Eating	0.8
Getting to or using the toilet	1.2
Has difficulty or needs assistance with:	
Going outside the home alone	4.2
Managing money	2.4
Preparing meals	2.4
Doing housework	3.2
Using the phone	1.2

Source: Brault, 2012.

For infants born with a disability, living with a disability is a long-term process for parents and child alike. Parents report that more than 289,000 children under age 3 have a disability, either a developmental delay or difficulty moving their arms or legs. In addition, about 465,000 children between the ages of 3 and 5 have a disability in which they experience a developmental delay or have difficulty walking, running, or playing (Brault, 2012).

Among persons who acquire disabilities later in life, through disease or accidents, the social significance of their disability can be seen in how they initially respond to their symptoms and diagnosis, how they view the immediate situation and their future, and how the illness and disability affect their lives. When confronted with a disability, most people adopt one of two strategies—avoidance or vigilance. Those who use the avoidance strategy deny

deinstitutionalization
the practice of rapidly discharging patients from mental hospitals into the community.

disability
a physical or mental impairment that substantially limits one or more major activities that a person would normally do at a given stage of life and that may result in stigmatization or discrimination against the person with a disability.

their condition in order to maintain hopeful images of the future and elude depression; for example, some individuals refuse to participate in rehabilitation following a traumatic injury because they want to pretend that the disability does not exist. By contrast, those using the vigilance strategy actively seek knowledge and treatment so that they can respond appropriately to the changes in their bodies (Weitz, 2013).

Sociological Perspectives on Disability

How do sociologists view disability? Those using the functionalist framework often apply Parsons's sick-role model, which is referred to as the medical model of disability. According to the medical model, people with disabilities become, in effect, chronic patients under the supervision of doctors and other medical personnel, subject to a doctor's orders or a program's rules and not to their own judgment. From this perspective, disability is deviance.

The deviance framework is also apparent in some symbolic interactionist perspectives. According to symbolic interactionists, people with a disability experience role ambiguity because many people equate disability with deviance. By labeling individuals with a disability as "deviant," other people can avoid them or treat them as outsiders. Society marginalizes people with a disability because they have lost old roles and statuses and are labeled as "disabled" persons (• Figure 18.20). According to the sociologist Eliot Freidson (1965), how people are labeled results from three factors: (1) their degree of responsibility for their impairment, (2) the apparent seriousness of their condition, and (3) the perceived legitimacy of the condition. Freidson concluded that the definitions of and expectations for people with a disability are socially constructed factors.

Finally, from a conflict perspective, persons with a disability are members of a subordinate group in conflict with persons in positions of power in the government, in

AP Images/Charles Dharapak

FIGURE 18.20 Disabilities are often the result of violent activity. Many veterans of the wars in Iraq and Afghanistan now suffer physical and emotional disabilities that may be with them for the rest of their lives.

the health care industry, and in the rehabilitation business, all of whom are trying to control their destinies. Those in positions of power have created policies and artificial barriers that keep people with disabilities in a subservient position (Asch, 1986; Hahn, 1987). Moreover, in a capitalist economy, disabilities are big business. When people with disabilities are defined as a social problem and public funds are spent to purchase goods and services for them, rehabilitation becomes a commodity that can be bought and sold by the medical–industrial complex. From this perspective, persons with a disability are objectified. They have an economic value as consumers of goods and services that will allegedly make them "better" people. Many persons with a disability endure the same struggle for resources faced by people of color, women, and older persons. Individuals who hold more than one of these ascribed statuses, combined with experiencing disability, are doubly or triply oppressed by capitalism.

Today, many working-age persons with a disability in the United States are unemployed. Most of them believe that they could and would work if offered the opportunity. However, even when persons with a severe disability are able to find jobs, they typically earn less than persons without a disability.

Employment, poverty, and disability are related. On the one hand, people may become economically disadvantaged as a result of chronic illness or disability. On the other hand, poor people are less likely to be educated and more likely to be malnourished and have inadequate access to health care—all of which contribute to risk of chronic illness, physical and mental disability, and the inability to participate in the labor force. As previously mentioned, the type of employment available to people with limited resources increases their chances of becoming disabled. They may work in hazardous places such as mines, factory assembly lines, and chemical plants, or in the construction industry, where the chance of becoming seriously disabled is much higher.

Looking Ahead: Health Care in the Future

Central questions regarding the future of health care are how to provide coverage for the largest number of people and how to do this without bankrupting the entire nation. Continued implementation of the Affordable Care Act will affect health care in this nation more than any other factor. Overall, our best hope for good medical care is a payment system that will result in the best outcomes for patients at reasonable costs and thus an overall transformation of the current U.S. health care system.

A key issue in the United States is how to prevent, reduce, and best treat epidemics that affect all Americans. If we do not have a specific illness or health condition, many of us do not see that disease as being "our problem." In the future, however, we must come to see problems such as HIV/AIDS as everyone's concern. The Obama administration has developed a national HIV/AIDS strategy to help reduce the number of people who become infected with HIV, to increase access to care and improve health outcomes for people living with HIV, and to reduce HIV-related health disparities with more and better community-level approaches that integrate HIV prevention and care with other social service needs. Along with meeting these goals, it will be important to reduce the stigma and the discrimination against people living with HIV. This HIV/AIDS national initiative is only one of many possible examples that show that in the future, we must have a coordinated national response to preventive health care and to the equitable distribution of health care services throughout the country.

Another key issue in contemporary health care is the role that advanced technologies play in the rising costs of medical care and their usefulness as major tools for diagnosis and treatment. Technology is a major stimulus for social change, and the health care systems in high-income nations such as the United States reflect the rapid rate of technological innovation that has occurred in the last few decades. In the future, advanced health care technologies will no doubt provide even more accurate and quicker diagnosis, effective treatment techniques, and increased life expectancy.

However, technology alone cannot solve many of the problems confronting us in health and health care delivery. In fact, some aspects of technological innovation may be dysfunctional for individuals and society. As we have seen, some technological "advances" raise new ethical concerns, such as the moral and legal issues surrounding the cloning of human life. Some "advances" also may fail: A new prescription drug may be found to cause side effects that are more serious than the illness that it was supposed to remedy. Whether advanced technology succeeds or fails in some areas, it will probably continue to increase the cost of health care in the future. As a result, the gap between the rich and the poor in the United States will contribute to inequalities of access to vital medical services. On a global basis, new technologies may lower the death rate in some low-income countries, but it will primarily be the wealthy in those nations who will have access to the level of health care that many people in higher-income countries take for granted.

In the developing nations of the world, preventive health care and more effective and efficient health care delivery are crucial as the world's population continues to increase and as some regions of the world are plagued by high rates of disease and poverty, a deadly combination in any setting. The concerns of the World Health Organization and other organizations must be heeded to prevent global pandemics: epidemics of infectious disease that spread through human populations across a wide region, country, continent, or the whole world. In the future it will be necessary to direct more money and attention toward preventing major health crises rather than trying to find some way to deal with them after they develop.

It goes without saying that health, illness, and health care will continue to change in the future. To a degree, health care in the future will be up to each of us. What measures will we take to safeguard ourselves against illness and disorders? How can we help others who are the victims of acute and chronic diseases or disabilities? Will we seek to have a voice in how our community, state, and nation deal with health care issues now and in the future?

CHAPTER REVIEW Q & A

LO1 What is the relationship between the social environment and health and illness?

According to the World Health Organization, health is a state of complete physical, mental, and social well-being. Illness refers to an interference with health; like health, illness is socially defined and may change over time and between cultures. In other words, health and illness are not only biological issues but also social issues. Studying health and health care issues around the world offers insights on illness and how political and economic forces shape health care in nations.

LO2 What is social epidemiology, and what key demographic factors are studied by social epidemiologists?

Social epidemiology is the study of the causes and distribution of health, disease, and impairment throughout a population. Typically, the target of the investigation is disease agents, the environment, and the human host (age, sex, race/ethnicity, physical condition, habits and customs, and lifestyle).

LO3 How did the profession of medicine emerge in the United States?

During the nineteenth century, medical schools were largely proprietary institutions, and their officials were often more interested in acquiring students than in enforcing standards. Gradually, the number of medical schools was reduced, and licensing laws were established to eliminate unqualified or irregular practitioners. Although medicine had been previously viewed more as an art than as a science, several significant discoveries during the nineteenth century in areas such as bacteriology and anesthesiology began to give medicine increasing credibility as a science.

LO4 How was U.S. health care paid for in the past, and how has the Affordable Care Act (Obamacare) changed this?

Throughout most of the past hundred years, medical care in the United States has been paid for on a fee-for-service basis. This approach to paying for medical services is expensive because few restrictions are placed on the fees that doctors, hospitals, and other medical providers can charge patients. Recently, there have been efforts at cost containment, and HMOs and managed care have produced both positive and negative results in the contemporary practice of medicine. Health maintenance organizations (HMOs) provide, for a set monthly fee, total care with an emphasis on prevention to avoid costly treatment later. Managed care is any system of cost containment that closely monitors and controls health care providers' decisions about medical procedures, diagnostic tests, and other services that should be provided to patients. The Affordable Care Act of 2010 includes measures to make insurance available to millions of persons who previously were uninsured or underinsured. However, this Act, also referred to as "Obamacare," includes cost-containment measures so that health plans will have incentives to compete and keep premiums low, more control of Medicare payments, and programs that penalize hospitals when patients are readmitted too frequently, among other cost-cutting endeavors.

LO5 How do other nations provide health services for their citizens?

Other nations have various ways in which they provide health care for their citizens. Some nations (such as Canada) have a universal health care system in which all citizens receive medical services paid for by tax revenues. Other nations (such as Great Britain) have a socialized health care system in which the government owns the medical care facilities and employs the physicians.

LO6 How has advanced medical technology changed the practice of medicine and the cost of health services?

Sociologists have identified specific social implications of some new medical technologies: (1) Advanced technologies create options for people and for society, but these options alter human relationships. (2) Advanced technologies increase the cost of medical care. (3) Advanced technologies such as cloning and stem cell research raise provocative questions about the very nature of life.

LO7 How do functionalist, conflict, symbolic interactionist, and postmodern approaches differ in their analysis of health and health care?

According to the functionalist approach, if society is to function as a stable system, it is important for people to be healthy and to contribute to their society. Consequently, sickness is viewed as a form of deviant behavior that must be controlled by society. Conflict theory tends to emphasize the political, economic, and social forces that affect health and the health care delivery system. Among these issues are the ability of all people to obtain health care; how race, class, and gender inequalities affect health and health care; power relations between doctors and other health care workers; the dominance of the medical model of health care; and the role of profit in the health care system. In studying health, symbolic interactionists focus on the fact that the meaning that social actors give their illness or disease will affect their self-concept and their relationships with others. Symbolic interactionists also examine medicalization—the process whereby nonmedical problems become defined and treated as illnesses or disorders. Postmodern theorists argue that doctors and the medical establishment have gained control over illness and patients at least partly because of the physicians' clinical gaze, which replaces all other systems of knowledge.

LO8 What is meant by the term *mental illness,* and why is it a difficult topic for sociological research?

Mental illness affects many people; however, it is a difficult topic for sociological research. An earlier social analyst, Thomas Szasz (1984), argued that mental illness is a myth. After decades of debate on this issue, social analysts are no closer to reaching a consensus than they were when Szasz originally introduced his ideas.

LO9 What is a disability, and what are some key sociological perspectives on disability?

Disability is a physical or health condition that stigmatizes or causes discrimination. In viewing disability, sociologists using the functionalist framework often apply the medical model of disability. According to the medical model, people with disabilities become chronic patients under the supervision of medical personnel, subject to a doctor's orders or a program's rules and not to their own judgment. According to symbolic interactionists, people with a disability experience role ambiguity because many people equate disability with deviance. From a conflict perspective, persons with a disability are members of a subordinate group in conflict with persons in positions of power in the government, the health care industry, and the rehabilitation business, all of whom are trying to control their destinies.

KEY TERMS

acute diseases 520
chronic diseases 520
deinstitutionalization 544
demedicalization 541
disability 544
drug 524
health 516

health care 516
health maintenance organization (HMO) 533
holistic medicine 537
infant mortality rate 518
life expectancy 517
managed care 533

medical–industrial complex 539
medicalization 540
medicine 516
sick role 538
social epidemiology 520
socialized medicine 535
universal health care 535

QUESTIONS for CRITICAL THINKING

1 Why is it important to explain the social, as well as the biological, aspects of health and illness in societies?

2 In what ways are race, class, and gender intertwined with physical and mental disorders?

3 How would functionalists, conflict theorists, and symbolic interactionists suggest that health care delivery might be improved in the United States?

4 Based on this chapter, how do you think illness and health care will be handled in the United States in the future? Are there things that we can learn from other nations regarding the delivery of health care? Why or why not?

ANSWERS to the SOCIOLOGY QUIZ

ON HEALTH, ILLNESS, AND HEALTH CARE

1 **False** The battle over passage of the historic health care reform law in 2010 was a reflection of how divided the United States is over universal health coverage. Debates continue over trying to end "Obamacare," as this law is now referred to by some.

2 **False** In the United States, more than half of HIV infections occur among gay and bisexual men and other men who have sex with men. However, globally, most cases of HIV infection are transmitted through heterosexual contact.

3 **False** The lower life expectancy of African Americans as a category is caused by a higher prevalence of life-threatening illnesses, such as cancer, heart disease, hypertension, and AIDS. However, it should be noted that African American males do have the highest death rates from homicide of any racial–ethnic category in the United States.

4	**True**	Unintentional injuries account for nearly half of all deaths among Americans ages 15–24, making unintentional injuries the most common cause of death in this age group. The majority of unintentional injury deaths in this age group result from motor vehicle accidents.
5	**False**	Most high-income, developed nations have some form of universal health coverage, which is either provided by the government or purchased by the government.
6	**False**	A number of government investigations have focused on rising health care payments and allegations of fraud in the health care delivery system. Billing fraud has been found in Medicare and Medicaid payments to physicians, hospitals, nursing homes, home health agencies, medical labs, and medical equipment manufacturers.
7	**False**	Until recently, chronic depression and other mental conditions were most often depicted as "female" problems. However, male depression has become more widely publicized through documentaries on individuals' lives and through advertising for antidepressants.
8	**True**	Overweight and obesity rates are alarmingly high in the United States, with two-thirds of U.S. adults being either overweight or obese.

Sources: Based on Kaiser Family Foundation, 2012; National Center for Health Statistics, 2012; and Weiss and Lonnquist, 2014.

POPULATION AND URBANIZATION

LEARNING OBJECTIVES

1 **Define** *demography* and explain the three processes that produce changes in populations.

2 **Explain** what is meant by population composition and describe how it is measured.

3 **Discuss** Malthusian, Marxist, and neo-Malthusian perspectives on population as well as demographic transition theory.

4 **Identify** the key points in international migration theories.

5 **Explain** how cities develop through preindustrial, industrial, and postindustrial stages.

6 **Compare** ecological/functionalist models with political economy/conflict models in their explanations of urban growth.

7 **Describe** what is meant by the experience of urban life and indicate how sociologists seek to explain this experience.

8 **Identify** the best-case and worst-case scenarios regarding population and urban growth in the twenty-first century, and discuss how some of the worst-case scenarios might be averted.

SOCIOLOGY & EVERYDAY LIFE

The Immigration Debate

After being born and reared in the Philippines, my mother wanted to give me a better life. So she sent me to live with my grandparents in Silicon Valley. It was 1993, and I was 12 years old. I loved America the moment I got here, and embraced the language, the culture, and the people. . . .

At 16, I rode my bike to the DMV [Department of Motor Vehicles] to get my driver's permit. I brought my green card with me. The woman at the DMV looked at it, leaned over and whispered, "This is fake. Don't come back here again."

I went home and confronted my grandfather. He confirmed it. That was the first time I realized I am an undocumented immigrant—what some people call an "illegal." I decided then that people could never doubt that I am an American. Speak English well. Write English well. Contribute to this society. If I worked hard enough, if I achieved enough, I felt I could earn what it means to be an American.

—Jose Antonio Vargas (2012), an award-winning journalist who informed the world of his undocumented immigrant status in the *New York Times* (see Vargas, 2011), describes

Jose Antonio Vargas

Justin Sullivan/Getty Images News/Getty Images

Immigration is one of the most politically and emotionally charged issues in the twenty-first century. Terms such as *illegal alien* and *undocumented immigrant* have been tossed around by the media and political candidates in the United States for decades, but concerns about the effects of immigration on cities have also escalated as major shifts in the global population have occurred.

Around the world, people move from one location to another for many reasons. Some individuals and families move for economic opportunities, while others move because they fear for their life. In the United States and many other high-income countries, there is a lack of consensus about the causes and consequences of immigration: Some are adamantly opposed to immigration; others believe that it is acceptable under certain circumstances. The issue of immigration has produced a divisive battle over immigration reform. A number of states have passed stringent immigration laws, causing extensive political strife and litigation. However, many people remain concerned that rapid population growth and the changing demographic characteristics of new arrivals will further harm the quality of life in the United States, and they demand more-restrictive immigration laws.

Clearly, immigration is an important factor in understanding the dynamics of population and urbanization;

however, immigration is only one factor associated with population growth and urban change. Birth rates and death rates are also important but have received relatively little attention. In this chapter we explore the dynamics of population and urbanization, with a focus on how birth rates, death rates, and migration affect growth and change in societies such as ours. Before reading on, test your knowledge about current migration and U.S. immigration issues by taking the "Sociology and Everyday Life" quiz.

 Define *demography* and explain the three processes that produce changes in populations.

Demography: The Study of Population

How large is the world's population? The world population has reached more than 7.2 billion people and continues to grow rapidly. Every year, the population increases by about 86,582,000 people when we calculate the natural increase (see • Figure 19.1). Births minus deaths equals the *natural increase* in population. If we break down the natural increase even further, 237,209 people are added to the

how he learned that he was in the United States illegally. Vargas founded Define American (see its website) to encourage others to use social media to support immigration reform.

How Much Do You Know About Migration and U.S. Immigration?

TRUE	FALSE		
T	F	1	All "unauthorized immigrants" in the United States entered the country illegally.
T	F	2	Most people migrate to another country because of "push" factors that make them unhappy in their country of origin, such as poor economic conditions, political unrest, or war.
T	F	3	U.S. immigrants are just as likely to have college degrees as are native-born Americans.
T	F	4	Immigrants increase unemployment and lower wages among native workers.
T	F	5	Most children living in undocumented immigrant families were born in the United States and are, therefore, U.S. citizens.
T	F	6	When a Gallup poll asked Americans if illegal immigrants mostly take jobs that Americans want, or if they mostly take low-paying jobs that Americans don't want, the majority of Americans indicated that they believe that most undocumented immigrants take low-paying jobs that Americans don't want.
T	F	7	To become a naturalized U.S. citizen, immigrants must pass a basic test on English and U.S. civics (history and government).
T	F	8	Foreign-born non–U.S. citizens who marry a U.S. citizen are automatically granted U.S. citizenship.

Answers can be found at the end of the chapter.

Births	143,341,000	392,714	273
minus deaths	−56,759,000	−155,505	−108
equals			
natural increase	86,582,000	237,209	165
	Each year	Each day	Each minute

FIGURE 19.1   Growth in the World's Population, 2014
Every single day, the world's population increases by more than 237,200 people.
Source: Population Reference Bureau, 2015.

population size, composition, and distribution. Many sociological studies use demographic analysis as a component of the research design because all aspects of social life are affected by demography. For example, an important relationship exists between population size and the availability of food, water, energy, and housing. Population size, composition, and distribution are also connected to issues such as poverty, racial and ethnic diversity, shifts in the age structure of society, and concerns about environmental degradation.

In the twenty-first century, nearly all of the world population growth is occurring in developing nations because of high birth rates combined with younger populations. For example, about nine out of ten people between the ages of ten and twenty-four live in less developed nations today. By contrast, birth rates

world's population each day, and 165 persons are added every minute (Population Reference Bureau, 2015).

What causes the population to grow so rapidly? This question is of interest to scholars who specialize in the study of *demography*—a subfield of sociology that examines

demography
a subfield of sociology that examines population size, composition, and distribution.

in developed nations are barely exceeding death rates because of a combination of low birth rates and much older populations (Population Reference Bureau, 2015).

Increases or decreases in population can have a powerful impact on the social, economic, and political structures of societies. As used by demographers, a *population* is a group of people who live in a specified geographic area. Changes in populations occur as a result of three processes: fertility (births), mortality (deaths), and migration.

Fertility

Fertility is the actual level of childbearing for an individual or a population. The level of fertility in a society is based on biological and social factors, the primary biological factor being the number of women of childbearing age (usually between ages 15 and 45). Other biological factors affecting fertility include the general health and level of nutrition of women of childbearing age. Social factors influencing the level of fertility include the roles available to women in a society and prevalent viewpoints regarding what constitutes the "ideal" family size.

Based on biological capability alone, most women could produce twenty or more children during their childbearing years. *Fecundity* is the potential number of children who could be born if every woman reproduced at her maximum biological capacity. Fertility rates are not as high as fecundity rates because people's biological capabilities are limited by social factors such as practicing voluntary abstinence and refraining from sexual intercourse until an older age, as well as by contraception, voluntary sterilization, abortion, and infanticide. Additional social factors affecting fertility include significant changes in the number of available partners for sex and/or marriage (as a result of war, for example), increases in the number of women of childbearing age in the workforce, and high rates of unemployment. In some countries, governmental policies also affect the fertility rate. For example, until recently the family-planning restrictions established by the People's Republic of China have allowed only one child per family in order to limit population growth in light of a population explosion that occurred between 1949 and 1976, when the population almost doubled. In 2014 these limits were relaxed for couples in which one or both partners come from single-child families themselves. Although it is too soon to tell how this change will affect the population in China, strong evidence indicates that the many years of the one-child policy have contributed to a rapidly aging population that will require more services and a shrinking workforce, both problematic for the future of the Chinese economy.

The most basic measure of fertility is the **crude birth rate**—the number of live births per 1,000 people in a population in a given year. The crude birth rate in the United States was 13.4 per 1,000 in 2014, as compared with an all-time high rate of 27 per 1,000 in 1947 (following World War II). This measure is referred to as a "crude" birth rate because it is based on the entire population and is not "refined" to

incorporate significant variables affecting fertility, such as age, marital status, religion, and race/ethnicity.

The United Nations divides countries into three categories based on fertility levels:

1. *Low-fertility countries*, where women are not having enough children to ensure that, on average, each woman is replaced by a daughter who survives to the age when she will be old enough to have children. Forty-two percent of the world's population lives in low-fertility countries, which include all countries in Europe except Iceland and Ireland; nineteen out of fifty-one countries in Asia; fourteen out of thirty-nine in the Americas; Mauritius and Tunisia in Africa; and Australia in Oceania.

2. *Intermediate-fertility countries*, where each woman is having, on average, between 1 and 1.5 daughters. Forty percent of the world's population lives in intermediate-fertility countries, such as India, the United States, Indonesia, Bangladesh, Mexico, and Egypt.

3. *High-fertility countries*, where the average woman has more than 1.5 daughters. Eighteen percent of the world's population lives in high-fertility countries, such as Pakistan, Nigeria, the Philippines, Ethiopia, the Democratic Republic of the Congo, the United Republic of Tanzania, Sudan, Kenya, Uganda, Iraq, Afghanistan, Ghana, Yemen, Mozambique, and Madagascar.

Obviously, high-fertility countries will potentially add the most to future population growth. However, it is difficult to estimate how much the world's population will *actually* increase over the next century because even small differences in fertility levels that are sustained over long periods of time can have a major effect on population size.

In most areas of the world, women are having fewer children. Women who have six or seven children tend to live in agricultural regions of the world, where children's labor is essential to the family's economic survival and child mortality rates are very high. For example, Nigeria has a crude birth rate of 46 per 1,000, as compared with 13 per 1,000 in the United States (*CIA World Factbook*, 2015). However, in Nigeria and some other African nations, families need to have many children in order to ensure that one or two will live to adulthood because of high rates of poverty, malnutrition, and disease.

Mortality

The primary cause of world population growth in recent years has been a decline in **mortality**—the incidence of death in a population. The simplest measure of mortality is the **crude death rate**—the number of deaths per 1,000 people in a population in a given year. The estimated world crude death rate for 2014 was 7.89 deaths per 1,000 population, which amounts to about 108 deaths per minute worldwide, or 1.8 deaths every second (*CIA World Factbook*, 2015).

TABLE 19.1

The Ten Leading Causes of Death in the United States, 1900 and 2014

Cause of Death—1900	Rank	Cause of Death—2014
Influenza/pneumonia	1	Heart disease
Tuberculosis	2	Cancer (malignant neoplasms)
Stomach/intestinal disease	3	Chronic lower respiratory diseases
Heart disease	4	Accidents (unintentional injuries)
Cerebral hemorrhage	5	Stroke (cerebrovascular diseases)
Kidney disease	6	Alzheimer's disease
Accidents	7	Diabetes mellitus
Cancer	8	Influenza and pneumonia
Diseases in early infancy	9	Nephritis, nephritic syndrome, and nephrosis
Diphtheria	10	Suicide (intentional self-harm)

Source: Kochanek et al., 2014.

Similarly, the U.S. crude death rate remains at about 8 deaths per 1,000 population (*CIA World Factbook,* 2015). In high-income, developed nations, such as the United States, mortality rates have declined dramatically as diseases such as malaria, polio, cholera, tetanus, typhoid, and measles have been virtually eliminated by vaccinations and improved sanitation and personal hygiene. The leading causes of deaths in most regions of the world are noncommunicable diseases such as cardiovascular diseases, cancer, diabetes, and chronic lung diseases. The ten leading causes of death in the United States in 1900 and 2014 are shown in • Table 19.1. In regions such as sub-Saharan Africa, infectious diseases remain the leading cause of death as mortality rates remain high as a result of HIV/AIDS.

Many children do not survive long enough to contract communicable diseases. On a global basis, large numbers of newborn infants do not live to see their first birthday. The measure of these deaths is referred to as the *infant mortality rate,* which is the number of deaths of infants under 1 year of age per 1,000 live births in a given year. The ten leading causes of infant death in the United States are congenital malformations, low birth weight, maternal complications, sudden infant death syndrome, unintentional injuries, cord and placental complications, bacterial sepsis, respiratory distress, circulatory system disease, and neonatal hemorrhage (Kochanek et al., 2014). The infant mortality rate is an important reflection of a society's

level of preventive (prenatal) medical care, maternal nutrition, childbirth procedures, and neonatal care for infants. Around the world, the infant mortality rate varies widely, as shown in the rates for Afghanistan: 117 deaths in the first year per 1,000 live births, compared to 1.8 deaths in the first year per 1,000 live births in Monaco.

In the United States the infant mortality rate has not changed significantly in recent years. In 2012 there were approximately 6.1 infant deaths per 1,000 live births. However, the mortality rate for African American (black) infants has remained 2.2 times as high as the rate for white American (non-Hispanic) infants: 12.4 deaths per 1,000 live births for black (African American) infants, as compared to 8.47 deaths for American Indian/Alaska Native infants, 5.33 deaths for white (non-Hispanic) infants, 5.29 for Hispanic infants, and 4.40 for Asian and Pacific Islander infants. Among other factors, differential levels of access to prenatal counseling and medical services are often reflected in the divergent infant mortality rates.

Life expectancy is an estimate of the average lifetime in years of people born in a specific year. For persons born in the United States in 2014, for example, life expectancy at birth was 79.6 years, with 77 years for males and 81.9 years for females (*CIA World Factbook,* 2015). However, disparities exist among racial groups and divergent educational categories. Studies in the second decade of the twenty-first century have found that adult men and women in the United States with fewer than twelve years of education have life expectancies that are about the same as adults in the 1950s and 1960s. When combined with race, the disparity grows even wider: White (non-Hispanic) women and men with 16 years or more of schooling have life expectancies far greater than African Americans with fewer than 12 years of education. Between white men and African American men, the difference is 14.2 additional years of life expectancy for white men. Between white women and African American women, the difference is 10.3 years more for white women than for African American women.

Migration

Migration is the movement of people from one geographic area to another for the purpose of changing residency. Migration affects the size and distribution of the population

fertility
the actual level of childbearing for an individual or a population.

crude birth rate
the number of live births per 1,000 people in a population in a given year.

mortality
the incidence of death in a population.

crude death rate
the number of deaths per 1,000 people in a population in a given year.

migration
the movement of people from one geographic area to another for the purpose of changing residency.

in a given area. *Distribution* refers to the physical location of people throughout a geographic area. In the United States, people are not evenly distributed throughout the country; many of us live in densely populated areas. *Density* is the number of people living in a specific geographic area. In urbanized areas, density may be measured by the number of people who live per room, per block, or per square mile.

Migration may be either international (movement between two nations) or internal (movement within national boundaries). Internal migration has occurred throughout U.S. history and has significantly changed the distribution of the population over time.

Migration involves two types of movement: immigration and emigration. *Immigration* is the movement of people into a geographic area to take up residency. Although immigration to the United States has continued steadily since the 1970s, as noted earlier, economic conditions in this century have reduced the flow of both legal and unauthorized immigration to the United States. Each year, more than one million people obtain legal permanent residence (LPR) in the United States. In 2013, for example, a total of 990,553 persons became legal permanent residents of this country, and the majority (54 percent) had already lived in the United States. About 66 percent were granted permanent resident status based on family relationship with a U.S. citizen or legal permanent resident of the United States. The leading countries of origin of new LPRs are Mexico (14 percent), China (7.2 percent), and India (6.9 percent) (Monger and Yankay, 2014).

Immigration rates are not an accurate reflection of the actual number of immigrants who enter a country. The U.S. Immigration and Naturalization Service records only legal immigration based on entry visas and change-of-immigration-status forms. Similarly, few records are maintained regarding *emigration*—the movement of people out of a geographic area to take up residency elsewhere. To determine the net migration in a geographic area, the number of people leaving that area to take up permanent or semipermanent residence elsewhere (emigrants) is subtracted from the number of people entering that area to take up residence there (immigrants), unless more people are moving out of the area than into it, in which case the mathematical process is reversed. It is estimated that the 2014 net immigration rate in the United States was 2.45 migrants per 1,000 population, which was down from previous years. After years of increases in immigration from Mexico, for example, net migration from Mexico fell to zero in 2012 because of economic problems and job loss in this country. However, by 2013, Mexican-born immigrants still accounted for about 28 percent of the 41.3 million total foreign-born residents of the United States. Today, persons from Mexico constitute the largest immigrant group in this country, followed by immigrants from India and China (migrationpolicy.org, 2015).

Why do people migrate? There are two key reasons why individuals and families migrate: People migrate either voluntarily or involuntarily. Voluntary migration is often related to pull factors. *Pull* factors at the international level, such as a democratic government, religious freedom, employment opportunities, or a more temperate climate, may draw voluntary immigrants into a nation. Within nations, people from large cities may be pulled to rural areas by lower crime rates, more space, and a lower cost of living. Some people are drawn by pull factors such as greater economic opportunities at their destination and are pushed by factors such as low wages and few employment opportunities in their previous place of residence. *Push* factors at the international level, such as political unrest, violence, war, famine, plagues, and natural disasters, may encourage people to leave one area and relocate elsewhere. Push factors in regional U.S. migration include unemployment, harsh weather conditions, a high cost of living, inadequate school systems, and high crime rates (• Figure 19.2).

Involuntary, or forced, migration usually occurs as a result of political oppression, such as when Jews fled Nazi Germany in the 1930s or when Afghans left their country to escape oppression there in the early 2000s. Slavery is the most striking example of involuntary migration; for example, the 10–20 million Africans forcibly transported to the Western Hemisphere prior to 1800 did not come by choice.

Where do we stand in the United States in the twenty-first century? A number of states, including Alabama, Arizona, Georgia, Indiana, South Carolina, and Utah, have passed stringent laws regarding illegal immigration. These laws have caused strife and legal contests throughout the nation. As an example, the original Arizona law required that immigrants carry necessary identity documents to show that they were legally in the United States, police were empowered to detain persons suspected of being in the country illegally, and employers would be legally sanctioned for hiring undocumented workers. However, in *United States of America v. Arizona* a district court judge blocked some controversial provisions of the law, and in 2011 Arizona Governor Jan Brewer appealed the case to the U.S. Supreme Court. In a 5–3 ruling, the Court ruled that the federal government has significant power to regulate immigration and struck down most of the key provisions of the Arizona law. However, the Court let stand the provision of the Arizona law involving police checks on people's immigration status while the police are enforcing other laws if "reasonable suspicion" exists that the person is in the United States illegally.

Members of Congress continue to propose immigration legislation to break the gridlock on this issue. Priorities of immigration bills often include one or more of these provisions:

1. *Securing the border:* The Department of Homeland Security would be required to monitor 100 percent of the southwest border with Mexico and intercept 90 percent of people trying to cross it illegally (• Figure 19.3).

2. *Path to citizenship:* Unauthorized immigrants who were in the United States before December 31, 2011, would be allowed to apply for temporary legal

FIGURE 19.2 Political unrest, violence, and war are "push" factors that encourage people to leave their country of origin. By contrast, job opportunities, such as construction work in the United States, are a major "pull" factor for people from low-income countries.

FIGURE 19.3 This fence between Mexico and the United States has been the subject of extensive media attention and public controversy. Although not a continuous fence, the barriers—combined with a "virtual fence" of sensors and cameras monitored by the U.S. Border Patrol—are designed to prevent illegal migration between the two countries.

status. After ten years, they could apply for a green card (a visa for legal permanent residents), and three years later they could apply to become U.S. citizens.

3. *Interior enforcement:* U.S. business owners would be required, within five years, to verify the legal status of new employees to ensure that they are not hiring undocumented immigrants. Homeland Security would also have to track all immigrants each time they enter and exit the United States.

4. *Immigration overhaul:* The legal immigration system would be overhauled, reducing the percentage of visas that go to immigrants based on family ties from 75 percent to 50 percent, with the other 50 percent based on occupational specializations in science, technology, math and engineering, or low-skilled jobs.

Initially it was thought by some political leaders that these immigration initiatives would also provide an opportunity for young undocumented immigrants who arrived in the United States as children, often referred to as "Dreamers," to become citizens in five years after they applied to do so. However, the federal DREAM Act proposed by the Obama administration did not pass. "DREAM" stood for Development, Relief and Education for Alien Minors.

FIGURE 19.4 How do media representations affect your views on legal and illegal immigration in the United States?

Some states passed their own version of immigration legislation using the term DREAM Act, causing confusion about which laws had passed. The state-level laws varied in what benefits they provided for students, including such things as the ability to pay in-state tuition rates, which are much lower than out-of-state tuition at most state universities (college.usatoday.com, 2015). As a result, many college students with undocumented immigrant status have been negatively affected by the failure of the federal DREAM Act bill, which was supposed to provide citizenship and educational benefits for them (●Figure 19.4).

Although our discussion has focused on how immigration affects the United States, many other nations face similar issues and often feel the need to act against migration to their country (see "Sociology in Global Perspective"). More nations are passing stringent laws to regulate immigration and make it easier to remove undocumented persons from within their nations' borders. Various factors, such as lack of jobs and poor economic conditions, fear of violence and terrorism, and concerns about how to provide an adequate social safety net for citizens, contribute to the belief that immigration is harmful to a nation's population and damages the existing social order.

LO2 **Explain** what is meant by population composition and describe how it is measured.

Population Composition

Changes in fertility, mortality, and migration affect the *population composition*—the biological and social characteristics of a population, including age, sex, race, marital status, education, occupation, income, and size of household.

One measure of population composition is the *sex ratio*—the number of males for every hundred females in a given population. A sex ratio of 100 indicates an equal number of males and females in the population. If the number is greater than 100, there are more males than females; if it is less than 100, there are more females than males. In the United States the sex ratio in 2014 was 97, which means there were 97 males per 100 females. Although approximately 124 males are conceived for every 100 females, male fetuses miscarry at a higher rate. At birth, there are 105 males for every 100 females, but because males at all ages generally have higher mortality rates than females, the sex ratio declines with age. In the 25–54 age category, the sex ratio in 2014 was 100, meaning that for every 100 men in that age category, there were 100 women. However, by age 65 and older, the sex ratio was 77, meaning that for every 100 women ages 65 and older, there were only 77 men in this age category.

For demographers, sex and age are significant population characteristics; they are key indicators of fertility and mortality rates. The age distribution of a population has a direct bearing on the demand for schooling, health, employment, housing, and pensions. The current distribution of a population can be depicted in a *population pyramid*—a graphic representation of the distribution of a population by sex and age. Population pyramids are a series of bar graphs divided into five-year age cohorts; the left side of the pyramid shows the number or percentage of males in each age bracket; the right side provides the same information for females. The age/sex distribution in the United States and other high-income nations does not have the appearance of a classic pyramid, but rather is more rectangular or barrel-shaped. By contrast, low-income nations, such as Mexico and Iran, which have high fertility and mortality rates, do fit the classic population pyramid. ● Figure 19.5 compares the demographic composition of Mexico, Iran, the United States, and France.

Population Growth in Global Context

What are the consequences of global population growth? Scholars do not agree on the answer to this question. Some biologists have warned that Earth is a finite ecosystem that cannot support the global population of 9.3 billion people predicted by 2050 or the population of more than 10 billion people predicted by the end of this century (UNFPA, 2011).

SOCIOLOGY in GLOBAL PERSPECTIVE

Problems People Like to Ignore: Global Diaspora and the Migrant Crisis

- "A tragedy of epic proportions" (United Nations statement, issued in April 2015, referring to refugees seeking but being refused asylum in Europe (nytimes.com/interactive, 2015)

- Transportation by human smugglers across the Mediterranean Sea cost African migrants from $400 to $700 per person per trip while Syrian migrants pay about $1,500. Vessels are overcrowded for maximum profit and often are not seaworthy. Once you're on the boat, there's no turning back (nytimes.com/interactive, 2015).

Are you familiar with the term *diaspora*? Here's a little background: Although it was originally used to describe the Jews exiled from Israel by the Babylonians in the fifth century B.C.E., many social scientists today use of the word *diaspora* to refer to a large group of people who possess a similar heritage or homeland but who have been forced to move, or exiled, from their country of origin and then try to relocate in other nations of the world. For example, the African diaspora describes communities that are descended from historical movements of people from Africa to regions such as North and South America, Europe, Asia, and the Middle East.

Contemporary human smugglers transport such refugees through various nations and on dangerous waterways so that they can flee such things as war and poverty in Africa and the Middle East. Many of today's refugees seek to reach Europe by way of the Mediterranean Sea, where at least 2,000 people have drowned in an effort to gain asylum (nytimes.com /interactive, 2015). Persons fortunate enough to live through the trip frequently find that countries such as Germany, Sweden, France, and Italy are not as receptive to them as these nations were in the past. Political officials believe that they simply cannot afford to provide asylum for the millions

of applicants who seek refuge in their country. For example, Italy had offered asylum to migrants rescued from the Mediterranean Sea but soon found itself overwhelmed by the number of migrants who wanted to stay. In one weekend alone, almost 6,000 migrants fleeing poverty and war showed up in Italy (*New York Times,* 2015). Likewise, hundreds of migrants set up a tent city in Paris until they were evacuated and their campsite bulldozed in an effort to control the swelling number of migrants seeking refuge there (Breeden, 2015).

What will happen? This is an important question, but the answer is unknown. Much like congressional political debates on immigration in the United States, spokespersons for European Union nations continue to argue about how to deal with the unprecedented flow of migrants into their various countries. Regardless of these endless debates, the key issue remains: How should nations deal with immigration brought about by extreme *push factors* such as poverty, persecution, and war, as well as *pull factors* such as a chance for safety, a better life, and more economic stability? What national and international policies might be implemented to deal with the global diaspora that continues to unfold in the twenty-first century?

REFLECT & ANALYZE

Are you familiar with policy debates and laws that deal with immigration in your own state and/or nation? What do you think might be done to protect the rights of all people—both permanent residents and individuals who enter a country legally or illegally? Is this a major policy dilemma of our era? Why or why not?

However, some economists have emphasized that free-market capitalism is capable of developing innovative ways to solve such problems. The debate is not a new one; for several centuries, strong opinions have been voiced about the effects of population growth on human welfare.

changes brought about by the Industrial Revolution in England, Malthus (1965/1798: 7) anonymously published *An Essay on the Principle of Population, As It Affects the Future Improvement of Society,* in which he argued that "the power of population is infinitely greater than the power of the earth to produce subsistence [food] for man."

 Discuss Malthusian, Marxist, and neo-Malthusian perspectives on population as well as demographic transition theory.

The Malthusian Perspective

English clergyman and economist Thomas Robert Malthus (1766–1834) was one of the first scholars to systematically study the effects of population. Displeased with societal

population composition
the biological and social characteristics of a population, including age, sex, race, marital status, education, occupation, income, and size of household.

sex ratio
the number of males for every hundred females in a given population.

population pyramid
a graphic representation of the distribution of a population by sex and age.

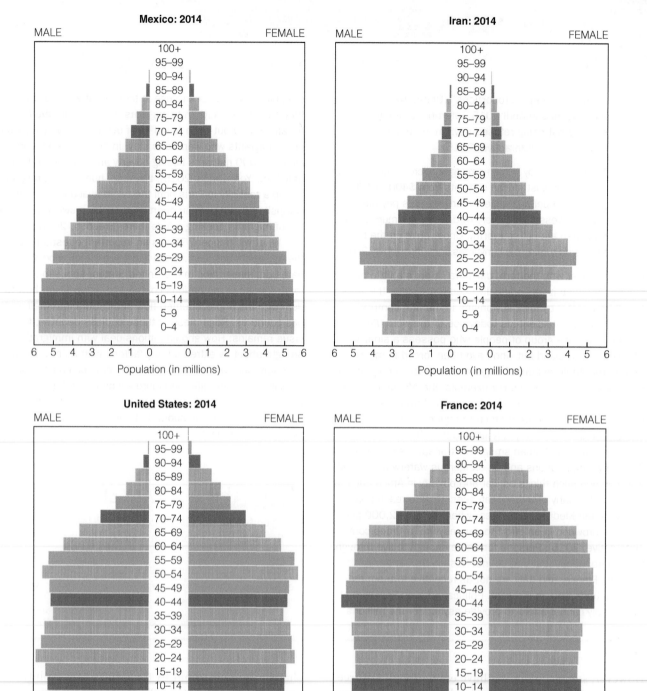

FIGURE 19.5 Population Pyramids for Mexico, Iran, the United States, and France, 2014

Source: CIA World Factbook 2015.

According to Malthus, the population, if left unchecked, would exceed the available food supply. He argued that the population would increase in a geometric (exponential) progression (2, 4, 8, 16 . . .) while the food supply would increase only by an arithmetic progression (1, 2, 3, 4 . . .). In other words, a *doubling effect* occurs: Two parents can have four children, sixteen grandchildren, and so on, but food production increases by only one acre at a time. Thus, population growth inevitably surpasses the food supply, and the lack of food ultimately ends population growth and perhaps eliminates the existing population (Weeks, 2012). Even in a best-case scenario, overpopulation results in poverty.

However, Malthus suggested that this disaster might be averted by either positive or preventive checks on population. *Positive checks* are mortality risks, such as famine, disease, and war; *preventive checks* are limits to fertility. For Malthus, the only acceptable preventive check was *moral restraint*; people should practice sexual abstinence before marriage and postpone marriage as long as possible in order to have only a few children.

Malthus has had a lasting impact on the field of population studies. Most demographers refer to his dire predictions when they examine the relationship between fertility and subsistence needs. Overpopulation is still a daunting problem that capitalism and technological advances thus far have not solved, especially in middle- and low-income nations with rapidly growing populations and very limited resources.

The Marxist Perspective

Among those who attacked the ideas of Malthus were Karl Marx and Frederick Engels. According to Marx and Engels, the food supply is not threatened by overpopulation; technologically, it is possible to produce the food and other goods needed to meet the demands of a growing population. Marx and Engels viewed poverty as a consequence of the exploitation of workers by the owners of the means of production. For example, they argued that England had poverty because the capitalists skimmed off some of the workers' wages as profits. The labor of the working classes was used by capitalists to earn profits, which, in turn, were used to purchase machinery that could replace the workers rather than supply food for all. From this perspective, overpopulation occurs because capitalists desire to have a surplus of workers (an industrial reserve army) in order to suppress wages and force workers concerned about losing their livelihoods to be more productive.

According to some contemporary economists, the greatest crisis today facing low-income nations is capital shortage, not food shortage. Through technological advances, agricultural production has reached the level at which it can meet the food needs of the world if food is distributed efficiently. *Capital shortage* refers to the lack of adequate money or property to maintain a business; it is a problem because the physical capital of the past no longer meets the needs of modern economic development. In the past, self-contained rural economies survived on local labor, using local materials to produce the capital needed for other laborers. For example, in a typical village a carpenter made the loom needed by the weaver to make cloth. Today, in the global economy the one-to-one exchange between the carpenter and the weaver is lost. With an antiquated, locally made loom, the weaver cannot compete against electronically controlled, mass-produced looms. Therefore, the village must purchase capital from the outside, using its own meager financial resources. In the process the complementary relationship between labor and capital is lost; modern technology brings with it steep costs and results in village noncompetitiveness and underemployment.

Marx and Engels made a significant contribution to the study of demography by suggesting that poverty, not overpopulation, is the most important issue with regard to food supply in a capitalist economy. Although Marx and Engels offer an interesting counterpoint to Malthus, some scholars argue that the Marxist perspective is self-limiting because it attributes the population problem solely to capitalism. In actuality, nations with socialist economies also have demographic trends similar to those in capitalist societies.

The Neo-Malthusian Perspective

More recently, *neo-Malthusians* (or "new Malthusians") have reemphasized the dangers of overpopulation. To neo-Malthusians, Earth is "a dying planet" with too many people and too little food, compounded by environmental degradation and overconsumption. The 1968 publication of Paul R. Ehrlich's *The Population Bomb* launched a worldwide discussion about the effects of overpopulation and rapid population growth. "The Population Bomb Revisited," published forty years later, reasserted the growing importance of the demographic element in the human predicament: "[t]he Earth's finite capacity to sustain human civilization" (Ehrlich and Ehrlich, 2009: 63). In other words, overpopulation and rapid population growth result in global environmental problems, ranging from global warming and rain-forest destruction to famine and vulnerability to epidemics (Ehrlich and Ehrlich, 2009). Unless significant changes are made, including improving the status of women, reducing racism and religious prejudice, reforming the agriculture system, and shrinking the growing gap between rich and poor, the consequences will be dire (Ehrlich and Ehrlich, 2009).

Early neo-Malthusians published birth control handbooks, and widespread acceptance of birth control eventually reduced the connection between people's sexual conduct and fertility (Weeks, 2012). Later neo-Malthusians have encouraged people to be part of the solution to the problem of overpopulation by having only one or two children in order to bring about ***zero population growth***—the point at which no population increase occurs from year to year because the number of births plus immigrants is equal to the number of deaths plus emigrants (Weeks, 2012).

Today, the Ehrlich proposal remains the same: "Adopt policies that gradually reduce birthrates and eventually start a global decline toward a human population size that is sustainable in the long run" (Ehrlich and Ehrlich, 2009).

Demographic Transition Theory

Some scholars who disagree with the neo-Malthusian viewpoint suggest that the theory of demographic transition offers a more accurate picture of future population

zero population growth
the point at which no population increase occurs from year to year.

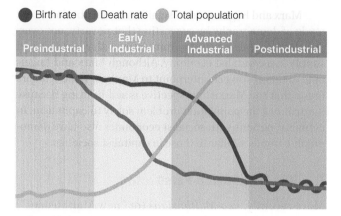

Birth rate ● Death rate ● Total population ●

| Preindustrial | Early Industrial | Advanced Industrial | Postindustrial |

FIGURE 19.6 **The Demographic Transition**

growth. ***Demographic transition*** is the process by which some societies have moved from high birth and death rates to relatively low birth and death rates as a result of technological development. Demographic transition is linked to four stages of economic development (see ● Figure 19.6):

- *Stage 1: Preindustrial societies.* Little population growth occurs because high birth rates are offset by high death rates. Food shortages, poor sanitation, and lack of adequate medical care contribute to high rates of infant and child mortality.

- *Stage 2: Early industrialization.* Significant population growth occurs because birth rates are relatively high whereas death rates decline. Improvements in health, sanitation, and nutrition produce a substantial decline in infant mortality rates. Overpopulation is likely to occur because more people are alive than the society has the ability to support.

- *Stage 3: Advanced industrialization and urbanization.* Very little population growth occurs because both birth rates and death rates are low. The birth rate declines as couples control their fertility through contraceptives and become less likely to adhere to religious directives against their use. Children are not viewed as an economic asset; they consume income rather than produce it. Societies in this stage attain zero population growth, but the actual number of births per year may still rise because of an increased number of women of childbearing age.

- *Stage 4: Postindustrialization.* Birth rates continue to decline as more women gain full-time employment and the cost of raising children continues to increase. The population grows very slowly, if at all, because the decrease in birth rates is coupled with a stable death rate.

Debate continues about whether this evolutionary model accurately explains the stages of population growth in all societies. Advocates note that demographic transition theory highlights the relationship between technological development and population growth, thus making Malthus's predictions obsolete. Scholars also point out that demographic transitions occur at a faster rate in now-low-income nations than they previously did in the nations that are already developed. For example, nations in the process of development have higher birth rates and death rates than the now-developed societies did when they were going through the transition. The death rates declined in the now-developed nations as a result of internal economic development—not, as is the case today, through improved methods of disease control (Weeks, 2012). Critics suggest that this theory best explains development in Western societies.

Other Perspectives on Population Change

In recent decades, other scholars have continued to develop theories about how and why changes in population growth patterns occur. Some have studied the relationship between economic development and a decline in fertility; others have focused on the process of secularization—the decline in the significance of the sacred in daily life—and how a change from believing that otherworldly powers are responsible for one's life to a sense of responsibility for one's own well-being is linked to a decline in fertility. Based on this premise, some analysts argue that the processes of industrialization and economic development are typically accompanied by secularization but that the relationship between these factors is complex when it comes to changes in fertility.

Shifting from the macrolevel to the microlevel, education and social psychological factors also play into the decisions that individuals make about how many children to have. Family planning information is more readily available to people with more years of formal education and may cause them to engage in decision making in accord with *rational choice theory,* which is based on the assumption that people make decisions based on a calculated cost–benefit analysis ("What do I gain and lose from a specific action?"). In low-income countries or other settings in which children are identified as an economic resource for their parents throughout life, fertility rates are higher than in higher-income countries. However, as modernization and urbanization occur in such societies, the positive economic effects of having more children may be offset by the cost of caring for those children and the lowered economic advantage gained from having children in an industrialized nation.

As demographers have reformulated the demographic transition theory, they have highlighted additional factors that are likely to be causes of fertility decline, and they have suggested that demographic transition is not just one process but rather a set of intertwined transitions. One is the epidemiological transition—the shift from deaths at younger ages because of acute, communicable diseases. Another is the fertility transition—the shift from natural fertility to controlled fertility, resulting in a decrease in the fertility rate. Other transitions include the migration transition, the urban transition, the age transition, and the family and household transition, which occur as a result of lower fertility, a longer life, an older age structure, and a predominantly urban residence.

A Brief Glimpse at International Migration Theories

Why do people relocate from one nation to another? Several major theories have been developed in an attempt to explain international migration. The *neoclassical economic approach* assumes that migration patterns occur based on geographic differences in the supply of and demand for labor. The United States and other high-income countries that have had growing economies and a limited supply of workers for certain types of jobs have paid higher wages than are available in areas with a less-developed economy and a large labor force. As a result, people move to gain higher wages and sometimes better living conditions. They may also take jobs in other countries so that they can send money to their families in their country of origin.

Unlike the neoclassical explanation of migration, which focuses on individual decision making, the *new households economics of migration approach* emphasizes the part that entire families or households play in the migration process. From this approach, Mexican workers' temporary migration to the United States would be examined not only from the perspective of the individual worker but also in terms of what the entire family gains from the process of having one or more migrant family members work in another country. By having a diversity of family income (originating from more than one source), the family is cushioned from the economic woes of the nation that most of the family members think of as "home."

Two conflict perspectives on migration add to our knowledge of why people migrate. Split-labor-market theory suggests that immigrants from low-income countries are often recruited for secondary labor market positions: dead-end jobs with low wages, unstable employment, and sometimes hazardous working conditions. By contrast, migrants from higher-income countries may migrate for primary-sector employment—jobs in which well-educated

FIGURE 19.7 Migrants from other high-income countries often move to the United States to gain primary-sector jobs, such as in the high-tech industry, where workers are paid high wages and receive generous benefit packages. How does this affect migration patterns in both the countries of origin and countries of destination?

workers are paid high wages and receive benefits such as health insurance and a retirement plan (● Figure 19.7). The global migration of some high-tech workers is an example of this process, whereas the migration of farm workers and construction helpers is an example of secondary labor market migration.

Finally, world systems theory (discussed later in this chapter) views migration as linked to the problems caused by capitalist development around the world (Massey et al., 1993). As the natural resources, land, and workforce in low-income countries with little or no industrialization have come under the influence of international markets, there has been a corresponding flow of migrants from those nations to the highly industrialized, high-income countries, especially those with which the poorer nations have had the most economic, political, or military contact.

After flows of migration commence, the pattern may continue because potential migrants have personal ties with relatives and friends who now live in the country of destination and can serve as a source of stability when the potential migrants relocate to the new country. Known as *network theory*, this approach suggests that once migration has begun, it takes on a life of its own and that the

demographic transition
the process by which some societies have moved from high birth and death rates to relatively low birth and death rates as a result of technological development.

migration pattern which ensues may be different from the original push or pull factors that produced the earlier migration. Another approach, *institutional theory,* suggests that migration may be fostered by groups—such as humanitarian aid organizations relocating refugees or smugglers bringing people into a country illegally—and that the actions of these groups may produce a larger stream of migrants than would otherwise be the case.

As you can see from these diverse approaches to explaining contemporary patterns of migration, the reasons that people migrate are numerous and complex, involving processes occurring at the individual, family, and societal levels.

LO5 **Explain** how cities develop through preindustrial, industrial, and postindustrial stages.

Urbanization in Global Perspective

Urban sociology is a subfield of sociology that examines social relationships and political and economic structures in the city. According to urban sociologists, a *city* is a relatively dense and permanent settlement of people who secure their livelihood primarily through nonagricultural activities. Although cities have existed for thousands of years, only about 3 percent of the world's population lived in cities 200 years ago, as compared with more than 50 percent today. Current estimates suggest that 69 percent of the global population will live in urban areas by 2050 (United Nations DESAPD, 2010).

Emergence and Evolution of the City

Cities are a relatively recent innovation when compared with the length of human existence. The earliest humans are believed to have emerged anywhere from 40,000 to 1,000,000 years ago, and permanent human settlements are believed to have begun first about 8000 B.C.E. However, some scholars date the development of the first city between 3500 and 3100 B.C.E., depending largely on whether a formal writing system is considered as a requisite for city life (Sjoberg, 1965; Weeks, 2012).

According to the sociologist Gideon Sjoberg (1965), three preconditions must be present in order for a city to develop:

1. *A favorable physical environment,* including climate and soil favorable to the development of plant and animal life and an adequate water supply to sustain both.
2. *An advanced technology* (for that era) that could produce a social surplus in both agricultural and nonagricultural goods.
3. *A well-developed social organization,* including a power structure, in order to provide social stability to the economic system.

Based on these prerequisites, Sjoberg places the first cities in the Middle Eastern region of Mesopotamia or in areas immediately adjacent to it at about 3500 B.C.E. However, not all scholars concur; some place the earliest city in Jericho (located in present-day Jordan) at about 8000 B.C.E., with a population of about 600 people (see Kenyon, 1957).

The earliest cities were not large by today's standards. The population of the larger Mesopotamian centers was between 5,000 and 10,000 (Sjoberg, 1965). The population of ancient Babylon (probably founded around 2200 B.C.E.) may have grown as large as 50,000 people; Athens may have held 80,000 people (Weeks, 2012). Four to five thousand years ago, cities with at least 50,000 people existed in the Middle East (in what today is Iraq and Egypt) and Asia (in what today is Pakistan and China), as well as in Europe. About 3,500 years ago, cities began to reach this size in Central and South America.

Preindustrial Cities

The largest preindustrial city was Rome; by 100 C.E., it may have had a population of 650,000. With the fall of the Roman Empire in 476 C.E., the nature of European cities changed. Seeking protection and survival, those persons who lived in urban settings typically did so in walled cities containing no more than 25,000 people. For the next 600 years, the urban population continued to live in walled enclaves, as competing warlords battled for power and territory during the "dark ages." Slowly, as trade increased, cities began to tear down their walls.

Preindustrial cities were limited in size by a number of factors. For one thing, crowded housing conditions and a lack of adequate sewage facilities increased the hazards from plagues and fires, and death rates were high. For another, food supplies were limited. In order to generate food for each city resident, at least fifty farmers had to work in the fields, and animal power was the only means of bringing food to the city. Once foodstuffs arrived in the city, there was no effective way to preserve them. Finally, migration to the city was difficult. Many people were in serf, slave, and caste systems whereby they were bound to the land. Those able to escape such restrictions still faced several weeks of travel to reach the city, thus making it physically and financially impossible for many people to become city dwellers.

In spite of these problems, many preindustrial cities had a sense of *community*—a set of social relationships operating within given spatial boundaries or locations that provided people with a sense of identity and a feeling of belonging. The cities were full of people from all walks of life, both rich and poor, and they felt a high degree of social integration. You will recall that Ferdinand Tönnies (1940/1887) described such a community as *Gemeinschaft*—a society in which social relationships are based on personal bonds of friendship and kinship and on intergenerational stability, such that people have a commitment to the entire group and feel a sense of togetherness. By contrast, industrial cities were characterized by Tönnies as *Gesellschaft*—societies exhibiting impersonal and specialized relationships, with little long-term commitment to the group or consensus on values. In *Gesellschaft* societies, even neighbors are "strangers" who perceive that they have little in common with one another.

Industrial Cities

The Industrial Revolution changed the nature of the city. Factories sprang up rapidly as production shifted from the primary, agricultural sector to the secondary, manufacturing sector. With the advent of factories came many new employment opportunities not available to people in rural areas. Emergent technology, including new forms of transportation and agricultural production, made it easier for people to leave the countryside and move to the city. Between 1700 and 1900, the population of many European cities mushroomed. For example, the population of London increased from 550,000 to almost 6.5 million. Although the Industrial Revolution did not start in the United States until the mid-nineteenth century, the effect was similar. Between 1870 and 1910, for example, the population of New York City grew by 500 percent. In fact, New York City became the first U.S. *metropolis*—one or more central cities and their surrounding suburbs that dominate the economic and cultural life of a region. Nations such as Japan and Russia, which became industrialized after England and the United States, experienced a delayed pattern of urbanization, but this process moved quickly once it began in those countries.

close to a central business district (• Figure 19.8). Technological advances in communication and transportation make it possible for middle- and upper-income individuals and families to have more work options and to live greater distances from the workplace; however, these options are not often available to people of color and those at the lower end of the class structure.

On a global basis, cities such as New York, London, and Tokyo appear to fit the model of the postindustrial city. These cities have experienced a rapid growth in knowledge-based industries such as financial services. London, Tokyo, and New York have—at least until recently—experienced an increase in the number of highly paid professional jobs, and more workers have been in high-income categories. Many people have benefited for a number of years from these high incomes and have created a lifestyle that is based on materialism and the gentrification of urban spaces. Meanwhile, those persons outside the growing professional categories have seen their own quality of life further deteriorate and their job opportunities become increasingly restricted to secondary labor markets in their respective "global" cities.

What is next for the postindustrial city? Various social analysts have proposed that cities are far from dead because of the many benefits they offer people. According to urban economist Edward Glaeser (2011), the future of nations and the world relies on cities that bring people together in a setting that is "healthier, greener, and richer" than urban myths would have us believe. From this perspective, incomes are higher in metropolitan areas; cities are more energy efficient than suburban areas, where people commute great

Postindustrial Cities

Since the 1950s, postindustrial cities have emerged in nations such as the United States as their economies have gradually shifted from secondary (manufacturing) production to tertiary (service and information-processing) production. Postindustrial cities increasingly rely on an economic structure that is based on scientific knowledge rather than industrial production, and, as a result, a class of professionals and technicians grows in size and influence. Postindustrial cities are dominated by "light" industry, such as software manufacturing; information-processing services, such as airline and hotel reservation services; educational complexes; medical centers; convention and entertainment centers; and retail trade centers and shopping malls. Most families do not live

FIGURE 19.8 Since the 1950s, postindustrial cities have emerged in which families do not live close to a central business district and more homes are located in suburban or outlying residential enclaves such as the neighborhood shown here.

distances, often in heavy traffic congestion with stop-and-go traffic, between work and home. Cities are also centers of consumption; however, this benefit primarily accrues to the wealthy. However, the poor fare better than many think because they have inexpensive mass transit, the ability to "cram into small apartments in the outer boroughs," and "plenty of entry-level service-sector jobs with wages that beat those in Ghana or Guatemala" (Glaeser, 2011). People in the middle-income category often have a harder time because of the costs of housing in good neighborhoods and a quality education for their children (Glaeser, 2011).

A second perspective on the postmodern city has been offered by John D. Kasarda, who coined the term *aerotropolis* to describe a new urban pattern in which cities are built around airports rather than airports being built around cities. An aerotropolis is a combination of giant airport, planned city, shipping facility, and business hub. According to this approach, the pattern of the twentieth century was city in the center, airport on the periphery. However, this pattern has shifted, with the airport now in the center and the city on the periphery, because of extensive growth in jet travel, 24/7 workdays, overnight shipping, and global business networks (Kasarda and Lindsay, 2011). Aerotropoli are now found in Seoul, Amsterdam, Dallas, Memphis, Washington, D.C., and other cities where globalization has forever changed the nature of urban life.

Perspectives on Urbanization and the Growth of Cities

Urban sociology follows in the tradition of early European sociological perspectives that compared social life with biological organisms or ecological processes. For example, Auguste Comte pointed out that cities are the "real organs" that make a society function. Emile Durkheim applied natural ecology to his analysis of *mechanical solidarity,* characterized by a simple division of labor and shared religious beliefs such as are found in small, agrarian societies, and *organic solidarity,* characterized by interdependence based on the elaborate division of labor found in large, urban societies. These early analyses became the foundation for ecological models/functionalist perspectives in urban sociology.

 Compare ecological/functionalist models with political economy/conflict models in their explanations of urban growth.

Functionalist Perspectives: Ecological Models

Functionalists examine the interrelations among the parts that make up the whole; therefore, in studying the growth of cities, they emphasize the life cycle of urban growth. Like the social philosophers and sociologists before him, the University of Chicago sociologist Robert Park (1915) based his analysis of the city on *human ecology*—the study of the relationship between people and their physical environment. According to Park (1936), economic competition produces certain regularities in land-use patterns and population distributions. Applying Park's idea to the study of urban land-use patterns, the sociologist Ernest W. Burgess (1925) developed the concentric zone model, an ideal construct that attempts to explain why some cities expand radially from a central business core.

The Concentric Zone Model Burgess's *concentric zone model* is a description of the process of urban growth that views the city as a series of circular areas or zones, each characterized by a different type of land use, which developed from a central core (see • Figure 19.9a). *Zone 1* is the central business district and cultural center. In *Zone 2,* houses formerly occupied by wealthy families are divided into rooms

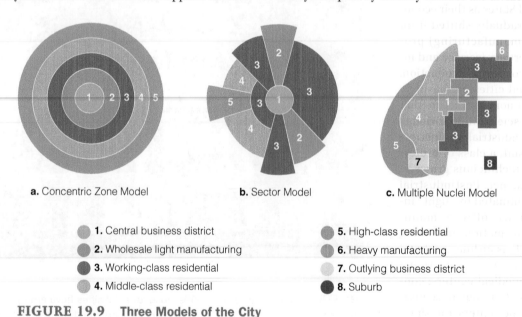

a. Concentric Zone Model **b.** Sector Model **c.** Multiple Nuclei Model

1. Central business district
2. Wholesale light manufacturing
3. Working-class residential
4. Middle-class residential
5. High-class residential
6. Heavy manufacturing
7. Outlying business district
8. Suburb

FIGURE 19.9 Three Models of the City
Source: Adapted from Harris and Ullman, 1945.

and rented to recent immigrants and poor persons; this zone also contains light manufacturing and marginal businesses (such as second-hand stores, pawnshops, and taverns). *Zone 3* contains working-class residences and shops and ethnic enclaves. *Zone 4* comprises homes for affluent families, single-family residences of white-collar workers, and shopping centers. *Zone 5* is a ring of small cities and towns populated by persons who commute to the central city to work and by wealthy people living on estates.

Two important ecological processes are involved in the concentric zone theory: invasion and succession. ***Invasion*** is the process by which a new category of people or type of land use arrives in an area previously occupied by another group or type of land use (McKenzie, 1925). For example, Burgess noted that recent immigrants and low-income individuals "invaded" Zone 2, formerly occupied by wealthy families. ***Succession*** is the process by which a new category of people or type of land use gradually predominates in an area formerly dominated by another group or type of land use (McKenzie, 1925). In Zone 2, for example, when some of the single-family residences were sold and subsequently divided into multiple housing units, the remaining single-family owners moved out because the "old" neighborhood had changed. As a result of their move, the process of invasion was complete, and succession had occurred.

Invasion and succession theoretically operate in an outward movement: Those who are unable to "move out" of the inner rings are those without upward social mobility, so the central zone ends up being primarily occupied by the poorest residents—except when gentrification occurs. ***Gentrification*** is the process by which members of the middle and upper-middle classes, especially whites, move into a central-city area and renovate existing properties (●Figure 19.10). Centrally located, naturally attractive areas are the most likely candidates for gentrification. To urban ecologists, gentrification is the solution to revitalizing the central city. To conflict theorists, however, gentrification creates additional hardships for the poor by depleting the amount of affordable housing available and by "pushing" them out of the area.

The concentric zone model demonstrates how economic and political forces play an important part in the location of groups and activities, and it shows how a large urban area can have internal differentiation. However, the model is most applicable to older cities that experienced high levels of immigration early in the twentieth century and to a few midwestern cities such as St. Louis. No city, including Chicago (on which the model is based), entirely conforms to this model.

FIGURE 19.10 These Brooklyn, New York, residences are in various stages of renovation as the Crown Heights area in which they are located continues to undergo gentrification. Do you believe that gentrification is the solution to revitalizing cities? Which categories of people are most likely to benefit from this process? Which are most likely to be disadvantaged as change occurs?

The Sector Model In an attempt to examine a wider range of settings, urban ecologist Homer Hoyt (1939) studied the configuration of 142 cities. Hoyt's *sector model* emphasizes the significance of terrain and the importance of transportation routes in the layout of cities. According to Hoyt, residences of a particular type and value tend to grow outward from the center of the city in wedge-shaped sectors, with the more-expensive residential neighborhoods located along the higher ground near lakes and rivers or along certain streets that stretch in one direction or another from the downtown area (see Figure 19.9b). By contrast, industrial areas tend to be located along river valleys and railroad lines. Middle-class residential zones exist on either side of the wealthier neighborhoods. Finally, lower-class residential areas occupy the remaining space, bordering the central business area and the industrial areas. Hoyt (1939) concluded that the sector model applied to cities such as Seattle, Minneapolis, San Francisco, Charleston (South Carolina), and Richmond (Virginia).

invasion
the process by which a new category of people or type of land use arrives in an area previously occupied by another group or type of land use.

succession
the process by which a new category of people or type of land use gradually predominates in an area formerly dominated by another group or type of land use.

gentrification
the process by which members of the middle and upper-middle classes, especially whites, move into a central-city area and renovate existing properties.

The Multiple Nuclei Model According to the *multiple nuclei model* developed by urban ecologists Chauncey Harris and Edward Ullman (1945), cities do not have one center from which all growth radiates but rather have numerous centers of development based on specific urban needs or activities (see Figure 19.9c). As cities began to grow rapidly, they annexed formerly outlying and independent townships that had been communities in their own right. In addition to the central business district, other nuclei developed around entities such as an educational institution, a medical complex, or a government center. Residential neighborhoods may exist close to or far away from these nuclei. A wealthy residential enclave may be located near a high-priced shopping center, for instance, whereas less-expensive housing must locate closer to industrial and transitional areas of town. This model may be applicable to cities such as Boston. However, critics suggest that it does not provide insights about the uniformity of land-use patterns among cities and relies on an after-the-fact explanation of why certain entities are located where they are.

Contemporary Urban Ecology Urban ecologist Amos Hawley (1950) revitalized the ecological tradition by linking it more closely with functionalism. According to Hawley, urban areas are complex and expanding social systems in which growth patterns are based on advances in transportation and communication. For example, commuter railways and automobiles led to the decentralization of city life and the movement of industry from the central city to the suburbs (Hawley, 1981).

Other urban ecologists have continued to refine the methodology used to study the urban environment. *Social area analysis* examines urban populations in terms of economic status, family status, and ethnic classification (Shevky and Bell, 1966). For example, middle- and upper-middle-class parents with school-age children tend to cluster together in "social areas" with a "good" school district; young single professionals may prefer to cluster in the central city for entertainment and nightlife.

The influence of human ecology on the field of urban sociology is still very strong today (see Frisbie and Kasarda, 1988). Contemporary research on European and North American urban patterns is often based on the assumption that spatial arrangements in cities conform to a common, most efficient design. However, some critics have noted that ecological models do not take into account the influence of powerful political and economic elites on the development process in urban areas (Feagin and Parker, 2002).

Conflict Perspectives: Political Economy Models

Conflict theorists argue that cities do not grow or decline by chance. Rather, they are the product of specific decisions made by members of the capitalist class and political elites. These far-reaching decisions regarding land use and urban development benefit the members of some groups at the expense of others (see Castells, 1977/1972). Karl Marx suggested that cities are the arenas in which the intertwined processes of class conflict and capital accumulation take place; class consciousness and worker revolt are more likely to develop when workers are concentrated in urban areas.

According to the sociologists Joe R. Feagin and Robert Parker (2002), three major themes prevail in political economy models of urban growth. First, both economic *and* political factors affect patterns of urban growth and decline. Economic factors include capitalistic investments in production, workers, workplaces, land, and buildings. Political factors include governmental protection of the right to own and dispose of privately held property as owners see fit and the role of government officials in promoting the interests of business elites and large corporations.

Second, urban space has both an exchange value and a use value. *Exchange value* refers to the profits that industrialists, developers, bankers, and others make from buying, selling, and developing land and buildings. By contrast, *use value* is the utility of space, land, and buildings for everyday life, family life, and neighborhood life. In other words, land has purposes other than simply for generating profits—for example, for homes, open spaces, and recreational areas (see • Figure 19.11). Today, class conflict exists over the use

Exchange Value
Profits from buying, selling, and developing urban land

Use Value
Utility of urban land, space, and buildings for everyday personal and community life

Examples of who profits
- Industrialists
- Developers
- Bankers
- Tax collectors

Examples of uses
- Affordable housing
- Open spaces
- Recreational areas
- Public services

FIGURE 19.11 **The Value of Urban Space**

of urban space, as is evident in battles over the rental costs, safety, and development of large-scale projects.

Third, both structure and agency are important in understanding how urban development takes place. *Structure* refers to institutions such as state bureaucracies and capital investment circuits that are involved in the urban development process. *Agency* refers to human actors, including developers, business elites, and activists protesting development, who are involved in decisions about land use.

Capitalism and Urban Growth in the United States

According to political economy models, urban growth is influenced by capital investment decisions, power and resource inequality, class and class conflict, and government subsidy programs. Members of the capitalist class choose corporate locations, decide on sites for shopping centers and factories, and spread the population that can afford to purchase homes into sprawling suburbs located exactly where the capitalists think they should be located (Feagin and Parker, 2002).

Today, a few hundred financial institutions and developers finance and construct most major and many smaller urban development projects around the country, including skyscrapers, shopping malls, and suburban housing projects. These decision makers set limits on the individual choices of the ordinary citizen with regard to real estate, just as they do with regard to other choices (Feagin and Parker, 2002). They can make housing more affordable or totally unaffordable for many people. Ultimately, their motivation rests not in benefiting the community but rather in making a profit; the cities that they produce reflect this mindset.

One of the major results of these urban development practices is *uneven development*—the tendency of some neighborhoods, cities, or regions to grow and prosper whereas others stagnate and decline. Conflict theorists argue that uneven development reflects inequalities of wealth and power in society. The problem not only affects areas in a state of decline but also produces external costs, even in "boom" areas, that are paid by the entire community. Among these costs are increased pollution, increased traffic congestion, and rising rates of crime and violence. According to the sociologist Mark Gottdiener (1985: 214), these costs are "intrinsic to the very core of capitalism, and those who profit the most from development are not called upon to remedy its side effects."

The Gated Community in the Capitalist Economy

The growth of *gated communities*—subdivisions or neighborhoods surrounded by barriers such as walls, fences, gates, or earth banks covered with bushes and shrubs, along with a secured entrance—is an example to many people of how developers, builders, and municipalities have encouraged an increasing division between public and private property in capitalist societies (• Figure 19.12). Many gated communities are created by developers who hope to increase their profits by offering potential residents a semblance of safety, privacy, and luxury that they might not have in nongated residential areas. Other gated communities have been developed after the fact in established neighborhoods by adding walls, gates, and sometimes security guard stations. In the past, for example, residents of elite residential enclaves, such as the River Oaks area of Houston or the "Old Enfield" area of Austin, Texas, were able to gain approval from the city to close certain streets and create cul-de-sacs, or to erect other barriers to discourage or prevent outsiders from driving through the neighborhood (Kendall, 2002). Gated communities for upper-middle-class and upper-class residents convey the idea of exclusivity and privilege, whereas such communities for middle- and lower-income residents typically focus on such features as safety for children and the ability to share amenities such as a "community" swimming pool or recreational center with other residents.

FIGURE 19.12 How do gated communities increase the division between public and private property in the United States?

Regardless of the social and economic reasons given for the development of gated communities, many analysts agree that these communities reflect a growing divide between public and private space in urban areas. According to a qualitative study by the anthropologist Setha Low (2003), gated communities do more than simply restrict access to the residents' homes: They also limit the use of public spaces, making it impossible for others to use the roads, parks, and open space contained within the enclosed community. Low (2003) refers to this phenomenon as the "fortressing of America."

Gender Regimes in Cities

Some feminist perspectives focus on urbanization as a reflection not only of the workings of the political economy but also of patriarchy. According to the sociologist Lynn M. Appleton (1995), different kinds of cities have different *gender regimes*—prevailing ideologies of how women and men should think, feel, and act; how access to social positions and control of resources should be managed; and how relationships between men and women should be conducted. The higher density and greater diversity found in central cities such as New York City serve as a challenge to the private patriarchy found in the home and workplace in lower-density, homogeneous areas such as suburbs and rural areas. *Private patriarchy* is based on a strongly gendered division of labor in the home, gender-segregated paid employment, and women's dependence on men's income.

At the same time, cities may foster *public patriarchy* in the form of women's increasing dependence on paid work and the government for income and their decreasing emotional interdependence with men. At this point, gender often intersects with class and race as a form of oppression because lower-income women of color often reside in central cities. Public patriarchy may be perpetuated by cities through policies that limit women's access to paid work and public transportation. However, such cities may also be a forum for challenging patriarchy; all residents who differ in marital status, paternity, sexual orientation, class, and/or race/ethnicity tend to live close to one another and may hold a common belief that both public and private patriarchy should be eliminated (Appleton, 1995).

LO7 **Describe** what is meant by the experience of urban life and indicate how sociologists seek to explain this experience.

Symbolic Interactionist Perspectives: The Experience of City Life

Symbolic interactionists examine the *experience* of urban life. How does city life affect the people who live in a city? Some analysts answer this question positively; others are cynical about the effects of urban living on the individual.

Simmel's View of City Life

According to the German sociologist Georg Simmel (1950/1902–1917), urban life is highly stimulating, and it shapes people's thoughts and actions. Urban residents are influenced by the quick pace of the city and the pervasiveness of economic relations in everyday life. Because of the intensity of urban life, people have no choice but to become somewhat insensitive to events and individuals around them. Many urban residents avoid emotional involvement with one another and try to ignore events taking place around them. Urbanites feel wary toward other people because most interactions in the city are economic rather than social. Simmel suggests that attributes such as punctuality and exactness are rewarded but that friendliness and warmth in interpersonal relations are viewed as personal weaknesses. Some people act reserved to cloak their deeper feelings of distrust or dislike toward others. However, Simmel did not view city life as completely negative; he also pointed out that urban living could have a liberating effect on people because they have opportunities for individualism and autonomy.

Urbanism as a Way of Life

Based on Simmel's observations on social relations in the city, the early Chicago School sociologist Louis Wirth (1938) suggested that urbanism is a "way of life." *Urbanism* refers to the distinctive social and psychological patterns of life typically found in the city. According to Wirth, the size, density, and heterogeneity of urban populations typically result in an elaborate division of labor and in spatial segregation of people by race/ethnicity, social class, religion, and/or lifestyle. In the city, primary-group ties are largely replaced by secondary relationships; social interaction is fragmented, impersonal, and often superficial. Even though people gain some degree of freedom and privacy by living in the city, they pay a price for their autonomy, losing the support and reassurance that come from primary-group ties.

From Wirth's perspective, people who live in urban areas are alienated, powerless, and lonely. A sense of community is obliterated and replaced by the "mass society"—a large-scale, highly institutionalized society in which individuality is supplanted by mass messages, faceless bureaucrats, and corporate interests.

Gans's Urban Villagers

In contrast to Wirth's gloomy assessment of urban life, the sociologist Herbert Gans (1982/1962) suggested that not everyone experiences the city in the same way (● Figure 19.13). Based on research in the west end of Boston in the late 1950s, Gans concluded that many residents develop strong loyalties and a sense of community in central-city areas that outsiders may view negatively. According to Gans, there are five major categories of adaptation among urban dwellers. *Cosmopolites* are students, artists, writers, musicians, entertainers, and professionals who choose to live in the city because they want to be close to its cultural facilities. *Unmarried people* and *childless couples* live in the city because they want to be close to work and entertainment. *Ethnic villagers* live in ethnically segregated neighborhoods; some are recent immigrants who feel most comfortable within their own group.

The *deprived* are poor individuals with dim future prospects; they have very limited education and few, if any, other resources. The *trapped* are urban dwellers who can find no escape from the city; this group includes persons left behind by the process of invasion and succession, downwardly mobile individuals who have lost their former position in society, older persons who have nowhere else to go, and individuals addicted to alcohol or other drugs. Gans concluded that the city is a pleasure and a challenge for some urban dwellers and a nightmare for others.

Cities and Persons with a Disability

Chapter 18 describes how disability rights advocates believe that structural barriers create a "disabling" environment for many people, particularly in large urban settings. Many cities have made their streets and sidewalks more user friendly for persons in wheelchairs and for individuals with visual disability by constructing concrete ramps with slide-proof surfaces at intersections or installing traffic lights with sounds designating when to "Walk." However, both urban and rural areas have a long way to go before many persons with disabilities will have the access to the things they need to become productive members of the community: educational and employment opportunities. Some persons with disabilities cannot navigate the streets and sidewalks of their communities, and some face obstacles getting into buildings that marginally, at best, meet the accessibility standards of the Americans with Disabilities Act; thus, many persons with a disability are unemployed.

Political scientist Harlan Hahn (1997: 177–178) traces the problem of lack of access to the beginnings of industrialism:

The rise of industrialism produced extensive changes in the lives of disabled as well as nondisabled people.

FIGURE 19.13 The people shown here enjoying a walk down East 6th Street, also known as Old Pecan Street, in Austin, Texas, may consider themselves to be "urban villagers" because they can enjoy small-town amenities such as little shops, neighborhood-like restaurants, and live music venues while living in a larger city. Do you consider yourself to be an urban villager?

Renato Granieri/Alamy

As factories replaced private dwellings as the primary sites of production, routines and architectural configurations were standardized to suit nondisabled workers. Both the design of worksites and of the products that were manufactured gave virtually no attention to the needs of people with disabilities. As a result, patterns of aversion and avoidance toward disabled persons were embedded in the construction of commodities, landscapes, and buildings that would remain for centuries. . . .

The social and economic changes fostered by industrialization may have been exacerbated by the accompanying process of urbanization. As workers increasingly moved from farms and rural villages to live near the institutions of mass production, the character of community life appeared to shift perceptibly. Deviant or atypical personal characteristics that may have gradually become familiar in a small community seemed bizarre or disturbing in an urban milieu.

As Hahn's statement suggests, historical patterns in the dynamics of industrial capitalism contributed to discrimination against persons with disabilities, and this legacy remains evident in contemporary cities. Structural barriers are further intensified when other people do not respond favorably toward persons with disabilities. A classic life experience was told by scholar and disability rights advocate Sally French (1999: 25–26), who described her experience as a person with a visual disability:

I have lived in the same house for 16 years and yet I cannot recognize my neighbors. I know nothing about them at all; which children belong to whom, who has come and gone, who is old or young, ill or well, black or white. . . . On moving to my present house I informed several neighbors that, because of my inability to recognize them, I would doubtless pass them by in the street without greeting them. One neighbor, who had previously seen me striding confidently down the road, refused to believe me, but the others said they understood and would talk to me if our paths crossed. For the first couple of weeks it worked and I was surprised how often we met, but after that their greetings rapidly decreased and then ceased altogether. Why this happened I am not sure, but I suspect that my lack of recognition strained the interaction and limited the social reward they received from the encounter.

The Concept Quick Review examines the multiple perspectives on urban growth and urban living.

Problems in Global Cities

Although people have lived in cities for thousands of years, the time is rapidly approaching when more people worldwide will live in or near a city than live in a rural area. In the middle-income and low-income regions of the world, Latin America is becoming the most urbanized: Four megacities—Mexico City, Buenos Aires, Lima, and Santiago—already contain more than half of the region's population and continue to grow rapidly. Soon, Rio de Janeiro and São Paulo are expected to have a combined population of about 35 million people living in a 350-mile-long megalopolis.

Rapid population growth will have a major effect on cities throughout the world. Essential services such as health, education, transportation, and sanitation are already strained in many cities, and the problem will only grow worse as the world's population moves upward toward a projected 10 billion or more by the end of the twenty-first century. In China alone it is estimated that 220 cities will have more

than 1 million people by 2025 and that more than 350 million rural residents will move to the cities from rural areas. Requirements for new and expanded infrastructure in these cities will be tremendous, particularly in the building of high-rise buildings, mass-transit systems, and other amenities that will be needed to support this colossal population shift.

Today, some social analysts look beyond the city proper, which is defined as a locality with legally fixed boundaries and an administratively recognized urban status that is usually characterized by some form of local government, to see the larger picture of what takes place in urban agglomerations. An *urban agglomeration* is defined as comprising the city or town proper and also the suburban fringe or thickly settled territory lying outside of, but adjacent to, the city boundaries, as a more accurate reflection of population composition and density in a given region. • Figure 19.14 shows the populations of the world's fifteen largest urban agglomerations.

Natural increases in population (higher birth rates than death rates) account for two-thirds of new urban growth. In recent years, fewer deaths and more births have occurred than demographers had anticipated. High fertility brings about booming populations and a corresponding strain on food and other resources; however, low fertility contributes to an aging population and stress on social services.

The other component of new urban growth is rural-to-urban migration. Some people move from rural areas to urban areas because they have been displaced from their land. Others move because they are looking for a better life. No matter what the reason, migration has caused rapid growth in cities in China, sub-Saharan Africa, India, Algeria, and Egypt. At the same time that the population is growing rapidly, the amount of farmland available for growing crops to feed people is decreasing. In Egypt, for example, land that

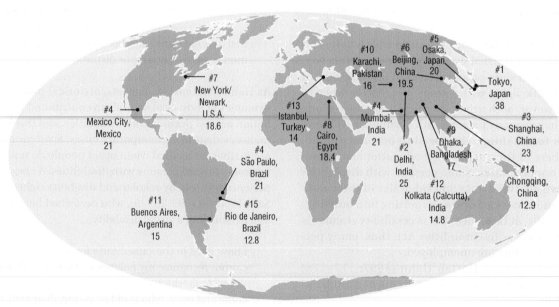

FIGURE 19.14   **The World's Fifteen Largest Agglomerations**

Note: Estimation of the 2014 population in millions (rounded).

Source: United Nations World Urbanization Prospects Report, 2014.

Perspectives on Urbanism and the Growth of Cities

Functionalist Perspectives: Ecological Models	Concentric zone model	Because of invasion, succession, and gentrification, cities are a series of circular zones, each characterized by a particular land use.
	Sector model	Cities consist of wedge-shaped sectors, based on terrain and transportation routes, with the most-expensive areas occupying the best terrain.
	Multiple nuclei model	Cities have more than one center of development, based on specific needs and activities.
Conflict Perspectives: Political Economy Models	Capitalism and urban growth	Members of the capitalist class choose locations for skyscrapers and housing projects, limiting individual choices by others.
	Gender regimes in cities	Different cities have different prevailing ideologies regarding access to social positions and resources for men and women.
	Gated communities in the capitalist economy	A belief in private ownership of property under capitalism promotes real-estate developments such as gated communities that contribute to a growing divide between public and private space in urban areas.
Symbolic Interactionist Perspectives: The Experience of City Life	Simmel's view of city life	Because of the intensity of city life, people become somewhat insensitive to individuals and events around them.
	Urbanism as a way of life	The size, density, and heterogeneity of urban population result in an elaborate division of labor and space.
	Gans's urban villagers	Five categories of adaptation occur among urban dwellers, ranging from cosmopolites to trapped city dwellers.
	Disability and city life	Cities provide unique challenges for persons with disabilities because urban areas often have barriers that make equal access difficult, despite rights that were gained under the Americans with Disabilities Act.

was previously used for growing crops is now used for petroleum refineries, food-processing plants, and other factories. Issues such as this contributed to a political uprising in 2011 and continued unrest in the following years.

Rapid global population growth in Latin America and other regions is producing a variety of urban problems, including overcrowding, environmental pollution, and the disappearance of farmland. In fact, many cities in middle- and low-income nations are quickly reaching the point at which food, housing, and basic public services are available to only a limited segment of the population.

As global urbanization has continued to increase, differences in urban areas based on economic development at the national level have become apparent. Some cities in what Immanuel Wallerstein's (1984) world systems theory describes as core nations are referred to as *global cities*—interconnected urban areas that are centers of political, economic, and cultural activity. New York City, London, Paris, Tokyo, Hong Kong, Los Angeles, Seoul, and Singapore are among the largest global cities today. These cities are the sites of new and innovative product development and marketing, and they are often the "command posts" for the world economy. But economic prosperity is not

shared equally by all of the people in the core-nation global cities. Sometimes the living conditions of workers in low-wage service-sector jobs or in assembly production jobs more closely resemble the living conditions of workers in semiperipheral nations than they resemble the conditions of middle-class workers in their own country.

Many African countries and some countries in South America and the Caribbean are *peripheral* nations, previously defined as nations that depend on core nations for capital, have little or no industrialization (other than what may be brought in by core nations), and have uneven patterns of urbanization. According to Wallerstein (1984), the wealthy in peripheral nations support the exploitation of poor workers by core-nation capitalists in return for maintaining their own wealth and position. Poverty is thus perpetuated, and the problems worsen because of the unprecedented population growth in these countries. Like peripheral nations, semiperipheral nations—such as India, Iran, and Mexico—are confronted with unprecedented population growth. In addition, a steady flow of rural migrants to large cities is creating enormous urban problems.

More recently, Wallerstein suggested that a new world system is emerging that will retain some basic features of the existing one, but it will not be based on capitalism as we know it. According to Wallerstein, the new system will be either hierarchical and exploitative or relatively democratic and relatively egalitarian. In either scenario, national and local governments will be faced with difficult choices, including how to deal with major fiscal crises and immigration. In the United States and throughout Europe, for example, a growing cry demands that governments "do something" about evicting "foreigners"; however, the effects of such an action may create great turmoil and economic instability in cities. Wallerstein's original theory that work migrates from core nations to workers in semi-peripheral and peripheral nations has shifted: Workers are now migrating to cities where they hope to find employment to support themselves and their families, and this process is not likely to stop in the near future.

Urban Problems in the United States

Even the most optimistic observers tend to agree that cities in the United States have problems brought on by years of neglect and deterioration. As we have seen in previous chapters, poverty, crime, racism, sexism, homelessness, inadequate public school systems, alcoholism and other drug abuse, gangs and guns, and other social problems are most visible and acute in urban settings. Issues of urban growth and development are intertwined with many of these problems (•Figure 19.15).

FIGURE 19.15 Despite an increase in telecommuting and more-diverse employment opportunities in the high-tech economy, our highways have grown increasingly congested. Can we implement measures to reduce the problems of urban congestion and environmental pollution, or will these problems grow worse with each passing year?

Divided Interests: Cities and Suburbs

In the twenty-first century the composition of the U.S. population in major cities is changing rapidly. The white population in cities is aging and declining in a number of cities, while the minority population is continuing to grow and disperse. According to the 2010 census, Hispanics (Latinos/as) represent the largest minority group in major U.S. cities. The Hispanic share of urban population rose in all of the largest 100 metropolitan areas to the extent that Hispanics now outnumber African Americans (blacks) as the largest minority group in major U.S. cities. The data show that Hispanics make up 26 percent of primary city (as compared to suburban) populations, while blacks make up 22 percent of city populations. Across all cities in 2010, white Americans account for 41 percent of residents in primary cities—continuing a downward trend from 53 percent in 1990 to 45 percent in 2000 and then reaching the current 41 percent (Frey, 2011).

The story of divided interests between cities and suburbs has continued for more than a century. Following World War II, a dramatic population shift occurred as thousands of families moved from cities to suburbs. Postwar suburban growth was fueled by aggressive land developers, inexpensive real estate and construction methods, better transportation, abundant energy, government subsidies such as liberal lending policies by federal agencies such as the Veterans Administration and the Federal Housing Authority, and racial stress in the cities. This pattern of postwar suburbanization and decades of white flight from cities produced a pattern in which the majority of today's white Americans who live in large urban areas are in the suburbs. By 2010, 78 percent of all white Americans who resided in large metro areas lived in suburbs, up from 74 percent in 1990 (Frey, 2011).

Suburbanization created a territorial division of interests between cities and suburban areas. City services and school districts have continued to languish for lack of funds. Affluent families living in "gentrified" properties in the city typically send their children to elite private schools, whereas the children of poor families living in racially segregated public housing projects attend underfunded (and often substandard) public schools. For decades the wealthy and the poor have lived in different spheres, even when they reside in close proximity in urban areas.

Race, Class, and Suburbs

Sharp racial and ethnic divisions between cities and suburbs have become more blurred in the 2010s; however, class lines remain more distinct, particularly in some wealthy suburban areas. In the past, most suburbs were predominantly white and middle class. Today, some upper-middle-class and upper-class suburbs still have a white majority, but the old image of "chocolate city and vanilla suburbs" is now obsolete (Frey, 2011: 2). According to the 2010 census, minorities represent 35 percent of suburban residents in the 100 largest metropolitan areas in the United States, similar to their share of the overall population. Hispanics make up 17 percent of suburbanites (as compared to 16 percent of the overall population); blacks make up 10 percent of suburbanites (as compared to 12 percent of the overall population) (Frey, 2011).

FIGURE 19.16 Class and other demographic factors are intertwined with suburban integration. Some suburbs, such as this one near Detroit, have become racially resegregated and provide few job opportunities and public services for residents who live there.

What most data do not reveal is how class and other demographic factors are intertwined with suburban integration. In some suburbs, people of color (especially African Americans) become resegregated (•Figure 19.16). An example is the Detroit metropolitan area, where Census Bureau data show a 25 percent drop in the city's population over the last decade and a corresponding increase in minority population in the suburbs. According to a study by historian and sociologist Thomas J. Sugrue (2011), many African Americans are moving into so-called second-hand suburbs in that city and others. Sugrue defines *second-hand suburbs* as "established communities with deteriorating housing stock that are falling out of favor with younger white homebuyers." He argues that if history holds true, these suburban areas will soon look like Detroit itself, with "resegregated schools, dwindling tax bases and decaying public services."

Although some analysts claim that the location of one's residence is a matter of personal choice, African Americans and other persons of color have sought the same things in suburban properties that white Americans have desired, namely safe streets and low crime rates, the best housing they can afford, quality schools for their children, and the amenities of life outside the hubbub of the center city. However, what African Americans have often found in areas such as Detroit are fewer job opportunities, poorer services, older houses, and run-down shopping districts (Sugrue, 2011). Fortunately, the overall picture is more positive in some suburbs located near Atlanta, Houston, Dallas, and Washington, where a sharp rise in black suburbanization has occurred as a result of economic progress among younger college-educated blacks (Frey, 2011).

However, Hispanics account for almost half (49 percent) of the overall growth in population in suburban areas in the 2000s. In the 100 largest metropolitan areas, Hispanics contributed more to the growth than any other racial and ethnic group in 49 of those areas. Hispanics had the largest gains in areas surrounding cities such as New York City, Houston, Miami, Los Angeles, and Riverside, California.

Beyond the Suburbs: Edge Cities and Exurban Areas

In the past, urban fringes (referred to as *edge cities*) developed beyond central cities and suburbs (Garreau, 1991). The Massachusetts Turnpike corridor west of Boston and the Perimeter area north of Atlanta are examples. Edge cities initially develop as residential areas; then retail establishments and office parks move into the area, creating the unincorporated edge city. Commuters from the edge city are able to travel around (rather than in and out of) the metropolitan region's center and can avoid its rush-hour traffic quagmires. Edge cities may not have a governing body or correspond to municipal boundaries; however, they drain taxes from central cities and older suburbs. Many businesses and industries have moved physical plants and tax dollars to these areas: Land is cheaper, and utility rates and property taxes are lower.

Eventually, edge cities in some states have become commuter towns or exurban areas with more structure. Fast-growing exurban areas in Kentucky, Georgia, Virginia, Missouri, Tennessee, Texas, North Carolina, Alabama, Oklahoma, and Pennsylvania have remained mostly white and have depended overwhelmingly on whites for growth in the twenty-first century (Frey, 2011). Of the 22 exurban areas, 16 are more than 75 percent white, with white Americans accounting for at least 80 percent of the population in 15 of the 20 areas. Examples include Spencer, Kentucky, in the Louisville/Jefferson County Kentucky–Indiana metro area; Dawsonville, Georgia, in the Atlanta metro area; New Kent, Virginia, in the Richmond metro area; and the Lincoln County, Missouri, metro area (Frey, 2011). According to a recent study, these exurban areas are a current reflection of what suburbia was in the past—"new housing, growth, and demographic detachment from the more urban portions of their metropolitan areas" (Frey, 2011: 11). And, to some extent, this includes the continuing financial woes of the cities and suburbs.

The Continuing Fiscal Crises of the Cities

The largest cities in the United States have faced periodic fiscal crises for many years; however, in the 2010s cities of all sizes and their adjoining suburbs are experiencing even greater financial problems partially linked to a major downturn in national and international economic trends. Economic recoveries in cities take at least two years longer than a national recovery, which means that financial problems brought about by the Great Recession may last for years to come. An example is the decline in the U.S. housing market, which affects city tax revenues: Housing prices fell by 4 percent during 2011, which reduced household wealth by $500 billion nationwide and further reduced local governments' property tax base (U.S. Conference of Mayors, 2012).

Why have national and international economic downturns hurt U.S. cities so drastically? What are cities doing about it? Cities have experienced extensive shortfalls in revenue because states have reduced the amount of money that they provide for cities, and the cities have had decreased revenue from sales taxes, corporate taxes, and personal income taxes. Funds from the federal government to states and cities have also been limited and are often earmarked for specific projects rather than for use in the general operating budget. These budget crises have forced states to cut funding to already cash-strapped cities. Vital services, including police, firefighting, and public works, have been cut drastically, and the slashing of city budgets and programs may continue in some areas for years to come. As cities lose revenue, officials must decide to lay off or furlough employees, charge higher fees for services, and cancel major projects such as street repairs or infrastructure improvements (building a new water treatment facility, for example).

A study by the Pew Charitable Trusts (2013) found that thirty cities in the most populous metropolitan areas of the United States faced more than $192 billion in unpaid commitments for pensions and other retiree benefits, such as health care, and that was as far back as 2009. The majority of the shortfall was in New York City, followed by Philadelphia; Portland, Oregon; and Chicago. It is a major concern not only for persons expecting to receive pensions or to be covered by health care benefits that these cities do not have the money needed to fully fund their pension plans, but it also is a major problem for everyone, particularly taxpayers. Although the pension systems and state insurance plans may be able to return to higher levels if the economy improves, the fiscal crisis of the cities will be deepened by the problem that this situation presents because it uniquely falls on cities to provide for many of their public-sector employees.

City officials continue to urge leaders at the state and federal levels to create new programs that will help cities meet their residents' needs. Demands are specifically being made for more federal aid through job-creation programs and other economic stimulus packages. Some analysts believe that inaction at the state and federal levels may create even greater financial chaos by forcing some cities into bankruptcy. Local officials emphasize that the state of America's cities continues to threaten the long-term national economic recovery (National League of Cities, 2010). It remains to be seen what the eventual effects of these continuing fiscal crises will be on various cities throughout the nation.

Rural Community Issues in the United States

Although most people think of the United States as highly urbanized, about 20 percent of the U.S. population resides in rural areas, identified as communities of 2,500 people or less by the U.S. Census Bureau (● Figure 19.17). Sociologists typically identify *rural communities* as small, sparsely settled areas that have a relatively homogeneous population of people who primarily engage in agriculture. However, rural communities today are more diverse than this definition suggests.

Unlike the standard migration patterns from rural to urban places in the past, recently more people have moved from large urban areas and suburbs into rural areas. Many of those leaving urban areas today want to escape the high cost of living, crime, traffic congestion, and environmental pollution that make daily life difficult. Technological advances make it easier for people to move to outlying rural areas and still be connected to urban centers if they need to be. The proliferation of computers, cell phones, commuter airlines, and highway systems has made previously remote areas seem much more accessible to many people. However, many recent immigrants to rural areas do not face some traditional problems experienced by long-term rural residents, particularly farmers, small-business owners, teachers, doctors, and other medical personnel in these rural communities.

For many people in rural areas who have made their livelihood through farming and other agricultural endeavors, recent decades have been very difficult, both financially and emotionally. Rural crises such as droughts, crop failures, and the loss of small businesses in the community have had a negative effect on many adults and their children. Like their urban counterparts, rural families have experienced problems of divorce, alcoholism, abuse, and other crises,

Carson Ganci/Design Pics/Newscom

FIGURE 19.17 About 20 percent of the U.S. population resides in rural areas. Are children's experiences different when they grow up in a rural setting? What are the benefits and limitations of living in a rural area?

but these issues have sometimes been exacerbated by such events as the loss of the family farm or business. Because home is also the center of work in farming families, the loss of the farm may also mean the loss of family and social life, and the loss of things dear to children such as their 4-H projects—often an animal that a child raises to show and sell. Some rural children and adolescents are also subject to injuries associated with farm work, such as from livestock kicks or crushing, from falling out of a tractor or a pickup, and from operating machinery designed for adults, that are not typically experienced by their urban counterparts.

Economic opportunities are limited in many rural areas, and average salaries are typically lower than in urban areas, based on the assumption that a family can live on less money in rural communities than in cities. An example is rural teachers, who earn substantially less than their urban and suburban counterparts. Some rural areas have lost many teachers and administrators to higher-paying districts.

Although many of the problems we have examined in this book are intensified in rural areas, one of the most

pressing is the availability of health services and doctors. Recently, some medical schools have established clinics and practices in outlying rural regions of the states in which they are located in an effort to increase the number of physicians available to rural residents. Typically, physicians who have just started to practice medicine have chosen to work in large urban centers with accessible high-tech medical facilities. Because of the pressing time constraints of tending to patients with life-threatening problems, the availability of community clinics and hospitals in rural areas may be a life-or-death matter for some residents. The loss of these facilities can have a devastating effect on people's health and life chances.

In addition to the movement of some urban dwellers to rural areas, two other factors have changed the face of rural America in some regions. One is the proliferation of superstores, such as Walmart, PetSmart, Lowe's, and Home Depot. In some cases these superstores have effectively put small businesses such as hardware stores and pet shops out of business because local merchants cannot meet the prices established by these large-volume discount chains. The development of superstores and outlet malls along the rural highways of this country has raised new concerns about environmental issues such as air and water pollution, and has brought about new questions regarding whether these stores benefit the rural communities where they are located.

A second factor that has changed the face of some rural areas (and is sometimes related to the growth of superstores and outlet malls) is an increase in tourism in rural America. The vast majority of adults in the United States have taken a trip to a rural destination, usually for leisure purposes, over the past few years. Tourism produces jobs; however, many of the positions are for food servers, retail clerks, and hospitality workers, which are often low-paying, seasonal jobs that have few benefits. Tourism may improve a community's tax base, but this does not occur when the outlet malls, hotels, and fast-food restaurants are located outside of the rural community's taxing authority, as frequently occurs when developers decide where to locate malls and other tourist amenities.

 LO8 **Identify** the best-case and worst-case scenarios regarding population and urban growth in the twenty-first century, and discuss how some of the worst-case scenarios might be averted.

Looking Ahead: Population and Urbanization in the Future

In the future, rapid population growth is inevitable. World population is projected to grow from more than 7.2 billion in 2014 to 9.5 billion in 2050, and 10.6 billion in 2085 (see • Figure 19.18). Nearly all of this growth in population is occurring in low-income nations, mostly in Africa and Asia. Higher population growth in low-income countries

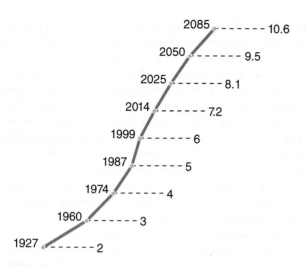

FIGURE 19.18   Increase in the World's Population in Billions of People

Source: United Nations World Urbanization Prospects: 2014 Revision. 2014.

stems from the higher total fertility rates—the average lifetime number of births per woman in a population. In sub-Saharan Africa, for example, women have an average of nearly 5 children in their lifetime. Worldwide, however, total fertility rates have declined significantly from nearly 5 in the 1950s to 2.5 in 2014 (*CIA World Factbook*, 2015).

According to the *World Urbanization Prospects: The 2011 Revision* (United Nations, 2012), the urban population worldwide will grow by 2.6 billion people between 2011 and 2050. In 2011 approximately 3.6 billion people resided in urban areas, and it is estimated that 6.3 billion will be urban dwellers by 2050. North America, Latin America and the Caribbean, and Europe, which are already highly urbanized, will continue the urbanization process. In 2014 about 53 percent of the world's population lived in urban areas, and it is estimated that by 2050 more than 84 percent of the world's population will live in urban areas.

In a worst-case scenario, by the 2020s or the 2030s central cities and nearby suburbs in the United States will have experienced bankruptcy exacerbated by sporadic race- and class-oriented violence. Sporadic urban violence brought about by mass killings in public locations such as schools and movie theaters, combined with random and terrorist attacks such as the Boston Marathon bombing, may also become more frequent occurrences, based on the nature of recent events. The infrastructure of cities will be beyond the possibility of repair. Families and businesses with the ability to do so will have long since moved to "new cities," where they may find that increased population growth has diminished the quality of life that they originally sought there. Areas that we currently think of as being relatively free from such problems will be characterized by depletion of natural resources and by greater air and water pollution.

By contrast, in a best-case scenario the problems brought about by rapid population growth in low-income nations will be remedied by new technologies that make goods readily available to people. International trade agreements

are removing trade barriers and making it possible for all nations to fully engage in global trade. People in low-income nations will benefit by gaining jobs and opportunities to purchase goods at lower prices. Of course, the opposite may also occur: People may be exploited as inexpensive labor, and their country's natural resources may be depleted as transnational corporations buy up raw materials without contributing to the long-term economic stability of the nation.

In the United States a best-case scenario for the future might include improvements in how taxes are collected and spent. Some analysts suggest that regional governments should be developed that would be responsible for water, wastewater (sewage), transportation, schools, and other public services over a wider area. Others believe that we should put more effort into smart growth and/or green movements to improve our cities. "Smart Growth" is the name given to development decisions that cities are making across the United States. Examples of Smart Growth planning include working to conserve resources, preserving natural lands and critical environmental areas, protecting water and air quality, and reusing already-developed land by reinvesting in existing infrastructure and reclaiming historic buildings. The following are Smart Growth principles (EPA.gov, 2013):

- Concentrate on mixed land use
- Take advantage of compact building design
- Create a range of housing opportunities and choices
- Create walkable neighborhoods
- Foster distinctive, attractive communities with a strong sense of place
- Preserve open spaces, farmland, natural beauty, and critical environmental areas
- Strengthen and direct development toward existing communities
- Provide a variety of transportation choices
- Make development decisions predictable, fair, and cost effective
- Encourage community and stakeholder collaboration in development decisions

Although some critics argue that cities that have tried Smart Growth have not brought about profound changes through their efforts, goals such as these seem desirable for bringing about a better future for our communities. At the macrolevel we may be able to do little about population and urbanization; however, at the microlevel we may be able to exercise some degree of control over our communities and our lives. Reclaiming public space for daily life would be an important start, along with making neighborhoods more sustainable and helping inhabitants feel that they have a vested interest in their own community. Hopefully, we will also have found a way to curb random violence and acts of terrorism that produce great harm and keep people living in a state of fear and consternation rather than moving forward for the betterment of themselves and future generations.

CHAPTER REVIEW Q & A

LO1 What is demography, and what are the three processes that produce population changes?

Demography is a subfield of sociology that examines population size, composition, and distribution. Populations change as the result of fertility (births), mortality (deaths), and migration.

LO2 What is meant by population composition, and how is it measured?

Changes in fertility, mortality, and migration affect the population composition—the biological and social characteristics of a population, including age, sex, race, marital status, education, occupation, income, and size of household. One measure of population composition is the sex ratio—the number of males for every hundred females in a given population.

LO3 What are the Malthusian, Marxist, and neo-Malthusian perspectives, and what are the components of demographic transition theory?

More than 200 years ago, Thomas Malthus warned that overpopulation would result in poverty, starvation, and other major problems that would limit the size of the population. According to Malthus, the population would increase geometrically while the food supply would increase only arithmetically, resulting in a critical food shortage and poverty. According to Karl Marx, poverty is the result of capitalist greed, not overpopulation. More recently, neo-Malthusians have reemphasized the dangers of overpopulation and encouraged zero population growth. Demographic transition theory links population growth to four stages of economic development: (1) the preindustrial stage, with high birth rates and death rates; (2) early industrialization, with relatively high birth rates and a decline in death rates; (3) advanced industrialization and urbanization, with low birth rates and death rates; and (4) postindustrialization, with additional decreases in the birth rate coupled with a stable death rate.

LO4 What are some of the key points in international migration theories?

The neoclassical economic approach assumes that migration patterns occur based on geographic differences in the supply of and demand for labor. The new households economics of migration approach emphasizes the part that families or households play in migration. Split-labor-market theory suggests that immigrants from low-income countries are often recruited for secondary labor market positions, whereas migrants from higher-income countries may migrate for primary-sector employment. World systems theory views migration as linked to the problems caused by capitalist development around the world.

LO5 How do cities develop through preindustrial, industrial, and postindustrial stages?

According to Gideon Sjoberg, three preconditions must be present in order for a city to develop: a favorable physical environment, an advanced technology, and a well-developed social organization. The largest preindustrial city was Rome; by 100 C.E., it may have had a population of 650,000. The Industrial Revolution changed the nature of the city. Emergent technology, including new forms of transportation and agricultural production, made it easier for people to leave the countryside and move to the city. Postindustrial cities increasingly rely on an economic structure that is based on scientific knowledge rather than industrial production.

LO6 How do ecological/functionalist models compare with political economy/conflict models of urban growth?

Functionalists view urban growth in terms of ecological models. The concentric zone model sees the city as a series of circular areas, each characterized by a different type of land use. The sector model describes urban growth in terms of terrain and transportation routes. The multiple nuclei model views cities as having numerous centers of development from which growth radiates. According to political economy models/conflict perspectives, urban growth is influenced by capital investment decisions, class and class conflict, and government subsidy programs. At the global level, capitalism also influences the development of cities in core, peripheral, and semiperipheral nations.

LO7 What is meant by the experience of urban life, and how do sociologists seek to explain this experience?

Symbolic interactionist perspectives focus on how people experience urban life. Some analysts view the urban experience positively; others believe that urban dwellers become insensitive to events and to people around them.

LO8 What are some best-case and worst-case scenarios regarding population and urban growth in the twenty-first century?

In a worst-case scenario, central cities and nearby suburbs in the United States will have experienced bankruptcy, and the infrastructure of cities will be beyond repair. Areas that we currently think of as being relatively free from such problems will be characterized by depletion of natural resources and by greater air and water pollution. In a best-case scenario, the problems brought about by rapid population growth in low-income nations will be remedied by new technologies that make goods readily available to people.

KEY TERMS

crude birth rate 556
crude death rate 556
demographic transition 564
demography 555
fertility 556

gentrification 569
invasion 569
migration 557
mortality 556
population composition 560

population pyramid 560
sex ratio 560
succession 569
zero population growth 563

QUESTIONS for CRITICAL THINKING

1 What impact does a relatively high rate of immigration have on culture in a nation? Does it affect personal identity? Why or why not?

2 If you were designing a study of growth patterns for the city where you live (or one you know well), which theoretical model(s) would provide the most useful framework for your analysis?

3 What do you think that everyday life in U.S. cities, suburbs, and rural areas will be like twenty years from the date you are reading this? Where would you prefer to live? What does your answer reflect about the future of U.S. cities?

ANSWERS to the SOCIOLOGY QUIZ

ON MIGRATION AND U.S. IMMIGRATION

1	False	Although the term *unauthorized immigrant* refers to a U.S. resident who is not a citizen of this country, who has not been admitted for permanent residence, or who does not have an authorized temporary status that permits longer-term residence and work, some "unauthorized immigrants" originally entered the country with valid visas but overstayed their visas' expiration or otherwise violated the terms of their admission.
2	False	Although "push" factors are important in migrating to another country, "pull" factors are even more important. Pull factors include better job opportunities, family or friends who already reside in the country of destination, and other favorable conditions that draw them to their new destination.
3	True	The percentage of U.S. immigrants (ages 25 and older) with at least a bachelor's degree (27 percent) is nearly identical to that of native-born U.S. adults (28 percent).
4	False	Research typically has not shown that immigrants increase unemployment in a region or lower wages among native workers. In fact, some studies have found that immigration produces modest gains in wages for native workers.
5	True	The majority (82 percent) of children living in undocumented immigrant families (with at least one undocumented immigrant parent) in the United States were born in this country and have U.S. citizenship status.

6	**True**	In a Gallup Organization survey, the majority (79 percent) of Americans said they believe that unauthorized immigrant workers mostly take low-paying jobs that Americans don't want.
7	**True**	To become a U.S. citizen, immigrants must go through an application process that includes taking a test on English and civics. To pass this test, applicants must demonstrate an ability to read, write, and speak simple words and phrases in English, and must demonstrate a knowledge and understanding of U.S. history and government.
8	**False**	A non–U.S. citizen who marries a U.S. citizen is *not* automatically granted U.S. citizenship. To become a U.S. citizen, a foreign-born spouse of a U.S. citizen must meet certain requirements (e.g., must have been married to and living with his or her U.S. citizen spouse for the past three years) and must go through the citizenship application process. The requirements are different for noncitizen spouses of military personnel who have died in service.

Sources: Based on Congressional Budget Office, 2011; Gallup Organization, 2011b; Gibson and Jung, 2006; Gryn and Larsen, 2010; Kandel, 2011; Passel and Cohn, 2011; Shierholz, 2010; and U.S. Citizenship and Immigration Services, 2011.

COLLECTIVE BEHAVIOR, SOCIAL MOVEMENTS, AND SOCIAL CHANGE

Bloomberg/Getty Images

LEARNING OBJECTIVES

1 **Define** collective behavior and list factors that contribute to it.

2 **Describe** the most common types of crowd behavior.

3 **Discuss** these explanations of crowd behavior: contagion theory, social unrest and circular reaction, convergence theory, and emergent norm theory.

4 **Explain** how mass behavior differs from other forms of collective behavior.

5 **Compare** fads and fashions, and evaluate the "trickle-down" approach as an explanation for the origin of contemporary fashions.

6 **Identify** and explain the major types of social movements.

7 **Compare** the following theories about conditions that are likely to produce social movements: relative deprivation theory, value-added theory, and resource mobilization theory.

8 **Describe** how research based on social constructionist theory—frame analysis, political opportunity theory, and new social movement theory—variously explains how people are drawn into social movements.

SOCIOLOGY & EVERYDAY LIFE

Collective Behavior and Environmental Issues

Growing up just outside of Houston, in a suburb containing the sixth largest refinery in the United States, the fossil fuel industry was all-encompassing in my childhood. . . . It wasn't until the summer before Sophomore year [of college], after having returned to Houston for three consecutive breaks only to become sick and unable to breathe properly from renewed exposure to the pollution, that I realized that I was safer away from home than in it. . . . Overnight, my merely ideological support for Bowdoin Climate Action's fossil fuel divestment campaign transformed into active participation. The dangerous realities of the fossil fuel industry lived too close to home—literally. To remain silent was no longer an option.

—College student ALLYSON GROSS (2015) describes why she has become an active participant in Bowdoin Climate Action, whose motto is "Divestment is the tactic, climate justice is the goal," because students want to encourage their school to disinvest from the fossil fuel industry that is directly linked to the refineries that make her physically ill.

Students on many U.S. college campuses, including these in Boston, have become actively involved in the divestment movement to demand that state legislators support bills that would require divestment by colleges, universities, and other institutional investors in the fossil fuel industry. Why do people believe that it is important to act collectively to get their ideas across to others?

While I was in college, I saw the power of students coming together to speak truth to power on a whole range of issues, both campus-specific and on a statewide level. . . . Our campus community was able to come together and organized multiple divestment campaigns. . . . By demanding accountability from the administration on multiple fronts, with multiple issues, we were able to build our power as students and unify the culture of shared community and support that already existed, while also providing a model for other campuses to move forward campaigns of their own.

A variety of demonstrations and protests are a part of daily life in the twenty-first century. Social media, the Internet, TV, and newspapers continually inform us about new or unresolved social, political, economic, or environmental concerns. As the chapter's opening narratives suggest, a key message of social movements is that people need to act collectively and often immediately to produce various kinds of social change, including reducing environmental degradation or encouraging divestment by colleges, universities and other institutional investors in the fossil fuel industry. For many activists, social change is essential for the improvement, or survival, of the entire planet.

What do you think of when you hear the term *social change*? For sociologists, **social change** is the alteration, modification, or transformation of public policy, culture, or social institutions over time. But social change does not occur of its own accord: Social change is usually brought about by collective behavior and social movements. In this chapter we will examine collective behavior, social movements, and social change from a sociological perspective. We will use the environment movement and activism directed at divestment in fossil fuels as examples of how people try to use mass

mobilization to bring about social transformation. Although the environmental movement, including the "Go-Green" campaign and various university divestment movements, has gained popularity in recent decades, earlier environmental movements started in the 1930s and culminated in the publication of *Silent Spring* (1962) by Rachel Carson, as discussed later in the chapter. Collective behavior and social movements may come and go, but core themes of concern often remain similar. Before reading on, test your knowledge about collective behavior and environmental issues by taking the "Sociology and Everyday Life" quiz.

 LO1 **Define** collective behavior and list factors that contribute to it.

Collective Behavior

Collective behavior is voluntary, often spontaneous activity that is engaged in by a large number of people and typically violates dominant-group norms and values. Unlike

—Sean Estelle's blog (2015) describes how students at University of California–San Diego organized to work for common causes, ranging from divestment in fossil fuels to campus workers' rights and better resources for black students on campus.

How Much Do You Know About Collective Behavior and Environmental Issues?

TRUE	FALSE		
T	F	1	Scientists are forecasting a global warming of between 2 and 11 degrees Fahrenheit over the next century.
T	F	2	The environmental movement in the United States started in the 1960s.
T	F	3	People who hold strong attitudes regarding the environment are very likely to be involved in social movements to protect the environment.
T	F	4	Environmental groups may engage in civil disobedience or use symbolic gestures to call attention to their issue.
T	F	5	People are most likely to believe rumors when no other information is readily available on a topic.
T	F	6	Influencing public opinion is a very important activity for many social movements.
T	F	7	Most social movements in the United States seek to improve society by changing some specific aspect of the social structure.
T	F	8	Sociologists have found that people in a community respond very similarly to natural disasters and to disasters caused by technological failures.

Answers can be found at the end of the chapter.

the *organizational behavior* found in corporations and voluntary associations (such as labor unions and environmental organizations), collective behavior lacks an official division of labor, hierarchy of authority, and established rules and procedures. Unlike *institutional behavior* (in education, religion, or politics, for example), it lacks institutionalized norms to govern behavior. Collective behavior can take various forms, including crowds, mobs, riots, panics, fads, fashions, and public opinion.

Early sociologists studied collective behavior because they lived in a world that was responding to the processes of modernization, including urbanization, industrialization, and the proletarianization of workers. Contemporary forms of collective behavior, such as the environmental and divestment movements and similar protests, are variations on the theme that originated during the transition from feudalism to capitalism and the rise of modernity in Europe. Some forms of collective behavior and social movements are directed toward public issues such as air pollution, water pollution, and the exploitation of workers in global sweatshops by transnational corporations.

Conditions for Collective Behavior

Collective behavior occurs as a result of some common influence or stimulus that produces a response from a collectivity. A *collectivity* is a number of people who act together and may mutually transcend, bypass, or subvert established institutional patterns and structures. Three major factors contribute to the likelihood that collective behavior will occur: (1) structural factors that increase the chances of people responding in a particular way, (2) timing, and (3) a breakdown in social control mechanisms and a corresponding feeling of normlessness.

Collective behavior often begins when there are structural factors that make people aware that a problem

social change
the alteration, modification, or transformation of public policy, culture, or social institutions over time.

collective behavior
voluntary, often spontaneous activity that is engaged in by a large number of people and typically violates dominant-group norms and values.

exists, and then a common stimulus further encourages them to respond to the problem in some specific manner. As previously mentioned, the publication of *Silent Spring* (1962) by former Fish and Wildlife Service biologist Rachel Carson is credited with triggering collective behavior directed at demanding a clean environment and questioning how much power that large corporations should have in the United States. Carson described the dangers of pesticides such as DDT, which was then being promoted by the chemical industry as the miracle that could give the United States the unchallenged position as food supplier to the world (Griswold, 2012). However, people, birds, and fish in the United States were experiencing health-related problems, and the purity of the waterways was being threatened. Many people were ready to acknowledge that problems existed. By writing *Silent Spring,* Carson made people aware of the hazards of chemicals in their foods and the destruction of wildlife. She also encouraged people to question the industries that they had entrusted with their lives and their resources, causing more people to demand accountability where pollution was occurring. These public outcries led to investigations throughout the United States as people demanded legal recognition of the right to a clean environment.

Finally, a breakdown in social control mechanisms is another powerful force in triggering collective behavior regarding environmental protection and degradation. During the 1970s, people in the "Love Canal" area of Niagara Falls, New York, became aware that their neighborhood and their children's school had been built over a canal where tons of poisonous waste had been dumped by a chemical company between 1930 and 1950. Over the next two decades an oily black substance began oozing into the homes in the area and killing the trees and grass on the lots; schoolchildren reported mysterious illnesses and feelings of malaise. Tests indicated that the dump site contained more than 200 different chemicals, many of which could cause cancer or other serious health problems. Upon learning this information, Lois Gibbs, a mother of one of the schoolchildren, began a grassroots campaign to force government officials to relocate community members injured by seepages from the chemical dump (• Figure 20.1). The collective behavior of neighborhood volunteers was not only successful in eventually bringing about social change but also inspired others to engage in collective behavior regarding environmental problems in their communities.

Similarly, the issue of global warming and the detrimental effects that the burning of fossil fuels might have on the climate encouraged activists to call people's

FIGURE 20.1   The Love Canal area of Niagara Falls, New York, has been the site of protests and other forms of collective behavior because of hazardous environmental pollution. Original protests in the 1970s, demanding a cleanup of the site, were followed in the 1990s by new protests, this time over the proposed resettlement of the area.

attention to the need for divestment as both a symbolic and a financial gesture because of the harm being created. In 1988 James Hansen of NASA provided testimony before the United States Senate Committee on Energy and Natural Resources, and NASA compiled a list (http://climate.nasa.gov/effects) documenting the effects of climate change—such as the loss of sea ice, accelerated sea level rise, and the presence of longer and more intense heat waves—on the global environment. Shortly after James Hansen gave his congressional testimony on the effects of global warming, Bill McKibben wrote a widely read book, *The End of Nature,* about climate change. McKibben also founded 350.org, a grassroots climate change movement that helped launched the fossil fuel divestment movement.

Dynamics of Collective Behavior

To better understand the dynamics of collective behavior, let's briefly examine several questions. First, how do people come to transcend, bypass, or subvert established institutional patterns and structures? Some environmental activists have found that they cannot get their point across unless they go outside established institutional patterns and organizations. For example, Lois Gibbs and other Love Canal residents initially tried to work within established means through the school administration and state health officials to clean up the problem. However, they quickly learned that their problems were not being solved through "official" channels. As the problem appeared to grow worse, "official" responses became more defensive and obscure. Accordingly, some residents began acting outside of established norms by holding protests and strikes (Gibbs, 1982). Some situations are more conducive to collective behavior than others. When people can communicate quickly and easily with one another, spontaneous behavior is more likely. When people are gathered together in one general location (whether lining the streets or assembled in a massive stadium), they are more likely to respond to a common stimulus.

Second, how do people's actions compare with their attitudes? People's attitudes (as expressed in public opinion surveys, for instance) are not always reflected in their political and social behavior. Issues pertaining to the environment are no exception. For example, people may indicate in survey research that they believe that the quality of the environment is very important, but the same people may not turn out on Election Day to support propositions that protect the environment or candidates who promise to focus on environmental issues. Likewise, individuals who indicate on a questionnaire that they are concerned about increases in ground-level ozone—the primary component of urban smog—often drive single-occupant, oversized vehicles that government studies have shown to be "gas guzzlers" that contribute to lowered air quality in urban areas. As a result, smog levels increase, contributing to human respiratory problems and dramatically reduced agricultural crop yields.

Third, why do people act collectively rather than singly? Many people believe that there is strength in numbers, whether they are attending a rock concert or protesting an injustice in society. Individuals may act as a collectivity when they believe it is the only way to fight those with greater power and resources. Collective behavior is not just the sum total of a large number of persons acting at the same time; rather, it reflects people's joint response to some common influence or stimulus.

Distinctions Regarding Collective Behavior

People engaging in collective behavior may be divided into crowds and masses. A **crowd** is a relatively large number of people who are in one another's immediate vicinity. Examples of crowds include the audience in a movie theater or people at a pep rally for a sporting event. By contrast, a **mass** is a number of people who share an interest in a specific idea or issue but who are not in one another's immediate vicinity. An example is the popularity of Facebook and Twitter. Through these forms of instantaneous communication, people express their views on everyday life and on larger social issues such as economic crises and environmental problems. Individuals who read what someone has posted and make comments in response usually share a common interest even if these individuals have not met in a face-to-face encounter and may not agree on certain issues.

Collective behavior may also be distinguished by the dominant emotion expressed. The *dominant emotion* refers to the "publicly expressed feeling perceived by participants

and observers as the most prominent in an episode of collective behavior" (Lofland, 1993: 72). Fear, hostility, and joy are three fundamental emotions found in collective behavior; however, grief, disgust, surprise, or shame may also predominate in some forms of collective behavior (Lofland, 1993). Can you think of other emotions that people might express when they engage in collective behavior?

 LO2 **Describe** the most common types of crowd behavior.

Types of Crowd Behavior

When we think of a crowd, many of us think of *aggregates*, previously defined as a collection of people who happen to be in the same place at the same time but who share little else in common. For example, think of thousands of people stranded in an airport when harsh weather or other conditions make it impossible for them to board flights and head for their destinations. Although stranded businesspeople and tourists are together in the airport, they do not necessarily share anything in common. Moreover, the presence of a relatively large number of people in the same location does not necessarily produce collective behavior. To help explain this phenomenon, early symbolic interactionist sociologist Herbert Blumer (1946) developed a typology that divides crowds into four categories: casual, conventional, expressive, and acting. Other scholars have added a fifth category, protest crowds.

Casual and Conventional Crowds *Casual crowds* are relatively large gatherings of people who happen to be in the same place at the same time; if they interact at all, it is only briefly. People in a shopping mall or a subway car are examples of casual crowds. Other than sharing a momentary interest, such as a clown's performance or a small child's fall, a casual crowd has nothing in common. The casual crowd plays no active part in the event—such as the child's fall—which likely would have occurred whether or not the crowd was present; the crowd simply observes.

Conventional crowds are made up of people who come together for a scheduled event and thus share a common focus. Examples include religious services, graduation ceremonies, concerts, and college lectures. Each of these events has pre-established schedules and norms. Because these events occur regularly, interaction among participants is much more likely; in turn, the events would not occur without the crowd, which is essential to the event (● Figure 20.2).

crowd
a relatively large number of people who are in one another's immediate vicinity.

mass
a number of people who share an interest in a specific idea or issue but who are not in one another's immediate vicinity.

FIGURE 20.2 Look closely at the bottom of this photo of Times Square in New York City, where millions of people have come together for a New Year's Eve celebration. Many people who do not know one other are standing in close proximity as they share a special occasion with other individuals who are experiencing similar emotions. According to sociologists, what kind of crowd does this constitute?

Expressive and Acting Crowds *Expressive crowds* provide opportunities for the expression of some strong emotion (such as joy, excitement, or grief). People release their pent-up emotions in conjunction with other persons experiencing similar emotions. Examples include worshippers at religious revival services; mourners lining the streets when a celebrity, a public official, or religious leader has died; and revelers assembled at Mardi Gras or on New Year's Eve at Times Square in New York.

Acting crowds are collectivities so intensely focused on a specific purpose or object that they may erupt into violent or destructive behavior. Mobs, riots, and panics are examples of acting crowds, but casual and conventional crowds may become acting crowds under some circumstances. A *mob* is a highly emotional crowd whose members engage in, or are ready to engage in, violence against a specific target—a person, a category of people, or physical property. Mob behavior in this country has included lynchings, fire bombings, effigy hangings, and hate crimes. Mob violence tends to dissipate relatively quickly once a target has been injured, killed, or destroyed. Sometimes, actions such as an effigy hanging are used symbolically by groups that are not otherwise violent. For example, Lois Gibbs and other Love Canal residents called attention to their problems with the chemical dump site by staging a protest in which they "burned in effigy" the governor and the health commissioner to emphasize their displeasure with the lack of response from these public officials.

Compared with mob actions, riots may be of somewhat longer duration. A *riot* is violent crowd behavior that is fueled by deep-seated emotions but not directed at one specific target. Riots are often triggered by fear, anger, and hostility; however, not all riots are caused by deep-seated hostility and hatred—people may be expressing joy and exuberance when rioting occurs. Examples include celebrations after sports victories that turn into riots when fans storm athletic courts or fields or when they go out and riot in the streets.

A *panic* is a form of crowd behavior that occurs when a large number of people react to a real or perceived threat with strong emotions and self-destructive behavior. The most common type of panic occurs when people seek to escape from a perceived danger, fearing that few (if any) of them will be able to get away from that danger. Examples include passengers on a sinking cruise ship or persons in a burning nightclub. Panics can also arise in response to larger social, financial, or political conditions that people believe are beyond their control—such as a major disruption in the economy. A "bank run" in which hundreds or thousands of customers seek to take out all of their money at the same time, fearing that the financial institution is becoming insolvent, is an example. Although panics are relatively rare, they receive massive media coverage because they provoke strong feelings of fear in readers and viewers, and the number of casualties may be large. An example was a tragic garment factory blaze in Bangladesh in which more than 100 workers died as fire gutted the large manufacturing warehouse and panic-stricken workers jammed a stairwell trying to escape their workplace.

Protest Crowds Sociologists Clark McPhail and Ronald T. Wohlstein (1983) added protest crowds to the four types of crowds identified by Blumer. *Protest crowds* engage in activities intended to achieve specific political goals. Examples include sit-ins, marches, boycotts, blockades, and strikes (• Figure 20.3). Some protests take the form of *civil disobedience*—nonviolent action that seeks to change a policy or law by refusing to comply with it. Acts of civil disobedience may become violent, as in a confrontation between protesters and police officers; in this case a protest crowd becomes an *acting crowd*. In the 1960s, African American students and sympathetic whites used sit-ins to call attention to racial injustice and demand social change. Some of these protests could escalate into violent confrontations even when violence was not the intent of the organizers. At the bottom line, protest crowds typically seek to change some aspect of the status quo. But some protest crowds do state specific policy demands, such as protesters who sought to call attention to the environmental impact of offshore drilling, particularly in the aftermath of the BP explosion at a Gulf of Mexico oil rig.

FIGURE 20.3 Protest crowds are often confronted by law enforcement officials who are responsible for maintaining order and safety in the community. Standoffs such the one shown here are not an unusual sight on a global basis as people seek to make their concerns known to others even as officials attempt to fulfill their duties.

In sum, people with many different issues take their concerns to the streets, as well as to mainstream and social media, to have their voices heard about problems that they believe need to be addressed.

LO3 **Discuss** these explanations of crowd behavior: contagion theory, social unrest and circular reaction, convergence theory, and emergent norm theory.

Explanations of Crowd Behavior

What causes people to act collectively? How do they determine what types of action to take? One of the earliest theorists to provide an answer to these questions was Gustave Le Bon, a French scholar who focused on crowd psychology in his contagion theory.

Contagion Theory *Contagion theory* focuses on the social–psychological aspects of collective behavior; it attempts to explain how moods, attitudes, and behavior are communicated rapidly and why they are accepted by others. Le Bon (1841–1931) argued that people are more likely to engage in antisocial behavior in a crowd because they are anonymous and feel invulnerable. Le Bon (1960/1895) suggested that a crowd takes on a life of its own that is larger than the beliefs or actions of any one person. Because of its anonymity, the crowd transforms individuals from rational beings into a single organism with a collective mind. In essence, Le Bon asserted that

emotions such as fear and hate are contagious in crowds because people experience a decline in personal responsibility; they will do things as a collectivity that they would never do when acting alone.

Le Bon's theory is still used by many people to explain crowd behavior. However, critics argue that the "collective mind" has not been documented by systematic studies.

Social Unrest and Circular Reaction

Sociologist Robert E. Park was the first U.S. sociologist to investigate crowd behavior. Park believed that Le Bon's analysis of collective behavior lacked several important elements. Intrigued that people could break away from the powerful hold of culture and their established routines to develop a new social order, Park added the concepts of social unrest and circular reaction to contagion theory. According to Park, social unrest is transmitted by a process of *circular reaction*—the interactive communication between persons such that the discontent of one person is communicated to another, who, in turn, reflects the discontent back to the first person (Park and Burgess, 1921).

Convergence Theory *Convergence theory* focuses on the shared emotions, goals, and beliefs that many people may bring to crowd behavior. Because of their individual characteristics, many people have a predisposition to participate in certain types of activities. From this perspective, people with similar attributes find a collectivity of like-minded persons with whom they can express their underlying personal tendencies. Although people may reveal their "true selves" in crowds, their behavior is not irrational; it is highly predictable to those who share similar emotions or beliefs.

Convergence theory has been applied to a wide array of conduct, from lynch mobs to environmental movements. In a now-classic study of lynching, social psychologist

mob
a highly emotional crowd whose members engage in, or are ready to engage in, violence against a specific target—a person, a category of people, or physical property.

riot
violent crowd behavior that is fueled by deep-seated emotions but not directed at one specific target.

panic
a form of crowd behavior that occurs when a large number of people react to a real or perceived threat with strong emotions and self-destructive behavior.

civil disobedience
nonviolent action that seeks to change a policy or law by refusing to comply with it.

Hadley Cantril (1941) found that the participants shared certain common attributes: They were poor and working-class whites who felt that their status was threatened by the presence of more-successful African Americans. Consequently, the characteristics of these individuals made them susceptible to joining a lynch mob even if they did not know the target of the lynching.

Convergence theory adds to our understanding of certain types of collective behavior by pointing out how individuals may have certain attributes—such as racial hatred or fear of environmental problems that directly threaten them—that initially bring them together. However, this theory does not explain how the attitudes and characteristics of individuals who take some collective action differ from those who do not.

Emergent Norm Theory Unlike contagion and convergence theories, *emergent norm theory* emphasizes the importance of social norms in shaping crowd behavior. Drawing on the symbolic interactionist perspective, the sociologists Ralph Turner and Lewis Killian (1993: 12) asserted that crowds develop their own definition of a situation and establish norms for behavior that fit the occasion:

> Some shared redefinition of right and wrong in a situation supplies the justification and coordinates the action in collective behavior. People do what they would not otherwise have done when they panic collectively, when they riot, when they engage in civil disobedience, or when they launch terrorist campaigns, because they find social support for the view that what they are doing is the right thing to do in the situation.

According to Turner and Killian (1993: 13), emergent norms occur when people define a new situation as highly unusual or see a long-standing situation in a new light.

Sociologists using the emergent norm approach seek to determine how individuals in a given collectivity develop an understanding of what is going on, how they construe these activities, and what type of norms are involved. For example, in a study of audience participation, the sociologist Steven E. Clayman (1993) found that members of an audience listening to a speech applaud promptly and independently but wait to coordinate their booing with other people; they do not wish to "boo" alone.

Some emergent norms are permissive—that is, they give people a shared conviction that they may disregard ordinary rules, such as waiting in line, taking turns, or treating a speaker courteously. Collective activity such as mass looting may be defined (by participants) as taking what rightfully belongs to them. In the aftermath of the 2010 Haiti earthquake, when relief aid was slow in coming, looting was commonplace, but so too was "mob justice" for those who were caught stealing other people's possessions.

Emergent norm theory points out that crowds are not irrational. Rather, new norms are developed in a rational way to fit the immediate situation. However, critics note that proponents of this perspective fail to specify exactly what constitutes a norm, how new ones emerge, and how they are so quickly disseminated and accepted by a wide variety of participants. One variation of this theory suggests that no single dominant norm is accepted by everyone in a crowd; instead, norms are specific to the various categories of actors rather than to the collectivity as a whole (Snow, Zurcher, and Peters, 1981). For example, in a study of football victory celebrations sociologists found that each week, behavioral patterns were changed in the postgame revelry, with some being modified, some added, and some deleted (Snow, Zurcher, and Peters, 1981).

LO4 **Explain** how mass behavior differs from other forms of collective behavior.

Mass Behavior

Not all collective behavior takes place in face-to-face collectivities. *Mass behavior* is collective behavior that takes place when people (who are often geographically separated from one another) respond to the same event in much the same way. For people to respond in the same way, they typically have common sources of information that provoke their collective behavior. The most frequent types of mass behavior are rumors, gossip, mass hysteria, public opinion, fashions, and fads. Under some circumstances, social movements constitute a form of mass behavior. However, we will examine social movements separately because they differ in some important ways from other types of dispersed collectivities.

Rumors and Gossip *Rumors* are unsubstantiated reports on an issue or a subject. Whereas a rumor may spread through an assembled collectivity, rumors may also be transmitted among people who are dispersed geographically, including people spreading rumors on Twitter, posting messages on Facebook, and texting or talking on the phone. Although rumors may initially contain a kernel of truth, they may be modified as they spread to serve the interests of those repeating them. Rumors thrive when tensions are high and when little authentic information is available on an issue of great concern.

Why do people believe rumors? People are often willing to give rumors credence when no opposing information is available. Once a rumor begins to circulate, it seldom stops unless compelling information comes to the forefront that either proves the rumor false or makes it obsolete. In contemporary societies with sophisticated technology, rumors come from a wide variety of sources and may be difficult to trace. Print media (newspapers and magazines) and electronic media (radio and television), cellular networks, satellite systems, the Internet, and social media aid the rapid movement of rumors around the globe. In addition, modern communications technology makes anonymity much easier. In a split second, messages (both factual and fictitious) can be disseminated to millions of people worldwide.

Whereas rumors deal with an issue or a subject, **gossip** refers to rumors about the personal lives of individuals. Charles Horton Cooley (1963/1909) viewed gossip as something that spread among a small group of individuals who personally knew the person who was the object of the rumor. Today, this is frequently not the case; many people enjoy gossiping about people whom they have never met. Tabloid newspapers and magazines, such as the *National Enquirer, People,* and *US Weekly,* along with television "news" programs, websites, Facebook, and Twitter, provide "inside" information on the lives of celebrities. We are constantly bombarded with information and gossip, much of which has not been checked for authenticity.

FIGURE 20.4   Although a spokesperson for CBS Radio stated to listeners that they were hearing a dramatization of a novel, the 1938 presentation of H. G. Wells's *The War of the Worlds,* as presented by Orson Welles and his Mercury Theatre, terrified untold numbers of people. Here Welles talks to interviewers the day after the event caused a nationwide panic.

Mass Hysteria and Panic   *Mass hysteria* is a form of dispersed collective behavior that occurs when a large number of people react with strong emotions and self-destructive behavior to a real or perceived threat. Does mass hysteria actually occur? Although the term has been widely used, many sociologists believe that this behavior is best described as a panic with a dispersed audience.

A classic example of mass hysteria or a panic with a widely dispersed audience was actor Orson Welles's 1938 Halloween eve radio dramatization of H. G. Wells's science fiction classic *The War of the Worlds.* A CBS radio dance-music program was interrupted suddenly by a news bulletin informing the audience that Martians had landed in New Jersey and were in the process of conquering Earth. Some listeners became extremely frightened even though an announcer had indicated before, during, and after the performance that the broadcast was a fictitious dramatization (● Figure 20.4). According to some reports, as many as 1 million of the estimated 10 million listeners believed that this astonishing event had occurred. Thousands were reported to have hidden in their storm cellars or to have gotten in their cars so that they could flee from the Martians (see Brown, 1954). In actuality, the program probably did not generate mass hysteria, but rather a panic among gullible listeners. Others switched stations to determine if the same "news" was

being broadcast elsewhere. When they discovered that it was not, they merely laughed at the joke being played on listeners by CBS. In 1988, on the fiftieth anniversary of the broadcast, a Portuguese radio station rebroadcast the program; once again, a panic ensued.

Today, panics such as this tend to occur among segments of the population but not the entire population. For example, in 2013 some gun enthusiasts became concerned that there was going to be a shortage of ammunition and started purchasing all the bullets they could find. The panic started following the reelection of President Barack Obama; the Newtown, Connecticut, school shootings and demands for more-stringent federal gun-control laws; and persistent rumors that the Department of Homeland Security and the federal government were stockpiling ammunition. Some people could not be persuaded that ammunition would still be available on store shelves when they needed it for their weapons.

mass behavior
collective behavior that takes place when people (who are often geographically separated from one another) respond to the same event in much the same way.

rumor
an unsubstantiated report on an issue or a subject.

gossip
rumors about the personal lives of individuals.

LO5 **Compare** fads and fashions, and evaluate the "trickle-down" approach as an explanation for the origin of contemporary fashions.

Fads and Fashions   As you will recall from Chapter 3, a *fad* is a temporary but widely copied activity enthusiastically followed by large numbers of people. Fads can be embraced by widely dispersed collectivities: TV, the Internet, and social media bring the latest fads—such as top memes, YouTube videos of cats or dogs doing funny things, zombie films and TV series like *The Walking Dead,* and Ben & Jerry's latest ice cream flavor—to the attention of audiences around the world.

Unlike fads, fashions tend to be longer lasting. In Chapter 3, *fashion* is defined as a currently valued style of behavior, thinking, or appearance. Fashion also applies to art, music, drama, literature, architecture, interior design, and automobiles, among other things. However, most sociological research on fashion has focused on clothing, especially women's apparel and how it may relate to culture and social customs in a specific era.

In preindustrial societies, clothing styles remained relatively unchanged. With the advent of industrialization, items of apparel became readily available at low prices because of mass production. Fashion became more important as people embraced the "modern" way of life and as advertising encouraged "conspicuous consumption."

Georg Simmel, Thorstein Veblen, and Pierre Bourdieu have all viewed fashion as a means of status differentiation among members of different social classes. Simmel (1957/1904) suggested a classic "trickle-down" theory (although he did not use those exact words) to describe the process by which members of the lower classes emulate the fashions of the upper class (• Figure 20.5). As the fashions descend through the status hierarchy, they are watered down and "vulgarized" so that they are no longer recognizable to members of the upper class, who then regard them as unfashionable and in bad taste (Davis, 1992). Veblen (1967/1899) asserted that fashion serves mainly to institutionalize conspicuous consumption among the wealthy. More than eighty years later, Bourdieu (1984) similarly (but more subtly) suggested that "matters of taste," including fashion sensibility, constitute a large share of the "cultural capital" possessed by members of the dominant class.

Herbert Blumer (1969) disagreed with the trickle-down approach, arguing that "collective selection" best explains fashion.

Blumer suggested that people in the middle and lower classes follow fashion because it is fashion, not because they desire to emulate members of the elite class. Blumer thus shifted the focus on fashion to collective mood, tastes, and choices: "Tastes are themselves a product of experience. . . . They are formed in the context of social interaction, responding to the definitions and affirmation given by others. People thrown into areas of common interaction and having similar runs of experience develop common tastes" (qtd. in Davis, 1992: 116). Perhaps one of the best refutations of the trickle-down approach is the way in which fashion today often originates among people in the lower social classes and is mimicked by the elites.

Public Opinion   *Public opinion* consists of the attitudes and beliefs communicated by ordinary citizens to decision makers. It is measured through polls and surveys, which use research methods such as interviews and questionnaires, as described in Chapter 2. Many people are not interested in all aspects of public policy but are concerned about issues that they believe are relevant to them. Even on a single topic, public opinion will vary widely based on race/ethnicity, religion, region, social class, education level, gender, age, and so on.

Scholars who examine public opinion are interested in the extent to which the public's attitudes are communicated to decision makers and the effect (if any) that public opinion has on policy making (Turner and Killian, 1993). Some political scientists argue that public opinion has a substantial effect on decisions at all levels of government; others

FIGURE 20.5   Georg Simmel suggested a "trickle-down" theory to describe the process by which the lower classes emulate the fashions of the upper class. Do runway shows such as this one in Paris provide evidence for such a theory?

strongly disagree. • Table 20.1 shows a public opinion survey of the top policy priorities of the U.S. public in 2015. As you will note, the environment is not listed among the ten items that the Pew Research Center (2015) survey identified as the public's top priorities. Slightly over half (51 percent) of Americans indicated that they believed the environment should be a top priority for the president and the Congress. Of course, other pressing issues such as terrorism, the economy, jobs, education, Social Security, and the budget deficit were on their minds when they were taking this survey.

Today, people attempt to influence elites, and vice versa. Consequently, a two-way process occurs with the dissemination of *propaganda*—information provided by individuals or groups that have a vested interest in furthering their own cause or damaging an opposing one. Although many of us think of propaganda in negative terms, the information provided can be correct and can have a positive effect on decision making.

Grassroots environmental activists have attempted to influence public opinion for many years. However, it is less clear that public opinion translates into action by either decision makers in government and industry or by individuals (such as a willingness to adopt a more ecologically sound lifestyle).

Initially, most grassroots environmental activists attempt to influence public opinion so that local decision makers will feel the necessity of correcting a specific problem through changes in public policy. Although activists usually do not start out seeking broader social change, they often move in that direction when they become aware of how widespread the problem is in the larger society or on a global basis. One of two types of social movements often develops at this point—one focuses on NIMBY ("not in my backyard"), whereas the other focuses on NIABY ("not in anyone's backyard").

Social Movements

Although collective behavior is short-lived and relatively unorganized, social movements are longer lasting, are more organized, and have specific goals. A *social movement* is an organized group that acts consciously to promote or resist change through collective action. Because social movements initially are not institutionalized and are outside the political mainstream, they offer "outsiders" an opportunity to have their voices heard.

Social movements are more likely to develop in industrialized societies than in preindustrial societies, where acceptance of traditional beliefs and practices makes such movements unlikely. Diversity and a lack of consensus (hallmarks of industrialized nations) contribute to demands for social change, and people who participate in social movements typically lack the power and other resources to bring about change without engaging in collective action. Social movements are most likely to spring up when people come to see their personal troubles as public issues that cannot be solved without a collective response. Although the government is most frequently the target of social movement activity, other organizations—such as schools, corporations, or financial institutions—are also the targets of social activism.

Social movements make democracy more available to excluded groups. Historically, people in the United States have worked at the grassroots level to bring about changes even when elites sought to discourage activism. For example, the civil rights movement brought into its ranks African Americans in the South who had never before been allowed to participate in politics. The women's suffrage movement gave voice to women who had been denied the right to vote. Similarly, a grassroots environmental movement gave the working-class residents of Love Canal a way to "fight city hall" and the chemical company. Most social movements rely on volunteers to carry out the work. Traditionally, women have been strongly represented in both the membership and the leadership of many grassroots movements.

TABLE 20.1

Top 15 Policy Priorities of the U.S. Public, 2015
Each percentage rating represents a top priority for the federal administration to tackle

Issue	Percentage
Terrorism	76
Economy	75
Jobs	67
Education	67
Social Security	66
Budget deficit	64
Health care costs	64
Medicare	61
Reducing crime	57
Poor and needy	55
Military	52
Immigration	52
Environment	51
Race relations	49
Moral breakdown	48

Sources: Pew Research Center, 2015.

public opinion
the attitudes and beliefs communicated by ordinary citizens to decision makers.

propaganda
information provided by individuals or groups that have a vested interest in furthering their own cause or damaging an opposing one.

social movement
an organized group that acts consciously to promote or resist change through collective action.

Other movements have grappled with issues that the sociologist Kai Erikson (1994) refers to as a "new species of trouble"—environmental problems that contaminate, pollute, befoul, taint, and scare human beings in new ways that produce uncanny fear (Erikson, 1991). The chaos that Erikson describes is the result of technological disasters caused by system failures, human error, faulty designs, and other problems that wreak havoc on people and things.

Social movements provide people who otherwise would not have the resources to enter the game of politics a chance to do so. We are most familiar with those movements that develop around public policy issues considered newsworthy by the media, ranging from abortion and women's rights to gun control and environmental justice. However, a number of other types of social movements exist as well.

FIGURE 20.6   Martin Luther King Jr., a leader of the civil rights movement in the 1950s and 1960s, advocated nonviolent protests that sometimes took the form of civil disobedience. Here he marches alongside his wife, Coretta Scott King, who for many years took over Dr. King's activities after he was assassinated.

 LO6   **Identify** and explain the major types of social movements.

Types of Social Movements

Social movements are difficult to classify; however, sociologists distinguish among movements on the basis of their *goals* and the *amount of change* they seek to produce. Some movements seek to change people, whereas others seek to change society.

Reform Movements   Grassroots environmental movements are an example of *reform movements,* which seek to improve society by changing some specific aspect of the social structure. Members of reform movements usually work within the existing system to attempt to change public policy so that it more adequately reflects their own value system. Examples of reform movements (in addition to the environmental movement) include labor movements, animal rights movements, antinuclear movements, Mothers Against Drunk Driving, and the disability rights movement.

Some social movements arise specifically to alter negative stereotypes and reduce stigma associated with specific categories of people (• Figure 20.6). Such social movements may not only bring about changes in societal attitudes and practices but also produce changes in participants' social emotions. For example, the civil rights and gay rights movements helped replace shame with pride.

Revolutionary Movements   Movements seeking to bring about a total change in society are referred to as *revolutionary movements.* These movements usually do not attempt to work within the existing system; rather, they aim to remake the system by replacing existing institutions with new ones. This was apparently the original goal of the Occupy Movement that started as Occupy Wall Street in 2011.

The goal of this movement at both location and international levels was to protest social and economic inequality and advocate change in how societies are organized and resources distributed. The movement emphasized reducing corporate greed and restructuring the global financial system. However, although the Occupy Movement introduced new language such as "the 1 percent" and "the 99 percent," it did not reach the impact of a major revolutionary movement. But some social analysts believe that this movement will bring about major changes in how presidential elections are run, in higher minimum wages, and in the U.S.-based environmental movement. After the Occupy Movement, environmental activists became highly visible as they protested the Keystone XL pipeline being constructed between the United States and Canada. Similarly, college students led the divestment movement to eliminate fossil-fuel assets from university investment funds (Levitin, 2015).

Revolutionary movements range from utopian groups seeking to establish an ideal society to radical terrorists who use fear tactics to intimidate those with whom they disagree ideologically. In the twenty-first century, people in Muslim countries around the world, including Tunisia, Egypt, and Iran, have participated in revolutionary movements and risen up against what they perceived to be tyrannical regimes (• Figure 20.7). For example, people in Tunisia gathered in 2013 to call for unity and support for the resistance movement against tyranny and injustices perpetrated by the former regime of Zine El Abidine Ben Ali, the ousted president of that nation who had banned all public meetings and engaged in other forms of oppression.

Movements based on terrorism often use tactics such as bombings, kidnappings, hostage taking, hijackings, and assassinations. A number of movements in the United

FIGURE 20.7 Revolutionary movements have taken place in Egypt and other Arab nations in recent years because of a strong belief that leaders are oppressive and governments are not benefiting the people.

States have engaged in terrorist activities or supported a policy of violence. However, the terrorist attacks in New York City and Washington, D.C., on September 11, 2001, and the events that followed those attacks proved to all of us that terrorism within this country can originate from the activities of revolutionary terrorists from outside the country as well. Some terrorist events—such as the Boston Marathon bombing in 2013—do not appear to be as fully organized or to include as many people as revolutionary movements are typically thought to involve.

Religious Movements Social movements that seek to produce radical change in individuals are typically based on spiritual or supernatural belief systems. Also referred to as *expressive movements, religious movements* are concerned with reforming or renewing people through "inner change." Fundamentalist religious groups seeking to convert nonbelievers to their belief system are an example of this type of movement. Some religious movements are *millenarian*—that is, they forecast that "the end is near" and assert that an immediate change in behavior is imperative. Examples include Hare Krishnas, the Unification Church, Scientology, and the Divine Light Mission, all of which tend to appeal to the psychological and social needs of young people seeking meaning in life that mainstream religions have not provided for them.

Alternative Movements Movements that seek limited change in some aspect of people's behavior are referred to as *alternative movements*. For example, early in the twentieth century the Women's Christian Temperance Union attempted to get people to abstain from drinking alcoholic beverages. Some analysts place "therapeutic social movements" such as Alcoholics Anonymous in this category; however, others do not because of their belief that people must change their lives completely in order to overcome alcohol abuse. More recently, a variety of "New Age" movements have directed people's behavior by emphasizing spiritual consciousness combined with a belief in reincarnation and astrology. Such practices as vegetarianism, meditation, and holistic medicine are often included in the self-improvement category. Some alternative movements have included the practice of yoga (usually without its traditional background in the Hindu religion) as a means by which the self can be liberated and union can be achieved with the supreme spirit or universal soul.

Resistance Movements Also referred to as *regressive movements, resistance movements* seek to prevent change or to undo change that has already occurred. Virtually all of the social movements previously discussed face resistance from one or more reactive movements that hold opposing viewpoints and want to foster public policies that reflect their own beliefs. Examples of resistance movements are groups organized to oppose same-sex marriage, abortion, and gun-control legislation.

Stages in Social Movements

Do all social movements go through similar stages? Not necessarily, but there appear to be identifiable stages in virtually all movements that are able to succeed beyond their initial phase of development (see • Figure 20.8).

In the *preliminary* (or *incipiency*) *stage*, widespread unrest is present as people begin to become aware of a problem. At this stage, leaders emerge to agitate others into taking

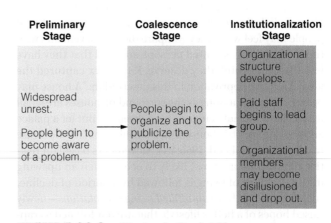

Preliminary Stage	Coalescence Stage	Institutionalization Stage
Widespread unrest. People begin to become aware of a problem.	People begin to organize and to publicize the problem.	Organizational structure develops. Paid staff begins to lead group. Organizational members may become disillusioned and drop out.

FIGURE 20.8 Stages in Social Movements

action. In the *coalescence stage,* people begin to organize and to publicize the problem. At this stage, some movements become formally organized at local and regional levels. In the *institutionalization* (or *bureaucratization*) *stage,* an organizational structure develops, and a paid staff (rather than volunteers) begins to lead the group. When the movement reaches this stage, the initial zeal and idealism of members may diminish as administrators take over management of the organization. Early grassroots supporters may become disillusioned and drop out; they may also start another movement to address some as-yet-unsolved aspect of the original problem. For example, some national environmental organizations—such as the Sierra Club, the National Audubon Society, and the National Parks and Conservation Association—that started as grassroots conservation movements are currently viewed by many people as being unresponsive to local environmental problems. As a result, new movements have arisen, such as the Go Green Movement to reduce problems in the environment and the economy.

LO7 **Compare** the following theories about conditions that are likely to produce social movements: relative deprivation theory, value-added theory, and resource mobilization theory.

Social Movement Theories

What conditions are most likely to produce social movements? Why are people drawn to these movements? Sociologists have developed several theories to answer these questions.

Relative Deprivation Theory

According to *relative deprivation theory,* people who are satisfied with their present condition are less likely to seek social change. Social movements arise as a response to people's perception that they have been deprived of what they consider to be their fair share. Thus, people who suffer relative deprivation are more likely to feel that change is necessary and to join a social movement in order to bring about that change. *Relative deprivation* refers to the discontent that people may feel when they compare their achievements with those of similarly situated persons and find that they have less than they think they deserve. Karl Marx captured the idea of relative deprivation in this description: "A house may be large or small; as long as the surrounding houses are small it satisfies all social demands for a dwelling. But let a palace arise beside the little house, and it shrinks from a little house to a hut" (qtd. in Ladd, 1966: 24). Movements based on relative deprivation are most likely to occur when an upswing in the standard of living is followed by a period of decline, such that people have *unfulfilled rising expectations*—newly raised hopes of a better lifestyle that are not fulfilled as rapidly as the people expected or are not realized at all.

Although most of us can relate to relative deprivation theory, it does not fully account for why people experience social discontent but fail to join a social movement. Even though discontent and feelings of deprivation may be necessary to produce certain types of social movements, they are not sufficient to bring movements into existence.

Value-Added Theory

The *value-added theory* developed by sociologist Neil Smelser (1963) is based on the assumption that certain conditions are necessary for the development of a social movement. Smelser called his theory the "value-added" approach based on the concept (borrowed from the field of economics) that each step in the production process adds something to the finished product. For example, in the process of converting iron ore into automobiles, each stage "adds value" to the final product (Smelser, 1963). Similarly, Smelser asserted, six conditions are necessary and sufficient to produce social movements when they combine or interact in a particular situation:

- *Structural conduciveness.* People must become aware of a significant problem and have the opportunity to engage in collective action. According to Smelser, movements are more likely to occur when a person, class, or agency can be singled out as the source of the problem; when channels for expressing grievances either are not available or fail; and when the aggrieved have a chance to communicate among themselves.

- *Structural strain.* When a society or community is unable to meet people's expectations that something should be done about a problem, strain occurs in the system. The ensuing tension and conflict contribute to the development of a social movement based on people's belief that the problem would not exist if authorities had done what they were supposed to do.

- *Spread of a generalized belief.* For a movement to develop, there must be a clear statement of the problem and a shared view of its cause, effects, and possible solution.

- *Precipitating factors.* To reinforce the existing generalized belief, an inciting incident or dramatic event must occur. With regard to technological disasters, some (including Love Canal) gradually emerge from a long-standing environmental threat, whereas others (including the meltdown of a nuclear power plant) involve a suddenly imposed problem.

- *Mobilization for action.* At this stage, leaders emerge to organize others and give them a sense of direction.

- *Social control factors.* If there is a high level of social control on the part of law enforcement officials, political leaders, and others, it becomes more difficult to develop a social movement or engage in certain types of collective action.

- Value-added theory takes into account the complexity of social movements and makes it possible to test

Smelser's assertions regarding the necessary and sufficient conditions that produce such movements. However, critics note that the approach is rooted in the functionalist tradition and views structural strains as disruptive to society.

Resource Mobilization Theory

Smelser's value-added theory tends to underemphasize the importance of resources in social movements. By contrast, *resource mobilization theory* focuses on the ability of members of a social movement to acquire resources and mobilize people in order to advance their cause. Resources include money, people's time and skills, access to the media, and material goods, such as property and equipment. Assistance from outsiders is essential for social movements. For example, reform movements are more likely to succeed when they gain the support of political and economic elites.

Resource mobilization theory is based on the assumption that participants in social movements are rational people. From this perspective, social movements are formed and dissolved, mobilized and deactivated, based on rational decisions about the goals of the group, available resources, and the cost of mobilization and collective action. Resource mobilization theory also assumes that participants must have some degree of economic and political resources to make the movement a success. In other words, widespread discontent alone cannot produce a social movement; adequate resources and motivated people are essential to any concerted social action.

In the twenty-first century, scholars continue to modify resource mobilization theory and to develop new approaches for investigating the diversity of movements. Emerging perspectives based on resource mobilization theory emphasize the ideology and legitimacy of movements as well as material resources.

Additional perspectives are also needed on social movements in other nations to determine how activists in those countries acquire resources and mobilize people to advance causes such as environmental protection (see "Sociology in Global Perspective").

LO8 **Describe** how research based on social constructionist theory—frame analysis, political opportunity theory, and new social movement theory—variously explains how people are drawn into social movements.

Social Constructionist Theory: Frame Analysis

Theories based on a symbolic interactionist perspective focus on the importance of the symbolic presentation of a problem to both participants and the general public. *Social constructionist theory* is based on the assumption that

FIGURE 20.9 How is the issue of immigration framed in these photos? Research based on frame analysis often investigates how social issues are framed and what names they are given.

a social movement is an interactive, symbolically defined, and negotiated process that involves participants, opponents, and bystanders.

Research based on this perspective often investigates how problems are framed and what names they are given (see ● Figure 20.9). This approach reflects the influence of the sociologist Erving Goffman's *Frame Analysis* (1974), in which he suggests that our interpretation of the particulars of events and activities is dependent on the framework from which we perceive them. According to Goffman (1974: 10), the purpose of frame analysis is "to try to isolate some of the basic frameworks of understanding available in our society for making sense out of events and to analyze the special vulnerabilities to which these frames of reference are subject." In other words, various "realities" may be simultaneously occurring among participants engaged in the same set of activities. When people come together in a social movement, they assign meanings to their activities in such a way that they build a framework for interacting and socially constructing their grievances so that they can more effectively voice them and know what resolution they want for these issues.

Old Environmental Pollution with New Social Pressures in China

- Up to 40 percent of China's rivers are seriously polluted after 75 billion tons of sewage and wastewater were discharged into them.

- Twenty percent of rivers are so polluted that their water quality was rated too toxic for people to come into contact with the rivers' water.

- Nearly 300 million rural residents lack access to drinking water.

- About two-thirds of the cities in China lack sufficient water. (*People's Daily Online*, 2012)

For many years, environmental problems in China have been in the news around the world. The Internet and social media have intensified and greatly sped up this coverage and made environmental activists worldwide more aware of the problems faced by people in that rapidly growing region of the world. For example, it is a well-known fact that cancer is now

Workers in plants in China and other emerging nations typically do not have the same legal protection as U.S. workers. If U.S. companies with factories in these nations become an active force in requiring higher environmental standards, will this improve the quality of life in those nations as well?

the leading cause of death in China, partly because of air pollution, water pollution, and other environmental contaminants, some of which may be attributed to factories related to production of goods for high-income nations such as the United States (Larsen, 2011).

What, if anything, is being done about this situation? Some social protests and the beginnings of social movements are now being found in various provinces of China. Here are a few examples:

- Approximately 12,000 protesters demonstrated in Dalian, China, in 2011 after a storm damaged a paraxylene (PX) factory. PX is a toxic chemical used to make polyester. The protesters were mobilized by cell phones and the Internet. Officials announced that the

© Imaginechina/Corbis

Sociologists have identified at least three ways in which grievances are framed. First, *diagnostic framing* identifies a problem and attributes blame or causality to some group or entity so that the social movement has a target for its actions. Second, *prognostic framing* pinpoints possible solutions or remedies, based on the target previously identified. Third, *motivational framing* provides a vocabulary of motives that compel people to take action. When successful framing occurs, the individual's vague dissatisfactions are turned into well-defined grievances, and people are compelled to join the movement in an effort to reduce or eliminate those grievances.

Beyond motivational framing, additional frame alignment processes are necessary in order to supply a continuing sense of urgency to the movement. *Frame alignment* is the linking together of interpretive orientations of individuals and social movement organizations so that there is congruence between individuals'

interests, beliefs, and values and the movement's ideologies, goals, and activities. Four distinct frame alignment processes occur in social movements: (1) *frame bridging* is the process by which movement organizations reach individuals who already share the same worldview as the organization, (2) *frame amplification* occurs when movements appeal to deeply held values and beliefs in the general population and link those to movement issues so that people's preexisting value commitments serve as a "hook" that can be used to recruit them, (3) *frame extension* occurs when movements enlarge the boundaries of an initial frame to incorporate other issues that appear to be of importance to potential participants, and (4) *frame transformation* refers to the process whereby the creation and maintenance of new values, beliefs, and meanings induce movement participation by redefining activities and events in such a manner that people believe they must become involved in collective action. Some or all of these frame alignment

plant would be closed, which was no small feat because $1.5 billion had been invested in a joint venture between a state-owned chemical company and a private developer (Economist.com, 2011).

- Apple announced in 2012 that it would allow independent environmental reviews of factories that supply parts for Apple products. Apple has faced rising criticism about toxic pollution and factory injuries of workers in suppliers' factories in China and other countries. Among the environmental problems cited are hazardous-waste leaks and the use of toxic chemicals that might create health risks not only for workers but also for neighboring communities. On its website, Apple stated that it insists that suppliers "provide safe working conditions, treat workers with dignity and respect, and use environmentally responsible manufacturing processes" (qtd. in Chu, 2012).

Are other Chinese activists and environmental movements trying to bring about environmental conservation in China? Some environmental activists are indeed attempting to highlight the causes and consequences of the various forms of pollution that are assaulting their nation. In the past, other social movement organizers in China found that they risked arrest and prosecution if they publicized their concerns and tried to mobilize others for causes. However, given the pressure that activists from China and other nations can bring to bear through communication on the Internet and Facebook, Twitter, and a growing number of other social media sites, it appears that Chinese officials are

concerned that another Arab Spring (the revolutionary wave of protests that occurred in Tunisia, Egypt, Yemen, Libya, and Syria in 2010–2011) might take place in China if sufficient numbers of people express their discontent about the current state of affairs in their nation.

What will the future hold for environmental protection in China? According to resource mobilization theory, widespread discontent alone cannot produce a social movement: Adequate resources and motivated people are essential for any concerned social action. Some analysts believe that environmental leaders will be able to mobilize people for change because cell phones, the Internet, and social media make it possible for people to organize quickly and demand governmental intervention regarding pressing problems. If companies such as Apple that do business in China become an active force in requiring environmental reviews of suppliers, this action might increase pressure for improvement by other Chinese suppliers and businesses. Even if this is just the tip of the iceberg, it is a start because many transnational companies currently have contract plants in China.

REFLECT & ANALYZE

Do you believe that U.S. companies can play a role in environmental movements in China? Should corporations also be concerned about environmental issues in the United States? Why or why not?

processes are used by social movements as they seek to define grievances and recruit participants.

Frame analysis provides new insights on how social movements emerge and grow when people are faced with problems such as technological disasters, about which greater ambiguity typically exists, and when people are attempting to "name" the problems associated with things such as nuclear or chemical contamination. However, frame analysis has been criticized for its "ideational biases" (McAdam, 1996). According to the sociologist Doug McAdam (1996), frame analyses of social movements have looked almost exclusively at ideas and their formal expression, whereas little attention has been paid to other significant factors, such as movement tactics, mobilizing structures, and changing political opportunities that influence the signifying work of movements. In this context, *political opportunity* means government structure, public policy, and political conditions that set the boundaries for change and political action. These boundaries are crucial

variables in explaining why various social movements have different outcomes.

Political Opportunity Theory

Why do social protests occur? According to political opportunity theorists, the origins of social protests cannot be explained solely by the fact that people possess a variety of grievances or that they have resources available for mobilization. Instead, social protests are directly related to the political opportunities that potential protesters and movement organizers believe exist within the political system at any given point in time. Political opportunity theory is based on the assumption that social protests that take place *outside* of mainstream political institutions are deeply intertwined with more-conventional political activities that take place *inside* these institutions. As used in this context, *opportunity* refers to "options for collective action, with chances and risks attached to them that depend on factors outside the mobilizing group" (Koopmans, 1999: 97). *Political opportunity theory* states that people will choose those options for collective action that are most readily available to them and those options that will produce the most favorable outcome for their cause.

What are some specific applications of political action theory? Urban sociologists and social movement analysts have found that those cities that provided opportunities for people's protests to be heard within urban governments were less likely to have extensive protests or riots in their communities because aggrieved people could use more-conventional means to make their claims known. By contrast, urban riots were more likely to occur when activists believed that all conventional routes to protest were blocked. Changes in demography, migration, and the political economy in the United States (factors that were seemingly external to the civil rights movement) all contributed to a belief on the part of African Americans in the late 1960s and early 1970s that they could organize collective action and that their claims regarding the need for racial justice might be more readily heard by government officials.

Political opportunity theory has grown in popularity among sociologists who study social movements because this approach highlights the interplay of opportunity, mobilization, and political influence in determining when certain types of behavior may occur. However, like other perspectives, this theory has certain limitations, including the fact that social movement organizations may not always be completely distinct from, or external to, the existing political system. For example, it is difficult to classify the Tea Party movement, which emerged in the aftermath of the election of President Barack Obama. Some supporters were outside the political mainstream and felt like they had no voice in what was happening in Washington. Keli Carender, who is credited with being one of the first Tea Party campaigners, complained that she tried to call her senators to urge them to vote against the $787 billion stimulus bill but constantly found that their mailboxes were full. As a result, she decided to protest against "porkulus"; in her words, "I basically thought to myself: 'I have two courses. I can give up, go home, crawl into bed and be really depressed and let it happen, or I can do something different, and I can find a new avenue to have my voice get out'" (qtd. in Zernike, 2010: A1).

By contrast, other active supporters of the Tea Party movement are players in the mainstream political process. An example is Sarah Palin, the former Alaska governor and Republican vice presidential candidate, who was a frequent spokesperson at Tea Party rallies across the United States. In political movements, social activists typically *create* their own opportunities rather than wait for them to emerge, and activists are often political entrepreneurs in their own right, much like the state and federal legislators and other governmental officials

whom they seek to influence on behalf of their social cause. Political opportunity theory calls our attention to how important the degree of openness of a political system is to the goals and tactics of persons who organize social movements.

New Social Movement Theory

New social movement theory looks at a diverse array of collective actions and the manner in which those actions are based on politics, ideology, and culture. It also incorporates factors of identity, including race, class, gender, and sexuality, as sources of collective action and social movements. Examples of "new social movements" include ecofeminism and environmental justice movements.

Ecofeminism emerged in the late 1970s and early 1980s out of the feminist, peace, and ecology movements. Prompted by the near-meltdown at the Three Mile Island nuclear power plant, ecofeminists established World Women in Defense of the Environment. *Ecofeminism* is based on the belief that patriarchy is a root cause of environmental problems. According to ecofeminists, patriarchy not only results in the domination of women by men but also contributes to a belief that nature is to be possessed and dominated, rather than treated as a partner.

Another "new social movement" focuses on environmental justice and the intersection of race and class in the environmental struggle. Sociologist Stella M. Capek (1993) investigated a contaminated landfill in the Carver Terrace neighborhood of Texarkana, Texas, and found that residents were able to mobilize for change and win a federal buyout and relocation by symbolically linking their issue to a larger *environmental justice* framework. Since the 1980s, the emerging environmental justice movement has focused on the issue of **environmental racism**—the belief that a disproportionate number of hazardous facilities (including industries such as waste disposal/treatment and chemical plants) are placed in low-income areas populated primarily by people of color (Bullard and Wright, 1992). (See • Figure 20.10.) These areas have been left out of most of the environmental cleanup that has taken place. Capek concludes that linking Carver Terrace with environmental justice led to it being designated as a cleanup site. She also views this as an important turning point in new social movements: "Carver Terrace is significant not only as a federal buyout and relocation of a minority community, but also as a marker of the emergence of environmental racism as a major new component of environmental social movements in the United States" (Capek, 1993: 21).

Sociologist Steven M. Buechler (2000: 11) has argued that theories pertaining to twenty-first-century social movements should be oriented toward the structural, macrolevel contexts in which movements arise. These theories

FIGURE 20.10 Referred to as "Cancer Alley," this area of Baton Rouge, Louisiana, is home to a predominantly African American population and also to many refineries that heavily pollute the region. Sociologists suggest that environmental racism is a significant problem in the United States and other nations. What do you think?

should incorporate both political and cultural dimensions of social activism:

> Social movements are historical products of the age of modernity. They arose as part of a sweeping social, political, and intellectual change that led a significant number of people to view society as a social construction that was susceptible to social reconstruction through concerted collective effort. Thus, from their inception, social movements have had a dual focus. Reflecting the political, they have always involved some form of challenge to prevailing forms of authority. Reflecting the cultural, they have always operated as symbolic laboratories in which reflexive actors pose questions of meaning, purpose, identity, and change.

As we have seen, social movements may be an important source of social change. Throughout this text we have examined a variety of social problems that have been the focus of one or more social movements. For this reason, many groups focus on preserving their gains while simultaneously fighting for changes that they believe are still

necessary. This chapter's Concept Quick Review summarizes the main theories of social movements.

Looking Ahead: Social Change in the Future

In this chapter we have focused on collective behavior and social movements as potential forces for social change in contemporary societies. For example, environmental problems like Love Canal in New York State and the Carver Terrace disaster in Texarkana produced social movements that were, to a large extent, responsible for passage of the Comprehensive Environmental Response, Compensation, and Liability Act, which led to the formation of the Agency for Toxic Substances and Disease Registry. Massive oil spills in the Gulf of Mexico and various oceans, as well as the collective

environmental racism
the belief that a disproportionate number of hazardous facilities (including industries such as waste disposal/treatment and chemical plants) are placed in low-income areas populated primarily by people of color.

Social Movement Theories

	Key Components
Relative Deprivation	People who are discontent when they compare their achievements with those of others consider themselves relatively deprived and join social movements in order to get what they view as their "fair share," especially when there is an upswing in the economy followed by a decline.
Value-Added	Certain conditions are necessary for a social movement to develop: (1) structural conduciveness, such that people are aware of a problem and have the opportunity to engage in collective action; (2) structural strain, such that society or the community cannot meet people's expectations for taking care of the problem; (3) growth and spread of a generalized belief about causes and effects of and possible solutions to the problem; (4) precipitating factors, or events that reinforce the beliefs; (5) mobilization of participants for action; and (6) social control factors, such that society comes to allow the movement to take action.
Resource Mobilization	A variety of resources (money, members, access to media, and material goods such as equipment) are necessary for a social movement; people participate only when they believe that the movement has access to these resources.
Social Constructionist Theory: Frame Analysis	Based on the assumption that social movements are an interactive, symbolically defined, and negotiated process involving participants, opponents, and bystanders, frame analysis is used to determine how people assign meaning to activities and processes in social movements.
Political Opportunity	People will choose the options for collective action (i.e., "opportunities") that are most readily available to them and those options that will produce the most favorable outcome for their cause.
New Social Movement	The focus is on sources of social movements, including politics, ideology, and culture. Race, class, gender, sexuality, and other sources of identity are also factors in movements such as ecofeminism and environmental justice.

behavior that followed these disasters, were responsible for new legislation and guidelines for offshore drilling and emergency preparedness by corporations and the federal government. Will current inequality-based protests generate long-term effects in our society? It is too soon to tell because a number of other factors also contribute to social change, including the physical environment, population trends, technological development, and social institutions.

The Physical Environment and Change

Changes in the physical environment often produce changes in the lives of people (● Figure 20.11). In turn, people can make dramatic changes in the physical environment, over which we have only limited control. Throughout history, natural disasters have taken their toll on individuals and societies. Major natural disasters—including tsunamis/hurricanes, floods, tornadoes, and earthquakes—can devastate an entire population.

In the twenty-first century, earthquakes have affected India, Pakistan, El Salvador, Iran, China, Italy, Haiti, Chile, New Zealand, Japan, Nepal, and the United States, among

Leonard Zhukovsky/Shutterstock.com

FIGURE 20.11 Destruction of the physical environment often produces dramatic fluctuations in the lives of people, a factor that can contribute to larger social change. What changes might natural disasters such as Hurricane Sandy bring about?

other nations. Hurricanes, tsunamis, and floods have devastated portions of Pakistan, Australia, and the United States. Tornadoes have plagued the United States, with one of the most devastating being the 2013 EF5 tornado that hit the city of Moore, south of Oklahoma City, with winds of 215 miles per hour that left a path of destruction more than a mile wide. Even comparatively "small" natural disasters change the lives of many people. As the sociologist Kai Erikson (1976, 1994) has suggested, the trauma that people experience from disasters may outweigh the actual loss of physical property—memories of such events can haunt people for many years.

Some natural disasters are exacerbated by human decisions. For example, floods are viewed as natural disasters, but excessive development may contribute to a flood's severity. As office buildings, shopping malls, industrial plants, residential areas, and highways are developed, less land remains as groundcover to absorb rainfall. When heavier-than-usual rains occur, flooding becomes inevitable; some regions of the United States—such as in and around New Orleans—had excessive water on the streets for days or even weeks after Hurricane Katrina, in 2005. Clearly, humans cannot control the rain, but human decisions can worsen the consequences. If Hurricane Katrina's first wave was the storm itself, the second wave was a *human-made disaster* resulting in part from decisions related to planning and budgetary priorities, allocation of funds for maintaining infrastructure, and the importance of emergency preparedness.

Infrastructure refers to a framework of systems, such as transportation and utilities, that makes it possible to have specific land uses (commercial, residential, and recreational, for example) and a built environment (buildings, houses, and highways) that support people's daily activities and the nation's economy. It takes money and commitment to make sure that the components of the infrastructure remain strong so that cities can withstand natural disasters and other concerns such as climate change. For example, consider that the city of Chicago has taken a proactive stance on dealing with future increases in temperature and climate conditions that will make Chicago's weather feel more like Baton Rouge, Louisiana, than a northern city. To cope with this change, Chicago city officials are already repaving alleyways with water-permeable materials and planting swamp oak and sweet gum trees from the South rather than the more indigenous white oak, which is the state tree of Illinois (Kaufman, 2011). Long-range planning such as this to cope with changes in the physical environment may seem far-fetched, but when the time comes, between 50 and 100 years from now, such endeavors may seem farsighted instead. In the words of one Chicago city official, "Cities adapt or they go away" (qtd. in Kaufman, 2011).

The changing environment is one of many reasons why experts are also concerned about availability of water in the future. Water is a finite resource that is necessary for both human survival and the production of goods. However, water is being wasted and polluted, and the supply of *potable* (drinkable) water is limited. People are causing—or at least contributing to—that problem.

People also contribute to changes in the Earth's physical condition. Through soil erosion and other degradation of grazing land, often at the hands of people, more than 25 billion tons of topsoil are lost annually. As people clear forests to create farmland and pastures and to acquire lumber and firewood, the Earth's tree cover continues to diminish. As millions of people drive motor vehicles, the amount of carbon dioxide in the environment continues to rise each year, contributing to global warming.

Just as people contribute to changes in the physical environment, human activities must also be adapted to changes in the environment. For example, we are being warned to stay out of the sunlight because of increases in ultraviolet rays, a cause of skin cancer, as a result of the accelerating depletion of the ozone layer. If the ozone warnings are accurate, the change in the physical environment will dramatically affect those who work or spend their leisure time outside.

Population and Change

Changes in population size, distribution, and composition affect the culture and social structure of a society and change the relationships among nations. As discussed in Chapter 19, the countries experiencing the most-rapid increases in population have a less developed infrastructure to deal with those changes. How will nations of the world deal with population growth as the global population continues to move toward eight billion? Only time will provide a response to this question.

In the United States a shift in population distribution from central cities to suburban and exurban areas has produced other dramatic changes. In some areas, central cities have experienced a shrinking tax base as middle-income and upper-middle-income residents and businesses have moved to suburban and outlying areas. In other areas, the largest metropolitan areas have become enclaves for the wealthiest transnational residents who can afford to own homes in many of the world's most expensive cities. Likewise, some suburban areas thrive with McMansion-style residential properties and security-gated communities, while other suburbs have run-down housing, high rates of poverty, and problems similar to the less-fortunate central cities, including decaying infrastructure, low-performing schools, inadequate transportation, and rising rates of crime, to name only a few concerns. The changing composition of the U.S. population has resulted in children from more-diverse cultural backgrounds entering school, producing a demand for new programs and changes in curricula. An increase in the number of single mothers and of women employed outside the household has created a need for more child care; an increase in the older population has created a need for services such as medical care and placed greater stress on programs such as Social Security. Population growth will affect many regions of the country and intensify existing social problems.

Technology and Change

Technology is an important force for change; in some ways, technological development has made our lives much easier. Advances in communication and transportation have made instantaneous worldwide communication possible but have also brought old belief systems and the status quo into question as never before. Today, we are increasingly moving information instead of people—and doing it instantly. Advances in science and medicine have made significant changes in people's lives in high-income countries.

Scientific advances will continue to affect our lives, from the foods we eat to our reproductive capabilities. Biotechnology is the process by which organisms or their components, such as enzymes, are used to make products such as yogurt, cheese, beer, and wine. Genetically modified products today include medicines and vaccines, as well as foods and food ingredients. Of course, these new technologies are not without controversy; they pose some risks (genomics.energy.gov, 2012). Advances in medicine have made it possible for those formerly unable to have children to procreate; women well beyond menopause are now able to become pregnant with the assistance of medical technology (• Figure 20.12). Advances in medicine have also increased the human life span, especially for white and middle- or upper-class individuals in high-income nations; medical advances have also contributed to the declining death rate in low-income nations, where birth rates have not yet been curbed.

Just as technology has brought about improvements in the quality and length of life for many, it has also created the potential for new disasters, ranging from global warfare to localized technological disasters at toxic waste sites. As the sociologist William Ogburn (1966) suggested, when a change in the material culture occurs in society, a period of *cultural lag* follows in which the nonmaterial (ideological) culture has not yet caught up with material development. The rate of technological advancement at the level of material culture today is mind-boggling. Many of us can

FIGURE 20.12 How might advances in medicine—such as the ability of postmenopausal women to bear children—create social change?

Cultura Creative/Alamy

never hope to understand technological advances in the areas of artificial intelligence, holography, virtual reality, biotechnology, cold fusion, and robotics.

One of the ironies of twenty-first-century high technology is the increased vulnerability that results from the increasing complexity of such systems. We have already seen this in situations ranging from jetliners used as terrorist weapons to identity theft and fraud on the Internet and the hacking of massive public and private databases.

Social Institutions and Change

Many changes occurred in the family, religion, education, the economy, and the political system during the twentieth century and early in the twenty-first century. The size and composition of families in the United States changed with the dramatic increase in the number of single-person and single-parent households. Changes in families produced changes in the socialization of children, many of whom spend large amounts of time playing video games, texting friends, posting their daily activities on Facebook or Twitter, or spending time in child-care facilities outside their own home. Although some political and religious leaders advocate a return to "traditional" family life, numerous scholars argue that such families never worked quite as well as some might wish to believe.

Public education changed dramatically in the United States during the last century. This country was one of the first to provide "universal" education for students regardless of their ability to pay. As a result, at least until recently, the United States has had one of the most highly educated populations in the world. Today, the United States still has one of the best public education systems in the world for the top 15 percent of the students, but it badly fails the bottom 25 percent. As the nature of the economy changes, schools almost inevitably will have to change, if for no other reason than demands from leaders in business and industry for an educated workforce that allows U.S. companies to compete in a global economic environment. Many business and political leaders believe that education is the single most important factor in the future of the United States. However, in difficult economic times when local, state, and federal budgets are strained, public education is one of the first institutions to undergo the axe as teachers are let go, school buildings are allowed to further decay, and students are not provided with the necessary physical setting and learning tools.

As we move further into the twenty-first century, we need new ways of conceptualizing social life at both the macrolevel and the microlevel. The sociological imagination helps us think about how personal troubles—regardless of our race, class, gender, age, sexual orientation, or physical abilities and disabilities—are intertwined with the public issues of our society and the global community of which we are a part. After one problem appears to be alleviated, additional problems may crop up, creating a negative chain reaction if we adopt a "business as usual" approach for dealing with these challenges (• Figure 20.13).

And, as we learned in the narratives that opened this chapter, vigilance and persistence are important when individuals or groups are trying to bring about positive change. After one problem appears to be alleviated, additional problems crop up, but we must rise to the challenge.

A Few Final Thoughts

In this text we have covered a substantial amount of material, examined different perspectives on a wide variety of social issues, and suggested different methods by which to deal with them. The purpose of this text is not to encourage you to take any particular point of view; rather, it is to provide different viewpoints that may be helpful to you and to society in dealing with the pressing issues of the twenty-first century. Possessing that understanding, we can hope that the future will be something we can all look forward to—producing a better way of life, not only for people in this country but worldwide as well.

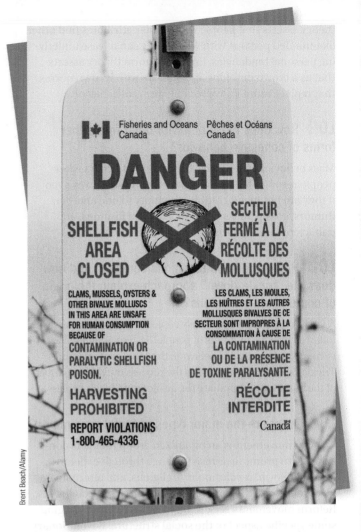

FIGURE 20.13 Pollution of lakes, rivers, and other bodies of water has an adverse effect on food supplies, air quality, and the entire environment. What influence does a "business as usual" approach have on environmental quality in your area?

CHAPTER REVIEW Q & A

LO1 What is collective behavior, and what factors contribute to it?

Social change—the alteration, modification, or transformation of public policy, culture, or social institutions over time—is usually brought about by *collective behavior,* which is defined as relatively spontaneous, unstructured activity that typically violates established social norms. Collective behavior occurs when some common influence or stimulus produces a response from a relatively large number of people.

LO2 What are the most common types of crowd behavior?

A crowd is a relatively large number of people in one another's immediate presence. Sociologist Herbert Blumer divided crowds into four categories: (1) casual crowds, (2) conventional crowds, (3) expressive crowds, and (4) acting crowds (including mobs, riots, and panics). A fifth type of crowd is a protest crowd.

LO3 What are some explanations of crowd behavior?

Social scientists have developed several theories to explain crowd behavior. Contagion theory asserts that a crowd takes on a life of its own as people are transformed from rational beings into part of an organism that acts on its own. A variation on this is social unrest and circular reaction—people express their discontent to others, who communicate back similar feelings, resulting in a conscious effort to engage in the crowd's behavior. Convergence theory asserts that people with similar attributes find other like-minded persons with whom they can release underlying personal tendencies. Emergent norm theory asserts that as a crowd develops, it comes up with its own norms that replace more-conventional norms of behavior.

LO4 How does mass behavior differ from other forms of collective behavior?

Mass behavior is collective behavior that occurs when people respond to the same event in the same way even if they are not in geographic proximity to one another. Rumors, gossip, mass hysteria, fads and fashions, and public opinion are forms of mass behavior.

LO5 How do fads and fashions compare, and how does the "trickle-down" approach explain the origin of contemporary fashions?

A fad is a temporary but widely copied activity enthusiastically followed by large numbers of people. Fashion is defined as a currently valued style of behavior, thinking, or appearance. Trickle-down theory describes the process by which members of the lower classes emulate the fashions of the upper class.

LO6 What are the major types of social movements?

A social movement is an organized group that acts consciously to promote or resist change through collective action. Reform, revolutionary, religious, and alternative movements are the major types identified by sociologists. Reform movements seek to improve society by changing some specific aspect of the social structure. Revolutionary movements seek to bring about a total change in society—sometimes by the use of terrorism. Religious movements seek to produce radical change in individuals based on spiritual or supernatural belief systems. Alternative movements seek limited change of some aspect of people's behavior. Resistance movements seek to prevent change or to undo change that has already occurred.

LO7 How do relative deprivation theory, value-added theory, and resource mobilization theory explain social movements?

Relative deprivation theory asserts that if people are discontent when they compare their accomplishments with those of others similarly situated, they are more likely to join a social movement than are people who are relatively content with their status. According to value-added theory, six conditions are required for a social movement: (1) a perceived problem, (2) a perception that the authorities are not resolving the problem, (3) a spread of the belief to an adequate number of people, (4) a precipitating incident, (5) mobilization of other people by leaders, and (6) a lack of social control. By contrast, resource mobilization theory asserts that successful social movements can occur only when they gain the support of political and economic elites, who provide access to the resources necessary to maintain the movement.

LO8 How does research based on social constructionist theory—frame analysis, political opportunity theory, and new social movement theory—explain the ways in which people are drawn into social movements?

Research based on social constructionist theory uses frame analysis, which often focuses on how people use social interactions to socially construct their grievances and to determine what they think should be done to resolve them. By contrast, research based on political opportunity theory identifies reasons why social protests are most likely to occur when potential protesters and movement organizers perceive that limited opportunities exist for them within the political system. Finally, research based on new social movement theory looks at various identity factors of potential protesters, including their race, class, gender, and sexuality, to help explain why collective action and social movements emerge (for example, people who believe that they have experienced environmental racism because of their race/ethnicity and class).

KEY TERMS

civil disobedience 590
collective behavior 586
crowd 589
environmental racism 602
gossip 593

mass 589
mass behavior 592
mob 590
panic 590
propaganda 595

public opinion 594
riot 590
rumor 592
social change 586
social movement 595

QUESTIONS for CRITICAL THINKING

1 What types of collective behavior in the United States do you believe are influenced by inequalities based on race/ethnicity, class, gender, age, or disabilities? Why?

2 Which of the four explanations of crowd behavior (contagion theory, social unrest and circular reaction, convergence theory, and emergent norm theory) do you believe best explains crowd behavior? Why?

....

3 In the text the environmental movement is analyzed in terms of the value-added theory. How would you analyze that movement under (a) the relative deprivation theory and (b) the resource mobilization theory?

....

4 Using the sociological imagination that you have gained in this course, what are some positive steps that you believe might be taken in the United States to make our society a better place for everyone? What types of collective behavior and/or social movements might be required in order to take those steps?

....

ANSWERS to the SOCIOLOGY QUIZ

ON COLLECTIVE BEHAVIOR AND ENVIRONMENTAL ISSUES

1	**True**	Scientists have made this prediction, and the Earth's average temperature has already increased about 1 degree Fahrenheit during the twentieth century. NASA scientists warn that small changes in temperature correspond to enormous changes in the environment.
2	**False**	The environmental movement in the United States is the result of more than 135 years of collective action. The first environmental organization, the American Forestry Association (now American Forests), originated in 1875.
3	**False**	Public opinion polls continue to show that the majority of people in the United States have favorable attitudes regarding protection of the environment; however, far fewer individuals are actively involved in social movements to further this cause.
4	**True**	Environmental groups have held sit-ins, marches, boycotts, and strikes, which sometimes take the form of civil disobedience. Others have hanged political leaders in effigy or held officials hostage. Still others have dressed as grizzly bears to block traffic in Yellowstone National Park or created a symbolic "crack" (made of plastic) on the Glen Canyon Dam on the Colorado River to denounce development in the area.
5	**True**	Rumors are most likely to emerge and circulate when people have very little information on a topic that is important to them. For example, rumors abound in times of technological disasters, when people are fearful and often willing to believe a worst-case scenario.
6	**True**	Many social movements, including grassroots environmental activism, attempt to influence public opinion so that local and national decision makers will feel obliged to correct a specific problem through changes in public policy.
7	**True**	Most social movements are reform movements that focus on improving society by changing some specific aspect of the social structure. Examples include environmental movements and the disability rights movement.
8	**False**	Most sociological studies have found that people respond differently to natural disasters, which usually occur very suddenly, than to technological disasters, which may occur gradually. One of the major differences is the communal bonding that tends to occur following natural disasters, as compared with the extreme social conflict that may follow technological disasters.

absolute poverty a level of economic deprivation that exists when people do not have the means to secure the most basic necessities of life.

achieved status a social position that a person assumes voluntarily as a result of personal choice, merit, or direct effort.

activity theory the proposition that people tend to shift gears in late middle age and find substitutes for previous statuses, roles, and activities.

acute diseases illnesses that strike suddenly and cause dramatic incapacitation and sometimes death.

affirmative action policies or procedures that are intended to promote equal opportunity for categories of people deemed to have been previously excluded from equality in education, employment, and other fields on the basis of characteristics such as race or ethnicity.

age stratification inequalities, differences, segregation, or conflict between age groups.

ageism prejudice and discrimination against people on the basis of age, particularly against older persons.

agents of socialization the persons, groups, or institutions that teach us what we need to know in order to participate in society.

aggregate a collection of people who happen to be in the same place at the same time but share little else in common.

aging the physical, psychological, and social processes associated with growing older.

agrarian societies societies that use the technology of large-scale farming, including animal-drawn or energy-powered plows and equipment, to produce their food supply.

alienation a feeling of powerlessness and estrangement from other people and from oneself.

animism the belief that plants, animals, or other elements of the natural world are endowed with spirits or life forces that have an impact on events in society.

anomie Emile Durkheim's designation for a condition in which social control becomes ineffective as a result of the loss of shared values and of a sense of purpose in society.

anticipatory socialization the process by which knowledge and skills are learned for future roles.

ascribed status a social position conferred at birth or received involuntarily later in life, based on attributes over which the individual has little or no control, such as race/ethnicity, age, and gender.

assimilation a process by which members of subordinate racial and ethnic groups become absorbed into the dominant culture.

authoritarian leaders leaders who make all major group decisions and assign tasks to members.

authoritarian personality a personality type characterized by excessive conformity, submissiveness to authority, intolerance, insecurity, a high level of superstition, and rigid, stereotypic thinking.

authoritarianism a political system controlled by rulers who deny popular participation in government.

authority power that people accept as legitimate rather than coercive.

B

beliefs the mental acceptance or conviction that certain things are true or real.

bilateral descent a system of tracing descent through both the mother's and father's sides of the family.

blended family a family consisting of a husband and wife, children from previous marriages, and children (if any) from the new marriage.

body consciousness how a person perceives and feels about his or her body.

bureaucracy an organizational model characterized by a hierarchy of authority, a clear division of labor, explicit rules and procedures, and impersonality in personnel matters.

bureaucratic personality a psychological construct that describes those workers who are more concerned with following correct procedures than they are with getting the job done correctly.

C

capitalism an economic system characterized by private ownership of the means of production, from which personal profits can be derived through market competition and without government intervention.

capitalist class (bourgeoisie) Karl Marx's term for those who own and control the means of production.

caste system a system of social inequality in which people's status is permanently determined at birth based on their parents' ascribed characteristics.

category a number of people who may never have met one another but share a similar characteristic, such as education level, age, race, or gender.

charismatic authority power legitimized on the basis of a leader's exceptional personal qualities or the demonstration of extraordinary insight and accomplishment that inspire loyalty and obedience from followers.

chronic diseases illnesses that are long term or lifelong and that develop gradually or are present from birth.

chronological age a person's age based on date of birth.

church a large, bureaucratically organized religious organization that tends to seek accommodation with the larger society in order to maintain some degree of control over it.

civil disobedience nonviolent action that seeks to change a policy or law by refusing to comply with it.

civil religion the set of beliefs, rituals, and symbols that makes sacred the values of the society and places the nation in the context of the ultimate system of meaning.

class conflict Karl Marx's term for the struggle between the capitalist class and the working class.

class system a type of stratification based on the ownership and control of resources and on the type of work that people do.

cohabitation a situation in which two people live together, and think of themselves as a couple, without being legally married.

cohort a group of people born within a specified period of time.

collective behavior voluntary, often spontaneous activity that is engaged in by a large number of people and typically violates dominant-group norms and values.

comparable worth (or *pay equity*) the belief that wages ought to reflect the worth of a job, not the gender or race of the worker.

conflict perspectives the sociological approach that views groups in society as engaged in a continuous power struggle for control of scarce resources.

conformity the process of maintaining or changing behavior to comply with the norms established by a society, subculture, or other group.

conglomerate a combination of businesses in different commercial areas, all of which are owned by one holding company.

content analysis the systematic examination of cultural artifacts or various forms of communication to extract thematic data and draw conclusions about social life.

contingent work part-time work, temporary work, or subcontracted work that offers advantages to employers but that can be detrimental to the welfare of workers.

control group in an experiment, the group that contains the subjects who are not exposed to the independent variable.

core nations according to world systems theory, nations that are dominant capitalist centers characterized by high levels of industrialization and urbanization.

corporate crime illegal acts committed by corporate employees on behalf of the corporation and with its support.

corporations large-scale organizations that have legal powers, such as the ability to enter into contracts and buy and sell property, separate from their individual owners.

correlation a relationship that exists when two variables are associated more frequently than could be expected by chance.

counterculture a group that strongly rejects dominant societal values and norms and seeks alternative lifestyles.

credentialism a process of social selection in which class advantage and social status are linked to the possession of academic qualifications.

crime behavior that violates criminal law and is punishable with fines, jail terms, and/or other negative sanctions.

criminal justice system the local, state, and federal agencies that enforce laws, adjudicate crimes, and treat and rehabilitate criminals.

criminology the systematic study of crime and the criminal justice system, including the police, courts, and prisons.

crossdresser a male who dresses as a woman or a female who dresses as a man but does not alter his or her genitalia.

crowd a relatively large number of people who are in one another's immediate vicinity.

crude birth rate the number of live births per 1,000 people in a population in a given year.

crude death rate the number of deaths per 1,000 people in a population in a given year.

cult (also known as *new religious movement* or *NRM*) a loosely organized religious group with practices and teachings outside the dominant cultural and religious traditions of a society.

cultural capital Pierre Bourdieu's term for people's social assets, including values, beliefs, attitudes, and competencies in language and culture.

cultural imperialism the extensive infusion of one nation's culture into other nations.

cultural lag William Ogburn's term for a gap between the technical development of a society (material culture) and its moral and legal institutions (nonmaterial culture).

cultural relativism the belief that the behaviors and customs of any culture must be viewed and analyzed by the culture's own standards.

cultural transmission the process by which children and recent immigrants become acquainted with the dominant cultural beliefs, values, norms, and accumulated knowledge of a society.

cultural universals customs and practices that occur across all societies.

culture the knowledge, language, values, customs, and material objects that are passed from person to person and from one generation to the next in a human group or society.

culture shock the disorientation that people feel when they encounter cultures radically different from their own and believe they cannot depend on their own taken-for-granted assumptions about life.

D

deinstitutionalization the practice of rapidly discharging patients from mental hospitals into the community.

demedicalization the process whereby a problem ceases to be defined as an illness or a disorder.

democracy a political system in which the people hold the ruling power either directly or through elected representatives.

democratic leaders leaders who encourage group discussion and decision making through consensus building.

democratic socialism an economic and political system that combines private ownership of some of the means of production, governmental distribution of some essential goods and services, and free elections.

demographic transition the process by which some societies have moved from high birth and death rates to relatively low birth and death rates as a result of technological development.

demography a subfield of sociology that examines population size, composition, and distribution.

denomination a large organized religion characterized by accommodation to society but frequently lacking in ability or intention to dominate society.

dependency theory the belief that global poverty can at least partially be attributed to the fact that the low-income countries have been exploited by the high-income countries.

dependent variable a variable that is assumed to depend on or be caused by the independent variable(s).

deviance any behavior, belief, or condition that violates significant social norms in the society or group in which it occurs.

differential association theory the proposition that people have a greater tendency to deviate from societal norms when they frequently associate with persons who are more favorable toward deviance than conformity.

disability a physical or mental impairment that substantially limits one or more major activities that a person would normally do at a given stage of life and that may result in stigmatization or discrimination against the person with a disability.

discrimination actions or practices of dominant-group members (or their representatives) that have a harmful effect on members of a subordinate group.

disengagement theory the proposition that older persons make a normal and healthy adjustment to aging when they detach themselves from their social roles and prepare for their eventual death.

division of labor how the various tasks of a society are divided up and performed.

domestic partnerships household partnerships in which an unmarried couple lives together in a committed, sexually intimate relationship and is granted some of the same rights and benefits as those accorded to married heterosexual couples.

dominant group a racial or ethnic group that has the greatest power and resources in a society.

dramaturgical analysis Erving Goffman's term for the study of social interaction that compares everyday life to a theatrical presentation.

drug any substance—other than food and water—that, when taken into the body, alters its functioning in some way.

dual-earner marriages marriages in which both spouses are in the labor force.

dyad a group composed of two members.

E

ecclesia a religious organization that is so integrated into the dominant culture that it claims as its membership all members of a society.

economy the social institution that ensures the maintenance of society through the production, distribution, and consumption of goods and services.

education the social institution responsible for the systematic transmission of knowledge, skills, and cultural values within a formally organized structure.

egalitarian family a family structure in which both partners share power and authority equally.

ego Sigmund Freud's term for the rational, reality-oriented component of personality that imposes restrictions on the innate pleasure-seeking drives of the id.

elder abuse physical abuse, psychological abuse, financial exploitation, and medical abuse or neglect of people age 65 or older.

elite model a view of society that sees power in political systems as being concentrated in the hands of a small group of elites whereas the masses are relatively powerless.

endogamy the practice of marrying within one's own group.

entitlements certain benefit payments made by the government.

environmental racism the belief that a disproportionate number of hazardous facilities (including industries such as waste disposal/treatment and chemical plants) are placed in low-income areas populated primarily by people of color.

ethnic group a collection of people distinguished, by others or by themselves, primarily on the basis of cultural or nationality characteristics.

ethnic pluralism the coexistence of a variety of distinct racial and ethnic groups within one society.

ethnocentrism the practice of judging all other cultures by one's own culture.

ethnography a detailed study of the life and activities of a group of people by researchers who may live with that group over a period of years.

ethnomethodology the study of the commonsense knowledge that people use to understand the situations in which they find themselves.

exogamy the practice of marrying outside one's own group.

experiment a carefully designed situation in which the researcher studies the impact of certain variables on subjects' attitudes or behavior.

experimental group in an experiment, the group that contains the subjects who are exposed to an independent variable (the experimental condition) to study its effect on them.

expressive leadership leadership that provides emotional support for members.

extended family a family unit composed of relatives in addition to parents and children who live in the same household.

F

face-saving behavior Erving Goffman's term for the strategies we use to rescue our performance when we experience a potential or actual loss of face.

faith a confident belief that cannot be proven or disproven but is accepted as true.

families relationships in which people live together with commitment, form an economic unit and care for any young, and consider their identity to be significantly attached to the group.

family of orientation the family into which a person is born and in which early socialization usually takes place.

family of procreation the family that a person forms by having, adopting, or otherwise creating children.

feminism the belief that men and women are equal and should be valued equally and have equal rights.

feminization of poverty the trend in which women are disproportionately represented among individuals living in poverty.

fertility the actual level of childbearing for an individual or a population.

field research the study of social life in its natural setting: observing and interviewing people where they live, work, and play.

folkways informal norms or everyday customs that may be violated without serious consequences within a particular culture.

formal education learning that takes place within an academic setting such as a school, which has a planned instructional process and teachers who convey specific knowledge, skills, and thinking processes to students.

formal organization a highly structured group formed for the purpose of completing certain tasks or achieving specific goals.

functional age observable individual attributes such as physical appearance, mobility, strength, coordination, and mental capacity that are used to assign people to age categories.

functionalist perspectives the sociological approach that views society as a stable, orderly system.

fundamentalism a traditional religious doctrine that is conservative, is typically opposed to modernity, and rejects "worldly pleasures" in favor of otherworldly spirituality.

G

Gemeinschaft (guh-MINE-shoft) a traditional society in which social relationships are based on personal bonds of friendship and kinship and on intergenerational stability.

gender the culturally and socially constructed differences between females and males found in the meanings, beliefs, and practices associated with "femininity" and "masculinity."

gender bias behavior that shows favoritism toward one gender over the other.

gender identity a person's perception of the self as female or male.

gender role the attitudes, behavior, and activities that are socially defined as appropriate for each sex and that are learned through the socialization process.

gender socialization the aspect of socialization that contains specific messages and practices concerning the nature of being female or male in a specific group or society.

gendered racism the interactive effect of racism and sexism on the exploitation of women of color.

generalized other George Herbert Mead's term for the child's awareness of the demands and expectations of the society as a whole or of the child's subculture.

genocide the deliberate, systematic killing of an entire people or nation.

gentrification the process by which members of the middle and upper-middle classes, especially whites, move into a central-city area and renovate existing properties.

gerontology the study of aging and older people.

Gesellschaft (guh-ZELL-shoft) a large, urban society in which social bonds are based on impersonal and specialized relationships, with little long-term commitment to the group or consensus on values.

global stratification the unequal distribution of wealth, power, and prestige on a global basis, resulting in people having vastly different lifestyles and life chances both within and among the nations of the world.

goal displacement a process that occurs in organizations when the rules become an end in themselves rather than a means to an end, and organizational survival becomes more important than achievement of goals.

gossip rumors about the personal lives of individuals.

government the formal organization that has the legal and political authority to regulate the relationships among members of a society and between the society and those outside its borders.

groupthink the process by which members of a cohesive group arrive at a decision that many individual members privately believe is unwise.

H

Hawthorne effect a phenomenon in which changes in a subject's behavior are caused by the researcher's presence or by the subject's awareness of being studied.

health a state of complete physical, mental, and social well-being.

health care any activity intended to improve health.

health maintenance organizations (HMOs) companies that provide, for a set monthly fee, total care with an emphasis on prevention to avoid costly treatment later.

hidden curriculum the transmission of cultural values and attitudes, such as conformity and obedience to authority, through implied demands found in the rules, routines, and regulations of schools.

high culture classical music, opera, ballet, live theater, and other activities usually patronized by elite audiences.

high-income countries (sometimes referred to as **industrial countries**) nations with highly industrialized economies; technologically advanced industrial, administrative, and service occupations; and relatively high levels of national and personal income.

holistic medicine an approach to health care that focuses on prevention of illness and disease and is aimed at treating the whole person—body and mind—rather than just the part or parts in which symptoms occur.

homogamy the pattern of individuals marrying those who have similar characteristics, such as race/ethnicity, religious background, age, education, or social class.

homophobia extreme prejudice and sometimes discriminatory actions directed at gays, lesbians, bisexuals, transgender persons, and others who are perceived as not being heterosexual.

horticultural societies societies based on technology that supports the cultivation of plants to provide food.

hospice an organization that provides a home-like facility or home-based care (or both) for people who are terminally ill.

hunting and gathering societies societies that use simple technology for hunting animals and gathering vegetation.

hypothesis in research studies, a tentative statement of the relationship between two or more concepts.

I

id Sigmund Freud's term for the component of personality that includes all of the individual's basic biological drives and needs that demand immediate gratification.

ideal type an abstract model that describes the recurring characteristics of some phenomenon.

illegitimate opportunity structures circumstances that provide an opportunity for people to acquire through illegitimate activities what they cannot achieve through legitimate channels.

impression management (presentation of self) Erving Goffman's term for people's efforts to present themselves to others in ways that are most favorable to their own interests or image.

income the economic gain derived from wages, salaries, income transfers (governmental aid), and ownership of property.

independent variable a variable that is presumed to cause or determine a dependent variable.

individual discrimination behavior consisting of one-on-one acts by members of the dominant group that harm members of the subordinate group or their property.

industrial societies societies based on technology that mechanizes production.

industrialization the process by which societies are transformed from dependence on agriculture and handmade products to an emphasis on manufacturing and related industries.

infant mortality rate the number of deaths of infants under 1 year of age per 1,000 live births in a given year.

informal education learning that occurs in a spontaneous, unplanned way.

informal side of a bureaucracy those aspects of participants' day-to-day activities and interactions that ignore, bypass, or do not correspond with the official rules and procedures of the bureaucracy.

ingroup a group to which a person belongs and with which the person feels a sense of identity.

institutional discrimination the day-to-day practices of organizations and institutions that have a harmful effect on members of subordinate groups.

instrumental leadership goal- or task-oriented leadership.

intergenerational mobility the social movement experienced by family members from one generation to the next.

interlocking corporate directorates members of the board of directors of one corporation who also sit on the board(s) of other corporations.

internal colonialism according to conflict theorists, a practice that occurs when members of a racial or ethnic group are conquered or colonized and forcibly placed under the economic and political control of the dominant group.

Internet crime illegal acts committed by criminals on the Internet, including FBI-related scams, identity theft, advance fee fraud, nonauction/nondelivery of merchandise, and overpayment fraud.

intersex person an individual who is born with a reproductive or sexual anatomy that does not correspond to typical definitions of male or female; in other words, the person's sexual differentiation is ambiguous.

interview a data collection encounter in which an interviewer asks the respondent questions and records the answers.

intragenerational mobility the social movement of individuals within their own lifetime.

invasion the process by which a new category of people or type of land use arrives in an area previously occupied by another group or type of land use.

iron law of oligarchy according to Robert Michels, the tendency of bureaucracies to be ruled by a few people.

J

job deskilling a reduction in the proficiency needed to perform a specific job that leads to a corresponding reduction in the wages for that job or in the use of nonhuman technologies to perform the work.

juvenile delinquency a violation of law or the commission of a status offense by young people.

K

kinship a social network of people based on common ancestry, marriage, or adoption.

L

labeling theory the proposition that deviance is a socially constructed process in which social control agencies designate certain people as deviants and they, in turn, come to accept the label placed upon them and begin to act accordingly.

labor union a group of employees who join together to bargain with an employer or a group of employers over wages, benefits, and working conditions.

laissez-faire leaders leaders who are only minimally involved in decision making and who encourage group members to make their own decisions.

language a set of symbols that expresses ideas and enables people to think and communicate with one another.

latent functions unintended functions that are hidden and remain unacknowledged by participants.

laws formal, standardized norms that have been enacted by legislatures and are enforced by formal sanctions.

leadership the ability to influence what goes on in a group or social system.

life chances Max Weber's term for the extent to which individuals have access to important societal resources such as food, clothing, shelter, education, and health care.

life expectancy a statistical estimate of the average number of years that a person born in a specific year will live.

looking-glass self Charles Horton Cooley's term for the way in which a person's sense of self is derived from the perceptions of others.

low-income countries (sometimes referred to as **underdeveloped countries**) primarily agrarian nations with little industrialization and low levels of national and personal income.

M

macrolevel analysis an approach that examines whole societies, large-scale social structures, and social systems instead of looking at important social dynamics in individuals' lives.

managed care any system of cost containment that closely monitors and controls health care providers' decisions about medical procedures, diagnostic tests, and other services that should be provided to patients.

manifest functions functions that are intended and/or overtly recognized by the participants in a social unit.

marginal jobs jobs that differ from the employment norms of the society in which they are located.

marriage a legally recognized and/or socially approved arrangement between two or more individuals that carries certain rights and obligations and usually involves sexual activity.

mass a number of people who share an interest in a specific idea or issue but who are not in one another's immediate vicinity.

mass behavior collective behavior that takes place when people (who are often geographically separated from one another) respond to the same event in much the same way.

mass education the practice of providing free, public schooling for wide segments of a nation's population.

mass media large-scale organizations that use print or electronic means (such as radio, television, film, and the Internet) to communicate with large numbers of people.

master status the most important status that a person occupies.

material culture the physical or tangible creations that members of a society make, use, and share.

matriarchal family a family structure in which authority is held by the eldest female (usually the mother).

matriarchy a hierarchical system of social organization in which cultural, political, and economic structures are controlled by women.

matrilineal descent a system of tracing descent through the mother's side of the family.

matrilocal residence the custom of a married couple living in the same household (or community) as the wife's parents.

mechanical solidarity Emile Durkheim's term for the social cohesion of preindustrial societies, in which there is minimal division of labor and people feel united by shared values and common social bonds.

medical–industrial complex local physicians, local hospitals, and global health-related industries such as insurance companies and pharmaceutical and medical supply companies.

medicalization the process whereby nonmedical problems become defined and treated as illnesses or disorders.

medicine an institutionalized system for the scientific diagnosis, treatment, and prevention of illness.

meritocracy a hierarchy in which all positions are rewarded based on people's ability and credentials.

microlevel analysis sociological theory and research that focus on small groups rather than on large-scale social structures.

middle-income countries (sometimes referred to as **developing countries**) nations with industrializing economies, particularly in urban areas, and moderate levels of national and personal income.

migration the movement of people from one geographic area to another for the purpose of changing residency.

militarism a societal focus on military ideals and an aggressive preparedness for war.

military–industrial complex the mutual interdependence of the military establishment and private military contractors.

mixed economy an economic system that combines elements of a market economy (capitalism) with elements of a command economy (socialism).

mob a highly emotional crowd whose members engage in, or are ready to engage in, violence against a specific target—a person, a category of people, or physical property.

modernization theory a perspective that links global inequality to different levels of economic development and suggests that low-income economies can move to middle- and high-income economies by achieving self-sustained economic growth.

monarchy a political system in which power resides in one person or family and is passed from generation to generation through lines of inheritance.

monogamy the practice or state of being married to one person at a time.

monotheism a belief in a single, supreme being or god who is responsible for significant events such as the creation of the world.

mores strongly held norms with moral and ethical connotations that may not be violated without serious consequences in a particular culture.

mortality the incidence of death in a population.

N

neolocal residence the custom of a married couple living in their own residence apart from both the husband's and the wife's parents.

network a web of social relationships that links one person with other people and, through them, with other people they know.

new international division of labor theory the perspective that views commodity production as being split into fragments that can be assigned to whatever part of the world can provide the most profitable combination of capital and labor.

nonmaterial culture the abstract or intangible human creations of society that influence people's behavior.

nonverbal communication the transfer of information between persons without the use of words.

norms established rules of behavior or standards of conduct.

nuclear family a family composed of one or two parents and their dependent children, all of whom live apart from other relatives.

O

occupational (white-collar) crime illegal activities committed by people in the course of their employment or financial affairs.

occupations categories of jobs that involve similar activities at different work sites.

official poverty line the income standard that is based on what the federal government considers to be the minimum amount of money required for living at a subsistence level.

oligopoly a condition that exists when several companies overwhelmingly control an entire industry.

organic solidarity Emile Durkheim's term for the social cohesion found in industrial (and perhaps postindustrial) societies, in which people perform very specialized tasks and feel united by their mutual dependence.

organized crime a business operation that supplies illegal goods and services for profit.

outgroup a group to which a person does not belong and toward which the person may feel a sense of competitiveness or hostility.

P

panic a form of crowd behavior that occurs when a large number of people react to a real or perceived threat with strong emotions and self-destructive behavior.

participant observation a research method in which researchers collect systematic observations while being part of the activities of the group being studied.

pastoral societies societies based on technology that supports the domestication of large animals to provide food.

patriarchal family a family structure in which authority is held by the eldest male (usually the father).

patriarchy a hierarchical system of social organization in which cultural, political, and economic structures are controlled by men.

patrilineal descent a system of tracing descent through the father's side of the family.

patrilocal residence the custom of a married couple living in the same household (or community) as the husband's parents.

pay gap the disparity between women's and men's earnings.

peer group a group of people who are linked by common interests, equal social position, and (usually) similar age.

peripheral nations according to world systems theory, nations that are dependent on core nations for capital, have little or no industrialization (other than what may be brought in by core nations), and have uneven patterns of urbanization.

personal space the immediate area surrounding a person that the person claims as private.

pink-collar occupations relatively low-paying, nonmanual, semiskilled positions primarily held by women.

pluralist model an analysis of political systems that views power as widely dispersed throughout many competing interest groups.

political action committees (PACs) organizations of special interest groups that solicit contributions from donors and fund campaigns to help elect (or defeat) candidates based on their stances on specific issues.

political crime illegal or unethical acts involving the usurpation of power by government officials or illegal/unethical acts perpetrated against the government by outsiders seeking to make a political statement, undermine the government, or overthrow it.

political party an organization whose purpose is to gain and hold legitimate control of government.

political socialization the process by which people learn political attitudes, values, and behavior.

political sociology the area of sociology that examines the nature and consequences of power within or between societies, as well as the social and political conflicts that lead to changes in the allocation of power.

politics the social institution through which power is acquired and exercised by some people and groups.

polyandry the concurrent marriage of one woman with two or more men.

polygamy the concurrent marriage of a person of one sex with two or more members of the opposite sex.

polygyny the concurrent marriage of one man with two or more women.

polytheism a belief in more than one god.

popular culture activities, products, and services that are assumed to appeal primarily to members of the middle and working classes.

population composition the biological and social characteristics of a population, including age, sex, race, marital status, education, occupation, income, and size of household.

population pyramid a graphic representation of the distribution of a population by sex and age.

positivism a term describing Auguste Comte's belief that the world can best be understood through scientific inquiry.

postindustrial societies societies in which technology supports a service- and information-based economy.

postmodern perspectives the sociological approach that attempts to explain social life in contemporary societies that are characterized by postindustrialization, consumerism, and global communications.

power according to Max Weber, the ability of persons or groups to achieve their goals despite opposition from others.

power elite C. Wright Mills's term for the group made up of leaders at the top of business, the executive branch of the federal government, and the military.

prejudice a negative attitude based on faulty generalizations about members of specific racial, ethnic, or other groups.

prestige the respect or regard that a person or status position is given by others.

primary deviance the initial act of rule breaking.

primary group a small, less specialized group in which members engage in face-to-face, emotion-based interactions over an extended period of time.

primary labor market the sector of the labor market that consists of high-paying jobs with good benefits that have some degree of security and the possibility of future advancement.

primary sector production the sector of the economy that extracts raw materials and natural resources from the environment.

primary sex characteristics the genitalia used in the reproductive process.

primary socialization the process of learning that begins at birth and occurs in the home and family.

probability sampling choosing participants for a study on the basis of specific characteristics, possibly including such factors as age, sex, race/ethnicity, and educational attainment.

profane the everyday, secular, or "worldly" aspects of life that we know through our senses.

professions high-status, knowledge-based occupations.

propaganda information provided by individuals or groups that have a vested interest in furthering their own cause or damaging an opposing one.

property crimes burglary (breaking into private property to commit a serious crime), motor vehicle theft, larceny-theft (theft of property worth $50 or more), and arson.

public opinion the attitudes and beliefs communicated by ordinary citizens to decision makers.

punishment any action designed to deprive a person of things of value (including liberty) because of some offense the person is thought to have committed.

Q

questionnaire a printed research instrument containing a series of items to which subjects respond.

R

race a category of people who have been singled out as inferior or superior, often on the basis of real or alleged physical characteristics such as skin color, hair texture, eye shape, or other subjectively selected attributes.

racial socialization the aspect of socialization that contains specific messages and practices concerning the nature of our racial or ethnic status as it relates to our identity, interpersonal relationships, and location in the social hierarchy.

racism a set of attitudes, beliefs, and practices that is used to justify the superior treatment of one racial or ethnic group and the inferior treatment of another racial or ethnic group.

random sampling a study approach in which every member of an entire population being studied has the same chance of being selected.

rational choice theory of deviance the proposition that deviant behavior occurs when a person weighs the costs and benefits of nonconventional or criminal behavior and determines that the benefits will outweigh the risks involved in such actions.

rationality the process by which traditional methods of social organization, characterized by informality and spontaneity, are gradually replaced by efficiently administered formal rules and procedures.

rational–legal authority power legitimized by law or written rules and procedures. Also referred to as *bureaucratic authority*.

reciprocal socialization the process by which the feelings, thoughts, appearance, and behavior of individuals who are undergoing socialization also have a direct influence on those agents of socialization who are attempting to influence them.

reference group a group that strongly influences a person's behavior and social attitudes, regardless of whether that individual is an actual member.

relative poverty a level of economic deprivation that exists when people may be able to afford basic necessities but are still unable to maintain an average standard of living.

religion a social institution composed of a unified system of beliefs, symbols, and rituals—based on some sacred or supernatural realm—that guides human behavior, gives meaning to life, and unites believers into a community.

representative democracy a form of democracy whereby citizens elect representatives to serve as bridges between themselves and the government.

research methods specific strategies or techniques for systematically conducting research.

resocialization the process of learning a new and different set of attitudes, values, and behaviors from those in one's background and previous experience.

respondents persons who provide data for analysis through interviews or questionnaires.

riot violent crowd behavior that is fueled by deep-seated emotions but not directed at one specific target.

rituals regularly repeated and carefully prescribed forms of behaviors that symbolize a cherished value or belief.

role a set of behavioral expectations associated with a given status.

role conflict a situation in which incompatible role demands are placed on a person by two or more statuses held at the same time.

role exit a situation in which people disengage from social roles that have been central to their self-identity.

role expectation a group's or society's definition of the way that a specific role ought to be played.

role performance how a person actually plays a role.

role strain a condition that occurs when incompatible demands are built into a single status that a person occupies.

role-taking the process by which a person mentally assumes the role of another person or group in order to understand the world from that person's or group's point of view.

routinization of charisma the process by which charismatic authority is succeeded by a bureaucracy controlled by a rationally established authority or by a combination of traditional and bureaucratic authority.

rumor an unsubstantiated report on an issue or a subject.

S

sacred those aspects of life that exist beyond the everyday, natural world that we cannot experience with our senses.

sanctions rewards for appropriate behavior or penalties for inappropriate behavior.

Sapir–Whorf hypothesis the proposition that language shapes the view of reality of its speakers.

scapegoat a person or group that is incapable of offering resistance to the hostility or aggression of others.

second shift Arlie Hochschild's term for the domestic work that employed women perform at home after they complete their workday on the job.

secondary analysis a research method in which researchers use existing material and analyze data that were originally collected by others.

secondary deviance the process that occurs when a person who has been labeled a deviant accepts that new identity and continues the deviant behavior.

secondary group a larger, more specialized group in which members engage in more-impersonal, goal-oriented relationships for a limited period of time.

secondary labor market the sector of the labor market that consists of low-paying jobs with few benefits and very little job security or possibility for future advancement.

secondary sector production the sector of the economy that processes raw materials (from the primary sector) into finished goods.

secondary sex characteristics the physical traits (other than reproductive organs) that identify an individual's sex.

secondary socialization the process of learning that takes place outside the home—in settings such as schools, religious organizations, and the workplace—and helps individuals learn how to act in appropriate ways in various situations.

sect a relatively small religious group that has broken away from another religious organization to renew what it views as the original version of the faith.

secularization the process by which religious beliefs, practices, and institutions lose their significance in society and nonreligious values, principles, and institutions take their place.

segregation the spatial and social separation of categories of people by race, ethnicity, class, gender, and/or religion.

self-concept the totality of our beliefs and feelings about ourselves.

self-fulfilling prophecy a situation in which a false belief or prediction produces behavior that makes the originally false belief come true.

semiperipheral nations according to world systems theory, nations that are more developed than peripheral nations but less developed than core nations.

sex the biological and anatomical differences between females and males.

sex ratio the number of males for every hundred females in a given population.

sexism the subordination of one sex, usually female, based on the assumed superiority of the other sex.

sexual orientation a person's preference for emotional–sexual relationships with members of the different sex (heterosexuality), the same sex (homosexuality), or both (bisexuality).

sexualization the act or processes whereby an individual or group is seen as sexual in nature or persons become aware of their sexuality.

shared monopoly a condition that exists when four or fewer companies supply 50 percent or more of a particular market.

sick role the set of patterned expectations that defines the norms and values appropriate for individuals who are sick and for those who interact with them.

significant others those persons whose care, affection, and approval are especially desired and who are most important in the development of the self.

simple supernaturalism the belief that supernatural forces affect people's lives either positively or negatively.

slavery an extreme form of stratification in which some people are owned or controlled by others for the purpose of economic or sexual exploitation.

small group a collectivity small enough for all members to be acquainted with one another and to interact simultaneously.

social bond theory the proposition that the probability of deviant behavior increases when a person's ties to society are weakened or broken.

social change the alteration, modification, or transformation of public policy, culture, or social institutions over time.

social construction of reality the process by which our perception of reality is shaped largely by the subjective meaning that we give to an experience.

social control systematic practices that social groups develop in order to encourage conformity to norms, rules, and laws and to discourage deviance.

social Darwinism Herbert Spencer's belief that those species of animals, including human beings, best adapted to their environment survive and prosper, whereas those poorly adapted die out.

social devaluation a situation in which a person or group is considered to have less social value than other individuals or groups.

social epidemiology the study of the causes and distribution of health, disease, and impairment throughout a population.

social facts Emile Durkheim's term for patterned ways of acting, thinking, and feeling that exist *outside* any one individual but that exert social control over each person.

social group a group that consists of two or more people who interact frequently and share a common identity and a feeling of interdependence.

social institution a set of organized beliefs and rules that establishes how a society will attempt to meet its basic social needs.

social interaction the process by which people act toward or respond to other people: the foundation for all relationships and groups in society.

social mobility the movement of individuals or groups from one level in a stratification system to another.

social movement an organized group that acts consciously to promote or resist change through collective action.

social script a "playbook" that "actors" use to guide their verbal replies and overall performance to achieve the desired goal of the conversation or fulfill the role they are playing.

social stratification the hierarchical arrangement of large social groups based on their control of basic resources.

social structure the complex framework of societal institutions (such as the economy, politics, and religion) and the social practices (such as rules and social roles) that make up a society and that organize and establish limits on people's behavior.

socialism an economic system characterized by public ownership of the means of production, the pursuit of collective goals, and centralized decision making.

socialization the lifelong process of social interaction through which individuals acquire a self-identity and the physical, mental, and social skills needed for survival in society.

socialized medicine a health care system in which the government owns the medical care facilities and employs the physicians.

society a large social grouping that shares the same geographical territory and is subject to the same political authority and dominant cultural expectations.

sociobiology the systematic study of "social behavior from a biological perspective."

socioeconomic status (SES) a combined measure that attempts to classify individuals, families, or households in terms of factors such as income, occupation, and education to determine class location.

sociological imagination C. Wright Mills's term for the ability to see the relationship between individual experiences and the larger society.

sociology the systematic study of human society and social interaction.

sociology of family the subdiscipline of sociology that attempts to describe and explain patterns of family life and variations in family structure.

special interest groups political coalitions made up of individuals or groups which share a specific interest that they wish to protect or advance with the help of the political system.

spirituality the relationship between the individual and something larger than oneself, such as a broader sense of connection with the surrounding world.

split labor market the division of the economy into two areas of employment, a primary sector or upper tier, composed of higher-paid (usually dominant-group) workers in more-secure jobs, and a secondary sector or lower tier, composed of lower-paid (often subordinate-group) workers in jobs with little security and hazardous working conditions.

state the political entity that possesses a legitimate monopoly over the use of force within its territory to achieve its goals.

status a socially defined position in a group or society characterized by certain expectations, rights, and duties.

status set all the statuses that a person occupies at a given time.

status symbol a material sign that informs others of a person's specific status.

stereotypes overgeneralizations about the appearance, behavior, or other characteristics of members of particular categories.

strain theory the proposition that people feel strain when they are exposed to cultural goals that they are unable to obtain because they do not have access to culturally approved means of achieving those goals.

subcontracting an agreement in which a corporation contracts with other (usually smaller) firms to provide specialized components, products, or services to the larger corporation.

subculture a category of people who share distinguishing attributes, beliefs, values, and/or norms that set them apart in some significant manner from the dominant culture.

subordinate group a group whose members, because of physical or cultural characteristics, are disadvantaged and subjected to unequal treatment and discrimination by the dominant group.

succession the process by which a new category of people or type of land use gradually predominates in an area formerly dominated by another group or type of land use.

superego Sigmund Freud's term for the conscience, consisting of the moral and ethical aspects of personality.

survey a poll in which the researcher gathers facts or attempts to determine the relationships among facts.

symbol anything that meaningfully represents something else.

symbolic interactionist perspectives the sociological approach that views society as the sum of the interactions of individuals and groups.

T

taboos mores so strong that their violation is considered to be extremely offensive and even unmentionable.

technology the knowledge, techniques, and tools that allow people to transform resources into usable forms, and the knowledge and skills required to use them after they are developed.

terrorism the calculated, unlawful use of physical force or threats of violence against persons or property in order to intimidate or coerce a government, organization, or individual for the purpose of gaining some political, religious, economic, or social objective.

tertiary deviance deviance that occurs when a person who has been labeled a deviant seeks to normalize the behavior by relabeling it as nondeviant.

tertiary sector production the sector of the economy that is involved in the provision of services rather than goods.

tertiary socialization the process of learning that takes place when adults move into new settings where they must accept certain ideas or engage in specific behaviors that are appropriate to that specific setting.

theism a belief in a god or gods who shape human affairs.

theory a set of logically interrelated statements that attempts to describe, explain, and (occasionally) predict social events.

theory of racial formation the idea that actions of the government substantially define racial and ethnic relations in the United States.

total institution Erving Goffman's term for a place where people are isolated from the rest of society for a set period of time and come under the control of the officials who run the institution.

totalitarianism   a political system in which the state seeks to regulate all aspects of people's public and private lives.

tracking   the practice of assigning students to specific curriculum groups and courses on the basis of their test scores, previous grades, or other criteria.

traditional authority   power that is legitimized on the basis of long-standing custom.

transcendent idealism   a belief in sacred principles of thought and conduct.

transgender person   an individual whose gender identity (self-identification as woman, man, neither, or both) does not match the person's assigned sex (identification by others as male, female, or intersex based on physical/genetic sex).

transnational corporations   large corporations that are headquartered in one or a few countries but sell and produce goods and services in many countries.

triad   a group composed of three members.

U

underclass   those who are poor, seldom employed, and caught in long-term deprivation that results from low levels of education and income and high rates of unemployment.

unemployment rate   the percentage of unemployed persons in the labor force actively seeking jobs.

universal health care   a health care system in which all citizens receive medical services paid for by tax revenues.

unstructured interview   an extended, open-ended interaction between an interviewer and an interviewee.

urbanization   the process by which an increasing proportion of a population lives in cities rather than in rural areas.

V

value contradictions   values that conflict with one another or are mutually exclusive.

values   collective ideas about what is right or wrong, good or bad, and desirable or undesirable in a particular culture.

victimless crimes   crimes involving a willing exchange of illegal goods or services among adults.

violent crime   actions—murder, rape, robbery, and aggravated assault—involving force or the threat of force against others.

W

war   organized, armed conflict between nations or distinct political and/or religious factions.

wealth   the value of all of a person's or family's economic assets, including income, personal property, and income-producing property.

welfare state   a state in which there is extensive government action to provide support and services to the citizens.

working class (proletariat)   Karl Marx's term for those who must sell their labor to the owners in order to earn enough money to survive.

world systems theory   a perspective that examines the role of capitalism, and particularly the transnational division of labor, in a truly global system held together by economic ties.

Z

zero population growth   the point at which no population increase occurs from year to year.

AARP. 2015. "Half of Older Workers Who Were Unemployed in Last Five Years Jobless Today, According to AARP Survey" (Mar. 30). Retrieved Apr. 18, 2015. Online: www.aarp.org/about-aarp/press-center/info-03-2015/older-workers-unemployment-survey.print.html

AARP.org. 2012. "Social Security and Older Minorities" (Nov. 30). Retrieved Jan. 13, 2013. Online: www.aarp.org/work/social-security/info-11-2012/social-security-older-minorities.html

AAUW (American Association of University Women). 1995. *How Schools Shortchange Girls/The AAUW Report: A Study of Major Findings on Girls and Education.* New York: Marlowe.

—. 2008. "Where the Girls Are: The Facts About Gender Equity in Education." Retrieved Apr. 2, 2010. Online: www.aauw.org/research/upload/whereGirlsARe.pdf

—. 2015. "The Simple Truth About the Gender Pay Gap." American Association of University Women. Retrieved July 17, 2015. Online: www.aauw.org/files/2015/02/The-Simple-Truth_Spring-2015.pdf

Abaya, Carol. 2011. "Welcome to the Sandwich Generation." Retrieved Mar. 20 2011. Online: www.sandwichgeneration.com

Adler, Tina. 2013. "Ageism: Alive and Kicking." *Observer, Association for Psychological Science* (Sept.). Retrieved Jan. 22, 2015. Online: www.psychologicalscience.org/index.php/publications/observer/2013/september-13/ageism-alive-and-kicking.html

Adorno, Theodor W., Else Frenkel-Brunswick, Daniel J. Levinson, and R. Nevitt Sanford. 1950. *The Authoritarian Personality.* New York: Harper & Row.

AFP. 2013. "Wombs for Rent: Commercial Surrogacy Big Business in India." *The Express Tribune with the International Herald Tribune* (Feb. 25). Retrieved July 7, 2013. Online: http://tribune.com.pk/story/512264/wombs-for-rent-commercial-surrogacy-big-business-in-india

AIDS.gov. 2013. "Global AIDS Overview: The Global HIV/AIDS Crisis Today." Retrieved Mar. 29, 2014. Online: http://aids.gov/federal-resources/around-the-world/global-aids-overview

Akers, Ronald L. 1998. *Social Learning and Social Structure: A General Theory of Crime and Deviance.* Boston: Northeastern University Press.

Albanese, Catherine L. 2013. *America: Religions and Religion* (5th ed.). Boston, MA: Cengage.

Albas, Daniel, and Cheryl Albas. 1988. "Aces and Bombers: The Post-Exam Impression Management Strategies of Students." *Symbolic Interaction,* 11 (Fall): 289–302.

—. 2011. "Aces and Bombers: The Post-Exam Impression Management Strategies of Students." *Symbolic Interaction,* 11 (2): 289–302. Retrieved Nov. 8, 2012. Online: http://onlinelibrary.wiley.com/doi/10.1525/si.1988.11.2.289/abstract

Albrecht, Gary L., Katherine D. Seelman, and Michael Bury (Eds.). 2001. *The Handbook of Disability Studies.* Thousand Oaks, CA: Sage.

Altemeyer, Bob. 1981. *Right-Wing Authoritarianism.* Winnipeg: University of Manitoba Press.

—. 1988. *Enemies of Freedom: Understanding Right-Wing Authoritarianism.* San Francisco: Jossey-Bass.

Alvarez, Erick. 2008. *Muscle Boys: Gay Gym Culture.* New York: Routledge.

Alvarez, Lizette. 2013. "Girl's Suicide Points to Rise in Apps Used by Cyberbullies." *New York Times* (Sept. 13). Retrieved Dec. 7, 2013. Online: www.nytimes.com/2013/09/14/us/suicide-of-girl-after-bullying-raises-worries-on-web-sites.html?_r=0

Alzheimer's Association. 2015. "2015 Alzheimer's Disease Facts and Figures." Retrieved Apr. 25, 2015. Online: www.alz.org/facts/downloads/facts_figures_2015.pdf

American Association of Community Colleges. 2013. "Community College Fact Sheet." Retrieved Apr. 14, 2013. Online: www.aacc.nche.edu/AboutCC/Documents/2013facts_print_revised.pdf

American Bar Association. 2012. "Your Privacy Online." Retrieved Dec. 26, 2014. Online: www.safeshopping.org/privacy.shtml

American Civil Liberties Union (ACLU). 2015. "Hate Speech on Campus." Retrieved Mar. 20, 2015. Online: www.aclu.org/free-speech/hate-speech-campus

American Foundation for Suicide Prevention. 2014. "Suicide Prevention: Facts and Figures." Retrieved Jan. 1, 2015. Online: www.afsp.org/understanding-suicide/facts-and-figures

American Indian College Fund. 2015. "Tribal Colleges and Universities Map." Retrieved Mar. 16, 2015. Online: www.collegefund.org/userfiles/2012TCUMapFinal.pdf

American Psychiatric Association. 2013. "DSM-5 Implementation and Support: What's New?" Retrieved May 26, 2013. Online: www.dsm5.org/Pages/Default.aspx

American Psychological Association. 2010. "Report of the APA Task Force on the Sexualization of Girls." Retrieved Feb. 22, 2013. Online: www.apa.org/pi/women/programs/girls/report-full.pdf

American Society for Aesthetic Plastic Surgery. 2014. "More Than 12 Billion Dollars Spent on Surgical and Nonsurgical Procedures in 2013." Retrieved Apr. 13, 2015. Online: www.surgery.org/media/news-releases/the-american-society-for-aesthetic-plastic-surgery-reports-americans-spent-largest-amount-on-cosmetic-surgery

American Sociological Association. 2008/1999. "Code of Ethics and Policies and Procedures of the ASA Committee on Professional Ethics." Washington, DC: American Sociological Association. Retrieved Jan. 1, 2015. Online: www.asanet.org/images/asa/docs/pdf/CodeofEthics.pdf

American Sociological Association (ASA) Section on Consumers and Consumption. 2014. "Mission." Retrieved Dec. 29, 2014. Online: https://asaconsumers.wordpress.com/about/mission

Americares.org. 2015. "More Aid Arriving To Help Nepal Survivors Face Monsoon Season and Aftershocks" (June 5). Retrieved June 7, 2015. Online: www.americares.org/who-we-are/newsroom/news/americares-responding-to-nepal-earthquake.html

Andersen, Margaret L., and Patricia Hill Collins (Eds.). 2010. *Race, Class, and Gender: An Anthology* (7th ed.). Belmont, CA: Wadsworth.

Angier, Natalie. 1993. "'Stopit!' She Said. 'No-more!'" *New York Times Book Review* (Apr. 25): 12.

Appleton, Lynn M. 1995. "The Gender Regimes in American Cities." In Judith A. Garber and Robyne S. Turner (Eds.), *Gender in Urban Research.* Thousand Oaks, CA: Sage, pp. 44–59.

Apuzzo, Matt, and Michael S. Schmidt. 2015. "U.S. Not Expected to Fault Officer in Ferguson Case." *New York Times* (Jan. 21). Retrieved Feb. 23, 2015. Online: www.nytimes.com/2015/01/22/us/justice-department-ferguson-civil-rights-darren-wilson.html

Ascani, Nathaniel. 2012. "Labeling Theory and the Effects of Sanctioning on Delinquent Peer Association: A New Approach to Sentencing Juveniles." *Perspective, University of New Hampshire* (Spring). Retrieved Feb. 22, 2015. Online: http://cola.unh.edu/sites/cola.unh.edu/files/student-journals/P12_Ascani.pdf

Asch, Adrienne. 1986. "Will Populism Empower Disabled People?" In Harry G. Boyle and Frank Reissman (Eds.), *The New Populism: The Power of Empowerment.* Philadelphia: Temple University Press, pp. 213–228.

—. 2001. "Critical Race Theory, Feminism, and Disability: Reflections on Social Justice and Personal Identity." *Ohio State Law Journal* (62). Retrieved May 14, 2011. Online: http://moritzlaw.osu.edu/lawjournal/issues/volume62/number1/asch.pdf

Asch, Solomon E. 1955. "Opinions and Social Pressure." *Scientific American,* 193 (5): 31–35.

—. 1956. "Studies of Independence and Conformity: A Minority of One Against a Unanimous Majority." *Psychological Monographs,* 70 (9) (Whole No. 416).

Ashe, Arthur R., Jr. 1988. *A Hard Road to Glory: A History of the African-American Athlete.* New York: Warner.

Associated Press. 2012. "Man Stabs 22 Children in China." New York Times.com (Dec. 14). Retrieved Dec. 27, 2012. Online: www .nytimes.com/2012/12/15/world/asia/man -stabs-22-children-in-china.html?_r=0

Aulette, Judy Root. 1994. *Changing Families.* Belmont, CA: Wadsworth.

Babbie, Earl. 2013. *The Practice of Social Research* (13th ed.). Belmont, CA: Cengage /Wadsworth.

Bacevich, Andrew J. 2013. *The New American Militarism: How Americans Are Seduced by War.* New York: Oxford University Press.

Ballantine, Jeanne H., and Floyd M. Hammack. 2012. *The Sociology of Education: A Systematic Analysis* (7th ed.). Upper Saddle River, NJ: Prentice-Hall/Pearson.

Barkan, Steven E. 2012. *Criminology.* Upper Saddle River, NJ: Prentice-Hall.

Bates, Daniel. 2011. "Globalization of Fat Stigma: Western Ideas of Beauty and Body Size Catching on in Developing Nations." Daily Mail.com (Apr. 14). Retrieved Apr. 13, 2015. Online: www.dailymail.co.uk/news /article-1372036/Globalisation-fat -stigma-Warped-ideas-beauty-body -size-born-West-exported-developing -nations.html

Baudrillard, Jean. 1983. *Simulations.* New York: Semiotext.

—. 1998. *The Consumer Society: Myths and Structures.* London: Sage (orig. pub. 1970).

Beaulieu, Catherine MJ. 2004. "Intercultural Study of Personal Space: A Case Study." *Journal of Applied Social Psychology,* 34 (4): 794–805.

Beck, Robyn. 2014. "Credit Card Debt—A Student's Story." YouTube (originally posted June 4, 2008). Retrieved Dec. 30, 2014. Online: www.youtube.com/watch ?v=7U6pmkTC8i0

Becker, Howard S. 1963. *Outsiders: Studies in the Sociology of Deviance.* New York: Free Press.

Belkin, Lisa. 2006. "Life's Work: The Best Part Comes in the Third Act." *New York Times* (July 2). Retrieved Mar. 25, 2007. Online: http://select.nytimes.com/search/restricted /article?res=FB0E10F83E540C718CDDA E0894DE404482

—. 2009. "Do Girls Have More Chores Than Boys?" *New York Times* (Oct. 5). *Motherlode: Adventures in Parenting.* Retrieved Feb. 5, 2012. Online: http://parenting.blogs .nytimes.com/2009/10/05/do-girls-have -more-chores-than-boys

Bellah, Robert N. 1967. "Civil Religion." *Daedalus,* 96: 1–21.

Berger, Peter. 1967. *The Sacred Canopy: Elements of a Sociological Theory of Religion.* New York: Doubleday.

Berger, Peter, and Hansfried Kellner. 1964. "Marriage and the Construction of Reality." *Diogenes,* 46: 1–32.

Berger, Peter, and Thomas Luckmann. 1967. *The Social Construction of Reality: A Treatise in the Sociology of Knowledge.* Garden City, NY: Anchor.

Bernard, Jessie. 1982. *The Future of Marriage.* New Haven, CT: Yale University Press (orig. pub. 1973).

Bernburg, Jon Gunnar, Marvin D. Krohn, and Craig J. Rivera. 2006. "Official Labeling, Criminal Embeddedness, and Subsequent Delinquency: A Longitudinal Test of Labeling Theory." *Journal of Research in Crime and Delinquency* (43). Retrieved Feb. 22, 2015. Online: www.uk.sagepub .com/tibbetts/study/articles/SectionVIII /Bernburg.pdf

Berrett, Dan. 2015. "Stunned by a Video, U. of Oklahoma Struggles to Talk About Race." *Chronicle of Higher Education* (Mar. 19). Retrieved Mar. 23, 2015. Online: http:// chronicle.com/article/Stunned -by-a-Video-U-of/228611/?cid=wb&utm _source=wb&utm_medium=en

Best, Joel. 1999. *Random Violence: How We Talk About New Crimes and New Victims.* Berkeley: University of California Press.

Biblarz, Arturo, R. Michael Brown, Dolores Noonan Biblarz, Mary Pilgram, and Brent F. Baldree. 1991. "Media Influence on Attitudes Toward Suicide." *Suicide and Life-Threatening Behavior,* 21 (4): 374–385.

Bishaw, Alemayehu, and Kayla Fontenot. 2014. "Poverty: 2012 and 2013: American Community Survey Briefs" (Sept.). Retrieved Mar. 1, 2015. Online: www.census .gov/content/dam/Census/library /publications/2014/acs/acsbr13-01.pdf

Bivens, Josh. 2014. "Another Measure of the Staggering Wage Gaps in the United States: Comparing Walton Family Wealth to Typical Households by Race and Ethnicity." Economic Policy Institute (Oct. 13). Retrieved Feb. 28, 2015. Online: www .epi.org/blog/measure-staggering-wage -gaps-united-states

Bizshifts-Trends. 2011. "Management Styles: U.S., Europe, Japan, China, India, Brazil, Russia. Retrieved Mar. 13, 2013. Online: http://bizshifts-trends.com/2011/01/10 /management-styles-u-s-europe-japan -china-india-brazil-russia

Bjornstrom, Eileen E., Robert L. Kaufman, Ruth D. Peterson, and Michael D. Slater. 2010. "Race and Ethnic Representations of Lawbreakers and Victims in Crime News: A National Study of Television Coverage." *Social Problems,* 57 (2): 269–293.

Blauner, Robert. 1972. *Racial Oppression in America.* New York: Harper & Row.

Bloom, Nicholas, James Liang, John Roberts, and Zhichun Jenny Ying. 2013. "Does Working from Home Work? Evidence from a Chinese Experiment." Stanford University Department of Economics (Feb. 22). Retrieved Mar. 10, 2013. Online: www .stanford.edu/~nbloom/WFH.pdf

Blumer, Herbert G. 1946. "Collective Behavior." In Alfred McClung Lee (Ed.), *A New Outline of the Principles of Sociology.* New York: Barnes & Noble, pp. 167–219.

—. 1969. *Symbolic Interactionism: Perspective and Method.* Englewood Cliffs, NJ: Prentice Hall.

Blynn, Jamie. 2013. "Inside the 2013 Presidential Inauguration." Seventeen.com (Jan. 22). Retrieved Mar. 18, 2013 . Online: www.seventeen.com/college /presidential-election-blog /inauguration-2013

Bonacich, Edna. 1972. "A Theory of Ethnic Antagonism: The Split Labor Market." *American Sociological Review,* 37: 547–549.

—. 1976. "Advanced Capitalism and Black–White Relations in the United States: A Split Labor Market Interpretation." *American Sociological Review,* 41: 34–51.

Bonanno, George A. 2009. "Grief Does Not Come in Stages and It's Not the Same for Everyone." *Psychology Today* (Oct. 26). Retrieved Mar. 25, 2011. Online: www .psychologytoday.com/blog/thriving-in -the-face-trauma/200910/grief-does-not -come-in-stages-and-its-not-the-same -everyone

—. 2010. *The Other Side of Sadness: What the New Science of Bereavement Tells Us About Life After Loss.* New York: Basic.

Bonvillain, Nancy. 2001. *Women & Men: Cultural Constructs of Gender* (3rd ed.). Upper Saddle River, NJ: Prentice Hall.

Bourdieu, Pierre. 1984. *Distinction: A Social Critique of the Judgement of Taste.* Trans. Richard Nice. Cambridge, MA: Harvard University Press.

Bourdieu, Pierre, and Jean-Claude Passeron. 1990. *Reproduction in Education, Society and Culture.* Newbury Park, CA: Sage.

Brault, Matthew W. 2012. "Americans with Disabilities: 2010." United States Census Bureau (July). Retrieved May 11, 2013. Online: www.census.gov/prod/2012pubs /p70-131.pdf

Braun, Ralph. 2009. "Disabled Workers: Employer Fears Are Groundless." *Bloomberg Businessweek* (Oct. 2). Retrieved Apr. 3, 2011. Online: www.businessweek .com/managing/content/oct2009 /ca2009102_029034.htm

Breault, K. D. 1986. "Suicide in America: A Test of Durkheim's Theory of Religious and

Family Integration, 1933–1980." *American Journal of Sociology*, 92 (3): 628–656.

Breeden, Aurelien. 2015. "Paris Police Clear Out Migrant Camp and Destroy Tents." *New York Times* (June 2). Retrieved June 16, 2015. Online: www.nytimes.com/2015/06/03/world/europe/paris-police-clear-out-migrant-camp-and-destroy-tents.html

Brewis, Alexandra A., Amber Wutich, Ashlan Falletta-Cowden, and Isa Rodriguez-Soto. 2011. "Body Norms and Fat Stigma in Global Perspective." *Current Anthropology* (Apr.). Retrieved Apr. 13, 2015. Online: www.jstor.org/stable/10.1086/659309#mainContent

Brown, E. Richard. 1979. *Rockefeller Medicine Men*. Berkeley: University of California Press.

Brown, Patrick R., Andy Alaszewski, Trish Swift, and Andy Nordin. 2011. "Actions Speak Louder Than Words: The Embodiment of Trust by Healthcare Professionals in Gynae-oncology." *Sociology of Health & Illness* 33 (2): 280–295.

Brown, Robert W. 1954. "Mass Phenomena." In Gardner Lindzey (Ed.), *Handbook of Social Psychology* (vol. 2). Reading, MA: Addison-Wesley, pp. 833–873.

Buechler, Steven M. 2000. *Social Movements in Advanced Capitalism: The Political Economy and Cultural Construction of Social Activism*. New York: Oxford University Press.

Bullard, Robert B., and Beverly H. Wright. 1992. "The Quest for Environmental Equity: Mobilizing the African-American Community for Social Change." In Riley E. Dunlap and Angela G. Mertig (Eds.), *American Environmentalism: The U.S. Environmental Movement, 1970–1990*. New York: Taylor & Francis, pp. 39–49.

Burawoy, Michael. 2005. "For Public Sociology: 2004 ASA Presidential Address." *American Sociological Association* (Feb.): 4–28.

Burgess, Ernest W. 1925. "The Growth of the City." In Robert E. Park and Ernest W. Burgess (Eds.), *The City*. Chicago: University of Chicago Press, pp. 47–62.

Burgess-Proctor, Amanda. 2006. "Intersections of Race, Class, Gender and Crime: Future Directions for Feminist Criminology." *Feminist Criminology*, 1: 27–47. Retrieved Dec. 24, 2012. Online: http://fcx.sagepub.com/content/1/1/27.full.pdf+html

Cacioppo, John T., and Louise C. Hawkley. 2003. "Social Isolation and Health, with an Emphasis on Underlying Mechanisms." *Perspectives in Biology and Medicine* (Summer). Retrieved Jan. 22, 2015. Online: http://psychology.uchicago.edu/people/faculty/cacioppo/jtcreprints/ch03.pdf

Campaign for Disability Employment. 2011. "What Can Employers Do?" Retrieved July 8, 2011. Online: www.whatcanyoudocampaign.org/blog/index.php/what-can-employers-do

Campus Times. 2008. "Survey Analyzes Students' Stress Levels" (Aug. 10). Retrieved Feb. 8, 2010. Online: www.campustimes.org/2.4981/survey-analyzes-students-stress-levels-1.492569

Cancian, Francesca M. 1992. "Feminist Science: Methodologies That Challenge Inequality." *Gender & Society*, 6 (4): 623–642.

Canetto, Silvia Sara. 1992. "She Died for Love and He for Glory: Gender Myths of Suicidal Behavior." *OMEGA*, 26 (1): 1–17.

Cantril, Hadley. 1941. *The Psychology of Social Movements*. New York: Wiley.

Capek, Stella M. 1993. "The 'Environmental Justice' Frame: A Conceptual Discussion and Application." *Social Problems*, 40 (1): 5–23.

Carson, Rachel. 1962. *Silent Spring*. Boston: Houghton Mifflin.

Castells, Manuel. 1977. *The Urban Question*. London: Edward Arnold (orig. pub. 1972 as *La Question Urbaine*, Paris).

Catalyst. 2015. "Women CEOs of the S&P 500." Retrieved Apr. 13, 2015. Online: www.catalyst.org/knowledge/women-ceos-sp-500

Cave, James. 2015. "How a Bar of Soap Got a Homeless Family Off the Street." Huffingtonpost.com (Jan. 29). Retrieved Feb. 1, 2015. Online: www.huffingtonpost.com/2015/01/22/pono-soap-honolulu-homeless_n_6527342.html

CBS News. 2007. "Outsourced 'Wombs-for-Rent' in India." Retrieved Feb. 9, 2008. Online: www.cbsnews.com/stories/2007/12/31/health/main3658750.shtml

cdc.gov. 2014. "American Indian and Alaska Native Death Rates Nearly 50 Percent Greater Than Those of Non-Hispanic Whites." Centers for Disease Control and Prevention (Apr. 22). Retrieved Mar. 10, 2015. Online: www.cdc.gov/media/releases/2014/p0422-natamerican-deathrate.html

Center for Responsive Politics. 2013. "Super PACs." OpenSecrets.org. Retrieved May 27, 2014. Online: www.opensecrets.org/pacs/superpacs.php

Chagnon, Napoleon A. 1992. *Yanomamo: The Last Days of Eden*. New York: Harcourt (rev. from 4th ed., *Yanomamo: The Fierce People*, published by Holt, Rinehart & Winston).

Chambliss, William J. 1973. "The Saints and the Roughnecks." *Society*, 11: 24–31.

Cherlin, A. J. 2010. "Demographic Trends in the United States: A Review of Research in the 2000s." *Journal of Marriage and Family*, 72: 403–419.

chiefexecutive.com. 2010. "Changing Management Practice for the 21st Century." Retrieved Mar. 8, 2010. Online: www.the-chiefexecutive.com/features/feature54442

Child Care Aware of America. 2014. "Child Care in America: 2014 State Fact Sheet." Retrieved Jan. 22, 2015. Online: http://usa.childcareaware.org/sites/default/files/19000000_state_fact_sheets_2014_v04.pdf

Child Trends Data Bank. 2014. "Teen Homicide, Suicide, and Firearm Deaths." Retrieved Jan. 2, 2015. Online: www.childtrends.org/wp-content/uploads/2014/07/70_Homicide_Suicide_Firearms.pdf

—. 2015. "Family Structure: Indicators on Children and Youth" (Mar.). Retrieved July 17, 2015. Online: www.childtrends.org/wp-content/uploads/2015/03/59_Family_Structure.pdf

Children's Defense Fund. 2013. "Child Poverty in America 2012: National Analysis." Retrieved Jan. 20, 2014. Online: www.childrensdefense.org/child-research-data-publications/data/child-poverty-in-america-2012.pdf

—. 2014. "The State of America's Children: 2014." Retrieved July 19, 2015. Online: www.childrensdefense.org/library/state-of-americas-children/2014-soac.pdf?utm_source=2014-SOAC-PDF&utm_medium=link&utm_campaign=2014-SOAC

—. 2015. *The State of America's Children 2014*. Retrieved Feb. 28, 2015. Online: www.childrensdefense.org/zzz-child-research-data-publications/data/2014-soac.pdf

Chokshi, Niraj. 2015. "The Economy's Bouncing Back. But Higher Education Funding Isn't." *Washington Post* (May 13). Retrieved May 30, 2015. Online: www.washingtonpost.com/blogs/govbeat/wp/2015/05/13/the-economys-bouncing-back-higher-education-funding-isnt

Chronicle of Higher Education. 2014. "Almanac: 2014–2015" (Aug. 22).

Chu, Kathy. 2012. "Apple Plans Environmental Audits of China Suppliers." *USA Today* (Feb. 20). Retrieved Feb. 17, 2012. Online: www.usatoday.com/tech/news/story/2012-02-20/apple-china-environmental-audits/53167970/1

CIA World Factbook. 2015a. "Country Comparison: Life Expectancy at Birth." Retrieved June 2, 2015. Online: www.cia.gov/library/publications/the-world-factbook/rankorder/2102rank.html

—. 2015b. "United States." Retrieved Apr. 19, 2015. Online: www.cia.gov/library/publications/the-world-factbook/geos/us.html

Clayman, Steven E. 1993. "Booing: The Anatomy of a Disaffiliative Response." *American Sociological Review*, 58 (1): 110–131.

Clinton, Hillary Rodham. 2011. "Empowering Women Helps Fuel Global Economic Growth." Bloomberg.com (Mar. 8). Retrieved Mar. 14, 2011. Online: www.bloomberg.com/news/print/2011-03-08/empowering-women-helps-global-growth-commentary-by-hillary-rodham-clinton.html

Cloward, Richard A., and Lloyd E. Ohlin. 1960. *Delinquency and Opportunity: A Theory of Delinquent Gangs*. New York: Free Press.

CNN.com. 2012a. "At a Glance: Supreme Court Decision on Arizona's Immigration Law." Retrieved May 14, 2013. Online: www.cnn

.com/interactive/2012/06/us/scotus .immigration/index.html?pos=canon

—. 2012b. "Knife Attack at Chinese School Wounds 22 Children" (Dec. 14). Retrieved Dec. 27, 2012. Online: www.cnn.com /2012/12/14/world/asia/china-knife -attack/index.html

Coakley, Jay. 2009. *Sports in Society: Issues and Controversies* (10th ed.). New York: McGraw-Hill.

Coleman, Richard P., and Lee Rainwater. 1978. *Social Standing in America: New Dimensions of Class.* New York: Basic.

Coleman-Jensen, Alisha, Christian Gregory, and Anita Singh. 2014. "Household Food Security in the United States in 2013." USDA (Sept.). Retrieved Feb. 28, 2015. Online: www.ers .usda.gov/media/1565415/err173.pdf

College Board. 2012. "2012 College-Bound Seniors: Total Group Profile Report." Retrieved Mar. 10, 2013. Online: http://media .collegeboard.com/digitalServices/pdf /research/TotalGroup-2012.pdf

—. 2014. "Trends in College Pricing: 2014." Retrieved May 29, 2015. Online: https:// secure-media.collegeboard.org /digitalServices/misc/trends/2014-trends -college-pricing-report-final.pdf

college.usatoday.com. 2015. "5 Facts You Need to Know About the DREAM Act" (Feb. 26). Retrieved June 13, 2015. Online: http:// college.usatoday.com/2015/02/26/5-facts -you-need-to-know-about-the-dream-act

Collins, Patricia Hill. 1990. *Black Feminist Thought: Knowledge, Consciousness, and the Politics of Empowerment.* London: Harper-Collins Academic.

—. 2000. *Black Feminist Thought: Knowledge, Consciousness, and the Politics of Empowerment* (2nd ed.). New York: Routledge.

Collins, Randall. 1982. *Sociological Insight: An Introduction to Non-Obvious Sociology.* New York: Oxford University Press.

Coltrane, Scott. 2010. "Fathering: Paradoxes, Contradictions, and Dilemmas." In Michael S. Kimmel and Michael A. Messner (Eds.), *Men's Lives* (8th ed.). Boston: Allyn & Bacon, pp. 432–449.

Commonwealth of Massachusetts. 2011. "Labor and Workforce Development: Job Search Tips." Retrieved Apr. 3, 2011. Online: www .mass.gov/?pageID=elwdterminal&L=5&L0 =Home&L1=Workers+and+Unions&L2 =Job+Seekers&L3=Special+Programs&L4 =Connecting+Disabled+Workers+and +Employers&sid=Elwd&b=terminalcontent &f=dcs_cc_services_job_search_tips _disabled&csid=Elwd

Congressional Budget Office. 2011. "A Description of the Immigrant Population: An Update." Retrieved Feb. 5, 2012. Online: www.cbo.gov/ftpdocs/121xx/doc12168 /06-02-Foreign-BornPopulation.pdf

Cooley, Charles Horton. 1963. *Social Organization: A Study of the Larger Mind.* New York: Schocken (orig. pub. 1909).

—. 1998. "The Social Self—the Meaning of 'I.'" In Hans-Joachim Schubert (Ed.), *On Self and Social Organization—Charles Horton Cooley.* Chicago: University of Chicago Press, pp. 155–175. Reprinted from Charles Horton Cooley, *Human Nature and the Social Order.* New York: Schocken, 1902.

Cooper, Elizabeth. 2011. "Citizen's Voice: Why Intelligent Design Doesn't Qualify as Science." Knoxnews.com (Apr. 9). Retrieved May 2, 2011. Online: www.knoxnews.com /news/2011/apr/09/why-intelligent -design-doesnt-qualify-as-science/?print=1

Copen, Casey E., Kimberly Daniels, and William D. Mosher. 2013. "First Premarital Cohabitation in the United States: 2006–2010 National Survey of Family Growth." National Health Statistics Report (Apr. 4). Centers for Disease Control and Prevention, U.S. Department of Health and Human Services. Retrieved Feb. 19, 2014. Online: www.cdc.gov/nchs/data/nhsr /nhsr064.pdf

Cormier, David. 2008. "Rhizomatic Education: Community as Curriculum." *Innovate: Journal of Online Education.* Retrieved Apr. 13, 2013. Online: http://davecormier.com /edblog/2008/06/03/rhizomatic-education -community-as-curriculum

Corr, Charles A., Clyde M. Nabe, and Donald M. Corr. 2003. *Death and Dying, Life and Living* (4th ed.). Pacific Grove, CA: Brooks/Cole.

Corr, William P. 2014. "Suicides and Suicide Attempts Among Active Component Members of the U.S. Armed Forces, 2010–2012." *MSMR: Medical Surveillance Monthly Report* (Oct.). Retrieved Jan. 2, 2015. Online: www. afhsc.mil/documents/pubs/msmrs/2014 /v21_n10.pdf#Page=14

Corsaro, William A. 2011. *The Sociology of Childhood* (3rd ed.). Thousand Oaks, CA: Pine Forge.

Cortese, Anthony J. 2004. *Provocateur: Images of Women and Minorities in Advertising* (2nd ed.). Latham, MD: Rowman and Littlefield.

Coser, Lewis A. 1956. *The Functions of Social Conflict.* Glencoe, IL: Free Press.

Cox, Oliver C. 1948. *Caste, Class, and Race.* Garden City, NY: Doubleday.

Cumming, Elaine C., and William E. Henry. 1961. *Growing Old: The Process of Disengagement.* New York: Basic.

Currid-Halkett, Elizabeth. 2014. "What People Buy Where." *New York Times* (Dec. 13). Retrieved Dec. 14, 2014. Online: www.nytimes .com/2014/12/14/opinion/sunday/what -people-buy-where.html?_r=0

Curtiss, Susan. 1977. *Genie: A Psycholinguistic Study of a Modern Day "Wild Child."* New York: Academic Press.

Cyrus, Virginia. 1993. *Experiencing Race, Class, and Gender in the United States.* Mountain View, CA: Mayfield.

Dallas Morning News. 2015. "Free-Speech Issues Arise in OU Case" (Mar. 12). Retrieved Mar. 24, 2015. Online: www.pressreader.com /usa/the-dallas-morning-news/20150312 /281797102469398/TextView

Daniels, Jessie. 2009. *Cyber Racism: White Supremacy Online and the New Attack on Civil Rights.* Lanham, MD: Rowman & Littlefield.

—. 2010. "Cyber Racism on College Campuses." *Race-Talk* (May 4). Retrieved Apr. 29, 2011. Online: http://blogs.alternet.org /speakeasy/2010/05/04/cyber-racism-on -college-campuses

Davidson, Janet, and Meda Chesney-Lind. 2009. "Discounting Women: Context Matters in Risk and Need Assessment." *Critical Criminology,* 17 (4): 221–245.

Davis, Fred. 1992. *Fashion, Culture, and Identity.* Chicago: University of Chicago Press.

Davis, Kingsley. 1940. "Extreme Social Isolation of a Child." *American Journal of Sociology,* 45 (4): 554–565.

Davis, Kingsley, and Wilbert Moore. 1945. "Some Principles of Stratification." *American Sociological Review,* 7 (April): 242–249.

Davis-Delano, Laurel R., April Pollock, and Jennifer Ellsworth Vose. 2009. "Apologetic Behavior Among Female Athletes: A New Questionnaire and Initial Results." *International Review for the Sociology of Sport,* 44: 131–150.

Death Penalty Information Center. 2015. "Facts About the Death Penalty" (Feb. 11). Retrieved Feb. 23, 2015. Online: www.deathpenaltyinfo .org/documents/FactSheet.pdf

Delgado, Richard. 1995. "Introduction." In Richard Delgado (Ed.), *Critical Race Theory: The Cutting Edge.* Philadelphia: Temple University Press, pp. xiii–xvi.

DelReal, Jose A. 2014. "Voter Turnout in 2014 Was the Lowest Since WWII" (Nov. 10). Retrieved May 15, 2015. Online: www .washingtonpost.com/blogs/post-politics /wp/2014/11/10/voter-turnout-in -2014-was-the-lowest-since-wwii

DeNavas-Walt, Carmen, Robert W. Cleveland, and Bruce H. Webster Jr. 2003. "Income in the United States: 2002." U.S. Census Bureau, Current Population Reports, P. 60–221. Washington, DC: U.S. Government Printing Office.

DeNavas-Walt, Carmen, and Bernadette D. Proctor. 2014. *Income and Poverty in the United States: 2013* (Sept.). Retrieved Feb. 21, 2015. Online: www.census.gov/content /dam/Census/library/publications/2014 /demo/p60-249.pdf

Derber, Charles. 1983. *The Pursuit of Attention: Power and Individualism in Everyday Life.* New York: Oxford University Press.

DeSilver, Drew. 2015. "Job Categories Where Union Membership Has Fallen Off Most" (Apr. 27). Retrieved May 13, 2015. Online: www.pewresearch.org/fact-tank /2015/04/27/union-membership

Dolan, Andy, and Martin Robinson. 2013. "Schoolgirl Is 'Trolled to Death': Parents'

Agony as Daughter, 14, 'Hangs Herself' After Horrific Abuse from Bullies on Website Ask.fm." *Daily Mail Online* (Aug. 5). Retrieved Dec. 7, 2013. Online: www .dailymail.co.uk/news/article-2384866 /Schoolgirl-Hannah-Smith-trolled-death -bullies-Ask-fm-website.html

Dollard, John, Neal E. Miller, Leonard W. Doob, O. H. Mowrer, and Robert R. Sears. 1939. *Frustration and Aggression.* New Haven, CT: Yale University Press.

Domhoff, G. William. 2002. *Who Rules America? Power and Politics* (4th ed.). New York: McGraw-Hill.

—. 2005. "The Class-Domination Theory of Power." Retrieved Mar. 23, 2013. Online: www2.ucsc.edu/whorulesamerica/power /class_domination.html

—. 2014. "Power in America: The Class-Domination Theory of Power." Retrieved May 27, 2014. Online: www2.ucsc.edu /whorulesamerica/power/class _domination.html

Dover, Elicia. 2011. "Gingrich Says Poor Children Have No Work Habits." abcnews .com (Dec. 1, 2011). Retrieved June 5, 2015. Online: http://abcnews.go.com/blogs /politics/2011/12/gingrich-says-poor -children-have-no-work-ethic

Driskell, Robyn Bateman, and Larry Lyon. 2002. "Are Virtual Communities True Communities? Examining the Environments and Elements of Community." *City & Community,* 1 (4): 1–18.

Du Bois, W. E. B. 1967. *The Philadelphia Negro: A Social Study.* New York: Schocken (orig. pub. 1899).

Dunbar, Polly. 2007. "Wombs to Rent: Childless British Couples Pay Indian Women to Carry Their Babies." *The Daily Mail* (Dec. 8). Retrieved Feb. 9, 2008. Online: www.dailymail .co.uk/pages/live/articles/news/worldnews .html?in_article_id=500601

Durkheim, Emile. 1933. *The Division of Labor in Society.* Trans. George Simpson. New York: Free Press (orig. pub. 1893).

—. 1956. *Education and Sociology.* Trans. Sherwood D. Fox. Glencoe, IL: Free Press.

—. 1964a. *The Rules of Sociological Method.* Trans. Sarah A. Solovay and John H. Mueller. New York: Free Press (orig. pub. 1895).

—. 1964b. *Suicide.* Trans. John A. Sparkling and George Simpson. New York: Free Press (orig. pub. 1897).

—. 1995. *The Elementary Forms of Religious Life.* Trans. Karen E. Fields. New York: Free Press (orig. pub. 1912).

Dye, Thomas R., Harmon Zeigler, and Louis Schubert. 2012. *The Irony of Democracy: An Uncommon Introduction to American Politics* (15th ed.). Boston: Wadsworth.

Early, Kevin E. 1992. *Religion and Suicide in the African-American Community.* Westport, CT: Greenwood.

Ebaugh, Helen Rose Fuchs. 1988. *Becoming an EX: The Process of Role Exit.* Chicago: University of Chicago Press.

Eckholm, Erik. 2009. "Gang Violence Grows on an Indian Reservation." *New York Times* (Dec. 14): A13.

—. 2013. "With Police in Schools, More Children in Court." *New York Times* (Apr. 12). Retrieved Apr. 13, 2013. Online: www.nytimes.com/2013/04/12/education /with-police-in-schools-more-children -in-court.html?_r=0

Economist. 2010. "The Net Generation, Unplugged." Monitor, *Economist* (Mar. 4). Retrieved Jan. 22, 2015. Online: www .economist.com/node/15582279?story _id=15582279

Economist.com. 2011. "Environmental Activism in China: Poison Protests—A Huge Demonstration Over a Chemical Factory Unnerves Officials" (Aug. 20). Retrieved Mar. 17, 2012. Online: www.economist .com/node/21526417

Edwards, Kathryn A., Anna Turner, and Alexander Hertel-Fernandez. 2011. *A Young Person's Guide to Social Security.* Economic Policy Institute. Retrieved Dec. 17, 2011. Online: www.epi.org

Egley, Arlen, Jr., and James C. Howell. 2012. "Highlights of the 2010 National Youth Gang Survey" (Apr.). U.S. Department of Justice, Office of Juvenile Justice and Delinquency Prevention. Retrieved Dec. 28, 2013. Online: www.ojjdp.gov/pubs/237542.pdf

Ehrenreich, Barbara. 2001. *Nickel and Dimed: On (Not) Getting by in America.* New York: Metropolitan.

—. 2011. *Nickel and Dimed: On (Not) Getting by in America* (10th anniversary edition). New York: Picador.

Ehrlich, Paul R., and Anne H. Ehrlich. 2009. "The Population Bomb Revisited." *Electronic Journal of Sustainable Development,* 1 (3). Retrieved May 22, 2011. Online: http:// fragette.free.fr/demography/The _Population_Bomb_Revisited.pdf

Eighner, Lars. 1993. *Travels with Lizbeth.* New York: St. Martin's.

Eisenstein, Zillah R. 1994. *The Color of Gender: Reimaging Democracy.* Berkeley: University of California Press.

Elster, Jon. 1989. *Nuts and Bolts for the Social Sciences.* Cambridge, England: Cambridge University Press.

Engels, Friedrich. 1970. *The Origins of the Family, Private Property, and the State.* New York: International (orig. pub. 1884).

Enloe, Cynthia H. 1987. "Feminists Thinking About War, Militarism, and Peace." In Beth H. Hess and Myra Marx Ferree (Eds.), *Analyzing Gender: A Handbook of Social Science Research.* Newbury Park, CA: Sage, pp. 526–547.

Ennis, Sharon, Merarys Rios-Vargas, and Nora G. Albert. 2011. "The Hispanic Population: 2010." 2010 Census Briefs (May). Retrieved Dec. 21, 2011. Online: www.census.gov /prod/cen2010/briefs/c2010br-04.pdf

Entwisle, Joe. 2015. "The Future Looks Bright for Hiring Workers with Disabilities" (Mar. 18). Retrieved May 14, 2015. Online: www .huffingtonpost.com/joe-entwisle/the -future-looks-bright-f_1_b_6889416.html

EPA.gov. 2013. "Creating Equitable, Healthy, and Sustainable Communities" (Feb.). Retrieved May 16, 2013. Online: www.epa.gov /smartgrowth/pdf/equitable-dev/equitable -development-report-508-011713b.pdf

Erikson, Erik H. 1963. *Childhood and Society.* New York: Norton.

Erikson, Kai T. 1962. "Notes on the Sociology of Deviance." *Social Problems,* 9: 307–314.

—. 1964. "Notes on the Sociology of Deviance." In Howard S. Becker (Ed.), *The Other Side: Perspectives on Deviance.* New York: Free Press, pp. 9–21.

—. 1976. *Everything in Its Path: Destruction of Community in the Buffalo Creek Flood.* New York: Simon & Schuster.

—. 1991. "A New Species of Trouble." In Stephen Robert Couch and J. Stephen Kroll-Smith (Eds.), *Communities at Risk: Collective Responses to Technological Hazards.* New York: Land, pp. 11–29.

—. 1994. *A New Species of Trouble: Explorations in Disaster, Trauma, and Community.* New York: Norton.

Essed, Philomena. 1991. *Understanding Everyday Racism.* Newbury Park, CA: Sage.

Estelle, Sean. 2015. "I'm in This Fight for the Long Haul: Will You Join Me?" wearepower shift.org (Apr. 1). Retrieved June 17, 2015. Online: www.wearepowershift.org/blogs /im-fight-long-haul-will-you-join-me

Etzioni, Amitai. 1975. *A Comparative Analysis of Complex Organizations: On Power, Involvement, and Their Correlates* (rev. ed.). New York: Free Press.

Farrigan, Tracey, and Timothy Parker. 2012. "The Concentration of Poverty Is a Growing Rural Problem." *Rural Economy & Population* (Dec. 5). Retrieved Mar. 3, 2013. Online: www.ers .usda.gov/amber-waves/2012-december /concentration-of-poverty.aspx

Fausto-Sterling, Anne. 1985. *Myths of Gender: Biological Theories About Women and Men.* New York: Basic.

Feagin, Joe R. 1991. "The Continuing Significance of Race: Antiblack Discrimination in Public Places." *American Sociological Review,* 56 (February): 101–116.

Feagin, Joe R., David B. Baker, and Clairece B. Feagin. 2006. *Social Problems: A Critical Power–Conflict Perspective* (6th ed.). Englewood Cliffs. NJ: Prentice Hall.

Feagin, Joe R., and Clairece Booher Feagin. 1994. *Social Problems: A Critical Power– Conflict Perspective* (4th ed.). Englewood Cliffs, NJ: Prentice Hall.

—. 2012. *Race and Ethnic Relations, Census Update* (9th ed.). Upper Saddle River, NJ: Pearson.

Feagin, Joe R., and Robert Parker. 2002. *Building American Cities: The Urban Real Estate Game* (2nd ed.). Hopkins, MN: Beard.

Feagin, Joe R., and Hernán Vera. 1995. *White Racism: The Basics.* New York: Routledge.

Federal Bureau of Investigation (FBI). 2013. *Crime in the United States 2012.* Federal Bureau of Investigation. Retrieved Dec. 31, 2013. Online: www.fbi.gov/about-us/cjis/ucr/crime-in-the-u.s/2012/crime-in-the-u.s.-2012/resource-pages/download-printable-files

—. 2014. *Crime in the United States: 2013.* Retrieved Feb. 14, 2015. Online: www.fbi.gov/about-us/cjis/ucr/crime-in-the-u.s/2013/crime-in-the-u.s.-2013/cius-home

Federal Interagency Forum on Child and Family Statistics. 2013. "America's Children: Key National Indicators of Well-Being, 2013." National Institute of Health. Retrieved Feb. 22, 2014. Online: www.nichd.nih.gov/publications/pubs/Documents/Americas_Children_2013_DRAFT.pdf

Federal Reserve Bulletin. 2014. "Changes in U.S. Family Finances from 2010 to 2013: Evidence from the Survey of Consumer Finances" (Sept.). Retrieved Feb. 27, 2015. Online: www.federalreserve.gov/pubs/bulletin/2014/pdf/scf14.pdf

Feiler, Bruce. 2013. "The Stories That Bind Us." *New York Times* (Mar. 15). Retrieved Apr. 1, 2013. Online: www.nytimes.com/2013/03/17/fashion/the-family-stories-that-bind-us-this-life.html?ref=fashion&_r=1&

Findlay, Deborah A., and Leslie J. Miller. 1994. "Through Medical Eyes: The Medicalization of Women's Bodies and Women's Lives." In B. Singh Bolaria and Harley D. Dickinson (Eds.), *Health, Illness, and Health Care in Canada* (2nd ed.). Toronto: Harcourt, pp. 276–306.

Firestone, Shulamith. 1970. *The Dialectic of Sex.* New York: Morrow.

Flexner, Abraham. 1910. *Medical Education in the United States and Canada.* New York: Carnegie Foundation.

Forbes. 2015. "World's Richest Billionaires." Forbes.com. Retrieved Feb. 21, 2015. Online: www.forbes.com/billionaires/list/#tab:overall

Ford, Zack. 2015. "Why Fixing 'Religious Liberty' Bills Is Good For LGBT Equality, but Not Nearly Good Enough." Thinkprogress.org (Apr. 3). Retrieved Apr. 6, 2015. Online: http://thinkprogress.org/lgbt/2015/04/03/3642280/fixing-religious-liberty-bills-good-lgbt-equality-not-good-enough

Foucault, Michel. 1979. *Discipline and Punish: The Birth of the Prison.* New York: Vintage.

—. 1988. *Madness and Civilization: A History of Insanity in the Age of Reason.* New York: Vintage (orig. pub. 1961).

—. 1994. *The Birth of the Clinic: An Archeology of Medical Perception.* New York: Vintage (orig. pub. 1963).

Franklin, John Hope. 1980. *From Slavery to Freedom: A History of Negro Americans.* New York: Vintage.

Freidson, Eliot. 1965. "Disability as Social Deviance." In Marvin B. Sussman (Ed.), *Sociology and Rehabilitation.* Washington, DC: American Sociology Association, pp. 71–99.

—. 1970. *Profession of Medicine.* New York: Dodd, Mead.

—. 1986. *Professional Powers.* Chicago: University of Chicago Press.

French, Sally. 1999. "The Wind Gets in My Way." In Mairian Corker and Sally French (Eds.), *Disability Discourse.* Buckingham, England: Open University Press, pp. 21–27.

Freud, Sigmund. 1924. *A General Introduction to Psychoanalysis* (2nd ed.). New York: Boni & Liveright.

Frey, William H. 2011. "Melting Pot Cities and Suburbs: Racial and Ethnic Changes in Metro America in the 2000s." Metropolitan Policy Program at Brookings. Retrieved Mar. 15, 2012. Online: www.brookings.edu/~/media/Files/rc/papers/2011/0504_census_ethnicity_frey/0504_census_ethnicity_frey.pdf

Friedman, Thomas L. 2005. "It's a Flat World, After All." *New York Times Magazine* (Apr. 3): 33ff.

—. 2007. *The World Is Flat 3.0: A Brief History of the Twenty-First Century.* New York: Picador.

Frisbie, W. Parker, and John D. Kasarda. 1988. "Spatial Processes." In Neil Smelser (Ed.), *The Handbook of Sociology.* Newbury Park, CA: Sage, pp. 629–666.

Frome, Pamela M., Corinne J. Alfeld, Jacquelynne S. Eccles, and Bonnie L. Barber. 2006. "Why Don't They Want a Male-Dominated Job: An Investigation of Young Women Who Changed Their Occupational Aspirations." *Educational Research and Evaluation,* 12 (4): 359–372.

Galbraith, John Kenneth. 1985. *The New Industrial State* (4th ed.). Boston: Houghton Mifflin.

Gallup Organization. 2011a. "Immigration." Retrieved Feb. 5, 2012. Online: www.gallup.com

—. 2011b. "Race Relations." Retrieved Dec. 27, 2011. Online: www.gallup.com/poll/1687/race-relations.aspx

Gambino, Richard. 1975. *Blood of My Blood.* New York: Doubleday/Anchor.

Gandara, Ricardo. 1995. "*Dichos de la Vida:* Homespun Proverbs Link Hispanic Culture's Past with the Present." *Austin American-Statesman* (Jan. 21): E1, E10.

Gans, Herbert. 1982. *The Urban Villagers: Group and Class in the Life of Italian Americans*

(updated and expanded ed.; orig. pub. 1962). New York: Free Press.

Garfinkel, Harold. 1967. *Studies in Ethnomethodology.* Englewood Cliffs, NJ: Prentice Hall.

Garreau, Joel. 1991. *Edge City: Life on the New Frontier.* New York: Doubleday.

Gates, Gary J. 2011. "How Many People Are Lesbian, Gay, Bisexual and Transgender?" Williams Institute at the University of California School of Law (Apr.). Retrieved Feb. 10, 2012. Online: http://williamsinstitute.law.ucla.edu/research/census-lgbt-demographics-studies/how-many-people-are-lesbian-gay-bisexual-and-transgender

Gates, Gary J., and Frank Newport. 2012. "Special Report: 3.4% of U.S. Adults Identify as LGBT" (Oct. 18). Retrieved Feb. 21, 2013. Online: www.gallup.com/poll/158066/special-report-adults-identify-lgbt.aspx

Gawande, Atul. 2002. *Complications: A Surgeon's Notes on an Imperfect Science.* New York: Picador.

Geertz, Clifford. 1966. "Religion as a Cultural System." In Michael Banton (Ed.), *Anthropological Approaches to the Study of Religion.* London: Tavistock, pp. 1–46.

genomics.energy.gov. 2007. "Human Genome Project Information: Minorities, Race, and Genomics." Retrieved Feb. 9, 2013. Online: www.ornl.gov/sci/techresources/Human_Genome/elsi/minorities.shtml

—. 2012. "Genetically Modified Food and Organisms." Retrieved Mar. 17, 2012. Online: www.ornl.gov/sci/techresources/Human_Genome/elsi/gmfood.shtml

Gerth, Hans H., and C. Wright Mills. 1946. *From Max Weber: Essays in Sociology.* New York: Oxford University Press.

Gibbs, Lois Marie, as told to Murray Levine. 1982. *Love Canal: My Story.* Albany: SUNY Press.

Gibson, Campbell, and Kay Jung. 2006. "Historical Census Statistics on the Foreign-Born Population of the United States: 1850 to 2000." U.S. Census Bureau. Retrieved Feb. 5, 2012. Online: www.census.gov/population/www/documentation/twps0081/twps0081.html#trends

Gilbert, Dennis. 2014. *The American Class Structure in an Age of Growing Inequality* (9th ed.). Thousand Oaks, CA: Sage.

Gill, Nikhila. 2012. "Extreme Poverty Drops Worldwide." *New York Times* (Mar. 4). Retrieved Feb. 3, 2013. Online: india.blogs.nytimes.com/2012/03/04/extreme-poverty-drops-worldwide/?pagewanted=print

Gilligan, Carol. 1982. *In a Different Voice: Psychological Theory and Women's Development.* Cambridge, MA: Harvard University Press.

Giuliano, Traci A., Kathryn E. Popp, and Jennifer L. Knight. 2000. "Footballs Versus Barbies: Childhood Play Activities as Predictors of Sports Participation by Women." *Sex Roles,* 42: 159–181.

Glaeser, Edward. 2011. *Triumph of the City*. New York: Penguin.

Glaser, Barney, and Anselm Strauss. 1967. *Discovery of Grounded Theory: Strategies for Qualitative Research*. Chicago: Aldine.

—. 1968. *Time for Dying*. Chicago: Aldine.

Glass, Jennifer. 2013. "It's About the Work, Not the Office." *New York Times* (Mar. 8): A23.

Global Terrorism Index. 2014. "Global Terrorism Index 2014: Measuring and Understanding the Impact of Terrorism." Institute for Economics and Peace. Retrieved May 15, 2015. Online: www.visionofhumanity.org /sites/default/files/Global%20Terrorism %20Index%20Report%202014_0.pdf

Godofsky, Jessica, Cliff Zukin, and Carl Van Horn. 2011. "Worktrends: Americans' Attitudes About Work, Employers, and Government" (May). John H. Heldrich Center for Workplace Development, Rutgers University. Retrieved July 3, 2013. Online: www .heldrich.rutgers.edu/sites/default/files /content/Work_Trends_May_2011.pdf

Goffman, Erving. 1956. "The Nature of Deference and Demeanor." *American Anthropologist*, 58: 473–502.

—. 1959. *The Presentation of Self in Everyday Life*. Garden City, NY: Doubleday.

—. 1961a. *Asylums: Essays on the Social Situation of Mental Patients and Other Inmates*. Chicago: Aldine.

—. 1961b. *Encounters: Two Studies in the Sociology of Interaction*. London: Routledge and Kegan Paul.

—. 1963a. *Behavior in Public Places: Notes on the Social Structure of Gatherings*. New York: Free Press.

—. 1963b. *Stigma: Notes on the Management of Spoiled Identity*. Englewood Cliffs, NJ: Prentice Hall.

—. 1967. *Interaction Ritual: Essays on Face to Face Behavior*. Garden City, NY: Anchor.

—. 1974. *Frame Analysis: An Essay on the Organization of Experience*. Boston: Northeastern University Press.

Gold, Matea, and Melanie Mason. 2012. "NRA Calls for Armed Guards in Schools to Prevent Gun Violence." *Los Angeles Times* (Dec. 22). Retrieved Dec. 27, 2012. Online: http://articles.latimes.com/print/2012 /dec/22/nation/la-na-guns-nra-20121222

Goolsby, Craig A. 2011. "Disaster Planning: The Scope and Nature of the Problem" (Mar. 18). Retrieved May 11, 2011. Online: http:// emedicine.medscape.com/article/765495 -overview.pdf

Gottdiener, Mark. 1985. *The Social Production of Urban Space*. Austin: University of Texas Press.

Gottfredson, Michael R., and Travis Hirschi. 1990. *A General Theory of Crime*. Palo Alto, CA: Stanford University Press.

Gough, Neil. 2015. "Lopsided Job Market Puts Strain on China." *New York Times* (Apr. 22): B1, B8.

Governors Highway Safety Association. 2011. "Mature Drivers." Retrieved Mar. 20, 2011. Online: www.ghsa.org/html/issues /olderdriver.html

Green, Erica L. 2011. "Prayer Service at City School Called Improper." *Baltimore Sun* (Mar. 13). Retrieved May 8, 2011. Online: http://articles.baltimoresun.com/2011 -03-13/news/bs-md-ci-msa-prayer -service-20110313_1_prayer-service-prayer -in-public-schools-voluntary-student-prayer

Greenfield, Kent. 2015. "The Limits of Free Speech." *Atlantic* (Mar. 13). Retrieved Mar. 24, 2015. Online: www.theatlantic.com /politics/archive/2015/03/the-limits-of -free-speech/387718

GreenParty.org. 2015. "Ten Key Values." Retrieved May 15, 2015. Online: www.gp.org /what-we-believe/10-key-values

Griswold, Eliza. 2012. "How 'Silent Spring' Ignited the Environmental Movement." *New York Times* (Sept. 21). Retrieved May 16, 2013. Online: www.nytimes .com/2012/09/23/magazine/how-silent -spring-ignited-the-environmental -movement.html?pagewanted=all&_r=0

Grogan, Sarah. 2007. *Body Image: Understanding Body Dissatisfaction in Men, Women and Children*. New York: Routledge.

Gross, Allyson. 2015. "A Personal Take: Why I Believe Bowdoin Should Divest from the Fossil Fuel Industry." wearepowershift.org (Feb. 9). Retrieved June 17, 2015. Online: www.wearepowershift.org/blogs/personal -take-why-i-believe-bowdoin-should-divest -fossil-fuel-industry-0

Grube, Nick. 2014. "Photos of Waikiki's Homeless Reveal What It's Like to Live on the Streets in Paradise." Huffingtonpost.com (June 19). Retrieved Feb. 1, 2015. Online: www.huffingtonpost.com/2014/06/19 /waikiki-homelessness_n_5501673.html

Gryn, Thomas A., and Luke J. Larsen. 2010. "Nativity Status and Citizenship in the United States: 2009." American Community Survey Briefs, U.S. Census Bureau. Retrieved Feb. 5, 2012. Online: www.census .gov/prod/2010pubs/acsbr09-16.pdf

Guehenno, Jean-Marie. 2015. "10 Wars to Watch in 2015." *Foreign Policy* (Jan. 2). Retrieved May 16, 2015. Online: http:// foreignpolicy.com/2015/01/02/10-wars -to-watch-in-2015

Guha, Ramachandra. 2004. "The Sociology of Suicide." Retrieved Dec. 20, 2007. Online: www.indiatogether.org/2004/aug/rgh -suicide.htm

Hagedorn, John. 2009. *A World of Gangs: Armed Young Men and Gangsta Culture*. Minneapolis: University of Minnesota Press.

Hahn, Harlan. 1987. "Civil Rights for Disabled Americans: The Foundation of a Political Agenda." In Alan Gartner and Tom Joe (Eds.), *Images of the Disabled, Disabling Images*. New York: Praeger, pp. 181–203.

—. 1997. "Advertising the Acceptably Employable Image." In Lennard J. Davis (Ed.), *The Disability Studies Reader*. New York: Routledge, pp. 172–186.

Hall, Edward. 1966. *The Hidden Dimension*. New York: Anchor/Doubleday.

Hall, Edward T. 1959. *The Silent Language*. New York: Anchor/Doubleday.

Hall, Judy A., Jason D. Carter, and Terrence G. Horgan. 2000. "Gender Differences in Nonverbal Communication of Emotion." In Agneta H. Fischer (Ed.), *Gender and Emotion: Social Psychological Perspectives*. New York: Cambridge University Press, pp. 97–117.

Hallinan, Maureen. 2005. "Should Your School Eliminate Tracking? The History of Tracking and Detracking in America's Schools." *Education Matters*, 1–2.

Harlow, Harry F., and Margaret Kuenne Harlow. 1962. "Social Deprivation in Monkeys." *Scientific American*, 207 (5): 137–146.

—. 1977. "Effects of Various Mother–Infant Relationships on Rhesus Monkey Behaviors." In Brian M. Foss (Ed.), *Determinants of Infant Behavior* (vol. 4). London: Methuen, pp. 15–36.

Harris, Chauncey D., and Edward L. Ullman. 1945. "The Nature of Cities." *Annals of the Academy of Political and Social Sciences* (November): 7–17.

Harris, Marvin. 1974. *Cows, Pigs, Wars, and Witches*. New York: Random House.

—. 1985. *Good to Eat: Riddles of Food and Culture*. New York: Simon & Schuster.

Harvard Magazine. 2014. "Faculty Tensions I: The Sanctity of the Classroom." Retrieved Jan. 19, 2015. Online: http://harvardmagazine.com /2014/11/harvard-professors-object-to -student-monitoring

Hatton, Erin, and Mary Nell Trautner. 2011. "Equal Opportunity Objectification? The Sexualization of Men and Women on the Cover of *Rolling Stone*." *Sexuality & Culture*, 15: 256–278.

Haun, Daniel B. M., and Michael Tomasello. 2011. "Conformity to Peer Pressure in Preschool Children." *Child Development*, 82 (6): 1759–1767.

Havighurst, Robert J., Bernice L. Neugarten, and Sheldon S. Tobin. 1968. "Patterns of Aging." In Bernice L. Neugarten (Ed.), *Middle Age and Aging*. Chicago: University of Chicago Press, pp. 161–172.

Haviland, William A., Harald E. L. Prins, Bunny McBride, and Dana Walrath. 2014. *Cultural Anthropology: The Human Challenge* (14th ed.). Boston, MA: Cengage.

Hawley, Amos. 1950. *Human Ecology*. New York: Ronald.

—. 1981. *Urban Society* (2nd ed.). New York: Wiley.

Hayes, Liz. 2011. "Black Officials Scarce, Despite Population Gains." *Valley News Dispatch* (Dec. 4). Retrieved Dec. 29, 2011. Online: www.pittsburghlive.com/x /valleynewsdispatch/s_770348.html

Healey, Joseph F. 2002. *Race, Ethnicity, Gender, and Class: The Sociology of Group Conflict and Change* (3rd ed.). Thousand Oaks, CA: Pine Forge.

Heard, Amy. 2011. "Point of View: Campus Kitchen Is Worth Being Excited About." Baylor Lariat.com (Feb. 8). Retrieved Feb. 11, 2011. Online: http://baylorlariat .com/2011/02/08/2310

Helpguide.org. 2015. "Adult Day Care Services." Retrieved Apr. 25, 2015. Online: www .helpguide.org/articles/caregiving/adult -day-care-services.htm

Heritage, John. 1984. *Garfinkel and Ethnomethodology.* Cambridge, MA: Polity.

Hermsen, Joan M., David Cotter, and Reeve Vanneman. 2011. "The End of the Gender Revolution? Gender Role Attitudes from 1977 to 2008." *American Journal of Sociology* 117 (1): 259–289.

Hesse-Biber, Sharlene, and Gregg Lee Carter. 2000. *Working Women in America: Split Dreams.* New York: Oxford University Press.

Hiebert, Paul G. 1983. *Cultural Anthropology* (2nd ed.). Grand Rapids, MI: Baker.

Hirschi, Travis. 1969. *Causes of Delinquency.* Berkeley: University of California Press.

Hixson, Lindsay, Bradford B. Hepler, and Myoung Ouk Kim. 2011. "The White Population: 2010." U.S. Census Bureau (Sept.). Retrieved July 16, 2015. Online: www.census.gov/prod /cen2010/briefs/c2010br-05.pdf

Hochschild, Arlie Russell. 1983. *The Managed Heart: Commercialization of Human Feeling.* Berkeley: University of California Press.

—. 1997. *The Time Bind: When Work Becomes Home and Home Becomes Work.* New York: Metropolitan.

—. 2003. *The Commercialization of Intimate Life: Notes from Home and Work.* Berkeley: University of California Press.

—. 2012. *The Outsourced Self: Intimate Life in Market Times.* New York: Metropolitan.

Hochschild, Arlie Russell, with Ann Machung. 1989. *The Second Shift: Working Parents and the Revolution at Home.* New York: Viking/ Penguin.

Holland, Dorothy C., and Margaret A. Eisenhart. 1990. *Educated in Romance: Women, Achievement, and College Culture.* Chicago: University of Chicago Press.

Holmes, Tamara E., and Yasmin Ghahremani. 2014. "Credit Card Debt Statistics." Credit cards.com (Sept. 23). Retrieved Dec. 30, 2014. Online: www.creditcards.com/credit-card -news/credit-card-debt-statistics-1276.php

Holyk, Greg. 2012. "Lackluster Popularity Dogs the Political Parties." ABCnews.com (Oct. 3). Retrieved Mar. 29, 2013. Online: http://abcnews .go.com/blogs/politics/2012/10/lackluster -popularity-dogs-the-political-parties

Hondagneu-Sotelo, Pierrette. 2007. *Doméstica: Immigrant Workers Cleaning and Caring in the Shadows of Affluence.* Berkeley: University of California Press.

Hooyman, Nancy, and H. Asuman Kiyak. 2011. *Social Gerontology: A Multidisciplinary Perspective* (9th ed.). Upper Saddle River, NJ: Pearson.

Howden, Lindsay M., and Julie A. Meyer. 2011. *Age and Sex Composition: 2010.* U.S. Census Bureau (May). Retrieved Mar. 3, 2013. Online: www.census.gov/prod/cen2010 /briefs/c2010br-03.pdf

Howie, John R. R. 2010. "A Final Word." *Blog-Abroad.* StudyAbroad.com. Retrieved Feb. 12, 2010. Online: www.studyabroad.com /blog-abroad/howie

Hoyt, Homer. 1939. *The Structure and Growth of Residential Neighborhoods in American Cities.* Washington, DC: Federal Housing Administration.

Huff, Mickey, and Andy Lee Roth, with Project Censored. 2012. *Censored 2013: The Top Censored Stories and Media Analysis of 2011–12.* New York: Seven Stories Press.

Hughes, Everett C. 1945. "Dilemmas and Contradictions of Status." *American Journal of Sociology,* 50: 353–359.

Humes, Karen R., Nicholas A. Jones, and Roberto R. Ramirez. 2011. "Overview of Race and Hispanic Origin: 2010." U.S. Census Bureau (Mar. 2011). Retrieved June 25, 2013. Online: www.census.gov/prod/cen2010 /briefs/c2010br-02.pdf

Hurtado, Aida. 1996. *The Color of Privilege: Three Blasphemies on Race and Feminism.* Ann Arbor: University of Michigan Press.

Hussey, Kristin, and John Leland. 2013. "After Boy's Suicide, Questions About Missed Signs." *New York Times* (Aug. 30): A13.

Institute for Women's Policy Research. 2015. "The Status of Women in the States: 2015— Employment and Earnings." Retrieved Apr. 27, 2015. Online: http:// statusofwomendata.org/app/uploads /2015/02/EE-CHAPTER-FINAL.pdf

Institute of International Education. 2014. "Open Doors Data." Retrieved Jan. 22, 2015. Online: www.iie.org/Research-and -Publications/Open-Doors/Data/US-Study -Abroad/Student-Profile/2000-13

International Federation of Red Cross and Red Crescent Societies. 2014. "World Disasters Report." Retrieved June 7, 2015. Online: www .ifrc.org/world-disasters-report-2014/data

International Foundation for Electoral Systems. 2011. "Gender Issues." Retrieved Mar. 12, 2011. Online: www.ifes.org /Content/Topics/Gender-Issues.aspx

International Labor Organization. 2013. "Global Employment Trends 2013." Retrieved Jan. 24, 2014. Online: www.ilo.org/wcmsp5/groups /public/—dgreports/—dcomm/—publ /documents/publication/wcms_202326.pdf

International Longevity Center. 2015. "Being Old and Healthy in Japan." Retrieved July 19, 2015. Online: www.ilcjapan.org /aging/doc/booklet_OHJ.pdf

International Monetary Fund. 2012. "World Economic Outlook Data Base." Retrieved Jan. 27, 2013. Online: www.imf.org/external /pubs/ft/weo/2012/02/weodata/index.aspx

Internet Crime Complaint Center. 2014. "2013 Internet Crime Report." Retrieved Feb. 14, 2015. Online: www.ic3.gov/media /annualreport/2013_IC3Report.pdf

—. 2015. "2014 Internet Crime Report." Online: www.ic3.gov/media/annualreport/2014 _IC3Report.pdf

Intersex Society of North America. 2015. "What Is Intersex?" Retrieved Apr. 5, 2015. Online: www.isna.org/faq/what_is_intersex

Irwin, Neil. 2015. "Why Less Educated Workers Are Losing Ground on Wages" (Apr. 23). Retrieved May 13, 2015. Online: www .nytimes.com/2015/04/22/upshot/why -workers-without-much-education-are -being-hammered.html?abt=0002&abg=1

James, Susan Donaldson. 2008. "Wild Child 'Genie': A Tortured Life." 6abc.COM (May 8). Retrieved Jan. 30, 2015. Online: http://6abc.com/archive/6130233

Janis, Irving. 1972. *Victims of Groupthink.* Boston: Houghton Mifflin.

—. 1989. *Crucial Decisions: Leadership in Policymaking and Crisis Management.* New York: Free Press.

Jiji, Kyodo. 2014. "Japan Population Drops for Third Year Straight; 25% Are Elderly." *Japan Times* (Apr. 15). Retrieved Apr. 23, 2015. Online: www.japantimes.co.jp /news/2014/04/15/national/japans -population-drops-for-third-straight-year -25-are-elderly/#.VTlexrl0wqw

Johnson, Jenna. 2010. "First Lady Pays Off on Challenge to Serve with GWU Commencement Speech." *Washington Post* (May 17). Retrieved Dec. 31, 2010. Online: www .washingtonpost.com/wp-dyn/content /article/2010/05/16/AR2010051601114.html

Kaiser Family Foundation. 2012. "The Global HIV/AIDS Epidemic: Fact Sheet" (Dec.). Retrieved May 12, 2013. Online: http:// kaiserfamilyfoundation.files.wordpress .com/2013/01/3030-17.pdf

—. 2015. "The Uninsured: A Primer—Key Facts About Health Insurance and the Uninsured in America Updated." (Jan. 13). Retrieved June 6, 2015. Online: http://kff.org/report -section/the-uninsured-a-primer-what -was-happening-to-insurance-coverage -leading-up-to-the-aca

Kaklenberg, Susan, and Michelle Hein. 2010. "Progression on Nickelodeon? Gender-Role Stereotypes in Toy Commercials." *Sex Roles,* 62 (11): 830–847.

Kandel, William A. 2011. "The U.S. Foreign-Born Population: Trends and Selected Characteristics." Congressional Research Service. Retrieved Feb. 5, 2011. Online: www.crs.gov

Kansas State University. 2010. "Timeline for Transition." Retrieved Dec. 22, 2010. Online: www.k-state.edu/parentsandfamily/resources/timeline.htm

Kanter, Rosabeth Moss. 1983. *The Change Masters: Innovation and Entrepreneurship in the American Corporation.* New York: Simon & Schuster.

—. 1985. "All That Is Entrepreneural Is Not Gold." *Wall Street Journal* (July 22): 18.

—. 1993. *Men and Women of the Corporation.* New York: Basic (orig. pub. 1977).

Karnowski, Steve. 2011. "Dalai Lama Says bin Laden's Killing Understandable." Associated Press (May 8). Retrieved May 8, 2011. Online: http://minnesota.publicradio.org/display/web/2011/05/08/dalai-lama-minnesota/?refid=0

Kasarda, John D., and Greg Lindsay. 2011. *Aerotropolis: The Way We'll Live Next.* New York: Farrar, Straus and Giroux.

Katzer, Jeffrey, Kenneth H. Cook, and Wayne W. Crouch. 1991. *Evaluating Information: A Guide for Users of Social Science Research.* New York: McGraw-Hill.

Kaufman, Leslie. 2011. "City Prepares for a Warm Long-Term Forecast." *New York Times* (May 23): A1, A14.

Kendall, Diana. 1980. Square Pegs in Round Holes: Non-Traditional Students in Medical Schools. Unpublished doctoral dissertation, Department of Sociology, the University of Texas at Austin.

—. 2002. *The Power of Good Deeds: Privileged Women and the Social Reproduction of the Upper Class.* Lanham, MD: Rowman & Littlefield.

—. 2008. *Members Only: Elite Clubs and the Process of Exclusion.* Lanham, MD: Rowman & Littlefield.

—. 2011. *Framing Class: Media Representations of Wealth and Poverty in the United States* (2nd ed.). Lanham, MD: Rowman & Littlefield.

Kenyon, Kathleen. 1957. *Digging Up Jericho.* London: Benn.

Kerkman, Maggie. 2011. "Educating Our Children: The Evolution of Home Schooling." Foxnews.com (Feb. 9). Retrieved Apr. 29, 2011. Online: www.foxnews.com/us/2011/02/09/educating-children-evolution-home-schooling

Kimmel, Michael S., and Michael A. Messner. 2012. *Men's Lives* (9th ed.). Upper Saddle River, NJ: Pearson.

King, Ryan D., Kecia R. Johnson, and Kelly McGeever. 2010. "Demography of the Legal Profession and Racial Disparities in Sentencing." *Law & Society Review,* 44 (1): 1–31.

Kitsuse, John I. 1980. "Coming Out All Over: Deviance and the Politics of Social Problems." *Social Problems,* 28: 1–13.

KITV.com. 2014. "Oahu Family Reacts to Homeless Decision." Retrieved Feb. 1, 2015. Online: www.kitv.com/news/oahu-family-reacts-to-homeless-decision/27142742

KJRH.com. 2013. "Oklahoma City–Area Hospitals Report Hundreds of Patients After Tornado" (May 21). Retrieved May 26, 2013. Online: www.kjrh.com/dpp/news/local_news/Oklahoma-City-area-hospitals-report-hundreds-of-patients-after-tornado-about-60-remain-hospitalized#

Kochanek, Kenneth D., Sherry L. Murphy, Jiaquan Xu, and Elizabeth Arias. 2014. "Mortality in the United States, 2013." U.S. Centers for Disease Control and Prevention (Dec.). Retrieved June 12, 2015. Online: www.cdc.gov/nchs/data/databriefs/db178.pdf

Kohl, Beth. 2007. "On Indian Surrogates." *The Huffington Post* (Oct. 30). Retrieved Feb. 9, 2008. Online: www.huffingtonpost.com/beth-kohl/on-indian-surrogates_b_70425.html

Kohlberg, Lawrence. 1969. "Stage and Sequence: The Cognitive–Developmental Approach to Socialization." In David A. Goslin (Ed.), *Handbook of Socialization Theory and Research.* Chicago: Rand McNally, pp. 347–480.

—. 1981. *The Philosophy of Moral Development: Moral Stages and the Idea of Justice,* vol. 1: *Essays on Moral Development.* San Francisco: Harper & Row.

Kohn, Melvin L., Atsushi Naoi, Carrie Schoenbach, Carmi Schooler, and Kazimierz M. Slomczynski. 1990. "Position in the Class Structure and Psychological Functioning in the United States, Japan, and Poland." *American Journal of Sociology,* 95: 964–1008.

Kohut, Andrew. 2015. "Are Americans Ready for Obama's 'Middle Class' Populism?" Pew Research Center (Feb. 19). Retrieved Feb. 26, 2015. Online: www.pewresearch.org/fact-tank/2015/02/19/are-americans-ready-for-obamas-middle-class-populism

Koopmans, Ruud. 1999. "Political. Opportunity. Structure. Some Splitting to Balance the Lumping." *Sociological Forum* (Mar.): 93–105.

Kovacs, M., and A. T. Beck. 1977. "The Wish to Live and the Wish to Die in Attempted Suicides." *Journal of Clinical Psychology,* 33: 361–365.

Kozol, Jonathan. 1991. *Savage Inequalities: Children in America's Schools.* New York: Crown.

Kristof, Nicholas D., and Sheryl WuDunn. 2009. *Half the Sky: Turning Oppression into Opportunity for Women Worldwide.* New York: Vintage.

Kruse, Kevin. 2012. "What Is Employee Engagement?" Forbes.com (June 22). Retrieved Mar. 10, 2013. Online: www.forbes.com/sites/kevinkruse/2012/06/22/employee-engagement-what-and-why

Kübler-Ross, Elisabeth. 1969. *On Death and Dying.* New York: Macmillan.

Kurtz, Lester. 1995. *Gods in the Global Village: The World's Religions in Sociological Perspective.* Thousand Oaks, CA: Sage.

Lachance-Grzela, Mylene, and Genevieve Bouchard. 2010. "Why Do Women Do the Lion's Share of Housework? A Decade of Research." *Sex Roles,* 63 (11–12): 767–780.

Ladd, E. C., Jr. 1966. *Negro Political Leadership in the South.* Ithaca, NY: Cornell University Press.

LaFrance, Marianne, and Marvin Hecht. 2000. "Gender and Smiling: A Meta-Analysis." In Agneta H. Fischer (Ed.), *Gender and Emotion: Social Psychological Perspectives.* New York: Cambridge University Press, pp. 118–142.

Lamanna, Mary Ann, Agnes Riedmann, and Susan D. Stewart. 2015. *Marriages, Families, and Relationships: Making Choices in a Diverse Society* (12th ed.). Belmont, CA: Cengage.

Lancet. 2012. "Suicide Mortality in India: A Nationally Representative Survey" (June). Retrieved Oct. 12, 2012. Online: www.thelancet.com/journals/lancet/article/PIIS0140-6736(12)60606-0/abstract

Landale, Nancy S., and Ralph S. Oropesa. 2007. "Hispanic Families: Stability and Change." *Annual Review of Sociology,* 33: 381–405.

Langhout, Regina D., and Cecily A. Mitchell. 2008. "Engaging Contexts: Drawing the Link Between Student and Teacher Experiences of the Hidden Curriculum." *Journal of Community & Applied Social Psychology,* 18 (6): 593–614.

Lapchick, Richard. 2014. "Racism Still Evident in Sports World." ESPN.com (Dec. 31). Retrieved Mar. 9, 2015. Online: http://espn.go.com/espn/print?id=12093538&type=story

Lapham, Lewis H. 1988. *Money and Class in America: Notes and Observations on Our Civil Religion.* New York: Weidenfeld & Nicolson.

Lareau, Annette. 2011. *Unequal Childhoods: Class, Race, and Family Life* (2nd ed.). Berkeley: University of California Press.

Larsen, Janet. 2011. "Cancer Now Leading Cause of Death in China." Earth Policy Institute (May 25). Retrieved Mar. 17, 2012. Online: www.earth-policy.org/plan_b_updates/2011/update96

Larson, Magali Sarfatti. 1977. *The Rise of Professionalism: A Sociological Analysis.* Berkeley: University of California Press.

Laumann, Edward O., John H. Gagnon, Robert T. Michael, and Stuart Michaels. 1994. *The Social Organization of Sexuality.* Chicago: University of Chicago Press.

Le Bon, Gustave. 1960. *The Crowd: A Study of the Popular Mind.* New York: Viking (orig. pub. 1895).

Leenaars, Antoon A. 1988. *Suicide Notes: Predictive Clues and Patterns.* New York: Human Sciences Press.

Lefrançois, Guy R. 1996. *The Lifespan* (5th ed.). Belmont, CA: Wadsworth.

Lemert, Charles. 1997. *Postmodernism Is Not What You Think.* Malden, MA: Blackwell.

Lemert, Edwin M. 1951. *Social Pathology.* New York: McGraw-Hill.

Lenhart, Amanda. 2015. "Teens, Social Media and Technology Overview 2015." Pew Research Center (Apr.). Retrieved Apr. 23, 2015. Online: www.pewinternet.org

/files/2015/04/PI_TeensandTech_Update2015_0409151.pdf

Lerner, Gerda. 1986. *The Creation of Patriarchy.* New York: Oxford University Press.

Levin, William C. 1988. "Age Stereotyping: College Student Evaluations." *Research on Aging,* 10 (1): 134–148.

Levine, Peter. 1992. *Ellis Island to Ebbets Field: Sport and the American Jewish Experience.* New York: Oxford University Press.

Levitin, Michael. 2015. "The Triumph of Occupy Wall Street." *Atlantic* (June 10). Retrieved June 23, 2015. Online: www.theatlantic.com/politics/archive/2015/06/the-triumph-of-occupy-wall-street/395408

Levy, Becca R. 2009. "Stereotype Embodiment: A Psychosocial Approach to Aging." *Current Directions in Psychological Science,* 18: 332–336.

Lewis, Jamie M., and Rose M. Kreider. 2015. "Remarriage in the United States." U.S. Census Bureau (Mar.). Retrieved May 24, 2015. Online: www.census.gov/content/dam/Census/library/publications/2015/acs/acs-30.pdf

Liebow, Elliot. 1993. *Tell Them Who I Am: The Lives of Homeless Women.* New York: Free Press.

Lipka, Sara. 2014. "Diversity: More Voices Call for Equity, Not Just Access." *Chronicle of Higher Education* (Aug. 22): 39.

Livingston, Gretchen. 2014. "Less Than Half of U.S. Kids Today Live in a 'Traditional' Family." Pew Research Center (Dec. 22). Retrieved May 17, 2015. Online: www.pewresearch.org/fact-tank/2014/12/22/less-than-half-of-u-s-kids-today-live-in-a-traditional-family

Lofland, John. 1993. "Collective Behavior: The Elementary Forms." In Russell L. Curtis, Jr., and Benigno E. Aguirre (Eds.), *Collective Behavior and Social Movements.* Boston: Allyn & Bacon, pp. 70–75.

Lorber, Judith. 1994. *Paradoxes of Gender.* New Haven, CT: Yale University Press.

Lorber, Judith (Ed.). 2005. *Gender Inequality: Feminist Theories and Politics* (3rd ed.). Los Angeles: Roxbury.

Low, Setha. 2003. *Behind the Gates: Life, Security, and the Pursuit of Happiness in Fortress America.* New York: Routledge.

Lowry, Erin. 2014. "College Students Still Face Crippling Credit Card Rates" (Aug. 25). Retrieved Dec. 16, 2014. Online: www.dailyfinance.com/2014/08/25/college-students-still-face-crippling-credit-card-interest-rates

Lundberg, Ferdinand. 1988. *The Rich and the Super-Rich: A Study in the Power of Money Today.* Secaucus, NJ: Lyle Stuart.

Luo, Michael, and Megan Thee-Brenan. 2009. "Emotional Havoc Wreaked on Workers and Families." *New York Times* (Dec. 15): A1, A26.

Lupton, Deborah. 1997. "Foucault and the Medicalisation Critique." In Alan Petersen and Robin Bunton (Eds.), *Foucault: Health and Medicine.* London: Routledge, pp. 94–110.

Lusca, Emanuel. 2008. "Race as a Social Construct." Anthropology.net (Oct. 1). Retrieved Feb. 9, 2013. Online: http://anthropology.net/2008/10/01/race-as-a-social-construct

Luscombe, Belinda. 2014. "There Is No Longer Any Such Thing as a Typical Family." *Time* (Sept. 4). Retrieved May 17, 2015. Online: http://time.com/3265733/nuclear-family-typical-society-parents-children-households-philip-cohen

Lyotard, Jean-Francois. 1984. *The Postmodern Condition.* Minneapolis: University of Minnesota Press.

Mack, Julie. 2010. "Charter Schools Split Along Racial Lines: New Study Finds Parents' Choices Accelerate Resegregation." Mlive.com (Feb. 15). Retrieved Apr. 25, 2011. Online: http://blog.mlive.com/kzgazette_impact/print.html?entry=/2010/02/charter_schools_split_along_ra.html

Mackin, Deborah. 2010. "BP Oil Spill, the Challenger and Columbia Explosions, and the Deaths on Mt. Everest: Key Decision Making Tragedies and What We Can Learn from Them." Retrieved Dec. 27, 2011. Online: www.newdirectionsconsulting.com/2010/06/bp-oil-spill-the-challenger-and-columbia-explosions-and-the-deaths-on-mt-everest-key-decision-making-tragedies-and-what-we-can-learn-from-them

Mahapatra, Rajesh. 2007. "Outsourced Jobs Take Toll on Indians' Health." *Austin American-Statesman* (Dec. 30): H1, H6.

Malinowski, Bronislaw. 1922. *Argonauts of the Western Pacific.* New York: Dutton.

Malthus, Thomas R. 1965. *An Essay on Population.* New York: Augustus Kelley (orig. pub. 1798).

Mangione, Jerre, and Ben Morreale. 1992. *La Storia: Five Centuries of the Italian American Experience.* New York: HarperPerennial.

Mann, Patricia S. 1994. *Micro-Politics: Agency in a Postfeminist Era.* Minneapolis: University of Minnesota Press.

Manning, Jennifer E. 2011. "Membership of the 112th Congress: A Profile." Congressional Research Service (Mar. 1). Retrieved Nov. 10, 2012. Online: www.senate.gov/reference/resources/pdf/R41647.pdf

Manning, Robert D. 2000. *Credit Card Nation: The Consequences of America's Addiction to Credit.* New York: Basic.

Marger, Martin N. 1994. *Race and Ethnic Relations: American and Global Perspectives.* Belmont, CA: Wadsworth.

—. 2015. *Race and Ethnic Relations: American and Global Perspectives* (10th ed.). Boston: Cengage Learning.

Martineau, Harriet. 1962. *Society in America* (edited, abridged). Garden City, NY: Doubleday (orig. pub. 1837).

Marx, Karl. 1967. *Capital: A Critique of Political Economy.* Ed. Friedrich Engels. New York: International (orig. pub. 1867).

Marx, Karl, and Friedrich Engels. 1967. *The Communist Manifesto.* New York: Pantheon (orig. pub. 1848).

—. 1970. *The German Ideology,* Part 1. Ed. C. J. Arthur. New York: International (orig. pub. 1845–1846).

Massey, Douglas J., G. Hugo Arango, A. Kowasuci, A. Pellegrino, and J. E. Taylor. 1993. "Theories of International Migration: A Review and Appraisal." *Population and Development Review,* 19: 431–466.

Mattingly, Beth, and Wendy A. Walsh. 2010. "Rural Families with a Child Abuse Report Are More Likely Headed by a Single Parent and Endure Economic and Family Stress." Carsey Institute Issue Brief no. 10 (Winter). Online: www.carseyinstitute.unh.edu/publications/FS-Mattingly-Childabuse.pdf

Mayo Clinic. 2011. "Infertility: Causes" (Sept. 9). Retrieved Jan. 5, 2012. Online: www.mayoclinic.com/health/infertility/DS00310/DSECTION=causes

McAdam, Doug. 1996. "Conceptual Origins, Current Problems, Future Directions." In Doug McAdam, John D. McCarthy, and Meyer N. Zald (Eds.), *Comparative Perspectives on Social Movements.* New York: Cambridge University Press, pp. 23–40.

McHale, Susan M., Ann C. Crouter, Ji-Yeon Kim, Linda M. Burton, Kelly D. Davis, Aryn M. Dotterer, and Dena P. Swanson. 2006. "Mothers' and Fathers' Racial Socialization in African American Families: Implications for Youth." *Child Development,* 77 (5): 1387–1402.

McKenzie, Roderick D. 1925. "The Ecological Approach to the Study of the Human Community." In Robert Park, Ernest Burgess, and Roderick D. McKenzie, *The City.* Chicago: University of Chicago Press.

McPhail, Clark, and Ronald T. Wohlstein. 1983. "Individual and Collective Behavior Within Gatherings, Demonstrations, and Riots." In Ralph H. Turner and James F. Short, Jr. (Eds.), *Annual Review of Sociology,* vol. 9. Palo Alto, CA: Annual Reviews, pp. 579–600.

McPherson, Barry D., James E. Curtis, and John W. Loy. 1989. *The Social Significance of Sport: An Introduction to the Sociology of Sport.* Champaign, IL: Human Kinetics.

Mead, George Herbert. 1934. *Mind, Self, and Society.* Chicago: University of Chicago Press.

Merton, Robert King. 1938. "Social Structure and Anomie." *American Sociological Review,* 3 (6): 672–682.

—. 1949. "Discrimination and the American Creed." In Robert M. MacIver (Ed.), *Discrimination and National Welfare.* New York: Harper & Row, pp. 99–126.

—. 1968. *Social Theory and Social Structure* (enlarged ed.). New York: Free Press.

Messenger, David. 2009. "Studies Show People More Stressed as Students Than at Other Stages of Life Due to Work, Relationships." Retrieved Feb. 8, 2010. Online: www.studlife.com/news/2009/10/30/studies-shows-people-more-stressed-as-students-than-at-other-stages-of-life-due-to-work-relationships

Meyer, Julie. 2012. *Centenarians: 2010.* U.S. Census Bureau (Dec.). Retrieved Mar. 3, 2013. Online: www.census.gov/prod/cen2010/reports/c2010sr-03.pdf

Michael, Robert T., John H. Gagnon, Edward O. Laumann, and Gina Kolata. 1994. *Sex in America.* Boston: Little, Brown.

Michels, Robert. 1949. *Political Parties.* Glencoe, IL: Free Press (orig. pub. 1911).

migrationpolicy.org. 2015. "Frequently Requested Statistics on Immigrants and Immigration in the United States." Migration Policy Institute (Feb. 26). Retrieved July 27, 2015. Online: www.migrationpolicy.org/article/frequently-requested-statistics-immigrants-and-immigration-united-states

Milgram, Stanley. 1963. "Behavioral Study of Obedience." *Journal of Abnormal and Social Psychology,* 67: 371–378.

—. 1967. "The Small World Problem." *Psychology Today,* 1: 61–67.

—. 1974. *Obedience to Authority.* New York: Harper & Row.

Miller, Casey, and Kate Swift. 1991. *Words and Women: New Language in New Times* (updated ed.). New York: HarperCollins.

Miller, Dan E. 1986. "Milgram Redux: Obedience and Disobedience in Authority Relations." In Norman K. Denzin (Ed.), *Studies in Symbolic Interaction.* Greenwich, CT: JAI, pp. 77–106.

Miller, Lisa. 2013. "The Retro Wife." *New York* (Mar. 25): 20–25, 78–79.

Miller, Seth. 2011. "The Biggest Addiction in College: Skipping Classes." EWU Admissions: Blogs for Future Eastern Eagles (Apr. 8). Retrieved Jan. 19, 2015. Online: http://sites.ewu.edu/admissions/2011/04/08/the-most-addictive-drug-in-college-skipping-class

Mills, C. Wright. 1956. *White Collar.* New York: Oxford University Press.

—. 1959a. *The Power Elite.* Fair Lawn, NJ: Oxford University Press.

—. 1959b. *The Sociological Imagination.* London: Oxford University Press.

—. 1976. *The Causes of World War Three.* Westport, CT: Greenwood.

Mills, Iain. 2010. "China Puts Healthcare Cart Before the Horse." *Asia Times Online* (Apr. 21). Retrieved May 13, 2011. Online: www.atimes.com/atimes/China/LD21Ad01.html

Minino, Arialdi M. 2013. "Death in the United States." NCHS Data Brief, No. 115 (Mar. 2013). U.S. Department of Health and Human Services. Retrieved May 7, 2013. Online: www.cdc.gov/nchs/data/databriefs/db115.pdf

Mollborn, Stefanie, and Jeff A. Dennis. 2011. "Investigating the Life Situations and Development of Teenage Mothers' Children: Evidence from the ECLS-B." *Population Research and Policy Review,* 31: 31–66.

Mollborn, Stefanie, and Peter J. Lovegrove. 2010. "How Teenage Fathers Matter for Children." *Journal of Family Issues,* 32: 3–30.

Monger, Randall, and James Yankay. 2014. "U.S. Lawful Permanent Residents: 2013." U.S. Department of Homeland Security (May). Retrieved June 12, 2015. Online: www.dhs.gov/sites/default/files/publications/ois_lpr_fr_2013.pdf

Mooney, Karen. 2011. "Raleigh Marchers Fight 'Re-segregation' Plan." abc.news.com (Feb. 12). Retrieved Apr. 25, 2011. Online: http://abcnews.go.com/Politics/raleigh-nc-marchers-fight-segregation-plan/story?id=12904201

Moore, Patricia, with C. P. Conn. 1985. *Disguised.* Waco, TX: Word.

Morgan, Sally. 2010. "Community College Changed My Life." BetterGrads.org (May 25). Retrieved Apr. 23, 2011. Online: http://bettergrads.org/blog/2010/05/25/why-college-part-7-%E2%80%93-community-college-changed-my-life

Morselli, Henry. 1975. *Suicide: An Essay on Comparative Moral Statistics.* New York: Arno (orig. pub. 1881).

MOWAA (Meals on Wheels Association of America). 2011. "Take Action: Volunteer." Retrieved June 15, 2011. Online: www.mowaa.org/page.aspx?pid=396

Muehlfried, Florian. 2007. "Sharing the Same Blood—Culture and Cuisine in the Republic of Georgia." *Anthropology of Food* (Dec.). Retrieved Oct. 13, 2012. Online: http://aof.revues.org/2342?&id=2342#text

Mueller, Anna S., Seth Abrutyn, and Cynthia Stockton. 2015. "Can Social Ties Be Harmful? Examining the Spread of Suicide in Early Adulthood." *Social Perspectives* (forthcoming). Retrieved Jan. 4, 2015. Online: www.memphis.edu/sociology/pdfs/mueller-abrutyn-stockton-sociological-perspectives.pdf

Murdock, George P. 1945. "The Common Denominator of Cultures." In Ralph Linton (Ed.), *The Science of Man in the World Crisis.* New York: Columbia University Press, pp. 123–142.

Murray, Jane, Rick Linden, and Diana Kendall. 2012. *Sociology in Our Times: The Essentials* (5th Canadian ed.). Toronto, CA: Nelson/Cengage.

Nagourney, Adam. 2015. "In U.C.L.A. Debate Over Jewish Student, Echoes on Campus of Old Biases." *New York Times* (Mar. 5). Retrieved Mar. 20, 2015. Online: www.nytimes.com/2015/03/06/us/debate-on-a-jewish-student-at-ucla.html

NASA. 2015. "Global Climate Change: The Current and Future Consequences of Global Change." Retrieved June 17, 2015. Online: http://climate.nasa.gov/effects

National Alliance for Public Charter Schools. 2013. "Back to School Tallies: Estimated Number of Public Charter Schools and Students, 2012–2013" (Jan.). Retrieved Apr. 11, 2013. Online: www.publiccharters.org/data/files/Publication_docs/NAPCS%202011-12%20New%20and%20Closed%20Charter%20Schools_20111206T125251.pdf

National Center for Educational Statistics. 2012a. *The Condition of Education 2012* (May). Retrieved Apr. 13, 2013. Online: http://nces.ed.gov/pubs2012/2012045.pdf

—. 2012b. *Indicators of School Crime and Safety: 2011* (Feb.). Retrieved Apr. 13, 2013. Online: www.bjs.gov/content/pub/pdf/iscs11.pdf

—. 2014. "Indicators of School Crime and Safety: 2013." Retrieved May 30, 2015. Online: https://nces.ed.gov/fastfacts/display.asp?id=49

National Center for Health Statistics. 2012. "Prevalence of Obesity in the United States, 2009–2010" (Jan.). Retrieved May 12, 2013. Online: www.cdc.gov/nchs/data/databriefs/db82.pdf

—. 2014. *Health, United States, 2013.* Retrieved Feb. 28, 2015. Online: www.cdc.gov/nchs/data/hus/hus13.pdf

National Center for Victims of Crime. 2011. "Elder Victimization." Retrieved June 28, 2013. Online: http://victimsofcrime.org/library/crime-information-and-statistics/elder-victimization

National Center on Elder Abuse. 2013. "What Is Elder Abuse?" Department of Health and Human Services. Retrieved June 28, 2013. Online: www.ncea.aoa.gov/faq/index.aspx

National Coalition for the Homeless. 2015. "State of the Homeless: 2014." Retrieved Feb. 1, 2015. Online: www.coalitionforthehomeless.org/wp-content/uploads/2014/03/StateoftheHomeless20141.pdf

National Collegiate Athletic Association (NCAA). 2012. "Estimated Probability of Competing in Athletics Beyond the High School Interscholastic Level." Retrieved Apr. 28, 2012. Online: www.ncaa.org/wps/wcm/connect/public/test/issues/recruiting/probability+of+going+pro

National Committee to Preserve Social Security and Medicare. 2015. "Social Security Is Important to Hispanic Americans." Retrieved Apr. 25, 2015. Online: www.ncpssm.org/SocialSecurity/HispanicAmericansandSS

National Education Association. 2015a. "The Condition of Education 2015." Retrieved May 29, 2015. Online: https://nces.ed.gov/programs/coe/indicator_coj.asp

—. 2015b. "Rankings & Estimates: Rankings of the States 2014 and Estimates of School Statistics 2015" (Mar.). Retrieved May 29, 2015. Online: www.nea.org/assets/docs/NEA_Rankings_And_Estimates-2015-03-11a.pdf

National Institute of Justice. 2011. "Transnational Organized Crime." Retrieved Feb. 9, 2011. Online: www.ojp.usdoj.gov/nij/topics/crime/transnational-organized-crime/welcome.htm

National League of Cities. 2010. "Significant Budget Shortfalls Could Mean More Job Losses" (May 24). National League of Cities. Retrieved June 10, 2010. Online: www.ncl.org.PRESSROOM/PRESSRELEASEITEMS/SoACJobsEcon5.10.aspx

National Marriage Project. 2010. "When Marriage Disappears: The New Middle America." Retrieved Apr. 18, 2011. Online: www.virginia.edu/marriageproject/pdfs/Union_11_12_10.pdf

National Survey of Sexual Health and Behavior. 2015. Indiana University. Retrieved May 27, 2015. Online: www.nationalsexstudy.indiana.edu

National Vital Statistics Report. 2014. "International Comparisons of Infant Mortality and Related Factors: United States and Europe, 2010." Retrieved June 16, 2015. Online: www.cdc.gov/nchs/data/nvsr/nvsr63/nvsr63_05.pdf

—. 2015. "Births: Final Data for 2013" (Jan. 15). Retrieved July 23, 2015. Online: www.cdc.gov/nchs/data/nvsr/nvsr64/nvsr64_01.pdf

National Women's Law Center. 2013. "Insecure and Unequal: Poverty and Income Among Women and Families 2000–2012." Retrieved Apr. 25, 2015. Online: www.nwlc.org/sites/default/files/pdfs/final_2013_nwlc_povertyreport.pdf

Navarrette, Ruben, Jr. 1997. "A Darker Shade of Crimson." In Diana Kendall (Ed.), *Race, Class, and Gender in a Diverse Society.* Boston: Allyn & Bacon, 1997: 274–279. Reprinted from Ruben Navarrette, Jr., *A Darker Shade of Crimson.* New York: Bantam, 1993.

NDTV.com. 2012. "Suicide Rates in India Are Highest in the 15–29 Age Group: Report." Retrieved Oct. 12, 2012. Online: www.ndtv.com/article/india/suicide-rates-in-india-are-highest-in-the-15-29-age-group-report-234986

New York Times. 2015. "Italy: Politicians Vow to Spurn Migrants." Nytimes.com (June 7). Retrieved June 16, 2015. Online: www.nytimes.com/2015/06/08/world/europe/italy-politicians-vow-to-spurn-migrants.html

Newport, Frank. 2011. "For First Time, Majority of Americans Favor Legal Gay Marriage." Gallup Organization (May 20). Retrieved Jan. 1, 2012. Online: www.gallup.com/poll/147662/first-time-majority-americans-favor-legal-gay-marriage.aspx

—. 2012. "In U.S., 77% Identify as Christian." Gallup.com (Dec. 24). Retrieved Apr. 16, 2013. Online: www.gallup.com/poll/159548/identify-christian.aspx

Ng, Christina. 2013. "Rebecca Sedwick Suicide Part of 'Cool to be Cruel' Cyber Culture."

ABCnews.com (Oct. 16). Retrieved Dec. 9, 2013. Online: http://abcnews.go.com/US/rebecca-sedwick-suicide-part-cool-cruel-cyber-culture/story?id=20588017

Nielsen. 2013. "The Teen Transition: Adolescents of Today, Adults of Tomorrow." Retrieved Jan. 18, 2015. Online: www.nielsen.com/us/en/insights/news/2013/the-teen-transition--adolescents-of-today--adults-of-tomorrow.html

NIH. 2012. "Regular Marijuana Use by Teens Continues to Be a Concern." National Institute of Health (Dec. 19). Retrieved May 7, 2013. Online: www.nih.gov/news/health/dec2012/nida-19.htm

Noel, Donald L. 1972. *The Origins of American Slavery and Racism.* Columbus, OH: Merrill.

Nolan, Patrick, and Gerhard E. Lenski. 2010. *Human Societies: An Introduction to Macrosociology* (11th ed.). New York: Oxford University Press.

NOW Foundation. 2015. "About 'Love Your Body Day.'" Retrieved Apr. 14, 2015. Online: http://now.org/now-foundation/love-your-body/love-your-body-whats-it-all-about/about

nytimes.com/interactive. 2015. "What's Behind the Surge in Refugees Crossing the Mediterranean Sea?" *New York Times* (May 21). Retrieved June 16, 2015. Online: www.nytimes.com/interactive/2015/04/20/world/europe/surge-in-refugees-crossing-the-mediterranean-sea-maps.html

Obama, Barack. 2015. "Remarks by the President at the 50th Anniversary of the Selma to Montgomery Marches." Whitehouse.gov (Mar. 7). Retrieved Mar. 8, 2015. Online: www.printfriendly.com/print?url=http%3A%2F%2Fwh.gov%2FiD3l4%23.VPyZeeM4qXI.printfriendly&title=Remarks+by+the+President+at+the+50th+Anniversary+of+the+Selma+to+Montgomery+Marches+%7C+The+White+House#

Obamacare Facts. 2015. "Obamacare Enrollment Numbers." Retrieved June 7, 2015. Online: http://obamacarefacts.com/sign-ups/obamacare-enrollment-numbers

Obesity Action Coalition. 2015. "Understanding Obesity." Retrieved July 17, 2015. Online: www.obesityaction.org/understanding-obesity

Ogburn, William F. 1966. *Social Change with Respect to Culture and Original Nature.* New York: Dell (orig. pub. 1922).

Ogden, Jane, and Talin Avades. 2011. "Being Homeless and the Use and Nonuse of Services: A Qualitative Study." *Journal of Community Psychology,* 39 (4): 499–505.

Omi, Michael, and Howard Winant. 1994. *Racial Formation in the United States: From the 1960s to the 1990s.* New York: Routledge.

—. 2013. *Racial Formation in the United States: From the 1960s to the 1990s* (3rd ed.). New York: Routledge.

Orfield, Gary, John Kucsera, and Genevieve Siegel-Hawley. 2012. "E Pluribus . . . Separation: Deepening Double Segregation for More Students" (Sept.). Retrieved Apr. 13, 2013. Online: http://civilrightsproject.ucla.edu/research/k-12-education/integration-and-diversity/mlk-national/e-pluribus . . . separation-deepening-double-segregation-for-more-students/orfield_epluribus_revised_omplete_2012.pdf

Ortman, Jennifer M., Victoria A. Velkoff, and Howard Hogan. 2014. "An Aging Nation: The Older Population in the United States" (May). U.S. Census Bureau. Retrieved Apr. 18, 2015. Online: www.census.gov/prod/2014pubs/p25-1140.pdf

Ott, Thomas. 2011. "Cleveland Students Hold Their Own with Voucher Students on State Tests." Cleveland.com (Feb. 22). Retrieved Apr. 27, 2011. Online: http://blog.cleveland.com/metro/2011/02/cleveland_students_hold_own_wi.html

Oxendine, Joseph B. 2003. *American Indian Sports Heritage* (rev. ed.). Lincoln: University of Nebraska Press.

Padilla, Felix M. 1993. *The Gang as an American Enterprise.* New Brunswick, NJ: Rutgers University Press.

Page, Charles H. 1946. "Bureaucracy's Other Face." *Social Forces,* 25 (October): 89–94.

Paloian, Andrea. 2015. "The Female/Athlete Paradox: Managing Traditional Views of Masculinity and Femininity." *Opus* (Steinhardt Department of Applied Psychology, New York University). Retrieved Apr. 11, 2015. Online: http://steinhardt.nyu.edu/opus/issues/2012/fall/female

Parenti, Michael. 1998. *America Besieged.* San Francisco: City Lights.

—. 2007. *Contrary Notions: The Michael Parenti Reader.* San Francisco: City Lights.

Park, Robert E. 1915. "The City: Suggestions for the Investigation of Human Behavior in the City." *American Journal of Sociology,* 20: 577–612.

—. 1928. "Human Migration and the Marginal Man." *American Journal of Sociology,* 33.

—. 1936. "Human Ecology." *American Journal of Sociology,* 42: 1–15.

Park, Robert E., and Ernest W. Burgess. 1921. *Human Ecology.* Chicago: University of Chicago Press.

Parker, Kim. 2011. "A Portrait of Stepfamilies." Pew Research Center (Jan. 13). Retrieved Jan. 1, 2012. Online: www.pewsocialtrends.org/2011/01/13/a-portrait-of-stepfamilies

—. 2013. "Modern Parenthood; Roles of Moms and Dads Converge as They Balance Work and Family." Pew Research Social and Demographic Trends (Mar. 14). Retrieved Feb. 11, 2014. Online: www.pewsocialtrends.org/2013/03/14/modern-parenthood-roles-of-moms-and-dads-converge-as-they-balance-work-and-family

Parker-Pope, Tara. 2011. "Fat Stigma Spreads Around the Globe." *New York Times* (Mar. 30). Retrieved Apr. 15, 2015. Online: http://well .blogs.nytimes.com/2011/03/30/spreading -fat-stigma-around-the-globe/?_r=0

Parsons, Talcott. 1951. *The Social System*. Glencoe, IL: Free Press.

—. 1955. "The American Family: Its Relations to Personality and to the Social Structure." In Talcott Parsons and Robert F. Bales (Eds.), *Family, Socialization and Interaction Process*. Glencoe, IL: Free Press, pp. 3–33.

—. 1960. "Toward a Healthy Maturity." *Journal of Health and Social Behavior*, 1: 163–173.

Passel, Jeffrey S., and D'Vera Cohn. 2011. "Unauthorized Immigrant Population: National and State Trends, 2010." Pew Hispanic Research Center. Retrieved Feb. 5, 2012. Online: www.pewhispanic.org

Passel, Jeffrey S., Wendy Wang, and Paul Taylor. 2010. "Marrying Out: One-in-Seven New U.S. Marriages Is Interracial or Interethnic." Pew Research Center (June 4). Retrieved Dec. 22, 2011. Online: http://pewresearch center.org

pbs.org. 2012. "Health Costs: How the U.S. Compares with Other Countries" (Oct. 22). Retrieved July 8, 2013. Online: www.pbs.org /newshour/rundown/2012/10/health-costs -how-the-us-compares-with-other-countries .html

Pearce, Diana. 1978. "The Feminization of Poverty: Women, Work, and Welfare." *Urban and Social Change Review*, 11 (1/2): 28–36.

Peck, Emily. 2015. "Proof That Working From Home Is Here to Stay: Even Yahoo Does It." Huffington Post Business (Mar. 18). Retrieved Apr. 18, 2015. Online: www .huffingtonpost.com/2015/03/18/the-future -is-happening-now-ok_n_6887998.html

Peitzman, Louis. 2013. "It Gets Better, Unless You're Fat." Buzzfeed.com (Oct. 10). Retrieved Apr. 2, 2015. Online: www.buzzfeed .com/louispeitzman/it-gets-better-unless -youre-fat?bffb#.ta4wp5rzw

People's Daily Online. 2012. "China's River Pollution a Threat to People's Lives" (Feb. 17). Retrieved Mar. 17, 2012. Online: http:// english.people.com.cn/90882/7732438.html

Perlstadt, Harry. 2007. "Applied Sociology." In Clifton D. Bryant and Dennis L. Peck (Eds.), *Handbook of 21st Century Sociology*, Vol. 2., pp. 341–353. Thousand Oaks, CA: Sage.

Peterson, Robert. 1992. *Only the Ball Was White: A History of Legendary Black Players and All-Black Professional Teams*. New York: Oxford University Press (orig. pub. 1970).

Pew Charitable Trusts. 2013. "Cities Squeezed by Pension and Retiree Health Care Shortfalls" (Mar.). Retrieved May 16, 2013. Online: www .pewstates.org/uploadedFiles/PCS_Assets /2013/Pew_city_pensions_brief.pdf

Pew Forum on Religion and Public Life. 2011. "Global Christianity: A Report on the Size and Distribution of the World's Christian Population" (Dec. 19). Retrieved July 8, 2013. Online: www.pewforum.org /Christian/Global-Christianity-exec.aspx

—. 2012a. "The Global Religious Landscape" (Dec. 16). Retrieved Apr. 19, 2013. Online: www.pewforum.org/uploadedFiles/Topics /Religious_Affiliation/globalReligion-full.pdf

—. 2012b. "'Nones' on the Rise" (Oct. 9). Retrieved Apr. 19, 2013. Online: www .pewforum.org/uploadedFiles/Topics /Religious_Affiliation/Unaffiliated /NonesOnTheRise-full.pdf

Pew Research Center. 2012a. "Hispanics of Puerto Rican Origin in the United States, 2012" (June 27). Retrieved Apr. 6, 2013. Online: www.pewhispanic.org/2012/06/27 /hispanics-of-puerto-rican-origin-in-the -united-states-2010

—. 2012b. "Majority of Public Agrees Labor Unions Needed" (Aug. 30). Retrieved Mar. 11, 2013. Online: www.pewresearch .org/daily-number/majority-of-public -agrees-labor-unions-needed

—. 2012c. "The Rise of Asian Americans" (July 12). Retrieved Apr. 6, 2013. Online: www .pewsocialtrends.org/files/2013/04/Asian -Americans-new-full-report-04-2013.pdf

—. 2013a. "Gay Marriage: Key Data Points from Pew Research." Retrieved Apr. 7, 2013. Online: www.pewresearch.org/2013/03/21 /gay-marriage-key-data-points-from-pew -research

—. 2013b. "Protecting the Environment Ranks in the Middle of Public's Priorities for 2013" (Apr. 22). Retrieved May 23, 2013. Online: www.pewresearch.org/daily-number /protecting-the-environment-ranks-in-the -middle-of-publics-priorities-for-2013

—. 2013c. "The State of the News Media: The Media and Campaign 2012." Retrieved May 27, 2014. Online: http://stateofthemedia. org/2013/special-reports-landing-page /the-media-and-campaign-2012

—. 2014. "Wealth Inequality Has Widened Along Racial, Ethnic Lines Since End of Great Recession" (Dec. 12). Retrieved Feb. 28, 2015. Online: www.pewresearch .org/fact-tank/2014/12/12/racial-wealth -gaps-great-recession

—. 2015a. "Changing Attitudes on Gay Marriage" (June 8). Retrieved July 23, 2015. Online: www.pewforum.org/2015/06/08 /graphics-slideshow-changing-attitudes -on-gay-marriage

—. 2015b. "Indian Americans." Retrieved Apr. 7, 2015. Online: www.pewsocialtrends.org /asianamericans-graphics/indians

—. 2015c. "Public Policy Priorities Reflect Changing Conditions at Home and Abroad" (Jan.). Retrieved June 20, 2015. Online: www.people-press.org/files/2015/01/01 -15-15-Policy-Priorities-Release.pdf

Pew Research Center Global Attitudes Project. 2011. "The American–Western European Values Gap" (Nov. 17). Retrieved Apr. 4, 2012. Online: www.pewglobal .org/2011/11/17/the-american-western -european-values-gap

Pew Research Center Internet, Science and Tech. 2015. "Mobile Access Shifts Social Media Use and Other Online Activities" (Apr. 9). Retrieved Apr. 11, 2015. Online: www.pewinternet.org/2015/04/09/teens -social-media-technology-2015

Pew Research Center Journalism and Media (PRCPEJ). 2015. "State of the News Media 2014." Retrieved July 20, 2015. Online: www .journalism.org/packages/state-of-the -news-media-2014

Pew Research Center on Religion and Public Life. 2014. "Global Religious Diversity" (Apr. 4). Retrieved June 1, 2015. Online: www.pewforum.org/2014/04/04/global -religious-diversity

Pew Research Center Social and Demographic Trends. 2013. "The Rise of Asian Americans" (Apr. 4). Original, June 19, 2012; updated edition, Apr. 4, 2013. Retrieved July 16, 2015. Online: www.pewsocialtrends.org/2012/06/19 /the-rise-of-asian-americans

Pew Research Center, 2014 Religious Landscape Study. 2015. "Christians Decline as Share of U.S. Population; Other Faiths and the Unaffiliated Are Growing" (May 7). Retrieved June 1, 2015. Online: www.pewforum.org /2015/05/12/americas-changing-religious -landscape/pr_15-05-12_rls-00

Phillips, John C. 1993. *Sociology of Sport*. Boston: Allyn & Bacon.

Piaget, Jean. 1954. *The Construction of Reality in the Child*. Trans. Margaret Cook. New York: Basic.

Picca, Leslie Houts, and Joe R. Feagin. 2007. *Two-Faced Racism: Whites in the Backstage and Frontstage*. New York: Routledge.

Pierre-Pierre, Garry. 1997. "Traditional Church's New Life." *New York Times* (Nov. 15): A11.

Pillsbury, George, and Julian Johannesen. 2012. "America Goes to the Polls 2012: A Report on Voter Turnout in the 2012 Presidential Election." Retrieved Mar. 24, 2013. Online: www.nonprofitvote.org/voter-turnout.html

Pines, Maya. 1981. "The Civilizing of Genie." *Psychology Today*, 15 (September): 28–29, 31–32, 34.

Plunkett Research. 2014. "Health Expenditures and Services in the U.S." (Nov. 11). Retrieved June 7, 2015. Online: www .plunkettresearch.com/trends-analysis /health-care-medical-business-market

Population Reference Bureau. 2015. "2014 World Population Data Sheet." Retrieved June 11, 2015. Online: www.prb.org/pdf14/2014 -world-population-data-sheet_eng.pdf

Postman, Joel. 2011. "MLK: From a Thing-Oriented Society to a Person-Oriented One." *Social Kapital*. Retrieved Apr. 20, 2011 . Online: http://socialized.tumblr.com/post /71619428/mlk-from-a-thing-oriented -society-to-a-person-oriented

Postman, Neil, and Steve Powers. 2008. *How to Watch TV News* (rev. ed.). New York: Penguin.

Powell, Michael. 2004. "Evolution Shares a Desk with 'Intelligent Design.'" *Washington Post* (Dec. 26): A1.

Prensky, Marc. 2001. "Digital Natives, Digital Immigrants." *On the Horizon* (MCB University Press, Oct. 2001). Retrieved Jan. 22, 2015. Online: www.marcprensky.com/writing /Prensky%20-%20Digital%20Natives,%20 Digital%20Immigrants%20-%20Part1.pdf

prideagenda.org. 2014. "New Report Highlights Eating Disorders Among Gay Men." Retrieved Apr. 6, 2015. Online: https:// prideagenda.org/news/2014-03-21-new -report-highlights-eating-disorders -among-gay-men

Project Censored. 2015. "Top 25 Most Censored Stories of 2013–2014." Retrieved May 16, 2016. Online: www.projectcensored.org /category/top-25-censored-stories-of-2014

ProQuest Statistical Abstract of the United States. 2015. "Section 2, Births, Deaths, Marriages, and Divorces." Retrieved July 23, 2015. Online: http://statabs.proquest.com.ezproxy .baylor.edu/ftv2/4c4e0000020f0f7f.pdf

Protess, Ben, and Jessica Silver-Greenberg, 2014. "Credit Suisse Pleads Guilty in Felony Case." *New York Times* (May 19). Retrieved Feb. 22, 2015. Online: http://dealbook .nytimes.com/2014/05/19/credit-suisse-set -to-plead-guilty-in-tax-evasion-case/?_r=0

Putnam, Robert D., and David E. Campbell. 2010. *American Grace: How Religion Divides and Unites Us*. New York: Simon & Schuster.

Quinney, Richard. 2001. *Critique of the Legal Order*. Piscataway, NJ: Transaction (orig. pub. 1974).

Reckless, Walter C. 1967. *The Crime Problem*. New York: Meredith.

Rehavi, M. Marit, and Sonia B. Starr. 2012. "Racial Disparity in Federal Criminal Charging and Its Sentencing Consequences." University of Michigan Law School, Empirical Legal Studies Center Paper #12-002. Retrieved May 13, 2013. Online: http://papers.ssrn .com/sol3/papers.cfm?abstract_id=1985377

Reich, Robert B. 2010. *Aftershock: The Next Economy and America's Future*. New York: Knopf.

Reilly, Corinne. 2014. "Navy: Store Guns of Sailors at Risk of Suicide." Military.com (Nov. 21). Retrieved Jan. 2, 2015. Online: www.military .com/daily-news/2014/11/21/navy-store -guns-of-sailors-at-risk-of-suicide .html?comp=700001075741&rank=4

Reiman, Jeffrey, and Paul Leighton. 2010. *The Rich Get Richer and the Poor Get Prison: Ideology, Class, and Criminal Justice*. Upper Saddle River, NJ: Pearson.

Resmovits, Joy. 2012. "Charter Schools Fall Short on Students with Disabilities." Huffingtonpost.com (June 19). Retrieved Apr. 13, 2013. Online: www.huffingtonpost .com/2012/06/19/charter-schools -disabilities-_n_1610744.html

Reuters News Service. 2008. "New mtvU and Associated Press Poll Shows How Stress, War, the Economy, and Other Factors Are Affecting College Students' Mental Health." Retrieved Feb. 8, 2010. Online: www .reuters.com/article/idUS173716+19-Mar -2008+PRN20080319

Rigler, David. 1993. "Letters: A Psychologist Portrayed in a Book About an Abused Child Speaks Out for the First Time in 22 Years." *New York Times Book Review* (June 13): 35.

Ritzer, George. 1997. *Postmodern Society Theory*. New York: McGraw-Hill.

—. 1998. *The McDonaldization Thesis*. London: Sage.

—. 1999. *Enchanting a Disenchanted World: Revolutionizing the Means of Consumption*. Thousand Oaks, CA: Pine Forge.

—. 2011. *Sociological Theory* (8th ed.). New York: McGraw-Hill.

—. 2014. *The McDonaldization of Society* (8th ed.). Thousand Oaks, CA: Sage.

Roberts, John. 2011. "Marine Corps Steps Up Suicide Prevention Efforts to Halt Deadly Trend" (Mar. 30). Retrieved Oct. 13, 2012. Online: www.foxnews.com/us/2011/03/30 /marine-corps-steps-suicide-prevention -efforts-halt-deadly-trend

Roberts, Sam. 2010. "Study Finds Cohabiting Doesn't Make a Union Last." *New York Times* (Mar. 2). Retrieved Feb. 11, 2012. Online: www.nytimes.com/2010/03/03 /us/03marry.html

Robertson, Campbell, Shaila Dewan, and Matt Apuzzo. 2015. "Ferguson Became Symbol, but Bias Knows No Border." *New York Times* (Mar. 7). Retrieved July 16, 2015. Online: www.nytimes.com/2015/03/08/us /ferguson-became-symbol-but-bias-knows -no-border.html

Robnett, Rachael D., and Joshue E. Susskind. 2010. "Who Cares About Being Gentle? The Impact of Social Identity and the Gender of One's Friends on Children's Display of Same-Gender Favoritism." *Sex Roles*, 64 (1/2): 90–102.

Roethlisberger, Fritz J., and William J. Dickson. 1939. *Management and the Worker*. Cambridge, MA: Harvard University Press.

Rollins, Judith. 1985. *Between Women: Domestics and Their Employers*. Philadelphia: Temple University Press.

Ropers, Richard H. 1991. *Persistent Poverty: The American Dream Turned Nightmare*. New York: Plenum.

Rose, Stephen J., and Heidi Hartman. 2008. "Still a Man's Labor Market: The Long-Term Earnings Gap." IWPR #C366 (Feb.). New York: Institute for Women's Policy Research.

Ross, Dorothy. 1991. *The Origins of American Social Science*. Cambridge, England: Cambridge University Press.

Ross, Sally R., and Kimberly J. Shinew. 2008. "Perspectives of Women College Athletes on Sport and Gender." *Sex Roles*, 58 (1): 40–57.

Rostow, Walt W. 1971. *The Stages of Economic Growth: A Non-Communist Manifesto* (2nd ed.). Cambridge: Cambridge University Press (orig. pub. 1960).

—. 1978. *The World Economy: History and Prospect*. Austin: University of Texas Press.

Rousseau, Ann Marie. 1981. *Shopping Bag Ladies: Homeless Women Speak About Their Lives*. New York: Pilgrim.

Ruth, Jennifer. 2015. "Remaking the Public U's Professoriate" (Apr. 2). Retrieved Apr. 28, 2015. Online: http://utotherescue.blogspot .com/2015/04/remaking-public-us -professoriate.html

Ryan, Camille. 2013. "Language Use in the United States: 2011." U.S. Census Bureau (Aug.). Retrieved Dec. 16, 2013. Online: www .census.gov/prod/2013pubs/acs-22.pdf

Rymer, Russ. 1993. *Genie: An Abused Child's Flight from Silence*. New York: HarperCollins.

Saad, Lydia. 2011. "Doctor-Assisted Suicide Is Moral Issue Dividing Americans Most." Gallup Organization (May 31). Retrieved Jan. 3, 2012. Online: www.gallup.com /poll/147842/Doctor-Assisted-Suicide -Moral-Issue-Dividing-Americans.aspx

—. 2015. "Beer Is Americans' Adult Beverage of Choice This Year." Gallup.com (June 23). Retrieved June 5, 2015. Online: www.gallup .com/poll/174074/beer-americans-adult -beverage-choice-year.aspx?version=print

Sadker, David, and Karen Zittleman. 2009. *Still Failing at Fairness: How Gender Bias Cheats Boys and Girls in Schools*. New York: Simon and Schuster.

Salcito, Jordan. 2014. "The Rebirth of Seriously Good Georgian Wine." Thedailybeast.com (Feb. 15). Retrieved Jan. 4, 2015. Online: www.thedailybeast.com/articles /2014/02/15/the-rebirth-of-seriously -good-georgian-wines.html

Sapir, Edward. 1961. *Culture, Language and Personality*. Berkeley: University of California Press.

SAT. 2014. "2014 College-Bound Seniors: Total Group Profile Report." Retrieved Apr. 27, 2015. Online: https://secure-media .collegeboard.org/digitalServices/pdf/sat /TotalGroup-2014.pdf

Saulny, Susan. 2011a. "Black? White? Asian? More Young Americans Choose All of the Above." *New York Times* (Jan. 29). Retrieved Mar. 7, 2011. Online: www.nytimes .com/2011/01/30/us/30mixed.html

—. 2011b. "In a Multiracial Nation, Many Ways to Tally." *New York Times* (Feb. 10): A1, A17.

Schaefer, Richard T., and William W. Zellner. 2010. *Extraordinary Groups: An Examination of Unconventional Lifestyles* (9th ed.). New York: Worth.

Scheiber, Noam. 2015. "In Test for Unions and Politicians, a Nationwide Protest on Pay" (Apr. 16). Retrieved May 13, 2015. Online: www.nytimes.com/2015/04/16/business/economy/in-test-for-unions-and-politicians-a-nationwide-protest-on-pay.html?_r=0

Schneider, Donna. 1995. *American Childhood: Risks and Realities.* New Brunswick, NJ: Rutgers University Press.

Schor, Juliet B. 1999. *The Overspent American: Upscaling, Downshifting, and the New Consumer.* New York: HarperPerennial.

Schulte, Brigid. 2014. "'The Second Shift' at 25: Q&A with Arlie Hochschild." *Washington Post* (Aug. 6). Retrieved Apr. 14, 2015. Online: www.washingtonpost.com/blogs/she-the-people/wp/2014/08/06/the-second-shift-at-25-q-a-with-arlie-hochschild

Schur, Edwin M. 1983. *Labeling Women Deviant: Gender, Stigma, and Social Control.* Philadelphia: Temple University Press.

Schwarz, John E., and Thomas J. Volgy. 1992. *The Forgotten Americans.* New York: Norton.

ScienceDaily. 2008. "Three Out of Four American Women Have Disordered Eating, Survey Suggests" (Apr. 22). Retrieved Jan. 30, 2012. Online: www.sciencedaily.com/releases/2008/04/080422202514.htm

—. 2011. "Educational Development Stunted by Teenage Fatherhood" (Mar. 30). Retrieved Feb. 12, 2012. Online: www.sciencedaily.com/releases/2011/03/110330094353.htm

Seaman, Andrew M. 2015. "Transgender People Face Discrimination in Health Care" (Mar. 13). Retrieved Apr. 5, 2015. Online: www.reuters.com/article/2015/03/13/us-transgender-healthcare-discrimination-idUSKBN0M928B20150313

Searcey, Dionne, and Robert Gebeloff. 2015. "Middle Class Shrinks Further as More Fall Out Instead of Climbing Up." *New York Times* (Jan. 25). Online: www.nytimes.com/2015/01/26/business/economy/middle-class-shrinks-further-as-more-fall-out-instead-of-climbing-up.html?_r=0

Seligman, Katherine. 2009. "Social Isolation a Significant Health Issue." *SFGATE* (Mar. 2). Retrieved Jan. 22, 2015. Online: www.sfgate.com/health/article/Social-isolation-a-significant-health-issue-3249234.php

Sengupta, Somini. 1997. "At Holidays, Test of Patience of Muslims." *New York Times* (Dec. 25): A12.

Senior, Jennifer. 2014. "Why Mom's Time Is Different from Dad's Time." *Wall Street Journal* (Jan. 24). Retrieved May 28, 2015. Online: www.wsj.com/articles/SB10001424052702304757004579335053525792432

Shapiro, Thomas, Tatjana Meschede, and Sam Osoro. 2013. "The Roots of the Widening Racial Wealth Gap: Explaining the Black–White Economic Division." Institute on Assets and Social Policy (Feb.). Retrieved Jan. 13, 2014. Online: http://iasp.brandeis.edu/pdfs/Author/shapiro-thomas-m/racialwealthgapbrief.pdf

Shawver, Lois. 1998. "Notes on Reading Foucault's *The Birth of the Clinic.*" Retrieved Oct. 2, 1999. Online: www.california.com/~rathbone/foucbc.htm

Shevky, Eshref, and Wendell Bell. 1966. *Social Area Analysis: Theory, Illustrative Application and Computational Procedures.* Westport, CT: Greenwood.

Shierholz, Heidi. 2010. "Immigration and Wages—Methodological Advancements Confirm Modest Gains for Native Workers." EPI Briefing Paper #255 (Feb. 4). Retrieved Feb. 5, 2012. Online: www.epi.org

Simmel, Georg. 1950. *The Sociology of Georg Simmel.* Trans. Kurt Wolff. Glencoe, IL: Free Press (orig. written in 1902–1917).

—. 1957. "Fashion." *American Journal of Sociology,* 62 (May 1957): 541–558. Orig. pub. 1904.

—. 1990. *The Philosophy of Money.* Ed. David Frisby. New York: Routledge (orig. pub. 1907).

Simon, David R. 2012. *Elite Deviance* (10th ed.). Upper Saddle River, NJ: Pearson.

Sjoberg, Gideon. 1965. *The Preindustrial City: Past and Present.* New York: Free Press.

Smelser, Neil J. 1963. *Theory of Collective Behavior.* New York: Free Press.

—. 1988. "Social Structure." In Neil J. Smelser (Ed.), *Handbook of Sociology.* Newbury Park, CA: Sage, pp. 103–129.

Smith, Aaron. 2014. "Older Adults and Technology Use." Pew Research Center (Apr. 3). Retrieved Apr. 25, 2015. Online: www.pewinternet.org/files/2014/04/PIP_Seniors-and-Tech-Use_040314.pdf

Smith, Adam. 1976. *An Inquiry into the Nature and Causes of the Wealth of Nations.* Ed. Roy H. Campbell and Andrew S. Skinner. Oxford, England: Clarendon (orig. pub. 1776).

Smith, Cooper. 2014. "The Surprising Facts About Who Shops Online and on Mobile." BusinessInsider.com (Dec. 10). Retrieved Dec. 30, 2014. Online: www.businessinsider.com/the-surprising-demographics-of-who-shops-online-and-on-mobile-2014-6

Smith, Huston. 1991. *The World's Religions.* San Francisco: HarperSanFrancisco.

Smith, Jessica C., and Carla Medalia. 2014. "Health Insurance Coverage in the United States: 2013" (Sept.). Retrieved July 7, 2015. Online: www.census.gov/content/dam/Census/library/publications/2014/demo/p60-250.pdf

Snow, David A., and Leon Anderson. 1991. "Researching the Homeless: The Characteristic Features and Virtues of the Case Study." In Joe R. Feagin, Anthony M. Orum, and Gideon Sjoberg (Eds.), *A Case for the Case Study.* Chapel Hill: University of North Carolina Press, pp. 148–173.

—. 1993. *Down on Their Luck: A Case Study of Homeless Street People.* Berkeley: University of California Press.

Snow, David A., Louis A. Zurcher, and Robert Peters. 1981. "Victory Celebrations as Theater: A Dramaturgical Approach to Crowd Behavior." *Symbolic Interaction,* 4 (1): 21–41.

Social Security Administration. 2015. "Monthly Statistical Snapshot, February 2015" (Mar.). Retrieved Apr. 25, 2015. Online: www.ssa.gov/policy/docs/quickfacts/stat_snapshot

Society for Human Resource Management. 2010. "Workers with Disabilities Face Steep Occupational Obstacles" (Dec. 8). Retrieved Apr. 2, 2011. Online: www.shrm.org/hrdisciplines/Diversity/Articles/Pages/FaceSteepOccupationalObstacles.aspx

—. 2012. "2012 Employee Job Satisfaction and Engagement: How Employees Are Dealing with Uncertainty." Retrieved July 3, 2013. Online: www.shrm.org/legalissues/stateandlocalresources/stateandlocalstatutesandregulations/documents/12-0537%202012_jobsatisfaction_fnl_online.pdf

Solem, Per Erik. 2008. "Age Changes in Subjective Work Ability." *International Journal of Aging and Later Life,* 3 (2): 43–70.

Stack, Steven. 1998. "Gender, Marriage, and Suicide Acceptability: A Comparative Analysis." *Sex Roles,* 38: 501–521.

Stack, Steven, and Ira Wasserman. 1995. "The Effect of Marriage, Family, and Religious Ties on African American Suicide Ideology." *Journal of Marriage and Family* (Feb.): 215–222.

Stainback, Kevin, and Donald Tomaskovic-Devey. 2012. *Documenting Discrimination: Racial and Gender Segregation in Private-Sector Employment Since the Civil Rights Act.* New York: Russell Sage Foundation.

Stark, Rodney. 2007. *Discovering God: The Origin of the Great Religions and the Evolution of Belief.* New York: HarperOne.

St. John, Warren. 2007. "A Laboratory for Getting Along." *New York Times* (Dec. 25): A1, A14.

Stoddard, Susan. 2014. "2014 Disability Statistics Annual Report." University of New Hampshire. Retrieved July 26, 2015. Online: www.disabilitycompendium.org/docs/default-source/2014-compendium/annual-report.pdf

Substance Abuse and Mental Health Services Administration (SAMHSA). 2014. "Nearly One in Five Adult Americans Experienced Mental Illness in 2013" (Nov.). Retrieved June 7, 2015. Online: www.samhsa.gov/newsroom/press-announcements/201411200815

Sugrue, Thomas J. 2011. "A Dream Still Deferred." *New York Times* (Mar. 26). Retrieved May 23, 2011. Online: www.nytimes.com/2011/03/27/opinion/27Sugrue.html?ref=michigan&pagewanted=print

Suhr, Jim. 2007. "Police: Elderly Driver in School Crash Was Bound for Driving School." *Chicago Tribune* (Jan. 30). Retrieved Mar. 25, 2007. Online: www.chicagotribune.com/news/local/illinois/chi-ap-il-carhitsschool,1,3832602.story?coll=chi-newsap_il-hed&ctrack=1&cset=true

Sullivan, Oriel. 2011. "Gender Deviance Neutralization Through Housework." *Journal of Family Theory & Review*, 3 (Mar.): 27–31.

Sumner, William G. 1959. *Folkways.* New York: Dover (orig. pub. 1906).

Sutherland, Edwin H. 1939. *Principles of Criminology.* Philadelphia: Lippincott.

—. 1949. *White Collar Crime.* New York: Dryden.

Swidler, Ann. 1986. "Culture in Action: Symbols and Strategies." *American Sociological Review*, 51 (April): 273–286.

Swisher, Kara. 2013. "Physically Together: Here's the Internal Yahoo No-Work-From-Home Memo for Remote Workers and Maybe More." *All Things D* (Feb. 22). Retrieved Mar. 8, 2013. Online: www.blogher.com/frame.php?url=allthingsd.com/20130222/physically-together-heres-the-internal-yahoo-no-work-from-home-memo-which-extends-beyond-remote-workers/?mod=obinsite#_=1362766877915&id=twitter-widget-0&lang=en&screen_name=karaswisher&show_count=false&show_screen_name=true&size=m

Szasz, Thomas S. 1984. *The Myth of Mental Illness: Foundations of a Theory of Personal Conduct.* New York: HarperCollins.

Takaki, Ronald. 1993. *A Different Mirror: A History of Multicultural America.* Boston: Little, Brown.

Tannen, Deborah. 1993. "Commencement Address, State University of New York at Binghamton." Reprinted in *Chronicle of Higher Education* (June 9): B5.

Tarbell, Ida M. 1925. *The History of Standard Oil Company.* New York: Macmillan (orig. pub. 1904).

Taylor, Paul. 2012. "The Growing Electoral Clout of Blacks in Drive by Turnout, Not Demographics." Pew Research Social and Demographic Trends (Dec. 26). Retrieved Mar. 13, 2013. Online: www.pewsocialtrends.org/files/2013/01/2012_Black_Voter_Project_revised_1-9.pdf

Taylor, Robert Joseph, Linda M. Chatters, and Sean Joe. 2011. "Religious Involvement and Suicidal Behavior Among African Americans and Black Caribbeans." *Journal of Nervous and Mental Disorders* (July): 478–486.

Taylor, Steve. 1982. *Durkheim and the Study of Suicide.* New York: St. Martin's.

Terkel, Studs. 1990. *Working: People Talk About What They Do All Day and How They Feel About What They Do.* New York: Ballantine (orig. pub. 1972).

thetaskforce.org. 2011. "Injustice at Every Turn: A Report on the National Transgender Discrimination Survey." Retrieved May 2, 2012. Online: www.thetaskforce.org/downloads/reports/reports/ntds_full.pdf

Tilly, Charles. 1975. *The Formation of National States in Western Europe.* Princeton, NJ: Princeton University Press.

TLC Channel. 2015. "Whitney's Video Diary." *My Big Fat Fabulous Life.* Retrieved July 17, 2015. Online: www.tlc.com/tv-shows/my-big-fat-fabulous-life/videos

Toft, Monica Duffy, Daniel Philpott, and Timothy Samuel Shah. 2011. "God's Partisans Are Back." *Chronicle Review* (Apr. 22): B4, B5.

Tönnies, Ferdinand. 1940. *Fundamental Concepts of Sociology* (Gemeinschaft *und* Gesellschaft). Trans. Charles P. Loomis. New York: American Book Company (orig. pub. 1887).

—. 1963. *Community and Society* (Gemeinschaft *and* Gesellschaft). New York: Harper & Row (orig. pub. 1887).

Toobin, Jeffrey. 2011. "Money Talks." *New Yorker* (Apr. 11). Retrieved Mar. 9, 2011. Online: www.newyorker.com/talk/comment/2011/04/11/110411taco_talk_toobin

Tracy, C. 1980. "Race, Crime and Social Policy: The Chinese in Oregon, 1871–1885." *Crime and Social Justice*, 14: 11–25.

Trounson, Rebecca. 2012. "Most Young People OK with Living at Home in Tough Times." *Los Angeles Times* (Mar. 16). Retrieved Apr. 1, 2013. Online: http://articles.latimes.com/2012/mar/16/local/la-me-0316-boomerang-20120316

Turkle, Sherry. 2011. *Alone Together: Why We Expect More from Technology and Less from Each Other.* New York: Basic.

Turner, Jonathan, Leonard Beeghley, and Charles H. Powers. 2007. *The Emergence of Sociological Theory* (6th ed.). Belmont, CA: Wadsworth.

Turner, Ralph H., and Lewis M. Killian. 1993. "The Field of Collective Behavior." In Russell L. Curtis Jr., and Benigno E. Aguirre (Eds.), *Collective Behavior and Social Movements.* Boston: Allyn & Bacon, pp. 5–20.

Truman, Jennifer L., and Lynn Langton. 2014. "Criminal Victimization, 2013." U.S. Department of Justice (Sept.). Retrieved Feb. 23, 2015. Online: www.bjs.gov/content/pub/pdf/cv13.pdf

UNAIDS. 2014. "Fact Sheet 2014: Global Statistics." Retrieved June 2, 2015. Online: www.unaids.org/sites/default/files/en/media/unaids/contentassets/documents/factsheet/2014/20140716_FactSheet_en.pdf

UNFPA (United Nations Population Fund). 2011. *The State of World Population 2011.* Retrieved Feb. 7, 2012. Online: http://foweb.unfpa.org/SWP2011/reports/EN-SWOP2011-FINAL.pdf

Unification Church News. 2011. "Who We Are." Retrieved May 10, 2011. Online: www.familyfed.org/about

United Nations. 1948. "The Universal Declaration of Human Rights." Retrieved Apr. 14, 2011. Online: www.un.org/en/documents/udhr/index.shtml

—. 1997. "Global Change and Sustainable Development: Critical Trends." United Nations Department for Policy Coordination and Sustainable Development. Posted online (Jan. 20).

—. 2012. "Highlights." *World Urbanization Prospects: The 2011 Revision.* Retrieved July 9, 2013. Online: http://esa.un.org/unup/pdf/WUP2011_Highlights.pdf

United Nations (Department of Economic and Social Affairs, Population Division [DESAPD]). *World Urbanization Prospects: The 2009 Revision.* Retrieved Feb. 7, 2012. Online: http://esa.un.org/unpd/wup/index.htm

United Nations Development Programme. 2011. *Human Development Report 2011: Sustainability and Equity: A Better Future for All.* Retrieved Mar. 4, 2012. Online: http://hdr.undp.org/en/media/HDR_2011_EN_Complete.pdf

—. 2013. *Human Development Report 2013: The Rise of the South, Human Progress in a Diverse World.* Retrieved July 8, 2013. Online: www.undp.org/content/dam/undp/library/corporate/HDR/2013GlobalHDR/English/HDR2013%20Report%20English.pdf

—. 2014. *Human Development Report 2014—Sustaining Human Progress: Reducing Vulnerabilities and Building Resilience.* Retrieved Mar. 6, 2015. Online: http://hdr.undp.org/sites/default/files/hdr14-report-en-1.pdf

—. 2015. "Human Development Index and Its Components." Retrieved June 2, 2015. Online: http://hdr.undp.org/en/content/table-1-human-development-index-and-its-components

United Nations Office of Drugs and Crime. 2014. "Transnational Organized Crime—the Globalized Illegal Economy." Retrieved May 14, 2014. Online: www.unodc.org/toc/en/crimes/organized-crime.html

United Nations World Urbanization Prospects: 2014 Revision. 2014. United Nations Economic and Social Affairs. Retrieved June 14, 2015. Online: http://esa.un.org/unpd/wup/Highlights/WUP2014-Highlights.pdf

United Press International. 2007. "Indian Women Carry Children for Foreigners." *United Press International.com* (Nov. 11). Retrieved Feb. 9, 2008. Online: www.upi.com/NewsTrack/Science/2007/11/11/indian_women_carry_children_for_foreigners/2909

University of Michigan. 2007. "Boys Mow Lawns, Girls Do Dishes: Are Parents Perpetuating the Chore Wars?" Institute for Social Research. Retrieved May 2, 2012. Online: www.isr.umich.edu/home/news/archive.html

Urbandictionary.com 2015. "Gay Fat." Retrieved Apr. 4, 2015. Online: www.urbandictionary.com/define.php?term=gay+fat

U.S. Army. 2012. "Army 2020: Generating Health & Discipline in the Force Ahead of the Strategic Reset: Report 2012." Retrieved Sept. 23, 2012. Online: http://usarmy.vo.llnwd.net/e2/c/downloads/232541.pdf

U.S. Bureau of Labor Statistics. 2013a. "The Employment Situation—February 2013."

Retrieved Mar. 10, 2013. Online: www.bls .gov/news.release/pdf/empsit.pdf

—. 2013b. "Employment Situation Summary" (June 7). Retrieved June 14, 2013. Online: www.bls.gov/news.release/empsit.nr0.htm

—. 2013c. "Major Work Stoppages in 2012." Retrieved Mar. 10, 2013. Online: www.bls .gov/news.release/pdf/wkstp.pdf

—. 2013d. "Union Members Summary." Retrieved Mar. 10, 2013. Online: http://data .bls.gov/cgi-bin/print.pl/news.release /union2.nr0.htm

—. 2013e. "Usual Weekly Earnings Summary" (Jan. 18). Retrieved Apr. 13, 2013. Online: http://data.bls.gov/cgi-bin/print.pl/news .release/wkyeng.nr0.htm

—. 2014a. "Fastest Growing Occupations: 2012–2022" (Jan. 8). Retrieved May 10, 2015. Online: www.bls.gov/ooh/fastest-growing.htm

—. 2014b. "Highlights of Women's Earnings in 2013." Retrieved Apr. 13, 2015. Online: www.bls.gov/opub/reports/cps/highlights -of-womens-earnings-in-2013.pdf

—. 2015a. "Economic News Release: Union Members Summary." Retrieved July 20, 2015. Online: www.bls.gov/news.release /union2.nr0.htm

—. 2015b. "Employment Characteristics of Families—2014" (Apr. 23). U.S. Department of Labor. Retrieved Apr. 27, 2015. Online: www.bls.gov/news.release/pdf/famee.pdf

—. 2015c. "The Employment Situation—Apr. 2015" (May 8). Retrieved May 11, 2015. Online: www.bls.gov/news.release/pdf /empsit.pdf

—. 2015d. "Employment Status of the Civilian Population 25 Years and Over by Educational Attainment" (May 8). Retrieved May 10, 2015. Online: www.bls.gov/news .release/empsit.t04.htm

—. 2015e. "Occupational Employment and Wages—May 2014" (Mar. 25). Retrieved Apr. 27, 2015. Online: www.bls.gov/news .release/archives/ocwage_03252015.pdf

—. 2015f. "Labor Force Statistics from the Current Population Survey: Table 11. Employed Persons by Detailed Occupation, Sex, Race, and Hispanic or Latino Ethnicity." Retrieved July 17, 2015. Online: www .bls.gov/cps/cpsaat11.pdf

U.S. Bureau of Labor Statistics Local Area Unemployment Statistics. 2015. "Unemployment Rates for States, March 2015." Retrieved Apr. 27, 2015. Online: www.bls .gov/web/laus/laumstrk.htm

U.S. Census Bureau. 2011a. "American Indian and Alaska Native Heritage Month: November 2011." U.S. Census Bureau News, Facts for Features. Retrieved Dec. 29, 2011. Online: www.census.gov/newsroom /releases/pdf/cb11ff-22_aian.pdf

—. 2011b. America's Families and Living Arrangements: 2011. Retrieved Jan. 4, 2012. Online: www.census.gov/population/www /socdemo/hh-fam/cps2011.html

—. 2011c. "Asian/Pacific American Heritage Month: May 2011." U.S. Census Bureau News, Facts for Features (May). Retrieved Dec. 29, 2011. Online: www.census.gov /newsroom/releases/archives/facts_for _features_special_editions/cb11-ff06.html

—. 2011d. "Census Bureau Releases Estimate of Same-Sex Married Couples." Retrieved Jan. 5, 2011. Online: http://2010.census.gov/news /releases/operations/cb11-cn181.html

—. 2011e. "Current Population Survey: Definitions and Explanations." Retrieved Apr. 14, 2011. Online: www.census.gov/population /www/cps/cpsdef.html

—. 2011f. Historical Income Tables: Families. Table F-3: Mean Income Received by Each Fifth and Top 5 Percent of Families, All Races: 1966 to 2010. Retrieved Dec. 12, 2011. Online: www.census.gov/hhes/www/income /data/historical/families/index.html

—. 2011g. Statistical Abstract of the United States: 2011 (130th edition). Washington, DC: U.S. Government Printing Office.

—. 2012a. "American Families and Living Arrangements: 2012." Retrieved Apr. 1, 2013. Online: www.census.gov/hhes /families/data/cps2012.html

—. 2012b. "Language Spoken at Home by Ability to Speak English for the Population 5 Years and Over." Retrieved Oct. 14, 2012. Online: http://factfinder2.census.gov /faces/tableservices/jsf/pages /productview.xhtml?pid=ACS_11 _1YR_B16001&prodType=table

—. 2015. "Living Arrangements of Children Under 18 Years: 2014." Retrieved May 18, 2015. Online: www.census.gov/hhes /families/data/cps2014C.html

U.S. Census Bureau American Indian and Alaska Native Heritage Month. 2014. "Facts for Features." Retrieved Mar. 28, 2015. Online: www.census.gov/content/dam /Census/newsroom/facts-for-features /2014/cb14ff-26_aian_heritage_month.pdf

U.S. Census Bureau Asian/Pacific American Heritage Month. 2014. "Facts for Features." Retrieved Mar. 28, 2015. Online: www.census .gov/content/dam/Census/newsroom/facts -for-features/2014/cb14-ff13_asian.pdf

U.S. Census Bureau, Current Population Survey, Annual Social and Economic Supplement. 2012. U.S. Census Bureau (Internet release date Dec. 2013). Retrieved Apr. 23, 2015. Online: www.census.gov/population /age/data/2012comp.html

U.S. Census Bureau Educational Attainment Table 3. 2014. "Detailed Years of Schooling Completed by People 25 Years and Over by Sex, Age Groups, Race and Hispanic Origin: 2014." Retrieved May 30, 2015. Online: www.census.gov/hhes/socdemo /education/data/cps/2014/tables.html

U.S. Census Bureau Facts for Features. 2014. "American Indian and Alaska Native Heritage Month: November 2014" (Nov. 12). Retrieved Feb. 28, 2015. Online: www

.census.gov/content/dam/Census /newsroom/facts-for-features/2014 /cb14ff-26_aian_heritage_month.pdf

U.S. Census Bureau, Families and Living Arrangements. 2014. Retrieved Apr. 24, 2015. Online: www.census.gov/hhes /families/data/cps2014.html

U.S. Census Bureau Hispanic Heritage Month. 2014. "Facts for Features." Retrieved Mar. 28, 2015. Online: www.census.gov/content /dam/Census/newsroom/facts-for -features/2014/cb14ff-22_hispanic.pdf

U.S. Census Bureau Historical Income Tables. 2015. "Table H.1: Income Limits for Each Fifth and Top Five Percent of All Households: 1967 to 2013." Retrieved July 7, 2015. Online: www.census.gov/hhes/www /income/data/historical/household

U.S. Census Bureau, Historical Poverty Tables—People. 2015. Retrieved Apr. 24, 2015. Online: www.census.gov/hhes/www /poverty/data/historical/people.html

U.S. Census Bureau Marital Status Data. 2015. "Figure MS-2, Median Age at First Marriage: 1890 to present." Retrieved July 213, 2015. Online: www.census.gov/hhes /families/files/graphics/MS-2.pdf

U.S. Census Bureau Newsroom. 2014. "As the Nation Ages, Seven States Become Younger." Retrieved Apr. 24, 2015. Online: www.census.gov/newsroom/press -releases/2014/cb14-118.html

U.S. Centers for Disease Control and Prevention. 2011a. "CDC Report: Mental Illness Surveillance Among Adults in the United States." Retrieved Jan. 31, 2012. Online: www.cdc.gov/mentalhealthsurveillance /fact_sheet.html

—. 2011b. "Preventing Teen Pregnancy in the US." Vital Signs. Retrieved Jan. 1, 2012. Online: www.cdc.gov/vitalsigns /TeenPregnancy/index.html

—. 2011c. "Sexually Transmitted Disease Surveillance: 2010." Retrieved Mar. 5, 2012. Online: www.cdc.gov/std/stats10 /surv2010.pdf

—. 2011d. "Teen Birth Rates Declined Again in 2009." Retrieved Jan. 1, 2012. Online: www .cdc.gov/features/dsteenpregnancy

—. 2012a. "African Americans and Sexually Transmitted Diseases" (Dec.). Retrieved May 8, 2013. Online: www.cdc.gov /nchhstp/newsroom/docs/AAs-and-STD -Fact-Sheet-042011.pdf

—. 2012b. "Binge Drinking." Retrieved Mar. 5, 2012. Online: www.cdc.gov/VitalSigns /pdf/2012-01-vitalsigns.pdf

—. 2012c. "HIV in the United States." Fact Sheet. Retrieved Jan. 29, 2012. Online: www .cdc.gov/hiv/resources/factsheets/us.htm

—. 2013a. "Adult Obesity: Data and Statistics." Retrieved Mar. 30, 2014. Online: www.cdc .gov/obesity/data/adult.html

—. 2013b. "HIV in the United States: At a Glance." Retrieved Mar. 30, 2014. Online:

www.cdc.gov/hiv/pdf/statistics_basics_factsheet.pdf

—. 2013c. "Secondhand Smoke (SHS) Facts." Retrieved July 8, 2013. Online: www.cdc.gov/tobacco/data_statistics/fact_sheets/secondhand_smoke/general_facts

—. 2014a. "2012 Sexually Transmitted Diseases Surveillance: Chlamydia." Retrieved Mar. 29, 2014. Online: www.cdc.gov/std/stats12/chlamydia.htm

—. 2014b. "2012 Sexually Transmitted Diseases Surveillance: Gonorrhea." Retrieved Mar. 29, 2014. Online: www.cdc.gov/std/stats12/gonorrhea.htm

—. 2014c. "2012 Sexually Transmitted Diseases Surveillance: Syphilis." Retrieved Mar. 29, 2014. Online: www.cdc.gov/std/stats12/syphilis.htm

—. 2014d. "Mortality in the United States, 2013." Retrieved Jan. 1, 2015. Online: www.cdc.gov/nchs/data/databriefs/db178.pdf

—. 2015a. "Alcohol Poisoning Deaths" (Jan.). Retrieved June 5, 2015. Online: www.cdc.gov/vitalsigns/pdf/2015-01-vitalsigns.pdf

—. 2015b. "Nursing Home Care." Retrieved Apr. 25, 2015. Online: www.cdc.gov/nchs/fastats/nursing-home-care.htm

U.S. Citizenship and Immigration Services. 2011. *A Guide to Naturalization.* Retrieved Feb. 6, 2012. Online: www.uscis.gov/files/article/M-476.pdf

U.S. Conference of Mayors. 2012. "U.S. Metro Economies: Key Findings" (Jan.). Retrieved Mar. 15, 2012. Online: http://usmayors.org/pressreleases/uploads/2012/MetroEconomiesKeyFindings_011812.pdf

—. 2014. "Hunger and Homelessness Survey: A Status Report on Hunger and Homelessness in America's Cities, A 25-City Survey" (Dec.). Retrieved Jan. 31, 2015. Online: www.usmayors.org/pressreleases/uploads/2014/1211-report-hh.pdf

U.S.Courts.gov. 2014. "U.S. Bankruptcy Courts—Bankruptcy Filings." Retrieved Dec. 30, 2014. Online: www.uscourts.gov/uscourts/Statistics/BankruptcyStatistics/BankruptcyFilings/2014/0914_f2.pdf

usda.gov. 2012. "Results of Community Facilities Program, FY 2009–2011: Healthcare, Education, and Public Service and Safety Investments Help Rural Communities Thrive." United States Department of Agriculture. Retrieved June 28, 2013. Online: www.rurdev.usda.gov/Reports/rdCFReportMay31-2012.pdf

U.S. Department of Health and Human Services. 2014a. "The AFCARS Report" (July). Retrieved July 23, 2015. Online: www.acf.hhs.gov/sites/default/files/cb/afcarsreport21.pdf

—. 2014b. "2013 National Healthcare Disparities Report" (May). Retrieved Mar. 10, 2015. Online: www.ahrq.gov/research/findings/nhqrdr/nhdr13/2013nhdr.pdf

U.S. Department of Health and Human Services, Children's Bureau. 2013. "Child Maltreatment 2012." Retrieved Jan. 22, 2015. Online: www.acf.hhs.gov/sites/default/files/cb/cm2012.pdf#page=11

U.S. Department of Homeland Security. 2014. "Immigrant Orphans Adopted by U.S. Citizens by Sex, Age, Region, and Country of Birth: 2013." Retrieved May 28, 2014. Online: http://statabs.proquest.com.ezproxy.baylor.edu/ftv2/4c4e0000020f0f50.pdf#43

U.S. Department of Housing and Urban Development. 2015a. "Affordable Housing." Retrieved Feb. 28, 2015. Online: http://portal.hud.gov/hudportal/HUD?src=/program_offices/comm_planning/affordablehousing

—. 2015b. "Fair Housing." Retrieved Apr. 6, 2015. Online: http://portal.hud.gov/hudportal/HUD?src=/program_offices/fair_housing_equal_opp/LGBT_Housing_Discrimination

U.S. Department of Labor. 2012. "Obama Administration Announces $500 Million in Community College Grants to Expand Job Training Through Local Employer Partnerships." Retrieved May 4, 2013. Online: www.dol.gov/opa/media/press/eta/ETA20121885.htm

—. 2013. "Disability Employment Policy Resources" (Feb.). Retrieved Mar. 9, 2013. Online: www.dol.gov/odep

—. 2015. "Current Disability Employment Statistics." Retrieved May 10, 2015. Online: www.dol.gov/odep

usgovernmentspending.com. 2015. "United States Total Spending Pie Chart." Retrieved May 15, 2015. Online: www.usgovernmentspending.com/united_states_total_spendingpie_chart

U.S. Office of Personnel Management. 2013. "Data Analysis and Documentation: Profile of Federal Civilian Non-Postal Employees." Retrieved Mar. 29, 2013. Online: www.opm.gov/policy-data-oversight/data-analysis-documentation/federal-employment-reports/reports-publications/profile-of-federal-civilian-non-postal-employees

U.S. State Department. 2014. "A Day in Your Life: Touched by Modern Slavery." Retrieved Mar. 1, 2015. Online: www.state.gov/documents/organization/233950.pdf

—. 2015a. "Independent States of the World." Fact Sheet (Dec. 30). Retrieved May 15, 2015. Online: www.state.gov/s/inr/rls/4250.htm

—. 2015b. "What Is Modern Slavery?" Retrieved Mar. 1, 2015. Online: www.state.gov/j/tip/what

Van Ausdale, Debra, and Joe R. Feagin. 2001. *The First R: How Children Learn Race and Racism.* Lanham, MD: Rowman & Littlefield.

Van Horn, Carl. 2013. "Changing Realities at Work Require Reforms in U.S. Workforce Policies" (Feb. 20). Conference on Long-Term Unemployment, National Association of State Workforce Agencies. Retrieved Mar. 10, 2013. Online: www.heldrich.rutgers.edu/sites/default/files/content/NASWA_Workforce_Policies_Van_Horn.pdf

Vargas, Jose Antonio. 2011. "My Life as an Undocumented Immigrant." *New York Times* (June 22). Retrieved Mar. 13, 2011. Online: www.nytimes.com/2011/06/26/magazine/my-life-as-an-undocumented-immigrant.html?pagewanted=all

—. 2012. "About Jose." defineamerican.com. Retrieved Mar. 14, 2012. Online: www.defineamerican.com/page/about/about-jose

Veblen, Thorstein. 1967. *The Theory of the Leisure Class.* New York: Viking (orig. pub. 1899).

Venkatesh, Sudhir Alladi. 2006. *Off the Books: The Underground Economy of the Urban Poor.* Cambridge: Harvard University Press.

—. 2013. *Floating City: A Rogue Sociologist Lost and Found in New York's Underground Economy.* New York: Penguin.

Vespa, Jonathan, Jamie M. Lewis, and Rose M. Kreider. 2013. "America's Families and Living Arrangements: 2012." U.S. Census Bureau (Aug.). Retrieved May 23, 2014. Online: www.census.gov/prod/2013pubs/p20-570.pdf

Vincent, Isabel. 2004. "Women Bear Brunt of 'Sandwich' Caregiving." *National Post* (Sept. 29). Retrieved Apr. 16, 2005. Online: www.albertacaregiversassociation.org/natpost-sandwichcaregiving.htm

VolunteerMatch.org. 2010. "Volunteer Spotlight: Christine French, the George Washington University" (June 16). Retrieved Dec. 29, 2010. Online: www.volunteermatch.org/volunteers/stories/spotlight.jsp?id=48

Wade, Lisa. 2012. "Women, Education, and Trends in Childlessness" (Sept. 24). Retrieved Apr. 8, 2013. Online: http://thesocietypages.org/socimages/2012/09/24/women-education-and-trends-in-childlessness

Wallace, Walter L. 1971. *The Logic of Science in Sociology.* New York: Aldine de Gruyter.

Wallerstein, Immanuel. 1979. *The Capitalist World-Economy.* Cambridge, England: Cambridge University Press.

—. 1984. *The Politics of the World Economy.* Cambridge, England: Cambridge University Press.

—. 2011. *The Modern World System I–IV.* Berkeley: University of California Press.

Walmart. 2014. "Our Story: Our Locations." Walmart.com. Retrieved Dec. 26, 2014. Online: http://corporate.walmart.com/our-story/our-business/locations

Walmart China. 2014. "Walmart China Factsheet." Retrieved Dec. 26, 2014. Online: www.wal-martchina.com/english/walmart

Warner, W. Lloyd, and Paul S. Lunt. 1941. *The Social Life of a Modern Community.* New Haven, CT: Yale University Press.

Watson, Tracey. 1987. "Women Athletes and Athletic Women: The Dilemmas and Contradictions of Managing Incongruent Identities." *Sociological Inquiry*, 57 (Fall): 431–446.

Weber, Max. 1968. *Economy and Society: An Outline of Interpretive Sociology.* Trans. G. Roth and G. Wittich. New York: Bedminster (orig. pub. 1922).

—. 1976. *The Protestant Ethic and the Spirit of Capitalism.* Trans. Talcott Parsons. Introduction by Anthony Giddens. New York: Scribner (orig. pub. 1904–1905).

Weeks, John R. 2012. *Population: An Introduction to Concepts and Issues* (11th ed.). Belmont, CA: Wadsworth.

Weigel, Russell H., and P. W. Howes. 1985. "Conceptions of Racial Prejudice: Symbolic Racism Revisited." *Journal of Social Issues*, 41: 124–132.

Wei-ming, Tu. 1995. "Confucianism." In Arvind Sharma (Ed.), *Our Religions.* San Francisco: HarperCollins, pp. 141–227.

Weiss, Gregory L., and Lynne E. Lonnquist. 2014. *Sociology of Health, Healing and Illness* (8th ed.). New York: Routledge.

Weitz, Rose. 2013. *The Sociology of Health, Illness, & Health Care: A Critical Approach* (6th ed.). Boston: Wadsworth/Cengage.

Wellman, Barry. 2001. "Physical Place and Cyberplace: The Rise of Personalized Networking." *International Journal of Urban and Regional Research*, 22 (2): 227–252.

Werhane, Veronica. 2011. "Food Trucks: Spreading Culture Through Food." Neontommy.com (Apr. 12). USC Annenberg School for Communication and Journalism. Retrieved Dec. 11, 2013. Online: www.neontommy.com/news /2011/04/food-and-food-trucks-using -food-spread-cultural-awareness

whitehouse.gov. 2011. "Strategy to Combat Transnational Organized Crime" (July). Retrieved Jan. 2, 2012. Online: www .whitehouse.gov/sites/default/files /microsites/2011-strategy-combat -transnational-organized-crime.pdf

—. 2012. "President Obama: 'Words Need to Lead to Action' on Gun Violence." Retrieved Dec. 27, 2012. Online: www .whitehouse.gov/blog/2012/12/19 /president-obama-words-need-lead -action-gun-violence

—. 2015. "Education: Knowledge and Skills for the Jobs of the Future—Race to the Top." Retrieved July 25, 2015. Online: www .whitehouse.gov/issues/education/k-12 /race-to-the-top

Whitesel, Jason. 2014. *Fat Gay Men: Girth, Mirth and the Politics of Stigma.* New York: New York University Press.

WHO Tobacco Facts. 2014. "World Health Organization Media Center: Tobacco" (May 2014). Retrieved Mar. 6, 2015. Online: www .who.int/mediacentre/factsheets/fs339/en

Whorf, Benjamin Lee. 1956. *Language, Thought and Reality.* Ed. John B. Carroll. Cambridge, MA: MIT Press.

Wikipedia.org. 2015. "List of Largest Companies by Revenue." Retrieved Apr. 10, 2015. Online: http://en.wikipedia.org/wiki/List _of_largest_companies_by_revenue

Wilson, David Sloan, and Edward O. Wilson. 2007. "Rethinking the Theoretical Foundation of Sociobiology." *Quarterly Review of Biology*, 82 (4): 327–348.

Wilson, William Julius. 1978. *The Declining Significance of Race: Blacks and Changing American Institutions.* Chicago: University of Chicago Press.

Wilson, William Julius, and Richard P. Taub. 2007. *There Goes the Neighborhood: Racial, Ethnic and Class Tension in Four Chicago Neighborhoods and Their Meaning for America.* New York: Vintage.

Wintour, Patrick. 2013. "UN Goals Must Aim to Eradicate Extreme Poverty." *Guardian* (Feb. 1). Retrieved Feb. 3, 2013. Online: www .guardian.co.uk/world/2013/feb/01/un-goals -extreme-poverty-david-cameron/print

Wirth, Louis. 1938. "Urbanism as a Way of Life." *American Journal of Sociology*, 40: 1–24.

World Bank. 2012. "World Development Indicators: 2012." Retrieved Jan. 14, 2013. Online: www.oecd-ilibrary.org/docserver /download/3011041ec020.pdf?expires=1359 239194&id=id&accname=guest&checksum =254AE45119DAF1066B9D9EEE031C65DD

—. 2014. "Updated Income Classifications." Online: http://data.worldbank.org /news/2015-country-classifications

—. 2015. "Country and Lending Groups." Retrieved Mar. 5, 2015. Online: http://data .worldbank.org/about/country-and -lending-groups#Low_income

World Health Organization. 2011a. "Disaster Risk Management for Health" (May). Retrieved May 11, 2011. Online: www.who.int /hac/events/drm_fact_sheet_

—. 2011b. "Frequently Asked Questions: What Is the WHO Definition of Health?" Retrieved May 12, 2011. Online: www.who .int/suggestions/faq/en

—. 2014a. "Preventing Suicide: A Global Imperative." Retrieved Jan. 1, 2015. Online: http://apps.who.int/iris/bitstream /10665/131056/1/9789241564779_eng .pdf?ua=1

—. 2014b. "World Health Statistics: 2014." Retrieved Mar. 6, 2015. Online: http://apps.who.int/iris /bitstream/10665/112738/1/9789240692671 _eng.pdf?ua=1

Wright, Erik Olin. 1978. "Race, Class, and Income Inequality." *American Journal of Sociology*, 83 (6): 1397.

—. 1979. *Class Structure and Income Determination.* New York: Academic Press.

—. 1985. *Class.* London: Verso.

—. 1997. *Class Counts: Comparative Studies in Class Analysis.* Cambridge, England: Cambridge University Press.

—. 2010. *Envisioning Real Utopias.* London: Verso.

Wyatt, Edward. 2014. "Weight-Loss Companies Charged with Fraud." *New York Times* (Jan. 7). Retrieved Feb. 6, 2015. Online: www.nytimes.com/2014/01/08/business /us-charges-4-companies-with-deception -in-weight-loss-products.html?_r=0

Yale News. 2013. "Students Map Diversity and Culture of New Haven Through Food Trucks." Yale University (Dec. 4). Retrieved Dec. 11, 2013. Online: http://news.yale.edu /2013/12/04/students-map-diversity-and -culture-new-haven-through-food-trucks

Yinger, J. Milton. 1960. "Contraculture and Subculture." *American Sociological Review*, 25 (October): 625–635.

—. 1982. *Countercultures: The Promise and Peril of a World Turned Upside Down.* New York: Free Press.

Yoder, Brian L. 2014. "Engineering by the Numbers." American Society for Engineering Education. Retrieved July 25, 2015. Online: www.asee.org/papers-and-publications /publications/14_11-47.pdf

Yong, Ed. 2011. "New Evidence That IQ Is Not Set in Stone." cbsnews.com (Apr. 26). Retrieved Apr. 28, 2011. Online: www .cbsnews.com/stories/2011/04/26/scitech /main20057536.shtml

Zellner, William M. 1978. Vehicular Suicide: In Search of Incidence. Unpublished M.A. thesis, Western Illinois University, Macomb. Quoted in Richard T. Schaefer and Robert P. Lamm. 1992. *Sociology* (4th ed.). New York: McGraw-Hill, pp. 54–55.

Zernike, Kate. 2010. "A Young and Unlikely Activist Who Got to the Tea Party Early." *New York Times* (Feb. 28): A1, A19.

Zernike, Kate, and Megan Thee-Brenan. 2010. "Discontent's Demography: Who Backs the Tea Party?" *New York Times* (Apr. 15): A1, A17.

Zetterberg, Hans L. 2002/1964. *Social Theory and Social Practice.* New York: Transaction.

Zuckerman, Mortimer B. 2011. "The Great Recession Goes on." *USNews* (Feb. 11). Retrieved Jan. 2, 2012. Online: www .usnews.com/opinion/mzuckerman /articles/2011/02/11/the-great-jobs -recession-goes-on

Zurbriggen, Eileen L., and Tomi-Ann Roberts (Eds.). 2012. *The Sexualization of Girls and Girlhood: Causes, Consequences, and Resistance.* New York: Oxford University Press.